THEORY OF
CONVEX STRUCTURES

North-Holland Mathematical Library

Board of Advisory Editors:

M. Artin, H. Bass, J. Eells, W. Feit, P.J. Freyd, F.W. Gehring,
H. Halberstam, L.V. Hörmander, J.H.B. Kemperman, H.A. Lauwerier,
W.A.J. Luxemburg, F.P. Peterson, I.M. Singer and A.C. Zaanen

VOLUME 50

NORTH-HOLLAND
AMSTERDAM • LONDON • NEW YORK • TOKYO

Theory of Convex Structures

M.L.J. VAN DE VEL
*Department of Mathematics and
Computer Science,
Free University
Amsterdam, The Netherlands*

1993
NORTH-HOLLAND
AMSTERDAM • LONDON • NEW YORK • TOKYO

ELSEVIER SCIENCE PUBLISHERS B.V.
Sara Burgerhartstraat 25
P.O. Box 211, 1000 AE Amsterdam, The Netherlands

Library of Congress Cataloging-in-Publication Data

Vel, M. L. J. van de, 1948-
　　Theory of convex structures / M.L.J. van de Vel
　　　p. cm. -- (North-Holland mathematical library; v. 50)
　　Includes bibliographical references and index.
　　ISBN 0-444-81505-8
　　1. Convex domains. I. Title. II. Series.
QA639.5.V45 1993
516' .08--dc20　　　　　　　　　　　　　　　　　　93-8207
　　　　　　　　　　　　　　　　　　　　　　　　　CIP

ISBN: 0 444 81505 8

© 1993 ELSEVIER SCIENCE PUBLISHERS B.V. All rights reserved.

No part of this publication may be reproduced, stored in a retrieval system or transmitted in any form or by any means, electronic, mechanical, photocopying, recording or otherwise, without the prior written permission of the publisher, Elsevier Science Publishers B.V., Copyright & Permissions Department, P.O. Box 521, 1000 AM Amsterdam, The Netherlands.

Special regulations for readers in the U.S.A. – This publication has been registered with the Copyright Clearance Center Inc. (CCC), Salem, Massachusetts. Information can be obtained from the CCC about conditions under which photocopies of parts of this publication may be made in the U.S.A. All other copyright questions, including photocopying outside of the U.S.A., should be referred to the publisher.

No responsibility is assumed by the publisher for any injury and/or damage to persons or property as a matter of products liability, negligence or otherwise, or from any use or operation of any methods, products, instructions or ideas contained in the material herein.

This book is printed on acid-free paper
Transferred to digital printing 2006.

voor Wendy
voor Wouter
voor Katleen

Introduction

This monograph presents the current state-of-the-art in the theory of convex structures. The notion of convexity covered here is considerably broader than the classic one; specifically, it is not restricted to the context of vector spaces. The classical concepts of *order-convex sets* (Birkhoff) and of *geodesically convex sets* (Menger) are directly inspired by intuition; they go back to the first half of this century. An axiomatic approach, started to develop from the early fifties on. I got attracted to it in the mid seventies. Ten years later, confronted with ever growing introductory sections to my papers on the subject, I decided to write a book that could serve as a general reference. This has resulted in the present volume, in which graphs appear side by side with Banach spaces, classical geometry with matroids, and ordered sets with metric spaces. A wide variety of results has been included (ranging for instance from the area of partition calculus to that of continuous selection). The tools involved are borrowed from areas ranging from discrete mathematics to infinite-dimensional topology.

One doesn't have to leave the context of vector spaces to encounter some interesting aspects of generalized convexity. For instance, H-convexity is a subsystem of the standard convexity, consisting of those convex sets which can be "generated" with the aid of a selected collection of linear functionals. The combinatorial properties of such convexities can be quite exotic, and the link with the theory of linear inequalities is obvious. Also, a standard convex set can be made into a convex structure of its own with the "relative" convexity, and it has some unusual features as well (non-affine geometry). Axiomatic convexity operates with a concept of "convexity preserving" functions. In case of standard convexity, this amounts to functions mapping line segments into line segments. Affine mappings share this property with perspective projection. For instance, a convex 4-gon is affinely isomorphic to a parallelogram iff it is a parallelogram itself, but from the general viewpoint of convexity preserving functions all convex 4-gons are isomorphic. This leads to still unsolved problems on classifying polytopes and on unicity of uniform structures relative to which the hull operator is uniformly continuous.

It is said that axiomatic convexity is a language, not a theory. There is considerable truth in the first half of this statement, but we firmly disagree with the second half. The convexity descending from a given mathematical structure usually contains essential information on its parent, which can sometimes even be reconstructed. This indeed allows to talk about variable structures in one and the same language. For instance, the classic conditions of Pasch and Peano for order geometries -- reformulated for interval operators -- are equivalent to, respectively, a separation property (S_4) and a hull property (JHC) of the induced convexity. When applied to matroids, the hull property characterizes projective spaces with a convexity of subspaces (points, lines, planes,..). When applied to lattices, the separation and hull properties are equivalent and characterize distributivity. On the other hand, a property of interval operators, known as modularity, characterizes modular lattices among all lattices, or modular graphs among all connected graphs. Trees are semilattices of which the usual convexity of all order-convex subsemilattices is modular, and median algebras are modular spaces with the separation property S_4 or with the hull property JHC. Modularity also holds for the geodesic interval operators, associated with the "sum" and "maximum" norms and some of their combinations in Euclidean spaces.

But one can say much more. A substantial amount of fairly general results has been developed, centered around concepts like separation properties (à la Kakutani), hull properties, interval operators, invariants of convex structures, compatible topologies with derived notions of weak topology or local convexity, etc.. To give a taste of the subject, let us briefly describe some of the results that have been obtained.

(1) A theorem on the rectilinear Steiner problem concerning graphs of minimal weight connecting a given set of points. The result states that a solution can be found within a set which is considerably smaller than the lattice which is usually considered. It is formulated in terms of median metric spaces (for instance, the plane with Manhattan metric) of any dimension. See Section I§6.

(2) A sharp lower bound on the number of Radon partitions in general convex structures. Its sharpness is illustrated by the standard convexity in Euclidean space. The result has been used to obtain an accurate estimate of Radon numbers in product spaces, a topic initiated by Eckhoff in a study of products of Euclidean spaces. The restriction to Euclidean factors turned out to be superfluous. See Section II§2.

(3) A compactness theorem for convex structures. The result has been used to develop some "infinite combinatorics". Particular applications involve extending Klee's countable intersection theorem, determination of the Helly number for symmetrically generated H-convexities, and embeddings into products of trees. See Section II§4.

(4) Embedding Bryant-Webster ("join") spaces as convex subspaces of vector spaces. This parallels -- and in part extends -- the traditional results on coordinatization of affine or projective spaces having Desargues' Property. See Section IV§1.

(5) A theorem on continuous selection. The assumptions are easy to state: a metric space equipped with a uniformly continuous hull operator, compact polytopes, connected convex sets, and Kakutani's separation property. The conclusion: each lower semicontinuous, compact-and-convex-valued function of a normal domain into this space has a single-valued, continuous selection. This includes the classical selection theorems of Michael (Fréchet spaces), Nadler (connected trees), a more recent result of Curtis on selection in spaces of arcs, and a result of Beer on approximate selections of upper semicontinuous multifunctions. The main auxiliary result on contractibility of convex open covers has some applications to fixed point theory as well. See Sections IV§3 and IV§6.

The open problem which is perhaps most intriguing concerns Tverberg's extension of the classical Radon Theorem. Although several proofs have been developed, and despite the purely combinatorial nature of the result, no counterpart is known in generalized convexity.

Introduction

Organisation

Chapter I deals with **general** (set-theoretic) **convexity theory.** The main types of examples are introduced early on, and are inspected regularly in the light of the theory developed so far. The important geometric conditions of join-hull commutativity and separation are introduced and studied in Sections 2 and 3. Section 4 gives an account of interval spaces, where the above properties are characterized in terms of the Pasch and Peano conditions, known from plane geometry. In Section 5 we develop a concept of base-point order and a related concept of completeness. Section 6 uses much of the theory developed so far on the class of modular spaces and its subclass of median spaces. Finally, Section 7 deals with a specialized class of so-called Bryant-Webster spaces, closely related with classic geometry.

Chapter II presents the **theory of convex invariants.** These are modeled after the well-known theorems of Helly, Carathéodory and Radon, and capture some combinatorial features of convexity. Sections 2 and 3 are especially devoted to the determination of the invariants in constructions like products and amalgamations. The Compactness Theorem of convexity and the related Compact Intersection Theorem are proved and exploited in Section 4, where some other convex invariants are considered as well. The last section deals with the intriguing Tverberg numbers.

Topological convexity theory is studied in Chapter III. Some of the main topics are: weak topology, local convexity, and (uniform) continuity of the hull operation. The first three sections are devoted to this. Separation properties are reconsidered in Section 4 from the viewpoint of an additional topology. Finally, Section 5 shows how to construct a reasonable topology from convexity data (using the base-point orders of Chapter I). This occasionally leads to uniqueness of the accompanying topology.

Chapter IV contains **miscellaneous results.** It is in part a continuation of the topological theory. Section 1 parallels the classic results on coordinatization of affine or projective spaces for the class of complete Bryant-Webster spaces without boundary points. Section 2 handles extremality and support properties. Section 3 is largely devoted to a proof of a theorem on continuous selection and its applications. Sections 4 and 5 present a theory of topological dimension in convex structures, and the relationship of such a dimension with some of the convex invariants. Finally, some results on fixed point theory (both combinatorial and topological) are included in Section 6.

Each chapter has a separate title page including a list of section headers, each followed by a list of key words. Each section opens with a brief description of the main concepts and results and is followed by a "topics section" including additional theory and examples, and by a notes section providing some bibliographic and background information. The first appearance of an item included in the index is written in *this font*. Reference to Chapter IV, Section 3, Theorem 5 is made by the string "IV§3.5". If the reference is made from within Chapter IV then we simply use "3.5". Proofs are ended by the mark ■

Some packages

The theory of convexity being somewhat inhomogeneous, the reader may wish to confine himself to certain parts of it. The following are a few suggestions.

General convexity. This package includes Chapter I §§ 1, 2, 3, 4; § 5 is a bit specialized. Reading can be continued with a study of the classical convex invariants (see next item), or extremality and support (Chapter IV §2), or with a study of general topological convexity theory (see the last item).

Convex invariants. This requires notions of (at least) Chapter I §§ 1, 7. The main body consists of Chapter II, of which §§ 3, 4 are somewhat specialized. The topological part of convex invariants theory requires notions of Chapter III §§ 1, 4, 5, with Chapter IV §§ 4, 5 as the main source of information.

Bryant-Webster (join) spaces. This subject comprises Chapter I § 1 (notions), §§ 2, 3, 4, § 5 (notions), and, essentially, § 7. Reading can be continued with Chapter III § 1 (notions) and § 5 (first half) for topological information. Conclude with Chapter IV § 1.

Modular and median spaces. This includes Chapter I §1 (notions), §§ 2, 3, 4 as prerequisites, and centers around §§ 5, 6. Topological information obtains from a selective reading of Chapter III §§ 1, 2. The main body of topological results can be found in III §§ 4, 5.

Topological convexity theory. This requires notions from Chapter I §§ 1, 2, 3. The main body of results is in Chapter III (where § 5 is a little specialized). One can continue with Chapter IV § 3 (continuous selection) or § 4 (dimension theory). Both appeal to a somewhat specialized knowledge in topology.

A generous index and abundant cross-referencing will hopefully be of help to pick up information from parts that have been skipped.

Although this monograph is addressed primarily to the researcher, I have not overlooked the educational aspects of the subject. The underlying concept of betweenness appeals to the intuition, and the theory makes for a natural contact with many branches of mathematics. A well-balanced, one-semester graduate course can be taught, based on (a selection of) the material in Chapter I §§ 1-4 and Chapter II §1. Sections I§7 and II§§2,5 can be used for additional material. I had the opportunity to present such a course at the University of Gent (Belgium) in the early spring of 1992.

Acknowledgements

Having spent seven years of my life to collect, organize, write and administer the material of this monograph, I add the last few lines to it with a feeling of relief. I wish to express my appreciation to J.-P. Doignon, M.A. Maurice, J. van Mill, and E.R. Verheul, who gave detailed comments on parts of the manuscript, and to H.-J. Bandelt and G. Sierksma for valuable suggestions and bibliographic remarks.

A considerable part of my research activities consisted of joined efforts with H.-J.

Bandelt, J. van Mill and E.R. Verheul. Through our cooperation, they deeply influenced my taste of the subject. The cooperation with Eric Verheul deserves special mentioning. During the four years that I directed his Ph.D. research, he rapidly evolved from a student to a colleague and friend. The results obtained in this period made me rewrite parts of the manuscript.

I was a regular speaker at the Amsterdam Topology Seminar, where I had the opportunity to try out parts of this monograph and to report on fresh results (even while the subject wasn't plain topology). The attention and the comments that I received were quite stimulating.

Finally, it is my pleasure to acknowledge the excellent computer facilities and printing equipment of the Department of Mathematics and Computer Science of the *Vrije Universiteit Amsterdam,* making it possible for me to produce high-quality camera-ready copy.

Amsterdam, March 10, 1993 M. van de Vel

>
> As for man, his days are like grass;
> he flourishes like a flower of the field;
> for the wind passes over it, and it is gone,
> and its place knows it no more.
> (Psalm 103:15-16)

Table of Contents

Introduction	vii
Table of Contents	xiii
List of frequent Symbols	xv
Chapter I. Abstract Convex Structures	1
§ 1. Basic Concepts	3
§ 2. The Hull Operator	31
§ 3. Half-spaces and Separation	49
§ 4. Interval Spaces	71
§ 5. Base-point Orders	91
§ 6. Modular Spaces	113
§ 7. Bryant-Webster Spaces	143
Chapter II. Convex Invariants	159
§ 1. Classical Convex Invariants	161
§ 2. Invariants and Product Spaces	181
§ 3. Invariants in other Constructions	205
§ 4. Infinite Combinatorics	219
§ 5. Tverberg Numbers	251
Chapter III. Topological Convex Structures	265
§ 1. Topology and Convexity on the same Set	267
§ 2. Continuity of the Hull Operator	287
§ 3. Uniform Convex Structures	303
§ 4. Topo-convex Separation	325
§ 5. Intrinsic Topology	349
Chapter IV. Miscellaneous	377
§ 1. Embedding Bryant-Webster Spaces into Vector Spaces	379
§ 2. Extremality, Pseudo-boundary and Pseudo-interior	405
§ 3. Continuous Selection	435
§ 4. Dimension Theory	453
§ 5. Dimension and Convex Invariants	469
§ 6. Fixed Points	487
Bibliography	507
Index of Terms	529

List of frequent Symbols

$aff(A)$	affine hull, 34, 151
$b, b(S)$	breadth, 171
$B(p,r)$	open disk at p of radius r, 22
$Bd(A)$	Topological boundary, 267
$c, c(X)$	Carathéodory number, 166
$\mathcal{C}(X)$	convexity of X, 3, 268
Cl	Kuratowski (topological) closure, 4, 267
$cl(A)$	generalized closure, 4
$co(A)$	convex hull, 3
$co^*(A)$	convex closure, 271
$\mathcal{D}_*$	$\mathcal{D} \setminus \{\emptyset\}$, 12
$\mathcal{D}_*$	convex hyperspace of a protopology $\mathcal{D}$, 12
$d, d(X)$	rank, 43, 227
$D(p,r)$	closed disk at p of radius r, 22
$\delta(A)$	pseudo boundary, 411
$\Delta(X)$	cone, 33
$dir, dir(X)$	directional degree, 231
$e, e(X)$	exchange (Sierksma) number, 166
$F_k(X)$	space with k-free convexity, 44
$FS_1,..,FS_4$	functional separation, 331
$\Gamma(S)$	space of all arcs, 318
$gen, gen(X)$	generating degree, 228
$h, h(X)$	Helly number, 166
$I(a,b), ab$	interval operator, 71
$Int(A)$	topological interior, 267
$\iota(A)$	pseudo interior, 411
$L(a), \downarrow(a)$	lower set, 6
$\lambda(n)$	superextension of free n-point set
$\Lambda(S)$	space of maximal arcs, 33
$\lambda(X, \mathcal{D})$	superextension w.r.t $\mathcal{D}$, 12
$\lambda(X)$	superextension w.r.t all (closed/convex) sets, 24, 280
$lin(A)$	linear hull, 34
$M(a,b,c)$	multimedian operator, 113
$m(a,b,c)$	median operator, 8
md	degree of minimal dependence, 222
$med(A)$	median stabilization, 130
$\mathbb{N}$	set of natural numbers; $0 \notin \mathbb{N}$
$NS_2,..,NS_4$	neighborhood separation, 327
p_k	k^{th} partition (Tverberg) number, 251
$pr(A)$	projective hull, 35
$\mathbb{R}$	set of real numbers
$r, r(X)$	Radon number (general), 166
r_n	Radon number (solid n-cube), 197
S^+	set of maximal linked systems containing S, 12
$S_1,..,S_4$	convex separation axioms, 53
$star(A, \mathcal{V})$	star w.r.t a cover, 304
$\mathcal{T}_{comp}(X)$	hyperspace of compact sets, 267
$\mathcal{TC}_{comp}(X)$	hyperspace of compact convex sets, 267
$\mathcal{TC}_*(X)$	hyperspace of convex closed sets, 267
$\mathcal{T}_{fin}(X)$	hyperspace of finite sets, 267
$\mathcal{T}_*(X)$	hyperspace of closed sets, 267
$\mathcal{T}, \mathcal{T}(X)$	topology of closed sets, 268
$U(a), \uparrow(a)$	upper set, 6
X_w	X with weak topology, 278
$\mathbb{Z}$	set of integers
$\#X$	number of elements, 161
$\mathbf{0}$	universal lower bound, origin, 33, 34
$\mathbf{1}$	universal upper bound, 33
2^X_*	collection of all non-empty subsets, 12
2^X_{fin}	collection of all finite subsets, 31
2^X	power set, 4
$\vee, \bigvee$	join in a (semi-)lattice, 6
$\wedge, \bigwedge$	meet in a (semi-)lattice, 6
$X^\wedge_x, \{..,\hat{x},..\}$	delete x, 167
$<A_1,...,A_n>$	hyperspace brackets, 12
$<n>$	$\{1,..,n\}$, 181
$\leq_b$	base-point order, 91
$a/b, A/B$	extension (ray), 54, 71, 145
$\lfloor x \rfloor$	largest integer $\leq x$, 235
$\lceil x \rceil$	smallest integer $\geq x$, 235
$X \multimap Y$	multivalued function, 287
$[a,b]$	order interval, 6
$a \circ b, A \circ B$	join (of points or sets), 145
$L \stackrel{\scriptscriptstyle\sim}{=} M$	perspectivity of lines, 379

CHAPTER I

ABSTRACT CONVEX STRUCTURES

1. Basic Concepts

Convex structures, hull operators, closure spaces, closure operators, bases and subbases, subspaces, products, CP and CC functions, disjoint sums, quotients, convex systems.

2. The Hull Operator

Betweenness, matroids, independence, Join-hull Commutativity, Cone-union Property, affine dimension.

3. Half-spaces and Separation

Half-spaces, screening, $S_1,..,S_4$, Polytope Screening Characterization, embedding in a Cantor cube, normal families.

4. Interval Spaces

Interval spaces, geometric interval spaces, subspaces and products, Pasch and Peano axioms, Characterization of S_3, S_4, and of Join-hull Commutativity.

5. Base-point Orders

Base-point (quasi-) orders, underlying graphs, discrete spaces, graphic spaces, gated sets, amalgamations, complete interval spaces, monotonely complete metric spaces.

6. Modular Spaces

Modularity and semimodularity, multimedian operators, Jordan-Hölder Theorem, characterization of median spaces among modular ones, metric completion of modular metric spaces, Steiner trees.

7. Bryant-Webster Spaces

Decomposability, Straightness and the Ramification Property of intervals, dense interval spaces, join and extension operator, Bryant-Webster spaces and a characterization, affine matroid, affine dimension.

1. Basic Concepts

>A convexity on a set is a family of subsets stable for intersection and for nested union. Each subset is included in a smallest convex set, the convex hull, and the hull of a finite set is called a polytope. Sets of this type determine the convexity. Various classes of examples have been selected for intensive use. Other convexities obtain from certain natural constructions, such as generation by subbases, or formation of subspaces, products, convex hyperspaces, quotients, etc..
>
>The appropriate notion of morphism is the one of convexity preserving (CP) function: convex sets in the range space are inverted to convex sets of the domain. Such functions arise in various constructions of convexities. For spaces derived from an algebraic structure, CP functions usually agree with the corresponding notion of homomorphism.
>
>Many concepts and results in convexity can actually be formulated in terms of convex systems, which will be studied very briefly. In a convex system, some sets do not have a convex hull; convex surfaces are perhaps the most relevant type of example.

1.1. Convex structures. It is typical of an axiomatic approach not to emphasize what an object *represents*, but rather how it *behaves*. In this spirit, we will not discuss the nature of a convex set; instead, we give a description of the properties of convex sets that we regard fundamental.

A family $\mathcal{C}$ of subsets of a set X is called a *convexity* (*alignment*) on X if

(C-1) The empty set $\varnothing$ and the universal set X are in $\mathcal{C}$.
(C-2) $\mathcal{C}$ is stable for intersections, that is, if $\mathcal{D} \subseteq \mathcal{C}$ is non-empty, then $\cap \mathcal{D}$ is in $\mathcal{C}$.
(C-3) $\mathcal{C}$ is stable for nested unions, that is, if $\mathcal{D} \subseteq \mathcal{C}$ is non-empty and totally ordered by inclusion, then $\cup \mathcal{D}$ is in $\mathcal{C}$.

The pair $(X, \mathcal{C})$ is called a *convex structure* (*convexity space, aligned space*, or briefly, a *space*). The members of $\mathcal{C}$ are called *convex sets* and their complements are called *concave sets*. We will frequently omit explicit reference to the name of the convexity, referring to the convexity of a space X by $\mathcal{C}(X)$.

By the axiom (C-1), a subset A of a convex structure X is included in at least one convex set, namely X. In regard to (C-2), A is included in a smallest convex set[1]

$$co(A) = \cap \{ C \mid A \subseteq C \in \mathcal{C} \},$$

the (*convex*) *hull of A*. A set of type $co(F)$, with F finite, is called a *polytope*. Note that the empty set is a polytope by (C-1). In order to discuss the role of the axiom (C-3), we will

---

1. Many authors prefer to use the name of the convexity -- as in "$\mathcal{C}(A)$" -- to describe the convex hull of A. We do not follow this custom since it forces us to fix a name for each convexity in consideration. We occasionally use a somewhat more descriptive name instead of *co*.

take a brief look at the structures obtained by deleting this axiom.

1.2. Closure Structures. Let X be a set and let $\mathcal{C}$ be a family of subsets of X. The pair $(X, \mathcal{C})$ is a *closure structure* (*closure space*) provided the following are true.

(**C-1**) The empty set $\varnothing$ and the universal set X are in $\mathcal{C}$.

(**C-2**) $\mathcal{C}$ is stable for intersections, that is, if $\mathcal{D} \subseteq \mathcal{C}$ is non-empty, then $\cap \mathcal{D}$ is in $\mathcal{C}$.

Members of $\mathcal{C}$ are called *closed sets* and $\mathcal{C}$ is called a *protopology* (*Moore family*) on X. Thus, a convex structure is a closure structure with an additional stability requirement involving nested unions.

A subset A of a closure space X is included in a smallest closed set

$$cl(A) = \cap \{ C \mid A \subseteq C \in \mathcal{C} \},$$

the *closure of* A. Dually, we say that $cl(A)$ is *spanned* by A. Let 2^X denote the *power set* of X, that is, the collection of all subsets of X. The function

$$cl: 2^X \to 2^X, \quad A \mapsto cl(A),$$

is known as the *closure operator of* $(X, \mathcal{C})$. It has the following properties.

(**CL-1**) *Monotone Law:* $A \subseteq B$ implies $cl(A) \subseteq cl(B)$.
(**CL-2**) *Extensive Law:* $A \subseteq cl(A)$.
(**CL-3**) *Idempotent Law:* $cl(cl(A)) = cl(A)$.
(**CL-4**) *Normalization Law:* $cl(\varnothing) = \varnothing$.

Conversely, an operator

$$cl: 2^X \to 2^X$$

with the properties (CL-1) to (CL-4) determines a protopology $\mathcal{C}$ on X consisting of all sets $C \subseteq X$ with $cl(C) = C$ and the closure operator of $\mathcal{C}$ equals cl. The argument is routine. Some authors exclude the normalization axiom (CL-4) from the definition of a closure operator; in this option, the empty set is not required to be closed. As this (minor) additional generality is of little use to us, we will always assume normalization. Note that the *Kuratowski closure* Cl of a topology has a property stronger than monotonicity, namely *additivity*:

$$Cl(A \cup B) = Cl(A) \cup Cl(B).$$

In addition to a convexity $\mathcal{C}$, we sometimes consider a topology $\mathcal{T}$ of closed sets on the same set X. This gives rise to a protopology $\mathcal{C} \cap \mathcal{T}$ and to a corresponding closure operator, which will be studied in Chapter III.

We close our list of definitions concerning set operators with the following one. A closure operator cl on a set X is *domain finite* (or: *algebraic*) provided for each set $A \subseteq X$ and for each point $p \in cl(A)$ there is a finite set $F \subseteq A$ with $p \in cl(F)$.

The first result describes the effect of imposing the third axiom of convexity on a protopology. Recall that a subset D of a partially ordered set is *up-directed* provided for each $d_1, d_2 \in D$ there is a third element $d \in D$ such that $d_1 \leq d$ and $d_2 \leq d$.

§1: Basic Concepts

1.3. Theorem. *The following are equivalent for a closure structure* $(X, \mathcal{C})$.

(1) $\mathcal{C}$ *is a convexity on* X.
(2) *The closure operator of* $\mathcal{C}$ *is domain finite.*
(3) $\mathcal{C}$ *is stable for up-directed union, that is, if* $\mathcal{D} \subseteq \mathcal{C}$ *is an up-directed set then* $\cup \mathcal{D} \in \mathcal{C}$.

Proof. To see that (1) $\Rightarrow$ (2), we verify by transfinite induction that

$$cl(S) = \cup \{ cl(F) \mid F \subseteq S, F \text{ finite} \}$$

for each set $S \subseteq X$. Since the result is valid for finite sets, let us assume that S is infinite and that the formula holds for sets of smaller cardinality than S. Well-order S in such a way that for each $s \in S$ the set $P(s)$ of points $< s$ has a cardinality strictly less than S. The sets $cl(P(s))$ for $s \in S$ form a chain in $\mathcal{C}$ by monotonicity, (CL-1). Its union C is a member of $\mathcal{C}$ by assumption on $\mathcal{D}$ and C includes the set $S = \cup_{s \in S} P(s)$ by extensiveness, (CL-2). Therefore $cl(S) \subseteq D \subseteq C$. As $cl(P(s)) \subseteq cl(S)$ for each $s \in S$ (monotonicity), we conclude that $C = cl(S)$. By the inductive assumption,

$$cl(P(s)) = \cup \{ cl(F) \mid F \subseteq P(s), F \text{ finite} \}.$$

Since each finite set in S is included in some $P(s)$, the assertion holds for S.

(2) $\Rightarrow$ (3). Let $\mathcal{D}$ be an up-directed family of convex sets and let $C = \cup \mathcal{D}$. If $F \subseteq C$ is finite then repeated application of the assumption on $\mathcal{D}$ yields a member D of $\mathcal{D}$ such that $F \subseteq D$. Hence $cl(F) \subseteq D \subseteq C$ and C is convex.

The implication (3) $\Rightarrow$ (1) is obvious. ■

1.4. Segments and arity. The condition of domain finiteness, cf. (2) of Theorem 1.3, illustrates the dominant role of polytopes in convexity theory: a set C is convex iff it includes $co(F)$ for each finite subset F of C. Informally, a convexity is "determined by its polytopes". Proposition 2.2 below gives a more accurate interpretation of this fact.

A polytope which can be spanned by n or less points (where $n \geq 0$) will be referred to as an n-*polytope*. The empty set is a 0-polytope. A 2-polytope $co\{a,b\}$ is alternatively called a *segment joining a and b*. A convex structure (or, its convexity) is *of arity* $\leq n$ provided its convex sets are precisely the sets C with the property that $co(F) \subseteq C$ for each subset F with $\#F \leq n$. Informally, a convexity of arity $\leq n$ is "determined by its n-polytopes". Convexities determined by their segments will be studied in Section 4.

1.5. Examples: first list. Our development of a general theory of convexity is supported by classes of examples, rather than individual *ad hoc* examples. Each class depends on a particular type of mathematical structure. The current list involves vector spaces, ordered spaces, semilattices (in particular, trees), lattices, metric spaces (in particular, normed vector spaces and graphs), and median algebras. Some other classes will be introduced at the proper moment.

1.5.1. Standard convexity. A *totally ordered field* consists of a field $\mathbb{K}$ and a total order $\leq$ on $\mathbb{K}$, such that

(OF-1) $a \leq b$ implies $x + a \leq x + b$;
(OF-2) $a, b \geq 0$ implies $ab \geq 0$.

This yields the familiar computation rules for (strict) inequality as valid in $\mathbb{Q}$ and $\mathbb{R}$. Let V be a vector space over a totally ordered field $\mathbb{K}$. A set $C \subseteq V$ is *convex* provided for all $x, y \in C$ and for each $t \in \mathbb{K}$ with $0 \leq t \leq 1$,
$$t \cdot x + (1-t) \cdot y \in C.$$
This is clearly a convexity of arity 2. The convex hull of a non-empty set A can be described directly as
$$co(A) = \{ \sum_{i=1}^{n} t_i \cdot a_i \mid n > 0; \, a_1, \ldots, a_n \in A ; \, \sum_{i=1}^{n} t_i = 1; \, t_1, \ldots, t_n \geq 0 \}.$$
The proof is routine. The resulting convexity is referred to as the *standard convexity of V*. Two alternative convexities are discussed in 2.7.

1.5.2. Order convexity. Let $(X, \leq)$ be a partially ordered set, briefly, a *poset*. A subset C of X is *order convex* provided $z \in C$ whenever $x \leq z \leq y$ and $x, y \in C$. The resulting *order convexity* is of arity two, as one can see from the hull formula
$$co(A) = \{ x \in X \mid \exists \, a_1, a_2 \in A : a_1 \leq x \leq a_2 \}.$$
Particular examples of convex sets are the closed intervals
$$[a,b] = \{ x \in X \mid a \leq x \leq b \}$$
and the (principal) *lower* and *upper sets* determined by a point $p \in X$:
$$L(p) = \downarrow(p) = (\leftarrow, p] = \{ x \mid x \leq p \};$$
$$U(p) = \uparrow(p) = [p, \rightarrow) = \{ x \mid p \leq x \}.$$
More generally, a lower (resp., upper) set in X is a subset A such that $L(p) \subseteq A$ (resp., $U(p) \subseteq A$) whenever $p \in A$. Such sets are order convex. As a matter of fact, the collection of all lower (upper) subsets of a poset X is another convexity, known as the *lower (upper) convexity of X*.

1.5.3. Convexity in a semilattice. The infimum (*meet*), resp., supremum (*join*), of two points x and y in a poset is denoted by $x \wedge y$, resp., $x \vee y$. A poset in which every pair of points has an infimum is a *meet semilattice*. Formally, this is a pair $(S, \wedge)$, where the operator
$$\wedge : S \times S \to S, \quad (x,y) \mapsto x \wedge y,$$
satisfies the idempotent law $x \wedge x = x$, the commutative law $x \wedge y = y \wedge x$, and the associative law $x \wedge (y \wedge z) = (x \wedge y) \wedge z$. A partial order is obtained by the prescription $x \leq y$ iff $x \wedge y = x$, and $x \wedge y$ is the infimum of x and y. The dual concept is a *join semilattice*; it is required that each pair of points has a supremum. A subset of a meet semilattice S is convex provided it is an order convex subsemilattice. It is easily seen that this *semilattice convexity* is of arity two. For a non-empty finite set $F \subseteq S$ the following hull formula is valid.

§1: Basic Concepts 7

$$co(F) = \{x \in S \mid \exists a \in F: \inf(F) \le x \le a\}.$$

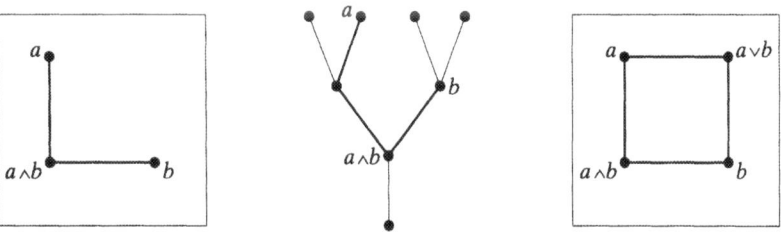

Fig. 1: (Semi)lattice convexity in a square and tree convexity: hull of a, b

An *(order) tree* is a meet semilattice such that the lower set of each point is totally ordered. The common definitions of a tree in graph theory or in topology are, in fact, instances of the above one. A *branch* emanating at a point p of a tree is a convex set C which is maximal with respect to the property that $p \notin C$. Such sets exist by Zorn's Lemma in combination with Theorem 1.3. By a *ramification point* is meant a point at which three or more branches emanate. Fig. 1 (middle) displays a typical segment of a tree.

All definitions can easily be adapted in case of join semilattices. We note that semilattices can be convexified in several other ways: see 1.23.3 for an example. Unless stated to the contrary, the above convexity is considered.

1.5.4. Convexity in a lattice. A poset which is at the same time a meet and a join semilattice is called a *lattice*. Formally, a lattice is a triple $(L, \wedge, \vee)$, where the operators

$$\wedge, \vee: L \times L \to L$$

both satisfy the idempotent, commutative and associative laws, in addition to the following absorptive laws:

$$x \wedge (x \vee y) = x; \quad x \vee (x \wedge y) = x.$$

A partial order on L, for which $x \wedge y$ and $x \vee y$ are the meet and the join, respectively, of x and y, is obtained by each of the prescriptions

$$x \le y \quad \text{iff} \quad x \wedge y = x; \quad x \le y \quad \text{iff} \quad x \vee y = y.$$

A subset of a lattice L is convex provided it is an order convex sublattice. For a nonempty finite set $F \subseteq L$ it follows that

$$co(F) = \{x \in L \mid \inf(F) \le x \le \sup(F)\}.$$

The resulting *lattice convexity* has the exceptional property that all polytopes are segments. It is of arity 2 (the previous observation is of no help to prove this).

As an important special case, a *cube* is a set of type

$$Q = \prod_{i \in I} L_i,$$

where I is an index set and each set L_i is totally ordered. If $\#I = n$ and $\#L_i > 1$ for each $i \in I$, then Q is an n *cube*. The *Cartesian order* of a product of posets $(X_i)_{i \in I}$ is defined as follows. If $u = (u_i)_{i \in I}$ and $v = (v_i)_{i \in I}$, where $u_i, v_i \in X_i$ for all $i \in I$, then

$$u \leq v \Leftrightarrow \forall i \in I: u_i \leq v_i.$$

This order produces a lattice structure on Q with the following meet and join operations.

$$x \wedge y = (\min\{x_i, y_i\})_{i \in I}; \quad x \vee y = (\max\{x_i, y_i\})_{i \in I}.$$

Fig. 1 displays a segment in the square $[0,1]^2$, equipped with a convexity of order convex subsemilattices (left) and with a convexity of order convex sublattices (right). There are many other ways to convexify a lattice; the one described here is regarded to be canonical.

1.5.5. Geodesic convexity. Let (X,d) be a metric space. The *triangle inequality* states that

$$\forall u, v, w \in X: d(u,w) \leq d(u,v) + d(v,w).$$

One of the best known concepts of general convexity is based on the following interpretation of equality in this expression. A point $x \in X$ is *(geodesically) between* the points $a, b \in X$ provided

$$d(a,x) + d(x,b) = d(a,b).$$

A set $C \subseteq X$ is *geodesically convex* provided each point in between two points of C is in C. The resulting convexity on X is known as the *geodesic convexity* of (X,d). It is clearly of arity two. The following types are of particular interest.

(i) X is a *normed space* and d is derived from the norm.
(ii) X is a *connected graph*[2] and d is the intrinsic metric of X.

In a connected graph a point is geodesically between two given points if it is on a geodesic (shortest path) joining these points. This may motivate the use of the term "geodesic convexity". In Section 4 we shall introduce the class of geometric interval spaces which properly generalizes the class of metric spaces.

1.5.6. Convexity in a median algebra. We now focus on a remarkable type of averaging operator on three arguments. A *median operator* on a set X is a function

$$m: X^3 \to X$$

satisfying the following properties.

(**M-1**)   *Absorption (Majority) Law:* $m(a,a,b) = a$.
(**M-2**)   *Symmetry Law:* if σ is any permutation of a, b, c, then

$$m(\sigma(a), \sigma(b), \sigma(c)) = m(a,b,c).$$

(**M-3**)   *Transitive Law:* $m(m(a,b,c),d,c) = m(a,m(b,c,d),c)$.

Assuming (M-2), the Transitive Law (M-3) expresses that the binary operator

$$(a,b) \mapsto a *_c b = m(a,b,c)$$

is associative for each $c \in X$. The point $m(a,b,c)$ is called the *median of* a, b, c, and the resulting pair (X,m) is a *median algebra*. The simplest type of example obtains from a

---

2.   We consider a *graph* as a set with a reflexive, symmetric relation describing the edges.

§1: Basic Concepts   9

totally ordered set, where the median of three points is the middle one. In a distributive lattice, the operator

$$m(a,b,c) = (a \wedge b) \vee (b \wedge c) \vee (c \wedge a)$$

defines a median. Note that by distributivity the right hand side is not altered if the operations $\wedge$ and $\vee$ are exchanged. Trees form a different class of examples. If a, b, c are elements of a tree then $a \wedge b$, $b \wedge c$, $c \wedge a$ are pairwise bounded from above and hence are comparable (in fact, at least two of them are equal). The median $m(a,b,c)$ is defined to be the largest element of this chain. See 1.22.

A subset C of a median algebra (X,m) is *convex* provided

$$m(C \times C \times X) \subseteq C.$$

Compare with the definition of an ideal in a ring. The resulting **median convexity** is easily seen to be of arity two. It will always be assumed on a median algebra. For a tree or a distributive lattice there is no conflict since, as we shall see later, the canonical proposals of a convexity agree. See Topic 1.22 or Section 6.

Other examples require more notions from convexity theory. We first describe a generating process involving the following characterization of polytopes.

1.6. Theorem. *The following are equivalent for a convex set C.*

(1)   *C is a polytope.*
(2)   *C is not the union of a non-empty up-directed family of proper convex subsets.*
(3)   *C is not the union of a non-empty chain of proper convex subsets.*

Proof. The implications (1) $\Rightarrow$ (2) $\Rightarrow$ (3) are clear. We verify that (3) $\Rightarrow$ (1). Let C be a convex set which is not a polytope. Let $\mathcal{S}$ be the family of all convex sets $D \subseteq C$ such that $C \neq co(D \cup F)$ for each finite set $F \subseteq C$. As C is not a polytope we have $\emptyset \in \mathcal{S}$. Let $\mathcal{M} \subseteq \mathcal{S}$ be a maximal chain (so $\mathcal{M} \neq \emptyset$), and consider the convex set $M_0 = \cup \mathcal{M}$. If $M_0 = C$, then we are done. So assume $M_0 \neq C$. If $M_0 \in \mathcal{S}$ then fix a point $p \in C \setminus M_0$ and let F be any finite set in C. We have $co(M_0 \cup \{p\} \cup F) \neq C$, and hence $co(M_0 \cup \{p\})$ is also in $\mathcal{S}$. But this allows us to enlarge the chain $\mathcal{M}$ in $\mathcal{S}$. So M_0 is not in $\mathcal{S}$, meaning that there is a finite subset F_0 of C such that $co(M_0 \cup F_0) = C$. Consider the following chain in $\mathcal{S}$, consisting of proper convex subsets of C.

$$\mathcal{N} = \{ co(M \cup F_0) \mid M \in \mathcal{M} \}.$$

Now $\cup \mathcal{N} = C$ since by (3), $\cup \mathcal{N}$ is a convex set including

$$\cup_{M \in \mathcal{M}} (M \cup F_0) = M_0 \cup F_0. \blacksquare$$

1.7. Base and subbase. A set can carry several convexities. For instance, a lattice L can be convexified as in 1.5.4, but one may also consider it as a meet- or a join semilattice and convexify it as in 1.5.3. Or, one may consider L as an ordered set and use 1.5.2. Yet another alternative is to take the collection of all sublattices as a convexity on L. A rough classification of convexities on a set X goes as follows. If $\mathcal{C}_1 \subseteq \mathcal{C}_2$, then $\mathcal{C}_1$ is *coarser than* C_2 and C_2 is *finer than* C_1. The finest of all convexities is the power set 2^X of X: the *free convexity*. The coarsest of all is $\{\emptyset, X\}$: the *coarse convexity*. The terms "free" and "coarse" will also be used with reference to the convex structures involved. If $(C_i)_{i \in I}$ is a family of convexities on X, then $\cap_{i \in I} C_i$ is also a convexity on X, commonly coarser than all C_i.

A collection $\mathcal{S}$ of sets in X is a *subbase of a convex structure* (X, C) (a *convex subbase*) provided $\mathcal{S} \subseteq C$ and C is the coarsest among all convexities that include $\mathcal{S}$. Alternatively, we say that $\mathcal{S}$ *generates the convexity* C. Every collection $\mathcal{S} \subseteq 2^X$ generates a convexity on X, namely the intersection of all convexities that include $\mathcal{S}$. Let us inspect this process of generation more closely.

A collection $\mathcal{B}$ of sets in X is a *base of a convex structure* (X, C) (a *convex base*) provided $\mathcal{B} \subseteq C$ and each member of C is the union of an up-directed subcollection of $\mathcal{B}$. In this situation, $\mathcal{B}$ is said to *generate the convexity* C. Note that each base is a subbase. In regard to Theorem 1.6(2) we conclude to the following.

1.7.1. Proposition. *Let $\mathcal{B} \subseteq C$, where C is a convexity on X. Then $\mathcal{B}$ is a base for (X, C) iff it contains all C-polytopes.* ∎

Let $(X, \mathcal{S})$ be a closure space. Then $\mathcal{S}$ is stable for intersections, and it contains the empty set and X. By adjoining up-directed unions, these properties persist. Indeed, let $C_i = \cup \mathcal{S}_i$ where, for each $i \in I$, the family $\mathcal{S}_i \subseteq \mathcal{S}$ is up-directed, and let $C = \cap_{i \in I} C_i$. Consider the family $\mathcal{S}_0$ of all sets of type $\cap_{i \in I} S(i)$, where

$$S: I \to \bigcup_{i \in I} \mathcal{S}_i$$

is a choice function (i.e., $S(i) \in \mathcal{S}_i$ for all $i \in I$). Then $\mathcal{S}_0$ is up-directed and $C = \cup \mathcal{S}_0$.

So we obtain a convexity C with $\mathcal{S}$ as a base. For $F \subseteq X$ finite we have $co(F) \subseteq cl(F)$ since $C \supseteq \mathcal{S}$. But $co(F)$ is in $\mathcal{S}$ by 1.7.1, yielding the following result.

1.7.2. Proposition. *If $(X, \mathcal{S})$ is a closure space then $\mathcal{S}$ generates a convexity on X with $cl(F) = co(F)$ for each finite set $F \subseteq X$.* ∎

In combinatorial convexity theory (as developed e.g., in Chapter II) there is a tendency to work on the level of closure structures rather than convex structures. Most results in the area relate with the behavior of polytopes only, and the axiom (C-3) seems superfluous (note that protopologies on *finite* sets satisfy (C-3) automatically). The previous result indicates that this is fake generality.

After closing a subbase $\mathcal{S}$ for intersections (and adding the empty set if it isn't there yet), we obtain a protopology which operates as a base for a convexity. This leads to the following useful description of generation by a subbase.

1.7.3. Proposition. *Let C be a convexity on X. Then $\mathcal{S} \subseteq C$ is a subbase for (X, C) iff each non-empty C-polytope is the intersection of a subfamily of $\mathcal{S}$.* ∎

In the situation of 1.7.3, the intermediate protopology $\mathcal{S}$ of X is said to be *generated*

§1: Basic Concepts

by a subbase $\mathcal{S}$, *and* $\mathcal{S}$ *is a subbase of the closure space* $(X, \mathcal{S})$.

1.8. Examples: second list. Natural subbases abound in the convexities that we met so far. The order convexity of a totally ordered set admits a subbase, consisting of the principal lower and upper sets. As to the usual convexity of a meet semilattice, the principal upper sets and their complements constitute a subbase. For a somewhat geometric example, the reader may check that the standard convexity of Euclidean space is generated by the subbase, consisting of all closed disks (Euclidean metric). We use generation by subbases to produce some new classes of examples: H-convexity, convex hyperspaces, and superextensions.

1.8.1. H-Convexity. Let V be a vector space over a totally ordered field $\mathbb{K}$ and let $\mathcal{F}$ be a collection of linear functionals $V \to \mathbb{K}$. Then the family

$$\mathcal{S} = \{ f^{-1}(\leftarrow, t] \mid t \in \mathbb{K}, f \in \mathcal{F} \}$$

generates a convexity $\mathcal{C}$ on V, coarser than the standard one. It is called an *H-convexity*.[3] By abuse of language, we also say that $\mathcal{F}$ *generates* $\mathcal{C}$. For an important special case, let $\mathcal{F}$ be *symmetric*, that is: $\mathcal{F}$ contains $-f$ whenever it contains f. Then $\mathcal{S}$ also contains all sets of type $f^{-1}[t, \to)$, with $t \in \mathbb{K}$ and $f \in \mathcal{F}$, and $\mathcal{C}$ is called a *symmetric H-convexity*. We usually

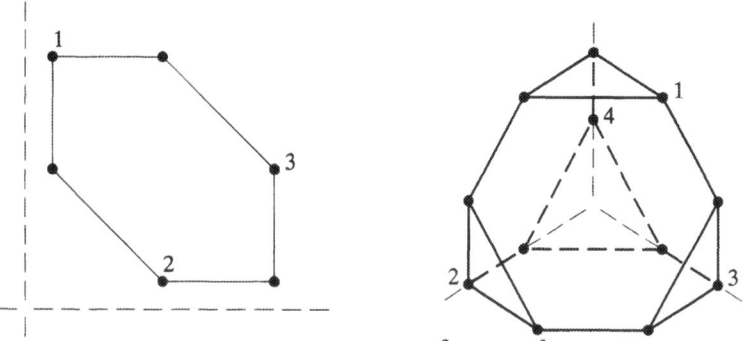

Fig. 2: H-convexity in $\mathbb{R}^2$ and in $\mathbb{R}^3$

omit one of f, $-f$, and say that $\mathcal{F}$ *symmetrically generates* the convexity $\mathcal{C}$. Fig. 2 displays a typical polytope of $\mathbb{R}^2$ and of $\mathbb{R}^3$ respectively, relative to the H-convexity which is symmetrically generated by the coordinate projections and their sum. In each situation, the numbered vertices form a spanning set.

1.8.2. Convex hyperspaces. Let X be a set, let $n \in \mathbb{N}$, and let $A_1, \ldots, A_n \subseteq X$. We use the following notation throughout.

---

3.   In the literature, the term is mostly used with reference to the *closure space* produced by the subbase $\mathcal{S}$.

$<A_1,..,A_n> = \{B \in 2^X \mid B \subseteq \cup_{i=1}^n A_i; B \cap A_i \neq \varnothing \text{ for } i=1,..,n\}$.

Such subsets of $2^X \setminus \{\varnothing\}$ occur in the next construction. Let $(X,\mathcal{S})$ be a closure structure. Then $\mathcal{S}$ is a base for a convexity $\mathcal{C}$ on X. As a matter of permanent notation, we let $\mathcal{S}_*$ denote $\mathcal{S} \setminus \{\varnothing\}$. In the same style, $2^X_* = 2^X \setminus \varnothing$. The sets of type

$<D> \cap \mathcal{S}_* = \{C \mid C \in \mathcal{S}_*; C \subseteq D; C \neq \varnothing\}$;
$<D,X> \cap \mathcal{S}_* = \{C \mid C \in \mathcal{S}_*; C \cap D \neq \varnothing\}$,

for $D \in \mathcal{S}$, generate the **Vietoris convexity** on the set $\mathcal{S}_*$, and the resulting convex structure is called a **convex hyperspace**. Throughout, cl denotes the closure operator of $\mathcal{S}_*$, and co denotes the convex hull operator of $\mathcal{C}$ and of the Vietoris convexity on $\mathcal{S}_*$ (the argument decides which one is meant). Let $C_1,..,C_n \in \mathcal{S}_*$. By 1.7.3, the polytope $co\{C_1,..,C_n\}$ is an intersection of subbasic sets, which fall into two classes. Of all subbasic sets of type $<D> \cap \mathcal{S}_*$ there is a smallest one, produced by

$D = cl(C_1 \cup .. \cup C_n)$.

Among the subbasic sets of type $<D,X> \cap \mathcal{S}_*$, we only need to consider the minimal ones, arising from sets of type

$D = co\{c_1,..,c_n\}$,

where $c_i \in C_i$ for $i=1,..,n$. Hence:

1.8.3. Hyperspace polytope formula. *If $C_1,..,C_n \in \mathcal{S}_*$, then $co\{C_1,..,C_n\}$ is the collection of all $C \in \mathcal{S}_*$ such that*

(i) $C \subseteq cl(\cup_{i=1}^n C_i)$;
(ii) *If $c_i \in C_i$ for $i=1,..,n$, then C meets $co\{c_1,..,c_n\}$.*   ∎

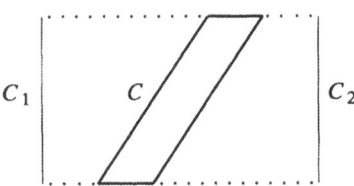

Fig. 3: Hyperspace convexity: $C \in co\{C_1,C_2\}$

In later chapters, $\mathcal{S}$ will be taken as the collection of all convex sets which are closed in an additional topology on X. Then $\mathcal{S}_*$ is a subcollection of the topological hyperspace of X. For the time being, one may consider the simpler case where $\mathcal{S} = \mathcal{C}$.

1.8.4. Superextensions. Let X be a set and let $\mathcal{S} \subseteq 2^X$. By a *linked system of sets* in $\mathcal{S}$ is meant a subcollection of $\mathcal{S}$ consisting of pairwise intersecting sets. Observe that by Zorn's Lemma each linked system in $\mathcal{S}$ extends to a maximal one. We let $\lambda(X,\mathcal{S})$ denote the set of all maximal linked systems (*mls's*) in $\mathcal{S}$. If $A \subseteq X$ then we put

$A^+ = \{m \in \lambda(X,\mathcal{S}) \mid A \text{ meets all members of } m\}$.

Note that if $S \in \mathcal{S}$, then S^+ consists of all $m \in \lambda(X,\mathcal{S})$ with $S \in m$. The collection

§1: Basic Concepts

$$\mathcal{S}^+ = \{ S^+ \mid S \in \mathcal{S} \}$$

generates a convexity on $\lambda(X, \mathcal{S})$. The resulting convex structure is called the *superextension of X relative to* $\mathcal{S}$. The family $\mathcal{S}$ is said to be T_1 provided

(i) for each $S \in \mathcal{S}$ and $x \in X \setminus S$ there is an $S' \in \mathcal{S}$ with $x \in S'$ and $S' \cap S = \emptyset$.
(ii) Each singleton in X is the intersection of a subfamily of $\mathcal{S}$.

The first statement is equivalent to the property that each collection of type

$$l(x) = \{ S \mid x \in S \in \mathcal{S} \}$$

is an mls in $\mathcal{S}$. The family $\mathcal{S}$ is T_1 iff the resulting function

$$l: X \to \lambda(X, \mathcal{S})$$

is injective. Note that if $\mathcal{S}$ is a topology on X, then "T_1" has the usual topological meaning. Here are a few useful results on superextensions.

1.8.5. Proposition. *Let $\mathcal{S} \subseteq 2^X$ and $S, S_1, S_2 \in \mathcal{S}$. Then the following are true.*

(1) $(X \setminus S)^+ = \lambda(X, \mathcal{S}) \setminus S^+$.
(2) $S_1 \cap S_2 = \emptyset \Leftrightarrow S_1^+ \cap S_2^+ = \emptyset$.
(3) *(under assumption of T_1)* $S_1 \cup S_2 = X \Leftrightarrow S_1^+ \cup S_2^+ = \lambda(X, \mathcal{S})$.
(4) *(under assumption of T_1)* $S_1 \subseteq S_2 \Leftrightarrow S_1^+ \subseteq S_2^+$.
(5) *(under assumption of T_1)* $l^{-1}(S^+) = S$.
(6) *If $F \subseteq \lambda(X, \mathcal{S})$ is a non-empty finite set, then $co(F)$ is the intersection of all sets S^+, where S is a common member of all mls's in F.* ∎

If $\mathcal{S}$ is an infinite family, a study of $\lambda(X, \mathcal{S})$ inevitably involves topological considerations (viz., the topology generated by the same subbase used to generate the convexity). Superextensions are instrumental in obtaining a counterpart in convexity of compactification in topology. See III§4. In this and the next chapter we will largely restrict to "combinatorial" superextensions, where X is finite.

1.9. Subspaces. Let $(X, \mathcal{C})$ be a convex structure and let Y be a subset of X. The family of sets

$$\mathcal{C} \mid Y = \{ C \cap Y \mid C \in \mathcal{C} \}$$

is easily seen to be a convexity on Y; it is called the *relative convexity of Y*. The resulting convex structure $(Y, \mathcal{C} \mid Y)$ is a *subspace* (*substructure*) of $(X, \mathcal{C})$, and the latter is an *extension* (or, a *superspace*) of $(Y, \mathcal{C} \mid Y)$. The members of $\mathcal{C} \mid Y$ are *relatively convex sets*. As a straightforward consequence of the definition we have:

1.9.1. Relative hull formula. *The hull operator co_Y of a subspace Y of X satisfies*

$$\forall A \subseteq Y: co_Y(A) = co(A) \cap X.$$ ∎

For instance, consider a semilattice S and a subsemilattice T of S. Relative to the convexity of S and of T described in 1.5.3, the latter is a subspace of the former. Similarly, if M is a sublattice of a lattice L, then M is a subspace of L with respect to the convexity in 1.5.4. For considerations concerning order convexity or geodesic convexity, see 4.19 and 4.20. For median algebras, consult 6.33.

1.10. Products and Joins. Let (X_i, C_i) for $i \in I$ be a family of convex structures, let X be the product of the sets X_i for $i \in I$, and let

$$\pi_i: X \to X_i$$

denote the ith projection. The *product convexity* C of X is the one generated by the subbase

$$\{\pi_i^{-1}(C_i) \mid C_i \in C_i; i \in I\}.$$

The resulting convex structure (X, C) is called the *product* of the spaces (X_i, C_i) for $i \in I$, and is denoted by

$$\prod_{i \in I} (X_i, C_i), \quad \text{or, if } I = \{1,..,n\}, \quad (X_1, C_1) \times .. \times (X_n, C_n).$$

For a better understanding of products, the following general concept is of use. Let $(C_i)_{i \in I}$ be a family of convexities on a set X. Its *join* is defined to be the convexity generated by $\cup_{i \in I} C_i$. For instance, the order convexity of a poset is the join of the lower and the upper convexity. The following fact is of importance.

1.10.1. Proposition. *The join of finitely many convexities C_i for $i=1,..,n$ on a set consists of all sets of type $\cap_{i=1}^n C_i$, where $C_i \in C_i$ for $i=1,..,n$.*

Proof. It suffices to show that these sets are stable under up-directed unions. Let $\mathcal{A}$ be an up-directed family of sets of type $A = \cap_{i=1}^n A_i$, where $A_i \in C_i$ for $i=1,..,n$. Let co_i be the hull operator of the ith convexity. Then $co_i(A) \subseteq A_i$ and hence $A = \cap_{i=1}^n co_i(A)$. For each $i=1,..,n$, the collection

$$\{co_i(A) \mid A \in \mathcal{A}\}$$

is up-directed and hence its union is in C_i. General considerations yield that

$$\cup \mathcal{A} \subseteq \bigcap_{i=1}^n \left(\bigcup_{A \in \mathcal{A}} co_i(A) \right).$$

Equality obtains since the index set is finite and the family $(co_i(A))_{A \in \mathcal{A}}$, occurring at the right, is up-directed for each i. Hence $\cup \mathcal{A}$ is convex. ∎

The convexity of a product space $\prod_{i \in I} (X_i, C_i)$ is the join of the convexities

$$\pi_i^{-1}(C_i) = \{\pi_i^{-1}(C) \mid C \in C_i\}$$

for $i \in I$. We conclude to the following.

1.10.2. Proposition. *In a product of finitely many convex structures (X_i, C_i) for $i=1,..,n$, all convex sets are of the product type*

$$C_1 \times C_2 \times .. \times C_n,$$

with $C_i \in C_i$ for $i=1,..,n$. ∎

If infinitely many factors are not coarse then the product space *must* include a convex non-product set, as the reader can easily verify. However, as each polytope is the

§1: Basic Concepts

intersection of subbasic sets, we find (with the above notation):

1.10.3. Product polytope formula. *Each polytope of a product space is a product set, and hence for each finite subset F of the product,*
$$co(F) = \prod_{i \in I} co(\pi_i F).$$
∎

For a different application of joins, consider an H-convexity $\mathcal{C}$ which is generated by a collection $\mathcal{F}$ of linear functionals. Then $\mathcal{C}$ is the join of convexities of type
$$\mathcal{C}(f) = \{ f^{-1}(S) \mid S \subseteq \mathbb{K} \text{ is a lower set} \},$$
where $\mathbb{K}$ is the totally ordered field involved, and $f \in \mathcal{F}$. If $\mathcal{F}$ is finite this leads to a simple description of the members of $\mathcal{C}$.

1.11. Types of morphisms. Let $f: X_1 \to X_2$ be a function between convex structures X_1 and X_2. Then f is said to be:

(i) a *convexity preserving function* (a *CP function*) provided for each convex set C in X_2, the set $f^{-1}(C)$ is convex in X_1.

(ii) a *convex-to-convex function* (a *CC function*) provided for each convex set C in X_1, the set $f(C)$ is convex in X_2.

Informally, a CP function *inverts* convex sets into convex sets, whereas a CC function *maps* convex sets to convex sets. The function f is an *isomorphism* (explicitly, a *CP isomorphism*) if it is a bijection and if both f and f^{-1} are CP. Equivalently, f is a CP and CC bijection. Finally, f is an *embedding of X_1 into X_2* if the restriction of f,
$$X_1 \to f(X_1),$$
(where the range set is furnished with the relative convexity) is an isomorphism. Composition of CP functions is CP and, plainly, there is a category consisting of all convex structures and of all CP functions. Here are a few simple facts.

1.12. Proposition. *Let X_1, X_2 be convex structures, let $\mathcal{S}$ be a subbase for X_2, and let $f: X_1 \to X_2$ be a function.*

(1) *f is CP iff $f(co(F)) \subseteq co(f(F))$ for each finite set $F \subseteq X_1$. If X_1 is of arity $\leq n$, then it suffices to consider sets F with $\#F \leq n$.*

(2) *f is CC iff $f(co(F)) \supseteq co(f(F))$ for each finite set $F \subseteq X_1$. If X_2 is of arity $\leq n$, then it suffices to consider sets F with $\#F \leq n$.*

(3) *f is CP iff $f^{-1}(S)$ is convex for each $S \in \mathcal{S}$.*

(4) *f is an embedding iff it is injective and $f^{-1}(\mathcal{S})$ is a subbase of X_1.*

(5) *If f is CP, CC and surjective, then the arity of X_2 is at most the arity of X_1.* ∎

1.13. Examples. Projection of a product onto its factors, and inclusion of a subspace in a superspace are both convexity preserving functions. The first one is convex-to-convex, as one can see from the polytope formula 1.10.3 and from Proposition 1.12(2). Another example obtains from the canonical injection in a superextension,

$$l : X \to \lambda(X, \delta),$$

defined for a T_1 family δ on X. In addition to the convexity of $\lambda(X, \delta)$, we consider the convexity on X generated by δ. Then l is an embedding by 1.8.5(5) and 1.12(4). Usually, $\lambda(X, \delta)$ is so much larger than X that the pleonasm "super(space)–extension" seems justified. See, for instance, Table II§4.2 following II§4.24.

Convexities are often derived from some mathematical structure: vector spaces, posets, (semi)lattices, median algebras, etc.. These structures have their own type of morphisms. Invariably, they lead to CP functions relative to the derived convexities as the list below may illustrate.

1.13.1. Order preserving functions. Let X and Y be posets, and let $f : X \to Y$. Then f is *order preserving* (*OP*) provided $x \leq x'$ implies $f(x) \leq f(x')$ for all $x, x' \in X$. It is clear that such functions are CP relative to the order convexities of domain and range. Since order convexity is unaffected by reversal of the order, CP functions of posets exist which are not OP. In general, an OP function need not be CC. However, a *surjective* OP function between *totally* ordered sets is CC, as one can easily verify.

1.13.2. Affine functions. Let V and W be vector spaces over a totally ordered field $\mathbb{K}$. A function $f : V \to W$ is *affine* provided that

$$f(t_1 \cdot v_1 + t_2 \cdot v_2) = t_1 \cdot f(v_1) + t_2 \cdot f(v_2) \quad (v_1, v_2 \in V; t_1, t_2 \in \mathbb{R}).$$

Equivalently, f is composed of a linear function $V \to W$ and a translation $W \to W$. Relative to the standard convexity of V and W, such a function satisfies the hypotheses in 1.12(1) and 1.12(2). Consequently it is both convexity preserving and convex-to-convex. The converse is not true. For instance, the composition of a linear functional with an order preserving map of $\mathbb{K}$ is CP but not necessarily CC; cf. 1.13.1. For related results and examples, consult 1.26 or 3.29.1.

1.13.3. Homomorphisms of semilattices. A function $f : S \to T$ of meet semilattices S and T is a (*semilattice*) *homomorphism* provided

$$f(x \wedge y) = f(x) \wedge f(y) \quad (x, y \in S).$$

Clearly, a homomorphism is order-preserving, and hence it is CP. Conversely, a CP function $f : S \to T$ of semilattices, which is order preserving, is a semilattice homomorphism. Indeed, if $a, b \in S$ then $f(a \wedge b) \leq f(a) \wedge f(b)$ since f is order preserving. As $a \wedge b \in co\{a, b\}$ and as f is CP, we find that $f(a \wedge b) \in co\{f(a), f(b)\}$, showing that $f(a) \wedge f(b) \leq f(a \wedge b)$.

1.13.4. Homomorphisms of lattices. A function $f : L \to M$ of lattices L and M is a (*lattice*) *homomorphism* provided

$$f(x \wedge y) = f(x) \wedge f(y), \quad f(x \vee y) = f(x) \vee f(y), \quad (x, y \in L);$$

it is an *anti-homomorphism* if the above formulas hold with the left hand "∧" and "∨" interchanged. Formally, anti-homomorphisms arise from inverting the order in the image

§1: Basic Concepts

lattice (this yields the so-called *opposite lattice*). Both a homomorphism and an anti-homomorphism are CP functions. The converse is false: consider the mapping $[0,1]^2 \to [0,1]^2$ which replaces the first coordinate x by $1-x$.

1.13.5. Homomorphisms of median algebras. Let (X, m_X) and (Y, m_Y) be median algebras. A function $f: X \to Y$ with the property

$$f(m_X(a,b,c)) = m_Y(f(a), f(b), f(c)) \quad (a, b, c \in X)$$

is called a *median preserving function* or an *MP function* for short. Alternatively, we say that f is a *homomorphism of median algebras*. It is easy to see that homomorphisms are CP. It will be shown in 6.11 that the converse is valid too and that surjective CP functions are CC.

1.13.6. Isometries. Let (X,d) and (Y, ρ) be metric spaces, and let $f: (X,d) \to (Y, \rho)$ be an isometry. Then f is a CP isomorphism of the corresponding geodesic convex structures.

1.13.7. Translations and homotheties. The following types of functions are CP isomorphisms of an H-convexity on a vector space V.

(i) Translations $V \to V$, i.e., functions of type $x \mapsto x + v$ for a fixed vector $v \in V$;
(ii) Positive homotheties $V \to V$.

A *homothety with center* $p \in V$ and *with coefficient* $c \in \mathbb{R} \setminus \{0\}$ is a function of type $x \mapsto c \cdot (x-p) + p$. The homothety is *positive* provided $c > 0$ and *negative* if $c < 0$. In case of symmetric H-convexity, a negative homothety is an isomorphism too.

> The remainder of this section includes a description of somewhat specialized concepts; it can be safely skipped at first reading.

1.14. Disjoint sums. Let $(X_i, \mathcal{C}_i)_{i \in I}$ be a family of convex structures. The disjoint sum X of the underlying sets $(X_i)_{i \in I}$ is defined and denoted as follows.

$$X = \sum_{i \in I} X_i = \bigcup_{i \in I} \{i\} \times X_i.$$

We consider the *(disjoint) sum convexity* $\mathcal{C}$ on X, consisting of all sets of type

$$\sum_{i \in I} C_i \quad (C_i \in \mathcal{C}_i, \, i \in I).$$

The resulting convex structure $(X, \mathcal{C})$ is called the *(disjoint) sum of* $(X_i, \mathcal{C}_i)_{i \in I}$. We extend this terminology to spaces isomorphic to the disjoint sum. For each i the function

$$X_i \to X, \quad x \mapsto (i, x),$$

is an embedding of the ith summand in the sum space.

1.15. Quotients. Let $(X, \mathcal{C})$ be a convex structure and let R be an equivalence relation on X. The quotient set X/R consists of all R-equivalence classes, and the quotient function $q: X \to X/R$ assigns to a point of X its R-equivalence class. A convexity $\mathcal{C}/R$ obtains on X/R as follows.

$$\mathcal{C}/R = \{ C \subseteq X/R \mid q^{-1}(C) \in \mathcal{C}(X) \}.$$

In words: the convex sets of X/R are the images of R-saturated convex sets of X. The resulting convex structure $(X/R, \mathcal{C}/R)$ is called a *quotient space of X;* the convexity $\mathcal{C}/R$ is called a *quotient convexity.* We extend this terminology to any space which is isomorphic to the genuine quotient.

A CP and CC surjection is a quotient function (up to isomorphism). A quotient map is convex-to-convex provided the saturation of each convex set in the domain is convex.

1.16. Examples. Homomorphic images of first-order structures tend to behave like quotients, and this observation largely extends to the corresponding convexity. Here are some illustrations.

1.16.1. Quotients of vector spaces. In the algebraic theory of vector spaces, quotients correspond with linear surjections. As we saw in 1.13.2, such functions are CP and CC with respect to the standard convexity, and hence they induce quotients of convex structures as well.

1.16.2. Quotients of semilattices. Let $f: S \to T$ be a surjective homomorphism of semilattices. Then f is CP by 1.13.3. To see that f induces a quotient of convex structures, let $C \subseteq T$ be such that $f^{-1}(C)$ is an order convex subsemilattice of S. Let $a, b \in C$ and take $a' \in f^{-1}(a)$, $b' \in f^{-1}(b)$. Then $a' \wedge b' \in f^{-1}(C)$ and $a \wedge b = f(a' \wedge b') \in C$, showing that C is a subsemilattice. If a, b are in C and $a \leq x \leq b$, then choose $b' \in f^{-1}(b)$, and let $x' \in f^{-1}(x)$ satisfy $x' \leq b'$ (one can replace x' by $x' \wedge b'$). Similarly, there is a point $a' \in f^{-1}(a)$ with $a' \leq x'$. As $f^{-1}(C)$ is order convex, we find that $x \in C$.

1.16.3. Quotients of lattices. It follows immediately from the previous argument that a surjective lattice homomorphism gives rise to a quotient of the corresponding convex structure.

1.16.4. A quotient of the 2–cell. Fig. 4 illustrates how a tree (the "character Y") can be obtained as a quotient of the two-cell with its standard convexity. In fact, by complicating the method, other trees can be obtained as well (cf. Topic III§3.26).

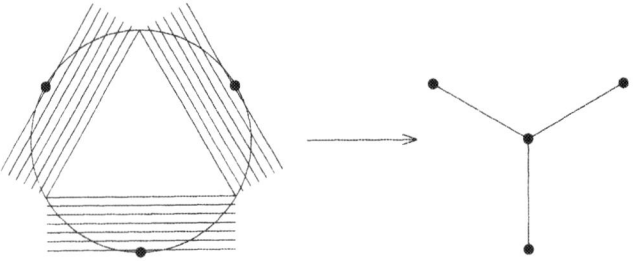

Fig. 4: A quotient of the two-cell

Some important examples of quotient spaces (cones, projective spaces) are

§1: Basic Concepts

described in the next section. For quotients of median algebras, see 1.27.2.

1.17. Convex systems. A *convex system* is a pair, consisting of a set X and a collection $\mathcal{C}$ of subsets of X (the convex sets of the system), which is stable for intersection and up-directed union, and which contains the empty set as a member. The collection $\mathcal{C}$ is the *partial convexity* of the system.

The difference with convex structures is simply that the universal set X need not be convex. However, this has a rather fundamental consequence: not every set has a convex hull. This leads to the next definition. A subset of a convex system is *admissible* provided it is included in some convex set of the system. Admissible sets constitute an up-directed family, which is the domain of the *hull operator* of the system.

If $(X, \mathcal{C})$ is a convex system and if $Y \subseteq X$, then the collection

$$\mathcal{C} \mid Y = \{ C \cap Y \mid C \in \mathcal{C} \}$$

is a partial convexity on Y, and $(Y, \mathcal{C} \mid Y)$ is a *subspace* (*subsystem*) of $(X, \mathcal{C})$. Observe that each convex set of a convex system is a genuine convex structure when regarded as a subspace. This leads to the adequate viewpoint that a convex system is just a system of mutually compatible convex structures -- hence the name. As a general rule, properties of convex systems are defined as properties valid on all convex subspaces. For instance, a convex system is *of arity* $\leq n$ provided each convex subspace is.

A convex system can be made into into a convex structure by adding the universal set to the partial convexity. The following, more subtle, method will be applied to a convex system $(X, \mathcal{C})$. A set $D \subseteq X$ is an *induced convex set* provided $D \cap C \in \mathcal{C}$ for all $C \in \mathcal{C}$.

1.17.1. *The collection of all induced convex sets of a convex system $(X, \mathcal{C})$ is a convexity on X including $\mathcal{C}$.* ∎

We refer to it as the *induced convexity* of the system.

1.17.2. *A set $D \subseteq X$ is an induced convex set iff $co(F) \subseteq D$ for each admissible finite set $F \subseteq X$.* ∎

1.18. Mappings of convex systems. Let X and Y be convex systems and let $f: X \to Y$ be a function. Then f is CP provided it maps admissible sets in X to admissible sets in Y and for each convex set $C \subseteq Y$ we have $co(A) \subseteq f^{-1}(C)$ whenever $A \subseteq f^{-1}(C)$ is an admissible set. The next result is in part inspired by Proposition 1.12.

1.18.1. *The following are equivalent for a function $f: X \to Y$ of convex systems.*

(i) *f is CP.*
(ii) *For each finite admissible set $F \subseteq X$, the set $f(F)$ is admissible in Y and $f(co_X(F)) \subseteq co_Y(f(F))$.*
(iii) *For each finite admissible set $F \subseteq X$, the set $f(F)$ is admissible in Y and f is a CP function with respect to the induced convexities of X and Y.* ∎

A function between convex systems is convex-to-convex (CC) provided it maps convex sets to convex sets. In particular, it maps admissible sets to admissible sets. An *isomorphism of convex systems* is a bijection which is CP both ways.

Let R be an equivalence relation on a convex system $(X, \mathcal{C})$ with a corresponding quotient function $q: X \to X/R$. We obtain a *quotient system* $(X/R, \mathcal{C}/R)$ as follows. Let $\mathcal{C}_{ind}$ be the induced convexity of X, and let $\mathcal{C}_{ind}/R$ be the corresponding quotient convexity with hull operator co_R. Then define $\mathcal{C}/R$ as the family of all sets $C \in \mathcal{C}_{ind}/R$ such that if F is a finite subset of C, then $F \subseteq co_R(q(G))$ for some admissible set $G \subseteq X$. We refer to an example in 1.31 to illustrate some finesses of this definition.

1.18.2. Proposition. *The collection $\mathcal{C}_{ind}/R$ is a partial convexity on X/R and $q: (X, \mathcal{C}) \to (X/R, \mathcal{C}/R)$ is a CP map. The induced convexity of $(X/R, \mathcal{C}/R)$ is precisely the quotient of the induced convexity of $(X, \mathcal{C})$. In formula,*

$$(\mathcal{C}/R)_{ind} = (\mathcal{C}_{ind})/R.$$

If q is even CC, then each admissible finite set in X/R is the image of an admissible subset of X.

Proof. By definition, sets of type $q(G)$, with $G \subseteq X$ admissible, are admissible for $\mathcal{C}/R$. As q is a CP map $(X, \mathcal{C}_{ind}) \to (X/R, \mathcal{C}_{ind}/R)$, we have $q(co(G)) \subseteq co_R(q(G))$ for admissible $G \subseteq X$. This shows that q is a CP map $(X, \mathcal{C}) \to (X/R, \mathcal{C}/R)$.

If $D \in (\mathcal{C}_{ind})/R$, then D meets each member C of $\mathcal{C}/R \subseteq (\mathcal{C}_{ind})/R$ in a member of $(\mathcal{C}_{ind})/R$. If $F \subseteq C \cap D$ is finite, then (as $F \subseteq C$) we have $F \subseteq co_R(q(G))$ for some admissible $G \subseteq X$. Therefore, $C \cap D \in \mathcal{C}/R$, showing that $D \subseteq (\mathcal{C}/R)_{ind}$. Conversely, let $D \in (\mathcal{C}/R)_{ind}$. If $G \subseteq q^{-1}(D)$ is admissible, then $q(G)$ is admissible and as D is an induced convex set, we have $co_R(q(G)) \subseteq D$. As q is a CP function with respect to the partial convexities $\mathcal{C}$ on X and $\mathcal{C}/R$ on X/R, we have $q(co(G)) \subseteq co_R(q(G))$. Hence $co(G) \subseteq q^{-1}(D)$. This shows that $q^{-1}(D) \in \mathcal{C}_{ind}$ and hence that $D \in (\mathcal{C}_{ind})/R$.

As to the last part, let $F \subseteq X/R$ be a finite admissible set. By definition, there is a finite admissible set $G \subseteq X$ such that $F \subseteq co_R(q(G))$. Then $q(co(G)) = co_R(q(G))$ since q is CP and CC. Hence there is a finite set $F' \subseteq co(G)$ with $q(F') = F$, and F' is admissible by definition. ∎

1.19. Examples

1.19.1. Wedge convexity. Let V be a vector space over a totally ordered field. A *(proper) wedge* at the origin $\mathbf{0}$ is a non-empty set $K \subseteq V$ such that $\mathbf{0} \notin K$ and

$$K + K \subseteq K; \quad \forall t > 0: tK \subseteq K.$$

If $p \in V$, then a wedge at p is a set of type $p + K$, where K is a wedge at the origin. Note that a wedge K at p is a standard convex set, and that a line passing through p cannot have points in K at both sides of p. The partial convexity of all wedges at p yields a convex system on V which we use in the next example.

§1: Basic Concepts   21

1.19.2. Convex surfaces. Let $S \subseteq \mathbb{R}^n$ be a *convex surface*, that is, S is the topological boundary of a closed convex set C with a non-empty interior. Fix a point $p \in Int(C)$. A convex system obtains by considering S as a subspace of the convex system of all wedges at p. Observe that this system depends on p. The differences are not essential, however, since all choices of p lead to isomorphic convex systems on S (cf. 1.31). It is often profitable to keep convex sets small, as the example below may illustrate.

1.19.3. Projective space. Let S^n be the unit n-sphere in $\mathbb{R}^{n+1}$, and let $q: S^n \to P^n$ be the quotient map which identifies antipodal pairs of points. If the sphere is partially convexified as a convex surface (using wedges at $\mathbf{0}$), then there are no saturated, induced convex sets in S^n other than subspheres centered at $\mathbf{0}$. However, if the partial convexity is restricted further to sets in which every two points are at angular distance $< \pi/2$ radials, then the quotient space is more interesting. For instance, a set C consisting of two antipodal spherical caps of diameter $< \pi/2$ radials is a saturated and induced convex set in S^n. Each component is a genuine convex set, and each admissible subset of C is included in one of them. Hence, the image set is convex. The resulting quotient space is a reasonable convex system on the real projective space of dimension n. See III§3.18.2 for more detailed results.

1.19.4. A general procedure. Let X be a convex structure and $Y \subseteq X$. There is a partial convexity on Y consisting of all sets $C \subseteq Y$ which are convex in X. Note that Y is, in general, not a subspace of X. The relatively convex sets of Y are induced convex sets of the system, but the converse is not always valid, as the reader may verify.

The question, whether a given convex system arises this way from a convexity (with suitable additional properties), is connected with several problems discussed in this monograph. We refer to the process of gated amalgamation (cf. §5), the convexification of cubical polyhedra of which the cubes carry the usual median convexity (cf. II§3.16) and to the problem of extending uniform convex systems to uniform convex structures (cf. III§3). The next example is also based on this general procedure.

1.19.5. Simplicial convexity. A *simplicial complex* consists of a set S of vertices and a set $\mathcal{S}$ of *simplices*, such that

(S-1) Each simplex is a finite subset of S.
(S-2) Each subset of a simplex is a simplex.
(S-3) The union of all simplices equals S.

The *realization of* $(S, \mathcal{S})$ is the set $|(S, \mathcal{S})|$ (or simply $|S|$) of all functions $x: S \to [0,1]$ such that the vertices with non-zero value (the *support* of x) form a simplex and $\sum_{s \in S} x(s) = 1$. We identify a vertex s with the function which is 1 on s and 0 elsewhere. Then S is a linear basis of the vector space of all finitely supported functions $S \to \mathbb{R}$.

We consider the partial convexity on $|S|$, consisting of all standard convex sets of V included in $|S|$. This is the *simplicial (partial) convexity of* $|S|$. For instance, each simplex $\sigma \in \mathcal{S}$ yields a convex set $|\sigma|$, and the subspace $|\sigma|$ of the system carries the standard convexity.

Let $(S, \mathcal{S})$ and $(T, \mathcal{T})$ be simplicial complexes. A function $\alpha: S \to T$ is *simplicial* provided $\alpha(\sigma) \in \mathcal{T}$ for each $\sigma \in \mathcal{S}$. Such a function extends in a unique way to a map

$$|\alpha|: |S| \to |T|$$

which is affine on each realized simplex. It is easily verified that $|\alpha|$ is CP and CC.

Further Topics.

1.20. Subspaces. Let T be a subsemilattice of a semilattice S. Show that the usual convexity of T equals the relative convexity, derived from the usual convexity of S. Formulate and prove a similar statement about lattices.

Note. Results of this type hold for other classes of spaces as well. For pospaces, see Section 4. For median algebras, consult Section 6. A corresponding statement on metric spaces is false; a (graphic) counterexample is given in 4.20.

1.21. Disk convexity (Lassak [1977]; in terms of closure spaces, Franchetti [1971]). Let (X,d) be a metric space, and let $\mathcal{C}$ be the convexity on X generated by the closed disks

$$D(p,r) = \{x \in X \mid d(p,x) \leq r\} \qquad (p \in X, r > 0).$$

1.21.1. Show that $\mathcal{C}$ is also generated by the open disks

$$B(p,x) = \{x \in X \mid d(p,x) < r\} \qquad (p \in X, r > 0).$$

1.21.2. Show that $a \in co\{a_1,..,a_n\}$ iff for all $x \in X$,

$$d(a,x) \leq \max_{i=1}^{n} d(a_i,x).$$

1.21.3. Let $(V, ||..||)$ be a normed vector space. Show that homotheties and translations of V are CP isomorphisms with respect to the disk convexity.

1.21.4. Let $(V, ||..||)$ be a normed vector space with a unit sphere

$$S = \{x \mid ||x|| = 1\}$$

and let $p \in S$. Then p is a *smooth point* of S (cf. e.g. Day [1962]) provided there is only one hyperplane tangent to S at p. Show that the disk convexity generated by $||..||$ is a symmetrically generated H-convexity. In fact, the functionals determining a hyperflat tangent at a smooth point of the sphere, form a set of generators.

A normed vector space (or, its norm) is called *smooth* provided all points of its unit sphere are smooth. If $||..||$ is a smooth norm on V then the corresponding disk convexity equals the standard convexity. For refinements of this result (which for the standard norm on $\mathbb{R}^n$ is attributed to Mazur) see Phelps [1960] or Giles and Gregory [1978].

§1: Basic Concepts

1.22. Median algebras and distributive lattices. Most results presented in this topic are either obtained in Section 6, or they can easily be derived from the results there.

1.22.1. Let X be a totally ordered set. If $a, b, c \in X$ are such that $a \leq b \leq c$, then we put $m(a,b,c) = b$. Verify that this prescription determines a median operator on X, and that the corresponding convexity is precisely the order convexity.

1.22.2. Let (X_i, m_i) for $i \in I$ be median algebras, and let $X = \prod_{i \in I} X_i$. Consider the following function $m: X^3 \to X$:

$$m(a,b,c) = (m_i(a_i,b_i,c_i))_{i \in I}, \qquad (a = (a_i)_{i \in I}, \, b = (b_i)_{i \in I}, \, c = (c_i)_{i \in I}).$$

Verify that m is a median operator on X, and that the resulting convexity on X is precisely the product of the median convexities, produced by m_i on X_i for $i \in I$.

1.22.3. (Birkhoff [1967]; see 6.12) Let L be a lattice. For $a, b, c \in L$ let

$$m(a,b,c) = (a \wedge b) \vee (b \wedge c) \vee (c \wedge a);$$
$$M(a,b,c) = (a \vee b) \wedge (b \vee c) \wedge (c \vee a).$$

As is well-known, $m \leq M$, and $m = M$ iff L is distributive. Assuming distributivity, show that (L,m) is a median algebra, and that the median convexity equals the usual lattice convexity of 1.5.4.

1.22.4. (Birkhoff and Kiss [1947]; Birkhoff [1967]; compare 5.3.4) Let (X,m) be a median algebra, and let $a, b \in X$ be such that $m(a,b,x) = x$ for all $x \in X$. Show that X is a distributive lattice with a and b as the smallest, resp., largest element under the operations

$$x \wedge y = m(a,x,y); \quad x \vee y = m(b,x,y).$$

In addition, the (usual) convexity of order convex lattices equals the median convexity.

1.22.5. If three elements a, b, c of a tree X are such that any two of them have a common upper bound, then $\{a,b,c\}$ is a chain. Consequently, each $a, b, c \in X$ there is a largest element $m(a,b,c)$ among $a \wedge b$, $b \wedge c$, $c \wedge a$. Show that (X,m) is a median algebra, and that the median convexity equals the original tree convexity. A simple argument is given after 6.10.

1.23. Extreme points. Let X be a convex structure. A point $p \in X$ is an *extreme point* provided $X \setminus \{p\}$ is convex. Extremality in a subset of X can be defined with the aid of relative convexity. Throughout, $ext(X)$ denotes the set of all extreme points of X.

1.23.1. (Soltan [1982]) Show that

$$ext(X) = \bigcap \{A \mid co(A) = X\}$$

and, if F is a finite subset of X, that all extreme points of $co(F)$ are in F.

1.23.2. Show that the extreme points of a tree are precisely its end points (see 5.3.3 for a definition in convexity terms).

1.23.3. Let $(S, \wedge)$ be a semilattice. By an *irreducible element* of S is meant a point q such that $x \wedge y = q$ implies $x = q$ or $y = q$. Contrasting with 1.5.3, consider the convexity of all *subsemilattices* of S. Show that the extreme points are exactly the irreducible elements of S.

1.23.4. (Jamison [1981b]) Show that the extreme points of a poset with the order convexity are precisely the maximal and minimal elements.

1.23.5. (van de Vel [1983d]) Let $\lambda(X)$ denote the superextension of a set X relative to the family of *all* subsets of X. Note that the natural function $X \to \lambda(X)$ embeds X as a subset of $\lambda(X)$. Show that if X is finite then the extreme points of $\lambda(X)$ are exactly the points of X.

1.23.6. (Smiley [1947]) A normed vector space (or, its norm) is *rotund* provided each point of the unit sphere is an extreme point of the unit disk. Show that the geodesic convexity and the standard convexity of a Banach space are equal iff the space is rotund.

1.24. Frattini-Neumann Intersection Theorem (compare Schmidt [1953]). Let X be a convex structure. Then $p \in X$ is called an *inner point* provided for each $A \subseteq X$ with $co(A) = X$ it is true that $co(A \setminus \{p\}) = X$. A non-inner point is called an *outer point*.

1.24.1. Show that the set of inner points in X is convex; specifically, it is the intersection of all maximal proper convex subsets of X.

1.24.2. Show that p is an outer point of X iff there is a convex set C with $p \notin C$ and $co(C \cup \{p\}) = X$. Deduce that a non-empty polytope is the convex hull of its outer points, and that the outer points of a standard convex set are exactly the extreme points. In the lattice $[0,1]^n$ the outer points are exactly the boundary points of this set in the topological space $\mathbb{R}^n$; for $n > 1$, *none* of these points is an extreme point.

1.24.3. Let G be a group. An element $g \in G$ is *absolutely dispensable* if it can be dropped from any set of generators of G. Show that the absolutely dispensable elements form a subgroup, equal to the intersection of all maximal proper subgroups in G (for finite groups this result was established in 1885 by G. Frattini, and it was extended to arbitrary groups in 1937 by B.H. Neumann).

1.25. Gluings and disjoint sums (Degreef [1982]; compare Sierksma [1981]; continued 2.25). Let $(X_i)_{i \in I}$ be a family of convex structures, let X be a set, and for each $i \in I$ let $f_i : X_i \to X$ be a function. There obviously exists a finest convexity on X turning each f_i into a CP function. If $X = \cup_{i \in I} X_i$ and if f_i is the inclusion of X_i into X then the resulting convex structure is called a *gluing* of the spaces X_i for $i \in I$.

1.25.1. If the sets X_i are disjoint, then the gluing corresponds with the disjoint sum. If all sets X_i are identical, then the gluing convexity is the join of the convexities $\mathcal{C}(X_i)$.

§1: Basic Concepts   25

1.25.2. A gluing $X = \cup_{i \in I} X_i$ is said to be *normal* provided for each $A \subseteq X$,
$$co(A) = \bigcup_{i \in I} co_i(A \cap X_i),$$
where co_i is the hull operator of X_i. Show that (i) a disjoint sum is a normal gluing, (ii) a gluing $X_1 \cup X_2$ is normal if each point of $X_1 \cap X_2$ is extreme in X_1 and in X_2, and (iii) the gluing of $[-1,1] \times [0,1]$ and $[0,2] \times [0,1]$ (standard convexity) is not normal.

1.26. CP functions and isomorphisms

1.26.1. Let $f_n: X \to \mathbb{R}$ for $n \in \mathbb{N}$ be a sequence of CP functions and let f be its pointwise limit. Show that f is CP. The same goes if the standard convexity of $\mathbb{R}$ is replaced by the lower or upper convexity.

1.26.2. Show that $C_*(X) \times C_*(Y)$ is isomorphic to $C_*(X \times Y)$.

1.26.3. (continued: 2.28 and 3.29) Let C be an H-convexity on $\mathbb{R}^n$. The unit sphere relative to the Euclidean norm is denoted by S^{n-1}. Each generating linear functional f of C can be represented by a some element $v \in S^{n-1}$, in the sense that $f(x) = v \cdot x$ (inner product) for all $x \in \mathbb{R}^n$. Let $H \subseteq S^{n-1}$ be the set of all points so obtained. Such a representation and notation is standard in real "H-convexity" -- hence the name. Show that if v is in the closure $\overline{H}$ of H in S^{n-1}, then the linear functional $x \mapsto v \cdot x$ is CP relative to C and to the lower convexity of $\mathbb{R}$. Hence the functionals corresponding with $\overline{H}$ generate the original H-convexity.

1.27. Quotients

1.27.1. (Burris [1971]) A *congruence* on a convex structure X is an equivalence relation E such that $co(E(A)) \subseteq E(co(A))$ for all $A \subseteq X$. Equivalently, the saturation of a convex set is convex. Prove that the congruences on X form a complete lattice.

1.27.2. Let X be a median algebra. This time, we use the term "congruence" in the usual algebraic sense: if a is congruent with a', then $m(a,b,c)$ is congruent with $m(a',b,c)$. It is clear that a median congruence satisfies Burris' condition. The converse holds as well; we advise to wait for Section 6 to prove this (cf. 6.33.2).

(Bandelt and Hedlíková [1983]) Let $C \subseteq X$ be convex, and consider the following binary relations in X.
$$\Theta[C] = \{(x,y) \in X^2 \mid \exists a,b \in C : x = m(a,x,y),\ y = m(b,x,y)\}.$$
$$\Xi[C] = \{(x,y) \in X^2 \mid \forall a,b \in C : m(a,b,x) \in C \Leftrightarrow m(a,b,y) \in C\}.$$
Show that $\Theta[C]$ and $\Xi[C]$ are the smallest, resp., the largest congruence on X for which C is a congruence class.

1.28. Inverse limits

1.28.1. Show that *inverse limits* exist in the category of convex structures. Specifically, let X_i for $i \in I$ be convex structures, where the index i ranges through a down-directed index set I, and let $f_{ij}: X_i \to X_j$ for $i \leq j$ in I be the morphisms of the system. The inverse limit X is the subspace of the product $\prod_{i \in I} X_i$, consisting of all points $(x_i)_{i \in I}$ with the property that $f_{ij}(x_i) = x_j$ for $i \leq j$. The canonical morphism $f_i: X \to X_i$ is the restriction of the ith projection.

1.28.2. Verify that the convexity of the inverse limit is generated by the sets
$$f_i^{-1}(C_i) \quad (i \in I; C_i \in \mathcal{C}(X_i)).$$

1.28.3. Show that in each of the following categories, the convexification of the inverse limit is the inverse limit of the convexifications:

(i) The category of vector spaces over a fixed ordered field and linear functions.
(ii) The category of meet semilattices and semilattice homomorphisms.
(iii) The category of lattices and lattice homomorphisms.
(iv) The category of median algebras and MP functions.

1.29. Direct limits

1.29.1. Show that *direct limits* exist in the category of convex structures. In fact, if $f_{ij}: X_i \to X_j$ for $i \leq j$ are the morphisms of the direct system, then the direct limit X is the quotient of the disjoint sum $\Sigma_{i \in I} X_i$ under the equivalence relation $\approx$, defined as follows.
$$(i, x_i) \approx (j, x_j) \text{ iff } \exists k \geq i, j : f_{ik}(x_i) = f_{jk}(x_j).$$
Prove that if all morphisms f_{ij} are CC, then the limit morphisms $X_i \to X$ are CC.

1.29.2. As is well-known, direct limits exist in each of the above described categories (i) - (iv). Show that in each case, the canonical convexifications lead to the correct direct limit in the category of convex structures.

1.30. Algebraic lattices

(compare Gierz et al [1980, Thm. 1§4.15]). Let L be a complete lattice with a largest and a smallest element. A point $x \in L$ is called a *compact element* if x is way-below itself. L is an *algebraic lattice* if each of its points is the supremum of compact elements.

1.30.1. Verify that a convexity on a set is an algebraic lattice, and that the compact elements are precisely the polytopes.

1.30.2. Show that each algebraic lattice L is isomorphic to a convexity on some set. Hint: let X be the set of compact elements of L and assign to $p \in L$ the set of compact elements below p.

Note. The convex hyperspace $\mathcal{C}_*(X) = \mathcal{C}(X) \setminus \{\emptyset\}$ of a convex structure X can be

§1: Basic Concepts 27

seen as a join semilattice under the operator
$$C_1 \vee C_2 = co(C_1 \cup C_2) \quad (C_1, C_2 \in \mathcal{C}_*(X)).$$
If X is a free space then (as the reader can easily verify) the Vietoris convexity of $\mathcal{C}_*(X)$ equals the convexity of order convex subsemilattices of $\mathcal{C}_*(X)$. In general, both kinds of convexities are firmly distinct.

1.31. Convex systems

1.31.1. Let $S \subseteq \mathbb{R}^n$ ($n > 0$) be a convex surface and let p, q be points of the interior convex set. Show that the partial convexities on S determined by the wedges at p resp., q, are isomorphic. Show that the gluing (cf. 1.25) of all convex subspaces of a convex surface S yields the induced convexity of S, and that this convexity has disconnected convex sets if S is compact.

1.31.2. Let $(X_i, \mathcal{C}_i)$ for $i \in I$ be convex systems, and let $X = \prod_{i \in I} X_i$. Define the *product* $(X, \mathcal{C})$ of these convex system as follows. Start with the product of the convexities induced on the factors, and restrict to those convex sets $C \subseteq X$ with the property that the projection of a finite subset of C to a factor is admissible. Verify that all coordinate projections are CP and that the induced convexity of the product is the product of the convexities, induced on the factors.

1.31.3. Let S^1 be the unit circle in the complex plane, let T denote the product of n copies of S^1, and let $q : \mathbb{R}^n \to T$ be defined by
$$q(x_1,..,x_n) = (e^{2\pi i x_1},..,e^{2\pi i x_n}).$$
The vector space $\mathbb{R}^n$ is equipped with the partial convexity, consisting of all standard convex sets C such that for each two points $x = (x_1,..,x_n)$, $x' = (x'_1,..,x'_n)$ in C,
$$\forall i = 1,.., n: \quad |x_i - x'_i| < \pi.$$
The n-dimensional torus T is equipped with the quotient convex system. Verify that the saturation of a convex set $C \subseteq \mathbb{R}^n$ equals $2\pi \mathbb{Z}^n + C$, which is an induced convex set. Conclude that the quotient map q is CP and CC.

1.31.4. Let $X \subseteq \mathbb{N} \times \mathbb{N}$ consist of all pairs (x,y) with $x \leq y$. On X we consider the partial convexity of all sets of type $C \times D \subseteq X$, where $C, D \subseteq \mathbb{N}$. Let $\pi_1 : X \to \mathbb{N}$ be the first projection and consider the corresponding quotient system on $\mathbb{N}$. Show that $\mathbb{N}$ is a genuine convex structure (with the free convexity) and that the image of any admissible subset of X is finite. In particular, the universal set $\mathbb{N}$ is not the convex hull of a set of type $q(A)$, where $A \subseteq X$ is admissible. Compare 1.18.

Notes on Section 1

The early papers on abstract convexity can be sorted roughly into two kinds. The first type deals with generalizations of particular problems, such as separation of convex sets (Ellis [1952]), extremality (Fan [1963]; Davies [1967]) or continuous selection (Michael [1959]). Papers of the second type are involved with a multi-purpose system of axioms. Several designs of such an axiom system are possible, each one expressing a particular point of view on convexity. For instance, Schmidt [1953] and Hammer [1955], [1963], [1963b], discuss the viewpoint of generalized topology which enters into convexity via the closure or hull operator. The arising of convexity from algebraic operations, and the related property of domain finiteness receive attention in Birkhoff and Frink [1948], Schmidt [1953], and Hammer [1963], with reference to a paper of Tarski [1930] on a metamathematical model for deduction. The viewpoint of combinatorial geometry originates in Levi [1951], where the relationship between Helly's and Radon's Theorem is discussed; cf. Section II§1. Finally, the viewpoint of ordered geometry enters via the axioms of a join geometry (Prenowitz [1946], Prenowitz and Jantosciak [1972]; cf. Section 7). For special interpretations, see Aumann [1971] (social contact relations, cf. 2.20) or Doignon and Falmagne [1985] (knowledge assessment).

The survey paper of Danzer, Grünbaum and Klee [1963] has considerably stimulated the investigations on abstract convexity. Two other papers should be credited with the same merit: Tverberg [1966] gave a deep and complicated extension of Radon's Theorem in $\mathbb{R}^n$ (cf. II§5). The search for a proof, expressible in terms of abstract convexity, still goes on. Eckhoff [1969] investigated Radon's Theorem on products of Euclidean spaces; cf. II§2. These papers accelerated the development of a general theory of convexity, the contours of which were shaped, among others, in the dissertations of Jamison [1974] and Sierksma [1976]. An extensive treatment of abstract convexity can be found in Soltan [1984]. Survey papers in particular areas are: Eckhoff [1979] (Radon numbers), Jamison [1982] (varieties, cf. 3.33.1), van de Vel [1984e] (convexity in semilattices), Edelman and Jamison [1985] (convex geometry), and Duchet [1987] (combinatorial convexity).

Closure spaces and the corresponding closure operators go back to Moore [1910]. They have a central place in lattice theory (cf. Birkhoff [1967]). The equivalence of stability under nested union and stability under up-directed union, Theorem 1.3, is actually a result from the theory of posets. See Cohn [1965, p. 33]. We regard domain finiteness as an essential part of the axiom system for convexity; see the discussion in 1.7. The characterization of polytopes, Theorem 1.6, has first been obtained by Klee [1972] for polytopes in real vector spaces and it was extended to abstract spaces (with a simpler proof) by Jamison [1974]. Part (3) (on unions of chains) appears in Birkhoff [1967].

Each example in 1.5 represents a natural notion of convex set, depending on some type of mathematical structure. A standard reference for convexity in real vector spaces is Valentine [1964]. Many concepts and results can be formulated and proved in a more

§1: Basic Concepts 29

general context of ordered fields. A further relaxing of the assumptions can be found e.g., in Doignon [1976] or in van Maaren's dissertation [1979]. The concept of order convexity is attributed to Birkhoff (1948 edition of [1967]). Convexity in general (semi)lattices has been studied by Varlet [1975], Jamison [1978], and van de Vel [1984], [1984c], [1984e]. The notion of convexity in metric spaces goes back to Menger [1928]. Several authors contributed to this area of convexity, e.g. Blumenthal [1953], Busemann [1955]. Particularly popular are the study of geodesic convexity in Banach spaces (cf. Boltyanskii and Soltan [1978]) and in graphs (cf. Mulder [1980]; Soltan [1983]). The convexity proposed here is not the only natural one. The notion of disk convexity (Lassak [1977], cf. 1.21) in a metric space is just one alternative. For yet another convexity in metric spaces, see Danzer [1961]. In connected graphs several path convexities have been considered by Bandelt [1989], Duchet [1988], Duchet and Meyniel [1983], Farber and Jamison [1986], and others. Median convexity is perhaps the least known -- but certainly one of the richest branches of abstract convexity. The interest in medians goes back to Birkhoff and Kiss [1947], Avann [1948] and Sholander [1952]. For a survey on median algebras see Bandelt and Hedlíková [1983]. More about this in Section 6.

The extension of a protopology to a convexity, 1.7.2, was considered by Sierksma [1984b]. The concept of a subbase is taken from Jamison [1974]. The subsequent examples given in 1.8 are somewhat more technical than the ones listed before. H-convexity is a well-established area of abstract convexity. See Boltyanskii and Soltan [1978] for a survey. Convex hyperspaces are an a-topological version of a construction given by the author in [1984d]. Superextensions originated in a context of compactifications in topology; they have also been introduced independently in a context of freely generated algebraic systems; cf. 6.33.4. The concept goes back to de Groot [1969] and developed later to an important example of (median) convexity. See van de Vel [1979], van Mill and van de Vel [1981].

Subspaces and products of convexities have been described by Jamison [1974], where the notion of a join was also introduced. For closure structures, a description of products appeared in Eckhoff's influential paper [1968]. Convexity preserving functions were first introduced by van Mill and Wattel [1978] in more restrictive circumstances, and redefined in the present way by van Mill and van de Vel [1978]. Convexity preserving and convex-to-convex functions have also been considered by Guay [1978]. CC functions are called "convex functions" by Kay [1977]. Quotients of convex structures were introduced by van Mill and van de Vel in [1986] in the context of uniform spaces; cf. Section III§3. Convex systems have been introduced by the author [1993]; the interesting problems are connected with the topological theory. See Chapters III§3 and IV§3.

2. The Hull Operator

A convex structure is completely determined by its hull operator, or even by its effect on finite sets. There is a simple axiom system characterizing the hull operator on the collection of finite sets. It can be interpretated in terms of betweenness, or alternatively, in terms of dependence, and it is used for a natural description of cone convexity and of convexity in spaces of arcs.

Convexities are often classified by the properties of the hull operator. This is the case, for instance, with matroids (alias independence structures), and with Join-hull Commutative (JHC) convexities. We also consider some related conditions, such as weak Join-hull Commutativity and the Cone-union Property (CUP). Among the main examples of matroids are the linear, affine, and projective convexities derived from vector spaces. The fact, that projective spaces over a field are JHC, leads to a general definition of projective spaces as JHC matroids (with convex singletons, and excluding free convexities). Some results on the rank function of a matroid are derived.

Domain finiteness of the hull operator, Theorem 1.3, turns out to be the main distinction between convexity and topology as it may appear from the next result. Throughout, 2^X_{fin}, resp. 2^X_n, denotes the collection of all finite subsets, resp., all subsets with at most n points, of a set X.

2.1. Proposition. *Let X be a convex structure and let $A \subseteq X$. Then*

(1) $\quad co(A) = \bigcup \{ co(F) \mid F \subseteq A ; F \in 2^X_{fin} \}$.

If X is n-ary, if $A_0 = A$ and $A_{k+1} = \bigcup \{ co(F) \mid F \subseteq A_k ; F \in 2^X_n \}$ for $k \in \mathbb{N}$, then

(2) $\quad co(A) = \bigcup_{k=0}^{\infty} A_k$.

Proof. Let C be the right hand set of (1), and let $F \subseteq C$ be finite. For each $x \in F$ there is a finite set $F_x \subseteq A$ such that $x \in co(F_x)$. We find that $co(F) \subseteq co(\bigcup_{x \in F} F_x) \subseteq C$ and C is convex by domain finiteness.

Let A_∞ denote the right hand set of (2), and let $F \subseteq A_\infty$ with $\#F \leq n$. The sequence A_k being increasing, there is $k \in \mathbb{N}$ such that $F \subseteq A_k$. Hence $co(F) \subseteq A_{k+1} \subseteq A_\infty$ and A_∞ is convex since X is of arity $\leq n$. ∎

The next result provides a quite useful characterization of hull operators among general set operators. The topics section contains an extension to convex systems.

2.2. Proposition. *Let X be a set and let*

$$h : 2^X_{fin} \to 2^X$$

be an operator satisfying the following conditions.

(H-1) $h(\emptyset) = \emptyset$.
(H-2) $F \subseteq h(F)$ *for each* $F \in 2^X_{fin}$.

(H-3) *For all $F, G \in 2_{fin}^X$, if $G \subseteq h(F)$ then $h(G) \subseteq h(F)$.*

Then there is precisely one convexity on X with a hull operator equal to h on 2_{fin}^X. Conversely, the hull operator of any convexity on X satisfies the conditions (H-1) to (H-3).

Proof. Suppose first that h is an operator as described above. A set $C \subseteq X$ is taken convex provided $h(F) \subseteq C$ for each finite set $F \subseteq C$. By (H-1), this yields a convexity $\mathcal{C}$ on X. Let co denote its hull operator. If $F \in 2_{fin}^X$, then $h(F) \subseteq co(F)$ by definition. On the other hand, the set $h(F)$ is convex by (H-3) and includes F by (H-2). As $co(F)$ is, by definition, the smallest convex set including F, we find that $co(F) = h(F)$. Uniqueness of $\mathcal{C}$ follows from domain finiteness, 1.3(2).

Conversely, the hull operation of any convex structure on X satisfies (H-1) by the normalization axiom (CL-4), (H-2) by extensiveness (CL-2) and (H-3) by monotonicity (CL-1) and idempotence (CL-3). ∎

The previous result is convenient to construct convexities which are difficult to describe directly. Two types of examples will be presented below. It will also be of use in considerations on completion and on compactification (see Section III§3).

2.3. Betweenness. A point which is in the convex hull of a finite set can be regarded as being between (the points of) this set, or, to depend on the set. The viewpoint of dependence will be described in 2.8 below. We interpret the conditions appearing in Proposition 2.2 as an axiom system for *betweenness* (or, after the Greec μετα ≈ between and θησις ≈ put down, *metathetism*[1]), by which we formally mean a binary relation involving points and finite subsets of a given set, subject to the conditions below.

(B-1) *Normalization:* No point is between the empty set.

(B-2) *Extensiveness:* If $x \in F$ then x is between F.

(B-3) *Transitive Law:* If x is between G and if each element of G is between F, then x is between F.

Note that, in the viewpoint of betweenness, CP functions correspond with "betweenness-preserving" functions (*metamorphisms*). A function f is betweenness-preserving provided that x is between F implies $f(x)$ is between $f(F)$.

2.4. Examples. The product formula 1.10.3 for polytopes states that betweenness in a product space is a product of betweenness relations given on the various factors. By the Relative Hull Formula 1.9.1, betweenness in a subspace is the restriction of betweenness in the superspace. The Hyperspace Polytope Formula (cf. 1.8.3) is a betweenness relation for closed sets of a closure space. Here are some new examples.

2.4.1. Cone over a convex structure. Let X be a space, let $x, x_1, ..., x_n \in X$, and let $t, t_1, ..., t_n \in [0,1]$. Consider the following prescription: (x,t) is *in between* the points $(x_1, t_1), ..., (x_n, t_n)$, provided

---
1. I am indebted to my daughter Kathleen for suggesting this term.

§2: The Hull Operator

(i) $\min_i t_i \leq t \leq \max_i t_i$, and
(ii) $x \in co\{x_i \mid t_i \leq t\}$.

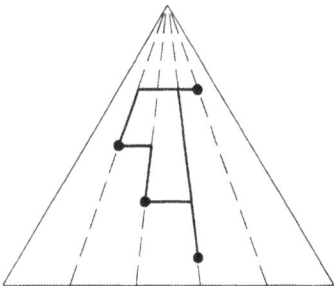 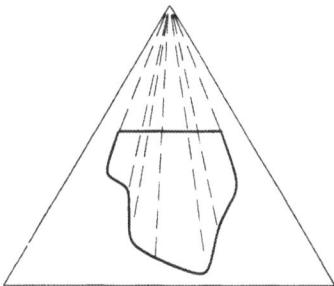

Fig. 1: Cone convexity: a polytope and a general convex set

It is clear that this prescription satisfies the axioms (B-1) to (B-3). Hence it produces a convex structure on $X \times [0,1]$. A convex set of this convexity meets each vertical fiber in a (perhaps empty) line segment, its horizontal section at level t corresponds with a convex set C_t of X, and C_t is increasing with t until a certain level t_0 is reached where $C_t = \emptyset$ for all $t > t_0$ or for all $t \geq t_0$.

Let $\Delta(X)$ be the quotient of $X \times [0,1]$, obtained by identifying $X \times \{1\}$ to one point. In regard to the previous description, the saturation of a convex set is convex. Therefore, in addition to being CP, the quotient function is convex-to-convex. The space $\Delta(X)$ is called the *cone over X*. Fig. 1 presents a typical polytope and a general convex set.

2.4.2. Spaces of arcs. Let $(S, \vee)$ be a join semilattice, and let $\Lambda(S)$ denote the set of all maximal chains (*arcs*) in S. We assume that S has a smallest element **0** (*lower universal bound*) and a largest element **1** (*upper universal bound*); briefly: S is *bounded*. All members of $\Lambda(S)$ join **0** with **1**.

Let $A, A_1, ..., A_n \in \Lambda(S)$. Then A is *in between* $A_1,...,A_n$ provided each element of A is of the form

$$a_1 \vee .. \vee a_n \quad (a_1 \in A_1,...,a_n \in A_n).$$

The collection of such points will be denoted by $A_1 \vee .. \vee A_n$, or simply $\vee_{i=1}^n A_i$. It is easily seen that the resulting betweenness relation satisfies the axioms (B-1) to (B-3). This yields a convexity on $\Lambda(S)$, and the resulting structure is the *space of arcs* of S.

If L is a bounded lattice, we can use each of the semilattices $(L, \vee)$ and $(L, \wedge)$ to produce a space of arcs. We denote them by $\Lambda(L, \vee)$, resp., $\Lambda(L, \wedge)$. The underlying set $\Lambda(L)$ is the same, and we can therefore produce a third convex structure $\Lambda(L, \vee, \wedge)$ on $\Lambda(L)$ which is the meet (intersection) of the previously considered convexities. The hull operators of these convexities will be denoted by, respectively, $co_\vee$, $co_\wedge$, and co.

2.5. Matroids. Since, by Proposition 2.2, convex structures are unambiguously determined by their hull operator (or even by its restriction to finite sets), it is possible to classify convex structures by the properties of this operator. As a first illustration, we consider the *Exchange Law:* If $A \subseteq X$ and if $p, q \in X \setminus co(A)$, then $p \in co(\{q\} \cup A)$ implies $q \in co(\{p\} \cup A)$.

Note that if $A = \emptyset$, then the sets $co\{p\}$ for $p \in X$ form a partition of X. A convex structure is called a *matroid* provided it satisfies the Exchange Law.

2.6. Theorem. *A CP and CC image of a matroid is a matroid.*

Proof. Let X, Y be convex structures of which X is a matroid, and let $f: X \to Y$ be a CP and CC surjection. Suppose $A \subseteq Y$, and let $p, q \in Y$ such that $p, q \notin co(A)$ and $p \in co(\{q\} \cup A)$. Fix a point $q' \in f^{-1}(q)$. Then

$$f(co(\{q'\} \cup f^{-1}(A))) = co(\{q\} \cup A)$$

since f is CP, CC, and surjective. Hence there is a point $p' \in f^{-1}(p)$ such that $p' \in co(\{q'\} \cup f^{-1}(A))$. As $p', q' \notin co(f^{-1}(A))$, the matroid property of X yields that $q' \in co(\{p'\} \cup f^{-1}(A))$. Consequently, $q \in co(\{p\} \cup A)$. ∎

2.7. Examples of Matroids. Below are three basic types of examples and an exotic one. The list is continued in 2.22.

2.7.1. Linear Matroids. Let V be a vector space over a field $\mathbb{K}$; the set V minus its origin $\mathbf{0}$ (a *punctured vector space*) will be denoted by V_0. A non-empty set $C \subseteq V$ is called a *linear set* provided for each pair of points $p, q \in C$ and for each $s, t \in \mathbb{K}$,

$$s \cdot p + t \cdot q \in C.$$

Equivalently,

$$\forall p_1,..,p_n \in C,\ \forall t_1,..,t_n \in \mathbb{K}: \sum_{i=1}^{n} t_i \cdot p_i \in C.$$

The vector $\sum_{i=1}^{n} t_i \cdot p_i$ is called a *linear combination* of the points $p_1,..,p_n$. The collection of all traces on V_0 of linear subsets of V is known as the *linear convexity* of V_0. We use *lin* instead of *co* to describe its hull operator. The resulting space is a matroid. In fact, if $p, q \notin lin(A)$ are non-zero vectors, then $p \in lin(\{q\} \cup A)$ iff there is a zero-valued linear combination of p, q and of some points of A, in which p and q occur with a non-zero coefficient. We call it a *linear matroid over* $\mathbb{K}$.

2.7.2. Affine matroids. Let V, $\mathbb{K}$ be as above. A set $C \subseteq V$ is called an *affine set* provided for each pair of points $p, q \in C$ and for each $t \in \mathbb{K}$,

$$t \cdot p + (1-t) \cdot q \in C.$$

Equivalently,

$$\forall p_1,..,p_n \in C,\ \forall t_1,..,t_n \in \mathbb{K}: \text{if } \sum_{i=1}^{n} t_i = 1 \text{ then } \sum_{i=1}^{n} t_i \cdot p_i \in C.$$

The vector $\sum_{i=1}^{n} t_i \cdot p_i$ is called an *affine combination* of the points $p_1,..,p_n$. This results into the *affine convexity* of V. Its hull operator is denoted by *aff*. If $p, q \notin aff(A)$, then

§2: The Hull Operator   35

$p \in \mathit{aff}(\{q\} \cup A)$ iff there is a zero-valued linear combination of p, q and of some points of A such that the coefficients sum up to zero and p, q appear with non-zero coefficient. The matroid property of the affine convexity follows at once. We refer to the resulting space as an *affine matroid over* $\mathbb{K}$.

2.7.3. Projective matroids. Consider the linear matroid in a vector space V over a field $\mathbb{K}$. The following is an equivalence relation on the punctured space V_0.

$$x \equiv y \Leftrightarrow \exists t \in \mathbb{K} \setminus \{0\}: tx = y.$$

Let P denote the quotient space. It is easily seen that the saturation of a linear set is linear. Hence the quotient function $V_0 \to P$ is CP and CC. By Theorem 2.6, P is a matroid. We refer to it as the *projective matroid over* $\mathbb{K}$. Its convex sets are called *projective sets*. Keeping to the style of the previous examples, we use the symbol pr to denote the hull operator of the projective convexity.

There is a standard way to embed the affine matroid of a vector space V into a projective matroid P over the same field $\mathbb{K}$. The idea is to embed V into the vector space $V \times \mathbb{K}$ by the affine map $v \mapsto (v, 1)$. The projective quotient P of $V \times \mathbb{K}$ does not identify points of $V \times \{1\}$.

2.7.4. Free spaces. On a set with the free convexity, the Exchange Law reduces to the situation where $p = q$. Hence a free space is a trivial example of a matroid.

2.8. Independence structures. In certain classes of convex structures, "betweenness" may be interpreted more accurately as "dependence". This is the case, for instance, with matroids. Let X be an arbitrary convex structure. A non-empty subset $F \subseteq X$ is called *convexly independent* provided $x \notin co(F \setminus \{x\})$ for each $x \in F$, and it is called *convexly dependent* otherwise. By domain finiteness, Theorem 1.3, it follows that a set is convexly independent iff each of its non-empty finite subsets is. A frequent procedure to obtain such sets is as follows. Let $p \in co(F)$, where F is finite. Then there is a minimal subset of F with p in its hull and each one is convexly independent.

The adjective "convexly" is usually omitted or is specified by the name of the convexity. For instance, in a vector space V over a field $\mathbb{K}$ we use "linear" resp., "affine" dependence in case of the linear resp., affine convexity. A set $F \subseteq V_0$ is *linearly independent* iff each zero-valued linear combination of elements of F is trivial (i.e. involves zero coefficients only). Note that the origin of V yields a linearly dependent singleton in the traditional sense. This corresponds with the fact that (properly speaking) the linear hull of $\emptyset$ is $\{0\}$; this phenomenon could have been captured by allowing non-normalized closure- or hull operators (cf. 1.2). A set $F \subseteq V$ is *affinely independent* provided each zero-valued linear combination in F, with coefficients summing up to zero, is trivial.

The important properties of the collection $\mathcal{E}$ of independent sets of the above matroids are as follows.

(IN-1) *Nondegeneracy Law:* All singletons are in $\mathcal{E}$.
(IN-2) *Finitary Law:* $E \in \mathcal{E}$ iff all non-empty finite subsets of E are in $\mathcal{E}$.

(IN-3) *Replacement Law:* If $A, B \in \mathcal{E}$ are finite and if $\#B > \#A$, then there is a point $b \in B \setminus A$ with $A \cup \{b\} \in \mathcal{E}$.

It appears that these laws are characteristic of matroids. Before proving this, let us complete our set of terminology. A pair $(X, \mathcal{E})$, consisting of a set X and a family $\mathcal{E} \subseteq 2^X$ of non-empty sets satisfying the previous laws, is an *independence structure*, and the members of $\mathcal{E}$ are *independent sets*. A point $p \in X$ *depends* on a set $A \subseteq X$ if $p \in A$ or if there is an independent set $B \subseteq A$ -- which can be taken finite -- such that $B \cup \{p\}$ is dependent (i.e., does not belong to $\mathcal{E}$). Finally, a *flat* is a set containing each point which depends on it.

2.9. Proposition (*Transitive Law of dependence*). *Let A be a finite subset of an independence structure. If x depends on A and if y depends on $A \cup \{x\}$, then y depends on A.*

Proof. We assume $x \notin A$ and $y \notin \{x\} \cup A$. As y depends on $A \cup \{x\}$, there is an independent set $F \subseteq A \cup \{x\}$ such that $F \cup \{y\}$ is dependent. Among all possible sets, we fix one with $\#F$ maximal. As x depends on A, there is an independent set $G \subseteq A$ with $G \cup \{x\}$ dependent. Again, we take a set of maximal cardinality. If $\#G < \#F$ then by the replacement law some point of $F \setminus G$ can be added to G without destroying its independence. However, this additional point cannot be x (since $G \cup \{x\}$ is dependent), nor can it be a point of A (this would contradict the maximality of $\#G$). Therefore, $\#G \geq \#F$. If y does not depend on A, then $G \cup \{y\}$ is independent. As $\#(G \cup \{y\}) > \#F$, by the replacement law, some point of $(G \cup \{y\}) \setminus F$ can be added to F without destroying its independence. This point cannot be y (since $F \cup \{y\}$ is dependent), nor can it be a point of A (this would contradict the maximality of $\#F$). ∎

We will use the following condition, which is somewhat stronger than (IN-3) (the sets involved need not be finite).

(IN-3′) *Strong Replacement Law:* For any two sets $A, B \in \mathcal{E}$ with $\#B > \#A$, there is a point $b \in B \setminus A$ with $A \cup \{b\} \in \mathcal{E}$.

2.10. Theorem

(1) *Let X be a matroid. Then the collection $\mathcal{E}$ of all independent subsets of X satisfies the Nondegeneracy-, Finitary-, and strong Replacement Laws, and hence it is an independence structure.*

(2) *Let $(X, \mathcal{E})$ be an independence structure. Then the flats of $(X, \mathcal{E})$ form a matroid. such that the convex hull of a set is the collection of all points depending on it.*

(3) *The above described transitions between matroids and independence structures are mutually inverse.*

Proof of (1). The Nondegeneracy and Finitary Laws are easy to verify (in fact, they are valid relative to *any* convex structure). As for the strong Replacement Law, let A, B be independent sets with $\#B > \#A$, and assume to the contrary that

§2: The Hull Operator 37

(4) $\{x\} \cup A$ is dependent for all $x \in B$.

Fix a point $b \in B$ for the moment. Then either $b \in co(A)$, or there is an $a \in A$ such that $a \in co(\{b\} \cup A \setminus \{a\})$. Note that $a \notin co(A \setminus \{a\})$ since A is independent. As X is a matroid, the second possibility yields

$$b \in co(\{a\} \cup A \setminus \{a\}) = co(A)$$

as well. We conclude that $B \subseteq co(A)$. If $\#A$ is infinite, then A, B can be rearranged (preserving (4)) such as to become finite. Indeed, for each $b \in B$ there is a finite set $F(b) \subseteq A$ with $b \in co(F(b))$ (domain finiteness, Theorem 1.3). As $\#2^A_{fin} = \#A$ when $\#A$ is infinite, we find that $\#B > \#2^A_{fin}$ and hence some finite set $F \subseteq A$ occurs as $F(b)$ for infinitely many $b \in B$. Then take this F instead of A and take the corresponding (infinite) subset of B instead of B. Then reduce B further to a finite size larger than $\#F$.

We assume from now on that $\#A < \#B < \infty$. Among all possible pairs A, B of this kind, satisfying (4), we consider one for which the number

$$n = \#B - \#(B \cap A)$$

is the smallest. Note that $n > 0$. Fix $b \in B \setminus A$ and let $F \subseteq A$ be minimal with respect to the property that $b \in co(F)$. Then $F \not\subseteq B$ since B is independent, and hence there is a point $a \in F \setminus B$. Now $b \notin co(F \setminus \{a\})$ by the minimality of F, and $a \notin co(F \setminus \{a\})$ by the independence of A. Hence by the Exchange Law and by the Monotone Law,

(5) $a \in co(\{b\} \cup F \setminus \{a\}) \subseteq co(\{b\} \cup A \setminus \{a\})$,

from which $b \notin co(A \setminus \{a\})$ follows (otherwise $co(\{b\} \cup A \setminus \{a\})$ equals $co(A \setminus \{a\})$ and contains a). This is a first step in establishing that the set $A' = \{b\} \cup A \setminus \{a\}$ is independent. Suppose that for some $a' \in A'$ different from b it is true that $a' \in co(A' \setminus \{a'\})$. Then by the Exchange Law,

$$b \in co(\{a'\} \cup (A \setminus \{a, a'\})) = co(A \setminus \{a\}),$$

which is a contradiction.

By (5) and (4), we have $co(A') = co(A \cup \{b\}) = co(A)$, and hence each point of B is dependent on A' as well. This establishes (4) for the sets A', B, whereas the independent set A' has one more point in common with B. Thus

$$\#B - \#(B \cap A') < n,$$

a contradiction with the definition of n.

Proof of (2). The Finitary Law can be used to show that a superset of a dependent set is again dependent. The Nondegeneracy Law will be used only in establishing (3).

We define an operator

$$h: 2^X_{fin} \to 2^X$$

as follows: $h(\emptyset) = \emptyset$ and for a non-empty finite set $F \subseteq X$, we let $h(F)$ be the set of all x depending on F. We aim at an application of Proposition 2.2.

Suppose $p \in h(\{p_1, ..., p_n\})$ and that for each $i = 1, ..., n$, the point p_i depends on the finite set F. Let $G_0 = F$ and for $k = 1, ..., n$ let

$G_k = F \cup \{p_1,..,p_k\} = G_{k-1} \cup \{p_k\}$.

Then p_k depends on G_{k-1} for each $k \leq n$, whereas p depends on $G_n = G_{n-1} \cup \{p_n\}$. The Transitive Law 2.9 then allows to eliminate the intermediate points $p_n,..,p_1$ in row. At the end of the process, we find that p depends on F.

This establishes condition (ii) of Proposition 2.2; the other conditions are trivial. By this proposition, there is a well-determined convex structure with h as a restricted hull operator: for $A \subseteq X$ non-empty,

$$co(A) = \bigcup \{h(F) \mid F \subseteq A \text{ finite}\}$$
$$= \{x \mid x \text{ depends on } A\},$$

as required. This also shows that the corresponding convex sets are exactly the flats. To see that X is a matroid, let $p, q \notin co(A)$ and $p \in co(\{q\} \cup A)$. Note that $p \in \{q\} \cup A$ yields $p = q$, which gives the desired result. So p depends properly on $\{q\} \cup A$, yielding an independent set $F \subseteq \{q\} \cup A$ with $\{p\} \cup F$ dependent. Let $G = \{p\} \cup F \setminus \{q\}$. Then G is independent (otherwise $p \in co(A)$), and $G \cup \{q\}$ is dependent, showing that $q \in co(\{p\} \cup A)$.

Proof of (3). We first start with a matroid $(X, \mathcal{M})$, we construct the derived collection $\mathcal{E}$ of convexly independent sets, and then we pass to the matroid $(X, \mathcal{M}_\mathcal{E})$. To show that $\mathcal{M} = \mathcal{M}_\mathcal{E}$, it suffices to compare hull operators on finite sets. For $\mathcal{M}_\mathcal{E}$ this operator will be denoted by h as before.

Let $F \subseteq X$ be non-empty finite. If $x \in co(F) \setminus F$, then there is minimal set $F_0 \subseteq F$ with $x \in co(F_0)$. Hence F_0 is independent and $F_0 \cup \{x\}$ is dependent; therefore, x depends on F, i.e., $x \in h(F)$. If $x \in h(F) \setminus F$, then there is an independent set $F_0 \subseteq F$ with $\{x\} \cup F_0$ dependent. We have either $x \in co(F_0)$, or $p \in co(\{x\} \cup F_0 \setminus \{p\})$ for some $p \in F_0$. Therefore, the Exchange Law for $(X, \mathcal{M})$ yields $x \in co(F_0)$. In both situations, we conclude that $x \in co(F)$.

Conversely, we start with an independence structure $(X, \mathcal{E}$, we derive a matroid $(X, \mathcal{M})$, and we construct a corresponding family $\mathcal{E}_\mathcal{M}$ of convexly independent sets (we insist upon using the adjective "convexly" in the latter case). It suffices to show that both independence structures have the same finite sets as members. First, let F be independent (i.e., $F \in \mathcal{E}$). By the construction of the hull operator h, we have $y \notin h(F \setminus \{y\})$ for each $y \in F$, and hence F is convexly independent. Next, consider a finite convexly independent set G. If $y \in G$ then $y \notin h(G \setminus \{y\})$, that is, y does not depend on $G \setminus \{y\}$). By the non-degeneracy law, G has at least some independent subsets (for instance, singletons). Consider a maximal one, say, H. If there is a point $y \in G \setminus H$ then $H \cup \{y\}$ is independent for otherwise y depends on $G \setminus \{y\}$. From this contradiction it follows that $H = G$. ∎

The Finitary Law (IN-2) implies that a nested union of independent sets is independent. By Zorn's Lemma there is a maximal independent set; such a set is called a *basis*. The strong Replacement Law (IN-3′) leads to the following fact of importance.

§2: The Hull Operator

2.11. Theorem. *In a matroid X, the hull of a basis equals X and all bases of X have the same cardinality.* ∎

The cardinality of one (and hence of each) basis is a parameter known as the *rank* of the matroid. Rank will be defined and studied for general convex structures in Chapter II.

2.12. Cone-union property; Join-hull commutativity. We now concentrate on some other classes of convex structures, specified by an additional condition on the hull operator. Consider the following statements concerning a convex structure X.

(CUP) If $C, C_1,.., C_n \subseteq X$ are convex sets with $C \subseteq \cup_{i=1}^n C_i$ and if $a \in X$, then
$$co(\{a\} \cup C) \subseteq \cup \{ co(\{a\} \cup C_i) \mid i = 1,..,n \}.$$

(JHC) If $C \subseteq X$ is a non-empty convex set and if $a \in X$, then
$$co(\{a\} \cup C) = \cup \{ co\{a,x\} \mid x \in C \}.$$

The labels "CUP" and "JHC" stand as abbreviations for *Cone-union Property* and *Join-hull Commutativity*.

2.13. Proposition

(1) *A convex structure is JHC iff for each non-empty finite set F and for each $a \notin co(F)$,*
$$co(\{a\} \cup F) \subseteq \cup \{ co\{a,x\} \mid x \in co(F) \}.$$

(2) *JHC implies CUP. Both properties are equivalent for convex structures with finite*[2] *polytopes.* ∎

A set of type $\cup_{x \in C} co\{a,x\}$ is sometimes called the join of the point a with the set C. The left side of the formula in (1) may be replaced by $co(\cup_{x \in F} co\{a,x\})$. The whole formula can now be paraphrased as follows. The hull of the join of a with F equals the join of a with the hull of F, which may explain the name "Join-hull Commutative".

2.14. Proposition. *If $A_1,..,A_n$ are non-empty convex sets in a JHC convex structure, then*
$$co(\bigcup_{i=1}^n A_i) = \cup \{ co\{a_1,..,a_n\} \mid \forall i = 1,..,n : a_i \in A_i \}.$$

Proof. It suffices to establish the result in case of a union of two non-empty polytopes, say: $A = co(F)$ and $B = co(G)$, with $\#F = n$, $\#G = m$ finite. The result is obvious if $n + m = 2$ (i.e. if $n = 1 = m$). Suppose $n + m > 2$ and let the result be valid for values $< n + m$. For $n = 1$ the result holds by the definition of JHC. Let $n > 1$ and let
$$x \in co(F \cup G) = co(co(F) \cup co(G)).$$
For a fixed $q \in F$ we use JHC to obtain a point y with

---

2. For an extension to spaces with compact polytopes, see III§4.7.

$y \in co(G \cup F \setminus \{q\})$; $x \in co\{q, y\}$.

The induction hypothesis yields two points a', b with

$a' \in co(F \setminus \{q\})$; $b \in co(G)$; $y \in co\{a', b\}$.

Hence $x \in co\{a', b, q\}$ and by JHC again we obtain a point a such that

$a \in co\{q, a'\} \subseteq co(F)$; $x \in co\{a, b\}$.   ■

This partially motivates the following additional concept. A convex structure X is **weakly Join-hull Commutative** provided for each pair of convex sets A, $B \subseteq X$ it is true that

$A \cap B \neq \varnothing$ implies $co(A \cup B) = \bigcup \{ co\{a, b\} \mid a \in A; b \in B \}$.

An equivalent definition is given in 4.23.

2.15. Proposition. *A JHC convex structure is weakly JHC, and a weakly JHC convex structure is of arity ≤ 2.*

Proof. The first part follows from the previous result. Let A be a set and assume that $co\{x, y\} \subseteq A$ whenever $x, y \in A$. We verify the property that $co(F) \subseteq A$ for all finite $F \subseteq A$ by induction on $n = \#F$. Suppose the property to be valid for cardinalities $< n$, and let the subset F of A have exactly n points, where $n > 2$. Fix two distinct points $p, q \in F$. Application of weak JHC yields

$co(F) = \bigcup \{ co\{u, v\} \mid u \in co\{p, q\}; v \in co(F \setminus \{p\}) \}$.

Both $co\{p, q\}$ and $co(F \setminus \{p\})$ are included in A by the inductive hypothesis. Therefore, each of the intervals $co\{u, v\}$ for $u \in co\{p, q\}$ and $v \in co(F \setminus \{p\})$ is included in A. It follows that $co(F) \subseteq A$.   ■

2.16. Theorem. *A CP and CC image of a JHC space is JHC.*

Proof. Let X, Y be JHC spaces and let $f : X \to Y$ be a CP and CC surjection. Let $p \in Y$ and let $C \subseteq Y$ be convex. Fix a point $p' \in f^{-1}(p)$. Then f maps the set $co(\{p'\} \cup f^{-1}(C))$ onto $co(\{p\} \cup C)$. As the first-named set is built with segments of type $co\{p', c\}$ for $c \in f^{-1}(C)$, it follows that its image is built with segments of type $co\{p, c\}$ for $c \in C$.   ■

See 2.26.1 for additional information.

2.17. Examples. We consider spaces of arcs, matroids and H-convexities. Some other relevant examples (vector spaces with standard convexity, posets) can be handled more conveniently with the geometric conditions studied in Section 4. We begin with convexity in spaces of arcs. For practical reasons, we consider a discrete version only; a topologically relevant approach will be given in Section III§3.

A partially ordered set X is *discrete* if it has no infinite bounded chains. For each non-maximal element $a \in X$, there exists a point a^+ such that $a < a^+$ and no point $x \in X$ satisfies $a < x < a^+$. In these circumstances, a^+ is said to *cover* the point a. A discrete poset X is (**upper**) **semimodular** provided for each pair of points $a \neq b$ covering a third point and bounded from above by a point d_0, there is a point $d \leq d_0$ covering both a and b. These concepts will be formulated and studied in greater detail in Section 6.

2.17.1. Proposition. *Let S be a bounded discrete semimodular join semilattice. Then the convex structure $\Lambda(S)$ is JHC.*

Proof. Consider the arcs
$$A, B_1,.., B_n \in \Lambda(S); \quad B \in co\{A, B_1,..,B_n\}.$$
Then $B \subseteq \vee_i B_i \vee A$ and hence each $b \in B$ can be written in the form
$$b = b_1 \vee .. \vee b_n \vee a,$$
where $b_i \in B_i$ and $a \in A$. By choosing each b_i maximal in B_i with the property $b \geq b_i$ we determine a function $f_i: B \to B_i$. Clearly, $b \leq b'$ in B implies $f_i(b) \leq f_i(b')$. This shows that the collection
$$C' = \{b_1 \vee .. \vee b_n \mid b \in B\}$$
is a chain. Let C be an extension of C' to a maximal chain in $B_1 \vee .. \vee B_n$. We verify that C is a maximal chain in S. Let $c < c'$ be successive elements of C. Then
$$c = b_1 \vee .. \vee b_n; \quad c' = b'_1 \vee .. \vee b'_n,$$
where $b_i, b'_i \in B_i$ for all i. We assume that the points b_i are taken maximal and that $b'_i \geq b_i$ is taken minimal. For at least one i there is a cover pair $b_i < b'_i$. Consider a chain
$$b_i = a_0 < a_1 < .. < a_k = c$$
of successive covers. Repeated application of modularity yields a chain
$$b'_i = a'_0 < a'_1 < .. < a'_k$$
of successive covers, such that $a'_k \leq c'$ and for each j, the point a'_j covers a_j. In particular, a'_k covers c. It follows that $c < a'_k$ and $a'_k = c \vee b'_i$, which is in $B_1 \vee .. \vee B_n$. Hence $a'_k = c'$ and c' covers c.

The arc C is in between $\{B_i \mid i = 1,..,n\}$ and B is in between C and A. ∎

2.17.2. Proposition

(1) *A linear matroid is Join-hull Commutative.*
(2) *An affine matroid is weakly Join-hull Commutative.*
(3) *A projective matroid over a field is Join-hull Commutative.*

Proof. Let V be a vector space. We consider the hull operator *lin* of the linear matroid V_0, and the affine hull operator *aff* of V. First, let $A, B \subseteq V$ be linear sets. Then $x \in aff(A \cup B)$ iff it is a linear combination of some points of A (say: with sum a) and of some points of B (with sum b say). Equivalently, $x = a + b$, where $a \in A$ and $b \in B$. This shows that x is a linear combination of a point of A with a point of B, establishing (1).

Second, let $A, B \subseteq V$ be affine sets meeting in a point p. By invariance under translation, without loss of generality, $p = 0$. Then (2) follows easily from (1). Finally, the projective matroid is JHC by (1) and Proposition 2.16. ∎

It is easily seen that in a vector space of dimension ≥ 2, the affine hull of a line L and a point $p \notin L$ is properly larger than the union of all lines joining p with the elements of L: the corresponding affine matroid is JHC in dimensions ≤ 1 only. The following is another example that fails to satisfy JHC.

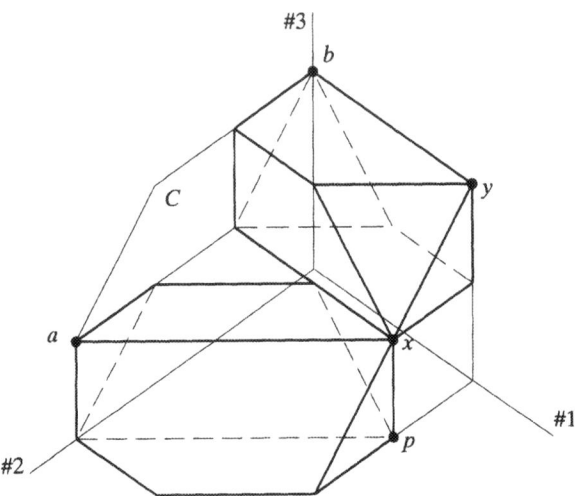

Fig. 2: Non-JHC H-convexity in $\mathbb{R}^3$

2.17.3. Symmetric H-convexity. Consider the symmetric H-convexity of $\mathbb{R}^3$, generated by the coordinate projections f_1, f_2, f_3 and their sum f_4. Let $C = co\{a,b\}$, where

$$a = (0, \tfrac{3}{4}, \tfrac{1}{4}); \quad b = (0,0,\tfrac{1}{2}),$$

and take $p = (\tfrac{1}{2}, \tfrac{1}{4}, 0)$. The set $D = \bigcup_{c \in C} co\{p,c\}$ is *not* convex. Assuming the contrary, we find that D includes the hull of $b \in C \subset D$ with the point

$$x = (\tfrac{1}{2}, \tfrac{1}{4}, \tfrac{1}{4}) \in co\{a,p\} \subset D.$$

However, consider the following point in $co\{b,x\}$:

$$y = (\tfrac{1}{2}, 0, \tfrac{1}{2}).$$

If $y \in co\{p,c\}$ for some $c \in C$, four inequalities arise, expressing that $f_i(c)$ is in between $f_i(a)$ and $f_i(b)$. The fact that $y \in co\{p,c\}$ leads to four inequalities expressing that $f_i(y)$ is in between $f_i(p)$ and $f_i(c)$. Solving for c gives

$$f_1(c) = 0; \quad f_2(c) = 0; \quad f_3(c) = \tfrac{1}{2}; \quad f_4(c) = 1,$$

contradicting that $f_1 + f_2 + f_3 = f_4$.

The result 2.17.2(3) on projective matroids over a field gives rise to the following definition. A *projective matroid (projective space)* is a Join-hull Commutative matroid in which all singletons are convex and which is not a free convexity[3].

---
3. We exclude free spaces from the definition of projective spaces to obtain a full agreement with the classical definition.

§2: The Hull Operator

2.18. Theorem. *Let X be a weakly JHC matroid and let $A, B \subseteq X$ be non-disjoint flats. If $d(C)$ denotes the rank of a flat C, then*

$$d(co(A \cup B)) + d(A \cap B) = d(A) + d(B). \qquad \text{(Rank Formula)}$$

If X is even JHC, then the above formula holds for disjoint flats as well.

Proof. We first show that if A, B are non-disjoint flats in a weakly JHC matroid and if $p \in A$, then

(*) $\qquad co(\{p\} \cup (A \cap B)) = A \cap co(\{p\} \cup B)$

To this end, fix $q \in A \cap B$. If x is in the right-hand set of (*) then by weak Join-hull Commutativity there exist $u \in co\{p,q\}$ and $v \in B$ with $x \in co\{u,v\}$. If $x \in co\{u\}$ then $x \in co\{p,q\}$ and if $v \in co\{u\}$ then $x \in co\{u,v\} \subseteq co\{u\} \subseteq co\{p,q\}$. In either case, x is in the left-hand set of (*). So assume $x, v \notin co\{u\}$. The matroid property then implies that $v \in co\{u,x\} \subseteq A$ and hence that $v \in A \cap B$. It follows that

$$x \in co(co\{p,q\} \cup (A \cap B)) = co(\{p\} \cup (A \cap B)).$$

This establishes the inclusion from right to left in (*). The opposite inclusion is trivial.

If X is JHC, then (*) holds for disjoint flats A, B as well. Indeed, if x is in the right-hand set, then $x \in co\{p,b\}$ for some $b \in B$. Repeat the previous argument with $u = p$ and $v = b$.

The rank formula can be derived as follows. If one of $d(A)$, $d(B)$ is infinite then so is $d(co(A \cup B))$ and the result is obvious. So assume $d(A), d(B) < \infty$. There is a sequence of points $(a_k)_{k=1}^n$, together with a sequence of flats $(B_k)_{k=0}^n$, such that

(i) $\quad \forall k \geq 0: B_0 = B$ and $B_{k+1} = co(B_k \cup \{a_{k+1}\})$;
(ii) $\quad \forall k \geq 0: a_{k+1} \in A \setminus B_k$;
(iii) $\quad A \subseteq B_n$.

Indeed, the prescriptions (i), (ii) indicate how to proceed by induction. Note that

$$a_k \notin co\{a_j \mid j < k\}$$

for all $k = 1,..,n$. A subset of a matroid with this property is easily seen to be independent (see Topic 2.23 below). As $d(A) < \infty$ it is clear why the process must end at some n. For all $k < n$ we find that $d(B_{k+1}) = d(B_k) + 1$ and

$$d(A \cap B_{k+1}) = d(co(\{a_{k+1}\} \cup (A \cap B_k))) = d(A \cap B_k) + 1.$$

Consequently,

$$d(A) = d(A \cap B_n) = d(A \cap B) + n; d\, co(A \cup B) = d(B_n) = d(B) + n.$$

Combining these results, we find

$$d(A \cap B) + d\, co(A \cup B) = d(A) - n + d(B) + n = d(A) + d(B). \qquad \blacksquare$$

The rank minus one of a projective set P is regarded as the **projective dimension** of P and is denoted by $dim(P)$. The **affine dimension** of an affine set A in a vector space equals its rank minus one and is also denoted by $dim(A)$. The dimension of the empty set is -1.

2.19. Proposition (*Dimension Formula*)

(1) If P_1, P_2 are projective sets in a projective space, then
$$\dim pr(P_1 \cup P_2) + \dim(P_1 \cap P_2) = \dim(P_1) + \dim(P_2).$$

(2) If A_1, A_2 are non-disjoint affine sets in a vector space, then
$$\dim aff(A_1 \cup A_2) + \dim(A_1 \cap A_2) = \dim(A_1) + \dim(A_2). \qquad \blacksquare$$

Further Topics

2.20. Convexity and social affinity (Aumann [1971]). Let S (for "society") and I (for "interests") be sets, and let $i: S \to 2^I$ be a function such that $i(x) \neq \emptyset$ for all $x \in S$. A person p has *social affinity* with a finite set of individuals $T \subseteq S$ provided his personal interests $i(p)$ are shared by the members of T:

$$i(p) \subseteq i(T) = \bigcup \{i(x) \mid x \in T\}.$$

Show that this prescription satisfies the axioms of betweenness. This leads to a convexity consisting of all sets $C \subseteq S$ which are stable in the sense that $p \in C$ whenever p has social affinity with a finite set of persons in C. Conversely, show that every convex structure arises from an "interest" function as described above. Hint: be interested in concave sets.

2.21. Convex systems and betweenness (compare Proposition 2.2). Let X be a set, let $\mathcal{A}$ be a family of finite sets in X with $\emptyset$ as a member, and let $h: \mathcal{A} \to 2^X$ be an operator with the following properties.

(H'-1) $h(\emptyset) = \emptyset$.
(H'-2) $F \subseteq h(F)$ for each $F \in \mathcal{A}$.
(H'-3) If $F \in \mathcal{A}$ and if $G \subseteq h(F)$ is a finite set, then $G \in \mathcal{A}$ and $h(G) \subseteq h(F)$.

Show that there is one and only one convex system on X with h as its partial hull operator and such that a set is admissible iff each of its finite subsets is in $\mathcal{A}$.

2.22. Matroids and independence structures

2.22.1. Show that the class of matroids is closed under the formation of subspaces. The square of the affine space $\mathbb{R}$ is *not* a matroid.

2.22.2. A set of edges of a connected graph G is called independent provided no circuit can be formed with them. Show that this determines an independence structure on the set of all non-trivial edges of G. Describe the bases of this matroid.

2.22.3. Let X be a set and $k > 0$. The k–*free convexity* of X consists of all subsets C of X such that $\#F < k$ or $C = X$. Show that the resulting space $F_k(X)$ is a matroid and that its rank equals k provided $k \leq \#X$. Note that $F_k(X)$ is a free space if $k \geq \#X$.

2.22.4. (compare Kay [1977b]) Let X be a convex structure such that for each convex set C, for each $c \in C$ and for each $x \in X \setminus C$ it is true that $xc \cap C = \{c\}$ (such spaces

§2: The Hull Operator   45

are called "affine structures" by Kay). Show that if two segments of X have more than one point in common, then these segments are equal. Deduce that if, in addition, X is weakly JHC, then it is a matroid.

2.23. Hyperflats and bases. Let X be a matroid.

2.23.1. Show that a set $A \subseteq X$ includes a basis for X iff $co(A) = X$, and (assuming A to be well-ordered) that A is independent iff no point of A depends on its predecessors.

2.23.2. A *hyperflat*[4] of a matroid is a maximal proper convex subset. Show that a set $H \subseteq X$ is a hyperflat iff there is a basis B of X and a point $b \in B$ such that $H = co(B \setminus \{b\})$.

2.24. Convex geometry. (Jamison [1980]; Edelman [1980]; for a survey, Edelman and Jamison [1985]) A convex structure is called a *convex geometry* (*anti-matroid*) provided it satisfies the following *Anti-Exchange Law*. If $A \subseteq X$ and if $p \neq q$ are in $X \setminus co(A)$, then $p \in co(\{q\} \cup A)$ implies $q \notin co(\{p\} \cup A)$.

2.24.1. Show that a space is a convex geometry iff each polytope is the hull of its extreme points. The outer points of a convex geometry are exactly the extreme points.

2.24.2. Show that the following are convex geometries: posets with the order convexity, or with the lower convexity (resp., the upper convexity), semilattices with the convexity of subsemilattices; trees; vector spaces over a totally ordered field with the standard convexity; any subspace of a convex geometry. The following are, in general, not a convex geometry: semilattices and lattices with the usual convexity (cf. 1.5), products, and cones (even over a convex geometry).

Problem. Is the property of being a convex geometry preserved by CP + CC images?

2.24.3. (Farber and Jamison [1986]; compare Soltan [1983]) A finite connected graph is a convex geometry iff it is a chordal graph. Hint. In a chordal graph, each cycle of length ≥ 4 admits a "chord" (an edge between two non-consecutive vertices), and a path is a geodesic iff it admits no chord. The extreme vertices are precisely the ones of which the set of neighbors induces a complete subgraph. Each non-extreme vertex lies on a chordless path connecting two extreme points.

2.25. Facets and normal gluings (for Euclidean spaces, Degreef [1982]; cf. 1.25). A non-empty subset Y of a convex structure X is called a *facet* of X provided it is convex and each segment of X meeting Y in more than one point is included in Y. Let X_1 and X_2 be two JHC spaces meeting in a common subspace $X_1 \cap X_2$ which is a facet of either space. Then the gluing $X_1 \cup X_2$ is normal.

---
4.   In affine or projective spaces, the term "hyperplane" is perhaps more standard.

2.26. Preservation of JHC and CUP

2.26.1. A CP and CC image of a weakly JHC convex structure is weakly JHC. The same holds for CUP.

2.26.2. Show that neither JHC or CUP are inherited by subspaces. For an example of the second, consider the following subspace of the 3–cube $[0,1]^3$ with the product (distributive lattice) convexity.

$$X = \{(x_1, x_2, x_3) \mid \exists i \in \{1,2,3\}: x_i = 0 \}.$$

The corner points of X yield a connected graph in an obvious way. The resulting space (which is also a subspace of the graphic 3–cube) is another counterexample. This violates a statement of Sierksma in [1982], that the CUP be hereditary on subspaces.

2.26.3. If X is JHC (resp., has the CUP), then so does its cone $\Delta(X)$. If X fails to be JHC, then $\Delta(X)$ is not weakly JHC.

2.26.4. (Sierksma [1982]; compare Valette [1982]) Products and disjoint sums of JHC convex structures are JHC. The same goes for CUP. A gluing of non-disjoint sets need *not* inherit JHC.

2.26.5. Deduce that the direct limit of convex structures in a system of CP + CC functions is JHC (resp., has the CUP) provided all spaces of the system are.

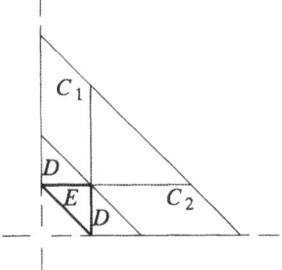

$C_1 = f_0^{-1}[\frac{1}{4}, 1] \cap f_1^{-1}[0, \frac{1}{4}];$
$C_2 = f_0^{-1}[\frac{1}{4}, 1] \cap f_2^{-1}[0, \frac{1}{4}];$
$D = f_0^{-1}[\frac{1}{4}, \frac{1}{2}] \cap f_1^{-1}[0, \rightarrow) \cap f_2^{-1}[0, \rightarrow).$

Fig. 3: Non-JHC hyperspace

2.26.6. (van de Vel [1983g]) Let $\mathbb{R}^2$ be equipped with the H-convexity symmetrically generated by the coordinate projections f_1, f_2 and their sum f_0. Prove that $\mathbb{R}^2$ is JHC under the given convexification (see also IV§5.7). Consider the convex subspace

$$X = \{x \mid 0 \le f_1(x) \le 1; 0 \le f_2(x) \le 1; \tfrac{1}{4} \le f_0(x) \le 1\}.$$

Let $\mathcal{S}$ be the Moore family of all closed convex subsets of X. Show that the convex hyperspace $\mathcal{S}_*$ is not JHC. Hint (Fig. 3). Let $E = D \cap C_1 \cap C_2$. Verify that $E \in co\{D, C_1, C_2\}$ and that there is no set $C \in co\{C_1, C_2\}$ with $E \in co\{D, C\}$.

2.27. JHC in semilattices

2.27.1. (Sholander [1954b]; compare Grätzer and Schmidt [1962]; see also 6.26) A (meet) semilattice S is said to be *(conditionally) distributive* provided each principal lower set is a distributive lattice. Show that in these circumstances there is a distributive lattice L and an embedding of (meet) semilattices $f: S \to L$ with the following properties.
(i) $f(a \vee b) = f(a) \vee f(b)$ whenever a, b are bounded from above.
(ii) If $f': S \to L'$ is an embedding into a distributive lattice L' with the above property, then there is a lattice embedding $i: L \to L'$ such that $i \circ f = f'$.
(iii) $f(S)$ is a lower subset of L.

2.27.2. Show that a distributive semilattice is JHC, and construct a semilattice which is not JHC.

Problem. Characterize Join-hull Commutativity in semilattices.

2.28. Affine isomorphisms (Murtha and Willard [1969]; continued III§5.26) Let V, W be vector spaces over a field $\mathbb{K}$ which does not admit automorphisms other than the identity (e.g., $\mathbb{K} = \mathbb{R}$). We regard these vector spaces as affine matroids. Show that if $dim(V) > 1$, then a bijective function $V \to W$ is a CP isomorphism iff it is an affine isomorphism.

This result actually shows that the terms "isomorphism of affine matroids" and "affine isomorphism" do not cause ambiguity for matroids on a vector space. The situation is different for subsets of vector spaces; cf. 3.29.

2.29. Projective geometry. Verify that projective spaces of dimension 2 (or: of rank 3) can be described by the following equivalent axiom system. A (synthetic) projective plane $\mathbb{P}$ consists of a set P of points, a disjoint set $\mathcal{L}$ (of "lines") and a relation ε of "incidence" between points and lines, such that

(P-1) If $a \neq b \in P$ then $a \, \varepsilon \, L$ and $b \, \varepsilon \, L$ holds for exactly one $L \in \mathcal{L}$. In words: two points are joined by exactly one line.
(P-2) If $L_1, L_2 \in \mathcal{L}$ then $p \, \varepsilon \, L_1$ and $p \, \varepsilon \, L_1$ holds for exactly one $p \in P$. In words: two lines meet in exactly one point.
(P-3) There exist four points in P, no three of which are on the same line.

The usual synthetic description of a projective space of dimension $n > 2$ amounts to specifying sets of "points", of "lines", of "planes", etc., with appropriate incidence properties. See Stevenson [1972] for an account of synthetic projective geometry.

2.30. Isomorphism of projective sets. Let P_1 and P_2 be two $(n-1)$-dimensional hyperflats in a projective matroid of n dimensions, $\mathbb{P}^n$. Verify that $P_1 \cap P_2$ is an $(n-2)$-dimensional flat P. Let $b \in \mathbb{P}^n \setminus (P_1 \cup P_2)$. Show that the set of all pairs (x_1, x_2), where $x_i \in P_i$ for $i = 1, 2$ are such that x_1, x_2 and b are collinear (i.e., projectively dependent), is the graph of a CP isomorphism $P_1 \to P_2$.

Notes on Section 2

Proposition 2.2 is taken from the author's paper [1983b]. It gives a general description of convex hull as a betweenness relation. A related axiom system for contact relations ("being in the closure of") was given by Aumann in [1971]. See 2.20.

The construction of a cone over a convex structure (cf. 2.4.1) is taken from the author's paper [1991]. The auxiliary convexity on $X \times [0,1]$ has been considered for $X = [0,1]$ by Jamison in [1974] under the name of "Krumm Alignment". Convexity in spaces of arcs (cf. 2.4.2) has been considered for the first time by Curtis in [1985]. However, Curtis' axioms of convexity are in terms of "convex combinations" rather than convex sets. The construction was adapted by the author in [1991] to the present setting of arcs in a semilattice. A discrete counterpart has been presented by Bandelt and van de Vel [1992]; semimodularity is defined in Birkhoff [1967].

The concept of a matroid and the related notion of independence go back to Whitney [1935]. The Exchange Law is also known as the *Mac Lane-Steinitz axiom*. The resulting theory is another well-established part of abstract convexity and of combinatorial mathematics: see Welsh [1976] or Bryant and Perfect [1980]. Most authors consider finite sets only. Infinite matroids were first studied by Rado [1949], where the axiom of "strong" replacement, (IN-3'), is derived from the much simpler axiom (IN-3) involving finite dependent sets; compare 2.10. Rado's argument involves "transversal" selection and is rather lengthy. The opposed notion of a convex geometry ("anti-matroid"; see 2.24) developed more recently to an intensively investigated branch of convexity. See Edelman and Jamison [1985] for a survey.

The condition of Join-hull Commutativity already occurred in the pioneering paper of Ellis [1952]. The term has been introduced by Kay and Womble in [1971], where Proposition 2.14 is obtained. The condition was independently studied by Calder [1971]. The Cone-union Property (CUP) has been introduced by Sierksma [1976]. It plays a role in the determination of convex invariants (see II§1 and II§2). Weak Join-hull Commutativity is introduced by Kay [1977b]. This paper contains Proposition 2.15. The counterexample in 2.17.3 appeared in van de Vel [1983g].

The rank formula 2.18 for intersecting flats is given by Kay [199*] for a class of convex structures including all weak JHC matroids.

3. Half-spaces and Separation

Half-spaces are two-sidedly convex sets. The existence of sufficiently many of them in a convex structure is required in a series of separation axioms S_1, S_2, S_3, S_4. The main result is the Polytope Screening Property, which states that S_4 (S_3) holds provided two disjoint polytopes (a polytope and a point outside) can be screened with convex sets. It can be used to handle many examples and constructions of the previous section: semilattices, products, quotients, cones, etc.; for lattices we are lead to the classical Stone Theorem characterizing distributivity. Products and subspaces behave well under the axioms S_1, S_2, S_3; the fourth axiom has a more subtle behavior. As a consequence, S_3 spaces can be characterized by means of embedding into a (Cantor) cube.

Other classes of examples, like standard convexity or order convexity, can be handled more conveniently with a different technique, developed in the next section.

3.1. Half-spaces. The concept of a half-space is a familiar one in vector spaces with standard convexity; it is often given a somewhat restricted meaning involving functionals. We propose a general, non-technical definition. Let X be a convex structure. A subset H of X is called a *half-space (hemispace, biconvex set)* provided H is both convex and concave. Note that $\varnothing$ and X are half-spaces in any convexity on X. All other half-spaces are called *non-trivial*. We present two elementary facts for later use.

3.1.1. *If $f: X \to Y$ is a CP function and if H is a half-space of Y, then $f^{-1}(H)$ is a half-space of X.* ∎

In particular, a half-space induces a relative half-space on a subspace. Note that in a free space every subset is a half-space. As the unit circle is a free subspace of the standard plane, it can be concluded that not all half-spaces of a subspace extend to half-spaces of the superspace.

3.1.2. *The half-spaces of a product $X \times Y$ are exactly the sets of type $H \times Y$, where H is a half-space of X, or of type $X \times H$, where H is a half-space of Y.*

Proof. All sets of a type described above are half-spaces: use 3.1.1 with the coordinate projections. By 1.10.2, all convex sets in $X \times Y$ are of the product type $C \times D$ with $C \subseteq X$ and $D \subseteq Y$ convex. The cases $C = \varnothing$ or $D = \varnothing$ being easy, suppose C and D are both proper non-empty subsets. Choose four points

$$x \in X \setminus C; \quad x' \in C; \quad y \in Y \setminus D; \quad y' \in D.$$

Then $(x',y), (x,y') \notin C \times D$, whereas $(x',y') \in C \times D$. By the product formula 1.10.3 for polytopes,

$$(x',y') \in co_X\{x',x\} \times co_Y\{y,y'\} = co\{(x',y),(x,y')\},$$

50   Chap. I: Abstract Convex Structures

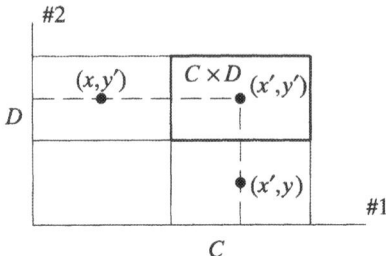

Fig. 1: $C \times D$ is not a half-space

showing that $C \times D$ is not a half-space. This leaves us with the cases $C = X$, resp., $D = Y$, where it can easily be shown that D, resp., C, is a half-space. ∎

3.2. Examples

3.2.1. H-convexity. Let V be a vector space over a totally ordered field $\mathbb{K}$, let $\mathcal{F}$ be a collection of linear functionals $V \to \mathbb{K}$, and let $\mathcal{C}$ denote the H-convexity which is symmetrically generated by $\mathcal{F}$. The following are half-spaces of the totally ordered set $\mathbb{K}$:

(*) $(\leftarrow, t)$; $(\leftarrow, t]$; $(t, \rightarrow)$; $[t, \rightarrow)$.

By definition, each $f \in \mathcal{F}$ is a CP function $V \to \mathbb{K}$, and hence by 3.1.1, the inverse image of a set in (*) is a half-space of $(V, \mathcal{C})$. Note that these half-spaces constitute a subbase.

There is a partial converse to this result. This time, let $V = \mathbb{R}^n$ and let $\mathcal{C}$ be a (not necessarily symmetric) H-convexity generated by a set of functionals $\mathcal{F}$. Let H be a nontrivial half-space of $\mathcal{C}$. As $\mathcal{C}$ is invariant under translation (cf. 1.13.7), we may assume that $\mathbf{0} \in H$. As $\mathcal{C}$ is invariant under positive homothety, the third axiom of convexity yields that the set $H_0 = \cap_{t>0} t \cdot H$ is another half-space of $\mathcal{C}$. Moreover, $\mathbf{0} \in H_0$. Let H' be a translated copy of $\mathbb{R}^n \setminus H$ containing $\mathbf{0}$, and let $H'_0 = \cap_{t>0} t \cdot H'$. One easily verifies[1] that $H_0 \cap H'_0$ is a proper linear subspace of $\mathbb{R}^n$. It determines a linear functional

$$f : \mathbb{R}^n \to \mathbb{R}$$

with $ker(f) = H_0 \cap H'_0$. Then

$$H_0 = f^{-1}(\leftarrow, 0]; \quad H'_0 = f^{-1}[0, \rightarrow)$$

(or vice versa) and it follows from invariance under translation that f is CP with respect to $\mathcal{C}$ and the order convexity of $\mathbb{R}$. Neither f nor $-f$ need to be in $\mathcal{F}$, but both can be adjoined without altering the convexity that is generated. For additional information on finitely generated H-convexity, see Topic 3.24.

3.2.2. Half-spaces in a semilattice. It can easily be verified that in a meet semilattice S with a convex subset C the collection

---

1.   Köthe [1960, p. 189]

§3: Half-Spaces and Separation   51

$$\uparrow(C) = \{y \in S \mid \exists x \in C : x \leq y\}$$

is a half-space. Conversely, any half-space $H \subseteq S$ is a lower or an upper set of S. Indeed, suppose that $a_1 \leq b_1$ and $b_2 \leq a_2$ are such that $a_i \in H$ and $b_i \notin H$ ($i = 1, 2$). Then $a = a_1 \wedge a_2 \in H$ and $b = a_1 \wedge b_2 \notin H$. However, $a \in [b, b_1]$ and hence $a \notin H$.

Note that the half-spaces of type $\uparrow(x)$ and $S \setminus \uparrow(x)$ for $x \in S$ generate the convexity of S (cf. the opening remarks in 1.8).

3.2.3. Half-spaces in a lattice. An *ideal* in a lattice L is a subset I such that
(i) $x \in I$ and $y \leq x$ imply $y \in I$;
(ii) $x, y \in I$ implies $x \vee y \in I$.

The ideal I is *prime* provided that $x \wedge y \in I$ implies x or y is in I. The notion of a (*prime*) *filter* or (*prime*) *dual ideal* is obtained if "$\leq$" is replaced by its inverse and "$\vee$" is replaced by "$\wedge$". By definition, the half-spaces of a lattice are exactly the prime ideals and the prime filters.

3.2.4. Superextensions. Let X be a convex structure. We use the abbreviation

$$\lambda(X) = \lambda(X, \mathcal{C}(X))$$

to denote the superextension of X relative to the family $\mathcal{C}(X)$ of all convex subsets. For any half-space $H \subseteq X$, by 1.8.5, (1) and (2), we have that

$$H^+ \cup (X \setminus H)^+ = \lambda(X); \quad H^+ \cap (X \setminus H)^+ = \emptyset,$$

whence H^+ is a half-space of $\lambda(X)$. Conversely, if the family $\mathcal{C}(X)$ is T_1, and if C^+ is a half-space of $\lambda(X)$, then C is a half-space of X since X embeds in $\lambda(X)$ (see the opening remarks in 1.13).

For superextensions relative to general subbases $\mathcal{S}$, the situation is more complicated and requires topological manipulations. A second problem is to decide whether every half-space of $\lambda(X, \mathcal{S})$ is of one of the types described above. If X is a finite free space, the answer is affirmative as the reader may verify. We will obtain a more general result for compact spaces in Chapter III.

3.2.5. The graphic n-cube. The product of n copies of the free space $\{0,1\}$ is called a *graphic n-cube*. By 3.1.2, the non-trivial half-spaces are exactly the $(n-1)$-faces of the cube. The term "graphic" will be given a precise meaning in Section 5.

3.3. Spaces of arcs. We next consider half-spaces in spaces of arcs in a discrete semimodular lattice. This requires some preparatory work.

3.3.1. Proposition. *Let L be a discrete semimodular lattice and let $C \subseteq L$ be a join subsemilattice.*
(1) *The collection of all arcs $A \in \Lambda(L)$ such that $A \subseteq C$ is convex in $\Lambda(L, \vee)$.*
(2) *If, in addition, C is order convex, then the collection of all $A \in \Lambda(L)$ such that $A \cap C \neq \emptyset$, is convex in $\Lambda(L, \vee)$.*

Proof. (1) is obvious. As to (2), by Join-hull commutativity (Proposition 2.17.1), it suffices to show that if $A_1, A_2 \in \Lambda(L, \vee)$ meet C, then so does $A \in co_\vee\{A_1, A_2\}$. Let $c_i \in A_i \cap C$ for $i = 1, 2$, and $c = c_1 \vee c_2$. Then $c \in C$ and, as C is order convex, we may replace each c_i with the largest point of A_i below c. Let $a \in A$ be the largest point with $a \leq c$, and let $a_i \in A_i$ be the largest point below a, $i = 1, 2$. As $A \in co_\vee\{A_1, A_2\}$, we have $a = a_1 \vee a_2$. We may assume that a is not the largest element of A (otherwise $a = c$ by the maximality of A and we are done). Consider the covering element a^+ of a in A, and let $a'_i \geq a_i$ in A_i be such that $a^+ = a'_1 \vee a'_2$. Observe that $a_i \leq c_i$ for $i = 1, 2$. We claim that at least one of these inequalities is an equality. Suppose the contrary. As $a < a^+$, we have e.g., $a_1 < a'_1$. Hence, the covering element a_1^+ of a_1 in A_1 satisfies $a_1^+ \vee a_2 = a^+$, and $a^+ \leq c$ would follow.

So we have e.g., $a_1 = c_1$. But then $c_1 \leq a \leq c$ and $a \in C$ follows since C is order convex. ∎

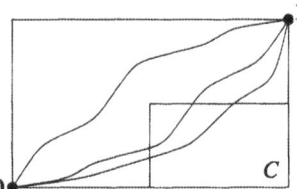

Fig. 2: Representing a half-space of $\Lambda(L, \vee)$

Observe that if $C \subseteq L$ is an order convex join subsemilattice and if $L \setminus C$ is a join subsemilattice, then the set

$$\mathcal{H} = \{ A \in \Lambda(L, \vee) \mid A \cap C \neq \emptyset \}$$

is a half-space of $\Lambda(L, \vee)$. Informally, we say that $\mathcal{H}$ is represented by the set C. The next result states that *every* half-space of $\Lambda(L, \vee)$ arises this way.

3.3.2. Proposition. *Let L be a bounded, discrete, distributive lattice. For each half-space $\mathcal{H} \subseteq \Lambda(L, \vee)$ there exists an order convex join subsemilattice $C \subseteq L$ such that $L \setminus C$ is a join subsemilattice and for each $A \in \Lambda(L)$,*

$$A \in \mathcal{H} \Leftrightarrow A \cap C \neq \emptyset.$$

Proof. We construct two sets as follows.

$$D = \bigcup \{A \mid A \in \Lambda(L) \setminus \mathcal{H}\};$$

$$C = L \setminus D = \{x \in L \mid \forall A \in \Lambda(L) : x \in A \Rightarrow A \in \mathcal{H}\}.$$

The set D is obviously a join subsemilattice. It is also evident that each arc meeting C is a member of $\mathcal{H}$. Assume next that $A \in \Lambda(L)$ is disjoint from C. For each $a \in A$ we take $A(a) \in \Lambda(L) \setminus \mathcal{H}$ pass through a. Now $A \subseteq \bigcup_{a \in A} A(a)$, and hence

$$A \in co_\vee \{A(a) \mid a \in A\}$$

We next show that C is order convex. Let $c_1 \leq c_2$ be in C and consider a point c

§3: Half-Spaces and Separation

between c_1, c_2. We verify that any arc $B \in \Lambda(L)$ passing through c is in $\mathcal{H}$. To this end, take two arcs $A_1, A_2 \in \Lambda(L)$ with $c_1 \in A_1$ and $c_2 \in A_2$. Define two sets

$$B_1 = \{x \in A_1 \mid x \leq c_1\} \cup (B \vee c_1);$$
$$B_2 = \{x \in B \mid x \leq c\} \cup (A_2 \vee c).$$

Both are arcs in L by virtue of semi-modularity. As B_i passes through c_i, we have $B_i \in \mathcal{H}$. Now $B \subseteq B_1 \subseteq B_2$, and à fortiori,

$$B \in co_\vee \{B_1, B_2\} \subseteq \mathcal{H}.$$

To complete the proof of the proposition, we show that C is a join subsemilattice. Let $c_1, c_2 \in C$ and $c = c_1 \vee c_2$. Let $B \in \Lambda(L)$ pass through c. Consider two arcs

$$B_1 = (B \wedge c_1) \cup (B \vee c_1);$$
$$B_2 = (B \wedge c_2) \cup (B \vee c_2).$$

Note that $c_i \in B_i$ and hence that $B_i \in \mathcal{H}$. For each point $b \in B$ we have either that $b \leq c$ (in which case $b = b \vee c_i \in B_i$ for both i), or $c \leq b$. In the latter case, we have by distributivity that

$$b = c \vee b = (c_1 \vee b) \vee (c_2 \vee b) \in B_1 \vee B_2.$$

We conclude that $B \in co\{B_1, B_2\} \subseteq \mathcal{H}$. ∎

The construction of C from $\mathcal{H}$ also leads to the following by-product.

3.3.3. Proposition. *Let L be a bounded, discrete, distributive lattice. A half-space $\mathcal{H} \subseteq \Lambda(L, \vee, \wedge)$ can be represented by an order convex sublattice $C \subseteq L$ such that $L \setminus C$ is a sublattice. Conversely, if $C \subseteq L$ is as described, then the set $\mathcal{H}$, consisting of all $A \in \Lambda(L, \vee, \wedge)$ meeting C, is a half-space.* ∎

3.4. Separation axioms. The following statements concerning a convex structure X will be considered in depth.

S$_1$:   all singletons in X are convex.
S$_2$:   if $x_1 \neq x_2 \in X$ then there is a half-space $H \subseteq X$ with

$$x_1 \in H; x_2 \notin H.$$

S$_3$:   if $C \subseteq X$ is convex and if $x \in X \setminus C$ then there is a half-space H of X with

$$C \subseteq H; x \notin H.$$

S$_4$:   if $C, D \subseteq X$ are disjoint convex sets then there is a half-space H of X with

$$C \subseteq H; D \subseteq X \setminus H.$$

In each of the statements S$_2$, S$_3$, S$_4$, we have a half-space H and two sets A, B with

$$A \subseteq H, B \subseteq X \setminus H.$$

In these circumstances we say that H *separates A from B,* or, that A and B are *in complementary half-spaces*. The properties S$_1$,..,S$_4$ are called *separation axioms*. If X satisfies the axiom S$_i$ then X is called an S$_i$ *convex structure* and $\mathcal{C}(X)$ is called an S$_i$ *convexity*. Clearly, S$_2$

implies S_1, and under assumption of S_1,

$S_4 \Rightarrow S_3 \Rightarrow S_2$.

Instead of S_1 we sometimes use the term *point-convex*. We next describe a related type of separation between sets.

3.5. Screening. Let X be an arbitrary set, and let A, $B \subseteq X$ be disjoint. We say that A, B are *screened* with the subsets C, D (in that order) of X provided

$A \subseteq C \setminus D; \quad B \subseteq D \setminus C; \quad C \cup D = X.$

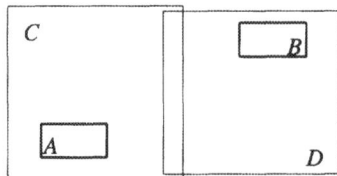

Fig. 3: Screening of A, B with C, D

Equivalently (Fig. 3),

$A \cap D = \emptyset = B \cap C; \quad C \cup D = X.$

Regarded this way, separating sets with a half-space H amounts to screening these sets with the convex pair H, $X \setminus H$. We now work towards a fundamental result on separation with half-spaces.

3.6. Extension of a convex set. Let X be a convex structure and let A, $B \subseteq X$ be non-empty. The *extension of A away from B* is the set

$A / B = \{x \mid co(\{x\} \cup B) \cap A \neq \emptyset\}.$

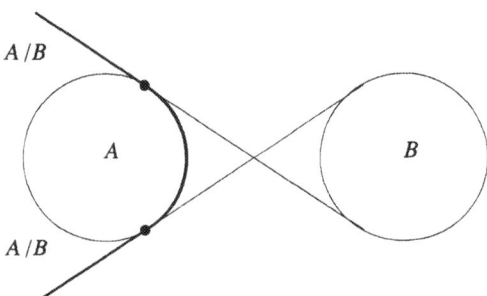

Fig. 4: Extension of convex sets

Observe that $A / B = \cup_{a \in A} \{a\} / B$. Fig. 4 gives an impression of A / B in the standard convexity of the plane.

§3: Half-Spaces and Separation   55

3.7. Lemma. *Let X be a convex structure, and let $A, B \subseteq X$ be non-empty convex sets. Then $A \mid B$ is convex provided for each pair of non-empty polytopes $P \subseteq A$, $Q \subseteq B$, the set $P \mid Q$ is convex.*

Proof. Let $x_1,..,x_n \in A \mid B$. Then there exist $a_1,..,a_n \in A$ with $a_i \in co(\{x_i\} \cup B)$ for $i = 1,..,n$. For each i we obtain a finite set $G_i \subseteq B$ with $a_i \in co(\{x_i\} \cup G_i)$. Taking $F = \{a_1,..,a_n\}$ and $G = \cup_{i=1}^{n} G_i$, we obtain $x_i \in \{a_i\} \mid co(G)$ and hence

$$x_1,..,x_n \in co(F) \mid co(G).$$

As $co(F) \mid co(G)$ is a convex set, it follows that $co\{x_1,..,x_n\} \subseteq A \mid B$. ∎

3.8. Theorem *(Polytope Screening Characterization)*

(1) *A convex structure is S_3, iff each polytope and each point in its complement can be screened with convex sets, iff for each point p and for each non-empty polytope P the extension $\{p\} \mid P$ is convex.*

(2) *A convex structure is S_4, iff each pair of disjoint polytopes can be screened with convex sets, iff for each pair of non-empty polytopes P and Q the extension $P \mid Q$ is convex.*

Proof. The argument proving (1) is similar to (but simpler than) the argument proving (2). We therefore concentrate on the second statement only. If X is an S_4 convex structure, then evidently each pair of disjoint polytopes can be screened with convex sets.

Assume that the latter statement holds. Let P and Q be non-empty polytopes, and let $x \in co(P \mid Q) \setminus (P \mid Q)$. Then $co(\{x\} \cup Q)$ is disjoint from P. Consider a screening of $co(\{x\} \cup Q)$ and P with convex sets C, D:

$$\{x\} \cup Q \subseteq X \setminus D; \quad P \subseteq X \setminus C; \quad C \cup D = X.$$

If $p \in P \mid Q$, then $co(\{p\} \cup Q)$ meets P, which implies that $p \in D$. Therefore $co(P \mid Q) \subseteq D$, contradicting that $x \notin D$.

Finally, assume that $P \mid Q$ is convex for all non-empty polytopes P, Q of X. By the previous lemma, we find that $A \mid B$ is convex whenever $A, B \subseteq X$ are non-empty and convex. Suppose A, B are disjoint convex sets in X. Then there exist convex sets $A_0 \supseteq A$ and $B_0 \supseteq B$, maximal with the property that $A_0 \cap B_0 = \emptyset$. If one of them is empty, then the other one equals X, and we have a separating half-space. Assume both sets are non-empty. If $x \in X \setminus B_0$, then $co(\{x\} \cup B_0) \cap A_0 \neq \emptyset$ by maximality. This shows that the convex set $A_0 \mid B_0$ includes $X \setminus B_0$. If $A_0 \cup B_0 \neq X$, then $(A_0 \mid B_0)$ is properly larger than A_0 and hence it meets B_0. However, the existence of a point $b \in B_0$ such that

$$co(\{b\} \cup B_0) \cap A_0 \neq \emptyset$$

contradicts the fact that B_0 is convex. ∎

It is important to have the order of objects right in part (1). Convexity of $P/\{p\}$ is, in general, not characteristic of S_3. In spaces of arity two, this property is in fact characteristic of S_4 (see the topics section or Section 4). As a consequence of 1.7.3, we have the following result.

3.9. Corollary. *A convex structure is S_3 iff it is generated by half-spaces.* ∎

The condition, that an extension set be convex, can be verified by using only subspaces which are polytopes. This directly leads to the following conclusion.

3.10. Corollary. *A convex structure is S_3 (resp., S_4) iff each polytope is.* ∎

3.11. Corollary. *Let $f : X \to Y$ be a convexity preserving and convex-to-convex surjection. If X is S_4 then so is Y.*

Proof. Let A, B be disjoint convex subsets of Y. Then $f^{-1}(A)$ and $f^{-1}(B)$ are disjoint and convex in X, and hence they can be screened with convex sets C, D:

$$f^{-1}(A) \cap D = \emptyset = f^{-1}(B) \cap C; \quad C \cup D = X.$$

Then $f(C)$, $f(D)$ are convex sets covering Y whereas

$$A \cap f(D) = \emptyset = B \cap f(C).$$

By Theorem 3.8, Y has to be S_4. ∎

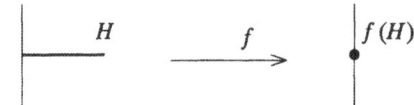

Fig. 5: CP and CC image of a half-space

We note that a CP and CC surjection need not map half-spaces to half-spaces. For a simple counterexample, consider a tree which looks like the character "T", and collapse one of its branches (Fig. 5). This results into a totally ordered space. An end segment of the collapsed branch is a half-space which maps to the "central" point of the quotient. Hence the definition of S_4 is of no use in proving Corollary 3.11.

3.12. Examples

3.12.1. Symmetric H-convexity. A symmetric H-convexity admits a subbase of half-spaces (see 3.2.1) and hence is S_3 by the last corollary. In general it is *not* S_4. Consider the symmetric H-convexity of $\mathbb{R}^3$, generated by the three coordinate projections and their sum. Let C be the unit 3-simplex with vertices

$$(0,0,0); \quad (0,0,1); \quad (0,1,0); \quad (1,0,0).$$

Let D be the translated copy $C + (½,-½,-⅔)$. Then C and D are convex sets of the given H-convexity and $C \cap D = \emptyset$ as one can verify. However, no plane in $\mathbb{R}^3$ parallel to one of the C-faces separates between C and D (cf. Fig. 6; the reader should check this). Apparently, this H-convexity is not S_4. In II§4.9, we will give evidence to the fact that all symmetric H-convexities in $\mathbb{R}^2$ are S_4.

3.12.2. Proposition. *A semilattice has the separation property S_4.*

§3: Half-Spaces and Separation

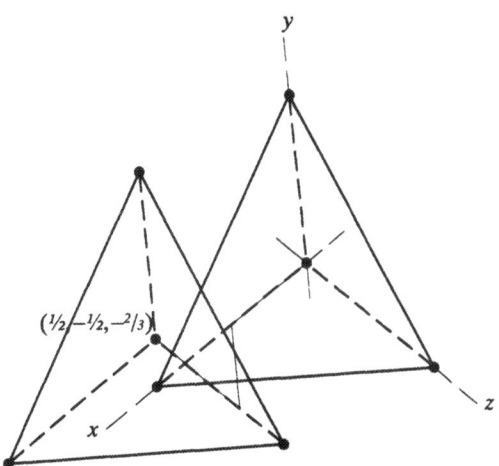

Fig. 6: An H-convexity which is not S_4

Proof. Let C, D be disjoint convex subsets of a semilattice S. We consider the sets $\downarrow(C)$ and $\downarrow(D)$ of all points below some element of C, resp., D. If
$$\downarrow(C) \cap D \neq \emptyset \neq C \cap \downarrow(D)$$
then there exist points $c_1, c_2 \in C$ and $d_1, d_2 \in D$ with $d_1 \leq c_1$ and $c_2 \leq d_2$. As C, D are subsemilattices of S we find that $c_1 \wedge c_2 \in C$ and $d_1 \wedge d_2 \in D$. As C, D are order convex, the inequalities
$$c_1 \wedge c_2 \leq c_1 \wedge d_2 \leq c_1; \quad d_1 \wedge d_2 \leq c_1 \wedge d_2 \leq d_2$$
yield that $c_1 \wedge d_2$ is common to C and D, a contradiction.

So, assume that C is disjoint from $\downarrow(D)$. This implies that D is disjoint from $\uparrow(C)$. The set $\uparrow(C)$ is easily seen to be a half-space; cf. 3.2.2. It includes C and is disjoint from D, as required. ∎

3.12.3. Proposition. *A lattice is S_4 iff it is distributive.*

Proof. Let L be a distributive lattice, and let $C, D \subseteq L$ be disjoint and convex. As in the previous example we find e.g., that
$$\downarrow(C) \cap D = \emptyset.$$
Note that $I = \downarrow(C)$ is an ideal of L. Let $I_0 \supseteq I$ be an ideal, maximal with the property of being disjoint from D. If I_0 is not prime then there exist $x, y \in L \setminus I_0$ such that $x \wedge y \in I_0$. But I_0 is included in the ideal
$$I_1 = \{u \mid \exists v \in I_0 : u \leq v \vee x\}$$
and $x \in I_1$. Then I_1, being properly larger than I_0, meets D. This yields two points

$u_1 \in D$, $v_1 \in I_0$ with $u_1 \leq v_1 \vee x$. Similarly, we obtain two points $u_2 \in D$ and $v_2 \in I_0$ with $u_2 \leq v_2 \vee y$. Distributivity of L implies that

$$u_1 \wedge u_2 \leq (v_1 \vee x) \wedge (v_2 \vee y) = (v_1 \wedge v_2) \vee (v_1 \wedge y) \vee (v_2 \wedge x) \vee (x \wedge y),$$

where $u_1 \wedge u_2 \in D$ and the right hand element is in I_0, a contradiction. We conclude that I_0 is a prime ideal -- hence a half-space, cf. 3.2.3 -- which separates C and D.

We finally show that S_2 implies distributivity (since L is evidently point-convex, this also shows that the axioms S_2, S_3, S_4 are equivalent for lattices).

$$m(a,b,c) = (a \wedge b) \vee (b \wedge c) \vee (c \wedge a);$$
$$M(a,b,c) = (a \vee b) \wedge (b \vee c) \wedge (c \vee a).$$

Then $m(a,b,c) \leq M(a,b,c)$ and

(*) $co\{a,b\} \cap co\{b,c\} \cap co\{c,a\} = \{x \mid m(a,b,c) \leq x \leq M(a,b,c)\}.$

Hence the left hand set is non-empty. Suppose it has two distinct elements u, v. As L is S_2 we find a half-space H with $u \in H$ and $v \notin H$. But $u \in H$ implies that two of a, b, c are in H, whereas $v \in L \setminus H$ implies that two of them are in $L \setminus H$. It follows from (*) that $m = M$ throughout, which is characteristic of distributive lattices (cf. 1.23.1). ∎

Example 3.12.3 is a reformulation of a classical result of M.H. Stone on distributive lattices. It requires the axiom of choice (AC), as does Theorem 3.8. This involvement of AC is a rather typical phenomenon in the context of separation properties beyond S_2.

We continue with some consequences of Theorem 3.8.

3.13. Corollary. *Let $(X, \mathcal{S})$ be a closure space such that every pair of disjoint closed sets, of which at least one is spanned by a finite set, can be screened with closed sets. Then the Vietoris convexity of the convex hyperspace $\mathcal{S}_*$ is S_4.*

Proof. Recall that $\mathcal{S}_* = \mathcal{S} \setminus \{\varnothing\}$. Let

$$\mathcal{A} = \{A_1,..,A_n\} \quad \text{and} \quad \mathcal{B} = \{B_1,..,B_m\}$$

be finite subsets of $\mathcal{S}_*$ such that $co(\mathcal{A}) \cap co(\mathcal{B}) = \varnothing$. Consider the closed sets

$$A = cl(\cup \mathcal{A}) \quad \text{and} \quad B = cl(\cup \mathcal{B}).$$

If $A \cap B$ meets all sets of type $cl\{a_1,..,a_n\}$ (with $a_i \in A_i$) and all sets of type $cl\{b_1,..,b_n\}$ (with $b_i \in B_i$), then by the Hyperspace Polytope Formula 1.8.3, $A \cap B \in co(\mathcal{A}) \cap co(\mathcal{B})$. Therefore we have, for instance, that

$$B \cap cl\{a_1,..,a_n\} = A \cap B \cap cl\{a_1,..,a_n\} = \varnothing.$$

Let B', C' be closed sets screening B and $cl\{a_1,..,a_n\}$:

$$B \subseteq X \setminus C'; \quad C \subseteq X \setminus B'; \quad B' \cup C' = X.$$

With the usual "bracket" notation for hyperspaces, we find that the sets

$$\mathcal{S}_* \cap <B'> \quad \text{and} \quad \mathcal{S}_* \cap <C',X>$$

form a screening of $\mathcal{A}$ and $\mathcal{B}$ with convex subsets of $\mathcal{S}_*$. ∎

§3: Half-Spaces and Separation   59

3.14. Corollary. *The cone of an* S_4 *convex structure is* S_4.

Proof. Let X be an S_4 convex structure. We recall that the cone $\Delta(X)$ of X is obtained as the quotient

$$X \times [0,1] \ / \ X \times \{1\},$$

where $X \times [0,1]$ carries the convexity determined by the polytopes

$$co\{(x_1,t_1),..,(x_n,t_n)\} = \{\ (x,t)\ |\ \min_i t_i \leq t \leq \max_i t_i;\ x \in co\{x_i\,|\,t_i {\leq} t\}\ \}.$$

The quotient function is CP by construction, and is CC because the saturation of a convex set is (obviously) convex. By Corollary 3.11 it suffices to show that the convexity of $X \times [0,1]$ is S_4. For $i = 1,..,m$ and $j = 1,..,n$ let $a_i = (x_i,t_i)$ and $b_j = (y_j,s_j)$ be points of $X \times [0,1]$ with

$$co\{a_1,..,a_m\} \cap co\{b_1,..,b_n\} = \emptyset.$$

We assume that $\max_i t_i \leq \max_j s_j$. Considering the maximal t_i–level, we have

$$co\{x_1,..,x_n\} \cap co\{y_j\,|\,s_j \leq \max t_i\} = \emptyset,$$

yielding a half-space H of X with

$$co\{x_1,..,x_n\} \subseteq H;\quad co\{y_j\ |\ s_j \leq \max t_i\} \subseteq X \setminus H.$$

Then $H \times [0, \max t_i]$ is a half-space and -- as the t-level section of a convex set decreases when t does -- we conclude that this half-space separates between $co\{a_1,..,a_n\}$ and $co\{b_1,..,b_n\}$. Theorem 3.8 gives the desired result. ■

3.15. Theorem

(1)   *A [convex] subspace of an* S_j *convex structure is* S_j *for* $j = 1, 2, 3, [4]$.
(2)   *The product of a family of* S_j *convex structures is* S_j *for* $j = 1, 2, 3, 4$.

Proof of (1): easy.

Proof of (2). The cases $j = 1, 2$ are left to the reader. For $j = 3$, let $X = \prod_{i \in I} X_i$ where each factor X_i is S_3. Let $F \subseteq X$ be finite and $x \notin co(F)$. If $\pi_i : X \to X_i$ is the ith projection, then (cf. 1.10.3)

$$co(F) = \prod_{i \in I} co(\pi_i(F)).$$

Hence $\pi_i(x) \notin co\,\pi_i(F)$ for some $i \in I$. Then separate with a half-space in X_i and invert it to X with π_i. The result follows from Theorem 3.8(1).

For S_4 convexities a similar argument works. ■

A non-convex subset of an S_4 convex structure need not be S_4: let $\mathbb{R}^4$ be equipped with the (4–fold) product convexity. The subspace with equation

$$x_1 + x_2 + x_3 = x_4$$

is CP isomorphic with the space $\mathbb{R}^3$, equipped with the symmetric H-convexity of Example 3.12.1. This space is S_3, but not S_4. For a simple example in the plane, see 3.28.

A *Cantor cube* is a cube of which all factors are two-point totally ordered sets. The standard type of example is $\{0,1\}^X$, where X is some index set. Cantor cubes are involved in some of the next results.

3.16. Lemma. *Let $\mathcal{F}$ be a collection of CP functionals $X \to \{0,1\}$, which*

(i) *is point-separating, that is: if $p \neq q$ are in X, then $f(p) \neq f(q)$ for some $f \in \mathcal{F}$, and*
(ii) *separates polytopes from points, that is: if $P \subseteq X$ is a non-empty polytope and if $p \notin P$ then $f(p) \notin f(P)$ for some $f \in \mathcal{F}$.*

In these circumstances, the function
$$F: X \to \{0,1\}^{\mathcal{F}}, \quad x \mapsto (f(x))_{f \in \mathcal{F}},$$
is an embedding of X into a Cantor cube.

Proof. Clearly, F is a CP function into $Q = \{0,1\}^{\mathcal{F}}$ which is injective by (i). Let $P \subseteq X$ be a non-empty polytope, and let $p \in X \setminus P$. By (ii), $f(p) \notin f(P)$ for some $f \in \mathcal{F}$, whence $F(p) \notin \prod_{f \in \mathcal{F}} f(P)$. The latter is a convex set of Q, and it follows that $F(P)$ is its trace on $F(X)$. We see that the convexity of $F(X)$, copied from X, has the same polytopes as the relative convexity, derived from the Cantor cube. Therefore, both convexities are the same. ∎

3.17. Corollary. *A point-convex space is S_3 iff it embeds in a Cantor cube.*

Proof. A Cantor cube is a product of free 2–point spaces, and it is S_4 by Theorem 3.15. As its singletons are convex, the cube is also S_3, and so is each of its subspaces. On the other hand, let X be an S_3 convex structure with convex singletons. For each non-trivial half-space H of X there is a CP function
$$f = f_H : X \to \{0,1\}$$
with $f(x) = 0$ for $x \in H$ and $f(x) = 1$ otherwise. The resulting collection of functionals is point-separating and it separates convex sets from points since X is point-convex and S_3. The result follows from the previous lemma. ∎

The previous results on embedding have several consequences. Let S be a meet semilattice. If $P \subseteq S$ is a polytope and if $p \notin P$, then (according to the argument in 3.12.2) either $p \notin \uparrow(P)$ or $P \cap \uparrow(p) = \varnothing$. Each of $\uparrow(P)$, $\uparrow(p)$ is a half-space of S. It follows that there is a semilattice homomorphism $f : S \to \{0,1\}$ separating p from P.

In a distributive lattice L, for each polytope P and each point $p \notin P$ there is a prime ideal including one of p, P and disjoint from the other. The function f, mapping the ideal to 0 and its complement to 1, is a lattice homomorphism.

Two elements x, y of a lattice L with universal bounds **0**, **1** are *complementary* provided $x \wedge y = \mathbf{0}$ and $x \vee y = \mathbf{1}$. In a distributive lattice, each element can have at most one complement. A function f as above of necessity maps one element of a complementary pair to 0 and the other one to 1. Recall that a *Boolean lattice* is a complemented distributive lattice. For instance, the Cantor cube $\{0,1\}^X$ is a Boolean lattice isomorphic to the power set of X. If the operation of taking complements is added to the structure, the term *Boolean*

§3: Half-Spaces and Separation   61

algebra is used. Derived concepts such as "Boolean subalgebra" are self-explaining. Combining the above observations with Lemma 3.16, we have:

3.18. Corollary.

(1) *A meet semilattice embeds as a subsemilattice of a Cantor cube.*
(2) *A distributive lattice embeds as a sublattice of a Cantor cube.*
(3) *A Boolean algebra embeds as a Boolean subalgebra of a power set.*

In either situation, a universal bound can be mapped to a universal bound of the cube (power set). ∎

A similar result can be obtained for median algebras; cf. 6.13. Part (3) of the above corollary is a version of the classic *Stone representation theorem*. The next result describes a combinatorial feature of the axiom S_4. The case $n = 2$ will be reconsidered in the next section to characterize S_4 for convexities of arity two.

3.19. Theorem. *Let X be an S_4 convex structure, let $p \in X$, and for $n \geq 2$ let*
$$F = \{a_1,..,a_n\}; \quad G = \{b_1,..,b_n\}$$
be subsets of X such that $b_i \in pa_i$ for each i. If $F_i = \{a_j \mid j \neq i\}$ and $G_i = \{b_j \mid j \neq i\}$, then
$$\bigcap_{i=1}^{n} co(F_i \cup \{b_i\}) \neq \emptyset \neq \bigcap_{i=1}^{n} co(G_i \cup \{a_i\}).$$

Proof. For $n = 2$ both formulae amount to the same. If $a_1b_2 \cap a_2b_1 = \emptyset$, we may consider a half-space H including a_1b_2 and disjoint from a_2b_1. But $p \in H$ implies $b_1 \in H$ and $p \notin H$ implies $b_2 \notin H$, contradiction.

We proceed by induction, assuming the results to be valid for sets with n points. Let $F = \{a_1,..,a_{n+1}\}$ and $G = \{b_1,..,b_{n+1}\}$. We use subscripts as in "$F_{i,j,k}$" to indicate the subset of F in which the points labeled with i, j, k are neglected. Consider a point $x \in \bigcap_{i=1}^{n} co(F_{i,n+1} \cup \{b_i\})$. Then
$$xa_{n+1} \subseteq \bigcap_{i=1}^{n} co(F_i \cup \{b_i\}).$$
We are finished with the first formula if xa_{n+1} meets $co(F_{n+1} \cup \{b_{n+1}\})$. Assuming this to be false, consider a half-space H including F_{n+1} and b_{n+1}, and such that $x, a_{n+1} \notin H$. Then $p \in H$ (otherwise $b_{n+1} \in pa_{n+1} \subseteq X \setminus H$) and consequently $b_n \in pa_n \subseteq H$. Therefore, $x \in co(F_{n,n+1} \cup \{b_n\}) \subseteq H$, a contradiction.

The second statement follows from a similar argument: take a point y in $\bigcap_{i=1}^{n} co(G_{i,n+1} \cup \{a_i\})$, and verify that yb_{n+1} meets $co(G_{n+1} \cup \{a_{n+1}\})$. ∎

This leads us to the following result on simultaneous separation.

3.20. Theorem. *Let X be a JHC and S_4 convex structure and let $C_1,.., C_n \subseteq X$ be convex sets with $\bigcap_{i=1}^{n} C_i = \emptyset$. Then there exist half-spaces $H_1,.., H_n$ in X with the following properties.*

$$C_i \subseteq H_i \ (i = 1,..,n); \quad \bigcap_{i=1}^{n} H_i = \emptyset; \quad \bigcup_{i=1}^{n} H_i = X.$$

Proof. Let $H_1 \supseteq C_1,..,H_n \supseteq C_n$ be maximal convex sets with the property that $\bigcap_{i=1}^{n} H_i = \emptyset$. Suppose H_1 is not a half-space. By S_4, there is a half-space $H \supseteq H_1$ disjoint from $\bigcap_{i=2}^{n} H_i$. Now H is properly larger than H_1, a contradiction.

Suppose $p \in X \setminus (\bigcup_{i=1}^{n} H_i)$. By maximality, the set $co(\{p\} \cup H_i)$ meets $\bigcap_{j \neq i} H_j$ in some point b_i. By JHC, there is a point $a_i \in H_i$ with $b_i \in pa_i$. By Theorem 3.19, the sets

$$P_i = co(\{b_j \mid j \neq i\} \cup \{a_i\}) \quad (i = 1,..,n)$$

have a point in common. However, we have $P_i \subseteq H_i$ for all i, a contradiction. ∎

An inspection of this proof shows that if X fails to be JHC, then half-spaces can still be constructed satisfying the first and second condition.

3.21. Normal families. A collection of sets $\mathcal{S} \subseteq 2^X$ is *normal* provided each pair of disjoint members of $\mathcal{S}$ can be screened with members of $\mathcal{S}$. For instance, if $\mathcal{S}$ is a convexity on X, then in regard to Theorem 3.8, normality is just the property S_4. If $\mathcal{S}$ is a topology, then normality reduces to the standard meaning.

3.21.1. Proposition. *A point-convex space is S_3 iff it can be generated by a normal T_1 subbase.*

Proof. The collection $\mathcal{H}$ of all half-spaces of a convex structure is a normal family, which generates the convexity under assumption of S_3, cf. Corollary 3.9. If the space is moreover S_1, then each singleton is the intersection of half-spaces and $\mathcal{H}$ is evidently T_1. Conversely, let $\mathcal{S}$ be a normal T_1 subbase for the convex structure X. If $P \subseteq X$ is a polytope and if $p \notin P$, then by Proposition 1.7.3, we have $P \subseteq S$ and $p \notin S$ for some $S \in \mathcal{S}$. By T_1 there exist $S' \in \mathcal{S}$ such that $S \cap S' = \emptyset$ and $p \in S'$. By normality, S and S' can be screened with members of $\mathcal{S}$. By Theorem 3.8, X is S_3. ∎

As an illustration, we consider a set X with a collection $\mathcal{S}$ of subsets. The superextension $\lambda(X, \mathcal{S})$ carries the natural convexity, generated by the family $\mathcal{S}^+$ of all sets of type

$$S^+ = \{m \in \lambda(X, \mathcal{S}) \mid S \in m\}$$

(cf. 1.8.4). Note that $\lambda(X, \mathcal{S})$ is point-convex. In general, $\lambda(X, \mathcal{S})$ has no particular separation property beyond S_1.

3.21.2. Proposition. *Let $\mathcal{S} \subseteq 2^X$ be a normal family. Then $\lambda(X, \mathcal{S})$ is S_4. In fact, each pair of disjoint polytopes of $\lambda(X, \mathcal{S})$ can be screened with subbasic sets.*

Proof. Let $F_1, F_2 \subseteq \lambda(X, \mathcal{S})$ be two non-empty finite sets spanning disjoint polytopes. For $i = 1, 2$, let $\mathcal{S}_i$ consist of all members of $\mathcal{S}$ common to all elements of F_i. If all sets in $\mathcal{S}_1$ meet all sets in $\mathcal{S}_2$, then $\mathcal{S}_1 \cup \mathcal{S}_2$ extends to a maximal linked system which is a common element of the polytopes. We conclude that there exist $S_i \in \mathcal{S}_i$ ($i = 1, 2$) such that $S_1 \cap S_2 = \emptyset$. Consider a screening with $T_1, T_2 \in \mathcal{S}$:

§3: Half-Spaces and Separation

$$S_1 \subseteq T_1 \setminus T_2; \quad S_2 \subseteq T_2 \setminus T_1; \quad T_1 \cup T_2 = X.$$

By 1.8.5 it follows that

$$F_1 \subseteq T_1^+ \setminus T_2^+; \quad F_2 \subseteq T_2^+ \setminus T_1^+; \quad T_1^+ \cup T_2^+ = \lambda(X, \delta).$$

Therefore, $co(F_1)$ and $co(F_2)$ are screened by T_1^+, T_2^+. By Theorem 3.8, $\lambda(X, \delta)$ is S_4. ∎

Further Topics

3.22. Separation and arity (Chepoi [1991])

3.22.1. Show that the following are equivalent for a convex structure X of arity n.

(i) X is S_4;
(ii) If a k-polytope P is disjoint from a l-polytope Q and if $k, l \leq n$, then P and Q can be screened with convex sets.
(iii) If P is an n-polytope and Q is an $(n-1)$-polytope, then P/Q is convex.

3.22.2. Let $n \geq 2$ and let X be the union of three disjoint sets $X_0, A, \{b\}$, where X_0 consists of $2n-2$ elements and A is a two-element set. Let $\mathcal{C}$ consist of all subsets of X of one of the following types. (i) All sets not containing b. (ii) All sets containing b and having at most $n-2$ elements in X_0. (iii) All sets containing b and including A. This determines a convexity on X of arity n, such that each $(n-1)$-polytope P and each convex set disjoint from P can be separated with complementary half-spaces, but $(X, \mathcal{C})$ is not S_4.

3.23. Expansion of convex structures (Mulder [1978]; Bandelt and Hedlíková [1983]; continued 6.38.3 and II§3.22).

Let $X = A \cup B$ be a convex structure. The *expansion* $E = E(X; A, B)$ *of* X *along* A *and* B is the subspace

$$A \times \{0\} \cup B \times \{1\}$$

of the product $X \times \{0, 1\}$ (where the second factor is given the discrete convexity). If $A \cap B \neq \emptyset$ then at least some points of X will be "split" in the expanded space. If A, B are both convex, then we use the term *convex expansion*. Verify the following statements.

3.23.1. The natural function $E \to X$, $(x, i) \mapsto x$, is CP.

3.23.2. If X is S_i, then so is E, for $i = 1, 2, 3$.

3.23.3. If E is a convex expansion, then the natural function is CC. Moreover, E is S_4, resp., of arity 2, iff X is.

3.23.4. Let A, B be half-spaces of X such that $X = A \cup B$ and if $a \in A$ and $b \in B$, then $A \cap B \cap co\{a, b\} \neq \emptyset$. Show that if X is JHC, then so is $E(X; A, B)$.

3.23.5. Let abc be a geometric simplex in $\mathbb{R}^2$, equipped with the standard convexity. Give a screening of ac and b with half-spaces such that the resulting convex expansion of abc is not JHC.

3.24. Copoints. Let X be a convex structure and let $p \in X$. A *copoint at* p is a convex set $C \subseteq X$ maximal with the property $p \notin C$. Conversely, p is an *attaching point of* C. For instance, a branch of a tree (cf. 1.5.3) is a copoint.

3.24.1. Note that the family of all copoints is an *intersectional subbase*, i.e., each convex set is the intersection of subbasic sets, and that it is the smallest subbase of this kind (compare Hammer [1955b], who uses the term *semispace* instead of copoint). Verify that a convex structure is S_3 iff each copoint is a half-space.

3.24.2. (Jamison [1980]) Show that a convex geometry is characterized by either of the following properties:
(i) For each copoint C at p the set $C \cup \{p\}$ is convex.
(ii) Each copoint has exactly one attaching point.

3.24.3. Let $\mathcal{C}$ be the join of finitely many convexities $\mathcal{C}_i$ on a set X. Show that all copoints of $(X, \mathcal{C})$ are in $\cup_i \mathcal{C}_i$.

3.24.4. Let V be a vector space over a totally ordered field $\mathbb{K}$, and let $\mathcal{F}$ be a finite collection of linear functionals $V \to \mathbb{K}$. Show that all half-spaces of the convexity generated by $\mathcal{F}$ are subbasic.

3.25. Posets. Throughout, X denotes a partially ordered set.

3.25.1. For each point $p \in X$ the following are half-spaces.

$$\downarrow(p) = \{x \mid x \leq p\}; \quad \downarrow(p) \setminus \{p\};$$
$$\uparrow(p) = \{x \mid p \leq x\}; \quad \uparrow(p) \setminus \{p\}.$$

Note that the only possible copoints at p are $X \setminus \downarrow(p)$ and $X \setminus \uparrow(p)$. What happens if p is extreme (i.e., is minimal or maximal; cf. 1.23.4)?

3.25.2. (Franklin [1962]) Verify the following property, much stronger than JHC:

$$\forall A \subseteq X: co(A) = \cup \{co\{a_1, a_2\} \mid a_1, a_2 \in A\}.$$

3.26. Separation properties in graphs

3.26.1. (compare Bandelt [1989]). Design three connected graphs G_1, G_2, G_3 such that G_i is S_i but not S_{i+1}, for $i = 1, 2, 3$.

3.26.2. (Bandelt [1989], Chepoi [1986]). If $a \neq b$ are neighbors in a connected graph, then V_a^b denotes the set of all vertices closer to a than to b. For a bipartite connected graph G the following are equivalent.
(i) All sets V_a^b, for $a \neq b$ neighbors in G, are convex.
(ii) G is S_2.
(iii) G is S_3.

§3: Half-Spaces and Separation

(iv) G embeds isometrically in a graphic (hyper)cube.

Hint for (iii) $\Rightarrow$ (iv). Note that in a bipartite graph each pair of neighbors can be separated by at most one pair of complementary half-spaces. Then adapt the argument proving Corollary 3.17.

3.26.3. Let $f: G \to G'$ be a function between connected graphs, of which G' is S_3. Then the following assertions are equivalent.

(i) f is EP and CP.
(ii) For each $a, b \in G$ and each geodesic $v_0, v_1, \ldots, v_n$ joining a and b, the image sequence $f(v_0), f(v_1), \ldots, f(v_n)$ (with duplicates removed) is a geodesic $f(a) \mapsto f(b)$ of G'.

3.26.4. Let G be a bipartite connected graph which is geodesically S_2 and let G' be an isometric subgraph. Show that G' is a subspace of G.

3.27. The Permutahedron (C. Reuter, communicated by Bandelt and van de Vel [1992]). Let Q^4 be the four-dimensional graphic cube. Show that $\Lambda(Q^4, \vee, \wedge)$ is not S_4. Hint. Represent a point of Q^4 as a subset of $\{1,2,3,4\}$. A set like $\{1,2\}$ is denoted by "12". Consider the following two sublattices.

$$C = \{\phi, 1, 3, 12, 13, 34, 123, 134, 1234\};$$
$$D = \{\phi, 2, 4, 23, 14, 24, 234, 124, 1234\}.$$

Note that $C \cap D = \{\phi, 1234\}$ and $C \cup D = Q^4$. Also, each point of C (resp., D) can be joined to the top and bottom elements by an arc in C (resp., D). Then the sets

$$\mathcal{C} = \{A \in \Lambda(Q^4, \vee, \wedge) \mid A \subseteq C\}; \quad \mathcal{D} = \{A \in \Lambda(Q^4, \vee, \wedge) \mid A \subseteq D\}$$

are non-empty, convex and disjoint. Now use Proposition 3.3.3.

There is an obvious 1-1 correspondence between maximal arcs in Q^n and permutations of an n-point set. For this reason, spaces of type $\Lambda(Q^n)$ are called *permutahedra*.

3.28. Preservation of separation

3.28.1. Design a 5-point subspace of $\mathbb{R}^2$ which is not S_4.

3.28.2. Problem (compare Corollary 3.11). Let $f: X \to Y$ be CP and CC, where X is S_2, resp., S_3. Must Y be S_2, resp., S_3?

3.28.3. For $i = 1, 2, 3, 4$, show that a disjoint sum of convex structures (cf. 1.25) is S_i iff all summands are.

3.28.4. Show that the direct limit of S_4 spaces and CP + CC functions is S_4. Beware: the quotient map of the disjoint sum is not CC.

3.28.5. Show that the inverse limit of S_i convex structures and CP functions is S_i for $i = 1, 2, 3$. Property S_4 is preserved under special circumstances only; see III§4.31.1.

3.29. CP Functions and isomorphisms

3.29.1. (compare 2.28 and 2.30). Let $n \geq 2$ and let H_1 and H_2 be two $(n-1)$-dimensional affine flats of $\mathbb{R}^n$ intersecting in an $(n-2)$-dimensional flat H. Let $b \in \mathbb{R}^n \setminus (H_1 \cup H_2)$. Observe that $b + H_1$ divides H_2 into two standard convex components; the same goes for $b + H_2$ with respect to H_1. Show that the set of all pairs (x_1, x_2) of points $x_i \in H_i$ collinear with b is the graph of a bijective function

$$f: H_1 \setminus (b+H_2) \to H_2 \setminus (b+H_1)$$

mapping each component of $H_1 \setminus (b+H_2)$ onto a component of $H_2 \setminus (b+H_1)$, in such a way that f is a CP isomorphism of corresponding components. Note that this isomorphism is *not* affine if $n > 1$. Hint. Use 2.30 to reduce the problem to $n = 2$ by considering two lines through b.

3.29.2. Use this technique to show that the following pairs of subspaces of $\mathbb{R}^2$ (with standard convexity) are CP isomorphic.

(i) $\{(x,y) \mid 0 \leq x, y; x+y < 1\}$ and $[0,\infty) \times [0,\infty)$ (cf. Fig. 7);
(ii) $[0,1] \times (0,1)$ and $\{(x,y) \mid 0 \leq x, y; x+y > 1\}$;
(iii) $[0,1] \times [0,1]$ and any convex 4-gon.

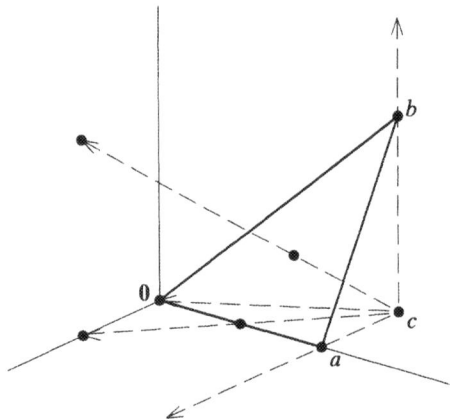

Fig. 7: Perspective isomorphism

None of these pairs are "affinely" isomorphic (an exception being made for the third type of pair, where the second space should not be a parallelogram). Compare IV§1.22.3.

3.29.3. Problem. Describe the CP isomorphism classes of polytopes in $\mathbb{R}^n$.

3.30. Separation and convex hyperspaces. Let X be a closure space with closed singletons, and let $\mathcal{D}_*(X)$ be its convex hyperspace. We consider the generated convexity $\mathcal{C}$ of X and the Vietoris convexity of $\mathcal{D}_*(X)$.

§3: Half-Spaces and Separation

3.30.1. Show that the function
$$X \to \mathcal{S}_*(X), \quad x \mapsto \{x\}$$
is an embedding.

3.30.2. Let X be finite (so, $\mathcal{S} = \mathcal{C}$) and S_2. Show that the half-spaces of $\mathcal{C}_*(X)$ are precisely the sets of type
$$<H> \cap \mathcal{C}_*(X) \text{ or } <H,X> \cap \mathcal{C}_*(X),$$
where $H \subseteq X$ is a half-space. An extension of this result to topological convex hyperspaces is given in Lemma IV§2.25.2)

3.30.3. (compare Corollary 3.13) Show that if $\mathcal{S}_*(X)$ is S_3 then X is S_4. Deduce that the separation properties S_3 and S_4 are equivalent in convex hyperspaces of type $\mathcal{C}_*(X)$ (case $\mathcal{S} = \mathcal{C}$).

3.30.4. (compare van de Vel [1983d], [1984d]) Suppose that each pair of disjoint closed sets, both spanned by a finite set, can be screened with closed sets. Show that for each finite set $F \subseteq X$ the following "cone function" is CP.
$$c_F : X \to \mathcal{S}_*(X), \quad c_F(x) = cl(\{x\} \cup F).$$

3.30.5. Reprove Theorem 3.8(2) with the aid of "cone functions". Hint. Consider sets of type
$$C(F) = \{x \mid c_F(x) \cap C \neq \emptyset\} \quad (C \subseteq X \text{ convex}, F \subseteq X \text{ finite}).$$

3.30.6. (compare van de Vel [1983]) Let X, Y be convex structures of which Y is S_4, and let $f : X \to Y$ be a CP function. Show that the following induced function of convex hyperspaces is CP.
$$\mathcal{C}_*(X) \to \mathcal{C}_*(Y), \quad C \mapsto co\, f(C).$$

3.31. Spaces of measurable functions. Let the set X be equipped with a σ–algebra, and let Y be a measure space. Two measurable functions $f, g : Y \to X$ are equivalent if they differ on a set of measure 0. Let $[f]$ denote the equivalence class of f, and let $M(Y,X)$ be the resulting set of equivalence classes. It is called a *space of measurable functions*. Usually Y is taken equal to $[0,1]$; cf. Bessaga and Pelczynski [1975].

A class $[f]$ is between $[f_1],..,[f_n]$ provided
$$f(y) \in \{f_1(y),...,f_n(y)\}$$
for all $y \in Y$ except on a set of zero measure. Show that this yields a betweenness relation on $M(Y,X)$, inducing a JHC and S_4 convexity.

3.32. Subsemilattice convexity (compare Jamison [1982]). Let S be a semilattice convexified by its subsemilattices; cf. 1.23.3, 2.24.2. Verify that S is JHC and S_4.

3.33. Separation and polytopes

3.33.1. Show that a convex structure is S_3 iff each finite subspace is. In other words, the class of S_3 spaces is a *variety* (Jamison [1982]), that is, a class of convex structures containing all isomorphic copies and subspaces of its members, and which is finitary in the sense that a space belongs to the class whenever all of its finite subspaces do.

3.33.2. Let $x \in co\{a_1,..,a_n\}$ in an S_2 space. Show that $\cap_{i=1}^n co\{x,a_i\} = \{x\}$.

3.34. Superextensions and extended CP mappings.
Let $\mathcal{S} \subseteq 2^X$ and $\mathcal{T} \subseteq 2^Y$ be T_1 families, of which $\mathcal{T}$ is normal.

3.34.1. Let $f: X \to Y$ be a function such that $f^{-1}(T) \in \mathcal{S}$ for all $T \in \mathcal{T}$. Then for each $m \in \lambda(X,\mathcal{S})$ there is a unique maximal linked system $\lambda(f)(m)$ in $\mathcal{T}$ which includes the family

$$\{T \in \mathcal{T} \mid f^{-1}(T) \in m\}.$$

Show that the resulting function

$$\lambda(f): \lambda(X,\mathcal{S}) \to \lambda(Y,\mathcal{T})$$

is a CP function making the diagram below commutative (l_X and l_Y are the canonical embeddings).

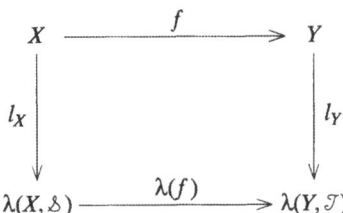

3.34.2. Show that the correspondence

$$(X,\mathcal{S}) \mapsto \lambda(X,\mathcal{S}); \quad f \mapsto \lambda(f)$$

is functorial on the appropriate categories. This provides a tool to extend point-convex S_4 spaces to special (median) convex structures in a functorial way. The condition "S_4" can be relaxed to "S_3" (cf. 3.36). We shall see later (cf. III§4.17) that, under the topological restriction of continuity, the induced function is the only one making the above diagram commutative. See 6.33.3 for finite superextensions.

3.35. Universal S_2 quotients and S_3 images.
The following yields a general method to produce convex structures with a separation property. Throughout, let X be a convex structure.

3.35.1. Show that there is an S_2 quotient space $\tilde{X}$ of X, together with a quotient function $q: X \to \tilde{X}$, such that the following "universal property" is valid. Every CP

§3: Half-Spaces and Separation

function $f: X \to Y$ to an S_2 convex structure Y factors in a unique way through q, that is, there is exactly one CP function $\tilde{f}: \tilde{X} \to Y$ with the property $\tilde{f} \circ q = f$.

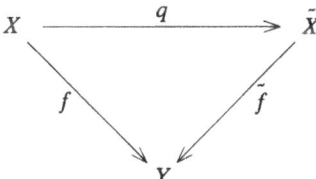

3.35.2. Show that there is a surjective CP function $g: X \to \hat{X}$ such that $\hat{X}$ is point-convex and S_3, and g satisfies a "universal" property (sensu part (1)) with respect to CP functions mapping X to a point-convex S_3 space. In general, g is *not* a quotient map.

3.35.3. Problem. Are there "universal" point-convex S_4 images?

3.36. S_3 spaces and superextensions. If X is a convex structure, then $\tilde{\lambda}(X)$ denotes the universal S_2 quotient of the superextension $\lambda(X) = \lambda(X, \mathcal{C}(X))$.

3.36.1. Show that if X is S_3 and point-convex, then the natural function
$$\tilde{i}_X: X \to \lambda(X) \to \tilde{\lambda}(X)$$
is an embedding. In particular, the universal S_3 image of a point-convex space X can be embedded in $\tilde{\lambda}(X)$.

3.36.2. Construct a functor $\tilde{\lambda}$ from the category of S_3, point-convex spaces to the category of median convex structures, such that for each CP function $f: X \to Y$,
$$\tilde{\lambda}(f) \circ \tilde{i}_X = \tilde{i}_Y \circ f.$$

3.37. Boolean lattices. Let L be a Boolean lattice. Show that a non-empty proper subset H of L is a half-space iff it is a maximal ideal or a maximal filter, iff it is an ideal or a filter containing precisely one element of each complementary pair in L. Deduce that a surjective homomorphism of Boolean lattices maps half-spaces to half-spaces.

Notes on Section 3

Half-spaces have been introduced under several names (hemispaces, semispaces, biconvex sets,..) and sometimes with slightly different meanings. We have adopted the definition given by the author in [1983d]. A description of half-spaces in semilattices and in lattices (cf. 3.2) belongs to the folklore. The representation of half-spaces in spaces of arcs, 3.3, is given by Bandelt and van de Vel [1992].

Separation properties have been considered by several authors. A common generalization of the Kakutani Separation Property S_4 in real vector spaces and of the Stone Theorem in distributive lattices was already given by Ellis in [1952]. For general convex structures, the axiom S_4 was considered by Kay and Womble [1971]. In a less known paper of Coelho [1970] an even finer list of separation properties has been considered. The labels $S_1,..,S_4$ were introduced by Jamison [1974] who gave a restricted version of the Polytope Screening Characterization in terms of screening with half-spaces; cf. Theorem 3.8(1), (2). Part (2) was obtained by van de Vel in [1983d] with a different method (see 3.30.5). The current formulation of the theorem and its proof are taken from Chepoi [1991]. Corollary 3.11 is taken from van Mill and van de Vel [1986]. The S_4 property of the Vietoris convexity, Corollary 3.13, is taken from the author's paper [1984d]. Separation in products and in subspaces was investigated by Jamison [1974] who gave Theorem 3.15(1) and (2), as well as Corollary 3.17.

Proposition 3.19 is related with a lemma of Tverberg [1966] on intersection and separation of linear subspaces of a vector space. It will be used in deriving Tverberg's Theorem; cf. II§5.9. Example 3.12.1 (that symmetric H-convexities need not be S_4) is taken from a paper of Bourgin [1952] on "restricted" separation of polytopes.

The process of convex expansions (see 3.23) was first considered for graphs by Mulder [1980]. Our description follows Bandelt and Hedlíková [1983] and is phrased in terms of general convex structures. Expansions by isometric subgraphs are considered by Chepoi [1986], who used them for a characterization of S_3, S_4 and of JHC in bipartite graphs (S_3 is handled in 3.26).

4. Interval Spaces

Interval operators provide a natural and frequent method of describing or constructing convex structures. We discuss a class of geometric interval operators, modeled after the properties of intervals derived from a metric (geodesic intervals). The class of geometric interval spaces is stable under products and relativation.

A Pasch-Peano (PP) space is an interval space satisfying the Pasch and Peano conditions, known from plane geometry. These properties imply geometricity. The convexity of a Pasch-Peano space is characterized by the separation property S_4 (Kakutani) and by Join-hull Commutativity (JHC). Examples are: vector spaces and spaces of arcs. The Sand-glass Property involves a different configuration and it characterizes S_3 under JHC. The main type of example is a poset.

4.1. Interval operators and convexity.
Let X be a set and let
$$I: X \times X \to 2^X$$
be a function with the following properties.

(I-1) *Extensive Law:* $a, b \in I(a,b)$.

(I-2) *Symmetry Law:* $I(a,b) = I(b,a)$.

Then I is called an *interval operator on* X, and $I(a,b)$ is the *interval between a and b*. The resulting pair (X,I) is called an *interval space*. For instance, the segment operator
$$(u,v) \mapsto co\{u,v\}$$
of a convex structure is an interval operator. Conversely, if (X,I) is an interval space, then the *interval convexity* $\mathcal{C}$, induced by I on X, is obtained as follows. A subset C of X is (*interval-*) *convex* provided $I(x,y) \subseteq C$ for all x, y in C. If co denotes the segment operator of $\mathcal{C}$, then
$$\forall a, b \in X : I(a,b) \subseteq co\{a,b\}.$$
The two operators need not be equal. There even are situations where $I(a,b)$ is a more natural function; see 4.16. One could say that an interval space contains more information than the induced convex structure.

In the sequel, we will freely use the terminology of interval spaces for convex structures and vice versa. In the former case, the terms should be applied to the segment operator, whereas in the latter case, the interval convexity should be considered. We will suppress the name of the interval operator if no ambiguity arises. The following shorthand notation is used.
$$ab = I(a,b); \quad a/b = \{x \mid a \in bx\}.$$
The last set is the *extension* (*ray*) at a away from b. This parallels the notation introduced in Section 3.

4.1.1. Proposition. *A convex structure is induced by an interval operator iff it is of arity* ≤ 2. ∎

4.1.2. Proposition. *The hull of a set A in an interval space is given by*
$$co(A) = \bigcup_{k=1}^{\infty} A_k,$$
where $A_0 = A$ *and (recursively)* $A_{k+1} = \bigcup \{I(a,a') \mid a, a' \in A_k\}$. ∎

The last result is a variant of Proposition 2.1.

4.2. Examples. The standard convexity of a vector space (1.5.1), the order convexity of a poset (1.5.2), and the geodesic convexity of a metric space (1.5.5) are examples of convex structures which can be defined in terms of a natural interval operator. The respective operators are defined as follows.

$$I(x,y) = \{t \cdot x + (1-t) \cdot y \mid 0 \leq t \leq 1\}; \qquad \text{(standard interval)}$$

$$I_\leq(x,y) = \begin{cases} \{x,y\}, & \text{if } x, y \text{ are incomparable;} \\ \{z \mid x \leq z \leq y\}, & \text{if } x \leq y; \end{cases} \qquad \text{(order interval)}$$

$$I_\rho(x,y) = \{z \mid \rho(x,z) + \rho(z,y) = \rho(x,y)\}. \qquad \text{(geodesic interval)}$$

Is is easily seen that standard intervals and order intervals are convex (and hence are equal to the corresponding segments). However, a geodesic interval need not be convex; see 4.16.

Let S be a semilattice and let the operator $I: S \times S \to 2^S$ be defined by

$$I_S(x,y) = \{z \in S \mid x \wedge y \leq z \leq x \text{ or } x \wedge y \leq z \leq y\}. \qquad \text{(semilattice interval)}$$

Note that $I_S(x,y)$ equals the segment between x, y of the semilattice convexity, defined in 1.5.3. Let $C \subseteq S$ be such that $I_S(x,y) \subseteq C$ for each $x, y \in C$. The above formula implies that $x \wedge y \in C$ whenever $x, y \in C$, so C is a subsemilattice. If $x \leq y$ are in C, then $I_S(x,y)$ equals the order segment and C is order convex.

If L is a lattice, then the following interval operators are used frequently:

$$I_L(x,y) = \{z \mid x \wedge y \leq z \leq x \vee y\} = [x \wedge y, x \vee y]; \qquad \text{(lattice interval)}$$

$$I_M(x,y) = \{z \mid (x \wedge z) \vee (y \wedge z) = z = (x \vee z) \wedge (y \vee z)\}. \qquad \text{(modular interval)}$$

The first one is easily seen to induce the lattice convexity, introduced in 1.5.4. We verify that the second operator (which has smaller intervals) yields the same convexity. To this end, note that

(i) $a \wedge b, a \vee b \in I_M(a,b)$,
(ii) $I_M(a,b) \subseteq [a \wedge b, a \vee b]$, and
(iii) $a \leq b$ implies $I_M(a,b) = [a,b]$.

To determine the convexity induced by I_M, we compute $co\{a,b\}$ in accordance with Proposition 4.1.2. Start with $I_M(a,b)$. For each pair $a', b' \in I_M(a,b)$, add $I_M(a',b')$. Repeat the process until the set stabilizes. Combining (i) and (iii), we see that the convex hull of

§4: Interval Spaces   73

a and b includes the order-interval $[a \wedge b, a \vee b]$. By (ii), no other points occur during the stabilization process. The term "modular interval" will be clarified later.

Finally, the convexity of a median algebra (M,m) is induced by the interval operator

$$I_m(x,y) = \{m(x,y,z) \mid z \in M\} = \{z \in M \mid m(x,y,z) = z\}. \quad \text{(median interval)}$$

As for the second equality, observe that $I_m(x,y)$ contains all points z with $z = m(x,y,z)$. Conversely, if $z \in I_m(x,y)$, say, $z = m(x,y,u)$, then

$$m(x,y,z) = m(m(u,x,y),x,y) = m(u,m(x,y,x),y) = m(u,x,y) = z$$

by the median axioms.

4.3. Subspaces and products of interval spaces. If (X,I) is an interval space, then a set $Y \subseteq X$ can be equipped with the *relative interval operator,* which is defined as follows.

$$I_Y(a,b) = I(a,b) \cap X \quad (a, b \in Y).$$

The resulting interval space (Y, I_Y) is a *subspace* of (X,I). It is easy to verify that if I is induced by a metric, then I_Y is induced by the relative metric of Y. If I is induced by order, then I_Y is induced by the relative order. In general, the relative interval operator induces a convexity which is coarser than the relative convexity. See 4.20.1 for an example where the two convexities are different.

The *product* of the interval spaces (X_j, I_j) for $j \in J$ is defined by the following interval operator on the product set $\prod_{j \in J} X_j$.

$$I(a,b) = \{(x_j)_{j \in J} \mid x_j \in I_j(a_j, b_j)\} \quad \text{(where } a = (a_j)_{j \in J}, b = (b_j)_{j \in J}\text{)}.$$

We collect some elementary observations on subspaces and products.

4.3.1. *In a product of finitely many interval spaces, the resulting interval convexity coincides with the product of the factor interval convexities.* ∎

One has to be careful with products of infinitely many factors: the induced convexity may be distinct from the product of the factor interval convexities. In fact, the product convexity can even fail to be of arity two; cf. 4.21.3.

4.3.2. *If the index collection J is finite and if for each $j \in J$ the interval operator I_j is derived from a metric ρ_j, then the product interval operator corresponds with the Manhattan (sum) metric ρ on X, defined as follows.*

$$\rho((a_j)_{j \in J}, (b_j)_{j \in J}) = \sum_{j \in J} \rho_j(a_j, b_j).$$
∎

For products of infinitely many metric spaces, see 4.20.2.

4.3.3. *If the jth interval operator I_j is induced by a median m_j for each $j \in J$, then the product interval operator is induced by the product median, defined as follows.*

$$m(x,y,z) = (m_j(x_j, y_j, z_j))_{j \in J} \quad \text{(where } x = (x_j)_{j \in J}; y = (y_j)_{j \in J}; z = (z_j)_{j \in J}\text{)}.$$
∎

4.4. Interval preserving functions. Let (X, I_X) and (Y, I_Y) be interval spaces. A function $f: X \to Y$ such that

(*) $\qquad f(I_X(a,b)) \subseteq I_Y(f(a), f(b)) \quad (a, b \in X)$

is called an *interval preserving function*, or, an *IP function* for short. The function f is *interval-to-interval* (briefly, an *II function*) provided equality holds in (*).

The inclusion of a subspace in a superspace and the projection of a product onto one of its factors are examples of interval preserving functions. By Proposition 1.12, (1) and (2), a CP (CP + CC) function between two convex structures is an IP (II) function relative to the respective segment operators; the opposite holds for convexities of arity two.

4.5. Geometric interval spaces. We now consider a set of additional axioms on an interval operator, capturing some quite useful properties. An interval operator I on a set X is *geometric* provided the following hold.

(G-1) *Idempotent Law:* $I(b,b) = \{b\}$ for all $b \in X$.
(G-2) *Monotone Law:* If $a, b, c \in X$ and $c \in I(a,b)$, then $I(a,c) \subseteq I(a,b)$.
(G-3) *Inversion Law:* If $a, b \in X$ and $c, d \in I(a,b)$, then $c \in I(a,d)$ implies $d \in I(c,b)$.

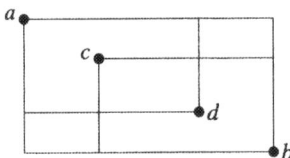

Fig. 1: Inversion Law: $c \in ad \Leftrightarrow d \in cb$

The last two properties together are equivalent to the following, single, condition.

(G-2&3) For all $a, b, c, d \in X$, if $c \in I(a,b)$ and $d \in I(c,b)$, then $d \in I(a,b)$ and $c \in I(a,d)$.

Observe that if I satisfies (G-2&3), then the sets of type $I(x,x)$ for $x \in X$ form a partition of X. In fact, if $c \in I(a,b)$ then $I(a,c) \cap I(c,b) = \{c\}$. A set with a geometric interval operator is called a *geometric interval space*.

4.6. Examples. We consider the geodesic interval operator of a metric space, the modular interval operator of a lattice, and the median interval operator in some detail. We leave it to the reader to verify that the interval operator of a poset is geometric.

4.6.1. Proposition. *The geodesic interval operator of a metric space is geometric.*

Proof. Let (X, ρ) be a metric space. The Idempotent Law (G-1) is evident. As to the Monotone Law of I_ρ, let $c \in I_\rho(a,b)$ and $x \in I_\rho(a,c)$. Then

$$\rho(a,b) = \rho(a,c) + \rho(c,b)$$
$$= \rho(a,x) + \rho(x,c) + \rho(c,b)$$

$$\leq \rho(a,x) + \rho(x,b)$$
$$\leq \rho(a,b),$$

whence equality holds throughout. As to the Inversion Law, let a, b, c, d be as in (G-3). Then

$$\rho(a,c) + \rho(c,b) \leq \rho(a,c) + \rho(c,d) + \rho(d,b) = \rho(a,d) + \rho(d,b) = \rho(a,b),$$

and equality holds throughout. ∎

Recall that a lattice L is *modular* provided $a \leq c$ implies $a \vee (b \wedge c) = (a \vee b) \wedge c$ for all a, b, $c \in L$. Equivalently, if $p \leq q$ are in L and if x, y, y' are such that

$$p \leq y' \leq y; \quad y \wedge x = p; \quad y \vee x = q,$$

then $y' = y$. This leads to another characterization of modularity in L, namely, that L

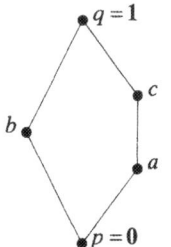

Fig. 2: Non-modular N_5 Fig. 3: Modular law: $a \vee (b \wedge c) = (a \vee b) \wedge c$ ($a \leq c$)

should not include a sublattice of type N_5 (see Fig 2). The next result may justify the term "modular interval operator".

4.6.2. Proposition. *The following are equivalent for a lattice L.*

(1) *The modular interval operator I_M is geometric.*
(2) *The modular interval operator I_M satisfies the Monotone Law, (G-2).*
(3) *For each a, b, $c \in L$, the following set is non-empty.*

$$M(a,b,c) = I_M(a,b) \cap I_M(b,c) \cap I_M(c,a).$$

(4) *The lattice L is modular.*

Proof. Recall the construction of the modular interval operator of a lattice,

$$I_M(x,y) = \{ z \mid (x \wedge z) \vee (y \wedge z) = z = (x \vee z) \wedge (y \vee z) \}.$$

By definition, (1) implies (2). We show that either (2) or (3) implies (4). Suppose L is not modular. Then L includes a sublattice as indicated in Fig. 2. Apparently, $q \in I_M(a,b)$ and $c \in I_M(q,b)$, but $c \notin I_M(a,b)$, conflicting with (2). Assume (3) and consider a point $z \in M(a,b,c)$. Then $b \leq z \leq c$, and hence $c = c \wedge q = (c \vee z) \wedge (z \vee a) = z$. Similarly, we have $b = b \vee p = (b \wedge z) \vee (z \wedge a) = z$. This yields $c = b$, a contradiction.

We next derive the implication (4) ⇒ (3). Assume L is modular and let a, b, $c \in L$. Consider the point $m = (a \vee b) \wedge (b \vee c) \wedge (c \vee a)$. By using the modular law we find

$$a \vee m = a \vee \big((b \vee c) \wedge ((c \vee a) \wedge (a \vee b))\big)$$
$$= \{a \vee (b \vee c)\} \wedge \{(c \vee a) \wedge (a \vee b)\}$$
$$= (c \vee a) \wedge (a \vee b).$$

Similar equalities hold for $b \vee m$ and $c \vee m$. If follows that $(a \vee m) \wedge (m \vee b) = m$, with similar formulas for all other combinations. For a dual formula consider the following computation:

$$(a \wedge m) \vee (b \wedge m) = \big((a \wedge m) \vee b\big) \wedge m$$
$$= \big((a \wedge (a \vee m) \wedge (m \vee b)) \vee b\big) \wedge m$$
$$= \big((a \wedge (m \vee b)) \vee b\big) \wedge m$$
$$= \big((a \vee b) \wedge (m \vee b)\big) \wedge m = m.$$

The first and fourth equality are applications of modularity, whereas the second equality follows from substituting $m = (a \vee m) \wedge (m \vee b)$. Similar formulas hold for other combinations. Therefore, $m \in I_M(a,b) \cap I_M(b,c) \cap I_M(c,a)$.

It remains to be shown that (4) implies (1). Let L be modular. The Idempotent Law $I_M(a,a) = \{a\}$ is evident. We next verify that I_M satisfies the Monotone Law. Let $b \in I_M(a,c)$ and $d \in I_M(a,b)$. Hence $a \wedge c \leq b$ and $a \wedge b \leq d$. We find that

$$d = (a \wedge d) \vee (d \wedge b)$$
$$= (a \wedge d) \vee \big((d \wedge ((a \wedge b) \vee (b \wedge c)))\big)$$
$$= (a \wedge d) \vee \big((a \wedge b) \vee (d \wedge b \wedge c)\big)$$
$$= (a \wedge d) \vee (a \wedge b) \vee (d \wedge b \wedge c)$$
$$\leq (a \wedge d) \vee (d \wedge c) \leq d.$$

The third equality involves modularity; the second last inequality involves the fact that $a \wedge b \leq d$. We conclude that $d = (a \wedge d) \vee (d \wedge c)$. Dually, $d = (a \vee d) \wedge (d \vee c)$, showing that $d \in I_M(a,c)$ as desired.

We next verify the intermediate result that if $c \in I_M(a,b)$, then

$$I_M(a,c) \cap I_M(c,b) = \{c\}.$$

We use the observation that I_M has smaller intervals than I_L. If $x \in I_M(a,c) \cap I_M(c,b)$, then $x \leq a \vee c$ and $x \leq b \vee c$, whence $x \leq c$. Similarly, $c \leq x$.

We finally deduce the Inversion Law. Let $a, b \in L$ and $c, d \in I_M(a,b)$, where $c \in I_M(a,d)$. The Monotone Law implies that $M(c,d,b) \subseteq M(a,d,b)$. The latter set equals $\{d\}$ since $d \in I_M(a,b)$, whereas the former set is non-empty by (3). We conclude that $d \in I_M(c,b)$. ∎

4.6.3. Proposition. *The median interval operator is geometric.*

Proof. The Idempotent Law of medians gives $I_m(a,a) = \{a\}$. As to the axiom (G-1&2), let $c \in I_m(a,b)$ and $d \in I_m(c,b)$. Then

$$m(a,d,c) = m(a,m(b,c,d),c) = m(m(a,b,c),d,c) = m(c,d,c) = c,$$

§4: Interval Spaces

showing that $c \in I_m(a,d)$. On the other hand,

$$m(a,d,b) = m(a,m(c,b,d),b) = m(m(a,c,b),d,b) = m(c,d,b) = d,$$

showing that $d \in I_m(a,b)$. ∎

In a distributive lattice L, the function $m: L^3 \to L$, defined by

$$m(x,y,z) = (x \wedge y) \vee (y \wedge z) \vee (z \wedge x),$$

yields a median operator (cf. Topic 1.22). In these circumstances, the interval operators I_L and I_m coincide. In 6.12 we will show that the interval operator I_L of a lattice is geometric if and only if the lattice is distributive. We just obtained (the easy) half of this result.

4.7. Theorem. *The segment operator of an S_3 convex structure with convex singletons is geometric.*

Proof. (G-1) follows from the fact that X is point-convex, whereas a segment operator satisfies (G-2) by definition. As to (G-3), let c, $d \in ab$, and assume that $c \in ad$ but $d \notin bc$. Separate d and bc with a half-space $H: d \in H$; b, $c \notin H$. Then $a \notin H$ (otherwise $c \in ad \subseteq H$) and hence we obtain $d \in ab \subseteq X \setminus H$, a contradiction. ∎

The interval operator of a poset or of a semilattice coincides with the induced segment operator. Both convexities are (at least) S_3, whence the order interval operator and the semilattice interval operator are geometric.

4.8. Theorem.

(1) A subspace of a geometric interval space is a geometric interval space.
(2) The product of a family of geometric interval spaces is a geometric interval space.
(3) The image of a geometric interval space under an II function is a geometric interval space. ∎

4.9. Peano and Pasch Property. The notation ab for an interval $I(a,b)$ and a/b for the related extension operator extends to subsets of an interval space as follows.

$$AB = \bigcup \{ab \mid a \in A; b \in B\} \quad \text{(join of } A, B\text{)};$$

$$A/B = \bigcup \{a/b \mid a \in A; b \in B\} \quad \text{(\textit{extension of } A \textit{ away from } B)}.$$

We note that the extension operator of sets, considered for convex structures in 3.6, is different if the convexity is of arity larger than two. The join and extension viewpoint will be reconsidered in Section 7 below. We use the extension operator of interval spaces in the next definitions, involving an interval space X. See Figs. 4 and 5.

Peano Property: For all a, b, $c \in X$ and $y \in bc$, $z \in ay$, there is a point $z' \in ab$ such that $z \in cz'$.

Pasch Property: For all p, a, $b \in X$ and $a' \in pa$, $b' \in pb$, the intervals ab' and $a'b$ intersect.

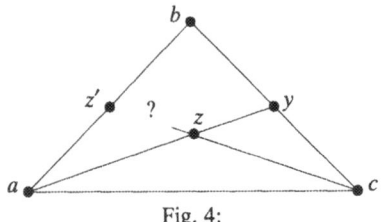

Fig. 4:
Peano axiom: $a(bc) = (ab)c$

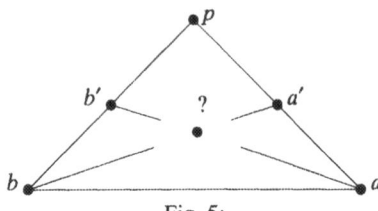

Fig. 5:
Pasch axiom: $b'/b \cap a'/a \neq \emptyset \rightarrow a'b \cap b'a \neq \emptyset$

Disregarding the formal difference between singletons and points for a moment, we arrive at constructions of type

$$a_1(a_2(..(a_{n-1}a_n)..)).$$

Such expressions are not particularly handsome, unless some computational rules are assumed. The Peano Property is such a rule; it states that the join operator is associative.

4.10. Proposition

(1) *An idempotent interval operator with the Peano Property (resp., the Pasch Property) satisfies the Monotone Law (resp., the Inversion Law). In particular, an interval space satisfying both conditions is geometric.*

(2) *In an idempotent interval space with the Peano Property, or in a connected graph with the Pasch Property, all intervals are convex.*

Proof of (1). Let $y \in ab$ and $z \in ax$. Then use the Peano Property to conclude that the interval bz extends to bz' for some $z' \in aa = \{a\}$, resp., use the Pasch Property to conclude that the intervals bz and $yy = \{y\}$ meet.

Proof of (2). We first consider a general interval space with the Peano Property. Let $u, v \in ab$ and $w \in uv$. The Peano Property yields a point $z \in ub$ such that $w \in az$. By the monotone property of intervals (cf. (1)), $z \in ub \subseteq ab$ and $w \in az \subseteq ab$.

Next, consider a connected graph with the Pasch Property. We verify by induction on the geodesic distance $\rho(u,v)$ that $uv \subseteq ab$ if $u,v \in ab$. The case $\rho(u,v) = 1$ being trivial, it suffices to show that each neighbor w of u in uv is in ab. If $\rho(a,w) \leq \rho(a,u)$ and $\rho(b,w) \leq \rho(b,u)$, then evidently w is in the metric interval ab. If $\rho(a,w) > \rho(a,u)$, then $u \in aw$ and the Pasch Property yields a point $z \in vw \cap ub$. If $z = v$ then $w \in uv \subseteq ub \subseteq ab$ by axiom (G-2). If $z \neq v$, then $\rho(u,z) < \rho(u,v)$. We have $z \in ub \subseteq ab$ by (G-2) and $w \in uz \subseteq ab$ by (G-3) and by inductive assumption. ∎

The last result fails even for finite metric spaces (cf. Topic 4.16.2). It turns out that the properties of Join-hull Commutativity, S_3, and S_4, can be characterized on interval spaces with the aid of some specific finite configurations of points.

4.11. Theorem. *A convex structure of arity two is JHC iff its segment operator satisfies the Peano Property.*

§4: Interval Spaces

Proof. Throughout, X is a convex structure of arity two. In terms of the join operator on sets, derived from the segment operator, this can be expressed as follows. A set $C \subseteq X$ is convex iff $CC = C$. Suppose first that X is JHC. Then $co\{a,b,c\} = a(bc)$ and $co\{a,b,c\} = (ab)c$ for a, b, $c \in X$, which yields the Peano Property.

Suppose next that X has the Peano Property, and let A, $B \subseteq X$ be convex. Combining the opening remark of this proof with the (evident) commutative rule and the assumed associative rule of joins (Peano Property), we find

$$(AB)(AB) = A(BA)B = A(AB)B = (AA)(BB) = AB,$$

whence AB is convex. ∎

Combining this result with Proposition 4.10(2), we conclude that an interval space with the Peano Property has a JHC convexity. Conversely, it is not possible to draw a conclusion about an interval operator producing a JHC convexity.

4.12. Theorem. *The following are equivalent for a convex structure X of arity two.*
(1) *X has the separation property S_4.*
(2) *Each pair of disjoint segments in X can be screened with convex sets.*
(3) *The segment operator of X has the Pasch Property.*
(4) *For each $p \in X$ and for each segment ab of X the set ab/p is convex.*

Proof. (1) evidently implies (2). Let X satisfy (2) and take p, a, b, a', b' as in the Pasch configuration (Fig. 5). If $a'b$ and ab' are disjoint, then there is a pair of convex sets C, D such that

$$a, b' \in C \setminus D; \quad a', b \in D \setminus C; \quad C \cup D = X.$$

Hence if $p \in C$ then $a' \in pa \subseteq C$, whereas if $p \in D$ then $b' \in pb \subseteq D$. Both yield a contradiction.

For a proof that (3) implies (4), consider c, $d \in ab/p$, say:

$$c' \in cp; \quad d' \in dp; \quad c',d' \in ab,$$

and let $x \in cd$. Two applications of the Pasch Property give that $px \cap c'd' \neq \emptyset$.

We finally show that (4) implies (1). Let $C, D \subseteq X$ be disjoint convex sets. Extend C to a convex set C_0, maximal with the property that $C_0 \cap D = \emptyset$. If C_0 is not a half-space, then there exist two points $p, q \in X \setminus C_0$ with $pq \cap C_0 \neq \emptyset$. By virtue of (4), C_0/p is a convex set properly larger than C_0, and by maximality it meets D, say, in d_0. Similarly, C_0/d_0 meets D in some point d_1. However, the segment $d_0 d_1$ is included in D and it meets C_0, a contradiction. ∎

An interval space satisfying the Pasch and Peano Properties will be called a *Pasch-Peano space,* or a *PP-space* for short. Observe that the distinction between a Pasch-Peano interval space and the induced S_4 JHC convex structure is quite formal. Indeed, the interval operator of a PP-space equals the segment operator, whereas an S_4 and JHC space is of arity two and its segment operator has the Pasch and Peano Properties.

An interval space X has the *Sand-glass Property* provided for each sextet

a, b, c, d, p, $v \in X$ such that $p \in ad \cap bc$ and $v \in ab$, there is a point $w \in cd$ such that $p \in vw$. See Fig. 6. We have the following result.

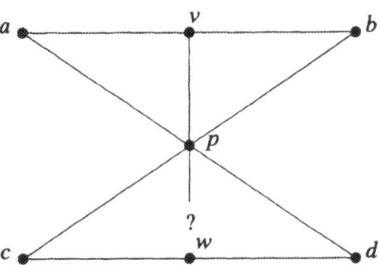

Fig. 6: Sand-glass Property

4.13. Theorem. *For a JHC convex structure X the following are equivalent:*

(1) X has the separation property S_3.
(2) X has the Sand-glass Property.

Proof that (1) implies (2). Let X be S_3, and let a, b, c, d, p, v be as above. If p is not in $co\{v,c,d\}$ then there is a half-space H with v, c, $d \in H$ and $p \notin H$. As $v \in ab$, one of a, b must be in H, say: $a \in H$. But then $p \in ad \subseteq H$, a contradiction. Now use JHC to obtain a point $w \in cd$ as required.

Proof that (2) implies (1). Let $C \subseteq X$ be convex and let $p \in X \setminus C$. By the third axiom of convexity, there is a convex set C_0 such that $C \subseteq C_0$; $p \notin C_0$, and maximal with these properties. Suppose C_0 is not a half-space. As X is of arity 2, there exist a, b in $X \setminus C_0$, together with a point v in $ab \cap C_0$. By maximality, p is in $co(\{b\} \cup C_0)$, and by JHC we find a point c in C_0 with $p \in bc$. Operating with a instead of b, we obtain a point d in C_0 with $p \in ad$. By the Sand-glass Property there is a point w in cd with $p \in vw$. However, v, $w \in C_0$ and hence $p \in C_0$, a contradiction. ∎

The equivalence between S_3 and the Sand-glass Property fails to hold for general spaces. See 4.24.4.

4.14. Examples

4.14.1. Proposition. *A vector space V over a totally ordered field $\mathbb{K}$ is a Pasch-Peano space.*

Proof that V has the Peano Property. Let a, b, $c \in V$; $y \in bc$; $z \in ay$. We must find a point $z' \in ab$ with $z \in cz'$. Now z is in the hull of a, b, c and hence it can be expressed as

$$z = r \cdot a + s \cdot b + t \cdot c,$$

with r, s, $t \geq 0$ and $r + s + t = 1$. If $t = 1$ then $z = c$ and z' is irrelevant. So let $r + s > 0$ and consider the point

§4: Interval Spaces

$$z' = \frac{r}{r+s}a + \frac{s}{r+s}b \in ab.$$

Then

$$(r+s)z' + (1-(r+s))c = r \cdot a + s \cdot b + t \cdot c = z,$$

showing that z' is as desired.[1]

Proof that V has the Pasch Property. Consider five points

$$a, b, p \in V; \quad a' \in pa; \quad b' \in pb$$

as in the Pasch configuration (Fig. 5). Let $s, t \in \mathbb{K}$ be such that

$$a' = s \cdot p + (1-s) \cdot a \quad (0 \le s \le 1); \quad b' = t \cdot p + (1-t) \cdot b \quad (0 \le t \le 1).$$

Define $u, v \in \mathbb{R}$ as follows.

$$u = \frac{t}{s+(1-s)t}; \quad v = \frac{(1-s)t}{s+(1-s)t}.$$

Then $0 \le u, v \le 1$ and $u \cdot a' + (1-u) \cdot b = v \cdot a + (1-v) \cdot b'$, yielding a common point of ab' and $a'b$. ∎

We may conclude that the standard convexity is JHC and S_4. For $\mathbb{K} = \mathbb{R}$, the separation property S_4 of V is a classical result attributed to S. Kakutani. For this reason, the axiom S_4 is also called the **Kakutani Separation Property**.

4.14.2. Proposition. *A poset is JHC and* S_3.

Proof that a poset is JHC. A poset X is of arity 2 and hence Theorem 4.11 applies. Consider the Pasch-Peano configuration as in Fig. 4: let

$$a, b, c \in X; \quad y \in cb; \quad z \in ay.$$

We are looking for a point z' with

$$z' \in ab; \quad z \in cz'.$$

If y or z are equal to one of the vertices of the segment in which they have been taken, then we are done. In the remaining cases membership of a segment arises from certain inequalities. Without loss of generality, we have $a \le z \le y$, together with one of the following:

(i) $b \le y \le c$: then $a \le z \le c$ and $z' = a$ is the desired point;
(ii) $c \le y \le b$: then $a \le z \le b$ and z' can be taken as z.

Proof that a poset is S_3. In the poset X we consider the Sand-glass configuration as in Fig. 6: let

$$a, b, c, d \in X; \quad p \in ad \cap bc; \quad v \in ab.$$

---

1. All figures in this monograph which involve a Peano or Pasch configuration have been drawn by a macro involving the formulae of this proof.

We look for a point w with

$$w \in cd; \quad p \in vw.$$

We assume that $v \neq a, b$. Without loss of generality, $a \leq v \leq b$ and $a \neq b$. Each of the "degenerate" cases $p = a, b, c, d$ is easily solved. This leaves us with a combination of the following possibilities.

(i) $a \leq p \leq d$ or (ii) $d \leq p \leq a$, and
(iii) $c \leq p \leq b$ or (iv) $b \leq p \leq c$.

Note that (ii) does not combine with (iv). As a sample we take (i) and (iv). Then $v \leq b \leq p \leq c$, so $w = c$ is the right point. The remaining two combinations are handled with the same ease. ∎

We note that a poset can fail to be S_4. Just take the vertices of the Pasch configuration (Fig. 5) without the desired intersection point, and design a suitable order on this 5-point set.

4.14.3. Proposition. *Let L be a bounded discrete semimodular lattice. Then its space of arcs $\Lambda(L, \vee)$ has the Kakutani Separation Property, S_4.*

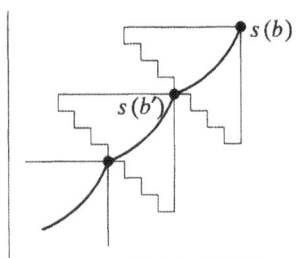

Fig. 7: Construction of an arc

Proof. By 2.17.1, it suffices to verify the Pasch Property. Let $A, A_1, A_2 \in \Lambda(L)$, and let $B_i \in co\{A, A_i\}$ for $i = 1, 2$. Consider the following homomorphism of join semilattices.

$$f: A_1 \times A_2 \times A \to L, \quad f(a_1, a_2, a) = a_1 \vee a_2 \vee a.$$

In addition, let

$$B'_1 = f^{-1}(B_1) \cap (A_1 \times \{\mathbf{1}\} \times A),$$
$$B'_2 = f^{-1}(B_2) \cap (\{\mathbf{1}\} \times A_2 \times A).$$

For $b \in B_1$ the set

$$f^{-1}(b) \cap A_1 \times \{\mathbf{1}\} \times A$$

is not empty since $B_1 \in co\{A, A_1\}$. Let $s(b)$ be its supremum. The collection $\{s(b) \mid b \in B_1\}$ is totally ordered. If $b' < b$ are consecutive elements of B_1, then

§4: Interval Spaces   83

$$[s(b'), s(b)] \setminus \{s(b')\} \subseteq f^{-1}(b).$$

We construct an arc C_1 as a compilation of maximal arcs in $[s(b'), s(b)]$ for successively increasing b', $b \in B_1$, and starting with a maximal arc joining **0** with $s(b)$, where $b > 0$ is the smallest non-trivial element of B_1. Note that $C_1 \subseteq B_1'$ joins $(\mathbf{0,0,0})$ to $(\mathbf{0,1,1})$ Similarly, we obtain an arc $C_2 \subseteq B_2'$, joining **0** to **1**. For each $a \in A$ we consider the level set

$$L(a) = \{(a_1, a_2, a) \mid (a_1, 0, a) \in C_1; (0, a_2, a) \in C_2\}.$$

Note that $L(a)$ is a non-empty rectangle. Suppose $a' < a$ are successive elements of A, and let (a_1, a_2, a) be the smallest element of $L(a)$. Then $(a_1, 0, a) \in C_1$, and its predecessor in C_1 is $(a_1, 0, a')$. This is so because successive elements of C_1 differ in only one coordinate; this cannot be a_1 by the minimality of (a_1, a_2, a). Similarly, the predecessor in C_2 of $(0, a_2, a)$ is $(0, a_2, a')$. This shows that $(a_1, a_2, a) \in L(a')$. It is evidently the largest element of $L(a')$. In each level set $L(a)$ we join the largest and the smallest element by a maximal arc. Compiling these arc pieces yields an arc C in $A_1 \times A_2 \times A$ joining $(\mathbf{0,0,0})$ to $(\mathbf{1,1,1})$. By the semimodularity of L, the image $f(C)$ of C is a maximal arc in L joining **0** to **1**.

If $c = (a_1, a_2, a) \in C$, then

$$(a_1, 0, a) \in B_1'; \quad (0, a_2, a) \in B_2',$$

whence $a_1 \vee a \in B_1, a_2 \vee a \in B_2$. We find that

$$f(c) = (a_1 \vee a) \vee a_2 \in B_1 \vee A_2; \quad f(c) = (a_2 \vee a) \vee a_1 \in B_2 \vee A_1.$$

We conclude that

$$f(C) \subseteq co\{A_1, B_2\} \cap co\{A_2, B_1\}.$$ ∎

An instructive aspect is that the half-spaces in spaces of arcs are not easily detected (compare Proposition 3.3.2): the Separation Theorems 3.8, 4.12 and 4.13 are *purely existential*. For an example involving $(\Lambda(L), \wedge, \vee)$, we refer to Topic 3.27. A somewhat surprising effect of separation properties in graphs is illustrated by the following result.

4.15. Proposition. *If a connected graph is* S_3, *then all of its intervals are convex.*

Proof. Let the connected graph G be S_3, and let $a \neq b \in G$. Throughout, I denotes the geodesic interval operator. The result is clear if $d(a,b) = 1$ since edges are always convex. We proceed by induction, assuming the result to be valid for points at a distance $< n$, where $n \geq 2$. Let $d(a,b) = n$. Take $u, v \in I(a,b)$ and $w \in I(u,v)$. We just have to show that $w \in I(a,b)$. The cases where u or v equal a or b are easily handled. So assume $u, v \neq a, b$. If w is on a geodesic joining u and b then composing with a geodesic from a to u yields a geodesic joining a and b via w, and we are done. We assume henceforth that

(1) $w \notin I(u,b)$

First, let w be a neighbor of u in $I(u,v)$. As $u \neq a$ we have $d(a,b) > d(u,b)$, whence by (1) and by the induction hypothesis, $w \notin ub$. As G is S_3 there is a half-space H such that (Fig. 8) $w \in H$; $u, b \notin H$. Then $v \in H$ (otherwise $w \in vu \subseteq G \setminus H$), and hence $a \in H$ (otherwise $v \in ab \subseteq G \setminus H$). Now, if $d(a,w) > d(a,u)$ then any geodesic joining a and u

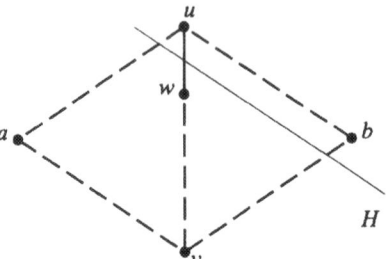

Fig. 8: Convex intervals in an S_3 graph

can be continued with the edge uw to a geodesic from a to w via u. However, $a, w \in H$ and H is geodesically convex, whence $u \in H$, a contradiction. We conclude that

(2) $\qquad d(a,w) \leq d(a,u)$.

By similar reasoning on b we find that

(3) $\qquad d(w,b) \leq d(u,b)$.

From (2) and (3) we obtain

$$d(a,b) \leq d(a,w) + d(w,b) \leq d(a,u) + d(u,b) = d(a,b),$$

and hence $w \in I(a,b)$.

To reach a general point $w \in I(u,v)$ we can proceed from u to w with one edge at the time, applying the previous result in each step. This gives $w \in I(a,b)$ in the end. ∎

It is not known whether this result extends to general metric spaces, or even whether it extends to finite metric spaces.

Further Topics

4.16. Intervals

4.16.1. Let $m, n \in \mathbb{N}$. The bipartite graph $K_{m,n}$ is defined on a vertex set which is the union of two disjoint sets A, B with $\#A = m$ and $\#B = n$. Non-trivial edges are drawn between a point of A and a point of B only. Show that the geodesic convex sets are: $\emptyset$, $K_{m,n}$, all singletons and all non-trivial edges. Therefore, no interval of diameter two is convex if $m, n \geq 2$ and $m + n \geq 5$. In these circumstances, by Proposition 4.10, the interval operator of $K_{m,n}$ satisfies neither the Peano nor the Pasch Property. A more involved example with non-convex geodesic intervals in normed spaces will be presented in 6.32.

4.16.2. Verify that the finite metric space, presented in Fig. 9 as a weighted graph, is a Peano space with a non-convex interval ab.

4.16.3. (Jamison [1974]) Let G be a group, and let $\mathcal{C}$ be the convexity on G consisting of all left cosets of subgroups, together with the empty set. Verify that $C \subseteq G$ is

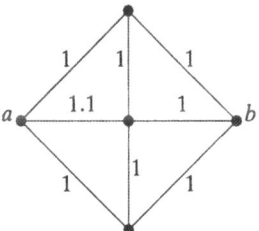

Fig. 9: A finite metric counterexample

convex iff for any three points $x,y,z \in C$ it is true that $co\{x,y,z\} \subseteq C$, and that this cannot be improved to pairs of points. Conclude that C is, in general, of arity three.

4.16.4. Problem. Is a (symmetric) H-convexity of arity ≤ 2 (cf. II§4.41)?

4.16.5. (for the standard plane, Ptolemaius, ± 150 A.C.; cf. Blumenthal [1953]) A metric space (X,d) (or, the metric d) is said to be *Ptolemaic* provided for each quadruple of points $p, q, r, s \in X$ the following inequality holds.

$$d(p,q){\cdot}d(r,s) + d(p,r){\cdot}d(q,s) \geq d(p,s){\cdot}d(q,r).$$

Verify that the standard Euclidean metric on $\mathbb{R}^n$ is Ptolemaic. Hint (Bottema et al [1968]): take one of the four points as the origin $\mathbf{0}$, say: p. Consider the inversion map $v : \mathbb{R}^n \setminus \{\mathbf{0}\} \to \mathbb{R}^n$, defined by

$$x \mapsto \frac{||q||{\cdot}||r||{\cdot}||s||}{||x||^2}{\cdot}x,$$

and observe that the distance of e.g., $v(q)$ and $v(r)$ equals $d(p,s){\cdot}d(q,r)$.

Finally, show that all geodesic intervals of a Ptolemaic space are convex.

4.17. Matroids. Let X be a matroid of arity 2. Show that if the segment function of X is geometric, then X is free. In particular, an S_3 matroid of arity two is free, and a projective matroid of rank ≥ 3 is never S_3. Note that the 2–free space $F_2(X)$ (with $\#X > 3$) is a matroid with a geometric segment function, but this space is neither free nor of arity two.

4.18. Vector lattices. Let V be a *vector lattice* (*Riesz space*), that is, V is a real vector space with a lattice order $\leq$ such that

(R-1) $0 \leq x$ implies $0 \leq t{\cdot}x$ and $v \leq v + x$ for all $v \in V$ and $t \geq 0$.
(R-2) $0 \leq x$ and $0 \leq y$ imply $0 \leq x + y$.

See Luxemburg and Zaanen [1971]. Consider the interval operator I_R, defined as follows (I denotes the standard line segment operator)

$$I_R(a,b) = I(a {\wedge} b, a) \cup I(a {\wedge} b, b).$$

Show that (V, I_R) is a geometric interval space. It is a PP space if $V = \mathbb{R}^2$ with the

Cartesian order. On the other hand, $(\mathbb{R}^3, I_R)$ is not a PP space.

4.19. Subspaces and products of posets

4.19.1. Show that the relative convexity on a subset of a poset is induced by the relative order. For additional information, see II§4.25.6.

4.19.2. For each $i \in I$ let $(X_i, \leq)$ be a poset, and let $X = \prod_{i \in I} X_i$ be equipped with the Cartesian order,

$$(x_i)_{i \in I} \leq (x'_i)_{i \in I} \text{ iff } x_i \leq x'_i \text{ for all } i \in I.$$

Note that if each factor poset is linearly ordered, then the Cartesian order makes X into a distributive lattice. Show that in this case, the product convexity of X is exactly the lattice convexity (cf. 1.5.4).

4.20. Subspaces and products of metric spaces

4.20.1. In general, the relative convexity on a metric subspace is not the geodesic convexity. For graphs, the situation is as follows. If G_1, G_2 are connected graphs then each isometric embedding $f: G_1 \to G_2$ is an EP and CP function, but not necessarily an

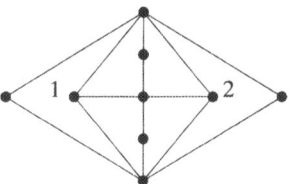

Fig. 10: Metric subspaces: a graphic counter-example

embedding of convex structures. Fig. 10 presents a bipartite counterexample. The dots labeled 1 and 2 are points of G_2 not in G_1. The outer 4-cycle is geodesically convex in G_1, but not relatively convex.

4.20.2. Let I be a countable infinite index set and for each $i \in I$ let d_i be a metric on X_i such that X_i is of diameter ≤ 1. Consider the following metric on the product set $X = \prod_{i \in I} X_i$.

$$d(x,x') = \sum_{i \in I} 2^{-i} \cdot d_i(x_i, x'_i).$$

Here, x_i and x'_i denote the ith coordinate of x, resp., x'. Show that the geodesic convexity of (X,d) is the product of the factor geodesic convexities.

4.20.3. Let the index set I be finite and let each G_i be a connected graph with intrinsic distance function d_i. On the product set G we consider the following graph structure: $(x_i)_{i=1}^n$ and $(x'_i)_{i=1}^n$ form an edge iff $x_i \neq x'_i$ for at most one i, and for each i the points x_i, x'_i form an edge in G_i. Show that the intrinsic metric of G is the Manhattan metric -- hence it

§4: Interval Spaces   87

produces the product convexity. The graph G is known as the *Cartesian product* of $G_1,..,G_n$.

4.21. Subspaces and products of interval spaces

4.21.1. The convexity of an interval subspace need not be the relative convexity. In fact, (cf. (Burris [1972]) every convex structure can be embedded in a convex structure of arity 2. Hint. Let $\hat{X}$ consist of all finite non-empty subsets of convex structure X, and consider the interval operator I on $\hat{X}$, defined by

$$I(G_1,G_2) = \{G \mid G \subseteq co(G_1 \cup G_2)\}.$$

Then X embeds in the interval space $\hat{X}$ by assigning to each point of X the corresponding singleton.

4.21.2. The convexity of a product of finitely many interval spaces is the product of the factor interval convexities. For an extension to spaces of higher arity, see II§4.41.3.

4.21.3. Let a, b, c be three distinct points. For each $n \geq 0$ consider the set

$$X_n = \{(a,k) \mid 0 \leq k \leq n\} \cup \{(b,k) \mid 0 \leq k \leq n\} \cup \{(c,k) \mid 0 \leq k \leq n\}.$$

For convenience, we say that (a,k) (resp., (b,k), (c,k)) is the predecessor of $(a,k+1)$ (resp., of $(b,k+1)$, $(c,k+1)$). A set $C \subseteq X_n$ is taken convex if it has the following property. Whenever two of $(a,k+1)$, $(b,k+1)$, $(c,k+1)$ are in C, then the predecessor of the third is in C. Verify that each X_n is of arity two, but not their product.

4.22. Ternary spaces (Hedlíková [1983]).
Consider a set X with a ternary relation abc satisfying the following requirements.

(T-1) abc implies cba.
(T-2) abc and acb iff $b = c$.
(T-3) abc and acd imply bcd.
(T-4) abc and acd imply abd.

The relation extends to an n–ary one in the following way. If $(x_i)_{i=1}^n$ are in X, then

$$x_1..x_n \Leftrightarrow i < j < k : x_i x_j x_k.$$

4.22.1. Let $I(a,b)$ be the set of all x satisfying axb. Verify that (X,I) is a geometric interval space. Conversely, each geometric interval space (X,I) is a ternary space under the ternary relation

$$abc \Leftrightarrow b \in I(a,c).$$

4.22.2. By a *ternary field* is meant a field $(F, +, \cdot)$, together with a ternary relation abc on F, subject to the following conditions.

(TF-1) $\forall a, b, c, x \in X$: abc implies $(a+x)(b+x)(c+x)$.
(TF-2) $\forall a, b, c, x \in X$: abc implies $(a \cdot x)(b \cdot x)(c \cdot x)$.

For instance, the rational or real field with the natural betweenness are ternary fields.

Each field can be made ternary by adding the trivial ternary relation, which consists of all triples with two equal entries at positions one and two or at positions two and three. Prove that if abc and bcd with $b \neq c$, then $abcd$. Furthermore:

(i) $0a1$ implies $0(1-a)1$;
(ii) $0a1$ and $0b1$ imply $0(a \cdot b)1$;
(iii) $0a1$ and $0b1$ imply $0a(a+b)$ provided at least one of a, b is not 1;

A field with a trivial ternary relation may illustrate the fact that the condition in (iii) cannot be omitted.

4.22.3. In a vector space over a ternary field, an interval operation is defined by

$$I(v,w) = \{ta + (1-t)b \mid 0t1\}.$$

Show that (V,I) is a Pasch-Peano space.

4.23. Characterizing weak JHC (cf. Kay [1977b]). Show that a convex structure X is weakly JHC iff for each convex set C and for each two points $b \in X$, $c \in C$, the hull of $\{b\} \cup C$ equals the join of bc and C. Compare with the definition of JHC.

Hint. Assume the above condition. First, derive the intermediate result that $(AB)C = A(BC)$ for each triple of convex sets A, B, C with $A \cap B \cap C \neq \emptyset$ (**weak Peano Property**). Next, deduce that $(AB)(AB) = AB$ for each pair of intersecting convex sets A, B (the second step is not quite as simple as suggested by Kay).

Problem. Is the weak Peano Property equivalent to weak JHC? Are the two conditions equivalent for matroids (as claimed by Kay)?

4.24. Separation and JHC in graphs

4.24.1. (Chepoi [1986]) A bipartite connected graph which is JHC and has convex intervals is S_3.

Hint. For each pair of points $a, b \in G$ consider the set

$$V_a^b = \{x \mid \rho(x,a) < \rho(x,b)\}$$

(cf. 3.26.2). Let $x_1, x_2 \in V_a^b$, and suppose that the vertex v following x_1 on a geodesic to x_2 is not in V_a^b. Then $x_1 \in co\{b,x_2,v\}$. Take a point $u \in vx_2$ with $x_1 \in bu$, and derive a contradiction.

Note. With considerable extra efforts, it is possible to prove that G is even S_4 in circumstances as above; cf. Chepoi's paper.

4.24.2. Verify that all cycles, and all connected graphs with at most four points, are PP-spaces. Observe that $K_{2,3}$ is a five-point connected graph with a non-convex geodesic interval.

4.24.3. (Bandelt, Chepoi, and van de Vel [1993]) Each polytope P in Euclidean space can be seen as a graph, the vertices of which are the extreme points of P, and the edges of which correspond with the 1–faces of P. Show that the five Platonic solids,

§4: Interval Spaces   89

viewed in this way as graphs, are PP spaces.[2] A *hyperoctahedron* of dimension $n > 0$ is defined as the complete graph K_{2n} minus a complete matching. Note that for $n > 1$ this is the graph of a polyhedron in $\mathbb{R}^n$. Show that each induced subgraph of a hyperoctahedron is JHC and S_4.

Remark. No example is known to us of a Euclidean polytope, the graph of which is not a Pasch-Peano space.

4.24.4. Inspect the graphs displayed in Fig. 11 for the Pasch, Peano and Sand-glass Properties. Hint: the first is Pasch, the fourth and fifth are Peano, and the third and fifth have the Sand-glass Property.

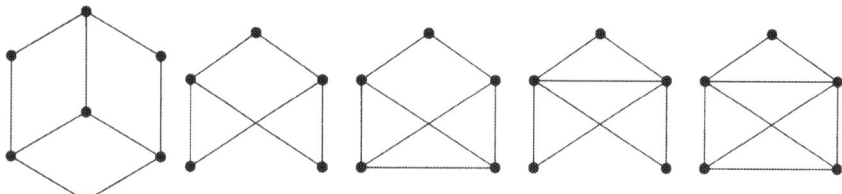

Fig. 11: On the Pasch, Peano and Sand-glass Property

4.25. Star-shaped sets. A subset S of an interval space is *star-shaped* at a point $p \in S$ provided

$$\forall x \in S : xp \subseteq S.$$

The *star-center of S* is the set of all points at which S is star-shaped. An interval convex set is star-shaped at each of its points. Relative to the geodesic interval operator, a metric disk $D(p,r)$ is star-shaped at p.

4.25.1. (Kołodziejczyk [1985]) Show that in a JHC convex structure the star-center of each set is convex (this is called *Brunn's Property*). Conversely, if the segment operator of a space has Brunn's Property, then the space is of arity 2.

4.25.2. (compare Kołodziejczyk [1985]) Show that $\mathbb{R}^n$ with one point removed has Brunn's property but is not JHC.

4.25.3. (Soltan [1979], Kołodziejczyk [1985]) Let X be point-convex. Show that X is JHC iff for each $S \subseteq X$ the star-center of S is the intersection of all convex sets, maximal with the property of being included in S.

4.25.4. Let $A \subseteq X$. Note that there is a smallest set $S \supseteq A$ which is star-shaped at each point of A. Verify that if X is JHC, then $S = co(A)$.

---

2. The Pasch and Peano axioms on the dodecahedron and icosahedron were verified with computer assistance. It should be possible to give readable proofs.

4.26. Separation and polytopes

4.26.1. Show that an interval space is S_4 iff each polytope on *three* vertices is S_4, and that a JHC convex structure is S_3 iff each polytope on *four* vertices is S_3. In fact, a convex structure of arity n is S_4 iff each $(2n-1)$–polytope is; cf. 3.22.1.

4.26.2. Let $p \in co\{a_1,..,a_n\}$ and for each $i = 1,..,n$ let $a'_i \in pa_i$. Show that if the space is S_3, then p is also a member of the shrunken polytope $co\{a'_1,..,a'_n\}$. This describes a process of "shrinking" a polytope. The reverse process of "swelling" a polytope need *not* preserve membership.

4.26.3. *Extended Sand-glass Property.* For each $i = 1,..,n$ let $p \in a_i b_i$, and let $a \in co\{a_1,..,a_n\}$. If the space is JHC and S_3, then there is a point $b \in co\{b_1,..,b_n\}$ such that $p \in ab$.

Notes on Section 4

Interval spaces have been introduced by Calder in [1971]. Most of the operators listed in 4.2 belong to the folklore; the same goes for the basic constructions of subspaces and products. The relation between the product of geodesic intervals and the Manhattan metric (cf. 4.3.1) is due to Lassak [1975b]; for normed spaces, the observation goes back to Soltan [1972].

Geometric interval spaces have been studied by Bandelt and Chepoi [1991], by Hedlíková [1983] ("ternary spaces"), and by Verheul [1993], who introduced the term. The characterization of modular lattices in terms of the geometric property of the modular interval operator I_M (cf. Proposition 4.6.2) goes back to Pitcher and Smiley [1942].

The characterization of Join-hull Commutativity by the associativity of the join, Theorem 4.11, is given by Calder [1971]. The characterization of S_3 under JHC (Sand-glass Property) is due to Chepoi [1986b], who gives a different proof. The corresponding characterization of S_4 by the Pasch Property is implicit in Ellis [1952]; it also appears in Chepoi [1986b] and (with an additional restriction) in van de Vel [1991]. Pasch-Peano spaces have been studied by Bandelt, Chepoi and van de Vel in [1993]. Example 4.24.3 (Platonic solids) is taken from this paper. The Pasch Property is involved in the axioms of join structures (more about this in Section 7).

The S_4 property of vector spaces, Example 4.14.1, is essentially Kakutani's result of [1937]. For Join-hull Commutativity of vector spaces, see Valentine [1964]. That posets are S_3 (Example 4.14.2) has been observed by Jamison in [1982]. A condition stronger than the Sand-glass Property was obtained on posets by Franklin [1962] (cf. 3.25). The Pasch Property of spaces of arcs was obtained by the author in [1991]. Its counterpart in discrete semimodular lattices (cf. Proposition 4.14.3) was obtained by Bandelt and van de Vel in [1992]. Proposition 4.15, on intervals being convex in graphs, has been obtained by Nieminen [1988]. It extends a result of Chepoi [1986] on bipartite graphs.

5. Base-point Orders

Interval operators can be used to construct quasi-orders on a convex structure in a natural way. These orders describe how a space looks like when seen from a particular base-point. This leads to a characterization of geometric interval spaces in terms of order. Each base-point order in a tree is a tree-order inducing the original tree convexity. Hence each base-point order can take the role of the original order. To some extent, the same goes for distributive lattices: each corner point leads to a different lattice structure with the same convexity.

Gated sets are subsets having a smallest element in each base-point order of the superspace. Such sets are convex and their gate maps satisfy several identities. Two geometric interval spaces can be amalgamated along a common, gated, convex subspace. The Pasch, Peano, and S_3 properties are preserved by this operation.

Completeness of interval spaces is defined in terms of base-point orders. This covers the traditional concepts of completeness in posets, in semilattices, and in lattices under the usual convexifications. In a vector space over a totally ordered field, completeness amounts to the requirement that the coefficient field be $\mathbb{R}$. A major result is that a segment bc of a complete S_4 convex structure is a lattice in the base-point order of b. A special kind of completeness is considered for the geodesic convexity in metric spaces, which is a mixture of metric and order completeness.

5.1. Base-point orders. Let (X,I) be an interval space and $b \in X$. We consider the following binary relation $\leq_b$ on X.

$u \leq_b v$ iff $I(b,u) \subseteq I(b,v)$ $(u, v \in X)$.

The relation $\leq_b$ is clearly reflexive and transitive; in other words: it is a quasi-order on X. We refer to it as the *base-point quasi-order at b*. This relation is a genuine partial order precisely if $I(b,u) = I(b,v)$ implies $u = v$ for all $u, v \in X$. Informally, $u <_b v$ means that u prevents b from seeing v (visibility order). Note that an interval preserving (IP) function $f: X \to Y$ can be characterized by the fact that for each base-point $b \in X$, if $u \leq_b v$ then $f(u) \leq_{f(b)} f(v)$. Briefly: IP functions are base-point order preserving.

Base-point (quasi-)orders can be considered for convex structures too; the segment operator is the canonical choice of an interval operator. The results of the previous sections assert that a segment operator has the Peano Property, resp., Pasch Property, iff for each base-point b and for each convex set C, the lower, resp., upper sets

$\downarrow_b(C) = \{ x \mid \exists c \in C : x \leq_b c \}$;
$\uparrow_b(C) = \{ x \mid \exists c \in C : c \leq_b x \}$

are interval convex. The Sand-glass Property implies that sets of type $\downarrow_b(p)$ are (interval) convex for each b, p. Under these circumstances, the b-infimum of a set A equals the b-infimum of $co(A)$ provided one of the named infima exists.

One of the features of geometric interval operators is that the corresponding base-point quasi-orders have some additional properties, thus allowing for some "order

5.2. Proposition. *The following are equivalent for an interval space X.*

(1) *The interval operator of X is geometric.*
(2) *For each $b \in X$ the relation $\leq_b$ is a partial order on X such that $\leq_b$ and $\leq_c$ are mutually inverse orders on the interval bc.*
(3) *For each $b \in X$ the relation $\leq_b$ is a partial order on X such that for all $u, v \in X$ with $u \leq_b v$ the following interval formula holds.*

$$uv = \{x \mid u \leq_b x \leq_b v\}.$$

Proof. (1) $\Rightarrow$ (2). We check the antisymmetry of $\leq_b$. Let $bc_1 = bc_2$. By (G-3), it follows that $c_2 \in c_1 c_1$, whence $c_1 = c_2$ by (G-1). If $u, v \in bc$ and $u \leq_b v$, then $u \in vb$ and (G-3) yields $v \in uc$, i.e., $v \leq_c w$.

(2) $\Rightarrow$ (3). Let $u \leq_b v$. If $x \in uv$ then $v \leq_v x \leq_v u \leq_v b$. In particular, we have $u, x \in bv$, whence $u \leq_b x \leq_b v$ by (2). Conversely, if $u \leq_b x \leq_b v$ then $u, x \in bv$. By (2) we find $x \leq_v u$, that is, $x \in uv$.

(3) $\Rightarrow$ (1). As to (G-1), note that $x \in bb$ expresses that $b \leq_b x \leq_b b$, which is solved by $x = b$ only. As to (G-2), let $c \in ba$. Then $c \leq_b a$. Hence $x \leq_b c$ implies $x \leq_b a$ for all x; in other words: $bc \subseteq ba$. To obtain (G-3), consider $c \in ba$ and $d \in ca$. We find $c \leq_b a$ and, by (3), $c \leq_b d \leq_b a$. The first of these inequalities yields $c \in bd$. ∎

In regard to Theorem 4.7, conditions (2) and (3) of the above result hold with respect to the base-point orders induced by the segment operator of a point-convex S_3 convex structure regardless of its arity.

5.3. Examples

5.3.1. Standard intervals. Let V be a vector space over a totally ordered field. V is equipped with the standard convexity. Let $b \in V$. The maximal chains in $(V, \leq_b)$ are precisely the half-lines of V emanating from b.

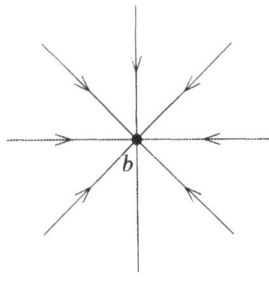

Fig. 1:
Base-point order in the standard plane

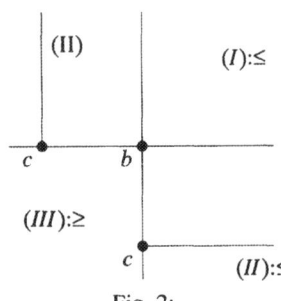

Fig. 2:
A semilattice base-point order

5.3.2. Semilattice intervals. Let S be a semilattice with a lower bound 0. Then the base-point order $\leq_0$ coincides with the original order of S. For other base-points b the corresponding order $\leq_b$ can be more complicated.

If $a \not\leq b$ then (cf. Fig. 2) $a \leq_b x$ iff $a \leq x$ and $a \wedge b = x \wedge b$. Hence on each of the sets of type

(I) $\uparrow(b) = \{x \mid x \geq b\}$;

(II) $\{x \mid b \wedge x = c\,; x \not\leq b\}$

(where c is a fixed constant), the base-point order $\leq_b$ coincides with the original order. If $a < b$ then $a \leq_b x$ iff $x \leq a$. Hence on sets of type

(III) $\downarrow(b) = \{x \mid x \leq b\}$,

the order $\leq_b$ is the inverse of the original order.

We now specialize to trees, where a lot more can be said on base-point orders.

5.3.3. Proposition. *If T is a tree, then for each $b \in T$ the base-point order $\leq_b$ is a tree order and the resulting semilattice convexity equals the original one.*

Proof that $(T, \leq_b)$ is a semilattice. Let $x_1, x_2 \in T$. We look for a point x such that

$$bx_1 \cap bx_2 = bx.$$

Such a point x is the b–infimum $x_1 \wedge_b x_2$ of x_1 and x_2. Observe that

(1) $bx_1 \cap bx_2 = \{u \mid b\wedge x_1 \leq u;\ b\wedge x_2 \leq u;\ u \leq b \text{ or } u \leq x_1;\ u \leq b \text{ or } u \leq x_2\}$.

As the lower set $L(b)$ is a chain, we may assume that $b \wedge x_1 \leq b \wedge x_2$.

Suppose first that $b \wedge x_1 < b \wedge x_2$ and let u be in the set (1). We have $b \wedge x_2 \leq u$ together with one of the following.

(2) $u \leq b \wedge x_1$; $u \leq b \wedge x_2$; $u \leq x_1 \wedge x_2$; $u \leq b$.

One solution is $u = b \wedge x_2$. All other solutions u satisfy $b \wedge x_2 < u$; hence they do not obey the first and second alternative in (2). The third one is also impossible, for it implies $b \wedge x_2 \leq b \wedge x_1$. So $u \leq b$ must hold, and it follows that

$$bx_1 \cap bx_2 = \{u \mid b\wedge x_2 \leq u \leq b\} = b(b\wedge x_2).$$

Suppose next that $b \wedge x_1 = b \wedge x_2$. Then (1) becomes

$$bx_1 \cap bx_2 = \{u \mid b\wedge x_1 \wedge x_2 \leq u;\ u \leq b \text{ or } u \leq x_1 \wedge x_2\} = b(x_1 \wedge x_2).$$

Summarizing, we find that

$$x_1 \wedge_b x_2 = \begin{cases} b \wedge x_1, & \text{if } b \wedge x_1 > b \wedge x_2; \\ b \wedge x_2, & \text{if } b \wedge x_1 < b \wedge x_2; \\ x_1 \wedge x_2, & \text{if } b \wedge x_1 = b \wedge x_2. \end{cases}$$

Proof that $(T, \leq_b)$ is an order tree. Let $x_1, x_2 \leq_b x$. If $b \wedge x_1 < b \wedge x_2$, then $x_1 \wedge_b x_2 = b \wedge x_2$, and we show that $b \wedge x_2 = x_2$, i.e. that $x_2 \leq b$. If not, then $x_2 \in bx$ implies $b \wedge x \leq x_2 \leq x$ and hence $b \wedge x_2 \leq b \wedge x \leq x_1$ (the last inequality comes from

$x_1 \in bx$). Hence $b \wedge x_2 \leq b \wedge x_1$, a contradiction.

If, on the other hand, $b \wedge x_1 = b \wedge x_2$, then $x_1 \wedge_b x_2 = x_1 \wedge x_2$. We show that x_1 and x_2 are comparable in the original order. If x_1, x_2 are both below x or are both below b (in the original order), then we are done. By assumption, x_i is either below x or below b. So assume $x_1 \leq x$ and $x_1 \nleq b$, whereas $x_2 \leq b$. Then $x_1 \wedge b < x_1 \leq x$ and $x_2 \wedge b = x_2$, yielding $x_2 < x_1$.

Proof that the convexities of $(T, \leq)$ and $(T, \leq_b)$ are the same. It suffices to show that each pair of points $x, y \in T$ leads to the same segment in either convexity. We use "co_b" to denote the convex hull operator of the tree $(T, \leq_b)$. Let $u \in xy$. Then $x \wedge y \leq u$ and either $u \leq x$ or $u \leq y$. It must be verified that

(3) $x \wedge_b y \leq_b u$; $u \leq_b x$ or $u \leq_b y$.

If $b \wedge x < b \wedge y$, then $x \wedge_b y = b \wedge y$. Also, $u \leq x$ or $u \leq y$ implies one of

$$b \wedge u \leq b \wedge x; \quad b \wedge u \leq b \wedge y,$$

and hence $b \wedge u \leq b \wedge y \leq b$, showing that $b \wedge y$ is in bu, i.e. $x \wedge_b y \leq_b u$. The second part of (3) will be derived simultaneously with the second case (under the common assumption that $b \wedge x \leq b \wedge y$).

If $b \wedge x = b \wedge y$, then the b-infimum of x, y equals the original infimum. To obtain $x \wedge y \leq_b u$, note that $u \leq x$ or $u \leq y$ implies that

$$b \wedge u \leq b \wedge x = b \wedge x \wedge y \leq x \wedge y,$$

whereas $x \wedge y \leq u$ by assumption. Hence $x \wedge y \in bu$, i.e. $x \wedge y \leq_b u$.

For the remainder of the proof (that $u \leq_b x$ or $u \leq_b y$) we only assume $b \wedge x \leq b \wedge y$. If $u \leq b$ then

$$b \wedge x = b \wedge x \wedge y \leq x \wedge y \leq u \leq b,$$

showing that $u \in bx$. If $u \leq x$, then we change the last inequality to "$u \leq x$" to obtain $u \in bx$. So assume that $u \leq y$. Then u and $b \wedge y$ are comparable in $(T, \leq)$ and $u \leq b \wedge y$ yields $u \leq b$, a case which we solved already. This leaves us with the case $b \wedge y < u \leq y$, expressing that $u \in by$.

So far we have shown that

$$xy \subseteq co_b\{x,y\}.$$

A proof of the opposite inclusion is left to the reader (but this part of the proof is, in fact, redundant by Theorem 6.5(2) below). ∎

The equivalence of all base-point orders suggests that a description of a tree can be given without specifying an order. This is indeed the case: see II§1.27.1. It also contributes to the following definition. An *end point* of a tree is a point which is maximal in some base-point order. See 5.26 for additional information on end points.

In a distributive lattice L, the interval operators I_M and I_L are both equal to the segment operator, as one can see directly from the definitions. This operator is geometric: use Proposition 4.6.2 or Theorem 4.7. Let $b \in L$ be such that $bb' = L$ for some $b' \in L$.

§5: Base-point Orders

Equivalently, $L = [b \wedge b', b \vee b']$. In these circumstances, $b \wedge b'$ and $b \vee b'$ are the universal bounds of L (usually denoted by **0** and **1**, respectively). This expresses that b and b' are complementary. Geometrically, such points can be regarded as opposite corner points of the lattice.

5.3.4. Proposition. *Let L be a distributive lattice and let $b \in L$.*

(1) *The partially ordered set $(L, \leq_b)$ yields a semilattice.*
(2) *If b has a complement, then $(L, \leq_b)$ yields a distributive lattice and the corresponding convexity equals the original one.*

Proof. We use the operator $m: L^3 \to L$ defined in 1.5.6. As to (1), it is not difficult to verify that the meet operator of $(L, \leq_b)$ is given by

$$x \wedge_b y = m(b, x, y)$$

(compare 1.22.4). Hence $\leq_b$ is a semilattice order on L.

As for (2), let b' be complement of b. By 5.2(2), the base-point orders $\leq_b$ and $\leq_{b'}$ are mutually inverse. It easily follows that

$$x \vee_b y = m(b', x, y).$$

Hence $\leq_b$ is a lattice order on L. Compare 1.22.4. It is possible to show directly that the resulting lattice is distributive. Alternatively, one can prove that the lattice convexities of $(L, \leq)$ and $(L, \leq_b)$ are identical. This goes again by verifying that

$$xy = co_b\{x, y\}$$

for all $x, y \in L$. We leave this to the reader. Then distributivity of $(L, \leq_b)$ follows from the fact that its convexity is S_4, cf. 3.12.3. ∎

This is another instance of an order-dependent structure, the convexity of which is not affected by a (suitable) change of the order. Once more, it suggests that a description of the convexity can be given which is independent of the lattice order. See II§1.27.

The operator m of a distributive lattice is a median operator (cf. Topic 1.22.4 or Section 6). Therefore, the first part of the previous proposition follows from the next result.

5.3.5. Proposition. *In a median algebra (X, m) each base-point order $\leq_b$ (where $b \in X$) is a semilattice order and the meet operator is given by*

$$x \wedge_b y = m(x, y, b).$$

Proof. We first verify that the operation $\wedge_b: X \times X \to X$ satisfies the axioms of a meet operation: reflexivity ($x \wedge_b x = x$) and symmetry ($x \wedge_b y = y \wedge_b x$) are obvious; the associative law holds by the axiom (M-3) of medians:

$$(x \wedge_b y) \wedge_b z = m(m(x, y, b), z, b) = m(x, m(y, z, b), b) = x \wedge_b (y \wedge_b z).$$

The meet operation $\wedge_b$ induces a partial order $\leq$ defined by

$$x \leq y \iff x \wedge_b y = x.$$

Therefore, $x \leq y$ holds iff $m(x,y,b) = x$, iff $x \in I_m(b,y)$. ∎

The proof of the following result involves the Relative Interval Formula and the Product Interval Formula (cf. 4.3) and is left to the reader.

5.4. Proposition

(1) *If Y is an interval subspace of the interval space X and if $b \in Y$, then the base-point quasi-order at b of Y is the trace of the base-point quasi-order at b of X.*

(2) *If X is the product of the interval spaces X_i for $i \in I$, if $b_i \in X_i$ for $i \in I$, and if $b = (b_i)_{i \in I}$, then the base-point quasi-order at b of X is the Cartesian product of the base-point quasi-orders at b_i of X_i.* ∎

Although the relative convexity on a subset of an interval space need not be induced by the relative interval operator, part (1) of the above result applies equally well to (the segment operator of) convex structures and their subspaces. This is because of the Relative Hull Formula, 1.9.1. Similarly, part (2) applies to products of convex structures, in regard to the Product Polytope Formula, 1.10.3.

5.5. Discrete spaces and graphs. An interval space is *discrete* provided for all base-point orders, each bounded totally (quasi-)ordered subcollection is finite. For a discrete space X, an *underlying graph* $G(X)$ obtains, the edges of which are the pairs of points $u, v \in X$ with $uv = \{u, v\}$.

5.6. Examples. Let us link the previous definition with a few familiar situations.

5.6.1. Let X be a poset with the usual interval function. The description of the base-point orders (see Topic 5.23.1) shows that X is discrete as a poset iff it is discrete as an interval space.

5.6.2. Let S be a semilattice with the (usual) interval function

$$I_S(a,b) = \{z \mid a \wedge b \leq z; z \leq a \text{ or } z \leq b\}.$$

I_S is a geometric interval operator, being the segment operator of an S_3 convexity. The description of the base-point orders in a semilattice shows that S is discrete as an interval space iff it is discrete as a poset. Note that $a \wedge b \in I_S(a,b)$. Therefore, two distinct points form an edge iff one of them covers the other. Hence $G(S)$ is the well-known covering graph of S.

5.6.3. Let L be a lattice with the modular interval function

$$I_M(a,b) = \{x \mid (a \wedge x) \vee (b \wedge x) = x = (a \vee x) \wedge (b \vee x)\}.$$

The description of base-point orders in a lattice shows that L is discrete as an interval space iff it is discrete as a poset. Observe that $a \wedge b \in I(a,b)$. Hence, two distinct points form an edge iff one of them covers the other, and $G(L)$ is the covering graph of L. The graphs in 5.6.2 and 5.6.3 are known as (unordered) *Hasse diagrams*.

§5: Base-point Orders

5.7. Proposition. *The underlying graph of a discrete geometric interval space is connected.*

Proof. Let X be a geometric interval space, let $a, b \in X$ and consider a maximal chain between a and b in $(X, \leq_a)$, say:

(*) $\quad a = x_0 <_a x_1 <_a \cdots <_a x_n = b.$

By Proposition 5.2(3), if $i < n$ and $x \in x_i x_{i+1} \setminus \{x_i, x_{i+1}\}$, then $x_i <_a x <_a x_{i+1}$. Therefore,

$$x_i x_{i+1} = \{x_i, x_{i+1}\},$$

and the chain (*) yields a path in X from a to b. ∎

In a discrete geometric interval space X the shortest-path (geodesic) metric gives rise to another interval operator, and X is said to be *graphic* provided both operators coincide. The next result gives a sufficient condition on a discrete space to be graphic. An interval space has the *Triangle Property* provided that for any three points a, b, c with the properties

$$ab \cap bc = \{b\}; \quad bc \cap ca = \{c\}; \quad ca \cap ab = \{a\},$$

all three intervals are edges whenever one of them is. For discrete geometric interval spaces, this property can be reformulated in terms of the base-point order $\leq_c$. If a, b form an edge and are not comparable in $\leq_c$, then the c-infimum of a, b exists and forms an edge with both a, b.

5.8. Proposition. *A discrete geometric interval space with the Triangle Property is graphic.*

Proof. Let X be as announced, and let ρ denote the metric of $G(X)$. We verify by induction on $n = \rho(a,b)$ that a point is geodesically between a and b iff it is in ab. For $n \leq 1$ there is nothing left to prove. Suppose the statement to be valid for distances $\leq n-1$, where $n \geq 2$. Consider a geodesic α joining b to a, and let a' be the point on α at distance $n-1$ of b. The inductive assumption implies that $a \notin a'b$. If $a' \notin ab$, then consider a maximal point $x \leq_b a, a'$. By the Triangle Property, we conclude that ax and $a'x$ are edges. As $x \in a'b$, the inductive assumption implies that

$$\rho(b,a) \leq \rho(b,x) + \rho(x,a) \leq n-1,$$

a contradiction. This shows that each geodesic joining a and b is in ab.

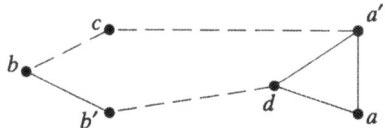

Fig. 3: Proof that Triangle Property ⇒ graphic

Conversely, let $c \in ab \setminus \{a,b\}$, and let $a' \geq_b c$ be a maximal element in ab distinct from a (Fig. 3). Then aa' is an edge. If $\rho(a',b) < n$ then a geodesic can be found in ab

98   Chap. I: Abstract Convex Structures

passing through a', c. Assume $\rho(a',b) \geq n$, and let b' be a neighbor of b on a geodesic joining a to b. Observe that $b' \in ab$ by the first half of the proof. We infer that $a \notin a'b'$ and $a' \notin ab'$. The first follows from the geometricity of the given interval operator; the second follows by inductive assumption. Consider a maximal element d below a, a' in the base-point order of b'. By the Triangle Property, both ad and $a'd$ are edges. As d is on a geodesic joining b' to a, we find that $\rho(d,b') = n-2$, and hence $d \in a'b$. But this implies $a' \in ad$ by geometricity, a contradiction. ∎

The characterization of modular lattices in terms of the interval operator I_M (cf. 4.6.2) can be used to see that such lattices have no non-trivial triples as described in the triangle property, and hence these spaces are always graphic. This fact will be generalized in the next section.

5.9. Gates. Let X be a geometric interval space, let $C \subseteq X$, and $b \in X$. A point $c \in C$ is the *gate of b in C* provided $c \in bx$ for each $x \in C$. In other words, a gate c of b in C is the smallest element of $(C, \leq_b)$. If $D \subseteq X$ is another set, then two points $c \in C$ and $d \in D$ are *mutual gates* of C and D provided c is the gate of d in C and d is the gate of c in D. If each point of X has a gate in C, then C is a *gated set within X,* and the resulting function $X \to C$, which assigns to a point of X its gate in C, is the *gate map* (*nearest point function*) of C.

5.10. Examples

5.10.1. Metric spaces. In a metric space (X, ρ), the gate c of b in C is a metric nearest point, that is: $\rho(b,c) \leq \rho(b,x)$ for all $x \in C$. The converse is not valid. For instance, the standard convexity of $\mathbb{R}^n$ is produced by the standard metric, and (as is well-known) each non-empty convex closed set admits metric nearest points. For $n > 1$, however, such sets rarely admit gates.

5.10.2. Normed spaces. Let $(V, ||.||)$ be a normed space. The following notion of orthogonality is natural:
$$x \perp y \iff ||x+y|| = ||x-y|| = ||x|| + ||y||.$$
If C is a subset of V, then $C^\perp$ denotes the set of all $x \in V$ which are orthogonal to all members of C. Let $L \subseteq V$ be a linear subspace. Then the following are equivalent.
(1) L has a gate map.
(2) $V = L + L^\perp$.

Indeed, let $p: V \to L$ be a gate map. Then
$$||x - p(x)|| + ||p(x) - c|| = ||x - c|| \quad (x \in V; c \in L).$$
Each element $c \in L$ can be written as a difference of two elements of L in the following way: $c = p(x) - (p(x) - c)$. We conclude that
$$||x - p(x)|| + ||c|| = ||x - p(x) + c||.$$
After replacing c with $-c$, we conclude that $x - p(x) \perp L$ and hence that

§5: Base-point Orders

$x = p(x) + (x - p(x)) \in L + L^\perp$.

Conversely, suppose $V = L + L^\perp$. Let $x \in V$, say: $x = x_L + x_\perp$, where $x_L \in L$ and $x_\perp \in L^\perp$. Then $x - x_L \in L^\perp$ and hence for each $c \in L$ we have

$$\|x - x_L\| + \|x_L - c\| = \|x - c\|,$$

showing that x_L is the gate of x in L.

It is not difficult to verify that the gate map $p: V \to L$ is linear iff $L^\perp$ is a linear space, iff there is a linear projection $p: V \to L$ satisfying $\|x\| = \|x - p(x)\| + \|p(x)\|$. Such maps are known as *L–projections* and the corresponding linear subspace is known as an *L–summand*.

5.10.3. Intervals in median algebras. In a median algebra (X, m), the point $p = m(a,b,c)$ is the gate of c in $I_m(a,b)$. First, $p \in I_m(a,b)$ by definition. According to 5.3.5, $m(a,b,c)$ is the meet $a \wedge_c b$ in $(X, \leq_c)$ of a, b. Let $x \in I_m(a,b)$. Then $x \leq_b a$ and hence $x \wedge_b c \leq_b a \wedge_b c = p$. Therefore, $x \wedge_c b \in pb$, which yields $p \in I_m(x \wedge_c b, c)$ by the geometric properties of I_m. This gives $p \leq_c x \wedge_c b \leq_c x$.

The following general observations will be used frequently.

5.11. Proposition. *Let X be a geometric interval space.*

(1) *If $D \subseteq X$ is a convex set containing b and meeting C, and if $b \in X$ has a gate c in the set $C \subseteq X$, then $c \in D$.*

(2) *(Screening Lemma) Let $c \in C$, $d \in D$ be mutual gates. Then each pair of convex sets screening c, d also screens C, D.*

(3) *(Transitive Law; Fig. 4) Let $D \subseteq C \subseteq X$ and let $b \in X$. If c is its gate in C, and let d be the gate of c in D, then d is the gate of b in D.*

(4) *(Idempotent Law; Fig. 5) Let c_1 be the gate in C_1 of some point in C_0, and let c_0 be the gate of c_1 in C_0. Then c_1 is the gate of c_0 in C_1.*

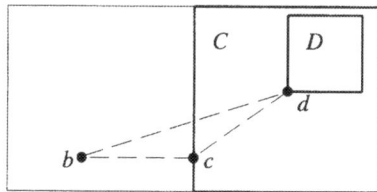
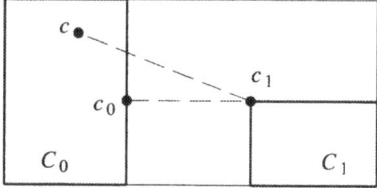

Fig. 4: Transitive rule of gates   Fig. 5: Idempotent rule of gates

Proof. (1) is evident. As to (2), let C', D' be convex sets screening c and d:

$$c \in C' \setminus D'; \quad d \in D' \setminus C'; \quad C' \cup D' = X.$$

If D' meets C then by Part (1), the gate c of d in C must be in D', contradiction. Similarly, C' cannot meet D. Therefore C, D are screened by C', D'.

(3). If $d' \in D \cap db$, then $d \in d'c \subseteq d'b$, whence $d = d'$.

(4). Assume $cc_1 \cap C_1 = \{c_1\}$. As c_0 is the minimum of C_0 from the viewpoint of

c_1, we obtain $c_0 c_1 \subseteq cc_1$ and $c_0 c_1$ meets C_0 and C_1 precisely in c_0, resp., c_1. ∎

The Transitive Law implies that if $C \subseteq X$ is gated within X, and if $D \subseteq C$ is gated within the subspace C, then D is gated within X and the associated gate map obtains as a composition of gate maps. On the other hand, the Idempotent Law implies that if C, D are gated sets in X with respective gate maps $p_C: X \to C$ and $p_D: X \to D$, then

$$p_C | D = p_C \circ p_D \circ p_C | D.$$

5.12. Proposition. *In a geometric interval space X, the following are true.*

(1) *Gated sets are convex.*
(2) *A finite collection of pairwise intersecting gated sets has a non-empty intersection.*
(3) *If $f: X \to Y$ is a surjective IP function, and if $C \subseteq X$ is gated, then $f(C) \subseteq Y$ is gated and f commutes with the respective gate maps.*
(4) *If X is S_3, then all gate maps of X are CP and CC.*

Proof. (1). Let $C \subseteq X$ be gated, let c_1, $c_2 \in C$, and let $x \in c_1 c_2$. Then x has a gate c in C. By construction, $c \in xc_1 \cap xc_2 = \{x\}$, whence $x \in C$.

(2). We argue by induction on the number of gated sets involved. For two sets, there is nothing left to prove. Suppose that the result is valid for $n \geq 2$ sets. Let C_i for $i = 1,..,n+1$ be pairwise intersecting gated sets. By assumption, there is a point $b \in \cap_{i=1}^n C_i$. Let c be its gate in C_{n+1}. For each $i \leq n$, the set C_i is convex, it contains b and it meets C_{n+1}. Hence $c \in C_i$ by 5.11.

(3). Let f and C be as announced, let $b \in X$, and let $c_0 \in C$ be its gate. If $d \in f(C)$, say: $d = f(c)$ with $c \in C$, then $c_0 \in bc$ and hence $f(c_0) \in f(b)d$.

(4). Let $C \subseteq X$ be a gated set with a gate map $p: X \to C$. Suppose $D \subseteq C$ is convex, u, $v \in p^{-1}(D)$ and $w \in uv$, but $w \notin p^{-1}(D)$. Now use S_3 to obtain a half-space H including D and excluding $p(w)$. The convex set $X \backslash H$ meets C (e.g., in $p(w)$) and consequently each point of $X \backslash H$ has a C-gate in $X \backslash H$. Therefore, u, $v \in H$. This gives $w \in uv \subseteq H$, a contradiction. The last part follows from (3) and (1). ∎

The next result gives an important method of constructing geometric interval spaces from smaller pieces.

5.13. Theorem (Amalgamation Theorem). *Let (X_i, I_i) for $i = 1, 2$ be geometric interval spaces such that $X_1 \cap X_2$ is a gated subset of X_1 and of X_2, on which the respective interval operators coincide. Then there is a unique geometric interval operator I on $X_1 \cup X_2$ subject to the following two conditions.*

(i) *I extends I_1 and I_2.*
(ii) *If $a \in X_1$ and $b \in X_2$, then $I(a,b)$ meets $X_1 \cap X_2$.*

If $p_i: X_i \to X_1 \cap X_2$ ($i = 1, 2$) are the gate maps, and if $a \in X_1$ and $b \in X_2$, then

(iii) *$I(a,b) = I_1(a, p_2(b)) \cup I_2(p_1(a), b)$.*

Moreover, X_1 and X_2 are gated, and the gate map $X_1 \cup X_2 \to X_i$ extends p_i for $i = 1, 2$.

§5: Base-point Orders   101

Proof. Let $X = X_1 \cup X_2$. Note that if $a \in X_1$ and $b \in X_1 \cap X_2$, then $p_2(b) = b$ and $p_1(a) \in I_1(a,b)$ by definition of a gate. Hence the prescriptions (i) and (iii) agree. The operator I, defined by (i) and (iii), is evidently an interval operator on X. We verify the following equalities (where $b_i \in X_i$ for $i = 1, 2$).

(1a) $I(b_1,b_2) \cap X_1 = I_1(b_1,p_2(b_2))$;
(1b) $I(b_1,b_2) \cap X_2 = I_2(p_1(b_1),b_2)$.

As to (1a), the right-to-left inclusion is evident. Let $u \in I(b_1,b_2) \cap X_1$ and suppose that

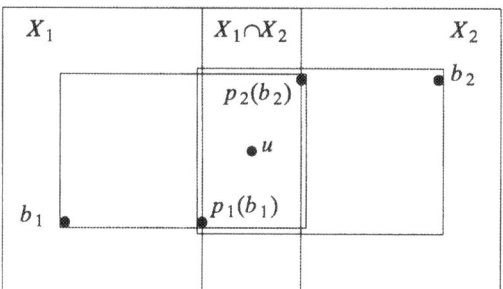

Fig. 6: Interval formula for amalgamations

$u \in I_2(p_1(b_1),b_2)$ (see Fig. 6). As $u \in X_1 \cap X_2$, we have $p_2(b_2) \in I_2(u,b_2)$ by definition of a gate. By the geometric property of I_2, we conclude that

$$u \in I_2(p_1(b_1),p_2(b_2)) = I_1(p_1(b_1),p_2(b_2)).$$

By the geometric property of I_1 the last set is included in $I(b_1,p_2(b_2))$.

Equality (1b) is obtained similarly. For a proof that I is geometric, let $a, b, u, v \in X$ and $u \in I(a,v)$, $v \in I(a,b)$. We have to show that

(2) $u \in I(a,b)$; $v \in I(u,b)$.

We may assume that $a \in X_1$ and $b \in X_2$. The situation where $v \in X_1$ or $u \in X_2$ can easily be reduced to the original interval spaces by using the formulae (1). So assume $u \in X_1$, $v \in X_2$. We have

(3) $u \in I_1(a,p_2(v))$

by the construction of I. By definition of a gate, $p_1(u) \in I_1(u,p_2(v))$. By (3) and the geometric property of I_1, we find $p_1(u) \in I_1(a,p_2(v))$. We obtain from (1b) that

(4) $p_1(u) \in I_2(p_1(a),p_2(v))$.

Next, we have $v \in I_2(p_1(a),b)$ by the construction of I and $p_2(v) \in I_2(p_1(a),v)$ by definition of a gate. By the geometric property of I_2, we conclude that

(5) $p_2(v) \in I_2(p_1(a),b)$,
(6) $v \in I_2(p_2(v),b)$.

By (5) and (1a) we find that $p_2(v) \in I_1(p_1(a),p_2(b))$. As $p_1(a) \in I_1(a,p_2(b))$ by definition of a gate, we conclude that

(7) $p_2(v) \in I_1(a, p_2(b))$.

By (4), (5), and the geometricity of I_2, we conclude that

(8) $p_2(v) \in I_1(p_1(u), b)$.

By (6), (8), and the geometricity of I_1, we find that $v \in I_2(p_1(u), b)$, whence $v \in I(u, b)$ by the construction of I. By (3), (7) and the geometricity of I_1 we finally obtain $u \in I_1(a, p_2(b))$ and hence $u \in I(a, b)$. This completes the proof that I is geometric (cf. (2)).

The statement concerning gate maps is a direct consequence of (1). If $a \in X_1$ and $b \in X_2$ then $I(a, b)$ meets $X_1 \cap X_2$, for instance, in the gate of a (or b) in $X_1 \cap X_2$. We are left with a proof of unicity. Suppose $\bar{I}$ is another geometric interval operator satisfying (i), (ii). with properties as described in the theorem. Let $a \in X_1$ and $b \in X_2$, and let $z \in \bar{I}(a, b) \cap X_1 \cap X_2$. Then

$$p_1(a) \in I_1(a, z) = \bar{I}(a, z) \subseteq \bar{I}(a, b).$$

It follows that $p_1(a)$ is the gate of a in X_2 relative to $(X, \bar{I})$. Similarly, $p_2(b)$ is the gate of b in X_1 relative to $(X, \bar{I})$. As $\bar{I}$ is geometric, the sets $I(a, p_2(b))$ and $I_2(p_1(a), b)$ are included in $\bar{I}(a, b)$. In other words, $I(a, b) \subseteq \bar{I}(a, b)$. Let $u \in \bar{I}(a, b)$, say $u \in X_1$. We showed earlier that $p_2(b)$ is the gate of b in X_1; in particular $p_2(b) \in \bar{I}(u, b)$. By the assumed geometric property of $\bar{I}$ we deduce that $u \in \bar{I}(a, p_2(b))$. Hence $u \in I(a, b)$. We conclude that $\bar{I} = I$. ∎

The resulting interval space is called the **gated amalgam** of X_1, X_2. Note that the latter are convex subsets of the amalgamation. Moreover, if a convexity on X is induced by a geometric interval operator and takes X_1 and X_2 as convex subspaces, then this convexity equals the original one. Among the properties that $X_1 \cup X_2$ inherits from its constituents X_1 and X_2 are the following.

5.14. Theorem. *The gated amalgam of two geometric interval spaces has the Peano Property (Pasch Property) iff both summands do.*

Proof. Let X_i for $i = 1, 2$ be geometric interval spaces with a commonly gated subspace $X_1 \cap X_2$. Hereafter, X denotes the resulting amalgam and $p_i: X \to X_i$ denotes the gate function. The "only if" statements are evident.

We first assume that both summands have the Peano Property. Let $a, b, c, x, y \in X$ with $x \in ba$ and $y \in cx$ (cf. Fig. 7). Taking into account the symmetric role of X_1 and X_2, it suffices to consider the following cases.

(1) None of a, b, c is in X_2. Use the fact that X_1 has the Peano Property.

(2) $a \in X_2, b, c \in X_1$.

(2a) If $x, y \in X_1$, then take $a' = p_1(a)$. The interval formula for amalgams gives $a' \in ax$ and by the geometric property of X's interval operator, we see that $x \in ba'$. Use (1) with reference to the configuration a', b, c, x, y to obtain a point $z \in ca'$ with $y \in bz$, and use the fact that $ca' \subseteq ca$.

§5: Base-point Orders   103

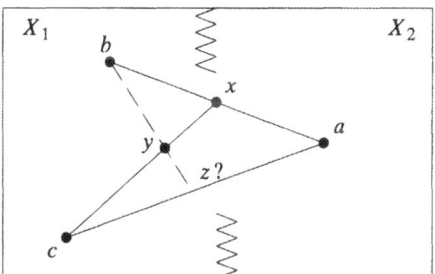

Fig. 7: Peano Property in gated amalgams

(2b) If $x \in X_2$ and $y \in X_1$, let $x' = p_1(x)$. Then $x' \in bx \subseteq ba$ whereas $x' \in yx$ yields $y \in x'c$. The configuration a,b,c,x',y brings us back to the previous situation.

(2c) If $x, y \in X_2$, then consider $b' = p_2(b)$. The interval formula yields $b' \in bx$ and hence $x \in b'a$. Use case (3a) below with reference to the configuration a,b',c,x,y to obtain a point $z \in ca$ with $y \in b'z$. As $b' \in by \subseteq bz$ we find $b'z \subseteq bz$.

(3) a,b are in X_2 and $c \in X_1$.

(3a) If $y \in X_2$, take $c' = p_1(c)$. We find $c' \in cy$ and hence $y \in c'x$. Apply (1) and use $c'a \subseteq ca$.

(3b) If $y \in X_1$, take $y' = p_1(y)$. Now $y' \in yx \subseteq cx$ and we use (3a) with reference to the configuration a,b,c,x,y'. This gives $z' \in ca$ with $y' \in bz'$. If $z' \notin X_1$ then consider $z'' = p_1(z')$. We have $z'' \in z'y'$ and hence $y' \in z''b$ on the one hand, whereas $z'' \in cz' \subseteq ca$ on the other hand. So without loss of generality, $z' \in X_1$. Use (4a) (with labels 1, 2 interchanged) to obtain $z \in cz' \subseteq ca$ with $y \in bz$.

(4) $a,c \in X_2$ and $b \in X_1$.

(4a) If $x,y \in X_2$, take $b' = p_2(b)$ and apply (1) (with labels 1, 2 interchanged) to obtain $z \in ca$ with $y \in zb'$. The interval formula for amalgams gives $b' \in bz$, and hence $y \in b'z \subseteq bz$.

(4b) If $x \in X_1$ and $y \in X_2$, then take $x' = p_2(x)$. Now $x' \in yx \subseteq cx$ and $x' \in ax \subseteq ab$. We can apply (4a) to the configuration a,b,c,x',y.

(4c) If $x,y \in X_1$, take $a' = p_1(a)$ and use (3a) (with labels 1, 2 interchanged) with reference to the configuration a',b,c,x,y to find $y' \in ca'$ with $y \in y'b$. Then use (4a) with reference to the configuration a,b,c,a',y' to obtain $z \in ca$ with $y' \in za$. Now $y \in y'b \subseteq zb$ as desired.

We next assume that both summands have the Pasch Property. Let $a,b,c \in X$ and $x \in ab$, $y \in ac$ (Fig. 8). Three major cases occur.

(1) $a,b,c \in X_1$: use the Pasch Property of X_1.

(2) $a \in X_2$ and $b,c \in X_1$. If $x,y \in X_1$, consider $a' = p_1(a)$ and apply (1). If $x \in X_1$ and $y \in X_2$, take $y' = p_1(y)$. The previous subcase applies to the resulting configuration. If $x,y \in X_2$, use $b' = p_1(b)$ and $c' = p_1(c)$ and apply case (1).

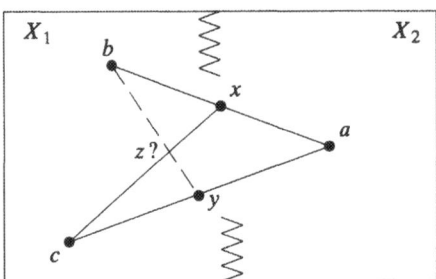

Fig. 8: Pasch Property in gated amalgams

(3) $a,b \in X_2$ and $c \in X_1$. If $y \in X_2$ then change c into $c' = p_1(c)$ and use (1). If $y \in X_1$, take $y' = p_1(y)$ and use the previous subcase. ∎

5.15. Theorem. *Let X_1 and X_2 be S_3 convex structures of arity two meeting in a non-empty gated subspace. Then there is one and only one S_3 convexity on $X = X_1 \cup X_2$ which is of arity two and takes X_1 and X_2 as convex subspaces. This convexity is derived from the gated amalgamation of the summands. Moreover, if $F_i \subseteq X_i$ for $i = 1, 2$ are sets with F_1 non-empty and if $p_i: X \to X_i$ ($i = 1, 2$) denotes the gate map, then*

$$co(F_1 \cup F_2) \cap X_1 = co(F_1 \cup p_2(F_2)).$$

Proof. By 4.7, the segment operator of each summand is geometric and the Amalgamation Theorem applies. Let $C \subseteq X$ be convex and $q \notin C$, say: $q \in X_1$. If $q \notin X_2$ and C is disjoint from X_1, then X_1, X_2 is a convex screening of q and C. Therefore, there is no loss of generality in assuming that C meets X_1. Observe that then $C \subseteq p_1^{-1}(C \cap X_1)$ (use the interval formula for amalgams). Take a convex screening D, E of q and C in X_1. If e.g., D is disjoint from X_2, then $E \supseteq X_1 \cap X_2$. Evidently, $E \cup X_2$ is a convex set of X. Now D and $E \cup X_2$ constitute a convex screening of q and C in X. The case where $E \cap X_2 = \emptyset$ is handled similarly. Suppose C, D both meet X_2. The sets $p_1^{-1}(D)$ and $p_1^{-1}(E)$ cover X_2 and by Proposition 5.12(4), these sets are convex. The gated interval formula gives that $D \cup p_1^{-1}(D)$ and $E \cup p_1^{-1}(E)$ are convex sets of X and they evidently form a screening of q and C.

If an S_3 convexity on X takes each X_i for a convex subspace, then its segment operator is geometric and it extends the respective segment operators of X_1, X_2. By the Amalgamation Theorem, it equals the canonical interval operator of X. If the convexity is, moreover, of arity two, it must be the original one.

As to the proposed hull formula, the inclusion from right to left is evidently valid -- even without S_3. Assume $x \notin co(F_1 \cup p_2(F_2))$ and fix a half-space H including $F_1 \cup p_2(F_2)$ but avoiding x. By 5.12(1), the points of $X_2 \setminus H$ have a gate in $X_2 \setminus H$. Hence, $F_2 \subseteq H$ and $x \notin co(F_1 \cup F_2)$. ∎

It is not known whether the related Sand-glass Property is preserved by gated

§5: Base-point Orders

amalgamation.

5.16. Complete interval spaces. A subset A of a poset X is *up-complete in X* provided each up-directed subset of A with an upper bound in X has a supremum in A. The set A is *down-complete in X* provided each down-directed subset of A with a lower bound in X has an infimum in A. Finally, A is *complete in X* provided it is both down- and up-complete in X. When $A = X$, the phrase "in X" will be omitted in either definition. For instance, the real line is complete. The above notions of completeness are *conditional* (the existence of a bound) and they are *relative* to a superset.

These concepts can be transferred to an interval space X as follows. A subset A of X is *(down/up-) complete in X* provided for each $b \in X$ the set A is (down/up-) complete in the poset $(X, \leq_b)$. Note that b is the smallest element in $(X, \leq_b)$: the notion of down-completeness is, in fact, unconditional. As a direct consequence of Proposition 5.4 and the Amalgamation Theorem 5.13, we have the following results.

5.17. Proposition

(1) *Let Y be an interval subspace of the interval space X. If Y is complete in X, then Y is (intrinsically) complete.*

(2) *If Y_i is a complete subset of X_i for each $i \in I$, then $\prod_{i \in I} Y_i$ is a complete subset of $\prod_{i \in I} X_i$.*

(3) *The gated amalgam of two complete geometric interval spaces is complete.* ∎

5.18. Examples. Completeness has an established meaning in (semi-)lattices and in metric spaces. A discussion of the latter is postponed to the end of this section. For semilattices with the usual interval operator, we obtain a full agreement between the established concept and the current one. In vector spaces, the requirement of completeness only affects the involved field. Here are the details.

5.18.1. Proposition. *A semilattice S is complete iff the interval space (S, I_S) is complete.*

Proof. By 5.3.2, S can be divided into a number of regions such that $\leq_b$ equals the given order or its inverse in each area. Elements of different areas have no common upper b-bound, and b is the only common lower bound. Hence an up-directed set resides in a single area and (disregarding the trivial case of a down-directed set of elements meeting pairwise in b) the same goes for down-directed sets. The result follows easily. ∎

5.18.2. Proposition. *A distributive lattice L is complete iff the interval space (L, I_L) is complete.*

Proof. First, note that for each $b \in L$ we have

(1) $\qquad x \leq_b y \iff y \wedge b \leq x \wedge b$ and $x \vee b \leq y \vee b$.

Hence in the areas $\downarrow(b)$ and $\uparrow(b)$, the new order $\leq_b$ coincides with the original order,

106   Chap. I: Abstract Convex Structures

resp., the inverse original order. It easily follows that if (L,I_L) is complete, then L is complete as a lattice.

As for the converse, we use the fact (Prop. 5.3.4, first part) that for each $b \in L$ the set of upper bounds relative to $\leq_b$ of an up-directed collection is down-directed when

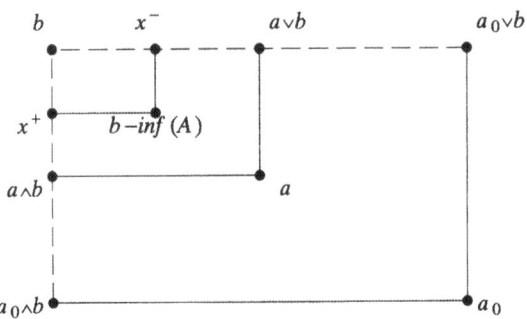

Fig. 9: Completeness in a distributive lattice

non-empty. Hence it suffices to establish down-completeness of $(L, \leq_b)$ for all $b \in L$. Let $A \subseteq L$ be b–down-directed. By (1), this leads to the following conclusions concerning the *given* order of L. The points $a \wedge b$, for $a \in A$, form an up-directed set, bounded from above by b. Let x^+ be the supremum. The points $a \vee b$, for $a \in A$, form a down-directed set, bounded from below by b. Let x^- be the infimum. Note that the mappings $x \mapsto x \wedge b$ and $x \mapsto x \vee b$ are the gate maps of L onto $\downarrow(b)$ and $\uparrow(b)$, respectively. These gate maps are CP by virtue of Proposition 5.12, and hence they preserve base-point orders.

The desired b-infimum of A is the point $p = m(a_0, x^+, x^-)$, where $a_0 \in A$ is a fixed, but arbitrary point. Clearly, p is an upper bound of A in the base-point order of a_0. Let c be another upper bound. Then $c \wedge b \geq a \wedge b$ for all $a \in A$, whence $c \wedge b \geq x^+$. This yields $c \wedge b \in x^+ b \subseteq x^+ x^-$ since $x^+ \leq b \leq x^-$. Similarly, $c \vee b \in x^+ x^-$. As p is the gate of a_0 in the interval $x^+ x^-$ (cf. 5.10.3), we conclude that $p \leq_a c$. Application of Proposition 5.2(2) completes the proof. ∎

5.18.3. Proposition. *Let V be a vector space over a totally ordered field $\mathbb{K}$. Then V is complete as a convex structure iff $\mathbb{K} \approx \mathbb{R}$ as totally ordered fields.*

Proof. A maximal chain in the base-point order $\leq_0$ is additively-, multiplicatively-, and orderly isomorphic to
$$\{x \in \mathbb{K} \mid x \geq 0\}.$$
Hence completeness of V is equivalent to completeness of $\mathbb{K}$. Up to isomorphism, $\mathbb{R}$ is the only complete ordered field.[1] ∎

The concepts of up- and down-completeness are not entirely independent, as is il-

---
1.   Birkhoff and Mac Lane [1958, Thm. IV,6 p. 98].

§5: Base-point Orders   107

lustrated by the next result.

5.19. Proposition. *Let X be a geometric interval space which is up-complete. Then a subset of X is down-complete in X iff it is up-complete in X.*

Proof. Let $C \subseteq X$ and assume first that C is up-complete in X. Let $b \in X$ and let $D \subseteq C$ be down-directed in $\leq_b$. We may assume that D has a largest element a; in particular, $D \subseteq ab$. By Theorem 5.2(2), D is an up-directed set for $\leq_a$ with b as an upper bound. Consequently the a-supremum d_0 of D exists and is in C. Note that $d_0 \leq_a b$, whence d_0 is in ab. By Theorem 5.2(2), d_0 is the b-infimum of D.

Suppose next that C is down-complete in X. Let $U \subseteq C$ be a b-up-directed set with an upper bound a. As X is up-complete, we find that the b-supremum of U exists. As above, we find that the b-supremum of U is the a-infimum of U, and as C is down-complete in X this point must be in C. ∎

Down-completeness need not imply up-completeness. Consider the set

$$[0,1) \cup \{1', 1''\},$$

(where $1'$, $1''$ denote distinct copies of 1) partially ordered with the natural order without a preference between $1'$ and $1''$). The resulting interval space is down-complete but not up-complete: in the base-point order of 0 (which equals the given order), the set $[0,1)$ is up-directed with two upper bounds $1'$, $1''$ and without a supremum.

The above result applies to the segment operator of a point-convex S_3 space. For S_4 spaces the following gives some additional information.

5.20. Theorem. *In an up-complete point-convex S_4 space, each base-point order is a semilattice order. In particular, each segment of type bc is a lattice under $\leq_b$.*

Proof. Let X be an up-complete, point-convex, S_4 space, let $b \in X$ and take $a_1, a_2 \in X$. We first show that the non-empty collection

$$A = \{x \mid x \leq_b a_1; x \leq_b a_2\}$$

is b-up-directed. To this end, take $u, v \in A$. If ua_1 and va_2 are disjoint, then there is half-space H of X with $u, a_1 \in H$ and $v, a_2 \notin H$. Then (for instance) $b \in H$, whence v is in $ba_1 \subseteq H$, a contradiction. Let $w \in ua_1 \cap va_2$. By Theorem 5.2(3), we find that

$$u \leq_b w \leq_b a_1; \quad v \leq_b w \leq_b a_2,$$

showing that w is a common upper bound of u and v. As X is up-complete, we conclude that the b-supremum of A exists. By definition, it is the b-infimum of a_1 and a_2.

The second part follows directly from Proposition 5.2(2). ∎

See 5.3.2 to 5.3.5 for examples of base-point orders which are semilattice orders. If V is a vector space over a totally ordered field and if $b \in V$, then two elements of V are either comparable or meet at b. Hence $\leq_b$ is a semilattice order. CP + CC functions between complete S_4 spaces need *not* induce semilattice homomorphisms on segments. For instance, consider the symmetric H-convexity in $\mathbb{R}^2$, generated by the two coordinate projections π_1, π_2 and their sum. The S_4 property can be verified directly (consult Chapter II for an argument involving Helly numbers). The segment $(0,0)(1,1)$ is just $[0,1]^2$. The restriction of $\pi_1 + \pi_2$ to this segment does not preserve the $(0,0)$–meet.

5.21. Monotonely complete metric spaces. Metric spaces allow for a different, traditional concept of completeness. Recall that a Cauchy sequence in a metric space (X,ρ) is a sequence of points $(x_n)_{n=1}^\infty$ in X with the property that for each $\varepsilon > 0$ there is $n_0 \in \mathbb{N}$ such that $\rho(x_m, x_n) < \varepsilon$ for $m, n > n_0$. A metric space (or, its metric) is *complete* if all Cauchy sequences converge. There is a different concept of completeness in metric spaces which combines this viewpoint with the one of base-point orders. To avoid confusion with the concept of a bounded subset of a poset, we will use the term "ρ–bounded" to mean "bounded with respect to the metric ρ".

Let $b \in X$. For convenience, a decreasing (increasing) sequence in $(X, \leq_b)$ will be called b–*decreasing* (b–*increasing*). It is easy to see that all b–decreasing sequences and all b–increasing, ρ–bounded sequences are Cauchy. The metric space (X, ρ) is *monotonely complete relative to b* (*monotonely b–complete*) provided each b–decreasing sequence and each ρ–bounded, b–increasing sequence in X converges.

It is clear that a complete metric space is monotonely b–complete for each base-point b. The converse is false. Indeed, for inner product spaces, the standard convexity coincides with the geodesic convexity. Such a space is monotonely complete relative to any base-point, but not necessarily complete as a metric space. In general, even unconditional completeness of (X, I_ρ) does not imply monotone completeness of (X, ρ); consider the set $\{0\} \cup \{x \mid 1 < x \leq 2\}$, equipped with the natural metric.

5.22. Proposition. *Let (X, ρ) be a metric space and let $b \in X$.*

(1) *If all b–decreasing sequences of X converge, then any down-directed net D in $(X, \leq_b)$ converges and its limit equals $b - \inf(D)$.*

(2) *If all ρ–bounded, b–increasing sequences of X converge, then any up-directed ρ–bounded net U in $(X, \leq_b)$ converges, and its limit equals $b - \sup(U)$.*

Hence, If (X, ρ) is monotonely b–complete, then (X, I_ρ) is complete.

Proof of (1). Let $r = \inf_{d \in D} \rho(b, d)$. We construct a b–decreasing sequence $(d_n)_{n=1}^\infty$ in D such that

$$\forall n \in \mathbb{N}: \rho(b, d_n) < r + 2^{-n}.$$

Take $d_1 \in D$ such that $\rho(b, d_1) < r + 2^{-1}$. Having obtained $d_n \leq .. \leq d_1$ in D such that $\rho(b, d_k) < r + 2^{-k}$ for $1 \leq k \leq n$, take $d \in D$ such that $\rho(b, d) < r + 2^{-(n+1)}$. As D is b–down-directed, there is a common lower bound $d_{n+1} \in D$ of d and d_n. Note that $\rho(b, d) = \rho(b, d_{n+1}) + \rho(d_{n+1}, d)$, whence $\rho(b, d_{n+1}) < r + 2^{-(n+1)}$. By assumption, the b–decreasing sequence $(d_n)_{n=1}^\infty$ converges to some point p.

We first show that the net D converges to p. Evidently, $\rho(b, p) = r$. Let $\varepsilon > 0$ and let $n \in \mathbb{N}$ be such that $2^{-n} < \varepsilon$ and $\rho(d_n, p) < \varepsilon$. For each $d \in D$ with $d \leq d_n$ we have

§5: Base-point Orders   109

$$r \le \rho(b,d) \le \rho(b,d) + \rho(d,d_n) = \rho(b,d_n) < r + \varepsilon.$$

We find that $\rho(d,d_n) < \varepsilon$, and hence that $\rho(p,d) < 2\varepsilon$.

We next show that $b - \inf(D) = p$. Let q be a lower bound of D in $(X, \le_b)$. Then

$$\rho(b,q) + \rho(q,d) = \rho(b,d)$$

for each $d \in D$. Passing to the limit $d \mapsto p$, the equality becomes

$$\rho(b,q) + \rho(q,p) = \rho(b,p).$$

Hence $q \le_b p$ and p is the largest lower bound of D in $(X, \le_b)$.

Proof of (2). With minor changes only, the previous argument can be repeated. ∎

In the next section we will show that metric completeness is equivalent to monotone completeness in so-called modular metric spaces.

Further Topics

5.23. Posets and semilattices

5.23.1. Let X be a poset and let $b \in X$. Show that if $x \ne y$, then $x \le_b y$ iff either $b \le x \le y$, or $y \le x \le b$, or $x = b$. Deduce that a poset is complete iff it is complete as a convex structure.

5.23.2. Let S be a meet semilattice with a lower bound and let $f: S \to Y$ be a surjective CP and CC function to a convex structure Y. Show that there is a semilattice structure on Y producing the given convexity, such that f is a homomorphism. Formulate and prove a similar result for distributive lattices.

5.24. Joins and meets of order convexities

5.24.1. (for finite sets, Edelman and Jamison [1985]) Show that the convexity of a poset is the join of all order convexities of linear extensions of the given order. Hint: let C be convex and $p \notin C$. Then either $\forall c \in C: p \not\le c$ or $\forall c \in C: c \not\le p$. In the first case, extend the given order to a total order in which all members of C come before p.

5.24.2. Let X be a geometric interval space. Show that the interval convexity of X is the meet (intersection) of poset convexities on X.

5.25. Superextensions.
Let $\lambda(X)$ be the superextension of X relative to the family of all subsets of X. Note that $\lambda(X)$ is a subset of the lattice 2^{2^X}.

5.25.1. (Verbeek [1972]) Show that if $a, b, c \in \lambda(X)$, then

$$(a \cap b) \cup (b \cap c) \cup (c \cap a) \in \lambda(X).$$

5.25.2. Show that if $a, b, c \in \lambda(X)$, then

$a \in bc \Leftrightarrow b \cap c \subseteq a \subseteq b \cup c \Leftrightarrow b \cap c \subseteq a \Leftrightarrow a \subseteq b \cup c.$

5.25.3. Show that the segment operator of $\lambda(X)$ has the Pasch and Peano Properties. Hint: if $p \in co\{a,b,c\}$, then $x = (a \cap b) \cup (b \cap p) \cup (a \cap p)$ is in ab and $p \in xc$.

5.25.4. Show that $\lambda(X)$ is (unconditionally) complete. Hint: if $(a_i)_{i \in I}$ is up-directed in the base-point order of b, then $a' = \cup_{i \in I} a_i \setminus b$ is a linked system and $b-\sup_{i \in I} a_i$ is the mls consisting of a', together with all $B \in b$ such that $a' \cup \{B\}$ is linked.

5.26. Extremality in trees (cf. 1.23). Let T be a tree.

5.26.1. Show that if a point $p \in T$ is maximal in some base-point order then it is maximal in all base-point orders, except for $\leq_p$.

5.26.2. Show that $ext(T)$ equals the set of end points of T, and that the non-end points of T are exactly the inner points (cf. 1.23; 1.24).

5.26.3. If T is absolutely complete, then $T = co(ext(T))$.

5.26.4. The *ramification order of a point* $p \in T$ is defined to be the number of branches emanating at p. Let us denote it by $ram(p)$. Note that p is a ramification point iff $ram(p) > 2$. Verify the following identity for an absolutely complete tree T:

$$\#ext(T) = 2 + \sum_p (ram(p)-2),$$

where p ranges over the ramification points of T.

5.27. Underlying graphs

5.27.1. Give an example of a finite semilattice that is not graphic. Note. A finite modular lattice is graphic; cf. 6.9 below

5.27.2. Let $f: X \to Y$ be an II function of interval spaces X and Y. Verify that f is an edge preserving (EP) function $G(X) \to G(Y)$. If X is a discrete space with the triangle property, then so is Y. We do not know whether Y is graphic if X is.

5.27.3. Show that $G(X \times Y) = G(X) \times G(Y)$.

5.28. More on gated amalgams

5.28.1. Show that the gated amalgam of complete geometric interval spaces is complete.

5.28.2. (compare van de Vel [1983e]) Let X_1, X_2 be disjoint geometric interval spaces, and let $C_i \subseteq X_i$ for $i = 1, 2$ be gated subsets. Let $f: C_1 \to C_2$ be an IP isomorphism. Show that there is a unique geometric interval operator on $X_1 \cup X_2$ extending the

§5: Base-point Orders   111

respective interval operators of X_1, X_2, such that f is the mutual gate map of C_1, C_2.

5.28.3. Let $X = X_1 \cup X_2$ and $Y = Y_1 \cup Y_2$ be gated amalgams, and for $i = 1, 2$ let $f_i: X_i \to Y_i$ be an IP function. We assume that f_1 and f_2 agree on the non-empty subspace $X_1 \cap X_2$. This determines a function $f: X \to Y$. Show that f is IP iff
$$f_1(X_1) \cap f_2(X_2) = f(X_1 \cap X_2).$$
Formulate and prove a version of this result for disjoint amalgamations (cf. part (1) of this topic).

5.29. Gated amalgams and metric spaces (Verheul [1993]). Let (X_i, ρ_i) for $i = 1, 2$ be metric spaces, such that $X_1 \cap X_2 \neq \emptyset$ and ρ_1, ρ_2 agree on $X_1 \cap X_2$. The following defines the *path metric* ρ on $X_1 \cup X_2$.
$$\rho(x_1, x_2) = \inf \{ \rho_1(x_1, c) + \rho_2(c, x_2) \mid c \in X_1 \cap X_2 \}.$$
Let $p_i: X_1 \cup X_2 \to X_i$ be a gate map, $i = 1, 2$.

5.29.1. Verify that ρ is the largest metric extending ρ_1 and ρ_2 and
$$\rho(a, b) = \rho_1(a, p_1(a)) + \rho_{12}(p_1(a), p_2(b)) + \rho_2(p_2(b), b)$$
(ρ_{12} denotes either of ρ_1, ρ_2). Moreover, if ρ_1 and ρ_2 are convex metrics, then so is ρ.

5.29.2. Show that the interval space induced by ρ is the gated amalgam of the interval spaces corresponding to ρ_i, $i = 1, 2$. In fact, ρ is the only metric extending ρ_1, ρ_2, and inducing this interval operator.

5.29.3. Conclude that the gated amalgam of two graphic spaces is graphic, and that the original graphs are isometric subgraphs.

5.30. Strong completeness (see III§5.12; compare Kay [1977b]). A subset C of an interval space X is *strongly complete in X* provided each family of intervals has a non-empty intersection with C whenever each non-empty finite subfamily has. When $C = X$ we drop the addition "in X". Show that a subset which is strongly complete in an interval space X is complete in X (in the usual sense). In a vector space over a totally ordered field, both concepts are equivalent.

5.30.1. (cf. Proposition 5.17(2)) Show that the product of a family of strongly complete interval spaces is a strongly complete interval space.

5.30.2. (cf. Proposition 5.17(3)) Show that the gated amalgam of strongly complete interval spaces is strongly complete.

Notes on Section 5

An early predecessor of the base-point order construction in median algebras can be found in Sholander [1954b] (see 5.3.5). For superextensions of topological spaces, base-point orders were introduced by van Mill [1977], and exploited by van de Vel [1979]. A related idea was developed by Ovchinnikov [1980] for lattices. The construction was extended to general (topological) convex structures by van de Vel [1983d]. This paper contains Theorem 4.7. Underlying graphs of finite convex structures were introduced by Mulder [1980] and independently by Jamison [1981b]. Proposition 5.7 on connectedness of underlying graphs is due to van de Vel [1983e]. The Triangle Property and the related Proposition 5.8 are taken from Bandelt and Chepoi [1991].

The notion of a gate (nearest point) has been considered by many authors. It was first used by van Mill and the author [1978b] in the context of superextensions. Gates were considered in median algebras by Isbell [1980], and in metric spaces by Dress and Scharlau [1987] and by Bandelt and Dress [1992]. The example on linear gate maps, 5.10.2, is taken from Verheul [1993].

Amalgamation (matching) of median algebras was described by Isbell [1980] and (in a different form) by the author in [1983e]. In its present form, the Amalgamation Theorem 5.13 is presented in Bandelt, van de Vel and Verheul [1993]. Amalgamation with disjoint constituents, 5.28.2, was presented by the author in [1983e] in case of median convexity. The result on extending CP maps from the constituents to the amalgam is new. Completeness of convex structures was introduced by van de Vel [1984] in the context of intrinsic topology; cf. Section III§5. Example 5.18.1 and Proposition 5.19 were given in the author's survey paper [1984e].

The notion of monotone completeness in metric spaces has been studied by van de Vel and Verheul in [1991] for modular spaces. This paper contains Proposition 5.22.

6. Modular Spaces

The characteristic property of a modular interval space is that the three intervals joining two out of three points have a non-empty intersection. Its members are called the medians of the triple, and they can be interpreted as maximal lower bounds of two points in the base-point order of the third one. Examples range from median algebras to Banach spaces of type L_1 and function spaces (with supremum norm).

The name "modularity" is derived from a characterization of modular lattices in terms of the properties of a certain interval operator and the Jordan-Hölder Theorem extends to all modular spaces. Median algebras are characterized as geometric interval spaces in which each triple has precisely one median, or as modular interval spaces with the separation property S_2, or as median stable subspaces of distributive lattices.

The completion of a modular metric space is modular. It can be obtained in two steps. First, add all limits of decreasing sequences; then, add all limits of increasing sequences. Only one base-point needs to be considered.

The median stabilization of a set in a median algebra is determined by a simple test involving pairs of intersecting half-spaces. An application is that Steiner trees connecting a given set of points in a median metric space can be found inside the median stabilization of the set.

6.1. Modular spaces. Let X be an interval space and let $a, b, c \in X$. In this section we will consider sets of type

$$M(a,b,c) = ab \cap bc \cap ca,$$

the points of which are called *medians* of a, b, c. A geometric interval space X (or, its interval operator) is *modular* provided the set of medians $M(a,b,c)$ is non-empty for all $a, b, c \in X$. The corresponding multivalued function M is the *multimedian operator of X*. Note that in a geometric interval space,

$$c \in M(a,b,c) \Leftrightarrow c \in ab \Leftrightarrow M(a,b,c) = \{c\}.$$

6.2. Examples. We already encountered the set $M(a,b,c)$ in a context of modular lattices; cf. 4.6.2. This type of example may explain the use of the term "modular":

6.2.1. Proposition. *The following are equivalent for a lattice L.*

(1)   *L is a modular lattice.*
(2)   *The interval space (L, I_M) is modular.*

In either situation, the point $(a \wedge b) \vee (b \wedge c) \vee (c \wedge a)$ is a median of $a, b, c \in L$.

Proof. As a modular interval operator is geometric by definition, the equivalence follows from Proposition 4.6.2. The additional statement was achieved in the course of proving 4.6.2. ∎

6.2.2. Proposition. *A median algebra (X,m) is a modular space with $M(a,b,c) = \{m(a,b,c)\}$ for all a, b, $c \in X$.*

Proof. We observed before (cf. 4.6.3) that a median algebra (X,m) induces a geometric interval operator I_m. By the symmetry of median operators, we have

$$m(a,b,c) \in I_m(a,b) \cap I_m(b,c) \cap I_m(c,a)$$

for all a, b, $c \in X$, showing that (X,I_m) is a modular interval space. Apparently, $m(a,b,c) \in M(a,b,c)$. Suppose $x \in M(a,b,c)$. Then all of the points $m(a,b,x)$, $m(b,c,x)$, $m(c,a,x)$ are equal to x, and hence

$$x = m(x,b,c) = m(m(x,a,b),c,b) = m(x,m(a,b,c),b).$$

This shows that $x \in I_m(b,m(a,b,c))$. Similarly, we obtain that $x \in I_m(a,m(a,b,c))$. As $m(a,b,c) \in I_m(a,b)$, we deduce from the geometric properties of I_m that $x = m(a,b,c)$. ∎

6.2.3. Banach spaces of type L_1. Consider a measure space $(X, \mathcal{A}, \mu)$, consisting of a set X, a σ-algebra $\mathcal{A}$ of measurable subsets of X, and a (σ-additive positive) measure $\mu: \mathcal{A} \to [0,\infty]$. The collection of absolutely integrable functions on X is a pseudo-normed space with pseudo-norm

$$||f|| = \int_X |f| \, d\mu.$$

If f is an integrable function, then $[f]$ denotes the class of measurable functions g which are almost everywhere equal to f, that is, $\mu(\{x \mid f(x) \neq g(x)\}) = 0$. Define

$$L_1(\mu) = L_1(X,\mathcal{A},\mu) = \{[f] \mid f \text{ is absolutely integrable}\}.$$

Such spaces are known as *Banach spaces of type L_1*. If μ is the *counting measure* on an index set I, then the associated Banach space is usually denoted by $\ell_1(I)$. If I is an n-point set, $\ell_1(I)$ is just $\mathbb{R}^n$, equipped with the *Manhattan (sum) norm:*

$$||(x_1, \cdots, x_n)|| = \sum_{i=1}^n |x_i|.$$

If f, g, $h: X \to \mathbb{R}$, then a new function $m(f,g,h): X \to \mathbb{R}$ obtains, which assigns to $x \in X$ the middle one of the three values $f(x)$, $g(x)$, $h(x)$. It is easily seen that if f, g, $h: X \to \mathbb{R}$ are absolutely integrable, then so is $m(f,g,h)$. Moreover, if f and f' are almost everywhere equal, then so are $m(f,g,h)$ and $m(f',g,h)$. So we arrive at a median quotient algebra $L_1(X,\mathcal{A},\mu)$. It is an elementary observation on real numbers a, b, c that

$$|a+b| = |a| + |b| \Leftrightarrow a = \text{middle one of } 0, a, a+b.$$

Hence, $||f+g|| = ||f|| + ||g||$ iff $f(x)$ is the middle one of 0, $f(x)$, $f(x)+g(x)$ almost everywhere. In terms of geodesic $(I_{||.||})$ and median (I_m) intervals, we obtain that $u \in I_{||.||}(0,v)$ iff $u \in I_m(0,v)$. As both the metric and the median of $L_1(\mu)$ are invariant under translation, we conclude that the corresponding interval operators coincide. In particular, the metric interval operator is modular.

6.2.4. Convex metric spaces with a "tri-spherical" intersection property. Let (X,ρ) be a metric space. A metric disk in X is a set of type

§6: Modular Spaces 115

$$D(c,r) = \{x \mid \rho(x,c) \le r\} \quad (c \in X, \, r \ge 0).$$

(this is sometimes called a "sphere"). The space X is said to have the *Tri-spherical Intersection Property* provided each triple of disks $D(c_i, r_i)$ for $i = 1, 2, 3$ in X with $r_i + r_j \ge \rho(c_i, c_j)$ has a non-empty intersection. Note that if $c_1, c_2, c_3 \in X$, then there is a unique triple of non-negative numbers r_1, r_2, r_3 such that if $i \ne j \in \{1,2,3\}$ then $r_i + r_j = \rho(c_i, c_j)$. In fact, if $\{i,j,k\} = \{1,2,3\}$, then

$$r_i = \tfrac{1}{2}(\rho(c_i, c_j) + \rho(c_i, c_k) - \rho(c_j, c_k)).$$

A metric ρ on X is said to be *convex* provided for each pair of points $a, b \in X$ and for each $r \in \mathbb{R}$ with $0 < r < \rho(a,b)$ there is a point $c \in I_\rho(a,b)$ with $\rho(a,c) = r$. It is by now evident that if a metric space satisfies the Tri-spherical Intersection Property, then its metric is convex and the corresponding interval operator is modular. Additional information is given in Topic 6.27.

The main type of example is a *function space with supremum norm*. Let X be a topological space and let $B(X)$ be the collection of all bounded, continuous functions $X \to \mathbb{R}$. The linear space $B(X)$ can be equipped with a (complete) norm defined as follows.

$$\|f\| = \sup \{|f(x)| \mid x \in X\}.$$

Let $f, g, h \in B(X)$ and let $r, s, t \ge 0$ be such that $|f(x) - g(x)| \le r + s$, $|g(x) - h(x)| \le s + t$, and $|h(x) - f(x)| \le t + r$ for all $x \in X$. The function

$$m: X \to \mathbb{R}, \quad x \mapsto m(x) = \max\{f(x) - r, g(x) - s, h(x) - t\}$$

is continuous and bounded, and $m \in D(f,r) \cap D(g,s) \cap D(h,t)$. Moreover, a metric derived from a norm is convex.

If X is a topologically discrete n-point space, then $B(X)$ corresponds with $\mathbb{R}^n$ equipped with the *Cartesian (maximum) norm:*

$$\|(x_1, \ldots, x_n)\| = \max_{i=1}^n |x_i|.$$

6.2.5. Products of metric spaces. Let (X_i, d_i) for $i = 1, 2$ be metric spaces. There are two (topologically equivalent) metrics on $X_1 \times X_2$ of current interest.

$$d_m(u,v) = \max\{d_1(u_1, v_1), d_2(u_2, v_2)\};$$
$$d_s(u,v) = d_1(u_1, v_1) + d_2(u_2, v_2)$$

(where $u = (u_1, u_2)$ and $v = (v_1, v_2)$). The first is the *Cartesian metric,* the second is the *Manhattan metric.* If both factors are modular metric spaces, then so is $X_1 \times X_2$ relative to each of these metrics. If the factor metrics are derived from a normed vector space, then so are the Cartesian and Manhattan metrics on the product. Repeated combination of the above constructions yields an abundance of modular norms on $\mathbb{R}^n$.

We next concentrate on discrete spaces. The definition of modularity in posets (cf. 2.17) can be extended as follows. A discrete interval space X is *semimodular* provided for each quadruple of points $a, b, c, d \in X$ such that $c, d \in ab$ are neighbors of a, there is a common neighbor of c, d in $bc \cap bd$. The following is a direct consequence of semimodularity. If a, b, c are such that $c \in ab$, and if $a' \in ab$ is a neighbor of a, then there is a neighbor $c' \in bc$ of c with $c' \in ba'$.

If X is a discrete modular space and if a, b, $c \in X$ are such that

$$ab \cap bc = \{b\}; \quad bc \cap ca = \{c\}; \quad ca \cap ab = \{a\},$$

then by modularity, $a = b = c$. Hence X has the triangle property, whence by Proposition 5.8, X is a graphic space. The relation between modularity and semimodularity is discussed for graphic spaces in the next result.

6.2.6. Proposition. *The following are equivalent for a connected graph G.*

(1) G is modular.
(2) G has no induced triangles, and each triple of pairwise intersecting metric disks of integer radius has a common point.
(3) G is bipartite and semimodular.

Proof. Throughout, ρ denotes the geodesic metric of G.

(1) $\Rightarrow$ (2). It is evident that the vertices of an induced K_3 cannot have a median. Let $a_i \in G$, and $r_i \geq 0$ be such that $D(a_i, r_i) \cap D(a_j, r_j) \neq \emptyset$ for $i \neq j$ among 1, 2, 3. Let $x \in M(a_1, a_2, a_3)$, and note that the inequality $r_i < \rho(a_i, x)$ cannot hold for two distinct i. So assume $r_i \geq \rho(a_i, x)$ for $i = 1, 2$. As $x \in a_3 a_i$ for $i \neq 3$, we have

$$r_3 + r_i - \rho(x, a_i) \geq \rho(a_3, x).$$

Let δ be the minimum of $\{r_1 - \rho(x, a_1)$ and $r_2 - \rho(x, a_2)\}$. Then $\delta \geq 0$ and

$$D(x, \delta) \subseteq D(a_1, r_1) \cap D(a_2, r_2).$$

As $r_3 + \delta \geq \rho(a_3, x)$, we also have $D(x, \delta) \cap D(a_3, r_3) \neq \emptyset$.

(2) $\Rightarrow$ (3). If G is not bipartite, then there is an odd circuit of minimal length. Take an edge ab in such a circuit, and take the point c of the circuit at equal distance n from a, b. This yields a triangle consisting of a, b, and any point of the set

$$D(a, 1) \cap D(b, 1) \cap D(c, n-1).$$

As to semimodularity, let a, b, c, d be such that $c \neq d$ are neighbors of a in ab, and let $n = \rho(a, b)$. Note that $n \geq 2$. The desired point is taken from the set

$$D(c, 1) \cap D(d, 1) \cap D(b, n-2).$$

(3) $\Rightarrow$ (1). We show that $M(a, b, c) \neq \emptyset$ by induction on the distance between a and b. If $\rho(a, b) = 1$, the result follows from bipartiteness. Suppose the result is valid for distances $< \rho(a, b)$. Let $b' \in ab$ be a neighbor of b and take $d \in M(a, b', c)$. If $d \in cb$ then $d \in M(a, b, c)$. So assume $d \notin cb$. As $d \in cb'$, we see that $b' \notin cb$ and as G is bipartite we conclude that $b \in cb'$. By virtue of semimodularity, there is a neighbor $d' \in dc$ of d with $d' \in bc$. We also have $d' \in dc \subseteq ac$. Finally, observe that, since $d \notin cb$,

$$\rho(c, d') + \rho(d', b) = \rho(c, b) < \rho(c, d) + \rho(d, b) = \rho(c, d') + \rho(d', d) + \rho(d, b),$$

whence $\rho(', b) \leq \rho(d, b)$. It follows that $d' \in db \subseteq ab$. ∎

Two critical examples of modular graphs are presented in Fig. 1.

§6: Modular Spaces

$K_{2,3}$ $K_{3,3}$ minus an edge

Fig. 1: Special modular graphs

6.3. Proposition. *Let X be a modular space, and let a, b, $c \in X$. Then $M(a,b,c)$ is precisely the set of maximal lower bounds of a, b in the poset $(X, \leq_c)$. Moreover, if $p \in ab$ then there is a point $q \in M(a,b,c)$ with $q \leq_c p$.*

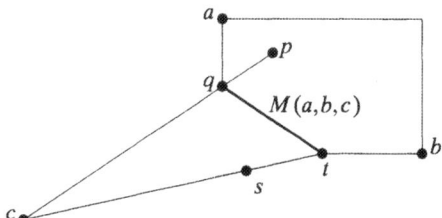

Fig. 2: Position of $M(a,b,c)$

Proof. Let $s \leq_c a,b$. By the geometric properties of the interval operator, we have $M(a,b,s) \subseteq M(a,b,c)$. Any element $t \in M(a,b,s)$ satisfies $s \leq_c t \leq_c a,b$. Conversely, if $t \in M(a,b,c)$ and $t \leq_c p \leq_c a,b$, then $p \in ta \cap tb = \{t\}$.

For a proof of the second part, let $s_a \in M(a,p,c)$ and $s_b \in M(b,p,c)$. As $s_b \in pb \subseteq s_a b$, we have $s_a s_b \subseteq s_a b \subseteq ab$, and hence $M(s_a,s_b,c) \subseteq M(a,b,c)$. Any point $q \in M(s_a,s_b,c)$ does the job. ∎

By Proposition 6.2.2 and the previous result, each pair of points in a median algebra has a **greatest** lower bound (infimum) in each base-point order; cf. Proposition 5.3.5.

Two pairs of points (a,b), (c,d) of an interval space X are *transposed* provided either a, $c \in bd$ and b, $d \in ac$, or a, $d \in bc$ and b, $c \in ad$. This relation is reflexive and symmetric; the relation obtained by transitive closure is known as *projectivity*. For instance, if a L is a lattice and if b, $c \in L$, then the pairs $(b, b \wedge c)$ and $(c \vee b, c)$ are transposed in (L, I_M).

In the next result, the expression "chain joining a, b" refers to a totally ordered subset C of ab in the base-point order of a, such that a, $b \in C$. The length of C is understood to be the number of "steps" in C, in other words: the number of elements minus one.

6.4. Theorem (*Jordan-Hölder Theorem*). *Let X be a modular space and let a, $b \in X$ be joined by a finite maximal chain. Then there is a bijection between the cover pairs of any two maximal chains connecting a and b, such that corresponding pairs are projective.*

In particular, all maximal chains joining a, b are of the same length.

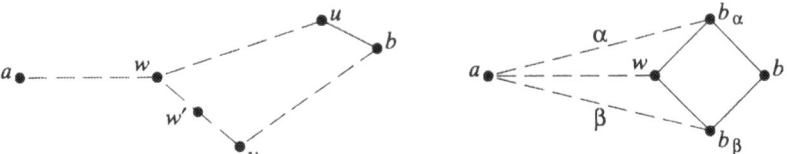

Fig. 3a: Proof on equal lengths   Fig. 3b: Proof on projective correspondence

Proof. We first verify by induction on $n \geq 1$ that if there is a maximal chain of length n between two points a, b of X, then all chains between a and b are finite and of length $\leq n$. For $n = 1$, the conclusion is evident. Assume the result to be valid for chains of length $n \geq 1$. Let α be a maximal chain of length $n + 1$ between a and b, and consider a finite chain β between a and b. Let u, resp., v, be the last element $\neq b$ in α, resp., β. Note that u, b are neighbors and that the part of α from a to u is a maximal chain of length n. If $v \in au$ then by inductive assumption, β cannot be longer than $n + 1$.

We assume $v \notin au$, or, equivalently, $u \notin vb$ (Fig. 3a). Let $w \in M(u,v,a)$. We verify that vw is an edge. Suppose $w' \in wv$ and $w' \neq w$. As bu is an edge, we have either $u \in w'b$ or $b \in w'u$. The former implies that w' is a lower bound of u and v in $\leq_c$ larger than w, contradicting Proposition 6.3. So, $b \in w'u$, whence $v \in w'b \subseteq w'u$. On the other hand, $w' \in vw \subseteq vu$, and it follows that $w' = v$. By inductive assumption, there is a maximal chain from a to u via w of length n. Composing the part from a to w with the edge to v we obtain a maximal chain from a to v via w of length at most n. By induction, the part of β up to v is of length $\leq n$, and β is of length $\leq n + 1$.

The statement on projective pairs is also obtained by induction on the length n of a maximal chain connecting two points (Fig. 3b). For $n = 1$ the result is evident. Suppose the statement is valid for $n \geq 1$, and let α, β be maximal chains of length $n+1$ between a and b. Consider the neighbors b_α and b_β of b in α and β, respectively. We may assume that these points are distinct. The previous argument shows that if $w \in M(a,b_\alpha,b_\beta)$ then the pairs

$$(b,b_\alpha); \quad (b_\beta,w); \quad (b,b_\beta); \quad (b_\alpha,w)$$

are cover pairs. In particular, the pairs (b,b_α) and (b_β,w) are transposed, as are the pairs (b,b_β) and (b_α,w). There is a maximal chain γ between a, w of length $n-1$. By adding b and b_α to γ, respectively by adding b and b_β to γ, we obtain two more maximal chains γ_α and γ_β respectively. Applying the inductive assumption twice, and considering the projective pairs constructed above, we obtain 1-1 correspondences of mutually projective pairs between α and γ_α, between γ_α and γ_β, and between γ_β and β. ∎

6.5. Theorem.

(1) *If X is a modular interval space and if $f: X \to Y$ is a surjective II function to an interval space Y, then Y is a modular space.*

(2) *Let I_1, I_2 be two geometric interval operators on a set X with corresponding*

multimedian operators M_1, M_2. If $M_1(a,b,c) \cap M_2(a,b,c) \neq \emptyset$ for all a, b, $c \in X$, then $I_1 = I_2$ and $M_1 = M_2$. In particular, if $I_1(a,b) \subseteq I_2(a,b)$ for all $a,b \in X$ then $I_1 = I_2$.

Proof. As to (1), Y is a geometric interval space by Theorem 4.8(3). Modularity of Y follows easily. We use the fact that if a multimedian operator M is derived from a (geometric) interval function I, then $x \in I(a,b)$ iff $M(a,b,x) = \{x\}$ iff $x \in M(a,b,x)$. Let $a,b \in X$, and let $x \in I_1(a,b)$. Then $M_1(a,b,x) = \{x\}$. By assumption, the sets $M_1(a,b,x)$, $M_2(a,b,x)$ meet, whence $x \in M_2(a,b,x)$. This gives $x \in I_2(a,b)$. We have shown that $I_1(a,b)$ is contained in $I_2(a,b)$. After reversing the role of the indices, we find that $I_1(a,b) = I_2(a,b)$. ∎

This result has several interesting consequences. First, we can handle the following type of example. A *valuation* on a lattice L is a function

$$v: L \to \mathbb{R}$$

with the properties listed below.

(VAL-1) $\forall x, y \in L : x < y$ implies $v(x) < v(y)$.
(VAL-2) $\forall x, y \in L : v(x \vee y) + v(x \wedge y) = v(x) + v(y)$.

The following formula determines a metric on a valuated lattice (L, v).

$$\rho(x,y) = v(x \vee y) - v(x \wedge y) \quad (x, y \in L).$$

For instance, let $L = \mathbb{N}$ be ordered by the prescription $a \leq b$ iff a divides b. Note that $a \wedge b$ and $a \vee b$ are the greatest common divisor, resp., the smallest common multiple of a, b. A valuation on $\mathbb{N}$ is defined by the following prescription.

$$v(n) = \log(n) \quad (n \in \mathbb{N}).$$

The lattice of convex sets in a JHC matroid is valuated by the rank function; cf. 2.18.

6.6. Corollary. *A valuated lattice is modular, and its modular and geodesic interval operators coincide.*

Proof. Let (L, v) be a valuated lattice which is not modular. Then the non-modular lattice N_5 embeds in L (cf. Fig. 2). This lattice has a largest element q, a smallest element p, and three more elements a, b, c, where $a < c$ and $b \wedge a = b \wedge c = p$; $b \vee a = b \vee c = q$. By (VAL-2), we see that $v(p) + v(q)$ equals both $v(b) + v(a)$ and $v(b) + v(c)$, whence $v(a) = v(c)$, contradicting (VAL-1).

To obtain the announced equality of interval operators, we require three metric formulas. First, we have $v(x) - v(x \wedge y) = v(x \vee y) - v(y)$ by definition. As $v(x \wedge y) \leq v(x)$ and $v(y) \leq v(x \vee y)$, the formula reads as

(1) $\quad \rho(x \wedge y, x) = \rho(y, x \vee y)$.

Secondly,

(2) $\quad x \leq z \leq y \quad$ implies $\quad \rho(x,z) + \rho(z,y) = \rho(x,y)$.

Indeed,

$$\rho(x,z) + \rho(z,y) = v(z) - v(x) + v(y) - v(z) = v(y) - v(x) = \rho(x,y).$$

Finally, as $\rho(x,y) = \rho(x \wedge y, x \vee y)$ and $x \wedge y \le x \le x \vee y$, we infer from (1) and (2) that

(3) $\quad \rho(x, x \wedge y) + \rho(x \wedge y, y) = \rho(x,y) = \rho(x, x \vee y) + \rho(x \vee y, y).$

Let I_M be the usual modular interval operator, and let I_ρ be the geodesic interval operator. Let $a,b \in L$ and $x \in I(a,b)$. Then $(a \wedge x) \vee (b \wedge x) = x$ and $a \wedge b \le x$. By the formulas (1), (2), (3) we obtain

$$\begin{aligned}\rho(a,b) &= \rho(a, a \wedge b) + \rho(a \wedge b, b) \\ &= \rho(a, a \wedge x) + \rho(a \wedge x, a \wedge b) + \rho(a \wedge b, b \wedge x) + \rho(b \wedge x, b) \\ &= \rho(a, a \wedge x) + \rho(a \wedge x, x) + \rho(x, b \wedge x) + \rho(b \wedge x, b) \\ &= \rho(a,x) + \rho(x,b).\end{aligned}$$

Therefore, $x \in I_\rho(a,b)$. Application of Theorem 6.5(2) gives the desired result. ∎

6.7. Corollary. *The segment operator of a modular space is geometric if and only if all intervals are convex.*

Proof. Segments being convex, axiom (G-1) is clearly fulfilled. Evidently,

$$M(a,b,c) \subseteq co(a,b) \cap co(b,c) \cap co(c,a).$$

We conclude from Theorem 6.5(2) that the segment operator of a modular space satisfies axiom (G-2) if and only if the interval operator equals the segment operator. ∎

For example, let L be a modular lattice. Then the operator I_M is modular, and the corresponding segment operator is given by

$$I_L(a,b) = [a \wedge b, a \vee b]$$

(cf. 4.2). In general, this set is properly larger than $I_M(a,b)$, in which case I_L cannot be geometric. We shall see later what it means for I_L to be geometric. As a direct consequence of 6.2.2 and 6.5(2), we arrive at the following result.

6.8. Corollary. *Let X be a modular space with a multimedian operator M, and let m be a median operator on X such that $m(a,b,c) \in M(a,b,c)$ for all $a,b,c \in X$. Then m and M coincide.* ∎

If X and Y are modular spaces with respective multimedian operators M_X, M_Y, then $f : X \to Y$ is *(multi) median preserving*, briefly, *MP*, provided

$$f(M_X(a,b,c)) \cap M_Y(f(a), f(b), f(c)) \ne \emptyset$$

for all $a, b, c \in X$. For median algebras, the condition reads as

$$f(m(a,b,c)) = m(f(a), f(b), f(c)),$$

i.e., f is a homomorphism of median algebras. Evidently, an interval preserving function of modular spaces is multimedian preserving and vice versa. The following result is a strengthening.

§6: Modular Spaces

6.9. Theorem. *Let X and Y be modular spaces with respective multimedian operators M_X, M_Y, and let $f: X \to Y$ be a surjective median preserving map. Then*
$$f(x_1 x_2) = f(x_1)f(x_2); \quad f(M_X(x_1,x_2,x_3)) = M_Y(f(x_1),f(x_2),f(x_3))$$
for all $x_1, x_2, x_3 \in X$. In particular, f maps convex subsets of X onto convex subsets of Y.

Proof. Let $a, b \in Y$ and $a' \in f^{-1}(a), b' \in f^{-1}(b)$. Define
$$I(a,b) = f(a'b').$$
We first verify that the definition of $I(a,b)$ does not depend on the choice of a', b' and that $I(a,b) \subseteq ab$. To this end, suppose $a'' \in f^{-1}(a)$, $b'' \in f^{-1}(b)$. Let $z \in f(a'b')$, say, $z = f(z')$ with $z' \in a'b'$. The following equalities are valid.

(1) $\quad \{z\} = \{f(z')\} = f(M_X(a',b',z'))$,

(2) $\quad M_Y(a,b,z) = M_Y(f(a'),f(b'),f(z'))$.

The sets on the right hand side of (1) and (2) meet. Hence the set $M_Y(f(a'),f(b'),f(z'))$ consists of z only. In particular, $z \in M_Y(a,b,z) \subseteq ab$, showing that

(3) $\quad I(a,b) \subseteq ab \quad (a,b \in Y)$.

Since f is MP, the sets $f(M_X(a'',b'',z'))$ and $M_Y(f(a''),f(b''),f(z'))$ have a non-empty intersection, whence, $z \in f(M_X(a'',b'',z'))$. Therefore, there exists a point $z'' \in M_X(a'',b'',z') \subseteq a''b''$ with $f(z'') = z$. This shows that $f(a'b') \subseteq f(a''b'')$; the other inclusion is obtained similarly.

It is clear that I is an interval operator. It is modular by Theorem 6.5(1). By (3), the modular interval operator I is comparable with the original one of Y. By Theorem 6.5(2), both operators are equal. In particular, $M = M_Y$. ∎

A noteworthy consequence of the last result is, that an IP bijection between modular spaces is an isomorphism. A subset A of a modular space X is **median stable** provided $M(u,v,w) \cap A \neq \emptyset$ for each triple of points $u, v, w \in A$. In case of a median algebra (X,m), this amounts to the condition that $m(A^3) \subseteq A$, that is: A is a subalgebra of X. Note that the image of a multimedian preserving function is median stable, and that a sublattice of a modular lattice is median stable. The following is a major result on modular spaces which provides further justification of our terminology.

6.10. Theorem. *The following are equivalent for a modular space X.*

(1) *The multimedian operator M of X is single-valued.*
(2) *X does not contain the graphic space $K_{2,3}$ as a median stable subspace.*
(3) *The interval operator of X has the Peano Property.*
(4) *The interval operator of X has the Pasch Property.*
(5) *All intervals of X are convex.*
(6) *Each interval ab of X is a distributive lattice in the base-point order $\leq_b$.*
(7) *There is a median operator m on X such that $m(u,v,w) \in M(u,v,w)$ for all $u, v, w \in X$.*

In either situation, the interval operator is induced by the canonical median operator and all non-empty polytopes are gated.

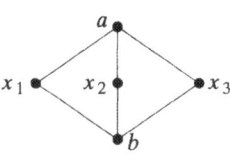

Fig. 4: $K_{2,3}$

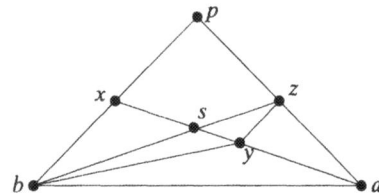

Fig. 5: Pasch implies Peano

Proof. (6) implies (2). Let $\{a,b,x_1,x_2,x_3\}$ be a median stable subset of X isomorphic with $K_{2,3}$. We let a, b correspond with the elements of $K_{2,3}$ of degree three (consult Fig. 4). Then $x_i \in ab$ for $i = 1, 2, 3$. Consequently, $M(x_i,x_j,a) = a$ and $M(x_i,x_j,b) = b$ for distinct i, j. This yields a five-point sublattice of $(ab, \leq_b)$ which is not distributive.

(2) implies (4). Let $a' \in ap$, $b' \in bp$, and assume that $a'b \cap b'a = \emptyset$. We construct a median stable subset isomorphic with $K_{2,3}$ in four steps as follows. First, we replace a' by a point of $M(a',b,p)$, and we replace b' by a point of $M(b',a,p)$. We still have a Pasch configuration without an intersection point, and with the additional conditions

$$a' \in pb; \quad b' \in pa.$$

Secondly, we replace a' by a point of $M(a',a,b)$ and we replace b' by a point of $M(b',a,b)$. The resulting Pasch configuration has the additional property that

$$a', b' \in ab.$$

Thirdly, we replace p by a member of $M(a',b',p)$. Our Pasch configuration now satisfies the additional property

$$p \in a'b'.$$

Fourthly, we replace a by a member of $M(a',b',a)$ and b by a member of $M(a',b',b)$. The Pasch configuration still hasn't an intersection point and it has the additional property

$$a, b \in a'b'.$$

We arrive at a configuration which is a median stable copy of $K_{2,3}$ in X (a', b' are of degree 3).

(4) implies (3) (cf. Fig. 5). Let a, b, $p \in X$ and $x \in pb$, $y \in ax$. Observe that if $x' \in M(y,b,x)$, then $x' \in pb$ and $y \in ax'$ by the geometric properties of X. We therefore assume without loss of generality that $x \in yb$. Take $z \in M(a,y,p)$. We have to verify that $y \in zb$. To this end, note that $z \in py$, whence by the Pasch Property we obtain a point $s \in zb \cap yx$. Now $s \in yx$ and $x \in yb$ imply that $x \in sb \subseteq zb$. On the other hand, $z \in ay$ and $y \in ax$ yields $y \in zx$. We conclude that $y \in zb$ as desired.

Next, (3) implies (5) by Proposition 4.10, whereas (5) implies (1) by the following argument. Let $u \neq v$ be in $M(a,b,c)$. Then $uc \subseteq ac$, $vc \subseteq bc$, and $uv \subseteq ab$ since ab is assumed convex. Hence $M(u,v,c) \subseteq M(a,b,c)$. However, a point of $M(u,v,c)$ is a

§6: Modular Spaces

common lower bound of u, v in $\leq_c$. This contradicts Proposition 6.3.

(1) implies (6). Suppose that the multimedian operator of X is single-valued. By virtue of Proposition 6.3, each interval is a lattice. If a *sublattice* of type $K_{2,3}$ occurs in an interval, then elementary base-point considerations yield that the multimedian of the three vertices of degree two contains both the top and bottom element of the sublattice. Apparently, such lattices do not occur and each interval is a distributive lattice.

(1) implies (7). Define a map $m: X^3 \to X$, as follows: $m(u,v,w)$ is the unique point of $uv \cap vw \cap wu$. Clearly, m satisfies the Idempotent and Symmetric Laws of a median operator. In regard to Proposition 6.3, the Transitive Law

$$(abc)dc = a(bcd)c$$

corresponds with the fact that, in the base-point order of c, the operation of taking the infimum of a, b, d is associative. The fact that (7) implies (1) follows from Corollary 6.8.

Observe that $x \in I(a,b)$ iff $M(a,b,x) = \{x\}$, and hence that the interval operator equals the median interval operator. By (3) and (4), X is JHC and S_4. Hence the Sand-glass Property is valid. By 6.8, each finite set of points has a meet in $(X, \leq_b)$. By virtue of the Sand-glass Property, the infimum of a non-empty finite set F equals the infimum of $co(F)$. We verify by induction on $\#F$ that this infimum is in $co(F)$. For $\#F \leq 2$, this follows from the fact that $m(a,b,c) \in I_m(a,c)$. If $\#F > 2$ and $p \in F$, then the b–infimum q of $F \setminus \{p\}$ is in $co(F \setminus \{p\})$ and

$$\inf{}_b(F) = q \wedge_b p \in I(p,q) \subseteq co(F).$$ ∎

As an illustration of the last result, let us discuss trees. Being a semilattice, a tree T is S_4 and of arity two. Therefore, the segment operator is geometric and has the Pasch Property. If $a, b, c \in T$, then $a \wedge b$, $b \wedge c$, $c \wedge a$ are pairwise bounded from above. Hence they are comparable. By the hull formula of semilattices, 1.5.3, the maximum is in $ab \cap bc \cap ca$, showing that T is modular. By (4) or by (5) of the last theorem, T is a median space and the above argument gives precise information about its median operator.

For additional characterizations of median spaces, consult Topic 6.29. The process of identifying median algebras with a class of interval spaces is summarized as follows.

6.11. Proposition. *The following transitions yield a 1-1 correspondence between median algebras and modular spaces with unique medians.*

(i) $\quad (X,m) \mapsto (X,I_m)$;

(ii) $\quad (X,I) \mapsto \{m(a,b,c)\} = I(a,b) \cap I(b,c) \cap I(c,a) \quad (a, b, c \in X)$.

Moreover, the convexity preserving functions between median convex structures are precisely the median preserving functions and surjective CP functions are CC.

Proof. As to the first part, we verified in 6.2.2 that

$$\{m(a,b,c)\} = I_m(a,b) \cap I_m(b,c) \cap I_m(c,a),$$

showing that application of (ii) after (i) returns the original median algebra. Conversely, use (ii) to define a median m, and use (i) afterwards to obtain the interval operator I_m. Then $m(a,b,c) \in I(a,b)$, whence by definition, $I_m(a,b) \subseteq I(a,b)$. By the incomparability of modular interval operators, Theorem 6.5(2), both operators are identical.

As to the second part, note that median convexity is of arity two, whence the adjectives "CP" and "IP" amount to the same. We observed before that IP functions between modular spaces are median-preserving and vice versa. The last part follows from Theorem 6.9. ∎

In the sequel we will use the term *median space* for a modular space with a single-valued multimedian operator. Since all intervals of a median space are convex, the distinction between the interval space and the induced convex structure is purely formal.

6.12. Corollary. *The following are equivalent for a modular lattice L.*

(1) *The interval operators I_M and I_L coincide.*
(2) *The lattice interval operator I_L is geometric.*
(3) *L with the lattice convexity is a median space.*
(4) *No sublattice of L is isomorphic with $K_{2,3}$.*
(5) *L is a distributive lattice.*

In either situation, the unique median of a, b, c $\in$ L is given by $(a \wedge b) \vee (b \wedge c) \vee (c \wedge a)$.

Proof. This goes very fast:

(1) $\Rightarrow$ (2): I_M is geometric.
(2) $\Rightarrow$ (3): I_L is a geometric -- hence modular -- interval operator with convex intervals. Then use 6.10(5).
(3) $\Rightarrow$ (4): there isn't even a median stable $K_{2,3}$.
(4) $\Rightarrow$ (5): standard.
(5) $\Rightarrow$ (1): a direct application of the Distributive Law.

The last part follows from Proposition 6.2.1. ∎

This leads to the following result, adding to Corollary 3.18.

6.13. Corollary. *The median spaces correspond with median stable subspaces of distributive lattices.*

Proof. A median stable subset of a distributive lattice is a median algebra by itself; the relative convexity agrees with the intrinsic median convexity (cf. Topic 6.33). On the other hand, the S_3 property of a median space X yields a collection $\mathcal{F}$ of homomorphisms $X \to \{0,1\}$ separating points from polytopes. According to Lemma 3.16, the function

$$X \to \{0,1\}^{\mathcal{F}}, \quad x \mapsto (f(x))_{f \in \mathcal{F}},$$

embeds X into a Cantor cube. As $\mathcal{F}$ consists of homomorphisms, the embedded copy is a subalgebra. ∎

§6: Modular Spaces   125

6.14. Theorem. *The gated amalgam X of two modular spaces X_1, X_2 is modular. If both summands are median spaces, then so is X.*

Proof. The first part is a consequence of the interval formula in 5.13. If both summands are median, Theorem 5.14 implies that the amalgam is a PP-space and hence median. ∎

The last result is an important instrument in constructing tailor-made examples of median spaces which are cubical polyhedra. We refer to Section IV§5. For additional information on median cubical polyhedra, see II§3.

> The remainder of this section is concerned exclusively with modular or median *metric* spaces.

In preparation to the next results, we develop some "metric calculus" in modular metric spaces. Additional results are described in Topics 6.31 and 6.36.

6.15. Lemma. *Let (X,ρ) be a modular metric space with a multimedian operator M, and let $a, a', b, c \in X$. Then for each point $d \in M(a,b,c)$ there is a point $d' \in M(a',b,c)$ such that $\rho(d,d') \le 2\rho(a,a')$.*

Proof. Let $d \in M(a,b,c)$. Then d is a maximal lower bound of a, b in $(X, \le_c)$. Consider a maximal lower bound e of d, a' with respect to this ordering. Then e is a common lower bound of a', b. By Proposition 6.3, there exists a maximal common lower bound $d' \ge_c e$ of these elements. Observe that $d' \in ea' \subseteq ca'$ and $d' \in eb \subseteq cb$. As $d' \in ba'$ by assumption, we see that $d' \in M(a',b,c)$. Hence,

$$\rho(a,d) = \tfrac{1}{2}(\rho(a,b) + \rho(a,c) - \rho(b,c));$$
$$\rho(a',d') = \tfrac{1}{2}(\rho(a',b) + \rho(a',c) - \rho(b,c)).$$

We have $e \in da'$ and $d' \in ea'$. Hence $d' \in da'$ and we obtain the following estimate.

$$\begin{aligned}\rho(d,d') &= \rho(a',d) - \rho(a',d') \\ &\le \rho(a',a) + \rho(a,d) - \rho(a',d') \\ &= \rho(a',a) + \tfrac{1}{2}(\rho(a,b) - \rho(a',b)) + \tfrac{1}{2}(\rho(a,c) - \rho(a',c)) \\ &\le 2\rho(a,a').\end{aligned}$$
∎

If a_1, a_2, a_3 are points of a metric space with metric ρ, and if $p \in M(a_1,a_2,a_3)$, then the real numbers $r_i = \rho(p,a_i)$ ($i = 1, 2, 3$) are the solution of three linear equations of type

$$r_i + r_j = \rho(a_i,a_j) \quad (i \ne j).$$

Consequently, each two points in $M(a_1,a_2,a_3)$ have the same distance to a_1. In modular graphs, owing to bipartiteness (cf. 6.2.6), it follows that each two medians of a triple are at even distance. We now derive the following remarkable property of medians in modular graphs.

6.16. Proposition. *Let G be a modular graph and let a, b, c ∈ G. Then each pair of points in $M(a,b,c)$ can be joined by a geodesic of which the vertices alternate between $M(a,b,c)$ and its complement.*

Proof. Let $m_1 \neq m_2$ be in $M(a,b,c)$ and let $a' \in M(a,m_1,m_2)$. If $\rho(m_1,m_2) = 2$ there is nothing left to be proved. Assume $\rho(m_1,m_2) > 2$. As

$$\rho(a',m_1) = \rho(a',m_2) = \tfrac{1}{2}\cdot\rho(m_1,m_2),$$

we can take a point $x \in m_1 a'$ distinct from m_1 and a'. Observe that $a' <_a x <_a m_1$, whence by elementary considerations,

$$M(a',b,c) \supseteq M(x,b,c) \supseteq M(m_1,b,c) = \{m_1\}.$$

Let $m_2 \in M(a',b,c)$. Applying Lemma 6.15 twice, we first obtain a point $y \in M(x,b,c)$ such that $\rho(m_2,y) \leq 2\cdot\rho(a',x)$, and we next find that the only point m_1 of $M(m_1,b,c)$ satisfies $\rho(y,m_1) \leq 2\cdot\rho(x,m_1)$. It follows that $y \in M(a,b,c)$ and $y \in m_1 m_2 \setminus \{m_1,m_2\}$. The result follows easily by induction. ∎

If A_1 and A_2 are bounded subsets of a metric space with metric ρ, then the *Hausdorff distance* of A_1 and A_2 is given by the formula

$$\rho_H(A_1,A_2) = \inf \{r \mid A_2 \subseteq B(A_1,r) \text{ and } A_1 \subseteq B(A_2,r)\},$$

where $B(A,r) = \{x \mid \exists\, a \in A : \rho(a,x) < r\}$. The Hausdorff distance yields a genuine metric on the set $\mathcal{T}_{bound}(Y)$ of all non-empty, bounded and closed subsets of a metric space Y. It is used in the next result.

6.17. Theorem. *The completion of a modular (resp., median) metric space is modular (resp., median).*

Proof. Assume first that X is a modular metric space, and let X^* be its metric completion. It follows from Lemma 6.15 that the operator

$$M : X^3 \to \mathcal{T}_{bound}(X)$$

(where the product is given the Manhattan metric) is uniformly continuous. As $\mathcal{T}_{bound}(X^*)$ is a complete metric space, $\overline{M}$ extends to a (uniformly) continuous operator[1]

$$M^* : X^* \times X^* \times X^* \to \mathcal{T}_{bound}(X^*).$$

Let $a, b, c \in X^*$ and fix three converging sequences $a_n \mapsto a$, $b_n \mapsto b$, and $c_n \mapsto c$ with $a_n, b_n, c_n \in X$ for all $n \in \mathbb{N}$. Consider a point $x \in M^*(a,b,c)$. There is a sequence of points $x_n \in M(a_n,b_n,c_n)$ for $n \in \mathbb{N}$, such that x_n converges to x. In each of the equations

$$\rho(a_n,b_n) = \rho(a_n,x_n) + \rho(x,b_n);$$
$$\rho(b_n,c_n) = \rho(b_n,x_n) + \rho(x,c_n);$$
$$\rho(c_n,a_n) = \rho(c_n,x_n) + \rho(x,a_n),$$

we take the limit as $n \mapsto \infty$. The resulting equalities express that x is a median of a, b, c.

---

1.   Engelking [1977, Theorem 8.3.10].

§6: Modular Spaces

Suppose next that X is a median metric space. Arguing as above, the median m of X extends to a uniformly continuous function

$$m^*: X^* \times X^* \times X^* \to X^*.$$

The axioms of a median operator are satisfied on a dense subset X of X^*, and hence they are satisfied throughout. By the first part, X^* has a multimedian operator[2], say M. Now M is (uniformly) continuous by Lemma 6.15, and as $m^*(u,v,w) \in M(u,v,w)$ on a dense collection of triples, this formula holds on all triples. Corollary 6.8 yields the desired result. ∎

Let X be a metric space, regarded as a subset of its metric completion X^*, and let $0 \in X$. Then X_0 denotes the set of all points $x \in X^*$ which obtain as the limit of a 0–decreasing sequence in X. Similarly, X^0 is the set of all $x \in X^*$ obtained as the limit of a 0–increasing sequence in X.

6.18. Corollary. *Let X be a modular metric space with a base-point $0 \in X$. Then the completion of X is given by $(X_0)^0$.*

Proof. Let ρ, resp., ρ^* denote the metric of X, resp., of its completion X^*. We need two auxiliary results.

6.18.1. Let $a, a', b \in X$, let $x \in ab$ and let $x' \in M(a',b,x)$. Then $\rho(x,x') \leq \rho(a,a')$.

This can be seen by the following computations.

$$\rho(x,x') = \tfrac{1}{2}(\rho(x,b) + \rho(x,a') - \rho(a',b))$$
$$= \tfrac{1}{2}(\rho(a,b) - \rho(x,a) + \rho(x,a') - \rho(a',b))$$
$$= \tfrac{1}{2}(\rho(a,b) - \rho(a',b) + \rho(x,a') - \rho(x,a))$$
$$\leq \rho(a,a').$$

The second statement involves the operator ∇, defined recursively on finite sequences by

$$\nabla(x) = \{x\}; \quad \nabla(x_1,x_2,...,x_{n+1}) = \bigcup \{xx_{n+1} \mid x \in \nabla(x_1,x_2,...,x_n)\}.$$

6.18.2. Let $(a_i)_{i=1}^n$ and $(b_i)_{i=1}^n$ be two sequences of points in X. Then

$$\rho_H(\nabla(a_1,...,a_n), \nabla(b_1,...,b_n)) \leq \sum_{i=1}^n \rho(a_i,b_i).$$

The formula is valid if $n=1$. Suppose $\rho_H(\nabla(a_1,...,a_n), \nabla(b_1,...,b_n)) \leq r$. Let $x \in \nabla(b_1,...,b_n,a_{n+1})$, say: $x \in ca_{n+1}$ for some $c \in \nabla(b_1,...,b_n)$. By assumption, there is a point $d \in \nabla(a_1,...,a_n)$ with $\rho(c,d) \leq r$. Note that

$$M(x,d,a_{n+1}) \subseteq da_{n+1} \subseteq \nabla(a_1,...,a_n,a_{n+1}).$$

---

2.   The multimedian operator of X^* is precisely the operator M^*, described in the first part of the proof. Verification of this fact (which is *not* needed here) takes a lot of work. See Topic 6.31.4.

Let $y \in M(x,d,a_{n+1})$. Then $y \in \nabla(a_1,..,a_{n+1})$, whereas $\rho(x,y) \le \rho(c,d)$ by the first auxiliary statement. It follows that the Hausdorff distance between the sets $\nabla(a_1,..,a_n,a_{n+1})$ and $\nabla(b_1,..,b_n,a_{n+1})$ is at most r. By the triangle inequality of ρ_H it suffices to prove that

$$\rho_H(\nabla(b_1,..,b_n,a_{n+1}),\nabla(b_1,..,b_n,b_{n+1})) \le \rho(a_{n+1},b_{n+1}).$$

Let $x \in \nabla(b_1,..,b_n,a_{n+1})$ and take a point $c \in \nabla(b_1,..,b_n)$ such that $x \in ca_{n+1}$. Take an arbitrary point $y \in M(b_{n+1},c,x)$. As before, one can deduce that $y \in \nabla(b_1,..,b_{n+1})$ and $\rho(x,y) \le \rho(a_{n+1},b_{n+1})$, completing the proof of the statement.

Let $x \in X^*$ and fix a sequence $(x_n)_{n=1}^{\infty}$ in X converging to x. Without loss of generality, $\rho^*(x,x_n) \le 2^{-n}$. We construct sequences $P^k = (p_n^k)_{n=k}^{\infty}$ in X for $k \in \mathbb{N}$, such that

(1:k) $p_k^k = x_k$.
(2:k) $p_n^k \in M(0, p_{n-1}^k, x_n)$ for $n > k$.
(3:k) $p_n^{k-1} \le_0 p_n^k$ for $n \ge k$.

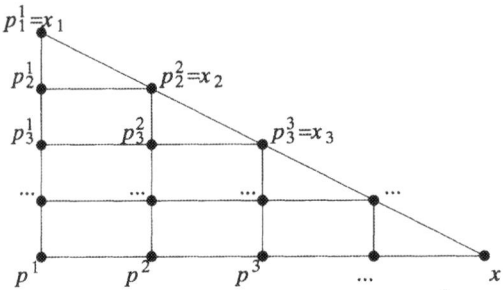

Fig 6: Construction of the sequences p_n^k

See Fig. 6. For $k = 1$, we take $p_1^1 = x_1$, and, recursively, $p_{n+1}^1 \in M(0, p_n^1, x_{n+1})$. Note that (1:1) and (2:1) are fulfilled; condition (3:1) is empty. Next, let $k > 1$ and suppose that the sequences P^l have been constructed for $1 \le l < k$ as in (1:l), (2:l), (3:l). We construct the sequence P^k as follows. Take $p_k^k = x_k$, and, recursively, $p_{n+1}^k \in M(p_{n+1}^{k-1}, p_n^k, x_{n+1})$. Note that $p_k^{k-1} \le_0 x_k = p_k^k$ by (2:k−1), so (3:k) is valid for the parameter value $n = k$. Let $n \ge k$. Assume that (2:k) and (3:k) are valid for all parameter values $\le n$. We have $p_{n+1}^{k-1} \le_0 p_n^{k-1}$; cf. (2:k−1). Combining with $p_n^{k-1} \le_0 p_n^k$ (inductive assumption) yields $p_{n+1}^{k-1} \le_0 p_n^k$. Combining instead with $p_n^{k-1} \le_0 x_n$ -- cf. (2:k−1) -- yields $p_{n+1}^{k-1} \le_0 x_n$. We now appeal to the general fact that, if $p \le_0 a$ and $p \le_0 b$, then $M(p,a,b) \subseteq M(0,a,b)$ and $p \le_0 p'$ for each $p' \in M(p,a,b)$. This yields

$p_{n+1}^k \in M(p_{n+1}^{k-1}, p_n^k, x_{n+1}) \subseteq M(0, p_n^k, x_{n+1})$ (cf. (2:k), parameter value $n+1$);

$p_{n+1}^{k-1} \le_0 p_{n+1}^k$ (cf. (3:k), parameter value $n+1$).

This completes the construction of the sequence P^k.

It follows from (1:k) and (2:k) that $p_n^k \in X$ ($n \ge k$). By (2:k), we conclude that each sequence $P^k = (p_n^k)_{n=k}^{\infty}$ is 0-decreasing. Hence it converges to a point p^k in X_0. By (1:k), $p_k^k = x_k \in \nabla(x_k)$. Inductively, if $n \ge k$ and $p_n^k \in \nabla(x_k,..,x_n)$, then, since $p_{n+1}^k \in p_n^k x_{n+1}$ by (2:k), we conclude that

§6: Modular Spaces 129

$$p^k_{n+1} \in \nabla(x_k, x_{k+1},..,x_{n+1}) \subseteq \nabla_{X^*}(x_k, x_{k+1},..,x_{n+1}).$$

The completion of a modular space being modular by Theorem 6.17, we can apply the formula in 6.18.2 (with one sequence constant and equal to x) to the effect that

$$\rho^*_H(x, \nabla_{X^*}(x_k, x_{k+1},..,x_n)) \leq \sum_{j=k}^{n} \rho^*(x, x_j) \leq \sum_{j=k}^{n} 2^{-j} \leq 2^{-(k-1)}.$$

In particular, $\rho(x, p^k_n) \leq 2^{-(k-1)}$ for $n \geq k$. Hence, $\rho^*(x, p^k) \leq 2^{-(k-1)}$ and $(p^k)_{k=1}^{\infty}$ converges to x. From formula (3:k) we deduce that $p^{k-1} \leq_0 p^k$. So, $(p^k)_{k=1}^{\infty}$ is a 0–increasing sequence in X_0. Therefore, this sequence converges in $(X_0)^0$, showing that $x \in (X_0)^0$. ∎

6.19. Corollary. *The following are equivalent for a modular metric space X.*
(1) X is a complete metric space.
(2) X is monotonely complete relative to some base-point order. ∎

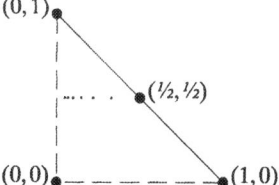

Fig. 7: Counterexample on monotone completeness

In contrast with Part (2), convergence of all decreasing sequences, or convergence of all metrically bounded increasing sequences in one base-point order does not imply a similar result for other base-points. Consider the following median subalgebra X of the plane (Manhattan norm, coordinate-wise median).

$$X = \{(x_1, x_2) \mid 0 < x_i < 1 \text{ for } i = 1, 2; x_1 + x_2 \leq 1\} \cup \{(0, 0)\}$$

(Fig. 7). All decreasing sequences converge relative to the base-point $(½, ½)$, but not relative to $(0,0)$. On the other hand, all metrically bounded, increasing sequences converge from the viewpoint of $(0,0)$, but not from the viewpoint of $(½, ½)$. Moreover, $(X^0)_0$ is not complete as it does not contain the points $(1,0)$ and $(0,1)$. This example also shows that the order of the operations in Corollary 6.18 cannot be altered.

Our final topic involves some new concepts. Throughout, we let $V(G)$, resp., $E(G)$, denote the vertex set, resp., the collection of non-trivial edges, of a graph G. Recall that a *weighted graph* is a finite graph G with a real number $w(e) > 0$ (the weight of e) attached to each edge $e \in E(G)$. The *weight of G* is the number

$$\Sigma_{e \in E(G)} w(e).$$

We consider weighted graphs G in a metric space (X, ρ), in the sense that $V(G) \subseteq X$ and the weight of an edge $e \in E(G)$ is the distance between the end points of e.

Let $C \subseteq X$ be a finite subset. A weighted graph G in X is said to *connect* C provided that G is connected and $C \subseteq V(G)$. By a *Steiner graph* of C is meant a weighted graph G

in (X, ρ) which connects C and has the least possible weight. It is clear that if G connects C and has a circuit, then removal of one edge from the circuit keeps G connected while reducing its weight. Therefore, a Steiner graph of C is a tree and we usually talk about a *Steiner tree*. It is also clear that each end point of a Steiner tree of C is a member of C (the converse is not true).

We will investigate the existence and location of Steiner trees in case of median metric spaces. The following additional terminology is needed. Let M be a median algebra and let $A \subseteq M$. Since the intersection of median stable subsets of M is evidently median stable, it makes sense to consider the smallest median stable subset of M which includes A. This set is called *median stabilization of A*, and it is denoted by $med(A)$.

6.20. Lemma. *Let M be a median space, let $A \subseteq M$ be finite, and let $p \in M$.*
(1) $p \in med(A)$ *iff for each pair of half-spaces $H_1, H_2 \subseteq M$ containing p,*
   $H_1 \cap H_2 \cap A \neq \emptyset.$
(2) *The set $med(A)$ is finite.*[3]

Proof of (1). The union of two convex sets of M is obviously median stable. Hence if H_1, H_2 are half-spaces containing p and if $H_1 \cap H_2$ does not meet A, then $(X \setminus H_1) \cup (X \setminus H_2)$ is a median stable set including A and not containing p. Conversely, suppose that each pair of half-spaces containing p meets on A. The family of convex sets

$$\mathcal{S}(p) = \{ co(H \cap A) \mid H \text{ a half-space containing } p \}$$

is finite since A is, and its members meet two by two. By 5.12, there is a point q common to all sets. If $q \neq p$, then some half-space H of M separates between these points: $p \in H$; $q \notin H$. But then $q \notin co(H \cap A)$. Consider the median space $med(A)$ and the relatively convex sets $D \cap med(A)$ for $D \in \mathcal{S}(p)$. Note that $co(H \cap A) \cap med(A)$ includes $H \cap A$, and hence that the restricted sets meet two by two. Once more, we conclude that there is at least one common point, this time in $med(A)$. However, as the unrestricted sets only have p in common, we conclude that $p \in med(A)$.

Proof of (2). To each $p \in med(A)$ we associate the family $\mathcal{S}(p)$ described above. If $p_1 \neq p_2$ are in $med(A)$ then consider a half-space $H \subseteq M$ with $p_1 \in H$ and $p_2 \notin H$. Clearly, $co(H \cap A) \in \mathcal{S}(p_1) \setminus \mathcal{S}(p_2)$, showing that the assignment $p \mapsto \mathcal{S}(p)$ is an injective function. Therefore, the number of points in $med(A)$ does not exceed $2^{2^{\#A}}$. ∎

For an extension of part (1) to infinite subsets, see II§4.25.7. For additional information on median stabilization, see 6.34. We are now able to derive the main result.

6.21. Theorem. *Let (X, ρ) be a median metric space and let $C \subseteq X$ be a finite set. Then there is a Steiner tree T of C such that $V(T)$ is part of the median stabilization of C.*

---

3. In contrast, there is an example of a modular space with a finite subset not included in a finite (multi)median stable set; cf. Mitschke-Wille [1973].

§6: Modular Spaces 131

Proof. It suffices to verify the result for finite median metric spaces. Indeed, assuming this to be done, consider any weighted graph G in X connecting C. By assumption, there is a Steiner tree T in the finite median space $med(V(G))$ such that $V(T) \subseteq med(C)$. The latter being independent of G, therefore, T is a Steiner tree of C in X.

We henceforth assume that X is finite. Among all Steiner trees of C in X, let T be one with #$med(V(T))$ minimal. Suppose $V(T) \not\subseteq med(C)$. By Lemma 6.20(1), there exist two half-spaces H_1, H_2 of X with (Fig. 8).

$$H_1 \cap H_2 \cap V(T) \neq \emptyset; \quad H_1 \cap H_2 \cap C = \emptyset.$$

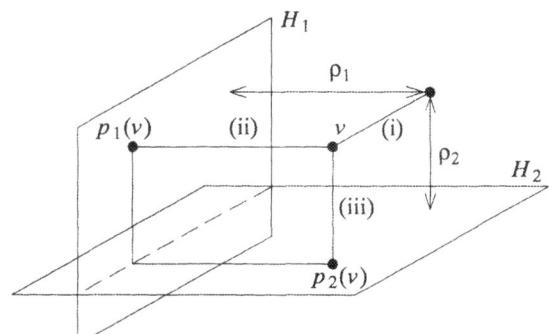

Fig. 8: Minimizing a Steiner tree

We assume that H_1 and H_2 are minimal with these properties. For each non-empty convex set $D \subseteq X$ and for each two points $a, b \in X$ with $b \in co(D \cup \{a\})$, we have $\rho(b,D) \leq \rho(a,D)$ (cf. Topic 6.31.2). In the present circumstances, this implies that for each $i = 1, 2$ there is a positive number ρ_i such that each point $x \in H_1 \cap H_2$ has the same distance ρ_i to $X \setminus H_i$. Let $p_i : X \to H_i$ be the gate map. If $x \in V(T) \cap H_1 \cap H_2$ has a T-neighbor x' in $X \setminus H_i$, we can add the vertex $p_i(x)$ to C and connect it with an edge to both x and x'. This does not increase the weight since $\rho(x,p_i(x)) + \rho(p_i(x),x') = \rho(x,x')$. We henceforth assume that an edge of T connecting a point u of $H_1 \cap H_2$ with a point in $X \setminus H_i$ is of type $up_i(u)$. Let there be n_i edges of this type in T. We note (cf. Topic 6.37.3) that $x \in p_1(x)p_2(x)$ for all $x \in med(V(T))$.

We produce two weighted graphs G_i ($i = 1, 2$) in X as follows. Replace each vertex $v \in V(T) \cap H_1 \cap H_2$ by its gate in H_i. Let $j \neq i$ in $\{1,2\}$. An edge vw of T is

(i) replaced by $p_i(v)p_i(w)$, if $v, w \in H_1 \cap H_2$;
(ii) dropped, if $v \in H_1 \cap H_2$ and $w \in X \setminus H_i$ (so $w = p_i(v)$);
(iii) replaced by $p_i(v)w$, if $v \in H_1 \cap H_2$ and $w \in X \setminus H_j$ (so $w = p_j(v)$);
(iv) kept the same, if $v, w \notin H_1 \cap H_2$.

In case (i), the edge will not get longer since a gate map is non-expansive (cf. Topic 6.31.1). In case (ii), we have a profit ρ_i, and this happens n_i times. In case (iii), the edge length increases with ρ_i and this happens n_j times. Making up the bill, we see that weight$(G_1) \leq$ weight$(T) + (n_1 - n_2) \cdot \rho_1$;

weight$(G_2) \leq$ weight$(T) + (n_2 - n_1) \cdot \rho_2$.

It follows that $n_1 = n_2$ and both G_1, G_2 are Steiner graphs. However, both are included in the median stable set $med(V(T)) \setminus (H_1 \cap H_2)$, contradiction. ∎

Steiner trees are of use in the design of electronic circuitry. In higher dimensions, they occur e.g., in problems on rectangular networks connecting consumers located in several buildings. A median metric is obtained by defining the distance of two locations in different buildings by the sum of the respective location heights and the horizontal (Manhattan) distance between the buildings. Consult Topic 6.31.3.

Further Topics.

6.22. Semimodularity

6.22.1. Let $(X, \leq)$ be a discrete poset. Show that the corresponding interval space is semimodular iff both $(X, \leq)$ and $(X, \geq)$ are semimodular posets.

6.22.2. (Birkhoff [1967]) A lattice L is modular iff the posets $(L, \leq)$ and $(L, \geq)$ are semimodular.

6.22.3. (Chepoi [1991]) Let G be a connected graph such that the geodesic convexity is semimodular and has the triangle property (such graphs are known as *quasi-modular graphs*). Show that a connected subset C of G is convex whenever each common neighbor of two distinct points of C is in C.

6.23. Alternative description of modularity (Bandelt, van de Vel and Verheul [1993]). Let X be a set and let $(a,b) \mapsto a \circ b$ be a multivalued function of X^2 to X, subject to the following conditions.

(1) $\forall a \in X: a \in a \circ a$.
(2) $\forall a, b \in X: a \circ (a \circ b) \subseteq a \circ b$.
(3) $\forall a, b, c \in X: b \in a \circ c$ implies $a \circ b \cap b \circ c = \{b\}$.
(4) $\forall a, b, c \in X: a \circ b \cap b \circ c \cap c \circ a \neq \emptyset$.

Show that $(a,b) \mapsto a \circ b$ is an interval operator and that the resulting interval space is modular. Conversely, the interval operator of a modular space satisfies (1)-(4).

6.24. Modular semilattices (Bandelt, van de Vel and Verheul [1993]). A meet semilattice S is called *modular* provided every principal ideal of S is a modular lattice and any three elements of S are bounded from above whenever each pair of them is. The first part of this condition states that S is *conditionally modular*.

6.24.1. Let S be a meet semilattice with the following interval operation.

§6: Modular Spaces   133

$$ab = \{\, x \vee y \mid x \in [a \wedge b, a], y \in [a \wedge b, b]\,\} \quad (a, b \in S).$$

Show that S is a modular interval space iff it is a modular semilattice. Hint: use the axiom system of 6.23. Conversely, let X be a modular interval space and let $0 \in X$ be such that $M(0,a,b)$ is a singleton for each $a, b \in X$ (a *neutral point*). Show that $(X, \leq_0)$ is a modular semilattice, and that its interval operator agrees with the one described above.

6.24.2. Verify that the interval function introduced above agrees with the standard one for modular lattices. (Nieminen [1979]) Let S be a conditionally modular semilattice. Show that the lattice of all ideals of S is modular. (Bandelt, van de Vel and Verheul [1993]) Conclude that a modular interval space with a neutral point embeds into a modular lattice, such that the neutral point is the least element.

6.24.3. An element a of a lattice L is *neutral* provided for each $x, y \in L$ the sublattice generated by a, x, y is distributive. Show that in a modular lattice an element a is neutral iff it is neutral in the sense of modular interval spaces.

6.24.4. Let A, B be disjoint sets. Consider the set S consisting of all points of $A \cup B$ and of $A \times B$, together with a point 0 not in one of the previous sets. A partial order is defined on S with 0 as the least element, and such that each element of type (a,b) is below a and below b. Verify that S is a modular semilattice. Note that if $\#A = m$ and $\#B = n$, then the graphic space S contains a (non-induced) copy of the graph $K_{m,n}$ as a median stable subspace.

6.25. Modular algebras (cf. Kolibiar and Marcisová [1974] and Hedlíková [1977]). Let $m: X^3 \to X$ be a function satisfying

(i) $\forall a, b \in X: m(a,b,b) = b$.
(ii) $\forall a, b, c, d \in X: m(m(a,d,c),b,c) = m(a,c,m(b,c,d))$.

The function m is a *modular operator* and the ternary algebra (X,m) is a *modular algebra*. For instance, the formula

$$m(a,b,c) = (a \wedge (b \vee c)) \vee (b \wedge c)$$

defines a modular operator on a modular lattice.

6.25.1. Verify the following identities for a modular operator m.

(i) $m(a,b,c) = m(a,c,b)$.
(ii) $m(a,a,b) = a$.
(iii) $m(a,m(a,b,c),m(d,b,c)) = m(a,b,c)$.
(iv) $m(m(a,b,c),a,c) = m(a,c,m(a,b,c)) = m(a,b,c)$.

6.25.2. (Hedlíková [1983]; cf. Bandelt, van de Vel, and Verheul [1993]). Let the interval operator I_m be defined by

$$I_m(a,b) = \{x \mid x = m(a,b,x)\} = \{m(a,b,x) \mid x \in X\}.$$

Show that (X,I_m) is a geometric interval space and $(I_m(a,b), \leq_a)$ is a modular lattice for all

$a, b \in X$. Observe that this yields the Jordan-Hölder Theorem (cf. 6.4) for this interval space. It is not known, however, whether (X, I_m) is modular (but see next topic). Prove that if a modular space (X, I) admits a modular operator m such that $m(a,b,c) \in M(a,b,c)$ for all $a, b, c \in X$, then (X, I) is a median space and m is its median operator.

6.25.3. (Bandelt, van de Vel, and Verheul [1993]). Let $0 \in X$ be such that $m(0,a,b) = m(a, 0,b)$. Show that $(X, \leq_0)$ is a modular semilattice and I_m agrees with the interval operator considered in 6.24.

6.26. Median semilattices (compare Sholander [1954b]; cf. 2.27.1). By a *median semilattice* is meant a (meet) semilattice such that each principal ideal is a distributive lattice (a *conditionally distributive semilattice*) and any three elements are bounded from above whenever each pair of them is. For instance, a tree is a median semilattice.

6.26.1. If X is a median space and if $b \in X$, then $(X, \leq_b)$ is a median semilattice. Note that $x \leq_b y$ iff $m(b,x,y) = x$ (Sholander's order).

6.26.2. Let S be a median semilattice and let $a, b, c \in S$. Then each pair among $a \wedge b$, $b \wedge c$, and $c \wedge a$ has an upper bound, and hence this triple has a supremum $m(a,b,c)$. Show that the resulting operation $m: S^3 \to S$ is a median. The median interval operator is given by the interval formula in 6.24.1.

6.26.3. Show that the transitions between median convex structures and median semilattices are mutually inverse. In particular, the construction of m in Part 2 does not depend on the choice of the base-point (compare 5.3.3 for trees): *one* base-point order suffices to reconstruct the entire median convexity!

6.27. Characterizing modular metric spaces (Verheul [1993]). Prove that the following are equivalent for a metric space (X, ρ).
(1) (X, I_ρ) is a modular interval space and ρ is a convex metric.
(2) Each triple of pairwise intersecting closed disks of (X, ρ) has a common point and ρ is a convex metric.
(3) (X, ρ) has the Tri-spherical Intersection Property.

6.28. Valuated lattices

6.28.1. If L is a lattice with 0, and if ρ is a metric on L such that $I_m = I_\rho$, then $v(x) = \rho(0,x)$ is a valuation on L.

6.28.2. Show that a measure algebra modulo zero sets yields a valuated lattice, where the valuation of an equivalence class is the measure of a representative.

6.28.3. (Verheul [1993]) Let $(V, \wedge, \vee, ||.||)$ be a vector lattice with an additional norm $||.||$ and with the usual modulus $|x| = x \vee -x$. Then $(V, \wedge, \vee, ||.||)$ is an L-space provided the following conditions hold.

§6: Modular Spaces

(i) $|x| \leq |y|$ implies $\|x\| \leq \|y\|$.
(ii) $\|x+y\| = \|x\| + \|y\|$ for positive x, y.

Spaces of type L_1 are prominent examples (the lattice structure is obtained from a maximal face F in the unit disk, determining a positive cone $\cup_{t \geq 0} t \cdot F$). Let $x^+ = x \vee 0$ and $x^- = x \wedge 0$. Show that the following is a valuation inducing the same metric as $\|.\|$.

$$v(x) = \|x^+\| - \|x^-\| \quad (x \in V).$$

6.29. More characterizations of median spaces. Show that the following are equivalent for an interval space (X, I).

(1) X is median.
(2) X is modular and its segment function is geometric.
(3) The interval function I of X is monotone (cf. (G-2)) and for each triple of points a, b, $c \in X$, the set $I(a,b) \cap I(b,c) \cap I(c,a)$ consists of one point.
(4) X is modular and its multimedian operator satisfies a "four-point transitive law":

$$M(M(a,b,c),d,c) = M(a,M(b,c,d),c).$$

Here, $M(A,b,c)$ denotes the union of all sets of type $M(a,b,c)$ for $a \in A$. Hint for (4). Use the forbidden structure $K_{2,3}$. Verheul [1993] has shown that a somewhat different form of the Transitive Law is valid in modular metric spaces.

6.30. Lattice intervals (Bandelt, van de Vel and Verheul [1993]). Show that in a modular space all intervals are (modular) lattices iff there is no subspace isomorphic to the graphic space $K_{3,3}$ minus an edge. In case of modular graphs, the term "subspace" may be replaced by "induced subgraph". There may be a non-induced subgraph isomorphic to $K_{3,3}$; cf. 6.24.4 above.

6.31. Modular and median metric spaces

6.31.1. Show that a gate map in a modular metric space is non-expansive.

6.31.2. If (X, ρ) is a median metric space, if $C \subseteq X$ is gated, and if $b \in co(C \cup \{a\})$ then $\rho(b,C) \leq \rho(a,C)$.

6.31.3. Show that the gated amalgam of two modular (median) metric spaces is a modular (median) metric space. Consult 5.29. Use this procedure to construct a (realistic) median metric on the Manhattan plane with several buildings erected.

6.31.4. (Verheul [199*]) By Theorem 6.17, the completion of a modular metric space X is modular. Part of the proof consists in showing that the multimedian operator of X extends in a unique way to a uniformly continuous operator M^* of the completion X^*. Verify that M^* is the multimedian operator of X^*.

6.31.5. (van de Vel and Verheul [1991]). A subset A of a modular interval space X is *multimedian with respect to a point* $0 \in X$ provided $M(0, a_1, a_2) \subseteq A$ for all $a_1, a_2 \in A$.

Note that a subset of a modular lattice with a universal lower bound **0** is multimedian w.r.t. **0** iff it is a meet subsemilattice; that a metric disk of a modular metric space is multimedian w.r.t. its center, and that an interval of a modular interval space is multimedian w.r.t. an endpoint. Extend Corollary 6.18 to subspaces of modular metric spaces which are multimedian with respect to a point.

6.32. Modular normed spaces

6.32.1. Verify that a 45° rotation of $\mathbb{R}^2$ induces an isomorphism of the Cartesian plane with the Manhattan plane.

6.32.2. Let V be a normed space and let e be an extreme point of its unit disk. Show that the geodesic interval $0e$ equals the line segment joining 0 with e. Conclude that if V is modular and if $H \subseteq V$ is a geodesically convex hyperplane not containing e, then the linear decomposition map $V \to H \times \mathbb{R}$, sending a vector x to the unique pair (h, t) with $x = h + t \cdot e$, is an isomorphism of normed spaces (sum norm on $H \times \mathbb{R}$).

6.32.3. (van de Vel and Verheul [1990]). Let $\mathbb{R}^n$ be given the Cartesian norm

$$\|x\| = \max\{|x_1|,\dots, |x_n|\} \quad (x = (x_1,\dots,x_n)).$$

Show that the Cartesian space $\mathbb{R}^n$ $(n \geq 3)$ has no geodesically convex bodies except $\mathbb{R}^n$ itself. Hint. One can use the theory of vector convexity (cf. III§1.25.3).

In fact (Boltyanskii and Soltan [1978]), the geodesic convexity of $(\mathbb{R}^3, \|..\|)$ consists of the following sets: $\emptyset$, all singletons, all 1-dimensional (ordinary) convex sets parallel to one of the main diagonals, and $\mathbb{R}^3$. Conclude that the line $\mathbb{R} \times \{0\} \times \{0\}$ is a median space which (i) is a median stable subset; (ii) is a subspace of $(\mathbb{R}^3, \|..\|)$ in the sense of interval spaces; (iii) is *not* a subspace in the sense of convex structures.

6.32.4. *Hanner's Problem.* Let $(V_i, \|.\|_i)$ for $i = 1, 2$ be normed spaces. Then the product space $V = V_1 \times V_2$ can be made into a normed space by either of the following constructions.

$$\|(v_1, v_2)\|_s = \|v_1\|_1 + \|v_2\|_2 \quad (v_1 \in V_1, v_2 \in V_2);$$
$$\|(v_1, v_2)\|_m = \max\{\|v_1\|_1, \|v_2\|_2\} \quad (v_1 \in V_1, v_2 \in V_2).$$

Show that if $(V_i, \|.\|_i)$ for $i = 1, 2$ are modular, then so are $(V, \|.\|_s)$ and $(V, \|.\|_m)$ (cf. 6.2.5).

Problem. Can every modular norm on $\mathbb{R}^n$ be obtained by a combination of the above constructions? Hanner [1956] has shown that the number of (non-isomorphic) modular norms on $\mathbb{R}^n$ is finite. Precise numbers are known for $n \leq 5$ only.

6.33. Median algebra

6.33.1. (compare Theorem 6.9) Let M, M' be median spaces and let $f: M \to M'$ be a CP function. Show that if $C \subseteq M$ is convex, then $f(C)$ is relatively convex in M'. In particular, the median and relative convexity of a median stable subset of a median

§6: Modular Spaces   137

algebra coincide. Topic 6.32.3 illustrates that a similar statement is not valid for modular spaces.

6.33.2. Let M be a median algebra, let Y be a convex structure, and let $f: M \to Y$ be a surjective CP and CC function. Show that there is a uniquely determined median operator on Y inducing the convexity of Y and turning f into an MP function.

6.33.3. Let X be a finite, S_4, and point-convex space embedded in $\lambda(X)$. Show that $\lambda(X) = med(X)$. Deduce that a CP function between finite, S_4, and point-convex spaces, extends in a unique way to a CP function of the corresponding superextensions.

6.33.4. Let M be a median convex structure and let $F \subseteq M$ be finite, say with r points. Throughout, $\lambda(r)$ denotes the superextension of the set $\{1,..,r\}$ with the free convexity. The result described in 6.33.3 justifies the interpretation of $\lambda(r)$ as the *free median algebra* on r points. Let $f: \{1,..,r\} \to F$ be a bijection. Show that the image of the canonical extension $\lambda(r) \to M$ of f is the median stabilization of F. Compare with a proof given in 6.20.

6.34. Median stable sets. The median stable subsets of median algebra M constitute a convexity $\mathcal{M}$ on M. The hull $med(Y)$ of a set $Y \subseteq M$ in this convexity is just the median subalgebra generated by Y and it can be obtained as follows.

$$med(Y) = \bigcup_{n=0}^{\infty} Y_n, \quad \text{where } Y_0 = Y \text{ and } \forall n: Y_{n+1} = m(Y_n^3).$$

6.34.1. Show that $\mathcal{M}$ is generated by sets of type $H_1 \cup H_2$, where H_1 and H_2 are half-spaces of the *original* convexity. Moreover, these generators are half-spaces of $\mathcal{M}$. In particular, $\mathcal{M}$ is S_3. Verify that $([0,1]^2, \mathcal{M})$ is not S_4 (standard median).

6.34.2. Let Y be a subset of M and let $p \in M$. Show that $p \in med(Y)$ iff there exist finite sets $F_1,..,F_n \subseteq Y$ with

(*) $\qquad \bigcap_{i=1}^{n} co(F_i) = \{p\}.$

In fact, the sets F_i can be chosen such as to intersect two by two. In a distributive lattice, the formula (*) is equivalent to the following equalities.

$$(\wedge F_1) \vee .. \vee (\wedge F_n) = p = (\vee F_1) \wedge .. \wedge (\vee F_n).$$

6.34.3. Let L be a distributive lattice, and let $X \subseteq L$. According to Monjardet [1975], a point $y \in L$ is **self-dual** (relative to X) if there is a polynomial expression of y in terms of the elements of X and the operations $\wedge$, $\vee$, returning the same value if "$\wedge$" and "$\vee$" are interchanged. Show that the polynomial $m(a,b,c)$ is self-dual, that the collection $M(X)$ of self-dual elements is median stable, and that $M(X) = med(X)$ (use Part (2)).

6.34.4. (Bandelt and van de Vel [1992b]) The sequence $(q_n)_{n=2}^{\infty}$ is defined by $q_2 = 2$ and $q_{n+1} = \lfloor 3q_n/2 \rfloor$. Let Y be a subset of the graphic d-dimensional cube Q^d, where $d \geq 4$. As above, let $Y_0 = Y$ and $Y_{n+1} = m(Y_n^3)$ for $n \geq 0$. Let n be the least integer with $d \leq q_n$. Prove that $med(Y) = Y_n$; in other words, each subset of Q^d, $d \geq 4$, stabilizes in at most $n \approx \log_{1.5} d$ steps. In contrast, each subset of Q^2 is median stable, and in Q^3 at most one step is required.

Hint. Represent Q^d as the power set of a d-point set. Verify that the median of three k-sets has at most $\lfloor 3k/2 \rfloor$ elements. Then consider the stabilization of the collection of all 2–sets. Prove that this is "the worst possible case".

6.35. Median graphs

6.35.1. (Bandelt, van de Vel and Verheul [1993]; compare Theorem 6.10). Show that a graph is median iff it is modular and it has no induced subgraph isomorphic to $K_{2,3}$.

6.35.2. (Bandelt and van de Vel [1987]; cf. IV§6.1). The median operator of a median graph G is edge preserving in each variable separately. Deduce that if $H \subseteq G$ is a connected set, then the median algebra $med(H)$ generated by H is also connected. Verify that each connected median stable subset of G is an isometric (median) subgraph and that its relative convexity equals the geodesic convexity.

6.35.3. (compare Bandelt [1983], Isbell [1980], van de Vel [1983e]) Show that a finite non-empty median graph G is a graphic cube iff the union of any two non-empty disjoint half-spaces equals G.

6.35.4. (Mulder [1980]) Let G be a finite non-empty median graph. Show that the number of non-trivial half-spaces of G is at least $2 \cdot \log_2 \#G$ and at most $2 \cdot (\#G - 1)$. Give examples illustrating the sharpness of these bounds.

6.36. Steiner points.
Let (X, ρ) be a metric space. A *Steiner point* of a triple $a, b, c \in X$ is a point minimizing the distance sum

$$\rho(a,x) + \rho(b,x) + \rho(c,x) \quad (x \in X).$$

6.36.1. (Verheul [1993]) Show that if (X, ρ) is modular, then the set of medians $M(a,b,c)$ consists exactly of the Steiner points of a, b, c.

6.36.2. (Avann [1961]) Show that a connected graph is median iff each triple of vertices has a unique Steiner point.

6.36.3. The restriction to graphic spaces is essential in the previous result. Relative to the Euclidean metric, each triple of points in the plane has a unique Steiner point.

6.37. Medians and gates.
Throughout, we consider a median space X.

6.37.1. (Kolibiar and Marcisová [1974]). Prove the *Five-point Transitive Law:*

$$m(m(a,b,u), m(a,b,v), w) = m(a,b,m(u,v,w))$$

(this law is used a.o. by Sholander [1954], Birkhoff [1967], and Isbell [1980]). Hint: A

§6: Modular Spaces   139

geometric argument involving half-spaces is far easier than an algebraic one.

6.37.2. Let $F \subseteq X$ be a non-empty finite set and let $b \in co(F)$. For $a \in F$, let $p_a: X \to ab$ denote the gate map. Show that if $C \subseteq co(F)$ is convex and contains b, then
$$C = \cap_{a \in F}\, p_a^{-1}(C \cap ab).$$

6.37.3. Let $C_1,..,C_n$ be convex sets and for each i let c_i be the gate of b in C_i. First, $b \in co(\cup_{i=1}^n C_i)$ iff $b \in co\{c_1,..,c_n\}$. Next, if $C = \cap_{i=1}^n C_i \neq \emptyset$, then the gate of b in C exists and equals the supremum (in $\leq_b$) of c_i for $i = 1,..,n$. Deduce that if a finite set F has a supremum s (as seen from a given base point) then $s \in co(F)$.

6.37.4. Let $C_1,..,C_n$ be convex sets. For each i let c_i be the gate of b in C_i. Show that the gate of b in $co(\cup_{i=1}^n C_i)$ exists and equals the infimum (in $\leq_b$) of c_i for $i = 1,..,n$. Note that this yields another proof of the fact that polytopes are gated (cf. 6.10). Deduce that if X is discrete and if c is the b-infimum of all neighbors of a in ab, then the segment ac is a graphic cube. Hint: use 6.35.3.

6.37.5. Let C, D be gated sets with gate maps p_C onto C and p_D onto D, such that $p_C(D) = C$ and $p_D(C) = D$. Then the restricted map $p_C|D,C$ is an isomorphism and $co(C \cup D)$ is the union of all intervals of type cd, where $c = p_C(d)$ and $d = p_D(c)$.

6.37.6. Let $f: X \to Y$ be a CP function of median spaces, and let $b \in X$ have a gate c in a convex set $C \subseteq X$. Show that $f(c)$ is the gate of $f(b)$ in $co(f(C))$. Deduce that in a median stable subspace, disjoint relative polytopes extend to disjoint polytopes.

6.37.7. The polytopes of X, together with the gate maps between pairs $P \subseteq Q$ of polytopes, constitute an inverse system and X can be embedded in the inverse limit.

6.38. Finite median spaces (van de Vel and Verheul [1990b]).

6.38.1. Let (X,ρ) be a finite median metric space. Show that, if $H \subseteq X$ is a half-space with a gate map $p: X \to H$, then the function
$$X \to H \times [0,\infty), \quad x \mapsto (p(x), \rho(x,H)),$$
is an isometric embedding (Manhattan metric on the product space).

6.38.2. Deduce that a finite median metric space embeds isometrically into $\mathbb{R}^n$ with Manhattan norm for some n. We note the following.

(i). It is shown by Verheul and the author [1990b] that a median metric space embeds isometrically into $\mathbb{R}^n$ iff it embeds algebraically into $\mathbb{R}^n$ (coordinate-wise median).

(ii). Assouad and Deza [1982] have shown that a metric space embeds isometrically into a space of type L_1 iff each of its finite subspaces does.

(iii). Verheul [1993] has shown that the median Banach spaces are precisely the spaces of type L_1.

6.38.3. Show that a convex expansion involving gated sets can be seen as a sequence of two gated amalgamations. Deduce that the expansion of a modular (median) space along gated convex sets is modular (median). Deduce that a finite graph is median iff it obtains from a one-point set by a series of convex expansions (cf. Mulder [1980]).

6.39. Medians in convex hyperspaces. Let M be a median space. Recall that the convex hyperspace $\mathcal{C}_*(M)$ consists of all non-empty convex sets in M.

6.39.1. (Bandelt and Hedlíková [1983]; compare van Mill and van de Vel [1978] or Nieminen [1978]). Show that the Vietoris convexity of $\mathcal{C}_*(M)$ is derived from the median operator m_* defined by

$$m_*(A,B,C) = m(A \times B \times C),$$

where m is M's median operator. Deduce that

$$m_*(A,B,C) = co(A \cup B) \cap co(B \cup C) \cap co(C \cup A).$$

6.39.2. Show that the collection $\mathcal{S}_*(M)$ of all gated convex sets of M is a median stable subset of $\mathcal{C}_*(M)$.

6.39.3. (compare van Mill and van de Vel [1978]). For $x \in M$ and $D \in \mathcal{S}_*(M)$ let $p(x,D)$ denote the gate of x in D. Show that p is CP in each variable separately; in general, it is not a CP function in two variables x,D.

6.39.4. (van Mill, communication) Deduce that for each $n \in \mathbb{N}$ there is a CP isomorphism of the superextension $\lambda(n+1)$ with one generating (extreme) point removed and the convex hyperspace of $\lambda(n)$.

Notes on Section 6

Instances of modular interval spaces were studied under different names and (almost) equivalent axioms by Isbell [1980] (media) and Hedlíková [1983] (taut media). The term "modular space" has been introduced by Bandelt, van de Vel and Verheul [1993]. Already in [1956], Hanner studied modular norms, and raised the (still unanswered) question whether all modular norms of $\mathbb{R}^n$ are obtained by the method of 6.2.5 (cf. Topic 6.32.4). Modularity in graphs was introduced by Howorka [1981]. The examples in 6.2.3 (on L_1 spaces) and in 6.2.4 (on the Tri-spherical Intersection Property) have been studied by Verheul [1993]. In the terminology of Aronszajn and Panitchpakdi [1956], the Tri-spherical Intersection Property corresponds with 4–hyperconvexity. For the modularity of valuated lattices (Corollary 6.6) see Birkhoff [1967, p. 42].

The Jordan-Hölder Theorem (cf. 6.4) was obtained for modular algebras (cf. 6.25) by Hedlíková [1983] and for modular interval spaces by Bandelt, van de Vel and Verheul in [1993]. In this paper, projective cover pairs have not been considered. An alternative proof of the Jordan-Hölder Theorem relies on the fact that the intervals of a modular space are so-called modular multilattices, to which Benado's [1955] version of the theorem

§6: Modular Spaces

applies. Isbell [1980] obtained the same result for his media by observing that all intervals are modular lattices. The incomparability of modular interval operators on the same set, Theorem 6.5, is due to Bandelt, van de Vel and Verheul [1993]. It was obtained for median spaces by the author in [1984]; a topological result in this direction is due to van Mill and the author [1978]. The equality of the geodesic interval operator and the modular interval operator of a valuated lattice, Corollary 6.6, goes back to Glivenko [1936].

A survey of the early theory of median spaces is given in Bandelt and Hedlíková [1983]; as for median graphs, Mulder [1980] is a good source. Birkhoff and Kiss [1947] discussed the median operator on distributive lattices and, in the presence of universal bounds **0**, **1**, they gave an axiom system in terms of the median. Median graphs were introduced by Avann [1961] (who called them "unique ternary distance graphs") and later by Nebeský [1970]. General median algebras were considered by Avann [1948], [1961], who called them "ternary distributive semilattices". Sholander [1952], [1954], [1954b] presented several axiom systems for median semilattices and trees.

Theorem 6.10 is taken from Bandelt, van de Vel and Verheul [1993]. Parts of it have been obtained before. The correspondence between median algebras and interval spaces with unique medians, (cf. (1) $\Leftrightarrow$ (7)) is due to Sholander [1954]. Convex sets in median algebras were studied by Nieminen [1978], who obtained the S_3 property, and by Isbell [1980], who also obtains JHC (cf. (1) $\Rightarrow$ (3)). The correspondence between median algebras and modular S_4 spaces (cf. (4) $\Leftrightarrow$ (7)) was obtained by the author in [1984].

Theorem 6.17, on the modularity of metric completions, was obtained by Verheul [199*]. It settles a problem raised by Aronszajn and Panitchpakdi [1956], whether the completion of a 4–hyperconvex metric space has the same property. The fact that such completions can be obtained with increasing and decreasing sequences in one base-point order only (cf. Corollary 6.18) was obtained by van de Vel and Verheul [1991].

Lemma 6.20 (on median stabilization) seems to be folklore; cf. Guénoche [1986], Barthelémy [1989]. The subsequent result on Steiner trees has recently been obtained by E. Verheul and the author. It extends and considerably improves the original result of Hanan [1966] on the "rectilinear" Steiner problem in the plane. Alas, this cannot change the fact that finding a Steiner tree is an NP-hard problem (Garey and Johnson [1977]).

7. Bryant-Webster Spaces

Interval operators are often regarded as a "join" operator. A frequent axiom system is the one for Pasch-Peano spaces. There is a different set of axioms for join operations, in which the join of two points is interpreted as the set of points which are strictly between the end points. This leads to the class of Bryant-Webster (BW) spaces, capturing most of the geometric properties of vector spaces (rather, convex subspaces thereof) without reference to linear algebra. The involved axiom system strongly resembles the axioms of ordered geometry. Among the exotic examples are the Moulton plane, totally densely ordered sets, and half-spheres.

The affine sets in a BW space without boundary yield a weakly Join-Hull Commutative matroid. The affine dimension formula relates the dimension of the intersection and the join of two flats.

We begin this section with a brief survey of a few special interval properties.

7.1. Some properties of intervals. Throughout, let X be an interval space. The space X is said to be *dense* provided for each pair of distinct points $a, b \in X$ the interval ab contains a point distinct from a, b. If X is geometric, this amounts to the statement that each base-point order $\leq_b$ is *dense:* if $u <_b v$ then there is a point w with $u <_b w <_b v$ (cf. Proposition 5.2).

An interval ab of X is *decomposable* provided for each $x \in ab$,

$$ax \cup xb = ab \text{ and } ax \cap xb = \{x\}.$$

Observe that if all intervals of X are decomposable, then the interval operator is geometric and all intervals are convex. In fact, it is even true then, that each base-point order is a total order on the lower set of any point. Consequently, if X is a complete and S_4 convex structure with decomposable segments, then by Proposition 5.20, $\leq_b$ is a tree-order. The same conclusion obtains from decomposability in combination with the *Ramification Property:* for all $b, c, d \in X$,

$$c \notin bd \text{ and } d \notin bc \text{ implies } bc \cap bd = \{b\}.$$

This means that if c, d are incomparable relative to $\leq_b$, then the base-point b is the only common lower bound of c, d. If X has decomposable segments and satisfies the Ramification Property, then $\leq_b$ is a tree order with at most one ramification point, namely b.

Finally, X is said to be *straight* provided the union of two intervals of X with more than one point in common is an interval. As in Section 4, all terminology involving interval spaces can be applied to convex structures via the segment operator.

The next results are used in discussing the axioms of convexity that lead to so-called Bryant-Webster spaces.

144   Chap. I: Abstract Convex Structures

7.2. Proposition. *A straight interval space with decomposable intervals has the Ramification Property.*

Proof. Suppose that $c \notin bd$ and $d \notin bc$, and consider a point $x \in bc \cap bd$ distinct from b. Straightness implies that there exist two points u, v with $bc \cup bd = uv$. Now $bc = bx \cup xc$ and $bd = bx \cup xd$. Hence the points u, v belong to one or more of the intervals bx, xc, xd. As we observed before, all intervals are convex. Therefore, u and v cannot occur both in the same interval, since this would imply that one of the points b, c, d is outside uv. If u and v are in $bx \cup xc$ then $d \in uv \subseteq bc$. Similarly, if u and v are in $bx \cup xd$ then $c \in bd$.

The only possibility left is that $u \in cx$ and $v \in dx$ (or vice versa). We have $b \in uv = ux \cup xv$, say, $b \in ux \subseteq cx$. Then

$$b \in cx \cap xb = \{x\}$$

whence $x = b$. ∎

7.3. Proposition. *In a dense interval space with the Ramification Property and with decomposable segments, the Peano Property implies the Pasch Property.*

Proof. By Proposition 4.10(2), all intervals are convex. Let $a' \in ap$ and $b' \in bp$. We have to show that the segments ab', $a'b$ intersect. Without loss of generality, the points a, b, p are convexly independent and $a' \neq a, p$; $b' \neq b, p$; a proof of the result for "degenerate" configurations is left to the reader. We claim that if $b' \in ax$ with $x \in a'b$, then $b' \in a'b$.

This can be seen as follows. Note that $b' \in ax \subseteq co\{a, a', b\}$. Use the Peano Property to extend bb' to a segment bb'' with $b'' \in aa' \subseteq ap$. Then $a' \in b''p$ (geometricity). Now $b' \in bb'' \cap bp$, and by the Ramification Property, either $b'' \in bp$ or $p \in bb''$. In the first case, a' is a common point of ap; bp, contradictory to the assumptions. In the second case, decomposability of $b''b$ yields that $p \in b''b'$. We see that

$$a' \in b''p \subseteq b''b' \subseteq b''b$$

and hence (geometricity) that $b' \in a'b$.

Continuing with the original proof, take any point $c \in ab'$ distinct from the end points a, b' (density). Note that if $c = a$ or $c = b$ then c is the desired intersection point. So let $c \neq b$. By the Peano Property, there is a point $d \in ap$ such that $c \in bd$. The segment ap decomposes as $ad \cup dp$. We distinguish between two possibilities (Fig. 1).

(i) $a' \in dp$ (left figure). By JHC, there is a point $c' \in a'b$ such that $c \in ac'$ As $c \neq a$, the Ramification Property yields that $c' \in ab'$ (whence c' is the desired point of intersection) or that $b' \in ac'$. In the second case, the claim shows that b' is the desired point of intersection.

(ii) $a' \in ad$, or, what amounts to the same, $d \in a'p$. By using JHC twice, we can extend the segment pc first to pd' with $d' \in a'b$ and next to pd'' with $d'' \in ab$. Now $d' \in cd'' \subseteq co\{a, c, b\}$ and hence bd' extends to a segment bc' with $c' \in ac$.

§7: Bryant-Webster Spaces 145

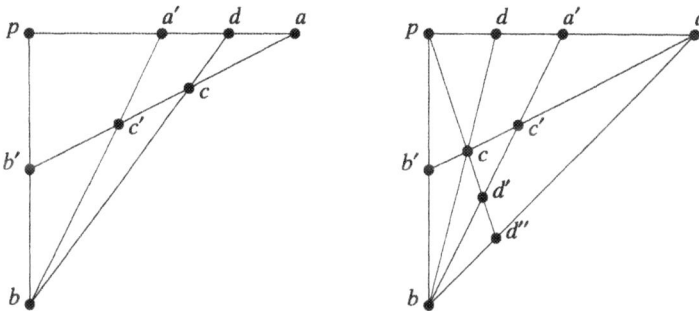

Fig. 1: The Peano Property implies the Pasch Property (two cases)

If $d' = b$ then $c \in pb$ and decomposability of the latter leads to one of the following possibilities.

— $b' \in pc$: As $a \notin bp \supseteq bb'$, the Ramification Property yields $b \in ab'$, showing that b is the desired point of intersection.

— $b' \in cb$: Then $c \in b'p$ and as $a' \notin pb'$ we obtain by the Ramification Property that $p \in b'a$. Hence $a' \in ap \subseteq ab'$, showing that a' is the desired intersection point.

If $d' \neq b$ then (Ramification) either $c' \in a'b$, whence c' is the desired intersection point, or $a' \in c'b$. The claim (with a, a' and b, b' interchanged and $x = c'$) yields that $a' \in b'a$ and a' is the desired point of intersection. ∎

A simpler proof is possible if the Ramification Property is replaced by the stronger condition of straightness. For a discussion on the sharpness of the above result, see 7.15.

7.4. Join operators. Let X be a set with an operator

$$\circ : X \times X \to 2^X,$$

subject to some of the axioms described below. The image of a pair $(a,b) \in X \times X$ is the *join* of the points a and b; it is denoted by $a \circ b$. The function $\circ$ is called a *join operator*. The related *extension operator* (*ray operator*) is constructed as follows.

$$/ : X \times X \to 2^X,$$
$$a/b = \{ x \mid a \in b \circ x \}.$$

The set a/b is called the *extension* (*ray*) *at a away from b*. If $A, B \subseteq X$, then we put

$$A \circ B = \bigcup \{ a \circ b \mid a \in A, b \in B \};$$
$$A/B = \bigcup \{ a/b \mid a \in A, b \in B \}.$$

The set $A \circ B$ is called the *join of the sets* A and B, and A/B is called the *extension of A away from B*. Expressions like $\{a\} \circ B$, or $\{a\}/B$, are abbreviated as $a \circ B$, or a/B. The following are properties which may hold for a join operator and its related extension operator. After a brief discussion, we shall make a choice.

(J-1)   Density: $a \circ b \neq \emptyset$.
(J'-1)   Closedness: $a \in a \circ b$.
(J-2a)   Idempotence of join: $a \circ a = \{a\}$.
(J-2b)   Idempotence of extension: $a/a = \{a\}$.
(J'-2)   Commutative Law: $a \circ b = b \circ a$.
(J-3)   Associative Law: $a \circ (b \circ c) = (a \circ b) \circ c$.
(J-4)   Pasch Property: $(a/b) \cap (c/d) \neq \emptyset$ implies $(a \circ d) \cap (b \circ c) \neq \emptyset$.
(J-5)   Ramification Property: $(a \circ b) \cap (a \circ c) \neq \emptyset$ implies $b = c$ or $b \in a \circ c$ or $c \in a \circ b$.
(J-6)   Extensibility: $a/b \neq \emptyset$.
(J-7)   Completeness: if $a \circ b = P \cup Q$ with P and Q non-empty, and if $a \circ x \cap y \circ b = \emptyset$ for all $x \in P$, $y \in Q$, then there is a point $c \in a \circ b$ with $P \supseteq a \circ c$; $Q \supseteq c \circ b$.

In Section 4 we took intervals as the join of points. The join spaces defined there correspond with join operations satisfying the axioms (J'-1), (J'-2), (J-3) and (J-4). Other combinations make sense as well. Here are a few consequences of some axiom subsets.

7.5. Proposition

(1) If (J-3) holds, then $A \circ (B \circ C) = (A \circ B) \circ C$.
(2) If (J-1), (J-2), (J-3), (J-4) hold, then $A \circ B = B \circ A$.
(3) Arithmetic Laws:
   (i)   If (J-3) holds, then $(A/B)/C = A/(B \circ C)$ and $A/(B/C) \subseteq (A \circ C)/B$.
   (ii) If (J-4) holds, then $A \circ (B/C) \subseteq (A \circ B)/C$.
   (iii) If (J-1) to (J-5) hold, then $(x \circ A)/(x \circ B) = (A/B) \cup (x \circ A)/B \cup A/(x \circ B)$.

Proof of (1): obvious.

Proof of (2). We just show that under the quoted axioms, commutativity (J'-2) holds. To see that $a \circ b \subseteq b \circ a$, let $z \in a \circ b$. By (J-1) there is a point $x \in z \circ b$. Then

$$b \in x/z \cap z/a,$$

and by (J-4), $x \circ a$ meets $z \circ z$. By (J-2a), $z \in x \circ a$. By definition and by (J-3),

$$z \in (z \circ b) \circ a = z \circ (b \circ a),$$

yielding a point $t \in b \circ a$ with $z \in z \circ t$, or: $t \in z/z$. Then $z = t$ by (J-2b), and indeed $z \in b \circ a$. The other inclusion is derived similarly.

Proof of (3). Except for the last formula, all proofs are straightforward. First, let $z \in (x \circ A)/(x \circ B)$. Then by commutativity and associativity, $x \circ (z \circ B) \cap (x \circ A) \neq \emptyset$, and (J-5) leads to one of the following conclusions. Either $z \circ B \cap A \neq \emptyset$ and hence $z \in A/B$. Or, $z \circ B \cap x \circ A \neq \emptyset$, whence $z \in (x \circ A)/B$. Or, $x \circ (z \circ B) \cap A \neq \emptyset$. Then, by commutativity and associativity, $z \in A/(x \circ B)$.

Conversely, let $z \in (A/B) \cup (x \circ A)/B \cup A/(x \circ B)$. If $z \in A/B$ then $z \circ B \cap A \neq \emptyset$. Hence $x \circ z \circ B \cap x \circ A \neq \emptyset$, yielding $z \in (x \circ A)/(x \circ B)$. If z is in the second or third summand, we proceed similarly, joining with x and applying idempotence of the join, (J-2a). ∎

§7: Bryant-Webster Spaces

Axioms (J-2) and (J'-2) imply that a or b is in $a \circ b$ iff $a = b$. Thus, assuming the axioms (J-1) to (J-4) (which yield (J'-2)), one should think of $a \circ b$ as an "open" interval between a and b; the operator $\circ$ is an *open join*. This suggests the following definition of an interval for open joins:

$$I(a,b) = \{a,b\} \cup a \circ b.$$

This interval operator on X in turn induces a convexity on X. In direct terms: $C \subseteq X$ is convex iff $a \circ b \subseteq C$ whenever $a, b \in C$. Briefly: $C \circ C \subseteq C$.

There is also a theory of *closed joins:* the most common axiom system is the one of Pasch-Peano spaces. See Section 4.

7.6. Join spaces. An *(open) join space* is a pair $(X, \circ)$, consisting of a set X and an operator

$$\circ : X \times X \to 2^X, \quad (a,b) \mapsto a \circ b,$$

satisfying the axioms (J-1) to (J-5). A point $p \in X$ such that $p/x = \emptyset$ for some $x \in X$ is called a *boundary point of X*. Note that X satisfies the axiom (J-6) exactly if it has an empty boundary. The axioms (J-6) and (J-7) will be considered later. See also 7.16 and 7.17.

Convex structures which arise from join spaces are characterized by certain properties of their segment operator, as the next two results will show.

7.7. Theorem. *Let $(X, \circ)$ be a join space. Then the corresponding convex structure is dense, straight, JHC, and its segments are decomposable. In particular, X is S_4. The convex hull of $a_1,..,a_n \in X$ is the union of all sets of type*

$$a_{i_1} \circ .. \circ a_{i_k}, \quad 1 \leq k \leq n.$$

Proof. We first verify the hull formula. For $a_1,..,a_n \in X$ let C be the union of sets as described above. Taking $k = 1$, we find that each a_i is in C. By commutativity and associativity, 7.5.1,(1) and (2), together with idempotence (J-2a), it follows that $C \circ C \subseteq C$. This shows that C is convex and hence that

$$co\{a_1,..,a_n\} \subseteq C.$$

On the other hand, $co\{a_1,..,a_n\}$ must include all sets of type $a_{i_1} \circ .. \circ a_{i_k}$ by construction of the convexity. This establishes the hull formula.

Join-hull commutativity is a direct consequence of the hull formula. The fact that X is dense follows directly from (J-1). We next verify that each segment ab of X ($a \neq b$) is decomposable. Let $c \in ab \setminus \{a,b\}$. The equality $ac \cap cb = \{c\}$ follows from S_2 (cf. Topic 3.33.2). To show that $ab \subseteq ac \cup cb$, consider x in ab. Without loss of generality, $x \in a \circ b$, and hence $a \in x/b$. On the other hand, $c \in a \circ b$, whence $a \in c/b$. By (J-4) we find that

$$(c \circ b) \cap (x \circ b) \neq \emptyset.$$

Application of (J-5) yields that either $c = x$, or $x \in c \circ b$, or $c \in x \circ b$. In the first and second case we are done. In the third case we have $b \in c/x$, whereas $b \in x/a$ since $x \in a \circ b$. By (J-4) we find that

$(c \circ a) \cap (x \circ x) \neq \emptyset$,

and (J-2) yields $x \in c \circ a$.

We finally verify the condition of straightness. For convenience, let us use the following terminology. A finite sequence of distinct points in a convex structure is a *convex succession* provided each point, except the first and the last, is in the segment between its left and right neighbor. We first verify that all points of a convex succession are in the convex hull of the first and last element. It suffices to consider a convex succession of type $(a_i)_{i=1}^4$. We have $a_2 \in a_1 \circ a_3$ and $a_3 \in a_2 \circ a_4$. Then

$$a_2 \in a_1 \circ (a_2 \circ a_4) = (a_1 \circ a_4) \circ a_2$$

by commutativity and associativity of join, yielding a point $t \in a_1 \circ a_4$ with $a_2 \in t \circ a_2$. Hence $a_2 = t$ and $a_2 \in a_1 \circ a_4$. One similarly shows that $a_3 \in a_1 \circ a_4$.

Next we handle the general straightness configuration. Suppose $x \neq y$ are in $ab \cap cd$. If one of a, b equals one of c, d then the result follows from (J-5). Appealing to the decomposability of segments, we may assume that $x \in ay \cap cy$ and $y \in xb \cap xd$ (cf.

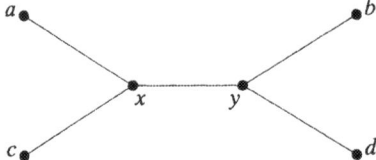

Fig. 2: Straightness in a join space

Fig. 2). If $a = x \neq c$ then we have a convex succession consisting of c, x, y, followed by b (provided $b \neq y$) or by d (provided $d \neq y$). If $d = y$ then $b \neq y$ and the union of the given segments is cb. If, instead, $b = y$, then $d \neq y$ and the union of the given segments is cd. If $b \neq y \neq$, then (J-5) settles the problem. Each of the other incidences $c = x$; $b = y$; $d = y$ can be handled similarly. If there are no such incidences, then (J-5) yields that either $a \in c \circ y$ (hence $a \in c \circ x$) or $c \in a \circ y$ (hence $c \in a \circ x$). Similarly, we have either $b \in y \circ d$ or $d \in y \circ b$. This yields a convex succession of six points, the first and last point of which determine the desired segment. ∎

Note that (J-4) leads directly to the Pasch Property, which is equivalent to S_4 by Theorem 4.12.

7.8. Theorem. *Let X be a dense, straight, and JHC convex structure with decomposable segments. Then the convexity of X is induced by a join operator $\circ$ on X, defined as follows.*

$$a \circ a = \{a\}; \quad a \circ b = ab \setminus \{a, b\} \quad (a \neq b).$$

Proof. Axiom (J-1) is a reformulation of density, and axiom (J-2) holds by construction of the join. The elaborate part of the proof is to verify (J-3): $a \circ (b \circ c) = (a \circ b) \circ c$. The various degenerate cases

§7: Bryant-Webster Spaces   149

$$a = b; a = c; b = c; b \in a \circ c; c \in a \circ b; a \in b \circ c$$

can be handled merely on the basis that X is dense and decomposable; this part is left to the reader. We henceforth assume that no such incidences occur. As the operation $\circ$ is commutative by construction, we will only show that $a \circ (b \circ c) \subseteq (a \circ b) \circ c$. Let x be in the left-hand set and let $y \in b \circ c$ be such that $x \in a \circ y$. As $x \in co\{a,b,c\}$, JHC yields a point $z \in ab$ such that $x \in zc$. Note that $x \neq a$ and hence that $y \neq x$.

By Proposition 7.2, X has the Ramification Property. If $x = z$, then this property implies $z = a$. But then $x \in a \circ y$ and (J-2) yield $a = y \in b \circ c$, which has been excluded. Therefore, $x \in z \circ c$.

If $z = a$ then the Ramification Property yields that either $c \in ay$, or $y \in ac$. In the first case we obtain a convex succession a, c, y, b, whence $c \in a \circ b$. In the second case, another application of the Ramification Property yields $c \in a \circ b$ or $b \in a \circ c$. If $z = b$, then by the decomposability of bc, the points x and y are comparable with respect to the orders $\leq_b$ and $\leq_c$. The Ramification Property then yields $x = y$. Since either $z = a$ or $z = b$ leads to an excluded incidence, we conclude that $z \in a \circ b$.

To see that $x \in a \circ z$, we examine two possibilities. If $x = c$ then the straightness property applied to the 4-tuple a, c, y, b implies that $c \in ab$, an excluded incidence. In case $x = z$, use a similar argument.

The Pasch axiom (J-4) now follows by virtue of Proposition 7.3 and (J-5) follows from the Ramification Property.   ∎

A *Bryant-Webster space* (briefly, a *BW space*) is a JHC, dense and straight convex structure with decomposable segments. It is clear that the transitions between join spaces and BW spaces, as described in Theorems 7.7 and 7.8 are mutually inverse. In regard to this intimate connection, we shall largely keep to the "convexity" viewpoint of Bryant-Webster spaces. Note that a convex structure is a BW space iff each polytope spanned by at most four points is.

7.9. Examples

7.9.1. Standard convexity. Let V be a vector space over a totally ordered field, and let C be a convex subset. Then C is a BW space with

$$a \circ b = \{t \cdot a + (1-t) \cdot b \mid 0 < t < 1\}$$

for $a, b \in V$.

7.9.2. Half-sphere. Consider the unit n-sphere in Euclidean $(n+1)$-space,

$$S^n = \{x \in \mathbb{R}^{n+1} \mid \|x\| = 1\},$$

and let $H \subseteq \mathbb{R}^{n+1}$ be a half-space, maximal with the property that $\mathbf{0} \notin H$ (i.e., H is a copoint at $\mathbf{0}$). Let $X_n = H \cap S^n$. A set $C \subseteq X^n$ is taken convex provided it is the trace of a wedge at $\mathbf{0}$. In fact, X_n is a subspace of the convex surface S^n (cf. Example 1.19.2).

Let $P \subseteq X_n$ be a polytope. If "*co*" denotes the standard hull operator in $\mathbb{R}^{n+1}$, then $co(P) \cap S^n = P$ by the definition of our convexity. As no antipodal pair occurs in X_n, we find that $0 \notin co(P)$. By virtue of the hull formula in 1.5.1 the set $co(P)$ is compact. Hence it is possible to find a hyperplane H missing both 0 and P and separating between them (this standard result is reproved in Chapter III). Let $S \subseteq S^n$ be the (minor) open hemisphere which is cut off by H and which includes K. Stereographic projection from the origin to H yields an isomorphism of the relative convexity of S and the standard convexity of an open disk in $\mathbb{R}^n$. As each polytope of X_n is isomorphic with a polytope of $\mathbb{R}^n$, we conclude that X_n is a BW space.

Sets of type X_n can be constructed recursively as follows.

$$X_0 = \{1\};$$
$$X_n = \{x \in S^n \mid x_{n+1} > 0\} \cup (X_{n-1} \times \{0\}).$$

The boundary of X_n is X_{n-1} ($n \geq 1$). The set X_n corresponds with real projective n-space, provided with an artificial cut in each line (circle) to make it ordered.

7.9.3. Moulton plane. In $\mathbb{R}^2$ we consider all "lines" given by one of the equations below.

$$x = c \qquad (c \in \mathbb{R} \text{ constant});$$
$$y = bx + c \qquad (b \geq 0, c \text{ constant});$$
$$y = \begin{cases} bx + c, & \text{if } x \geq 0 \\ \tfrac{1}{2}bx + c, & \text{if } x < 0 \end{cases} \quad (b < 0, c \text{ constant}).$$

The "lines" of negative slope are bent like refracted light rays at the y-axis. All other lines are of a standard type. This so-called *Moulton plane* is turned into a BW space as follows. First, note that every two distinct points a, b are joined by exactly one of the above

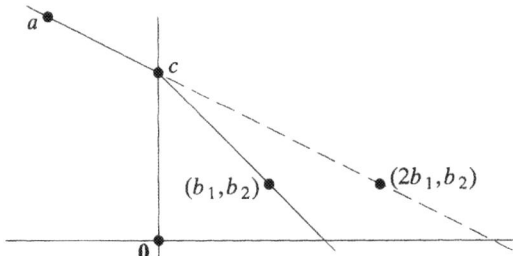

Fig. 3: A line in the Moulton plane

listed lines. We only verify this in case $a = (a_1, a_2)$ and $b = (b_1, b_2)$ are such that

$$a_1 < 0; b_1 > 0; b_2 \leq a_2.$$

Let L be the Euclidean line joining a with $(2b_1, b_2)$, and let c be its intersection with the y-axis. See Fig. 3. The desired line is the union of two standard half-lines: the intersection of L with the half-plane $x \leq 0$, and the half-line at c which passes through b.

Each line of the Moulton plane carries a natural total order, and we define the join of two points $a \neq b$ to be the part of the unique line through a, b which is strictly in

§7: Bryant-Webster Spaces

between these points. The verification that this is indeed a BW space is left to the reader (cf. Topic 7.20 for an approach via so-called line spaces).

7.9.4. Totally ordered sets. A densely totally ordered set is a BW space. Its boundary consists of the end points, if any.

7.10. Affine subsets. A subset A of a join space X is called an **affine set** (**flat**, **linear variety**) provided $A \circ A \subseteq A$ and $A/A \subseteq A$. Note that all singletons are flats by (J-2). As a direct consequence of the arithmetic formulae (i), (ii) in 7.5.1(3), if $C \subseteq X$ is convex then C/C is an affine set. It is easy to see that if X satisfies the axiom (J-6) of extensibility, then A is a flat iff $A/A \subseteq A$.

The collection of all affine sets in a BW space X is a convexity, coarser than the original one. Thus it makes sense to consider the **affine hull** of a set $A \subseteq X$, which we denote with $\mathit{aff}(A)$. By an earlier remark, it follows that $\mathit{aff}(A) = co(A)/co(A)$.

7.11. Proposition. *The following are valid in a BW space X.*

(1) (*Affine Hull Formula*) *The set $\mathit{aff}\{x_1,..,x_n\}$ is the union of all sets of type*

 (i) $x_{i_1} \circ .. \circ x_{i_k} / x_{i_{k+1}} \circ .. \circ x_{i_l}$, *where $i_1,..,i_k,..,i_l \in \{1,..,n\}$ are distinct.*

 (ii) $x_{i_1} \circ .. \circ x_{i_m}$, *where $i_1,..,i_m \in \{1,..,n\}$.*

(2) *If $A \subseteq X$ is a non-empty affine set and if $x \in X$, then*

$$\mathit{aff}(\{x\} \cup A) = A \cup (x \circ A)/A \cup (A/x).$$

(3) *If A, B are non-disjoint affine sets in a BW space without boundary, then*

$$\mathit{aff}(A \cup B) = A/B.$$

Proof of (1). Use the polytope formula in 7.7, together with the last of the Arithmetic Laws in 7.5.1(3).

Proof of (2). We have $\mathit{aff}(\{x\} \cup A) = (x \circ A)/(x \circ A)$ since the latter is affine and includes both x and A. By the third Arithmetic Law in 7.5.1(3) we find that

$$\mathit{aff}(\{x\} \cup A) = A/A \cup (x \circ A)/A \cup A/(x \circ A).$$

Note that $A/A = A$ since A is affine. Then $z \in A/(x \circ A)$ iff $z \circ x \circ A \cap A \neq \emptyset$, iff $z \circ x$ meets $A/A = A$, iff $z \in A/x$, showing that $A/(x \circ A) = A/x$.

Proof of (3). By definition, A/B is included in $\mathit{aff}(A \cup B)$. The fact that $A, B \subseteq A/B$ follows easily from (J-6) by considering any point of $A \cap B$. It remains to be verified that A/B is an affine set. First, this set is convex since

$$(A/B) \circ (A/B) \subseteq ((A/B) \circ A)/B$$
$$\subseteq ((A \circ A)/B)/B$$
$$= (A/B)/B$$
$$= A/(B \circ B)$$

$$= A/B,$$

by the Arithmetic Laws 7.5.1(3) and by the fact that A, B are convex. We use the above fact, together with the Arithmetic Laws, to see that

$$(A/B)/(A/B) \subseteq ((A/B) \circ B)/A$$
$$\subseteq ((A/B) \circ (A/B))/A$$
$$= (A/B)/A$$
$$= A/(A \circ B)$$
$$= (A/A) \circ B$$
$$= A/B.$$ ∎

The following result is in part an extension of Proposition 2.17.2(2).

7.12. Theorem. *The family of all affine sets in a BW space is a matroid. If the space has no boundary points, then this matroid is weakly JHC.*

Proof. Let X be a BW space, let $A \subseteq X$ be an affine set, and let p, $q \notin $ aff (A) such that $p \in$ aff $(\{q\} \cup A)$, We show that $q \in$ aff $(\{p\} \cup A)$. By 7.11(2),

$$\textit{aff } (\{q\} \cup A) = A \cup (q \circ A)/A \cup A/q.$$

As p is not in the third summand, we are left with two possibilities.
(i) $p \in (q \circ A)/A \Leftrightarrow p \circ A \cap q \circ A \neq \emptyset \Leftrightarrow q \in (p \circ A)/A$.
(ii) $p \in A/q \Leftrightarrow p \circ q \cap A \neq \emptyset \Leftrightarrow q \in A/p$.

Both $(p \circ A)/A$ and A/q are included in *aff* $(\{p\} \cup A)$.

As for the final part of the result, consider two affine sets A, B with a non-empty intersection. By Proposition 7.11(3) we find that

$$\textit{aff } (A \cup B) = A/B \subseteq \bigcup \{ \textit{aff}\{a, b\} \mid a \in A; b \in B \},$$

from which weak join-hull commutativity follows directly. ∎

7.13. Dimension of BW spaces. Let X be a BW space, and let $\mathcal{A}$ be the matroid convexity, consisting of all affine subsets of X. By Theorem 2.10 and the Basis law, 2.11, all bases of the **affine matroid** $(X, \mathcal{A})$ have the same cardinality: the rank of the underlying independence structure. A basis of $(X, \mathcal{A})$ is also called an **affine basis of** X. The **affine dimension** of a Bryant-Webster space X is defined to be one less than the rank of the matroid $(X, \mathcal{A})$. Explicitly: X is of affine dimension n (where $n < \infty$) provided there is a collection of $n + 1$ affinely independent points, the affine hull of which is X. Note that the empty space has dimension -1.

As a consequence of Theorem 7.12 and Proposition 2.18, we have the following result.

7.14. Proposition (*Affine Dimension Formula*). *If X is a BW space without boundary and if A, B are non-disjoint affine sets, then*

§7: Bryant-Webster Spaces

$$dim\ aff\ (A \cup B) + dim\ (A \cap B) = dim\ (A) + dim\ (B).$$ ∎

The formula is not valid for disjoint flats, as two parallel lines in Euclidean plane may illustrate. The result also fails to hold for BW spaces with boundary points. For instance, consider the convex subspace

$$X = (B \times \mathbb{R}) \cup \{(p, 0)\}$$

of $\mathbb{R}^3$, where B is the open unit disk of $\mathbb{R}^2$ and p is a point in its planar boundary. Consider two lines in $B \times \{0\}$ meeting in p. Multiplying with $\mathbb{R}$ yields two planes of X meeting in p only. The affine hull of their union equals X.

Further Topics

7.15. Independence of certain properties

7.15.1. Show that the segment operator of a matroid is straight and has the Ramification Property. No non-trivial segment of a matroid is decomposable.

7.15.2. Show that a JHC convex structure is a convex geometry iff all points of a convex succession are in the convex hull of the first and last element. Design a poset (hence a JHC convex geometry) which is not straight.

7.15.3. Consider the subset X of the complex plane, consisting of all points $e^{2\pi i \alpha}$ with $\alpha \in [0, 1/6) \cup [2/6, 3/6) \cup [4/6, 5/6)$. The set is equipped with the geodesic convexity, induced by the angular distance. Note that there are no diametric pairs in X. Show that X is dense and complete, that its intervals are decomposable, that X has the Ramification Property and that X is JHC (hence S_4). However, X is not straight.

7.15.4. Let $\mathbb{P}^2$ be the two-dimensional real projective space. Then $\mathbb{P}^2$ is JHC, dense, straight, and it has the Ramification Property, but it is not decomposable.
Problem. Find such an example which, in addition, is S_4.

7.15.5. Verify that a tree has the Ramification Property iff it has no ramification points. Therefore, a tree which looks like a character "T" is dense, decomposable, JHC and S_4, but it does not have the Ramification Property. Note that this space is a CP + CC image of a BW space.

7.15.6. The following examples satisfy all but one hypothesis (as indicated between brackets) of Proposition 7.3, yet they all fail to be S_4. (i) (density) the circumference of a triangle in the standard plane. (ii) (JHC) the plane with one point removed. (iii) (decomposability) the projective plane. No example is known to us showing that the conclusion of 7.3 is false without the Ramification Property.

7.16. Join spaces without boundary. Let X be a BW space of dimension > 1 satisfying the axiom (J-6) of extensibility. Show that for any pair of non-trivial segments $co\{a,b\}$, $co\{a',b'\}$ there is a CP isomorphism $f: co\{a,b\} \to co\{a',b'\}$ with $f(a) = a'$ and $f(b) = b'$. What happens if dimension equals one?

7.17. Completeness of join spaces

7.17.1. Show that a join space is complete (cf. axiom (J-7)) iff it the corresponding BW space is a complete interval space. Deduce that a BW space without boundary and of dimension > 1 is complete provided it has *one* non-trivial segment which is complete with respect to the order of an end point.

7.17.2. (compare Bryant and Webster [1973]) Show that a BW space is complete iff for each convex set A and for each $a \in A$, $b \notin A$, there exists a point $c \in co(a,b)$ such that $a \circ c \subseteq A$ and $b \circ c \cap A = \emptyset$.

7.18. Affine and projective dimension. Let V be a vector space over a field $\mathbb{K}$ and consider the projective space $\mathbb{P} = \mathbb{P}(V \times \mathbb{K})$.

7.18.1. Show that the affine matroid V embeds in $\mathbb{P}$ by the mapping

$V \to \mathbb{P}, \quad v \mapsto$ equivalence class of $(v, 1)$,

7.18.2. Show that, under this embedding, $dim\, A = dim\, pr(A)$ for each affine set $A \subseteq V$.

7.18.3. Conclude that if $A_1,..,A_n$ are finite-dimensional affine subspaces of V having a point in common, then $pr(\cap_{i=1}^n A_i) = \cap_{i=1}^n pr(A_i)$.

7.19. Base-point orders and lines

7.19.1. (compare van de Vel [1982]) Let X be a JHC convex structure which is S_3 and has the Ramification Property. Show that if $C \subseteq X$ is a maximal chain in $\leq_b$, then $C \setminus \{b\}$ is a convex set. In general, C is not convex; consider the geodesic convexity of a 6–cycle.

7.19.2. (cf. Levi [1951]) Show that in an interval space with decomposable intervals, each base-point quasi-order is a genuine partial order.

7.19.3. (Kay [1977]) The *line* spanning two points $a \neq b$ in an interval space X is the set $L(a,b)$ of all $x \in X$ such that one of a, b, x is in the interval between the other two. Let X be a straight interval space with decomposable intervals. Show that that two lines meet in at most one point and (consequently) that two distinct points are on exactly one line. Note that X has the Ramification Property and that each line is a convex set, the relative convexity of which is induced by a total order.

§7: Bryant-Webster Spaces

7.20. Line spaces (Doignon [1976]). Let E be a set with a family of subsets (to be regarded as "lines"). Each line L has a total order $<_L$. We say that a is between b and c provided there is a line L containing the three points, such that $b <_L a <_L c$ or $c <_L a <_L b$ in the total order of L. We assume the following axioms.

(L-1) Two distinct points belong to a unique line.
(L-2) Each line has at least two points.
(L-3) The total order of each line is dense.
(L-4) If $a, b, c, d, e \in E$ are such that d is between b, c and e is between a, d, then the line joining c, e meets the closed segment $[a,b]$ ("Pasch-Peano axiom").

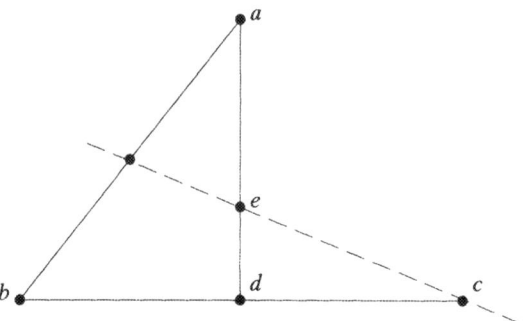

Fig. 4: Pasch-Peano axiom

Structures of this kind are called *line spaces*. As the direction of the order on each line is irrelevant, one usually considers a description of order in terms of "betweenness". After writing out (L-3) this way, the above list becomes part of the axiom system for *ordered geometry* (see Coxeter [1961]). The line spaces of Cantwell [1974] constitute a subclass of line spaces in which lines are order isomorphic to $\mathbb{R}$. See IV§1.2 and IV§1.23.

7.20.1. Show that in (L-4) the intersection with the open segment is non-empty in case a, b, c are not collinear.

7.20.2. Derive the "Pasch axiom": if p, a, a', b, b' are such that a' is in between p, a and b' is in between p, b, then the segments $[a',b]$ and $[a,b']$ meet.

7.20.3. (compare Precup [1981]) Design a natural 1-1 correspondence between line spaces and BW spaces.

7.20.4. Show that the Moulton plane is a BW space.

7.21. Sylvester's problem of collinear points (Coxeter [1961]). Let X be a join space and let $p_1,...,p_n \in X$ be a non-collinear set of points, that is: $aff\{p_1,...,p_n\}$ has affine dimension > 1. Show that there is a line (a one-dimensional affine subspace) in X passing through precisely *two* of the given points p_i.

Hint. Without loss of generality, X is two-dimensional. Some set of three p_i's is affinely independent. Assume p_1 is one of them. Construct a line L containing only p_1, such that L meets at least one of the lines $\mathit{aff}\{p_i,p_j\}$ in a point distinct from p_1. Such intersection points divide L into a number of segments. Let $p_1 \circ a$ be one of them, where $a \in M \cap L$ for $M = \mathit{aff}\{p_2,p_3\}$. If M is not the desired line (say: $p_4 \in M$), then renumber p_2, p_3, p_4 if necessary and assume that $p_2 \in p_3 \circ a$. The desired line is $N = \mathit{aff}\{p_3,p_1\}$.

7.22. Gauss' construction of parallel rays (Coxeter [1961]). Let X be a join space and let $a \neq b \in X$. Recall that the set a/b is a ray at a; we now call this an *open ray*. If a is added to it, then the resulting set is called a *closed ray*. An *angle* at a consists of two open rays R_1, R_2 at a. Such an angle is cut by another ray R at a provided R meets the join of some (any) point of R_1 with some (any) point of R_2 (this won't work for a stretched angle, where the two rays cover a whole line). Let L be a line in a join space X and let $a \in X \setminus L$. A ray R_0 at a is *parallel* to L provided (Fig. 5)

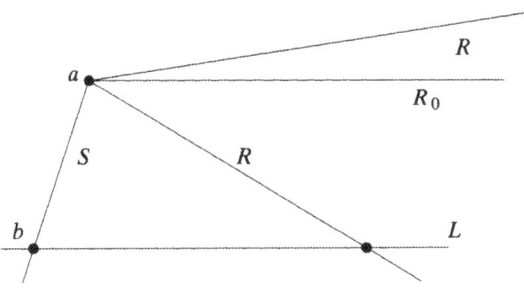

Fig. 5: Parallel ray R_0

(i) R_0 is disjoint from L.
(ii)   There is a point $b \in L$ (yielding a ray S at a and an angle α at a formed with R_0) such that each ray at a cutting α meets L, and each ray at a forming an angle with S cut by R_0 does not meet L.

7.22.1. Show that if X is complete, without boundary, and two-dimensional, then at each point a not on a line L there is a ray parallel to L. Note that the opposite ray is, in general, not parallel to L. If it happens to be parallel as well, then we talk about *parallel lines* of the join space. There can be at most one line parallel to a given line and passing through a given point.

7.22.2. Deduce that if X is complete, without boundary, and of dimension > 1, then all open segments $a \circ b$ ($a \neq b$), all open rays, and all lines of X are CP isomorphic (compare 7.16). More information will be given in Section IV§1.

§7: Bryant-Webster Spaces

7.23. Distance geometry (compare Blumenthal [1953]; continued IV§1.25).

7.23.1. Recall that a metric ρ is *convex* provided for each pair of points a, b there is a point $c \neq a, b$ with

$$\rho(a,b) = \rho(a,c) + \rho(c,b).$$

Show that a metric is convex iff the the corresponding geodesic interval operator is dense.

7.23.2. A metric space is said to have the *Two-triples Property* provided for each set of four distinct points in X, if two (out of four) triples are convexly dependent, then so are the remaining ones. Verify that in a metric space with this property, all metric intervals are convex, and that the Two-triples Property can be stated equivalently by replacing "segment operator" by "geodesic interval operator". Show next that for a metric space (X, ρ) the following are equivalent.

(i) (X, ρ) has the Two-triples Property.
(ii) (X, ρ) is decomposable and straight.

7.23.3. A metric is called *externally convex* provided for each two distinct points a, b there is a point c different from b such that b is metrically between a, c. Note that this corresponds with the extensibility property of join structures. By a *metric line* is meant a convex subspace which is isometric with the real line. Show that if a metric space has the Two-triples Property relative to a convex, externally convex, and complete metric, then each pair of distinct points is joined by a unique metric line.

7.23.4. A metric space has the *weak Euclidean Four-point Property* provided each quadruple of distinct points with a dependent triple can be isometrically embedded in $\mathbb{R}^2$. Note that this implies the two-triples property. Show that if a metric space X has the weak Euclidean Four-point Property relative to a complete, convex, and externally convex metric, then X is a Bryant-Webster space.

Notes on Section 7

It is natural to investigate conditions on an interval operator which highlight certain features of the standard operator. The Ramification Property was first considered by Calder [1971] and Bean [1974]. The concepts of straightness and of decomposability are taken from Kay [1977]. In fact, decomposable intervals occurs implicitly in Levi [1951].

Join operators form a quite extensive branch of abstract convexity, going back to Prenowitz's papers [1946] and [1961]). These spaces link up with the theory of "ordered" geometry as developed by Pasch, Peano, Veblen, and others (Coxeter [1961]). In ordered geometry, the fundamental notion is a ternary relation of betweenness. Prenowitz and Jantosciak [1972] consider the join operation as the atomic concept. Their axioms (almost) correspond with the ones for Pasch-Peano spaces (cf. §4) and are more general than the ones of Bryant and Webster [1969] resp., [1972], which correspond with (J-1) to

(J-5), resp., to (J-6). Extensible BW spaces correspond with Kay's [1977] "regular convex structures". There are still other combinations of axioms that received attention in the literature; see Degreef [1981] for a survey. The dependence of commutativity (J-2′) on some of the other axioms (cf. 7.5.1(2)) was discovered by Bryant and Webster [1972]. For results concerning the independence of join axioms, see Bryant [1974] or Gottwald [1980]. Completeness of join structures was introduced by Bryant and Webster [1973]. Their definition is equivalent to axiom (J-7); cf. 7.17.

Theorems 7.7 and 7.8, characterizing join spaces in terms of convex structures, are based on the work of Calder [1971] and R. Hammer [1977]. Comparable work for line spaces (cf. 7.20) has been done by Whitfield and Yong [1981] and by Kay [1977b]. The relationship of join spaces with Cantwell's line spaces [1974] (see 7.20) was studied by Doignon [1976]. The Moulton plane (cf. 7.9.3) is a classical example of a non-Desarguesian affine plane (cf. Stevenson [1972] and IV§1).

The hull formula 7.11 for intersecting affine sets, the matroid property of affine sets, Theorem 7.12, and the resulting definition of dimension for join structures (cf. 7.13) are taken from Bryant and Webster [1977]. The Affine Dimension Formula in 7.14 is also considered by Bryant and Webster [1977] and, in case of line spaces, by Cantwell and Kay [1978]. A similar result for a class of so-called exchange spaces (spaces defined by a certain type of join operator, with flats satisfying the Exchange Law) was given earlier by Prenowitz and Jantosciak [1972].

CHAPTER **II**

CONVEX INVARIANTS

1. Classical Convex Invariants

Dependence, classical invariants: Helly-, Carathéodory-, Radon- and exchange number, Theorems of Helly, Carathéodory and Radon in $\mathbb{R}^n$ (more generally, in Bryant-Webster spaces), inequalities of Levi, Sierksma and Eckhoff-Jamison.

2. Invariants and Product Spaces

Helly number of products, Exchange function, Carathéodory number of products, Radon function, Radon number of products, the Radon function of $\mathbb{R}^n$, Radon number in products of totally ordered sets.

3. Invariants in other Constructions

Helly number in spaces of arcs, classical invariants of disjoint sums and gated amalgams, cubical polyhedra and geometric realization.

4. Infinite Combinatorics

Cantor cube, Compactness Theorem and Compact Intersection Theorem, Helly number of H-convexity, Countable Intersection Property, rank, generating degree and directional degree.

5. Tverberg Numbers

Tverberg (partition) numbers, full independence, Tverberg Theorem in $\mathbb{R}^n$, directional degree at a point and the Eckhoff conjecture.

1. Classical Convex Invariants

The numbers of Helly, Carathéodory and Radon are a central theme in abstract convexity. Each of them is defined as a degree of independence, tolerated by a convex structure, and is invariant under isomorphism. Comparing with affine independence leads to the classic theorems on Euclidean space, bearing the names of the above quoted mathematicians. In fact, these results are valid for all Bryant-Webster (BW) spaces.

In general convex structures the mutual relationship between these invariants is more complicated. The inequalities of Levi, Sierksma, and Eckhoff-Jamison describe the relative range of the invariants. The second one involves the so-called exchange number, which is defined as a modification of the Carathéodory number.

1.1. Types of dependence. Let X be a convex structure, and let $F \subseteq X$ be a non-empty finite set. Throughout, #F will denote the number of elements in F. The following notions will be studied in depth throughout this chapter.

The set F is *Helly dependent* (or, *H-dependent*) provided

$$\bigcap_{a \in F} co(F \setminus \{a\}) \neq \varnothing,$$

and it is *Helly (H-) independent* otherwise.

The set F is *Carathéodory dependent* (or, *C-dependent*) provided

$$co(F) \subseteq \bigcup_{a \in F} co(F \setminus \{a\}),$$

and it is *Carathéodory (C-) independent* otherwise.

The set F is *Radon dependent* (or, *R-dependent*) if there is a partition $\{F_1, F_2\}$ of F such that

$$co(F_1) \cap co(F_2) \neq \varnothing.$$

In these circumstances, $\{F_1, F_2\}$ is called a *Radon partition* of F. If no such partitions exist, then F is *Radon (R-) independent*.

The set F is called *exchange dependent* (or, *E-dependent*), provided for each $p \in F$,

$$co(F \setminus \{p\}) \subseteq \bigcup \{ co(F \setminus \{a\}) \mid a \in F; a \neq p \},$$

and it is called *exchange (E-) independent* otherwise.

Note that singletons are dependent in any of the four senses described above. A property (or, a class) $\mathcal{P}$ of sets is called *hereditary* provided each non-empty subset of a set with property $\mathcal{P}$ also has this property. It can be deduced (cf. 1.16) that H- and R-independence are hereditary, whereas C- and E-independence are not, unless the convex structure satisfies the Cone-union Property (CUP). For a comparison of the above introduced notions of independence with convex independence (cf. I§2.5) see 4.12. We first derive a few results of a geometric flavor, involving some concepts of Section I§7.

1.2. Proposition. *Let X be a join-hull commutative (JHC) space and let $F \subseteq X$ be a finite set.*

(1) *If X has the Ramification Property, and if F is R-independent, then for each pair of subsets $F_1, F_2 \subseteq F$,*
$$co(F_1) \cap co(F_2) = co(F_1 \cap F_2).$$

(2) **(Subdivision Property)** *If X has decomposable segments, and F has at least two points, then for all $x \in co(F)$,*
$$co(F) = \bigcup_{a \in F} co(\{x\} \cup F \setminus \{a\}).$$

Proof of (1). If $F_1 \cap F_2 = \emptyset$ then the result follows from R-independence. We proceed by induction on $\#(F_1 \cap F_2)$. Let $a \in F_1 \cap F_2$ be fixed and consider the polytopes
$$P_i = co(F_i \setminus \{a\}) \quad (i = 1, 2).$$
Note that if (for instance) $P_1 = \emptyset$, then $F_1 = \{a\} \subseteq F_2$, which gives the desired result. So assume $P_i \neq \emptyset$ for $i = 1, 2$. As X is JHC we have
$$co(F_i) = \bigcup \{ax \mid x \in P_i\}.$$
This leads to the equality of sets
$$co(F_1) \cap co(F_2) = \bigcup \{ax_1 \cap ax_2 \mid x_1 \in P_1; x_2 \in P_2\}.$$
Let $x \in co(F_1) \cap co(F_2)$. There are points $x_i \in P_i$ for $i = 1, 2$ such that $x \in ax_1 \cap ax_2$. If $x_1 \in ax_2$ then by the inductive hypothesis,
$$x_1 \in co(F_1 \setminus \{a\}) \cap co(F_2) = co(F_1 \cap F_2 \setminus \{a\}),$$
which yields
$$x \in ax_1 \subseteq co(F_1 \cap F_2).$$
A similar argument works if $x_2 \in ax_1$. In the remaining case, the Ramification Property implies that $x = a$, yielding $x \in co(F_1 \cap F_2)$ once again. This establishes one half of the desired equality; the other half is trivial.

Proof of (2). Let $\#F = n$. If $n = 2$, then the required result corresponds with the fact that segments are decomposable. If $n > 2$, we proceed by induction. Let $x \in co(F)$ and fix $a_0 \in F$. As X is JHC we find a point x' in $co(F \setminus \{a_0\})$ with $x \in a_0 x'$. Take $y \in co(F)$ and let $y' \in co(F \setminus \{a_0\})$ be such that $y \in a_0 y'$. By the induction hypothesis there is a point $a_1 \in F \setminus \{a_0\}$ with
$$y' \in co(\{x'\} \cup F \setminus \{a_1, a_0\}).$$
See Fig. 1. Hence,
$$y \in a_0 y' \subseteq co(a_0 x' \cup co(F \setminus \{a_1, a_0\})).$$
Now $a_0 x' = a_0 x \cup xx'$. By JHC, the right hand set above is included in the union of the

§1: Classical Convex Invariants 163

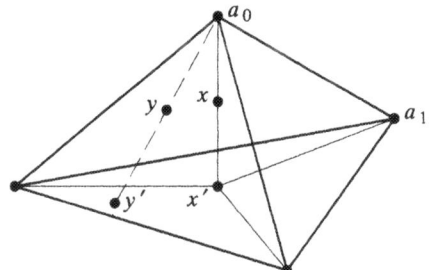

Fig. 1: Subdivision Property

two sets described below:

$$co(\{a_0,x\} \cup F \setminus \{a_1,a_0\}) = co(\{x\} \cup F \setminus \{a_1\});$$
$$co(\{x,x'\} \cup F \setminus \{a_1,a_0\}) \subseteq co(\{x\} \cup F \setminus \{a_0\}).$$

This shows that

$$y \in \bigcup_{a \in F} co(\{x\} \cup F \setminus \{a\}),$$

establishing one half of the desired equality. The other half is trivial. ■

1.3. Proposition. *For a non-empty finite subset of a convex structure X the following are true.*

(1) *R-dependence implies H-dependence.*
(2) *If X is JHC and has the Ramification Property, then R-dependence is equivalent to H-dependence.*
(3) *If X has the CUP, then E-dependence implies C-dependence.*
(4) *If X is JHC and has decomposable segments, then H-dependence implies E-dependence.*

Proof of (1). Let $F \subseteq X$ be R-dependent. Consider a Radon partition $\{F_1, F_2\}$ of F and a point $p \in co(F_1) \cap co(F_2)$. For each $a \in F$ we have either $F_1 \subseteq F \setminus \{a\}$ or $F_2 \subseteq F \setminus \{a\}$. So $p \in co(F \setminus \{a\})$ and F is H-dependent.

Proof of (2). Let $F \subseteq X$ be an R-independent set. Repeated application of 1.2(1) yields

$$\bigcap_{a \in F} co(F \setminus \{a\}) = co\left(\bigcap_{a \in F} F \setminus \{a\}\right) = co(\emptyset) = \emptyset,$$

showing that F is H-independent.

Proof of (3). Let X have the CUP, and let $F \subseteq X$ be E-dependent. Then $F \neq \emptyset$ and we fix $p \in F$. By E-dependence we have

$$co(F \setminus \{p\}) \subseteq \bigcup \{co(F \setminus \{a\}) \mid a \in F \setminus \{p\}\}.$$

By CUP, we have

$$co(F) = co(\{p\} \cup (F \setminus \{p\}))$$
$$\subseteq \bigcup \{co(\{p\} \cup co(F \setminus \{a\})) \mid a \in F \setminus \{p\}\}$$
$$= \bigcup \{co(F \setminus \{a\}) \mid a \in F \setminus \{p\}\},$$

whence F is C-dependent.

Proof of (4). Let $F \subseteq X$ be H-dependent and consider a point
$$x \in \bigcap_{a \in F} co(F \setminus \{a\}).$$
Then for each $p \in F$, the Subdivision Property 1.2(2) yields that
$$co(F \setminus \{p\}) = \bigcup \{co(\{x\} \cup F \setminus \{a,p\}) \mid a \in F \setminus \{p\}\}$$
$$\subseteq \bigcup \{co(F \setminus \{a\}) \mid a \in F \setminus \{p\}\}.$$
Hence F is exchangeable. ∎

1.4. Examples. We consider independence in vector spaces with standard convexity, semilattices with the convexity of order convex subsemilattices, lattices with the convexity of order convex subsemilattices, and free spaces.

1.4.1. Proposition. *In a vector space over a totally ordered field, the notions of H-, C-, R-, and E-dependence are all equivalent to affine dependence.*

Proof. The standard convexity of a vector space V is JHC (cf. I§4.14.1) and its segments are evidently decomposable. Hence by Proposition 1.3, (1), (4), (3),

R-dependent ⇒ H-dependent ⇒ E-dependent ⇒ C-dependent.

We note that V has the Ramification Property, and hence that the first implication is already an equivalence. We won't use this now. Instead, we establish two more implications involving affine dependence, which "close the circle".

First, affine dependence implies R-dependence. Indeed, let $F = \{a_1,..,a_n\} \subseteq V$ be affinely dependent. There exist scalars $t_1,..,t_n$ such that
$$\sum_{i=1}^{n} t_i a_i = 0; \quad \sum_{i=1}^{n} t_i = 0,$$
where at least some t_i are non-zero. Let
$$F_1 = \{a_i \mid t_i > 0\}; \quad F_2 = \{a_i \mid t_i \leq 0\}.$$
Then $\{F_1, F_2\}$ is a partition of F and
$$s_1 = \sum_{a_i \in F_1} t_i > 0; \quad s_2 = \sum_{a_i \in F_2} t_i < 0.$$
We find that the following point is common to $co(F_1)$ and $co(F_2)$.
$$\sum_{a_i \in F_1} \frac{t_i}{s_1} a_i = \sum_{a_i \in F_2} \frac{t_i}{s_2} a_i.$$

Next, C-dependence implies affine dependence. Indeed, let $F = \{a_1,..,a_n\} \subseteq V$ be

§1: Classical Convex Invariants

C-dependent. The point
$$p = \sum_{i=1}^{n} \frac{1}{n} a_i$$
is in the hull of $F \setminus \{a_j\}$ for some j. Therefore, p can be expressed as $\sum_{i \neq j} t_i a_i$, with $\sum_{i \neq j} t_i = 1$. Solving for a_j, we find
$$a_j = \sum_{i \neq j} (nt_i - 1) a_i; \quad \sum_{i \neq j} (nt_i - 1) = 1,$$
showing that a_j is an affine combination of a_i, $i \neq j$. ∎

Recall that a subset $\{a_1,..,a_n\}$ of a semilattice S is *reducible* provided there is a $j \in \{1,..,n\}$ such that $\wedge_{i \neq j} a_i = \wedge_{i=1}^{n} a_i$. In this situation, a_j is said to be *redundant*. If none of the points $a_1,..,a_n$ is redundant, then the set is *irreducible*. This leads to another notion of "dependence", adapted to semilattices, and which fits into the implication scheme below.

1.4.2. Proposition. *For any finite subset of a semilattice,*

E-dependent $\Rightarrow$ H-dependent $\Rightarrow$ C-dependent $\Rightarrow$ reducible.

Together with Proposition 1.3(1), these are the only relations valid for all semilattices (cf. 1.25 below). A finite set $F \neq \varnothing$ is H-dependent iff

(1) $\quad \wedge F \setminus \{a\} = \wedge F = \wedge F \setminus \{b\},$

for some $a \neq b$ in F, and that F is R-dependent iff there is a partition $\{F_1, F_2\}$ of F, together with points $a_1 \in F_1$, $a_2 \in F_2$, such that

(2) $\quad \wedge F_1 \leq a_2; \quad \wedge F_2 \leq a_1.$

The other types of dependence are not so easily translated.

Proof of Proposition 1.4.2. Throughout, F denotes the set $\{a_1,..,a_n\}$ and $F_i = F \setminus \{a_i\}$. First, let F be exchangeable. Then for each i,
$$\wedge F_i \in co(F_i) \subseteq \bigcup_{j \neq i} co(F_j),$$
yielding an index j distinct from i such that $\wedge F_j \leq \wedge F_i$. Consider a minimal element of type $\wedge F_i$. By (1), F is H-dependent.

Next, let F be H-dependent, and let $i \neq j$ be such that $\wedge F_i = \wedge F_j$. If $x \in co(F)$ then $\wedge F \leq x \leq a_k$ for some k distinct from i or from j, say: $k \neq i$. Then
$$\wedge F_i \leq x \leq a_k; \quad a_k \in F_i,$$
whence $x \in co(F_i)$. It follows that F is C-dependent.

Finally, let F be C-dependent. Then
$$\wedge F \in co(F) \subseteq \bigcup_{i=1}^{n} co(F_i),$$
and hence $\wedge F_i \leq \wedge F$ for some i, showing that a_i is redundant. ∎

1.4.3. Convexity in modular spaces. In a modular space (X,I) each finite collection F with $\#F \geq 3$ is Helly dependent. Indeed, let $a, b, c \in F$ be distinct. By definition, the set of medians

$$M(a,b,c) = I(a,b) \cap I(b,c) \cap I(c,a)$$

is non-empty. For each $p \in F$ it is clear that $M(a,b,c) \subseteq co(F \setminus \{p\})$.

1.4.4. Free convexity. Let the set X be equipped with the free convexity, and let $F \subseteq X$ be non-empty and finite. Then the following are easy to verify.

(i) F is C-dependent iff $\#F > 1$.
(ii) F is E-dependent iff $\#F > 2$.
(iii) F is always H- and R-independent.

1.4.5. Lattices. In the lattice $[0,1]^2$ we consider the set F, consisting of the points $\mathbf{0} = (0,0)$, $p = (1,0)$ and $q = (\frac{1}{2},1)$. Then

$co(F \setminus \{q\}) \subseteq \bigcup \{F \setminus \{a\} \mid a \in F; a \neq q\};$
$co(F \setminus \{p\}) \not\subseteq \bigcup \{F \setminus \{a\} \mid a \in F; a \neq p\}.$

[Diagram: square with q at top, $\mathbf{0}$ at bottom-left, p at bottom-right; upper-left region labeled $co(F \setminus \{p\})$, upper-right region labeled $co(F \setminus \{\mathbf{0}\})$, bottom edge labeled $\leftarrow co(F \setminus \{q\}) \rightarrow$]

This illustrates that some point of a finite set may be "exchangeable" with the remaining points, whereas some other point isn't. The set F is not exchangeable. For a general lattice no implications are known other than the one in Proposition 1.3(1). Some different types of relations are valid: see 1.14. For distributive lattices we will be able to give complete information later on (cf. 4.37).

1.5. Some invariants. A convex structure X gives rise to four numbers

$$h(X), c(X), r(X), e(X) \in \{0,1,2,..,\infty\}$$

determined by the following prescriptions. For each n with $0 \leq n < \infty$,

$h(X) \leq n$ iff each finite set $F \subseteq X$ with $\#F > n$ is H-dependent;
$c(X) \leq n$ "   C-dependent;
$r(X) \leq n$ "   R-dependent;
$e(X) \leq n$ "   E-dependent.

We use the following terminology:[1]

---
1. Many authors define the Radon number to be one unit larger (" .. first n such that each set with *at least* n points has a Radon partition .. "). This is motivated by the traditional interests in Radon partitions. The present definition is not only completely similar to the other ones, but it also makes *all* of the formulas below more aesthetical.

§1: Classical Convex Invariants

$h(X)$:   the *Helly number* of X;
$c(X)$:   the *Carathéodory number* of X;
$r(X)$:   the *Radon number* of X;
$e(X)$:   the *exchange number* (or, *Sierksma number*) of X.

Let f be a (class) function defined on the class of all convex structures, and ranging into the set $\{0,1,2,..,\infty\}$. Then f is called a *convex invariant* provided that isomorphic convex structures have equal f-values. Obviously each of the above defined functions h, c, r, e is a convex invariant. Such functions allow for a classification of convex structures according to their combinatorial properties. The functions h, c, r go back to traditional topics in the combinatorial geometry of Euclidean spaces, and they are therefore called *classical* invariants. In the sequel we will often omit reference to the convex structure, using the abbreviations $h = h(X)$, etc..

1.6. The use of weighted sets. Let X be a set and $w: X \to \mathbb{N}$ a function[2]. For $x \in X$ we regard $w(x)$ as the *weight* (or, *multiplicity*) of x, and the pair (X,w) is called a *weighted set*. Such pairs obtain, for instance, via the indexing of a set X with a (surjective) function $f: I \to X$. We regard i as a "name" of $f(i)$, and the number of names of a given point is its weight. Conversely, a weighted set (X,w) can be transformed into an indexing $f: I \to X$ by letting I consist of all points of type (i,x), where $x \in X$ and $1 \le i \le w(x)$, and by defining $f(i,x) = x$.

If X is a set of weight n and if x is one of its elements, then $X_x^\wedge$ denotes the set of weight $n-1$, obtained by decreasing the weight of x with one unit, and dropping the point if it has no weight left. In the sequel, we will freely choose between the weight - or index viewpoint, according to what is locally most convenient. Explicit reference to weight or indexing will often be omitted, and weighted (indexed) sets will mostly be treated like ordinary sets. A few deviating terms may serve as an indicator. For instance, weight (sum of the point weights) is used instead of cardinality, and weighted partitions are used (essentially, partitions of the index set) instead of ordinary partitions.

In this way, the various notions of dependence can be carried over in a natural way to weighted sets. Clearly, such sets are dependent in any of the four senses above whenever some point weighs more than one. Allowing weighted sets does not alter the value of the invariants. Nevertheless, they are a labour saving device in many arguments.

The first result provides an alternative description of h and of c.

1.7. Theorem. *Let X be a convex structure and let $n < \infty$.*

(1) $h \le n$ *iff each finite collection of convex sets in X meeting n by n has a non-empty intersection.*

(2) $c \le n$ *iff for each set $A \subseteq X$ and $p \in co(A)$ there is a subset F of A with $\#F \le n$ and $p \in co(F)$.*

---

2.   We assume that the set $\mathbb{N}$ of natural numbers excludes zero.

Proof of (1). Suppose first that each finite collection of convex sets meeting n by n has a non-empty intersection. Apply this to a family of sets of type $co(F_i)$, $i = 1,..,m$, where $m > n$ and
$$F = \{a_1,..,a_m\}; \quad F_i = F \setminus \{a_i\},$$
to see that F is H-dependent.

Conversely, assume $h \leq n$, and let $\mathcal{D}$ be a finite collection of sets meeting n by n. We may assume that $\#\mathcal{D} = m > n$, and that the result is valid for families with less than m members. We put
$$\mathcal{D} = \{D_1,..,D_m\}; \quad \mathcal{D}_i = \mathcal{D} \setminus \{D_i\}.$$
By the inductive hypothesis, $\cap \mathcal{D}_i \neq \emptyset$, and hence the members of $\mathcal{D}$ meet with $m-1$ at the time. For each $i = 1,..,m$ fix a point $d_i \in \cap \mathcal{D}_i$. These points constitute a weighted set F of weight $m > n$. We have $co(F_{d_i}^\wedge) \subseteq D_i$, and as $h \leq n$ we obtain
$$\emptyset \neq \bigcap_{i=1}^m co(F_{d_i}^\wedge) \subseteq \bigcap_{i=1}^m D_i.$$

Proof of (2). Let $c \leq n$ and $p \in co(A)$. By domain finiteness there is a finite set $F \subseteq A$ with $p \in co(F)$. Among all possible sets F we choose the one with the smallest cardinality. If $\#F > n$ we find that
$$co(F) \subseteq \bigcup\{co(F_a^\wedge) \mid a \in F\},$$
showing that $p \in co(F_a^\wedge)$ for some $a \in F$, a contradiction.

Conversely, let $c > n$. Then there is a set $A \subseteq X$ with $\#A > n$ and
$$co(A) \not\subseteq \bigcup\{co(A_a^\wedge) \mid a \in A\}.$$
Hence some point of $co(A)$ is in the hull of no proper subset of A. ∎

Combining Proposition 1.4.1 with the above theorem, we obtain the following result, which is classical for real vector spaces.

1.8. Theorem. *Let V be a vector space of dimension $n < \infty$ over a totally ordered field.*

(1) *(Helly Theorem) If $\mathcal{D}$ is a finite collection of convex sets in V meeting $n+1$ by $n+1$, then the members of $\mathcal{D}$ have a common point. Moreover, some finite collection of convex sets meeting n by n has an empty intersection.*

(2) *(Carathéodory Theorem) If $A \subseteq V$ and $p \in co(A)$ then there is a set $F \subseteq A$ with $\#F \leq n+1$ and $p \in co(F)$. Moreover, there is a set A and a point $p \in co(A)$ such that $p \notin co(F)$ for each set $F \subseteq A$ with $\#F \leq n$.*

(3) *(Radon Theorem) If $F \subseteq V$ is a finite set with at least $n+2$ points then there is a partition $\{F_1, F_2\}$ of F with $co(F_1) \cap co(F_2) \neq \emptyset$. If $\#F \leq n+1$ then such a partition need not exist.* ∎

The theorem states in by now familiar terms that the classical invariants of V are equal to the dimension of V plus one. By 1.4.1, the same holds for the exchange number.

§1: Classical Convex Invariants   169

A generalization to Bryant-Webster (BW) spaces is given in 1.12. The next result describes the basic relations between the invariants in general spaces.

1.9. Theorem. *The following hold for all convex structures.*

(1) *(Levi inequality)* $h \leq r$.
(2) *(Sierksma inequalities)* $e-1 \leq c \leq max \{h, e-1\}$.
(3) *(Eckhoff-Jamison inequality)* $r \leq c(h-1)+1$ *if* $h \neq 1$ *or* $c < \infty$.

Proof of (1). Just use Proposition 1.3(1).

Proof of (2). To establish $e-1 \leq c$, assume $c < \infty$ and let $F \subseteq X$ be a finite set with $\#F > c+1$. Take $p \in F$. Then $F_p^\wedge$ has more than c elements and hence

$$co(F_p^\wedge) \subseteq \bigcup \{co(F_{pa}^{\wedge\wedge\wedge}) \mid a \in F; a \neq p\} \subseteq \bigcup \{co(F_a^\wedge) \mid a \in F; a \neq p\},$$

showing that F is exchangeable.

To establish $c \leq max \{h, e-1\}$, assume h and e to be finite, and let n be the maximum of h, $e-1$. Let F be a finite subset of X with more than n points. Then F is H-dependent, and we find a point

$$p \in \bigcap \{co(F_a^\wedge) \mid a \in F\}.$$

Consider the weighted set $F \cup \{p\}$ (if p happens to be in F, its weight is increased with one unit). Now, $F \cup \{p\}$ has weight $> e$ and hence it is E-dependent:

$$co(F) \subseteq \bigcup_{a \in F} co(\{p\} \cup F_a^\wedge) = \bigcup_{a \in F} co(F_a^\wedge).$$

It follows that F is C-dependent.

Proof of (3). First, note that the combined assumption of $h = 1$ and $c = \infty$ makes the right hand expression undefined (see also Topic 1.17.2). In the remaining case, we may assume that $c(h-1) < \infty$. Let $F \subseteq X$ be a finite set such that $\#F > c(h-1)+1$ and let $p \in F$. We first show that the sets $co(F_p^\wedge)$ and $co(F \setminus A)$ for $p \notin A \subseteq F$ and $\#A \leq c$, meet h by h. To this end, take h sets among the above ones. If $co(F_p^\wedge)$ is selected then the remaining sets are of type $co(F \setminus A_i)$ for $i = 1,..,h-1$. Now $\#\bigcup_{i=1}^{h-1} A_i \leq c(h-1)$, and hence there is at least one point $q \neq p$ of F left uncovered by the sets A_i. Then q is common to all selected sets. If $co(F_p^\wedge)$ is not selected, then all chosen sets contain p.

We conclude from Theorem 1.7(1) that there is a point

$$x \in co(F_p^\wedge) \cap \bigcap \{co(F \setminus A) \mid p \notin A \subseteq F; \#A \leq c\}.$$

By Theorem 1.7(2) there is a set $A \subseteq F_p^\wedge$ with $\#A \leq c$ and $x \in co(A)$. This gives a Radon partition $\{A, F \setminus A\}$ of F.   ∎

A modification of the Sierksma inequalities is given in 1.18. For a discussion on the sharpness of these inequalities, see 3.18. We next investigate the behavior of convex invariants under the formation of CP images and subspaces.

1.10. Theorem. *For a surjective CP function $f: X \to Y$ the following are true.*
(1) $h(X) \geq h(Y)$ and $r(X) \geq r(Y)$.
(2)   *If f is also CC then $c(X) \geq c(Y)$ and $e(X) \geq e(Y)$.*

Proof of (1). Let $F = \{a_1,..,a_n\}$ be a weighted subset of Y and, for each i, let $b_i \in X$ be such that $f(b_i) = a_i$. Let G be the weighted set $\{b_1,..,b_n\}$. Note that $f(G_{b_i}^\wedge) \subseteq F_{a_i}^\wedge$. The function f being CP, it maps $co(G_{b_i}^\wedge)$ into $co(F_{a_i}^\wedge)$. Hence, if $\cap_{i=1}^n co(G_{b_i}^\wedge)$ is non-empty then so is $\cap_{i=1}^n co(F_{a_i}^\wedge)$, showing that F is Helly-dependent provided G is. The proof of $h(X) \geq h(Y)$ is now routine, and a similar argument can be given for the inequality involving the Radon number.

Proof of (2). Take F, G as above, and assume G is C-dependent:
$$co(G) \subseteq \bigcup_{i=1}^n co(G_{b_i}^\wedge).$$
As f is CP and CC, I§1.11 yields
$$f(co(G)) = co(F); \quad f(co(G_{b_i}^\wedge)) = co(F_{a_i}^\wedge).$$
It easily follows that
$$co(F) \subseteq \bigcup_{i=1}^n co(F_{a_i}^\wedge),$$
i.e., F is C-dependent.

A similar argument works for the exchange number, and (2) follows immediately. ∎

The additional assumption in (2) is necessary. Consider an infinite free convex structure $X \neq \emptyset$; then $c(X) = 1$ (see Example 1.4.4). The space Y can be any convex structure with $\#Y \leq \#X$ and with a high c-value.

1.11. Theorem. *If X is a subspace of Y then the following are true.*
(1) $c(X) \leq c(Y)$ and $e(X) \leq e(Y)$.
(2)   *If X is convex in Y, then $h(X) \leq h(Y)$ and $r(X) \leq r(Y)$.* ∎

The unit circle is a free subspace of $\mathbb{R}^2$. Hence its Helly and Radon number are infinite by 1.4.4, whereas on the superspace $\mathbb{R}^2$ these invariants are equal to 3. A more striking example is given in Topic 1.22.2. We next determine the invariants for several types of examples.

1.12. Theorem. *In a Bryant-Webster space, the notions of H-, C-, R-, and E-dependence are equivalent to affine dependence. In particular, if the affine dimension is n, then*
$$h = c = r = e = n+1.$$

Proof. Let X be a BW space. Then X is JHC and has decomposable segments. By Proposition 1.3 we therefore have

R-dependence $\Rightarrow$ H-dependence $\Rightarrow$ E-dependence $\Rightarrow$ C-dependence.

§1: Classical Convex Invariants   171

We first show that C-dependence implies affine dependence. Let $x_0, x_1, .., x_m$ be C-dependent and let

$$p \in x_0 \circ x_1 \circ .. \circ x_m.$$

By C-dependence and the hull formula of Theorem I§7.7, there are indices $i_1, .., i_k$ ($k \leq m$) such that

$$p \in x_{i_1} \circ .. \circ x_{i_k}.$$

Without loss of generality, 0 is not among these indices. Then $p \in x_0 \circ y$ for some y in $x_1 \circ .. \circ x_m$, whence

$$x_0 \in p/y \subseteq x_{i_1} \circ .. \circ x_{i_k} \,/\, x_1 \circ .. \circ x_m.$$

The right hand set is part of the affine hull of the points $x_1, .., x_m$ (Affine Hull Formula, I§7.11(1)). It follows that x_0 depends affinely on the other points.

We next show that affine dependence implies R-dependence. Let $F = \{x_0, .., x_m\}$ be affinely dependent, say: x_0 depends on the other points. By the Affine Hull Formula I§7.11 we are lead to one of the following possibilities.

(i) $x_0 \in x_{i_1} \circ .. \circ x_{i_r}$, where $i_1, .., i_r$ are among $0, 1, .., m$. Then x_0 is in the convex hull of $x_{i_1}, ..., x_{i_r}$, and F is R-dependent.

(ii) $x_0 \in x_{i_1} \circ .. \circ x_{i_r} / x_{i_{r+1}} \circ .. \circ x_{i_t}$, where all involved indices are distinct. Consider the sets

$$A = \{x_0, x_{i_{r+1}}, ..., x_{i_t}\}; \quad B = \{x_{i_1}, ..., x_{i_r}\}.$$

We have $co(A) \cap co(B) \neq \emptyset$, and F is R-dependent. ∎

Recall that the **breadth** of a semilattice S is the number $b(S) \in \{0, 1, .., \infty\}$ satisfying the following rule for each $n < \infty$:

$b(S) \leq n$ iff each finite set $F \subseteq S$ with $\#F > n$ is reducible.

Note that $b(S) = 0$ is equivalent to $S = \emptyset$. This important invariant from the theory of semilattices is now compared with the convex invariants.

1.13. Proposition. *The following inequalities are valid in a semilattice.*

(1) $h-1 \leq b \leq c \leq h \leq e \leq c+1$.
(2) $b \leq r \leq 2b$.

Proof of (1). The inequalities $b \leq c \leq h \leq e$ follow from Proposition 1.4.2 and $e \leq c+1$ is a Sierksma inequality, 1.9(2). To show that $h-1 \leq b$, we may assume that b is finite. Let F be a finite set with $\#F > b+1$. Then there exist $u \neq v$ in F with $\wedge F = \wedge(F_{uv}^{\wedge\wedge})$, and hence the infima of $F_u^\wedge$ and of $F_v^\wedge$ are equal. As we verified in 1.4.2, this means that F is H-dependent.

Proof of (2). We have $h \leq r$ by the Levi inequality, and hence $b \leq r$ by (1). Next, let F have more than $2b$ points. Take a set $F'_1 \subseteq F$ with

172   Chap. II: Convex Invariants

$\#F'_1 \leq b;\quad \wedge F'_1 = \wedge F.$

Now $F \setminus F'_1$ has more than b points, yielding a set $F_2 \subsetneq F \setminus F'_1$ with

$\#F_2 \leq b;\quad \wedge F_2 = \wedge (F \setminus F'_1).$

Let v_1 in $F \setminus (F'_1 \cup F_2)$ and let $F_1 = F \setminus F_2$. Then $v_1 \in F_1$ and for any $v_2 \in F_2$,

$\wedge F_1 \leq v_2;\quad \wedge F_2 \leq v_1.$

Hence (as we verified in 1.4.2) F is R-dependent. ∎

For a discussion on the sharpness of this proposition, see 1.25.

1.14. Proposition. *The following (in)equalities are valid in a lattice.*

(1) $h \leq 2$.
(2) $e - 1 \leq c \leq e$; *specifically* $c = e$ *if* $e < 2$ *and* $c = e - 1$ *if* $e > 2$.
(3) $b \leq e - 1$.

Proof of (1). As Helly independence is hereditary, it suffices to investigate sets with at most three points. For each triple of points u, v, w we have

$$m(u,v,w) \in co\{u,v\} \cap co\{v,w\} \cap co\{w,u\},$$

where m is the function defined in I§3.12.3. Hence $h \leq 2$.

Proof of (2). The Sierksma inequalities imply that

$$e - 1 \leq c \leq \max\{h, e-1\} \leq \max\{2, e-1\}.$$

Hence if $e > 2$ then $c = e - 1$, and if $e = 2$ then $1 \leq c \leq 2$. The case $e < 2$ is handled as follows. The statements $e = 0$ and $e = 1$ are equivalent to the lattice being empty, resp., a singleton (compare 1.17 on low values of the invariants, applied to an S_1 convexity), which in turn implies $c = 0$ and $c = 1$, respectively. Summarizing, $c \in \{e-1, e\}$.

Proof of (3). Let $k \geq e$ and consider k points $x_1,..,x_k$ of the lattice. Define

$$y_0 = \bigwedge_{i=1}^{k} x_i;\quad y_j = \bigwedge_{i \neq j} x_i\ (j = 1,..,k);\quad z = \bigvee_{i \neq 0} y_i.$$

Then $\wedge_{j \neq 0} y_j \leq z \leq \vee_{j \neq 0} y_j$, and since $\{y_0, y_1,..,y_k\}$ is E-dependent, we find an index $i \neq 0$ with $\wedge_{j \neq i} y_j \leq z \leq \vee_{j \neq i} y_j$. This yields $z \leq \vee_{j \neq i} y_j \leq x_i$ and hence $\wedge_{j \neq i} x_j = y_i \leq z \leq x_i$, showing that x_i is a redundant point of $\{x_1,..,x_k\}$. ∎

For $e = 2$ it is not possible to determine c more accurately; the lattices $\{0,1\}$ and $[0,1]$ have $e = 2$, and $c = 1$, 2, respectively. As we shall see later (cf. 4.37), the value of b relative to e can be determined exactly for distributive lattices.

1.15. Proposition. *A convex structure X is JHC in each of the following situations.*

(1) $c(X) \leq 2$.
(2) $h(X) \leq 2$ *and distinct points in X can be screened with convex sets.*

Proof. Let F be a non-empty finite set, let $a \notin co(F)$, and let $x \in co(F \cup \{a\})$. We must show that $x \in pa$ for some $p \in co(F)$. Without loss of generality, $x \notin co(F)$. If

§1: Classical Convex Invariants   173

$c(X) \leq 2$, we obtain a two-point subset of $F \cup \{a\}$ with x in its hull. At least one of them must be a. For the remainder of the proof, we assume that $h(X) \leq 2$. The convex sets

$$co(F),\ xv\ (v \in F)$$

meet two by two, and hence they have a point p in common. If $x \notin pa$, then use $h(X) \leq 2$ once more to obtain a point $q \in pa \cap ax \cap xp$. Note that $q \neq x$. By assumption, there exist convex sets C, D such that $q \notin D, x \notin C, C \cup D = X$. Then $q \in ax \cap px$ yields that $a, p \notin D$, whereas $x \in co(F \cup \{a\})$ yields that $v \notin D$ for some $v \in F$. However, $p \in vx \subseteq D$, contradiction. ∎

Part (2) of the above result implies that an S_2 space X of Helly number 2 is of arity two. This is an essential step before the results of Section I§6 can be applied: X is a median space. Conversely, a median space is S_4 and JHC. In Section IV§5, JHC is derived from the combined assumption of $c \leq 3$ and $e \leq 3$ under topological conditions.

Further Topics

1.16. Heredity of independence. We consider the problem whether a subset of an independent set (in any of the four senses introduced in this section) is independent.

1.16.1. Verify that a subset of an H- (resp., R-) independent set is again H- (resp., R-) independent. In spaces with the CUP, a similar property holds for C- and E-independence. If F denotes one of H, C, R, E, then heredity of F-independence implies that the corresponding invariant is $\leq n$ iff each $(n+1)$–set is F-dependent.

1.16.2. For $0 < m < n$, let $F_m(n)$ be the m–free convex structure on an n–point set (cf. I§2.22). Show that the Carathéodory number of $F_m(n)$ equals m. On the other hand, each 2–point set is C-dependent whenever $m > 2$. Hence Carathéodory independence is not hereditary. Note that the Helly- and Radon number of this space are equal to m; compare 5.16.1.

1.16.3. (Duchet [1988]) First, we recursively define a label to be either 0, or to be a symbol of type (τ_1, τ_2), where τ_1 and τ_2 are labels. The complexity $\gamma(\tau)$ of a label τ is the number of left-handed brackets "(" that occur in it; formally,

$$\gamma(0) = 0;\quad \gamma((\tau_1, \tau_2)) = \gamma(\tau_1) + \gamma(\tau_2) + 1.$$

Next, let I be an interval operator on a set X, and let $A \subseteq X$. For each label τ define

$$A^\tau = \begin{cases} A, & \text{if } \tau = 0; \\ I(A^{\tau_1} \times A^{\tau_2}), & \text{if } \tau = (\tau_1, \tau_2). \end{cases}$$

Note that $co(A) = \cup_\tau A^\tau$. The A–weight of $x \in co(A)$ is defined to be the minimal complexity of a label τ with $x \in A^\tau$, and the A–weight of a finite set $B \subseteq co(A)$ is the sum of its point-weights.

Now suppose that each $(n+1)$–subset of X is C-dependent, and let $A \subseteq X$ be a finite set with $\#A \geq n+1$. Show that for each $p \in co(A)$, if $B \subseteq co(A)$ is a finite set with n points such that $p \in co(B)$, and if B is of minimal possible A–weight, then $B \subseteq A$.

1.16.4. (Duchet op. cit.; compare part (1)) Deduce that in a space X of arity two the following is true: $c(X) \leq n$ (where $n < \infty$) iff each $(n+1)$-set in X is C-dependent.

1.17. Low values of the invariants. Let X be a convex structure.

1.17.1. Let f denote any of the invariants introduced in this section. Then
$$f(X) = 0 \Leftrightarrow X = \emptyset.$$

1.17.2. $h(X) \leq 1$ iff $r(X) \leq 1$ iff each finite collection of non-empty convex sets has a common point. Note that this can be combined with $c = \infty$.

1.17.3. $c(X) \leq 1$ iff each union of convex sets is convex (*Union Property*). Note that by the third axiom of convexity, the inequality $c \leq 1$ is equivalent with the convexity being a topology (of closed sets).

1.17.4. $e(X) \leq 1$ iff X has the coarse convexity.

1.17.5. A convex structure X has the *weak Union Property* (Jamison [1981]) provided the union of any two intersecting convex sets in X is convex. Equivalently, for each point p of X, any two distinct copoints at p are disjoint. Let X be of arity two. Show that $e(X) \leq 2$ iff X has the weak Union Property. Observe that the k-free space $F_k(n)$ $(1 < k < n)$ satisfies $e = 2$ and fails to have the weak Union Property.

1.18. Special Eckhoff-Jamison inequality (compare Sierksma [1976]; continued 3.18). Show that the following so-called *Special Eckhoff-Jamison inequality* is valid for all convex structures.
$$r \leq (\max\{c,e\}-1)(h-1)+2.$$

1.19. Totally ordered sets and trees

1.19.1. For a totally ordered set X with more than one point, show that
$$h = r = e = 2; \quad c \leq 2; \quad c = 2 \text{ iff } \#X \geq 3.$$

1.19.2. Let X be a tree with more than two points. Then
$$h = c = e = 2; \quad 2 \leq r \leq 3,$$
where $r = 3$ iff X has a ramification point. Deduce that each subspace of a tree is JHC.

1.20. Posets

1.20.1. (compare Calder [1971] and Jamison [1981b]) For a partially ordered set with more than one point, show that

§1: Classical Convex Invariants   175

$$c = 2; \quad 2 \leq e \leq 3; \quad h = r.$$

Verify that $e = 2$ holds in a poset iff two incomparable elements with a common lower (upper) bound are always maximal (minimal).

1.20.2. (Jamison [1981b]; compare Bean [1974]) The *width* of a poset is the largest possible number of elements in a set of mutually incomparable points (an antichain). Show that the following inequalities are sharp on the class of posets which are not an antichain.

$$width + 1 \leq h \leq 2 \cdot width$$

(for an antichain, the first inequality should be changed into $width = h$). Conclude that $h \leq 2$ iff the poset is linearly ordered.

1.21. Convex invariants and cliques. A *free set* in a convex structure is a convex subset which is relatively free. A *clique* is a maximal free set, and the *clique number*, is the largest possible number of points in a free set. For (connected) graphs this corresponds with the usual notions.

1.21.1. (Jamison [1981b]) Show that the Helly number of a finite convex geometry equals the clique number. In case of connected graphs, this result covers the class of chordal graphs (cf. I§2.24.3).

1.21.2. The class $\mathcal{D}$ of *dismantlable finite graphs* is defined recursively as follows.
(a)   The one-point graph is in $\mathcal{D}$.
(b) $G \in \mathcal{D}$ provided there is a vertex $v \in G$ such that $G \setminus \{v\} \in \mathcal{D}$ and the vertex v can be "dismantled" in the sense that there exists a vertex $w \neq v$ such that each neighbor of v is a neighbor of w.
(c)   No graph is in $\mathcal{D}$ but those described in (a) and (b).

In the situation of (b), v is said to be *dominated* by w; note that v, w are neighbors. Show that a finite connected graph, which is a convex geometry, is dismantlable.

A *Helly graph* is defined by the property that each family of pairwise intersecting metric disks of integer radius has a non-empty intersection provided the disks meet two by two. Verify that a finite Helly graph is dismantlable.

1.21.3. (Bandelt and Mulder [1990]) Let G be a dismantlable graph. Show that the Helly number of G equals the clique number. Hint. Among all Helly independent sets with n vertices, choose F such that the sum of all distances $d(p,q)$ for $p, q \in F$ is minimal. Let $u \in G$ be dominated by v. Without loss of generality, $u \in F$. For each $p \neq u$ in F take a proper neighbor of u on a geodesic $p \mapsto u$, and let F' be the resulting set of neighbors. If two members of F' coincide, then replace u with this vertex in F and derive a contradiction with the minimality of F. If all members of F' are distinct, then consider two possibilities. If $v \in F'$, then $F' \cup \{u\}$ is a clique. If $v \notin F'$, then either $F' \cup \{u\}$ is a clique, or there is a $p \in F'$ such that $\{u,v\} \cup F' \setminus \{p\}$ is a clique.

1.21.4. (Doignon [1973]; Jamison [1981b]; continued 1.22.3) Show that the relative standard convexity of $\mathbb{Z}^n$ (the set of *Gaussian integers*) has Helly number 2^n.

1.21.5. (Bell [1977]) If a finite collection of half-spaces of $\mathbb{R}^n$ has no intersection with $\mathbb{Z}^n$, then the same is true for a subcollection of at most 2^n half-spaces.

1.21.6. (Bandelt and Pesch [1989]) Show that the Radon number of a Helly graph equals its clique number provided the latter is ≥ 3. Hint. If G is a Helly graph, then each vertex v which is maximal in some base-point order, is dominated by some neighbor, and $G \setminus \{v\}$ is a Helly graph. Let $X \subseteq G$ be Radon independent. Without loss of generality, X consists precisely of the points which are maximal in some base-point order. Show by induction on the number of vertices of G that there is a bijection f of X with some clique of G such that $f(x_1), f(x_2) \in x_1 x_2$ for all $x_1, x_2 \in X$.

Conclude that $\mathbb{Z}^n$ with the Cartesian metric has a Radon number 2^n (cf. Changat and Vijayakumar [1992]).

1.22. Radon number

1.22.1. Let Y be a subspace of X such that disjoint Y–polytopes extend to disjoint X–polytopes. Show that $r(Y) \leq r(X)$. Verify that the above assumption is fulfilled for each of the following pairs.

(i)   A median algebra (X, m) and a median stable subspace Y (cf. I§6.10).
(ii) A semilattice and a subsemilattice.
(iii) A product $X \times \{0,1\}$ versus a convex expansion of X.
(iv) A connected, S_4, induced subgraph of a graphic hypercube (Hint: first, verify that the graph is a subspace of the cube. This works even under S_3. Then proceed with the hint in I§3.26.2).

1.22.2. (compare Jamison [1981b]) Let X be a convex structure. Let $Z = X \times \mathbb{N}$ and consider the convexity on Z for which the convex hull of a finite set $F = \{(x_1, i_1), .., (x_n, i_n)\}$ is given by

$$F \cup \{(x, i) \mid x \in co\{x_k \mid i > i_k\}\}.$$

Then $r(Z) = r(X)$. Let $Z_n = X \times [1, n]$ for $n \in \mathbb{N}$. If X has a finite Radon number $r \geq 2$, and if there exists a minimally R-dependent subset of X with $r + 1$ points, then

$$\forall n \in \mathbb{N}: r(Z_n) > r(Z).$$

Conclude that there exist convex structures of any prescribed finite Radon number ≥ 2, and including a cofinal family of subspaces with a strictly larger Radon number.

1.22.3. (Onn [1990]) For $n \geq 1$, consider $\mathbb{Z}^n$, the set of Gaussian integers, equipped with the relative standard convexity. Show that

$$2 \cdot r(\mathbb{Z}^n) \leq r(\mathbb{Z}^{n+1}); \quad r(\mathbb{Z}^n) \leq n \cdot (2^n - 1) + 2.$$

Hint. If a set $A \subseteq \mathbb{Z}^n$ has no Radon partition, then $A \times \{0,1\} \subseteq \mathbb{Z}^{n+1}$ has no Radon

§1: Classical Convex Invariants   177

partition either. As for the second inequality, use 1.21.4 in combination with the special Eckhoff-Jamison inequality 1.18.

In addition, prove that $r(\mathbb{Z}^2) = 5$. Deduce that $r(\mathbb{Z}^n) \geq 5 \cdot 2^{n-2}$. Onn's paper [1990] contains further information on the computational complexity of Radon partitions in $\mathbb{Z}^n$.

1.22.4. (Levi [1951]; compare Proposition I§7.3) Let X be a JHC and straight convex structure with decomposable segments. Show that if each polytope spanned by three vertices has a Radon number ≤ 3, then X is S_4.

1.22.5. (van de Vel [1983e]) Let $\lambda(r)$ be the superextension of the free space with $r < \infty$ points. Note that the r extreme points of $\lambda(r)$ are Radon independent. Show that a median convex structure X has a Radon number $\geq r$ iff $\lambda(r)$ embeds in X and (consequently) that $\lambda(r)$ has Radon number r.

1.23. Subsemilattices as convex sets (Jamison [1982]). The *depth of a poset* is the supreme length of a chain. Let S be a semilattice equipped with the convexity of all subsemilattices; cf. I§1.23.3. Show that $c(S) = b(S)$ and that $h(S) = depth(S)$.

1.24. Equal union of sets

1.24.1. (Lindström [1972], Tverberg [1971][3]; continued: 5.17.3). Let $\mathcal{A}$ be a collection of m subsets of $\{1,2,..,n\}$, where $m > n$. Prove that there exist two disjoint and non-empty subcollections $\mathcal{A}_1, \mathcal{A}_2$ of $\mathcal{A}$, such that $\cup \mathcal{A}_1 = \cup \mathcal{A}_2$. Hint: construct an incidence matrix and apply Radon's Theorem for Euclidean space.

1.24.2. Let the meet semilattice $S = \{0,1\}^n$ with the convexity of subsemilattices (as in 1.23) or with the convexity of order convex subsemilattices (as usual). Conclude that $r(S) = n$.

1.25. Examples on semilattices

1.25.1. Consider the meet semilattice $S = \{0,1\}^n$ (Cartesian order). Let $1 \leq k \leq \lfloor n/2 \rfloor$, and let

$N_k = \{x \in S \mid \text{at least } k \text{ coordinates of } x \text{ are } 0\}$.

Then the quotient semilattice S_k / N_k satisfies

$b(S_k) = k;\quad r(S_k) = 2k-1$.

1.25.2. Let $S = [0,1]^2$ with a vertical line segment erected at $(\frac{1}{2},\frac{1}{2})$. Regard S as a subsemilattice of the 3-cube (usual order). Show that $b(S) = 2$ and $r(S) = 4$.

---

3.   This result has been proved *originally* by Lindström; the hint is based on an argument presented later by Tverberg.

1.25.3. Problem (compare Proposition 1.13). Is the inequality $r \leq 2b$ sharp for all values of b?

1.25.4. Show that none of the inequalities in Proposition 1.13(1) can be replaced by an equality: give examples $S_1,..., S_5$ of (finite) semilattices according to the following demands (an invariant f takes the value f_i on S_i).

(S_1) $b_1 < c_1 = h_1 < e_1$; (S_2) $b_2 = c_2 = h_2 < e_2$;
(S_3) $b_3 = c_3 < h_3 = e_3$; (S_4) $b_4 < c_4 = h_4 = e_4$;
(S_5) $b_5 = c_5 = h_5 = e_5$.

1.25.5. Show that a chain can be defined as a semilattice of Radon number 2 and that a tree can be defined as a semilattice of exchange number 2 or, equivalently, as a semilattice of Helly number 2.

1.26. A semi-duality between c and h (Jamison [1975]). Let X be a convex structure, and let $p \in X$. The collection of all copoints at p (cf. I§3.24) is denoted by $\Sigma(X,p)$. As a basis for a convexity on this collection, take all sets of type

$$A^* = \{ S \mid S \in \Sigma(X,p); A \subseteq S \}.$$

Let $0 \leq n < \infty$. Then $c(X) \leq n$ iff $h(\Sigma(X,p)) \leq n$ for all $p \in X$. If X is S_3 then $c(\Sigma(X,p)) \leq h(X)$ for all $p \in X$.

1.27. Some characterizations

1.27.1. A space is derived from a tree iff it is S_2 and $e \leq 2$; $h \leq 2$.

1.27.2. A space is derived from a total order iff it is S_2 and $e \leq 2$; $r \leq 2$.

1.27.3. A segment is derived from a distributive lattice iff it is S_2 and $h \leq 2$.

1.27.4. A space is derived from a median algebra iff it is S_2 and $h \leq 2$.

1.28. Plane cosector convexity (P.C. Hammer [1965]). Let $0 < \theta \leq \pi$. A *half-open sector* at $p \in \mathbb{R}^2$ is a subset of $\mathbb{R}^2$ lying between two rays at p forming an angle θ, and including one of the rays (with p). A *half-open cosector* is the complement of such a set. Consider the protopology $\mathcal{C}(\theta)$ formed on the plane by the intersections of half-open cosectors with a fixed angle θ. Note the special case $\theta = \pi$: the cosectors are precisely the copoints for the standard convexity, and it follows from I§3.24.1 that the resulting protopology equals the standard convexity. Show that $\mathcal{C}(\theta)$ is, in fact, a convexity in the plane, and that the Carathéodory number of $(\mathbb{R}^2, \mathcal{C}(\theta))$ is the largest integer $< \lfloor 4\pi/\theta \rfloor$.

1.29. Disk convexity in Euclidean spaces. Consider the Manhattan norm on $\mathbb{R}^n$,

§1: Classical Convex Invariants 179

$$||x|| = \sum_{i=1}^{n} |x_i| \quad (\text{where } x = (x_i)_{i \in I}).$$

Show that the Helly number of the disk convexity, generated by this norm, equals $n+1$ for all n, except for $n = 2$, where it equals 2.

Notes on Section 1

The theory of convex invariants has grown out of the classical results of Helly [1923], Carathéodory [1907], and Radon [1921] on intersection or on union of convex sets in Euclidean space. These results are summarized in Theorem 1.8. The first general definition of h and r was given by Levi [1951], who gave part (1) of Proposition 1.3. This result is, in fact, the proof of Helly's Theorem by Radon, reformulated in an abstract environment. A general theory of convex invariants was first developed by Kay and Womble [1971]. The terms "H-", "C-" and "R-dependence" are taken from Soltan [1976]. They were first used for geodesic convexity by Lassak [1975]. The exchange number was introduced by Sierksma in [1975].

The Subdivision Property 1.2(2) was considered as an axiom by Levi [1951]. Parts (2), (3) and (4) of Proposition 1.3 were obtained by R. Hammer [1977], who used them to prove the equality of the classical invariants of a BW space and dimension plus one, Theorem 1.12. The fact that these invariants do not exceed the dimension by more than one was obtained before by Bryant and Webster [1969]. Part (1) of Theorem 1.7 is due to Calder [1971]; it is also found in Sierksma [1975] and in Berge and Duchet [1975]. The statements involved in this result are often used as an alternative definition of h or of c. Part (1) of Theorem 1.9 is due to Levi [1951]. The inequalities (2) of this theorem are taken from Sierksma's [1976] dissertation. Part (3) of the result was obtained by Eckhoff and Jamison, and was communicated by Sierksma in [1976]. Kay and Womble [1971] obtained a slightly weaker version of the Eckhoff-Jamison inequality before. For a survey of inequalities with many examples on their sharpness, see Sierksma [1977].

The inequalities of Propositions 1.13 and 1.14 for semilattices, resp., lattices, are new. Results of this kind illustrate that the convex invariants provide a delicate measure of the combinatorial behavior of convex structures, and suggest that it should be possible to characterize certain classes of convex structures by the (relative) values of their invariants. For instance, the results given in 1.27 are in terms of separation and invariants, and relate with known characterizations of (finite) trees and totally ordered sets. For these and other characterizations, see Duchet's survey paper [1987].

2. Invariants and Product Spaces

Computing the convex invariants of a product space by hand is often complicated. In this section some techniques and results are presented leading to sharp bounds or even to the determination of the product invariants. The main tools are provided by the so-called exchange function and the Radon function, which are defined for general convex structures.

The Carathéodory and Sierksma number of a product are determined in tandem: to compute one of them, both invariants must be known on each factor. For instance, the product's Carathéodory number depends on the number of factors with an exchange number not exceeding the Carathéodory number. On the other hand, the product exchange number depends on the maximal value of either invariant on each factor.

It appears that the Radon function of Euclidean space realizes the lower bound predicted by the general theory. A table of such bounds is computed. For products with two factors, sharp lower and upper bounds exist on the Radon number. When more factors occur, a somewhat complicated upper bound obtains, which is slightly better than the one deduced from the two-factor formula. A product of two Euclidean spaces has the Radon number one would expect from the product dimension, but the Radon number is deficient for products with three or more Euclidean (or other) factors. For products of linearly ordered spaces a quite accurate determination of the Radon number is possible.

The easiest problem concerning product spaces is to determine the Helly number, which will be of our concern first. Throughout, if a factor is labeled with an index i and if f is a convex invariant then f_i denotes the value of this factor. A plain "f" refers to the value of the product. As a matter of permanent notation, we let $<m>$ denote the set $\{1,..,m\}$.

2.1. Theorem. *Let X_i for $i \in I$ be a family of non-empty convex structures and let X be the product. Then the Helly number of X is the supremum of the factor Helly numbers,*

$$h = \sup\{h_i \mid i \in I\}.$$

Proof. Let $\pi_i: X \to X_i$ denote the ith projection. As CP surjections do not raise the Helly number, Theorem 1.10, it follows that $h \geq h_i$ for all $i \in I$.

To establish the opposite inequality we may assume that the number $n = \max_{i \in I} h_i$ is finite. Let $F = \{x_1,...,x_m\} \subseteq X$ be a weighted set, where $m > n$, and let $F_k = F_{x_k}^\wedge$. By the product formula I§1.10.3 for polytopes,

$$co(F_k) = \prod_{i \in I} co(\pi_i F_k) = \prod_{i \in I} \pi_i co(F_k).$$

We regard $\pi_i(F)$ as a set indexed by $<m>$. Then $m > h_i$ implies that $\cap_{k=1}^m \pi_i co(F_k) \neq \emptyset$ for all $i \in I$. Hence $\cap_{k=1}^m co(F_k) \neq \emptyset$, showing that F is H-dependent. ∎

All of the remaining invariants will be studied for products with finitely many factors only. The next result explains why.

2.2. Proposition. *Let X_i for $i \in I$ be an infinite collection of convex structures. Then the product space $\prod_{i \in I} X_i$ has the following properties.*
(1) *If each X_i has a pair of incomparable convex sets, then $c = e = \infty$.*
(2) *If each X_i admits a non-trivial half-space, then $r = \infty$.*

Proof. We first consider the case where $X_i = \{0,1\}$ is equipped with the free convexity and $I = \mathbb{N}$. Let $a(i) \in \{0,1\}^{\mathbb{N}}$ be defined by

$$a(i)_n = \begin{cases} 0, & \text{if } n \neq i; \\ 1, & \text{if } n = i. \end{cases}$$

Any finite collection of $a(i)$'s is both C- and E-dependent, showing that $c = e = \infty$. Note that $h = 2$ by the previous theorem. Hence the equality $e = \infty$ can also be derived from the fact that $c = \infty$ in combination with the Sierksma inequalities 1.9(2). On the other hand, let p be a prime number and let $b(p) \in \{0,1\}^{\mathbb{N}}$ be defined by

$$b(p)_n = 1 \quad \Leftrightarrow \quad p \text{ divides } n.$$

Each finite collection of $b(p)$'s is R-independent, whence $r = \infty$.

We now consider the general case. If X_i admits two incomparable convex sets there is a two-point free subspace of X_i. This allows to embed $\{0,1\}^{\mathbb{N}}$ in the product X, and the result for c and e follows from Theorem 1.11(1) and the previous argument.

Assume next that each factor X_i admits a non-trivial half-space H_i. We define a CP function as follows.

$$f_i : X_i \to \{0,1\}, \quad f_i(x) = 1 \Leftrightarrow x \in H_i.$$

After fixing a countable infinite set $\{i_n \mid n \in \mathbb{N}\}$, we obtain a CP surjection

$$f : X \to \prod_{n \in \mathbb{N}} X_{i_n} \to \{0,1\}^{\mathbb{N}},$$

where the first arrow is the projection, and the second arrow is the product function formed with f_{i_n} for $n \in \mathbb{N}$. By Theorem 1.10, f does not raise the Radon number, whence $r = \infty$ by a previous argument. ∎

We first pay attention to products with two factors. The following result gives a complete description of the Carathéodory number in terms of the Carathéodory and exchange numbers of the factors. Note that the product of a space X with a non-empty coarse space has the same Carathéodory number as X; we therefore confine ourselves to non-coarse factor spaces.

2.3. Proposition. *Let X_1, X_2 be non-coarse convex structures. Then the Carathéodory number of $X_1 \times X_2$ is determined as follows.*
(1) *If $c_i < e_i$ for $i = 1, 2$, then $c = c_1 + c_2$.*
(2) *If $c_i < e_i$ for exactly one of $i = 1, 2$, then $c = c_1 + c_2 - 1$.*

§2: Invariants and Product Spaces

(3) *If $c_i \geq e_i$ for $i = 1, 2$, then $c = c_1 + c_2 - 2$.*

Proof. Observe that $c_i > 0$ and $e_i > 1$ by 1.17. We verify three statements in advance. Throughout, X is the product space and $\pi_i : X \to X_i$ denotes the ith projection.

2.3.1. *In all circumstances, $c \leq c_1 + c_2$.*

Proof. Let $A \subseteq X$, and let $x \in co(A)$. We regard $\pi_1(A)$ and $\pi_2(A)$ as being indexed by A. This yields two subsets F_1, F_2 of A of weight $\#F_i \leq c_i$ such that $\pi_i(x) \in co(\pi_i(F_i))$ for $i = 1, 2$. Now $\#(F_1 \cup F_2) \leq c_1 + c_2$ and

$$x \in co(\pi_1 F_1) \times co(\pi_2 F_2) \subseteq co(F_1 \cup F_2).$$ ∎

2.3.2. *If $c_i \geq e_i$ for at least one of $i = 1, 2$, then $c \leq c_1 + c_2 - 1$.*

Proof. Let us assume that $c_2 \geq e_2$. Let $A \subseteq X$ and let $x \in co(A)$. As in 2.3.1 there are finite sets $F_1, F_2 \subseteq A$ with $\#F_i \leq c_i$ and

$$\pi_i(x) \in co(\pi_i F_i) \quad (i = 1, 2).$$

Hence $x \in co(F_1 \cup F_2)$. If one of the inequalities $\#F_i \leq c_i$ is strict, or if F_1 meets F_2, then $\#(F_1 \cup F_2) < c_1 + c_2$. Assume $\#F_i = c_i$ and $F_1 \cap F_2 = \emptyset$. Consider a point $a \in F_1$. As $\#(F_2 \cup \{a\}) > c_2 \geq e_2$, we can exchange a with some point $a' \neq a$ in F_2:

$$\pi_2(x) \in co(\pi_2(\{a\} \cup F_2 \setminus \{a'\})).$$

Define $F'_2 = \{a\} \cup F_2 \setminus \{a'\}$. It appears that x is in the hull of the set $F_1 \cup F'_2 \subseteq A$ and that $F_1 \cup F'_2$ has at most $c_1 + c_2 - 1$ points. ∎

2.3.3. *For $i = 1, 2$, let $F_i \subseteq X_i$ be a set with a distinguished point p_i, such that $\#F_i > 1$ and*

$$co(F_i) \not\subseteq \bigcup \{co(F_i \setminus \{q\}) \mid q \in F_i; q \neq p_i\}.$$

Then the following subset of X is C-independent (Fig. 1):

$$F = (\{p_1\} \times (F_2 \setminus \{p_2\})) \cup ((F_1 \setminus \{p_1\}) \times \{p_2\}).$$

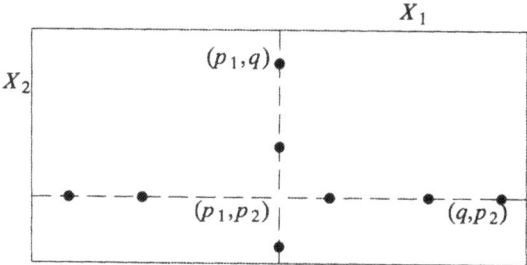

Fig. 1: C-dependent set in a product

Proof. For $i = 1, 2$, consider a point x_i in $co(F_i)$ but not in the sets $co(F_i \setminus \{q\})$ for $q \in F_i$, $q \neq p_i$. Let $x = (x_1, x_2)$. Note that $\pi_i(F) = F_i$ since $\#F_i > 1$, whence x is in $co(F)$.

If $a \in F$, say, $a = (p_1, q)$ with $q \neq p_2$, then $\pi_2(F \setminus \{a\}) = F_2 \setminus \{q\}$. Consequently, the set $co(\pi_2(F \setminus \{a\}))$ does not contain x_2. Similarly, if $a = (q, p_2)$ with $q \neq p_1$, then the set $co(\pi_1(F \setminus \{a\}))$ does not contain x_1. In either case we conclude that $x \notin co(F \setminus \{a\})$ and hence that F is C-independent.

Proof of (1). Assume $e_i > c_i$ for $i = 1, 2$. For each i there is a non-exchangeable set $F_i \subseteq X_i$ of cardinality $> c_i$. This yields a point $p_i \in F_i$ together with a point

$$x_i \in co(F_i \setminus \{p_i\}) \setminus \bigcup \{ co(F_i \setminus \{q\}) \mid q \in F_i; q \neq p_i \}$$

for $i = 1, 2$. By 2.3.3 it follows that the set

$$F = (\{p_1\} \times (F_2 \setminus \{p_2\})) \cup ((F_1 \setminus \{p_1\}) \times \{p_2\})$$

is C-independent. As F has at least $c_1 + c_2$ points, it follows that $c \geq c_1 + c_2$. In combination with 2.3.1, this completes the proof of part (1).

Proof of (2). Let $c_1 < e_1$ and $c_2 \geq e_2$. We first assume that c_2 is finite. There is a non-exchangeable set $F_1 \subseteq X_1$ with at least $c_1 + 1$ points and a C-independent set $F_2 \subseteq X_2$ with c_2 points. Note that $c_2 > 1$ since $e_2 > 1$. For a suitable choice of $p_i \in F_i$, By 2.3.3, the set

$$F = (\{p_1\} \times (F_2 \setminus \{p_2\})) \cup ((F_1 \setminus \{p_1\}) \times \{p_2\}).$$

is C-independent. Now F has at least $c_1 + c_2 - 1$ points and hence $c \geq c_1 + c_2 - 1$. Combining with 2.3.2 yields part (2).

If $c_2 = \infty$ we can repeat the above argument with c_2 replaced by an arbitrarily large number.

Proof of (3). Assume $c_1 \geq e_1$ and $c_2 \geq e_2$, so $c_1, c_2 > 1$. Let $A \subseteq X$ and $x \in co(A)$. We look for a subset of A with no more than $c_1 + c_2 - 2$ points and with x in its convex hull. We may assume that c_1 and c_2 are finite. By 2.3.2 there is a finite set $F \subseteq A$ of cardinality $\leq c_1 + c_2 - 1$, such that $x \in co(F)$. After projecting to each factor space, we obtain $F_1, F_2 \subseteq F$ with $\#F_i \leq c_i$ and

$$\pi_i(x) \in co(\pi_i F_i).$$

Without loss of generality, $F = F_1 \cup F_2$. Note that if $\#F \leq c_1 + c_2 - 2$, then we have achieved our goal. In the remaining case, one of the inequalities $\#F_i \leq c_i$ ($i = 1, 2$) is an equality (say: $\#F_1 = c_1$) and F_1 meets F_2 in at most one point. Hence $F_2 \not\subseteq F_1$ since $c_2 > 1$. As $e_1 \leq c_1$, any $p_2 \in F_2 \setminus F_1$ can be used in exchange of some $q_1 \in F_1$:

$$\pi_1(x) \in co(\pi_1(\{p_2\} \cup F_1 \setminus \{q_1\})).$$

Then the role of the set F_1 can also be taken by $F'_1 = \{p_2\} \cup F_1 \setminus \{q_1\}$. Note that F'_1 and F_2 have the point p_2 in common. This shows two things. First, it may be assumed without loss of generality that $F_1 \cap F_2$ has *exactly* one point, hence $\#F_2 = c_2$. Second, if this point isn't q_1, then F'_1 and F_2 have **two** points in common and we are done. The only possibility left is that q_1 is the unique point in $F_1 \cap F_2$.

We now consider the symmetric process. Take any $p_1 \in F_1 \setminus F_2$ and exchange with some $q_2 \in F_2$. Again, we are done if q_2 wasn't common to F_1, F_2. In the

§2: Invariants and Product Spaces   185

remaining case we find that $q_1 = q_2$ (= q, say). With F'_2 constructed from F_2 as above, it follows that

$$\pi_1(x) \in co(\pi_1 F'_1); \quad \pi_2(x) \in co(\pi_2 F'_2).$$

Hence x is in the hull of the set $F'_1 \cup F'_2$ which has $c_1 + c_2 - 2$ points since q is missing.

To verify the opposite inequality $c_1 + c_2 - 2 \leq c$, we just consider C-independent sets $F_i \subseteq X_i$ with c_i points for $i = 1, 2$, and proceed with 2.3.3 to obtain a C-independent set in X of cardinality $c_1 + c_2 - 2$. When c_1 or c_2 is infinite we can repeat the argument with an arbitrarily large number. This finishes the proof of part (3). ∎

To handle products with more than two factors, we must keep control of the exchange number of a product. Dealing with this problem first, we introduce the following function to measure the exchange phenomenon.

2.4. The exchange function. Let X be a convex structure and let $F = \{a_1,...,a_n\}$ be an indexed set. As before, we let F_i denote the set $F_{a_i}^{\wedge}$. Let $p \in X$ and $x \in co(F)$. The set

$$E_n(x;p,F) = \{ i \mid 1 \leq i \leq n; x \in co(\{p\} \cup F_i) \}$$

describes the points in F that can be exchanged with p without uncovering x. The *exchange function*

$$\varepsilon = \varepsilon_X : \mathbb{N} \to \mathbb{N} \cup \{0\}$$

of X is defined as follows:

$$\varepsilon(n) = \min \{ \#E_n(x;p,F) \mid \#F = n; x \in co(F); p \in X \}$$

(weighted cardinality). If $\#X \geq n$ then the minimum is attained for an unweighted set, as the reader can verify. Note that $\varepsilon(n) > 0$ iff for each n-point set F and for each $p \in X$ the set $F \cup \{p\}$ is exchangeable. Hence

$$e(X) \leq n \Leftrightarrow \forall k \geq n : \varepsilon(k) > 0.$$

The following auxiliary result is needed.

2.5. Lemma. *Let X be a convex structure of Carathéodory number $c < \infty$. Then the exchange function of X is strictly increasing on $[c, \infty)$.*

Proof. Let $n \geq c$ and let $F \subseteq X$ be of weight $n+1$, say: $a_0,...,a_n$ represents the weighted list of its points. Then F is C-dependent and hence $co(F) \subseteq \cup_{i=0}^{n} co(F_i)$. Let $p \in X$ and $x \in co(F)$. Then one of the points of F -- say a_0 -- is redundant in the sense that $x \in co(F_0)$. A fortiori, x is in $co(\{p\} \cup F_0)$, and hence $0 \in E_{n+1}(x;p,F)$. For any weighted subset[1] G of F of weight n and with $x \in co(G)$, it is evidently true that

$$E_n(x;p,G) \subseteq E_{n+1}(x;p,F).$$

Taking $G = F_0$, we find that the inclusion is proper because the right-hand set contains 0

---

1. It is understood that the weight of a point in the subset does not exceed its weight in the superset.

whereas the left-hand set doesn't. Therefore
$$\varepsilon(n) \leq \#E_n(x;p,F_0) < \#E_{n+1}(x;p,F)$$
and the desired inequality follows by taking the minimum at the right. ∎

As a consequence of this result, note that $\varepsilon_X(k) > 0$ for all $k \geq c(X) + 1$, confirming the Sierksma inequality $e \leq c + 1$ (cf. 1.9(2)).

2.6. Proposition. *Let X_1, X_2 be non-coarse convex structures with product X. Then*
$$e = \max\{c_1, e_1\} + \max\{c_2, e_2\} - 1.$$

Proof. Put $m_i = \max\{c_i, e_i\}$. We first show that $e \leq m_1 + m_2 - 1$. Without loss of generality, m_1 and m_2 are finite. Let $F \subseteq X$ be a weighted set of weight $n \geq m_1 + m_2$, and let $p \in F$ have coordinates (p_1, p_2). To see that F is E-dependent, take $x = (x_1, x_2)$ in $co(F_p^\wedge)$. We let ε_i denote X_i's exchange function. Then $\varepsilon_i(k) > 0$ for $k \geq e_i$, and by Lemma 2.5, $\varepsilon_i(k)$ is strictly increasing for $k \geq c_i$. Hence,
$$\varepsilon_i(n-1) \geq \varepsilon_i(m_i) + (n-1) - m_i \geq n - m_i.$$
Consequently, the sets $E_{n-1}(x_i; p_i, \pi_i(F_p^\wedge))$ for $i = 1, 2$ -- indexed via $F_p^\wedge$ -- each have at least $n - m_i$ points, and hence there must be a point $a \in F_p^\wedge$ common to both. This gives
$$x_i \in co(\pi_i(\{p\} \cup F_a^\wedge)) = co(\pi_i(F_a^\wedge))$$
for $i = 1, 2$, and hence that $x \in co(F_a^\wedge)$ for a suitable $a \neq p$ in F.

To obtain the opposite inequality we first establish the following result, comparable with 2.3.3.

2.6.1. *For $i = 1, 2$, let $F_i \subseteq X_i$ be a set of weight > 1 and with a distinguished point p_i, such that*
$$co(F_i) \not\subseteq \cup\{co(F_i \setminus \{q\}) \mid q \in F_i; q \neq p_i\}.$$
Then the following subset of X is E-independent:
$$F = (F_1 \times \{p_2\}) \cup (\{p_1\} \times F_2).$$

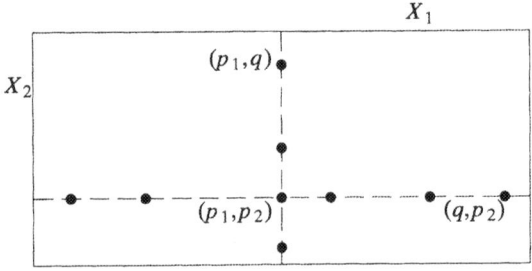

Fig. 2: E-dependent set in a product

Proof. For $i = 1, 2$, consider a point x_i in $co(F_i)$ but not in the sets $co(F_i \setminus \{q\})$ for

§2: Invariants and Product Spaces   187

$q \in F_i$, $q \neq p_i$. Let $x = (x_1, x_2)$ and $p = (p_1, p_2)$. Now $\pi_i(F \setminus \{p\}) = F_i$ since $\#F_i > 1$, and it follows directly that $x \in co(F \setminus \{p\})$. Furthermore, if $a \neq p$ is in F, then (Fig. 2) one of the following occurs.

(i) If $a = (p_1, q)$, then $\pi_1(F \setminus \{a\}) = F_1$ and $\pi_2(F \setminus \{a\}) = F_2 \setminus \{q\}$.
(ii) If $a = (q, p_2)$, then $\pi_1(F \setminus \{a\}) = F_1 \setminus \{q\}$ and $\pi_2(F \setminus \{a\}) = F_2$.

In either case it follows that

$$x \in co(F \setminus \{p\}) \setminus \cup \{ co(F \setminus \{a\}) \mid a \in F; a \neq p \}.$$

This allows to derive the following inequalities.

(i) $e_1 + e_2 - 1 \leq e$;   (ii) $c_1 + e_2 - 1 \leq e$;
(iii) $e_1 + c_2 - 1 \leq e$;   (iv) $c_1 + c_2 - 1 \leq e$.

All four proofs invoke 2.6.1 in about the same way. For each $i = 1, 2$, consider an E-, resp., C-independent subset F_i of X_i with at least e_i, resp., c_i points. The former leads to a point $p_i \in F_i$ with

$$co(F_i \setminus \{p_i\}) \setminus \cup \{ co(F_i \setminus \{q\}) \mid q \in F_i; q \neq p_i \} \neq \emptyset,$$

whereas for the latter kind of set, we have

$$co(F_i) \setminus \cup \{ co(F_i \setminus \{q\}) \mid q \in F_i \} \neq \emptyset,$$

in which case the choice of p_i is indifferent. In either case, $\#F = \#F_1 + \#F_2 - 1$, and F is E-independent. If one of the invariants e_i, c_i is infinite, it should be replaced with an arbitrarily large number in an otherwise identical argument.

In applying 2.6.1, we assumed that $\#F_i > 1$. In case of an E-independent set F_i, this can be achieved because a non-coarse convex structure has an exchange number ≥ 2 (cf. 1.17). In case of a C-independent set F_i, it is possible that $c_i = 1$. But then $e_i > c_i$, and it suffices to handle the inequality where this c_i is replaced by e_i. ∎

We are now in a position to deal with products of several factors.

2.7. Theorem. *Let $n \geq 2$ and let $X_1, \ldots, X_n$ be non-coarse convex structures. Then the following inequalities are valid for the product.*

(1) $e = \sum_{i=1}^{n} \max \{c_i, e_i\} - n + 1$.
(2) *If exactly k of the factors X_i satisfy $e_i \leq c_i$, then $c = \sum_{i=1}^{n} c_i - k$.*

Proof. We first verify that $c < e$ on a product of two non-coarse convex structures Y_1, Y_2. This amounts to the following case study: $c_1 < e_1$ and $c_2 < e_2$. By the Sierksma inequality 1.9(2), we find $c_i = e_i - 1$, whence by Propositions 2.3(1) and 2.6,

$$c = c_1 + c_2 = e_1 + e_2 - 2 = e - 1.$$

$c_1 \geq e_1$ and $c_2 < e_2$. From 2.3(2) and 2.6 we find, similarly, that

$$c = c_1 + c_2 - 1 = c_1 + e_2 - 2 = e - 1.$$

$c_1 \geq e_1$ and $c_2 \geq e_2$. Now $e_i > 1$ since X_i is not coarse (see 1.17), and hence none of

c_1, c_2 equals 1. By 2.3(3) and 2.6 this yields

$$c = c_1 + c_2 - 2 = e - 1.$$

The statements of the theorem are established simultaneously by induction on the number n of factors. For $n = 2$, just use Propositions 2.3 and 2.6. Suppose that the result holds for n factors, $n \geq 2$, and let $X = \prod_{i=1}^{n+1} X_i$. The product of the first n factors is X' and has invariants c', e'. Then $c' < e'$ and we are lead to consider two possibilities. If $c_{n+1} < e_{n+1}$, then

$$c = c' + c_{n+1} = \sum_{i=1}^{n+1} c_i - k,$$

where k is the number of factors X_i with $e_i \leq c_i$ (which are all among $X_1,..,X_n$ now). If $c_{n+1} \geq e_{n+1}$, then

$$c = c' + c_{n+1} - 1 = \left(\sum_{i=1}^{n} c_i\right) - k' + c_{n+1} - 1 = \left(\sum_{i=1}^{n+1} c_i\right) - k,$$

where $k' = k - 1$ is the number of X_i, $i \leq n$, with $e_i \leq c_i$. In either case we obtain

$$e = \max\{c', e'\} + \max\{c_{n+1}, e_{n+1}\} - 1 = \sum_{i=1}^{n+1} \max\{c_i, e_i\} - n. \quad \blacksquare$$

2.8. Examples

2.8.1. Products of trees. Let $n \geq 2$ and let $X_1,..,X_n$ be trees, each with at least three points. The invariants of X_i have the following values (see 1.19).

$$h_i = 2; \quad c_i = 2; \quad e_i = 2.$$

The invariants of $\prod_{i=1}^{n} X_i$ are

$$h = 2; \quad c = n; \quad e = n+1.$$

2.8.2. Products of Euclidean spaces. Let $n \geq 2$, let $d_1,..,d_n > 0$ and for each $i \leq n$, let $\mathbb{R}^{d_i}$ be equipped with the standard convexity. The product space $\mathbb{R}^{d_1} \times .. \times \mathbb{R}^{d_n}$ is of dimension $d = d_1 + .. + d_n$ and its invariants are

$$h = \max\{d_1,..,d_n\} + 1; \quad c = d; \quad e = d+1.$$

It is also possible to compute h with the aid of Theorem 4.7 below.

2.8.3. Products of free spaces. Let $n \geq 2$ and let $\#X_i \geq 2$ for $i = 1,..,n$. Each X_i is taken with the free convexity; so $h_i = r_i = \#X_i$, $c_i = 1$, and $e_i = 2$. The invariants of the product space are

$$h = \max\{\#X_1,..,\#X_n\}; \quad c = n; \quad e = n+1.$$

2.8.4. Products of pospaces. Let $n \geq 2$ and for each $i = 1,..,n$, let X_i be a pospace with more than one point. Then by the results in 1.20, the invariants of the product space satisfy the equalities

§2: Invariants and Product Spaces

$$c = 2n - k; \quad e = \sum_{i=1}^{n} e_i - n + 1,$$

where k (with $0 \le k \le n$) is the number of factors with exchange number equal to two.

2.9. The Radon function. In order to estimate the Radon number of a product, we introduce the following function. Let X be a convex structure. For a finite weighted set F in X, the collection of all (weighted) Radon partitions will be denoted by $R(F)$. Define

$$\rho_X(m) = \min \{ \#R(F) \mid \#F = m \}.$$

This determines the so-called *Radon function* of X,

$$\rho_X: \mathbb{N} \to \mathbb{N} \cup \{0\}.$$

For $m \le \#X$ it is easily seen that the minimum of $\#R(F)$ for $\#F = m$ is attained on an unweighted m-set. The next result is fundamental.

2.10. Lemma (*Extension Lemma*). *Let $1 \le k \le m$ and for each k-set $K \subseteq <m>$ let $\phi(K) \subsetneq K$. Then there are at least*

$$\sum_{i=0}^{m-k} \binom{m-1}{i}$$

many distinct sets $L \subseteq <m>$ such that there is a k-set K with $L \cap K = \phi(K)$.

Proof. Let $L(\phi, m, k)$ denote the number of sets $L \subseteq <m>$ extending some $\phi(K)$ in the sense that $L \cap K = \phi(K)$. Clearly,

$$L(\phi, 1, 1) = 1; \quad L(\phi, m, 1) = 2^{m-1},$$

showing that the lemma holds if $m = k = 1$ or if $k = 1$ and m is arbitrary. We proceed by induction on $m + k$, and we assume $k \ge 2$. We have a function

$$\phi: 2_k^{<m>} \to 2^{<m>},$$

where $2_k^{<m>}$ is the collection of all k-sets in $<m>$. Two new functions arise from ϕ:

$$\phi': 2_k^{<m-1>} \to 2^{<m-1>}, \quad \phi'(K) = \phi(K);$$

$$\phi'': 2_{k-1}^{<m-1>} \to 2^{<m-1>}, \quad \phi''(K) = \phi(K \cup \{m\}) \setminus \{m\}.$$

The collection of all sets extending some $\phi(K)$ can be split into two parts.

(i) The sets $L \subseteq <m>$ extending some $\phi(K)$ with $m \notin K$. Sets L of this type yield $L(\phi', m-1, k)$ many distinct traces $L \cap <m-1>$, each trace extending to two admissible L's. This gives a total of $2L(\phi', m-1, k)$ many sets.

(ii) The sets $L \subseteq <m>$ extending only some $\phi(K)$ with $m \in K$. For such an L, the set $L \cap <m-1>$ extends some $\phi''(K)$ with $K \subseteq <m-1>$ a $(k-1)$-set, but it extends no $\phi'(K)$ for a k-set $K \subseteq <m-1>$. This counts for at least

$$L(\phi'', m-1, k-1) - L(\phi', m-1, k)$$

distinct candidates in $<m-1>$. Each of these is easily seen to be the trace of an admissible set in $<m>$ (if $L' \subseteq <m>$ extends $\phi''(K)$ for some $K \subseteq <m-1>$, then just add m to L' in case $m \in \phi(K \cup \{m\})$ and don't add m otherwise).

This leads to the following estimate.

$$\begin{aligned}
L(\phi,m,k) &\geq 2L(\phi',m-1,k) + L(\phi'',m-1,k-1) - L(\phi',m-1,k) \\
&= L(\phi',m-1,k) + L(\phi'',m-1,k-1) \\
&\geq \sum_{i=0}^{m-k-1}\binom{m-2}{i} + \sum_{i=0}^{m-k}\binom{m-2}{i} \\
&= \sum_{i=0}^{m-k}\binom{m-1}{i}.
\end{aligned}$$

We used the inductive hypothesis and a well-known combinatorial identity. ∎

Sharpness of the previous lemma follows either from Example 2.12 or from the method suggested in 2.25. The following is a naive estimate of the number of extensions. Fix *one* k-set $K \subseteq \, <m>$. Any subset of $<m> \setminus K$ can be added to $\phi(K)$, yielding 2^{m-k} distinct extensions. One can easily verify that the Extension Lemma gives a much better result.

2.11. Corollary. *Let $r < \infty$ be the Radon number of a non-empty space X. Then the Radon function ρ of X satisfies the following inequalities.*

$$\sum_{i=0}^{m-r-1}\binom{m-1}{i} \leq \rho(m) \leq 2^{m-1}-1 \quad (m \in \mathbb{N}).$$

Proof. For $m \leq r$ there is a Radon independent set with m points, and hence $\rho(m) = 0$, in agreement with the above estimates. Suppose $m > r$ and let $F \subseteq X$ be a set indexed by $<m>$. We recall that partitions of F correspond with partitions of the index set. For each $(r+1)$-set $K \subseteq \, <m>$ there is a Radon partition of K of which we fix one member, $\phi(K)$. Any extension L of $\phi(K)$ yields an extended partition $\{L, <m> \setminus L\}$, and this leads to at least

$$\sum_{i=0}^{m-r-1}\binom{m-1}{i}$$

many Radon partitions of F, establishing the left-hand inequality. The upper bound is the total number of partitions of the set $<m>$. ∎

The relevant part of Corollary 2.11 is the lower bound. Its sharpness is illustrated by the next result. A subset F of a matroid of finite rank d is said to be *in general position* provided each subset of F with at most d points is independent. In case of the affine space $\mathbb{R}^n$, let $F = \{p_1,...,p_m\}$ and consider the associated matrix

$$M(F) = \begin{bmatrix} p_{11} & p_{21} & \cdots & p_{m1} \\ \vdots & \vdots & & \vdots \\ p_{1n} & p_{2n} & \cdots & p_{mn} \\ 1 & 1 & \cdots & 1 \end{bmatrix}$$

The ith column represents the n coordinates of p_i with an additional $(n+1)^{\text{th}}$ entry, 1. By

§2: Invariants and Product Spaces   191

definition, F is in general position iff each $(n+1) \times (n+1)$ submatrix of $M(F)$ is invertible.

2.12. Proposition. *The Radon function ρ of $\mathbb{R}^n$ with standard convexity satisfies*

$$\rho(m) = \sum_{i=0}^{m-n-2} \binom{m-1}{i}.$$

Moreover, an m-set in general position has exactly $\rho(m)$ Radon partitions.

Proof. Let $F = \{p_1,...,p_m\}$ be a subset of $\mathbb{R}^n$ in general position. Note that for $m \leq n+1$ the set F is affinely independent and hence R-independent, establishing the result. So assume $m > n+1$. Consider the following system of $n+1$ equalities in m unknowns.

$$(2.12.1) \quad \begin{cases} \sum_{i=1}^{m} x_i \cdot p_i = \mathbf{0}. \\ \sum_{i=1}^{m} x_i = 0. \end{cases}$$

The system's matrix is $M(F)$. Each non-trivial solution of 2.12.1 gives rise to a Radon partition $\{F_1, F_2\}$, where

$$F_1 = \{p_i \mid x_i > 0\}; \quad F_2 = \{p_i \mid x_i \leq 0\}.$$

See 1.4.1. Conversely, if $\{F_1, F_2\}$ is a Radon partition of F with $p \in co(F_1) \cap co(F_2)$ then p is a convex combination of points in F_1 as well as of points in F_2, leading to a non-trivial solution of 2.12.1. Let $V \subseteq \mathbb{R}^m$ be the linear space of all solutions of 2.12.1, and let $H_i \subseteq V$ be the hyperplane determined by $x_i = 0$. As F is in general position, we find that each square $(n+1)$-submatrix of $M(F)$ is of rank $n+1$, whence

$$d = dim \ V = m - n - 1,$$

and for each $k \leq d$,

$$(2.12.2) \quad dim \ (H_{i_1} \cap .. \cap H_{i_k}) = d - k.$$

Let $\mathcal{O}$ be the collection of all components of $V \setminus \cup_{i=1}^{m} H_i$. Each member of $\mathcal{O}$ is an open cone to which the origin is adherent. We construct a function

$$f : \mathcal{O} \rightarrow R(F)$$

as follows. For $O \in \mathcal{O}$ fix a point $(s_1,...,s_m) \in O$ and let F_1, F_2 be determined as above. Note that $s_i = 0$ does not occur. This gives a Radon partition of F depending only on the *sign* of the points s_i, in other words, depending on O only. To see that f is surjective, consider a Radon partition $\{F_1, F_2\}$ and a corresponding solution $s = (s_1,...,s_m) \neq \mathbf{0}$ of 2.12.1. We assume that $s_i \geq 0$ for $p_i \in F_1$ and $s_i \leq 0$ otherwise. Put

$$N = \{i \mid s_i = 0\}.$$

Then $\#N < m-n-1 = d$ since the points p_i are in general position, and by 2.12.2 the restricted coordinate projections

$$\pi_i: V \subseteq \mathbb{R}^m \to \mathbb{R}$$

for $i \in N$ are linearly independent. Hence the set

$$P = \{ x \mid \pi_i(x) > 0 \text{ for } i \in N \text{ and } p_i \in F_1; \; \pi_i(x) < 0 \text{ for } i \in N \text{ and } p_i \in F_2 \}$$

is non-empty. Choose a point $s^* \in P$ close enough to the origin in order that the open disk of radius $\|s^*\|$ around s does not cross any of the hyperplanes H_i with $i \notin N$. Then $s + s^*$ is another solution of 2.12.1 with non-zero coordinates only, and we have

$$F_1 = \{ p_i \mid \pi_i(s + s^*) > 0 \}; \quad F_2 = \{ p_i \mid \pi_i(s + s^*) < 0 \}.$$

We next note that the function f is exactly "2-to-1": only pairs of type $O, -O \in \mathcal{O}$ give rise to the same Radon partition. We conclude that $\#\mathcal{O} = 2 \cdot \#R(F)$. For a counting of $\mathcal{O}$ we consider V to be any d-dimensional real vector space, and we let $H_1,..,H_m$ be hyperplanes through the origin satisfying 2.12.2 for each $k \leq d$.

2.12.3. *The number of components of $V \setminus \cup_{i=1}^{m} H_i$ equals*

$$2 \cdot \sum_{j=0}^{d-1} \binom{m-1}{j}.$$

Indeed, if $d = 1$ or $m = 1$ then there are exactly two components as predicted. We assume $m, d \geq 2$, and we let $C(V; H_1,..,H_m)$ be the number of components of $V \setminus \cup_{i=1}^{m} H_i$. Let $V' = H_m$ and $H'_i = H_i \cap V'$ for $i < m$. By 2.12.2 and by the inductive assumption, the sets H'_i for $i = 1,..,m-1$ are (relative) hyperplanes dividing V' into $C(V'; H'_1,..,H'_{m-1})$ many components. On the other hand, the V–hyperplanes $H_1,..,H_{m-1}$ divide V into $C(V; H_1,..,H_{m-1})$ many components and $C(V'; H'_1,..,H'_{m-1})$ many of them are cut in halves by H_m. Therefore,

$$C(V; H_1,..,H_m) = C(V; H_1,..,H_{m-1}) + C(V'; H'_1,..,H'_{m-1}),$$

and the desired result follows by recursion.

According to Corollary 2.11, our set F in general position has the least possible number of Radon partitions, completing the proof. ∎

The values of the Radon function ρ_n of Euclidean n–space are given in Table 2.1. Note the following identities.

$$\rho_n(m) + \rho_{n+1}(m) = \rho_{n+1}(m+1); \quad \rho_1(m) + m = 2^{m-1}.$$

As a consequence, all values in the table can be computed from the left upper corner to the right lower corner with additions only.

2.13. Theorem. *Let X_1, X_2 be non-empty convex structures. Then the Radon number of the product space satisfies*

$$\max \{ r_1, r_2 \} \leq r \leq r_1 + r_2 - 1.$$

Proof. The first inequality is easily obtained as in 2.1. As for the second inequality, let F be a finite subset of the product with $r_1 + r_2$ points, and let π_i be the ith projection. We regard $\pi_i(F)$ as a set indexed by F, and we let $R_i(F)$ denote the set of all partitions of F which cause an indexed Radon partition of $\pi_i(F)$. If F were Radon-independent, then

§2: Invariants and Product Spaces 193

Table 2.1:

The Radon function of $\mathbb{R}^n$, $\rho_n(m) = \sum_{i=0}^{m-n-2} \binom{m-1}{i}$

m	$\rho_1(m)$	$\rho_2(m)$	$\rho_3(m)$	$\rho_4(m)$	$\rho_5(m)$	$\rho_6(m)$	$\rho_7(m)$	$\rho_8(m)$	$\rho_9(m)$
3	1								
4	4	1							
5	11	5	1						
6	26	16	6	1					
7	57	42	22	7	1				
8	120	99	64	29	8	1			
9	247	219	163	93	37	9	1		
10	502	466	382	256	130	46	10	1	
11	1013	968	848	638	386	176	56	11	1
12	2036	1981	1816	1486	1024	562	232	67	12
13	4083	4017	3797	3302	2510	1586	794	299	79
14	8178	8100	7814	7099	5812	4096	2380	1093	378
15	16369	16278	15914	14913	12911	9908	6476	3473	1471

$R_1(F) \cap R_2(F) = \emptyset$, and hence

$$2^{r_1+r_2-1} - 1 \geq \#(R_1(F) \cup R_2(F)) = \#R_1(F) + \#R_2(F).$$

Filling in the lower bounds of $\#R_i(F)$ as obtained in Corollary 2.11 yields

$$\#R_1(F) + \#R_2(F) \geq \sum_{i=0}^{r_2-1} \binom{r_1+r_2-1}{i} + \sum_{i=0}^{r_1-1} \binom{r_1+r_2-1}{i}$$

$$= \sum_{i=0}^{r_1+r_2-1} \binom{r_1+r_2-1}{i}$$

$$= 2^{r_1+r_2-1},$$

which is a contradiction. ∎

The predicted upper bound is reached in products of Euclidean spaces.

2.14. Proposition. *Let $n_1, n_2 \geq 1$ and let $\mathbb{R}^{n_1}$ and $\mathbb{R}^{n_2}$ be equipped with the standard convexity. Then*

$$r(\mathbb{R}^{n_1} \times \mathbb{R}^{n_2}) = n_1 + n_2 + 1.$$

Proof. For $m = n_1 + n_2 + 1$ we will construct two m–sets $F_i \subseteq \mathbb{R}^{n_i}$, $i = 1, 2$, having no common Radon partition. We start with some general considerations.

Let V be a d–dimensional real vector space, $d > 0$, and let $f_1,...,f_m: V \to \mathbb{R}$, $m \geq d + 2$, be linear functionals with the following properties.

(i) Each d of $f_1,...,f_m$ are linearly independent.
(ii) Each $d+1$ of $f_1,...,f_m$ are affinely independent.

(iii) $\sum_{i=1}^{m} f_i = \mathbf{0}$.

For any $d > 0$ and $m \geq d + 1$ one can easily create examples. By (ii) we have an embedding

$$(f_1,..,f_m): V \to \mathbb{R}^m,$$

and f_i corresponds with the restricted i^{th} coordinate projection π_i. Let V' be the subspace of $\mathbb{R}^m$ defined by

$$V^{\perp} \cap \{x \mid \Sigma \pi_i(x) = 0\}.$$

Here, $V^{\perp}$ denotes the orthogonal complement of V. The dimension d' of V' equals $m-d-1$. Note that $d' > 0$. We let $f'_1,..,f'_m$ denote the restriction to V' of the respective coordinate projections $\pi_1,..,\pi_m$. Then V can be recovered from V' as follows.

$$V = V'^{\perp} \cap \{x \mid \Sigma \pi_i(x) = 0\}.$$

We first show that the role of V and V' is entirely symmetric. It is clear that $\sum_{i=1}^{m} f'_i = 0$. To show that each $d'+1$ of $f'_1,..,f'_m$ are affinely independent, suppose $\sum_{i=1}^{d'+1} a_i f'_i = \mathbf{0}$ and $\sum_{i=1}^{d'+1} a_i = 0$. Then the point

$$a = (a_1,..,a_{m-d},0,..,0)$$

(with d zeros added) is in V and belongs to $\cap_{i=m-d+1}^{m} \ker \pi_i$, whence $a = \mathbf{0}$ by (ii). To show that each d' of $f'_1,..,f'_m$ are linearly independent, suppose $b \in \cap_{i=1}^{d'} \ker f'_i$. As $b \in V'$,

$$\sum_{j=1}^{m} \pi_j(b) f_j = \mathbf{0}; \quad \sum_{j=1}^{m} \pi_j(b) = 0.$$

Also, $\pi_j(b) = 0$ for $j = 1,..,d'$. After dropping the terms with these indices we are left with a relation between $d+1$ functionals $f_{m-d},..,f_m$ which are affinely independent by (iii). Hence the remaining coefficients $\pi_j(b)$ are also zero, showing that $\cap_{i=1}^{d'} \ker f'_i = \{\mathbf{0}\}$. In other words, the functionals $f'_1,..,f'_{d'}$ are linearly independent.

Let $e_1,..,e_n$ be the standard basis of $\mathbb{R}^m$ and let $\tau': \mathbb{R}^m \to V'$ be the orthogonal projection. Each Radon partition of $F' = \{\tau'(e_1),..,\tau'(e_m)\}$ leads to equalities of type

$$\sum_{i=1}^{m} a_i \tau'(e_i) = \mathbf{0}; \quad \sum_{i=1}^{m} a_i = 0.$$

with $a = (a_1,..,a_m) \neq \mathbf{0}$. Then

$$a \in \ker \tau' \cap \{x \mid \Sigma \pi_i(x) = 0\} = V$$

and by an argument as above, no more than $d-1$ coordinates of a can be zero. Reversing the roles, we let $\tau: \mathbb{R}^m \to V$ be the orthogonal projection. Each Radon partition of $F = \{\tau(e_1),..,\tau(e_m)\}$ induces a point $b \neq \mathbf{0}$ in V' with at most $d'-1$ zero coordinates. Hence there is always a common index i with $a_i, b_i \neq 0$.

Assume that the weighted sets

$$F = \{\tau(e_1),..,\tau(e_m)\}; \quad F' = \{\tau'(e_1),..,\tau'(e_m)\},$$

have a common Radon partition, and consider the corresponding points $b \in V'$; $a \in V$.

§2: Invariants and Product Spaces

Then a_i, b_i have the same sign for each i and we find that

$$\sum_{i=1}^{m} a_i b_i > 0.$$

However, this implies that $a \in V$ is not orthogonal to $b \in V'$, a contradiction.

The result follows by taking $m = n_1 + n_2 + 1$ and $d = n_1$. Then $d' = n_2$ and $m \geq d + 2$, $d' + 2$ since $n_1, n_2 \geq 1$. ∎

One can verify that the sets F and F' are in general position in V and V', respectively. The sharpness of the lower bound in Theorem 2.13 can be deduced from 2.24. Theorem 2.17 below can also be used to establish sharpness for most pairs of factor Radon numbers. We now concentrate on products of $n \geq 2$ convex structures. Repeated application of Theorem 2.13 yields an upper bound of type

$$r \leq r_1 + .. + r_n - (n-1).$$

However, it is possible to do somewhat better with the aid of the following result.

2.15. Theorem. *Let $n \geq 2$ and let $X_1, .., X_n$ be non-empty convex structures with product X. If ρ_i denotes the Radon function of X_i, then the Radon function ρ of X satisfies*

$$\rho(m) \geq \sum_{i=1}^{n} \rho_i(m) - (n-1)(2^{m-1} - 1).$$

Proof. We first reconsider the case $n = 2$. Let $F \subseteq X$ be an m-set. Its projections to X_i, $i = 1, 2$, are regarded as sets indexed by F. As in 2.13, we let $R_i(F)$ denote the set of all Radon (weighted) partitions of the ith projection. Then

$$\#R_1(F) + \#R_2(F) - \#(R_1(F) \cap R_2(F)) = \#(R_1(F) \cup R_2(F)) \leq 2^{m-1} - 1,$$

where $R_1(F) \cap R_2(F) = R(F)$, and hence

$$\#R(F) \geq \#R_1(F) + \#R_2(F) - (2^{m-1} - 1)$$
$$\geq \rho_1(m) + \rho_2(m) - (2^{m-1} - 1).$$

Taking the minimum on the left over all m-sets $F \subseteq X$ gives the desired result for $n = 2$. Products with more factors can now be handled by induction. ∎

Corollary 2.11 and Theorem 2.15 lead to the following conclusion.

Let $k \geq 2$, let $X = \prod_{i=1}^{k} X_i$, and let ρ_n denote the Radon function of Euclidean space $\mathbb{R}^n$ as tabulated in Table 2.1. With the abbreviations $r = r(X)$ and $r_i = r(X_i)$ ($i = 1, .., n$),

$$r \leq \max \left\{ m \mid \sum_{i=1}^{k} \rho_{r_i - 1}(m) \leq (k-1)(2^{m-1} - 1) \right\}.$$

The resulting estimates of r are displayed in Table 2.2, where k factors are considered, $3 \leq k \leq 5$, and where the factor Radon numbers r_i are arranged in increasing order. For $k = 3$, the table is limited to $\Sigma r_i \leq 13$; this bound is increased to 15 resp., 17 in case $k = 4$ resp., 5. For $k = 3$ and for low values of r_i only are the tabulated values the same as the ones predicted before. All other estimates are one or more units sharper. For values marked with an asterisk, see notes on this section.

Table 2.2:
Estimated Radon numbers for products with 3, 4, or 5 factors

$k = 3$		$k = 4$		$k = 5$	
(r_1,r_2,r_3)	$r \leq$	(r_1,r_2,r_3,r_4)	$r \leq$	(r_1,r_2,r_3,r_4,r_5)	$r \leq$
2, 2, 2	4	2, 2, 2, 2	5*	2, 2, 2, 2, 2	5
2, 2, 3	5	2, 2, 2, 3	5	2, 2, 2, 2, 3	6*
2, 2, 4	6	2, 2, 2, 4	6	2, 2, 2, 2, 4	7*
2, 2, 5	7	2, 2, 2, 5	7	2, 2, 2, 2, 5	7
2, 2, 6	8	2, 2, 2, 6	8	2, 2, 2, 2, 6	8
2, 2, 7	9	2, 2, 2, 7	9	2, 2, 2, 2, 7	9
2, 2, 8	10	2, 2, 2, 8	10	2, 2, 2, 2, 8	10
2, 2, 9	11	2, 2, 2, 9	11	2, 2, 2, 2, 9	11
2, 3, 3	6	2, 2, 3, 3	6	2, 2, 2, 3, 3	6
2, 3, 4	7	2, 2, 3, 4	7	2, 2, 2, 3, 4	7
2, 3, 5	8	2, 2, 3, 5	8	2, 2, 2, 3, 5	8
2, 3, 6	9	2, 2, 3, 6	9	2, 2, 2, 3, 6	9
2, 3, 7	10	2, 2, 3, 7	10	2, 2, 2, 3, 7	10
2, 3, 8	11	2, 2, 3, 8	11	2, 2, 2, 3, 8	11
2, 4, 4	8	2, 2, 4, 4	8	2, 2, 2, 4, 4	8
2, 4, 5	9	2, 2, 4, 5	9	2, 2, 2, 4, 5	9
2, 4, 6	10	2, 2, 4, 6	10	2, 2, 2, 4, 6	10
2, 4, 7	11	2, 2, 4, 7	11	2, 2, 2, 4, 7	11
2, 5, 5	10	2, 2, 5, 5	10	2, 2, 2, 5, 5	10
2, 5, 6	11	2, 2, 5, 6	11	2, 2, 2, 5, 6	11
3, 3, 3	7*	2, 3, 3, 3	7	2, 2, 3, 3, 3	7
3, 3, 4	7	2, 3, 3, 4	7	2, 2, 3, 3, 4	8
3, 3, 5	8	2, 3, 3, 5	8	2, 2, 3, 3, 5	8
3, 3, 6	9	2, 3, 3, 6	9	2, 2, 3, 3, 6	9
3, 3, 7	10	2, 3, 3, 7	10	2, 2, 3, 3, 7	10
3, 4, 4	8	2, 3, 4, 4	8	2, 2, 3, 4, 4	8
3, 4, 5	9	2, 3, 4, 5	9	2, 2, 3, 4, 5	9
3, 4, 6	10	2, 3, 4, 6	10	2, 2, 3, 4, 6	10
3, 5, 5	10	2, 3, 5, 5	10	2, 2, 3, 5, 5	10
4, 4, 4	9	2, 4, 4, 4	9	2, 2, 4, 4, 4	9
4, 4, 5	10*	2, 4, 4, 5	10	2, 2, 4, 4, 5	10
		3, 3, 3, 3	7	2, 3, 3, 3, 3	7
		3, 3, 3, 4	8	2, 3, 3, 3, 4	8
		3, 3, 3, 5	9	2, 3, 3, 3, 5	9
		3, 3, 3, 6	10	2, 3, 3, 3, 6	10
		3, 3, 4, 4	9*	2, 3, 3, 4, 4	9
		3, 3, 4, 5	9	2, 3, 3, 4, 5	9
		3, 4, 4, 4	9	2, 3, 4, 4, 4	9
				3, 3, 3, 3, 3	8
				3, 3, 3, 3, 4	8
				3, 3, 3, 3, 5	9
				3, 3, 3, 4, 4	9

§2: Invariants and Product Spaces

We now consider products of totally ordered sets. The following combinatorial lemma is required.

2.16. Lemma. *Let the integer k satisfy $0 < k < \tfrac{1}{2}m$. There exists an injective function*
$$\phi: 2_k^{<m>} \to 2_{k+1}^{<m>}$$
such that $A \subseteq \phi(A)$ for all $A \in 2_k^{<m>}$.

Proof. Consider a bipartite graph with a vertex set consisting of all k- or $(k+1)$-sets, and an edge between each two comparable vertices. This graph is bipartite with colors $V_1 = 2_k^{<m>}$ and $V_2 = 2_{k+1}^{<m>}$. There are $\#\mathcal{S} \cdot (m-k)$ edges going out of the set $\mathcal{S} \subseteq V_1$ and at most $k+1$ of them can end at a same vertex. It follows that the total number of proper neighbors of $\mathcal{S}$ is at least
$$\#\mathcal{S} \cdot \frac{m-k}{k+1} \geq \#\mathcal{S}.$$
The result then follows from Hall's Kopplungssatz.[2] ∎

2.17. Theorem. *Let $n \geq 2$ and let $X_1,..,X_n$ be totally ordered sets with product X. Then the Radon number of X satisfies*
$$r \leq \max \{ m \mid \binom{m}{\lfloor \tfrac{1}{2}m \rfloor} \leq 2n \}.$$
Equality holds if each factor X_i is infinite.

Proof. We introduce the permanent abbreviation
$$r_n = \max \{ m \mid \binom{m}{\lfloor \tfrac{1}{2}m \rfloor} \leq 2n \}.$$
To see that $r \leq r_n$, let $F \subseteq X$ be a Radon independent set with m points. For each $A \subseteq F$ with $\#A = \lfloor \tfrac{1}{2}m \rfloor$ we have
$$co(A) \cap co(F \setminus A) = \varnothing,$$
and hence for some $i \leq n$ there is a half-space $H_i \subseteq X_i$ with

(*) $\quad \pi_i(A) \subseteq H_i; \quad \pi_i(F \setminus A) \subseteq X_i \setminus H_i.$

Here, π_i denotes the ith projection. A half-space in a totally ordered set is a lower or an upper subset. Hence a given index i can occur in (*) for at most two sets A. This yields
$$\binom{m}{\lfloor \tfrac{1}{2}m \rfloor} \leq 2n.$$

To show that $r \geq r_n$, an R-independent subset of X must be found with $m = r_n$ points. The strategy is to construct n permutations $\sigma_1,..,\sigma_n$ of $<m>$ such that for each k with $1 \leq k \leq \lfloor \tfrac{1}{2}m \rfloor$ the following holds.

---

2. Bondy and Murty [1976, Thm. 5.2].

(2.17.1$_k$) If $A \subseteq \ <m>$ is a k-set then there is an $i \in \{1,..,n\}$ such that $\sigma_i(A)$ is a lower or an upper subset of $<m>$.

For each $j = 1,..,n$, fix a totally ordered sequence

(**) $x_j(1) < x_j(2) < .. < x_j(m)$

in X_j. The points x^i of an R-independent set in X are located as follows.

$$x^i = (x_1(\sigma_1(i)),..,x_n(\sigma_n(i))).$$

Note that the jth coordinate x^i_j equals the $\sigma_j(i)$th element of the coordinate sequence (**), to the effect that the jth coordinate projection of $F = \{x^i \mid i = 1,..,n\}$ corresponds with the jth permutation. For a given 2-partition of F, consider the smaller set with $k \leq \lfloor \tfrac{1}{2}m \rfloor$ elements. For some j, this set projects to a lower or an upper subset of the jth coordinate sequence, and the resulting partition of the jth coordinate sequence is not Radon. Hence F has no Radon partition.

The permutations will be constructed by downward induction on k, starting at $k = \lfloor \tfrac{1}{2}m \rfloor$. We let $q = \lfloor \tfrac{1}{2}\binom{m}{k} \rfloor$. There is a sequence[3]

$$\{A_1, B_1\},..,\{A_q, B_q\}$$

with q disjoint pairs of disjoint k-sets in $<m>$. As $m = r_n$ we find that $q \leq n$. Also, there is at most one $\lfloor \tfrac{1}{2}m \rfloor$-set C not occurring in the sequence above (in this situation, $q < n$).

We start with q (resp., $q+1$) provisional permutations $\sigma_1,..,\sigma_q$ (σ_{q+1}) of $<m>$, such that σ_i maps A_i and B_i respectively to a lower and an upper set of $<m>$ (and σ_{q+1} maps C to any "end" of $<m>$). This yields (2.17.1$_k$) for $k = \lfloor \tfrac{1}{2}m \rfloor$.

Suppose (2.17.1$_{k+1}$) holds for $k+1 \leq \lfloor \tfrac{1}{2}m \rfloor$. We use 2.16: each k-set D is included in the $(k+1)$-set $\phi(D)$. Now σ_i maps $\phi(D)$ to a lower or upper end of $<m>$, and by suitable rearrangement within $\sigma_i\phi(D)$ we can manage that D is mapped to an initial part of the lower end or to a final part of the upper end. This leads to a modified permutation σ_i, but the previously established properties are not affected.

The induction ends at $k = 1$, and leaves us with q (or $q+1$) permutations of $<m>$ satisfying (2.17.1$_k$) for all k. If q (or $q+1$) $< n$, then adjoin some extra permutations until the required amount of n functions is attained. ∎

The above result applies to products of real lines, and it is therefore of interest to compare the values in Table 2.3 with the ones of Table 2.2: r_n is a somewhat better bound if the number of factors is large. An extension of Theorem 2.17 to products of trees is given in 4.40. The column headed by $\log_2(2n+2)$ anticipates on Topics 2.23 and 4.31, where the Radon number of the graphic n-cube is determined.

---

3.   This is a result of Baranyai [1975].

§2: Invariants and Product Spaces

Table 2.3:
Radon number of the solid and the graphic n-cube

n	r_n	$\log_2 2\cdot(n+1)$	n	r_n	$\log_2 2\cdot(n+1)$
1	2	2	231-461	11	8,9
2	3	2	462-857	12	9,10
3-4	4	3	858-1715	13	10,11
5-9	5	3,4	1716-3217	14	11,12
10-17	6	4,5	10^4	16	14
18-34	7	5,6	10^5	20	17
35-62	8	6	10^6	23	20
63-125	9	7	10^7	26	24
126-230	10	7,8	10^8	30	27

Further Topics

2.18. Example. For $n \geq 1$ consider the convex structure $C(n)$ on $\{0,1,...,n\}$ with, as convex sets, all subsets C satisfying either $\#C \geq n$, or $0 \in C$.

2.18.1. Verify that $e = n+1$ and $h = c = r = n$.

2.18.2. Let $X = C(n) \times C(m)$ and $Y = C(n) \times F(m)$, where $m \geq 2$ and $F(m)$ is the standard m-point free space. Verify that

$$h(X) = \max\{n,m\}; \quad c(X) = n+m-2; \quad e(X) = n+m-1;$$
$$h(Y) = \max\{n,m\}; \quad c(Y) = n; \quad e(Y) = n+1.$$

For products of type $F(n) \times F(m)$, see 2.24.

2.19. Example (Sierksma [1976]). Let $X = \cup_{n=1}^{\infty} X_n$, where $\#X_n = n$ and the sets X_n are pairwise disjoint. Let

$$\mathcal{C}(X) = \{C \mid \exists n : C \subset X_n \text{ or } C = X\}.$$

Show that $c(X) = \infty$ and $e(X) = 2$ (compare Lemma 2.5). Verify that X's exchange function is 0 for $n = 1$ and is 1 for $n > 1$.

2.20. Graph of the exchange function (Sierksma [1976])

2.20.1. Show that if $e < \infty$ then $\varepsilon(e) = 1$, and that $1 \leq \varepsilon(n) \leq n$ for $n \geq e$.

2.20.2. If X has the CUP then $\varepsilon_X(n) > 0$ implies $\varepsilon_X(n+1) > \varepsilon_X(n)$.

2.20.3. For two non-empty sets A, B in a space X, the intermediate cone of B over A is the convex set

$C(B;A) = \bigcap \{co(\{b\} \cup A) \mid b \in B\}$.

Note that $C(B;A) = co(A)$ whenever $B \cap A \neq \emptyset$. Then X has the *Intermediacy Property* ("Continuity property" in Sierksma's terminology) provided for each finite set $A \neq \emptyset$ and for each convex set B with $B \cap A \neq \emptyset \neq B \setminus A$, it is true that $C(B \setminus A; A) = co(A)$.

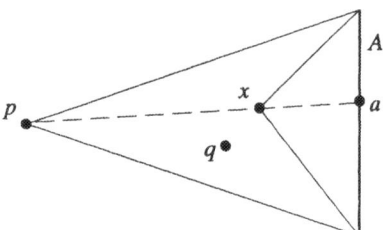

Fig. 3: Intermediacy Property

Equivalently (Fig. 3), if $A \subseteq X$ is finite and non-empty, if $a \in A$, if $p \notin A$ and if $q \in co(A \cup \{p\}) \setminus co(A)$, then there is a point $x \in pa \setminus A$ with $q \notin co(A \cup \{x\})$. Observe that join spaces and H-convexities have this property. Show that if X has the Intermediacy Property, then ε_X is progressing with at most unary steps: $\varepsilon_X(n+1) \leq \varepsilon_X(n) + 1$.

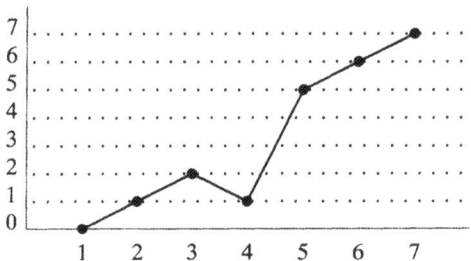

Fig. 4: Declining exchange function of $F_4(\mathbb{N})$

2.20.4. Consider the k-free space $X = F_k(\mathbb{N})$, where $k > 1$. Verify that

$$\varepsilon_X(n) = \begin{cases} n-1, & \text{if } 1 \leq n < k; \\ 1, & \text{if } n = k; \\ n, & \text{if } n > k. \end{cases}$$

For $k = 4$, the graph of ε_X is sketched in Fig. 4. Note that $F_k(\mathbb{N})$ does not have the CUP.

2.21. Relations between e, c

2.21.1. (compare Lassak [1982]) Let X be a convex structure with $c < \infty$. Then the following are equivalent:

(i) each $(c+1)$-set in X is exchangeable.

§2: Invariants and Product Spaces

(ii) $e(X) \le c(X)$.

2.21.2. (compare 2.20.2) Show that $c \le e$ for spaces having the CUP.

2.22. Exchange and Radon function

2.22.1. If $f: X \to Y$ is a CP surjection, then $\rho_X \le \rho_Y$. If f is CC moreover, then $\varepsilon_X \le \varepsilon_Y$.

2.22.2. If X is a subspace of Y, then $\varepsilon_X \ge \varepsilon_Y$. If X is moreover convex, then $\rho_X \ge \rho_Y$.

2.22.3. Show that for each finite set $F \ne \varnothing$ in Euclidean space, $\#R(F) = \rho(\#F)$ iff F is in general position.

2.22.4. Problem. Describe ρ_X in case X is a semilattice (low values expected) and in case X is a lattice (high values expected).

2.23. Free convexity and Radon numbers. Show that the Radon function of the free r-point space $F(r)$ satisfies

$$\rho(m) = 2^{m-1} - 2^{r-1} \quad (m \ge r).$$

Deduce that if $X = \prod_{i=1}^{n} F(r_i)$, then the Radon number r of X satisfies

$$r \le \log_2 \left(\sum_{i=1}^{n} 2^{r_i - 1} - n + 1 \right) + 1.$$

For $X = \{0,1\}^n$ -- the graphic n-cube -- this becomes

$$r \le \log_2(n+1) + 1.$$

Verify that $\log_2 2(n+1) < r_n$ for all $n \ge 2$, where r_n is the general bound presented in Theorem 2.15 (cf. Table 2.3). See 4.31 for the sharpness of the above result.

2.24. Lower bound of r in a product (Eckhoff [1968]). Let $0 < m \le \#X$ and $0 \le n \le \#Y$. Show that the Radon number of the m-free matroid $F_m(X)$ (cf. I§2.22.3) equals m and that

$$r(F_m(X) \times F_n(Y)) = \max\{m, n\}.$$

In particular, the lower bound in Theorem 2.13 is sharp.

2.25. Alternating partitions (Sierksma [1976]). For each k-set $K \subseteq <m>$, arrange the elements in increasing order $i_1 < i_2 < .. < i_k$, and put

$$\phi(K) = \{ i_{2l-1} \mid 1 \le l \le \lceil \tfrac{1}{2}k \rceil \}.$$

ϕ is the *alternating set function*. For $k \ge 3$, show that the number of distinct extensions of the sets $\phi(K)$ is minimal, namely

$$\sum_{i=0}^{m-k} \binom{m-1}{i} \quad \text{(see Lemma 2.10).}$$

2.26. A sharper product formula for r (for Euclidean factors, Eckhoff [1969]). Consider the following statement about a space Y.

(*) If $F \subseteq Y$ is a finite Radon independent subset, and if $G \subseteq Y$ is such that $co(F) \subseteq co(G)$, then $\#F \leq \#G$.

Every Bryant-Webster space, every matroid, and every tree has this property; most (semi)lattices and most pospaces do not. For yet another counterexample, consider an H-convexity on $\mathbb{R}^d$ ($d > 1$) which is symmetrically generated by finitely many linear functionals.

Let X be the product of k spaces each satisfying condition (*) above. Show that

$$r \leq \max \{ m \mid \sum_{i=1}^{k} \rho_{r_i-1}(m) \leq (k-1)(2^{m-1}-1) + m - \sum_{i=1}^{n} r_i \}.$$

Hint: let F be a finite subset of the product space such that $\#F$ is larger than the above indicated maximum. The projection of F to the ith factor is regarded as a set indexed by F, and we let $R'_i(F)$ be the number of its indexed Radon partitions other than the ones of type "1 versus $\#F - 1$". Then estimate $\sum_{k=1}^{m} R'_i(F)$ from below.

Verify that the values in Table 2.2, which are marked with an asterisk, can be decreased with one unit if all factors satisfy (*). Eckhoff op. cit. has shown that many of the (adapted) values in Table 2.2 are sharp for products of Euclidean spaces.

Notes on Section 2

The Helly number of a product with two factors was first determined by Sierksma [1976]. The general result, Theorem 2.1, is described in Soltan [1976]. Proposition 2.2 is new; it is a set-theoretic counterpart of certain results in dimension theory of topological convex structures (see IV§5).

The material on the Carathéodory and exchange numbers in products is largely based on Sierksma's treatment [1976], from which we took the definition of the exchange function. See also Sierksma [1984]. The key to the evaluation of these invariants on products with more than two factors is Lemma 2.5. This is a mild strengthening of a result of Sierksma op. cit., who showed that the exchange function is strictly increasing on $[e, \infty)$ if X has the CUP (see 2.20), and consequently obtained a more restricted version of Proposition 2.6 and the corresponding Theorem 2.7 on general products. These results were obtained by Lassak [1982] and Soltan [1981] with somewhat more elaborate methods.

The Radon function was first used by Eckhoff [1969] in his study of Euclidean products. For general spaces, the function was introduced by Sierksma [1976]. The Extension Lemma, 2.10, and the bounds on the number of Radon partitions, given in Corollary 2.11, were obtained by Boland and Sierksma [1974]; cf. Sierksma [1976]. The short proof of 2.10 is taken from Lindquist and Sierksma [1981]. Proposition 2.12 -- on Euclidean space realizing the Boland-Sierksma lower bound for the Radon function -- has been considered by Eckhoff [1969] (of course, without the above interpretation), and is one of the remarkable results in abstract convexity.

§2: Invariants and Product Spaces

The Radon function has been used by Sierksma [1976] to obtain an upper bound for the Radon number of a product with two factors; the corresponding Theorem 2.13 improves an upper bound obtained earlier by Eckhoff [1968] with one unit. The sharpness of the adjusted bound is discussed in some detail by Eckhoff [1969], who gives Example 2.14 showing that the upper bound is realized by the product of two Euclidean spaces. The argument given here is new, and is less complicated than Eckhoff's original one. The product formula for many factors, Theorem 2.15, is given by Sierksma [1982c], and is almost identical with the special formula obtained by Eckhoff [1969] for Euclidean products (see Topic 2.26). As a consequence, the values given in Table II can hardly be improved in Euclidean spaces: only the values marked with an asterisk can be decreased with one unit.

The formula in 2.17 for the Radon number of products of totally ordered sets is another major achievement of the theory and is due to Eckhoff [1969]. A gap in the original proof was filled by Cochand and Duchet [1983] with the aid of Baranyai's [1975] result on paired sets. Eckhoff's Theorem will be of use in determining the Radon number of median continua with great precision. See Chapter IV.

3. Invariants in other Constructions

In this section we concentrate on constructions other than product spaces, and we determine or estimate the invariants. In the space of arcs over a discrete distributive lattice, the Helly number equals the breadth of the lattice. For disjoint sums, the numbers of Helly, Carathéodory, Radon and Sierksma are determined by the corresponding invariants of the summands. In gated amalgams we are able to determine numbers of Helly, Carathéodory, and Sierksma; the Radon number is determined up to one.

The result on the Helly number of amalgams is the starting point for a brief excursion in polyhedra theory. It is shown that if a compact connected cubical polyhedron carries a median convexity in which cubes are convex subspaces isomorphic with standard median cubes, then its vertex graph is median, and the relative convexity equals the geodesic convexity. Conversely, each finite median graph gives rise to a cubical polyhedron with a median convexity as above by a process called realization. The Carathéodory and Sierksma numbers of a realization are equal to the corresponding invariants of the underlying median graph. This is not true for the Radon number.

In order to determine the Helly number in spaces of arcs, the following auxiliary result on the breadth $b(L)$ of a lattice L is of use (cf. 1.13).

3.1. Lemma. *Let L be a lattice of breadth n and let $C_0,..,C_n$ be order convex meet subsemilattices of L such that for each $i = 0,1,..,n$ there is a chain in L meeting all but the ith set C_i. Then there is a chain meeting all sets C_i.*

Proof. For each $i = 0,..,n$ let $A_i \subseteq L$ be a chain consisting of points $x_j^i \in C_j$, one for each $j \neq i$. For a fixed j we let

$$x_j = \wedge_{i \neq j} x_j^i.$$

Note that $x_j \in C_j$ since the latter is a meet subsemilattice. The breadth of L is the same when described in terms of "$\wedge$" or "$\vee$". Hence, as $b(L) \leq n$ there is an index k with

(1) $\qquad \vee_{j \neq k} x_j = x_0 \vee x_1 \vee .. \vee x_n.$

Assume $x_0^k \in C_0$ to be the largest member of the chain A_k (in particular, we assume that $k \neq 0$). We verify that $\vee_{j=0}^n x_j \in C_0$. For $j \neq k$ we have $x_j \leq x_j^k \leq x_0^k$, and hence $\vee_{j \neq k} x_j \leq x_0^k$. On the other hand, $x_0 \leq \vee_{j \neq k} x_j$ since $k \neq 0$. As C_0 is order convex, we conclude that $\vee_{j \neq k} x_j \in C_0$. Equality (1) then establishes the claim.

Next, consider the chain A_0. Without loss of generality,

(2) $\qquad x_n^0 \leq x_{n-1}^0 \leq .. \leq x_1^0.$

By definition, we have an increasing sequence

(3) $\qquad x_n \leq x_n \vee x_{n-1} \leq .. \leq \vee_{j \geq l} x_j \leq .. \leq \vee_{j=1}^n x_j.$

By (2) we have for each $j \geq l$ that $x_j \leq x_j^0 \leq x_l^0$, and hence that $x_l \leq \vee_{j \geq l} x_j \leq x_l^0$. It follows that $\vee_{j\geq l} x_j \in C_l$ since the last set is order convex.

So far this proves that the points in the chain (3) are successively in $C_n, C_{n-1},..,C_1$. Adding the element $\vee_{j=0}^{n} x_j$ of C_0 completes the proof. ∎

3.2. Theorem. *Let L be a bounded, discrete, distributive lattice. Then the spaces*

$$\Lambda(L,\wedge),\ \Lambda(L,\vee),\ \Lambda(L,\vee,\wedge)$$

have the same Helly number, namely the breadth of L.

Proof. Let $h_\vee$ resp., $h_\wedge$ denote the Helly number of $\Lambda(L,\vee)$ and $\Lambda(L,\wedge)$, respectively. The Helly number of $\Lambda(L,\vee,\wedge)$ will be denoted by h and b is the breadth of L. We will show that

$$b \leq h \leq h_\vee \leq b.$$

The result about $h_\wedge$ follows for reasons of symmetry.

The inequality $h \leq h_\vee$ follows directly from the fact that the convexity of $\Lambda(L,\vee,\wedge)$ is coarser that the one of $\Lambda(L,\vee)$. To see that $h_\vee \leq b$, suppose that $b \leq n$, where $n < \infty$. Let $A_0,..,A_n \in \Lambda(L)$ and assume

$$\bigcap_{i=0}^{n} co\{A_j \mid j \neq i\} = \emptyset.$$

By Theorem I§3.20, there exist half-spaces $\mathcal{H}_i$ ($i = 0,1,..,n$) with

(1) $co\{A_j \mid j \neq i\} \subseteq \mathcal{H}_i$.
(2) $\bigcap_{i=0}^{n} \mathcal{H}_i = \emptyset$.

By Theorem I§3.3.2, for each $i = 0,1,..,n$ there is an order convex subsemilattice $C_i \subseteq L$, representing $\mathcal{H}_i$, $i = 0,1,..,n$. By (1), it follows that $A_j \in \mathcal{H}_i$ for $j \neq i$, and hence that A_j meets all C_i for $j \neq i$. According to Lemma 3.1, there is a chain meeting all sets C_i. An extension of this chain to an arc connecting the universal bounds yields a member of $\bigcap_{i=0}^{n} \mathcal{H}_i$, contradicting (2). This shows that $h_\vee \leq n$. We finally verify that $b \leq h$. Let $b \geq n$, where $n < \infty$. Then L includes a graphic n-cube as a convex sublattice; cf. Theorem 4.19(4). It therefore suffices to prove the result if L is a graphic n-cube. The cube is represented as the power set of $\mathbb{Z}/n\mathbb{Z}$. For each $i = 1,..,n$ consider the linear order

$$i < i+1 < .. < i+n-1 \quad (\text{mod } n).$$

The arc A_i consists of all lower sets of this order. Except for the ith arc, none of A_j contains a "point" between $\{i\}$ and $\{0,1,..,n-1\} \setminus \{i+n-1\}$. The complement of the lattice segment

$$\{S \subseteq L \mid i \in S,\ i+n-1 \notin S\}$$

is a sublattice. Therefore, no arc $A \in co\{A_j \mid j \neq i\}$ contains a set between $\{i\}$ and $\{0,1,..,n-1\} \setminus \{i+n-1\}$. Consequently, the set $\bigcap_{i=0}^{n} co\{A_j \mid j \neq i\}$ is empty, showing that $h \geq n$. ∎

The next results deal with disjoint sums. The following notation will be used

§3: Invariants in other Constructions 207

throughout the proofs of these results. If F is a subset of the sum space and if i is the index of the ith summand X_i, then F_i denotes $F \cap X_i$ ($i = 1, 2$). If f is a convex invariant, then f_i denotes its value on the ith summand.

3.3. Theorem. *Let $X_1,...,X_n$ be convex structures with disjoint sum $X = \Sigma_{i=1}^{n} X_i$. Then the respective Helly numbers satisfy*

$$h(X) = \sum_{i=1}^{n} h(X_i).$$

Proof. It suffices to consider two summands with a finite Helly number. Let $F \subseteq X$ and $\#F > h_1 + h_2$. Then $\#F_1 > h_1$ or $\#F_2 > h_2$, showing that one of the sets $\cap_{a \in F_1} co(F_1 \setminus \{a\})$ or $\cap_{a \in F_2} co(F_2 \setminus \{a\})$ is non-empty. In either case we find that

$$\bigcap_{a \in F} co(F \setminus a) \neq \emptyset.$$

On the other hand there exist sets $F_i \subseteq X_i$ such that $\#F_i = h_i$ and $\cap_{a \in F_i} co(F_i \setminus \{a\}) = \emptyset$ ($i = 1, 2$). Then $F = F_1 \cup F_2$ has $h_1 + h_2$ points and the sets $co(F \setminus \{a\})$ for $a \in A$ have no point in common. ∎

3.4. Theorem. *Let $X_1,...,X_n$ be convex structures with disjoint sum $X = \Sigma_{i=1}^{n} X_i$. Then the respective Radon numbers satisfy*

$$r(X) = \sum_{i=1}^{n} r(X_i).$$

Proof. It suffices to consider two summands with a finite Radon number. If $F \subseteq X$ and $\#F > r_1 + r_2$, then, for instance, $F_1 > r_1$, and we can add F_2 to one of the members of a Radon partition of F_1. On the other hand, for each i there is a set $F_i \subseteq X_i$ without Radon partition, such that $\#F_i = r_i$. If $\{A, B\}$ is a partition of $F_1 \cup F_2$, then the fact that

$$co(A) = co(A_1) \cup co(A_2); \quad co(B) = co(B_1) \cup co(B_2),$$

can be used to see that this is not a Radon partition. ∎

3.5. Theorem. *Let $X_1,...,X_n$ be convex structures with disjoint sum $X = \Sigma_{i=1}^{n} X_i$. Then the respective Carathéodory numbers satisfy*

$$c(X) = \max\{c(X_i) \mid i = 1,...,n\}.$$

Proof. We consider two summands X_1 and X_2 of finite Carathéodory number. Since X_1 and X_2 are subspaces, the number $k = \max\{c_1,c_2\}$ does not exceed c. Conversely, suppose $F \subseteq X$ has more than k points, and let $x \in co(F)$. It must be shown that x is in the hull of a subset with at most k points. We have, for instance, $x \in co(F_1)$. If $\#F_1 \leq k$ we are done. If $\#F_1 > k$, then there is a subset G of F with $\#G \leq c_1$ and $x \in co(G)$, as desired. ∎

3.6. Theorem. *Let $X_1,...,X_n$ ($n \geq 2$) be non-empty convex structures with disjoint sum $X = \sum_{i=1}^{n} X_i$. Then*

$$e(X) = c(X) + 1.$$

Proof. As before, we consider two summands. We first show that $e \leq c + 1$, assuming c to be finite. Let $F \subseteq X$ have more than $c + 1$ points, and let p be one of them. As usual, $F_q^\wedge$ denotes $F \setminus \{q\}$ for $q \in F$. Take $x \in co(F_p^\wedge)$; we may assume that $x \in X_1$. It must be shown that $x \in co(F_a^\wedge)$ for some $a \neq p$ in F. If $F_p^\wedge$ meets X_2, then just take a in the intersection. Suppose $F_p^\wedge \subseteq X_1$. As $F_p^\wedge$ has at least $c + 1$ points, the previous theorem yields that $x \in co(F_{pa}^{\wedge\wedge})$ for some $a \in F_p^\wedge$.

We now verify that $e \geq c + 1$. Without loss of generality, $c_2 \geq c_1$. Take $p \in X_1$ and let $G \subseteq X_2$ be a C-independent set with c_2 points. If this number happens to be infinite, we use an arbitrarily large, finite number instead. There is a point $x \in co(G)$ not in $co(G_a^\wedge)$ for $a \in G$. Let $F = G \cup \{p\}$. For each $a \in F_p^\wedge$ we find that

$$co(F_a^\wedge) = co(\{p\} \cup G_a^\wedge) = co(\{p\}) \cup co(G_a^\wedge),$$

and none of these sets contains x. ∎

We now switch to gated amalgams. This is a rather different method of globalizing the information contained in two subspaces, and it is restricted to interval spaces meeting in a common, gated, convex subspace. See Section I§5.13.

3.7. Theorem. *Let X be the gated amalgam of S_3 spaces X_1, X_2 of arity two. Then*

$$c(X) = \max\{c(X_1), c(X_2)\},$$

unless X_1, X_2 are free convex structures with more than one point, which meet in one point. In this situation, the Carathéodory number is one larger.

Proof. Let $p_i: X \to X_i$ be the gate map ($i = 1, 2$). We adopt the notation of the previous proofs concerning the invariants. The inequalities $c_i \leq c$ ($i = 1, 2$) are clear. Let $n \geq 1$, let $c_i \leq n$, and let $F \subseteq X$ be a finite set with more than n points. If $F \subseteq X_i$ for some i we are done. Otherwise, we decompose F as $F_1 \cup F_2$, with $F_i \subseteq X_i$ (points common to $X_1 \cap X_2$ are distributed randomly over F_1, F_2 to maintain the symmetry of the data). We regard $F_1 \cup p_1(F_2)$ as a weighted set, indexed by $F_1 \cup F_2$. Its total weight exceeds n and hence

$$co(F_1 \cup p_1(F_2)) = \bigcup_{a \in F_1} co(F_1 \setminus \{a\} \cup p_1(F_2)) \cup \bigcup_{a \in F_2} co(F_1 \cup p_1(F_2 \setminus \{a\})).$$

By the hull formula in I§5.15, we have for each $a \in F_2$ that

$$co(F_1 \cup p_1(F_2 \setminus \{a\})) = co(F_1 \cup F_2 \setminus \{a\}) \cap X_1.$$

If F_1 has more than one point, then for each $a \in F_1$,

$$co(F_1 \setminus \{a\} \cup p_1(F_2)) = co(F_1 \setminus \{a\} \cup F_2) \cap X_1.$$

So we are left with a problem if F_1 has only one point, in which case some points of

§3: Invariants in other Constructions 209

$co(p_1(F_2))$ may still be uncovered. We operate the same way on X_2. Again, we have a problem if $\#F_2$ has only one point, and the problem only involves the points of $co(p_2(F_1))$. In the first case, these points are in $co(F) \cap X_2$, and if $\#F_2 > 1$ we are done. By arguing the same way in the second case, we see that the covering problem persists only if both F_1 and F_2 consist of one point. But this implies $n = 1$, meaning that both summands are free spaces which, in order to have a gated intersection, meet in one point. In this situation, the conclusion of the theorem is obviously valid. ∎

The situation is slightly more regular in case of the exchange number:

3.8. Theorem. *Let X be the gated amalgam of S_3 spaces X_1, X_2 of arity two. Then*
$$e(X) = \max\{e(X_1), e(X_2)\}.$$

Proof. We adopt the notation of the previous proof. The inequalities $e_i \leq e$ for $i = 1, 2$ are obvious. We verify that if $n \geq 1$ and $\max\{e_1, e_2\} \leq n$, then $e \leq n$. Let $F \subseteq X$ have more than n points, and let $q \in F$. We assume that $q \in X_2$, and we decompose F into $F_1 \cup F_2$ with $F_1 \subseteq X_1$, $q \in F_2 \subseteq X_2$, where the points of F in $X_1 \cap X_2$ (other than q if the case may be) are randomly distributed over F_1 and F_2.

Application of the hull formula shows that $co(F \setminus \{q\}) \cap X_1$ is covered by the sets $co(F \setminus \{a\})$ for $a \in F \setminus \{q\}$ provided $\#F_1 > 1$, and, if F_1 is a singleton, that the only problem lies in covering points of $co(F \setminus \{q\})$ which are in $co(p_1(F_2)) \subseteq X_2$. Similarly, the set $co(F \setminus \{q\}) \cap X_2$ is covered by the sets $co(F \setminus \{a\})$ for $a \in F \setminus \{q\}$, unless $F_2 = \{q\}$, and the problem only concerns the covering of points in $co(p_2(F_1)) \subseteq X_1$. Therefore, the problem is persistent only if F_1 and F_2 both consist of one point. This implies $n = 1$, meaning that both parts of the amalgam singletons. But even in this case, the proposed formula works. ∎

3.9. Theorem. *Let X be the gated amalgam of S_3 spaces X_1, X_2. Then*
$$h(X) = \max\{h(X_1), h(X_2)\}.$$

Proof. We adopt the notation of the previous proof. As X_i is a convex subspace of X, it is evident that $h_i \leq h$ for both i. We show that for all $n \geq 1$, if $h_i \leq n$ for $i = 1, 2$ then $h \leq n$. Note that a Helly number one belongs to a one-point space, so we are done if $n = 1$. Let $F \subseteq X$ consist of more than n points. If F is included in one of X_1, X_2, then it is a Helly dependent set. In the remaining case, we have e.g that $\#(F \cap X_1) \geq 2$. We define $F_1 = F \cap X_1$ and $F_2 = F \setminus F_1$. Application of the hull formula of Theorem I§5.15 gives

$$\bigcap_{a \in F} co(F \setminus \{a\}) \supseteq \bigcap_{a \in F_1} co(F_1 \setminus \{a\} \cup p_1(F_2)) \cap \bigcap_{a \in F_2} co(F_1 \cup p_1(F_2 \setminus \{a\})).$$

We regard $F_1 \cup p_1(F_2)$ as a weighted set indexed by $F_1 \cup F_2$. Then the right-hand side is non-empty since $h_1 \leq n$. ∎

We finally consider the Radon number of an amalgamation. Observe that the next result is valid for general amalgams.

3.10. Theorem. *Let X be the gated amalgam of two geometric interval spaces X_1, X_2. Then*

$$\max\{r(X_1), r(X_2)\} \le r(X) \le \max\{r(X_1), r(X_2)\} + 1.$$

Proof. The first inequality follows from the fact that X_1 and X_2 are convex subspaces. If $r_i = \infty$ for some $i = 1, 2$ there is nothing left to prove. We assume that $n = \max\{r_1, r_2\} + 1$ is finite. Let $F \subseteq X$ be a finite set with $\#F > n$. If $\#(F \cap X_1) \le 1$ then $\#(F \cap X_2) \ge n > r_2$, and hence $F \cap X_2$ has a Radon partition. We assume that the set $F_1 = F \cap X_1$ has at least two points. Take a point $a \in F_1$, let $F_2 = F \setminus F_1$, and let $p_2: X \to X_2$ be the gate map. We regard $p_2(F_2) \cup (F_1 \setminus \{a\})$ as a weighted set of weight $\ge n > r_1$. This leads to a Radon partition $\{P, Q\}$ of weighted sets.

If both P and Q meet $F_1 \setminus \{a\}$, then a Radon partition of $F \setminus \{a\}$ obtains as follows. Replace each point of $P \cap p_2(F_2)$ and of $Q \cap p_2(F_2)$ by its pre-image in F_2. This changes the sets P and Q into the sets P' and Q' respectively. As p_2 is a gate map we have

$$P \subseteq co(P'); \quad Q \subseteq co(Q'),$$

whence $co(P) \cap co(Q) \ne \emptyset$.

Suppose next that $P \cap (F_1 \setminus \{a\}) = \emptyset$, and construct the sets P' and Q' from P and Q as above. This time we have

$$P \subseteq co(P' \cup \{a\})$$

since $a \in X_1$, and we have $Q \subseteq co(Q')$ as before. Once more, we obtain a Radon partition $P' \cup \{a\}$, Q', this time of F. ∎

In combination with Theorem I§5.15, the result on Helly numbers shows once more that the gated amalgam of median spaces is median (compare Theorem I§6.14). This fact is often used with solid cubes $[0,1]^n$ as the basic building stones. Repeated amalgamation then leads to *cubical polyhedra*. The product convexity of a solid cube will be referred to as the *standard median convexity*.

On a related topic, a *vertex* of a cubical polyhedron is a corner point of some cube of the complex. The set of vertices can be turned into a graph -- the *underlying graph* of the polyhedron -- by the following prescription. Two distinct vertices form an edge provided they occur as the end points of an edge in some cube of the complex. We wish to determine the Helly number of this graph (seen as a subspace of the polyhedron) in case the complex is equipped with a median convexity. We require two auxiliary results.

3.11. Lemma. *Let X be an S_2 space of Helly number ≤ 2, let $a, b \in X$, and let C be a gated subset of X with a gate function $p: X \to C$. If C meets the segment ab, then*

$$ab \cap C = p(a)p(b).$$

Proof. Recall that, if D is a convex set meeting C and if $x \in D$, then $p(x) \in D \cap C$. In the present situation, this gives $p(a), p(b) \in ab$. The inclusion from right to left now follows from the convexness of ab and of C.

§3: Invariants in other Constructions

On the other hand, gate functions are CP and CC, whence $p(co(A)) = co(p(A))$ for any set $A \subseteq X$. If $x \in C$, then evidently $p(x) = x$. Combining these observations for a point $x \in ab \cap C$ yields

$$x = p(x) \in p(ab) = p(a)p(b),$$

thereby establishing the opposite inclusion. ∎

3.12. Lemma. *Let X be an S_2 space of Helly number ≤ 2, let $a_1, a_2 \in X$, and let $b, c \in a_1 a_2$. If $p_i : X \to a_i b$ denotes the nearest point projection and if $p_i(c) = c_i$ for $i = 1, 2$, then the function (p_1, p_2) induces an isomorphism*

$$cb \to c_1 b \times c_2 b.$$

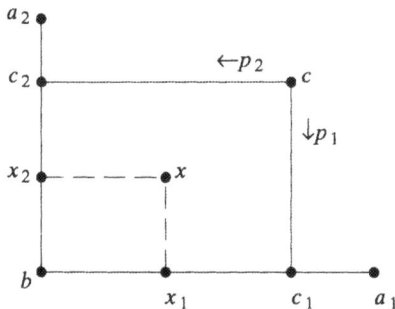

Fig. 1: Product of intervals $c_1 b$ and $c_2 b$

Proof. Let $x_i \in c_i b$ for $i = 1, 2$. The five sets

$$p_1^{-1}(a_1 x_1); \ p_1^{-1}(x_1 b); \ p_2^{-1}(a_2 x_2); \ p_2^{-1}(x_2 b); \ bc$$

meet two by two and hence the intersection $bc \cap p_1^{-1}(x_1) \cap p_2^{-1}(x_2)$ is non-empty. This shows that the function p is surjective.

To see that p is injective, consider a second point y in $p_1^{-1}(x_1) \cap p_2^{-1}(x_2)$, and let $H \subseteq X$ be a half-space with $x \in H$, $y \notin H$. If $b \in H$ then by Proposition I§5.11, we see that x_1 and x_2 are in H. The half-space $X \setminus H$ meets bc in y and hence it contains c. As $c \in a_1 a_2$, we find that e.g., $a_1 \in X \setminus H$. But then $c_1 \in X \setminus H$ and Proposition I§5.11 shows that $p_1(y) \in X \setminus H$, a contradiction. If $b \in X \setminus H$, then a contradiction obtains from similar reasoning. ∎

Although the next results deal with compact connected polyhedra, we do not really use topology: compactness can be replaced by the assumption that the polyhedron has finitely many cubes, and connectedness can be replaced by the assumption that any two vertices can be joined with a path of edges.

3.13. Theorem. *Let P be a compact, connected, cubical polyhedron with a median convexity in which each cube, equipped with its standard median convexity, is a convex subspace. Let G be the underlying graph of P. Then the following are true.*

(1) *G is a median graph.*

(2) *The geodesic convexity of G is precisely the relative convexity, inherited from P.*

(3) *The graphic cubes of G are precisely the intersection of cubes in P with G.*

Proof of (1) and (2). Let $a_1, a_2, a \in G$. We show that the gate of a into the P-segment $a_1 a_2$ is a vertex of G. This goes in three stages. The first one is rather subtle.

Step I. Suppose $a_1 \neq a_2$ form an edge. Let $p: P \to a_1 a_2$ denote the gate function and assume that $p(a) \neq a_1, a_2$. As P is connected, there must be a path of edges connect-

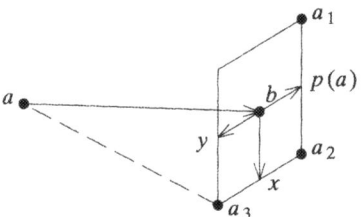

Fig. 2: Median stability of vertices

ing a to the given edge. We consider a path with a minimal number of edges. Let the edge before $a_1 a_2$ be $a_2 a_3$. Note that $a_1 a_2 \cap a_2 a_3 = \{a_2\}$ and hence that $a_2 \in a_1 a_3$ since P has Helly number two. Suppose for the moment that the union of these two line segments is convex. Its relative convexity is median and hence it obtains from the amalgamation of the two line segments. By the gated polytope formula I§5.15, we find that the segment, joining $p(a)$ with a's gate x in $a_2 a_3$, equals $p(a) a_2 \cup a_2 x$. It follows that the convex sets $ap(a)$ and ax both meet $p(a)x$ in a single point, contradicting that the Helly number is ≤ 2.

Consider a point $c \in a_1 a_3 \setminus (a_1 a_2 \cup a_2 a_3)$. By Lemma 3.12, $a_2 c$ is isomorphic to the product $a_2 c_1 \times a_2 c_3$, where c_i denotes the gate of c in $a_2 a_i$ ($i = 1, 3$). This product is non-trivial in the sense that $c_i \neq a_2$ for $i = 1, 3$. Indeed, if (for instance) $c_1 = a_2$, then $a_2 \in ca_1$, whence by the geometric property of the median segment operator (cf. I§6), $c \in a_2 a_3$. Each non-empty intersection of this standard median square with a cube of the polyhedron must be a subsquare (being a convex subset), and the whole product is covered with finitely many of them. Hence, there must be a cube Q such that $p(a) \in Q$ and such that the trace of Q on our square has a non-empty (intrinsic) interior. This implies that $a_1 a_2$ is an edge of Q, and (as the product is non-trivial) that $a_2 a_3$ is an edge of Q. Now we rely on the Transitive Law of gates; cf. I§5.11. The gate of a in $a_1 a_2$ obtains by first gating a into Q, (giving a point b, say) and then gating b into the line segment. Arguing in the standard median cube Q, we see that b must be on the hyperplane perpendicular to the edge $a_1 a_2$ at $p(a)$. But then the gate y of b on the edge opposite to $a_1 a_2$ in the face spanned by a_1, a_2, a_3 is also an interior one. The Transitive Law of gates is used once more to conclude that y is the gate of a in this edge. This leads us to a configuration with a shorter path, a contradiction.

Step II. We now assume that a_1, a_2 span a cube Q of P, and we operate by induction on the dimension $n \geq 1$ of Q, the case $n = 1$ being handled in the previous step. Let

§3: Invariants in other Constructions   213

$p: P \to Q$ be the gate map and let $Q' \subseteq Q$ be the smallest possible cube of P containing $p(a)$. If $dim(Q') = 0$ then we are done. If $dim(Q') > 0$, note that $p(a)$ is at the same time the gate of a into the cube Q', and that this cube is spanned by two vertices of P. This conflicts with the inductive assumption, unless $Q' = Q$. Fix any $(n-1)$–face Q_0 of Q and note that $n-1 \geq 1$. The Transitive Law of gate maps states that the composed projection $P \to Q \to Q_0$ equals the projection $p_0: P \to Q_0$. Evidently, an inner point of Q maps to an inner point of Q_0, so that under the composed mapping, a goes to a non-vertex. Under the second mapping, however, a maps to a vertex by the inductive assumption.

Step III. We finally consider the general case. Let $p: P \to a_1 a_2$ be the projection and let Q be the smallest cube of P containing $p(a)$. By using the gate map $q: P \to Q$, we see (Lemma 3.11) that

$$p(a) \in a_1 a_2 \cap Q = q(a_1)q(a_2).$$

By Step II, the points $q(a_1)$, $q(a_2)$ are vertices. Hence $q(a_1)q(a_2) = Q$ by the minimality of this cube. It appears that $Q \subseteq a_1 a_2$ and as a direct consequence, we have $p(a) = q(a)$. The latter is a vertex by Step II.

We have shown so far that the vertex set G of P is a median stable subset. In particular, the relative convexity of G inherited from P coincides with the convexity, derived from the restricted median. Each non-trivial edge of G comes from an edge of P and is therefore a two-point relatively convex set. Conversely, if a_1, a_2 is a two-point relatively convex set of G, then -- as all convex sets of P are connected -- there is a point

$$b \in a_1 a_2 \setminus \{a_1, a_2\}.$$

As in the proof of Step III, the smallest cube Q of P containing b is included in $a_1 a_2$, and hence the only vertices of this cube are a_1 and a_2: the cube is an edge. By Proposition I§5.8, it follows that the canonical convexity of G is precisely the geodesic convexity; in particular, G is a median graph.

Proof of (3). The solid cube of P inducing a graphic cube Q of G will be denoted by $|Q|$. By definition, every cube of P meets G in a graphic cube. Conversely, let Q be a graphic cube of G of dimension n. By the previous results, each edge of G corresponds to an edge of P. Proceeding by induction on $n \geq 2$, we assume that all $(n-1)$–faces of Q correspond to an $(n-1)$–cube of P. Let Q_1^i, Q_2^i be the ith pair of opposite $(n-1)$–faces of Q, $i = 1,..,n$. Consider a vertex x_1 of Q_1^i, and let $x_2 \in Q_2^i$ be the vertex joined to x_1 by an edge. If x is a point on this edge distinct from the end points, then the sets xx_j ($j = 1, 2$) form a pair of relative half-spaces of $x_1 x_2$. As the gate map of $x_1 x_2$ is CP, this leads us to two half-spaces H_1^i, H_2^i of P such that

$$Q_1^i \subseteq H_1^i \setminus H_2^i; \quad Q_2^i \subseteq H_2^i \setminus H_1^i; \quad H_1^i \cap H_2^i \neq \emptyset.$$

Consider the family $\mathcal{F}$, consisting of the following convex sets of P:

$$co_P(Q), H_j^i \quad (j = 1, 2, \ i = 1,..,n).$$

The members of $\mathcal{F}$ meet pairwise. As far as $co_P(Q)$ is involved, this is evident. Two sets of type H_1^i, H_2^i meet by construction. Two sets with a distinct superscript i meet in a vertex. As the Helly number of P is two, there is a point a common to all members of $\mathcal{F}$. Observe that $a \in co_P(Q)$, but $a \notin |Q'|$ if Q' is a proper face of Q. Let $|Q_0|$ be the smallest cube of P containing a, and consider two diametrical vertices v_1, v_2 of the graphic cube Q. The elementary "calculus" of hull operators yields that

$$co_P(Q) = co_P\{v_1, v_2\} = co_P(co_G\{v_1, v_2\}).$$

Hence $|Q_0|$ meets the segment $v_1 v_2$ of P in a. By using the gate map $p: P \to |Q_0|$, we conclude (Lemma 3.11) that

$$a \in |Q_0| \cap v_1 v_2 = p(v_1) p(v_2).$$

Now $p(v_i)$ is a vertex (Step I), whence $Q_0 = co_G\{p(v_1), p(v_2)\}$ by the minimal choice of this cube. Consequently (Step II),

$$Q_0 \subseteq co_P\{v_1, v_2\} \cap G = co_G\{v_1, v_2\} = Q.$$

Since a is in no solid cube extending a proper face of Q, we conclude that $Q_0 = Q$. In particular, Q extends to a solid cube of P. ∎

Part (3) of the last result states that the cubes of the underlying graph correspond with the cubes of P; informally, P "has no missing cubes". Fig. 3 below depicts a (non-median) polyhedron with missing cubes.

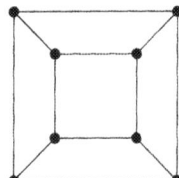

Fig. 3: A planar complex with a missing square and a missing 3–cube.

If G is the underlying graph of a cubical polyhedron P and if $C \subseteq G$, then $|C|$ denotes the union of all solid cubes $|Q|$ of P, where $Q \subseteq C$ is a graphic cube of G. Sets of this type are involved in the next results.

3.14. Lemma. *Let P be compact connected cubical polyhedron with a median convexity such that each cube of P occurs as a convex set of which the relative convexity equals the standard median convexity. If C is a geodesically convex subset of the underlying graph, then $|C|$ is a convex set of P.*

Proof. Let G be the underlying graph. As G is a subspace of P (Theorem 3.13), we have $|C| \subseteq co_P(C)$. Suppose $x \in co_P(C) \setminus |C|$, let $Q \subseteq G$ be the smallest cube such that $x \in |Q|$, and let $q: P \to |Q|$ be the gate map. Then $Q \not\subseteq C$ and, again since G is a subspace, $Q \not\subseteq co_P(C)$. Hence there is a half-space $H \subseteq P$ with $Q \not\subseteq H$ and $C \subseteq H$. Note that $x \in H$. Let $Q' \subseteq Q$ be the maximal face included in H. As $|Q'|$ is a relative half-space of $|Q|$, we obtain a half-space $H' = q^{-1}(|Q'|)$ of P. Then $P \setminus H'$ is a half-space containing x and hence it intersects C, say in v. However, the gate of v in Q is in the strip

§3: Invariants in other Constructions 215

$|Q| \cap H \setminus H'$, which contains no vertices of Q. This contradicts the fact that G is median stable. ∎

We now consider the reverse problem of extending a median graph to a cubical polyhedron, and of extending its convexity to a median convexity of the polyhedron.

3.15. Realization. Let G be a finite graph. For each induced graphic cube $Q \subseteq G$ we consider a solid cube $|Q|$ of the same dimension, such that for each pair of graphic cubes Q_1, Q_2 of G,
$$|Q_1 \cap Q_2| = |Q_1| \cap |Q_2|.$$
This leads to a compact cubical polyhedron $|G| = \cup_Q |Q|$ with G as its underlying graph. It is called the *realization of G*.

Although the details of this realization procedure are immaterial for the results below, we give a brief description of how it can be done. For each $n \geq 0$, let V_n be the set of all graphic n-cubes of the graph G. In particular, V_0 is the vertex set and V_1 is the edge set. The resulting set $\cup_n V_n$ is embedded as an affinely independent subset of a real vector space. For each $v \in V_0$ we let $|\{v\}| = \{v\}$. Having defined the realization of each $Q \in V_n$, let $Q \in V_{n+1}$ and let $|Q|$ be the union of all line segments joining Q with the points of $|Q'|$ for each n-face Q' of Q. Observe that there is a natural PL (piece-wise linear) isomorphism of $|Q_n|$ with $[0,1]^n$ for each n-cube Q_n of G.

3.16. Theorem. *Let G be a median graph. Then the realization $|G|$ carries one and only one median convexity such that*

(i) *Each realized cube is a convex set and its relative convexity is the standard median convexity.*
(ii) *G is a subspace of $|G|$.*

Proof. The theorem is obviously valid if G is a graphic cube. We proceed by induction on the number of cubes needed to cover G. As G is not a cube itself, there is a pair of non-trivial, non-disjoint half-spaces $H_1, H_2 \subseteq G$ covering G (cf. I§6.35.3). If Q is a graphic cube meeting H_1 and H_2, then (Helly number 2 property) $H_1 \cap H_2 \cap Q \neq \emptyset$. Applying the graphic cube characterization once more, we conclude that Q is included in one of the half-spaces.

We have shown so far that $|H_1| \cup |H_2| = |G|$. Moreover, each of H_1, H_2 is a median graph coverable with less cubes than G. By assumption, each of them can be realized as a cubical polyhedron with a median convexity as required by the theorem. Observe that
$$|H_1| \cap |H_2| = |H_1 \cap H_2|; \quad |H_1| \cup |H_2| = |G|.$$
Combining the Amalgamation Theorem I§5.13 and Theorem I§6.14, there is a median convexity on $|G|$ extending the convexity of $|H_1|$, $|H_2|$, and G. Since each cube of G is in one of the canonical half-spaces, this convexity is as required.

To see that the convexity is unique, observe that if a median convexity on $|G|$ is as required, then the sets $|H_1|$ and $|H_2|$ are convex; cf. Lemma 3.14. The relative convexity on these subsets satisfies the requirements of the theorem and hence it is unique. Another application of the Amalgamation Theorem I§5.13 yields that the convexity of G is unique. ∎

The passage from a median graph to its realization is yet another construction in convexity and the effect on the invariants is worth considering. It is clear that the graphic n-cube and the corresponding solid n-cube have the same Carathéodory number if $n \neq 1$ and they have the same exchange number for all n. As a consequence of Theorems 3.7 and 3.8, we find

3.17. Theorem. *Let G be a finite median graph. Then $e(|G|) = e(G)$ and, if G is not reduced to an edge, $c(|G|) = c(G)$.* ∎

It appears from 2.23 that the Radon number of $\{0,1\}^n$ is strictly smaller than the Radon number of its realization $[0,1]^n$ for each $n \geq 2$.

Further Topics

3.18. Sharpness of the inequalities between h, r, c, e

3.18.1. *Levi inequality $h \leq r$* (compare Hammer [1977]). Let $h, r \in \{2, 3, .., \infty\}$ be such that $h \leq r$. Show that there is a convex structure (even one which is JHC and S_4) with h and r as its Helly- resp., Radon number. Hint: for $r < \infty$, the superextension $\lambda(r)$ of a free r-point set has Radon number r; cf. 4.22.1. Consider a disjoint sum of such a space with a number of singleton spaces.

3.18.2. *Sierksma inequalities $e - 1 \leq c \leq \max\{h, e - 1\}$.*
Case $h \leq e - 1$ (Sierksma [1981]). Note that $c = e - 1$ in the present circumstances. Use products of Euclidean spaces to obtain sharpness of the above (in)equalities.
Case $h > e - 1$ (compare Kołodziejczyk [1987]). Note that $c \leq h$ in this case. Let e, c, h be integers such that $2 \leq e \leq c \leq h$. Consider three mutually disjoint sets X_e, X_c, X_h with e, c and h points respectively. In X_e we consider a "distinguished" point $\mathbf{0}$. Let $X = X_e \cup X_c \cup X_h$. A proper subset C of X is convex iff one of the following alternatives holds.
(i) C is a proper subset of X_e, and if $\#C \geq e - 1$ then $\mathbf{0} \in C$.
(ii) C is a proper subset of X_c.
(iii) $C \subseteq X_h$.
Observe that if $e = 2$ then (i) reduces to $C = \emptyset$ or $C = \{\mathbf{0}\}$ and that for $n = e - 1$ the space $C(n)$, described in 2.18, is isomorphic with the subspace X_e of X. Show that the numbers e, c, h are the exchange-, Carathéodory- and Helly number of X, respectively.

3.18.3. *Special Eckhoff-Jamison inequality, $r \leq (\max\{e, c\} - 1)(h - 1) + 2$* (Sierksma [1981]; cf. 1.18). Use disjoint sums and products of Euclidean spaces to produce

§3: Invariants in other Constructions   217

examples with $r = (c-1)(h-1)+2$ and $c \le 3$.

3.18.4. Problems. Is the Eckhoff-Jamison inequality sharp? Explicitly, let $h, c, r \in \{1,..,\infty\}$ be such that $h \ne 1$ or $c \ne \infty$ and $r \le c(h-1)+1$. Is there a convex structure with these numbers as (canonically assigned) invariants? Can these convex structures be chosen such as to be S_4? JHC? If one assumes S_4 and $h = 2$ (which implies JHC; cf. Section I§6), the answer turns out to be negative (cf. 4.20.2). Is Sierksma's modification of this inequality sharp?

3.19. Disjoint sums and unions

3.19.1. Let $C_1,..,C_n$ be convex subsets of a convex structure X. Verify the following (in)equalities.

$$c(\bigcup_{i=1}^{n} C_i) = \max\{c(C_i) \mid i = 1,..,n\};$$

$$\max\{h(C_i) \mid i = 1,..,n\} \le h(\bigcup_{i=1}^{n} C_i) \le \sum_{i=1}^{n} h(C_i);$$

$$\max\{r(C_i) \mid i = 1,..,n\} \le r(\bigcup_{i=1}^{n} C_i) \le \sum_{i=1}^{n} r(C_i).$$

The inequalities at the right are sharp; they are even valid for non-convex C_i.

3.19.2. In the previous situation, assume that $h(X) = 2$, $n \ge 2$ and that the convex sets C_i meet. Show that the Helly number of $\cup_{i=1}^{n} C_i$ is at most n.

3.19.3. Let $Q^n = \{0,1\}^n$ and let $\{C_1,..,C_n\}$ be the set of faces through **0**. Show that the Helly number of $\cup_{i=1}^{n} C_i$ is exactly n (establishing sharpness of the previous result). Verify that the exchange number of a 2–face of Q^3 is 3, whereas the exchange number of the union of all faces of Q^3 passing through **0** equals 4. Similar results are valid for *solid* cubes.

3.20. Convex hyperspaces (compare van de Vel [1983g]). Let X be an S_1 convex structure and let $\mathcal{C}_*(X)$ denote its convex hyperspace. Show that

$$h(X) = h(\mathcal{C}_*(X)); \quad r(X) \le r(\mathcal{C}_*(X)),$$

and that the last inequality is sharp. Hint: use $X = \lambda(n)$ for $1 \le n < \infty$.

3.21. Cones

3.21.1. Verify the following equalities.

$$h(\Delta(X)) = h(X)+1; \quad r(\Delta(X)) = r(X)+1; \quad e(\Delta(X)) = e(X)+1.$$

3.21.2. If X satisfies the CUP, then $c(X) \le c(\Delta(X)) \le c(X)+1$. The first inequality is valid without CUP as well.

Problem. Can the second inequality be replaced by an equality?

3.22. Convex expansion (cf. I§3.23). Let E be a convex expansion of X, where $\#X > 1$. Show that $h(E) = h(X)$ and if f denotes one of c, r, e, then

$$f(X) \le f(E) \le f(X) + 1.$$

3.23. Medians and cubes

3.23.1. Let $Q = [0,1]^n$ and let $\mathcal{C}$ be a median convexity on Q, subject to the following conditions.

(i) Q is the hull of its corner points.
(ii) Each edge of Q is a convex set and its relative convexity is the natural one.

Show that $(Q, \mathcal{C})$ is isomorphic with the standard median cube.

3.23.2. Let T be a finite tree. Show that there is a median convexity on $[0,1]^2$ in which a copy of the realization $|T|$ occurs as a convex set. Hint. Use the amalgamation process to realize the ramifications of T successively by attaching a few standard median squares. Note that at each stage the resulting polyhedron is a topological copy of $[0,1]^2$.

Notes on Section 3

The equality of the Helly number of a space of arcs and the breadth of the discrete lattice underneath (cf. Theorem 3.2) has been obtained by Bandelt and van de Vel in [1992]. The proof is a modification of a result by the author [1988] concerning spaces of arcs over compact connected lattices.

The results on convex invariants of disjoint sums (Theorems 3.3 to 3.6) are taken from Sierksma [1981]. The combination of these results with the product theorems of the previous section lead Sierksma to a discussion on the sharpness of the various inequalities between the classical invariants. The problem on the sharpness of the (special) Eckhoff-Jamison inequality -- cf. 3.18 -- has been particularly popular in the late seventies. In spite of considerable efforts, the problem seems to have been abandoned in a rather poor state of solution.

The results on gated amalgams (Theorems 3.7 to 3.10) are taken from Bandelt, Chepoi, and van de Vel [1993]. A discussion of median convexity on median polyhedra was first presented by the author in [1983e]. One of the conditions used in this paper (concerning the realization of a graphic convex set) turned out to be redundant; see Lemma 3.14. The improved result, Theorem 3.16, is taken from the author's paper [1990]. Verheul [1993] has studied median metrics on cubical polyhedra.

4. Infinite Combinatorics

In this section we highlight some aspects of classical and other invariants which partially go beyond ordinary finite combinatorics. The departing result is the Compactness Theorem, which states that a convexity on a set X induces a compact subset of the Cantor cube 2^X. This leads to the so-called Compact Intersection Theorem, which allows to deduce that certain compact families of convex sets have a non-empty intersection. This fact has numerous consequences: for low Helly numbers, weak separation properties imply stronger ones; the Helly number of H-convexities can be determined with relative ease; and finally, a countable intersection theorem can be derived for convex sets in a space satisfying rather weak assumptions on separation and Helly independence.

There are other convex invariants measuring combinatorial phenomena. For instance, the rank measures the amount of variation among the convex sets. It has briefly been met as a fundamental parameter in a study of matroids (cf. I§2.5) and it is an upper bound of all classical invariants. The generating degree dominates the rank, and frequently both are equal. It is defined as a measure of the least possible complexity of a subbase. Thus, it controls the generating of a convexity, as it is the case with another invariant: the directional degree. Applications involve the determination of Radon numbers in spaces of low Helly number, and a computation of the exchange number in median spaces.

4.1. The Cantor cube. The symbol 2^X has so far been used to describe the power set of X. If one regards 2 as shorthand notation for $\{0,1\}$ then 2^X refers to the X-fold product of a 2-point set, which is called a *Cantor cube*. The formal correspondence between 2^X and $\{0,1\}^X$ is given by

$A \subseteq X \Leftrightarrow$ characteristic function of A.

A Cantor cube is given the product topology ($\{0,1\}$ is taken with the discrete topology). The resulting space is Hausdorff, and is compact by the Tychonov Theorem.[1] This topology can be transferred to 2^X as follows. For $A, B \subseteq X$ let

$\mathbb{E}(A,B) = \{Y \in 2^X \mid A \subseteq Y; B \cap Y = \emptyset\}$.

(the symbol $\mathbb{E}$ stands as a contraction of I -- from inclusion -- and E -- from exclusion). The family of all sets of type

$\mathbb{E}(F,G)$ (F, G finite),

is an open base for the topology of 2^X. Note that $\mathbb{E}(F,G)$ is at the same time a compact set. The resulting topology on 2^X is known as the *inclusion-exclusion topology*.

4.2. Theorem (*Compactness Theorem*). *A convexity on X is a compact subset of 2^X, relative to the inclusion-exclusion topology.*

---
1. Engelking [1977, 3.2.4].

Proof. Let $A \subseteq X$ be non-convex. By domain finiteness (cf. Theorem I§1.3(2)) we obtain a finite set $F \subseteq A$ and a point $p \in co(F) \setminus A$. Then $\mathbb{E}(F, \{p\})$ is a basic neighborhood of $A \in 2^X$, and no member of it can be convex. Hence the convex sets of X form a closed collection in 2^X. ∎

In spite of its short proof, this result is fundamental. It lies at the basis of many other results of this section. The first one involves the following concept. A *copoint at* p is a convex set C, maximal with the property that $p \notin C$ (cf. I§3.24).

4.3. Theorem. *Let X be a convex structure.*

(1) *The collection of all half-spaces in X is a compact subset of 2^X, relative to the inclusion-exclusion topology.*

(2) *The closure in 2^X of a subbase includes all copoints of X.*

Proof. As usual, $\mathcal{C}(X)$ denotes the convexity of X. Then $\mathcal{C}(X)$ is compact as a subspace of 2^X, Theorem 4.2. On the other hand, the operation of taking complements,

$$c : 2^X \to 2^X, \quad A \mapsto X \setminus A,$$

is easily seen to be continuous. Hence $c(\mathcal{C}(X)) \cap \mathcal{C}(X)$ is compact, and this is exactly the collection of all half-spaces of X.

As to the second part, let $\mathcal{S}$ be a convex subbase of X, let $P \subseteq X$ be a copoint at $p \in X$, that is: P is a convex set, $p \notin P$, and P is maximal with these properties. For each finite set $F \subseteq P$ the set $\mathbb{E}(F, \{p\})$ meets $\mathcal{S}$ since $\mathcal{S}$ is a subbase. In this way, we obtain a down-directed family of non-empty compact sets of type $\mathbb{E}(F, \{p\}) \cap \mathcal{S}$, where $F \subseteq P$ is finite. A common member of the sets $\mathbb{E}(F, \{p\})$ includes P and is disjoint from p. Therefore, $P \in \overline{\mathcal{S}}$. ∎

We will first investigate a few deeper properties involving the Helly number. The first one is concerned with the following problem. Let X have Helly number $h < \infty$ and let there be given a *compact* family of convex sets in X meeting h by h. Is it true that the whole family has a non-empty intersection (compare Theorem 1.7)? For a direct approach one may wish to prove that a compact family with an empty intersection admits a finite subfamily with this property. However, there is a simple counterexample: let X be an infinite free convex structure and consider the following compact subcollection of 2^X.

$$\{X \setminus \{p\} \mid p \in X\} \cup \{X\}.$$

Yet, the original question has an affirmative answer.

4.4. Theorem (*Compact Intersection Theorem*). *Let X be a convex structure of Helly number $h < \infty$, and let $\mathcal{S}$ be a family of convex sets, compact in 2^X. If $\cap \mathcal{S} = \emptyset$, then some subfamily of at most h sets from $\mathcal{S}$ has an empty intersection.*

Proof. Without loss of generality, $\mathcal{S}$ does not contain the empty set. Consider a decreasing chain $(\mathcal{S}_i)_{i \in I}$ of compact families $\mathcal{S}_i \subseteq \mathcal{S}$, each with an empty intersection, and let $\mathcal{S}_\infty = \cap_{i \in I} \mathcal{S}_i$. Assume that the point p is common to all members of $\mathcal{S}_\infty$. Then

§4: Infinite Combinatorics   221

(with the notation of 4.1) we have

$$\mathcal{S}_\infty \subseteq \mathbb{E}(\{p\},\emptyset),$$

where the latter is a basic open set. Hence by compactness of the families $\mathcal{S}_i$ there is an $i \in I$ with

$$\mathcal{S}_i \subseteq \mathbb{E}(\{p\},\emptyset),$$

and it follows that p is a common point of $\mathcal{S}_i$, a contradiction. We conclude that the family of all compact subcollections of $\mathcal{S}$ having an empty intersection is (downward) inductive, and by Zorn's lemma it has a minimal member $\mathcal{S}_0$.

Suppose $\mathcal{S}_0$ to be infinite; in particular, there is a cluster point $D_0 \in \mathcal{S}_0$. Then $D_0 \neq \emptyset$ and we fix a point $x_1 \in D_0$. Proceeding by induction, suppose $x_1,..,x_n$ have already been chosen in D_0 such that if $1 \leq i < j \leq n$ and if

$$\mathcal{S}_i = \{ D \in \mathcal{S}_0 \mid x_i \notin D \},$$

then x_j is in each member of $\mathcal{S}_i$. Let

$$\mathcal{C} = \mathcal{S}_1 \cup .. \cup \mathcal{S}_n \cup \{D_0\}.$$

Clearly, $\mathcal{C}$ is closed in $\mathcal{S}_0$, and D_0 is an isolated member of it, since the family

$$\{Y \subseteq X \mid x_i \in Y \text{ for } i=1,..,n\}$$

is an open set of 2^X meeting $\mathcal{C}$ in D_0 only. This shows that $\mathcal{C}$ is properly included in $\mathcal{S}_0$, and we conclude from the minimality of the latter that $\cap \mathcal{C} \neq \emptyset$. Let x_{n+1} be a point in this intersection. Note that x_{n+1} is in D_0 and is in all members of $\mathcal{S}_i$ for $i < n+1$.

This inductive process can go on forever, but we stop at $n = h+1$. Let

$$P_i = co\{x_1,..,\hat{x_i},..,x_{h+1}\}$$

(the hat refers to an element that has to be omitted from the list). and let $D \in \mathcal{S}_0$. We show that $P_i \subseteq D$ for some i. If D contains all x_i then we are done. So assume that D misses some of the points x_i, and let i be the first index with this property. By the construction of $\mathcal{S}_i$ and of x_j we have $D \in \mathcal{S}_i$ and $x_j \in D$ for $j > i$, whereas $x_j \in D$ for $j < i$ by the minimal choice of i. Therefore, D contains all x_j except for $j = i$, whence $P_i \subseteq D$.

We conclude that $P = \cap_{i=1}^{h+1} P_i$ is empty, being included in $\cap \mathcal{S}_0$. However, this implies that the set

$$\{x_1,..,x_{h+1}\}$$

is Helly independent, a contradiction.

We have proved so far that the minimal family $\mathcal{S}_0$ must be finite. By Theorem 1.7(1), it can have no more than h members.   ∎

This result enables us to determine the Helly number by means of subbases.

4.5. Corollary. *Let $\mathcal{S}$ be a compact subbase of a convex structure of finite Helly number h, and let $n < \infty$. Then the following assertions are equivalent.*
(1) $h \leq n$.
(2)   *If $m \geq n$ and $S_1,..,S_m \in \mathcal{S}$ meet n by n, then $\cap_{i=1}^m S_i \neq \emptyset$.*

Proof. The implication (1) $\Rightarrow$ (2) is a particular case of Theorem 1.7(1). We assume (2) and derive (1) as follows. Let $m > n$ and let $F = \{a_1,..,a_m\}$ have m points. For each i we let $\mathcal{S}_i$ be the collection of all $S \in \mathcal{S}$ such that $a_j \in S$ for all $j \neq i$. Then $\mathcal{S}_i$ is a closed subcollection of the compact set $\mathcal{S}$ and by the property S_3,

$$\cap \mathcal{S}_i = co(F \setminus \{a_i\}).$$

Suppose $\cap_{i=1}^m co(F \setminus \{a_i\}) = \emptyset$. Then the compact collection $\mathcal{S}_1 \cup .. \cup \mathcal{S}_m$ has an empty intersection. As h is finite we conclude from Theorem 4.4 that some finite subcollection $\mathcal{S}_0$ has an empty intersection. By (2), we may assume that $\mathcal{S}_0$ has no more than n members. Each of these includes one of the sets $co(F \setminus \{a_i\})$, and as $m > n$, they do have some vertex a_j in common.

From this contradiction we conclude that the sets $co(F \setminus \{a_i\})$ for $i = 1,..,m$ have a common point, whence F is Helly dependent. ∎

In regard to Corollary 4.3 and Theorem I§3.9, the corollary applies to the set of half-spaces of an S_3 convex structure. For further information, consult 4.26.

4.6. Minimal dependence of functionals. Let V be a vector space over a totally ordered field $\mathbb{K}$. A weighted set $\{f_1,..,f_n\}$ of linear functionals $V \to \mathbb{K}$ is called *minimally (linearly) dependent* provided

(MD-1) $\{f_1,..,f_n\}$ is linearly dependent.

(MD-2) For each $i = 1,..,n$ the indexed collection $\{f_1,..,\hat{f_i},..,f_n\}$ is linearly independent.

For instance, $\{f,f\}$ is minimally dependent (at least, if $f \neq 0$), but $\{f,f,g\}$ is not. Under the assumption of (MD-2), statement (MD-1) is equivalent with the following apparently stronger statement.

(MD-1′) For each i the functional f_i is dependent of f_j, $j \neq i$.

If $\mathcal{F}$ is a collection of linear functionals $V \to \mathbb{K}$, then the *degree of minimal dependence of* $\mathcal{F}$ is the supremum $md(\mathcal{F})$ of all $n < \infty$ such that there exists a minimally dependent sequence of length n in $\mathcal{F}$. In other words, we have for each $n < \infty$ that

4.6.1. $md(\mathcal{F}) \leq n$ iff for each $m > n$ and for each sequence $f_1,..,f_m$ of linearly dependent functionals in $\mathcal{F}$ there is an i such that $f_1,..,\hat{f_i},..,f_m$ are linearly dependent. ∎

We need the following general result on linearly independent functionals.

4.6.2. Let $m \geq n$ and let $f_1,..,f_m$ be linear functionals of a vector space V into its coefficient field $\mathbb{K}$. Then the following are equivalent.
(i) At least n of the functions f_i are linearly independent.
(ii) The image of the linear function $f = (f_1,..,f_m): V \to \mathbb{K}^m$ is at least n-dimensional.

Moreover, if $\mathbb{K}$ is totally ordered then $f_1,..,f_m$ are linearly independent iff for each $i = 1,..,m$ and for each choice of $t_i \in \mathbb{K}$, if E_i is a subset of $\mathbb{K}$ of type

$$(t_i,\infty); \quad [t_i,\infty); \quad (\infty,t_i); \quad (\infty,t_i],$$

§4: Infinite Combinatorics   223

then $\cap_{i=1}^m f_i^{-1}(E_i) \neq \emptyset$.

Proof. The equivalence of (i) and (ii) is standard. As to the final part, let $\pi_i: \mathbb{K}^m \to \mathbb{K}$ denote the projection to the ith factor. With $E_i \subseteq \mathbb{R}$ as above and $H_i = \pi_i^{-1}(E_i)$, we have

$$\bigcap_{i=1}^n f_i^{-1}(E_i) = f^{-1}(\bigcap_{i=1}^n H_i).$$

If $f_1,\ldots,f_m$ are linearly independent, then by (ii) we find that f is surjective and hence the above inverse image is non-empty.

Conversely, take E_i successively equal to $(0,\to)$ and $(\leftarrow,0)$ for each i. This gives 2^m points of $\mathbb{K}^m$ which are mutually different by the corresponding sequence of coordinate signs. A straightforward induction argument yields that f is surjective. ∎

We arrive at the following description of the Helly number for H-convexities.

4.7. Theorem. *Let V be a finite-dimensional vector space over the totally ordered field $\mathbb{K}$, and let $\mathcal{C}$ be the H-convexity on V generated symmetrically by a set $\mathcal{F}$ of linear functionals. If $\mathcal{F}$ is finite or if $\mathbb{K} = \mathbb{R}$, then $h(V,\mathcal{C}) = md(\mathcal{F})$.*

Proof. The convexity $\mathcal{C}$ is S_3 by Example I§3.12.1. As V is finite-dimensional, it follows from Theorem 1.8 that the **standard** Helly number of V is finite. As the standard convexity is finer than $\mathcal{C}$, it follows from Theorem 1.10 that $(V,\mathcal{C})$ has a finite Helly number h.

We assume first that $\mathcal{F}$ is finite. By I§3.24.4, all half-spaces of $\mathcal{C}$ are of type $f^{-1}(K)$, where $K \subseteq \mathbb{K}$ is a half-space and $f \in \mathcal{F}$. Hence by Corollary 4.5 we obtain the following intermediate result.

(*) $h \leq n$ iff for each $m > n$, for each sequence $f_1,\ldots,f_m$ in $\mathcal{F}$, and for each collection of non-empty half-spaces of type

$$H_i = f_i^{-1}(K_i),$$

where K_i is a half-space in $\mathbb{K}$ and $i = 1,\ldots,m$, it is true that $\cap_{i=1}^m H_i \neq \emptyset$ whenever these half-spaces meet n by n.

We first show that $md \leq h$. Let $h \leq n < \infty$. Take $m > n$ and let the sequence $f_1,\ldots,f_m$ in $\mathcal{F}$ be linearly dependent. By 4.6.2, for each $i = 1,\ldots,m$ there is a half-space H_i of a type as above, such that $\cap_{i=1}^m H_i = \emptyset$. By (*) there exist $i_1,\ldots,i_n$ among $1,\ldots,m$ such that $\cap_{k=1}^n H_{i_k} = \emptyset$. By 4.6.2 again, we conclude that f_{i_k}, for $k = 1,\ldots,n$, are linearly dependent. Taking $i \in \{1,\ldots,m\} \setminus \{i_1,\ldots,i_n\}$, we arrive at a linearly dependent sequence $f_1,\ldots,\hat{f_i},\ldots,f_m$. By 4.6.1, it follows that $md(\mathcal{F}) \leq n$.

We next verify that $h \leq md(\mathcal{F})$. Let $md(\mathcal{F}) \leq n < \infty$, let $m > n$, and let $H_1,\ldots,H_m$ be non-trivial half-spaces of the convexity $\mathcal{C}$ meeting n by n. Our purpose is to show that in these circumstances each $n+1$ of the selected half-spaces have a point in common; repetition of this argument with n replaced by $n+1$, etc., yields that the intersection of all m half-spaces is non-empty. For this restricted purpose, it suffices to consider $m = n+1$.

Recall that each H_i comes from a non-trivial $\mathbb{K}$–half-space, say: $H_i = f_i^{-1}(K_i)$, where $f_i \in \mathcal{F}$. One of the following two possibilities is in order; cf. (MD-1′), (MD-2).

Case I: For each i, the functional f_i is dependent of the other ones, and for some i the sequence f_j, $j \neq i$ is dependent.

This condition implies that the image W of the linear function
$$f = (f_1,..,f_{n+1}): V \to \mathbb{K}^{n+1}$$
is at most $(n-1)$-dimensional. If $\pi_i: \mathbb{K}^{n+1} \to \mathbb{K}$ denotes the ith projection, and if $L_i = \pi_i^{-1}(K_i) \cap W$, then $H_i = f^{-1}(L_i)$. By assumption, the (standard) convex sets $L_1,..,L_{n+1}$ of the linear space W meet n by n. As the standard Helly number of W is at most n, we find that $\cap_{i=1}^{n+1} L_i = \varnothing$ and hence that
$$\bigcap_{i=1}^{n+1} H_i = f^{-1}\left(\bigcap_{i=1}^{n+1} L_i\right) \neq \varnothing.$$

Case II: For some i, the functional f_i is independent of the remaining ones.

We take $i = n+1$ for convenience. Consider a point $x \in \cap_{j=1}^{n} f_j^{-1}(K_j)$. Since $\cap_{j=1}^{n} \ker(f_j) \not\subseteq \ker(f_{n+1})$, we find that $x + \cap_{j=1}^{n} \ker(f_j)$ meets each coset of $\ker(f_{n+1})$. Take a coset $u + \ker(f_{n+1})$ with $f_{n+1}(u) \in K_{n+1}$. Any point in the intersection
$$x + \bigcap_{j=1}^{n} \ker(f_j) \cap (u + \ker(f_{n+1}))$$
settles the problem.

As for the second part of the theorem, we assume that $\mathcal{F}$ is arbitrary. Note that if $md(\mathcal{F}) > n$, then there is an indexed set of $n+1$ minimally dependent functionals of $\mathcal{F}$, generating a convexity coarser than $\mathcal{C}$ and of Helly number $> n$. See the first part of the proof. Hence the original Helly number satisfies $h > n$, establishing that $md(\mathcal{F}) \leq h$.

To obtain the opposite inequality, assume $\mathbb{K} = \mathbb{R}$. For each finite subset $\mathcal{b} \subseteq \mathcal{F}$ let $co_{\mathcal{b}}$ denote the corresponding hull operation. Then for each $A \subseteq V$ we have
$$co(A) = \bigcap \{ co_{\mathcal{b}}(A) \mid \mathcal{b} \subseteq \mathcal{F} \text{ finite } \}.$$
The case where $\mathcal{F}$ does not separate the points of V requires working with linear quotients and is left to the reader. We assume henceforth that $\mathcal{F}$ separates points, and hence that for each finite set $F \subseteq V$ and for each sufficiently large but finite subfamily $\mathcal{b} \subseteq \mathcal{F}$, the set $co_{\mathcal{b}}(F)$ is compact in the natural topology of V.

Let $md(\mathcal{F}) \leq n$, and let $F = \{p_1,..,p_m\}$ be a set with $m > n$ points. As usual, F_i denotes F with p_i removed. For $\mathcal{b} \subseteq \mathcal{F}$, we define $P(i, \mathcal{b}) = co_{\mathcal{b}}(F_i)$. As $md(\mathcal{b}) \leq n$, we have $\cap_{i=1}^{m} P(i, \mathcal{b}) \neq \varnothing$ for all finite $\mathcal{b}$. For sufficiently large but finite set $\mathcal{b}$, this yields a down-directed family of compact sets, which has a non-empty intersection, equal to $\cap_{i=1}^{m} co(F_i)$. This shows that $h \leq n$, and hence that $h \leq md(\mathcal{F})$. ∎

4.8. Corollary. *Let $n > 0$ and let X be $\mathbb{R}^n$, equipped with a point-convex, symmetric H-convexity. Then $h(X) = 2$ iff X is symmetrically generated by a family of functionals which is a dual basis of $\mathbb{R}^n$.*

§4: Infinite Combinatorics   225

Proof. Suppose first that $\mathcal{F} = \{f_1,..,f_n\}$ is a basis for the dual space of $\mathbb{R}^n$. The only dependent weighted sets, formed with the members of $\mathcal{F}$, are the ones in which some f_i is repeated. If a third element occurs, then the indexed set minus this element is still dependent. Hence $md(\mathcal{F}) \leq 2$, and equality follows by considering the pair f_1, f_1.

As for the converse, note that a generating collection of functionals includes a dual basis since X is S_1. Let $\mathcal{F}$ be a symmetrically generating collection of functionals, among which are the members $f_1,..,f_n$ of a dual linear basis. If $f \in \mathcal{F}$ then there is a minimal set $\mathcal{b} \subseteq \mathcal{F}$ on which f depends. We assume $f \neq 0$. Then the indexed set $\mathcal{b} \cup \{f\}$ is minimally dependent and as $md(\mathcal{F}) \leq 2$, we conclude that $\#\mathcal{b} = 1$. Hence f is a multiple of some f_i and is redundant in the symmetric generating process. ∎

4.9. Theorem. *The following are true for any convex structure X.*
(1)   *If $h(X) \leq 3$, and if X is S_3 then X is S_4.*
(2)   *If $h(X) \leq 2$, and if X is S_2 then X is S_4.*

Proof of (1). Let X be S_3 and of Helly number ≤ 3. Suppose C, D are disjoint convex sets. We may assume that these sets are non-empty. Consider the following families:

$$\mathcal{H}(C) = \{H \mid H \subseteq X \text{ a half-space}, C \subseteq H\};$$
$$\mathcal{H}(D) = \{H \mid H \subseteq X \text{ a half-space},\}.$$

By virtue of Theorem 4.3(1), either family is compact in 2^X, whereas by S_3,

$$\cap \mathcal{H}(C) = C; \quad \cap \mathcal{H}(D) = D.$$

Consequently, the compact family $\mathcal{H}(C) \cup \mathcal{H}(D)$ has an empty intersection. By the Compact intersection Theorem 4.4, there are three or less members of it with no common point. These sets do not all come from $\mathcal{H}(C)$, nor do they all come from $\mathcal{H}(D)$. So assume that one is from $\mathcal{H}(C)$, say: H_1, and that the remaining ones are from $\mathcal{H}(D)$, say: H_2 and H_3 (if there is no third set then we can always take $H_3 = X$). Then

$$C \subseteq H_1; \quad D \subseteq H_2 \cap H_3; \quad H_1 \cap (H_2 \cap H_3) = \emptyset,$$

whence H_1 is the desired half-space. ∎

Proof of (2). In regard to (1) it suffices to show that S_2 implies S_3 if $h \leq 2$. Let C be a non-empty convex set and $p \notin C$. Consider the compact family $\mathcal{H}(p)$ of all half-spaces H of X with $p \in H$. The property S_2 yields $\cap \mathcal{H}(p) \cap C = \emptyset$. By Theorem 4.4, there are two disjoint sets in $\mathcal{H}(p) \cup \{C\}$. They are necessarily of type C, and $H \in \mathcal{H}(p)$, and H yields the desired half-space. ∎

This result implies, for instance, that all symmetrically generated H-convexities in the plane are S_4. Sharpness follows from the existence of S_3 convexities of Helly number 4 which are not S_4. Indeed, the example given in I§3.12.1 (an H-convexity on $\mathbb{R}^3$, symmetrically generated by the coordinate projections and their sum) is S_3 but not S_4 and its Helly number equals 4 (cf. 4.33.2).

4.10. Sigma-finite Helly number. A convex structure X is said to have a *sigma-finite Helly number* provided there is a sequence $(X_n)_{n \in \mathbb{N}}$ of subspaces of X such that $\cup_{n \in \mathbb{N}} X_n = X$ and each X_n has a finite Helly number.

The simplest example is a countable-dimensional vector space over a totally ordered field. More generally, a BW space of countable dimension has a sigma-finite Helly number by Theorem 1.12.

Recall (cf. I§1.23) that $ext(X)$ denotes the set of all *extreme points* of X, that is, points x with $X \setminus \{x\}$ convex.

4.11. Theorem (*Countable Intersection Theorem, I*). *Let X be an S_3 convex structure of sigma-finite Helly number, such that for each half-space H in X there is a countable set $A \subseteq H$ with $H \setminus ext(X) \subseteq co(A)$. Then each collection of convex sets with an empty intersection has a countable subcollection with an empty intersection.*

Proof. Let $X = \cup_{n=1}^{\infty} X_n$, where each X_n has a finite Helly number. If C is a convex set of X included in $ext(X)$, then C is countable. Indeed, as $C \cap X_n$ is a relatively convex set of X_n, by Theorem 1.11 we obtain that $h(C \cap X_n) < \infty$. Moreover, each point of C is extreme in X and hence C is free. Therefore $C \cap X_n$ is finite for each n.

As X is S_3, each convex set is an intersection of half-spaces. Hence it suffices to verify the result for families $\mathcal{H}$ consisting entirely of half-spaces, and such that $\cap \mathcal{H} = \emptyset$. Suppose first that $(H_n)_{n=1}^{\infty}$ is a sequence in $\mathcal{H}$ with $\cap_{n=1}^{\infty} H_n \subseteq ext(X)$. The intersection is a countable set $\{x_k \mid k \in \mathbb{N}\}$. For each $k \in \mathbb{N}$ there is an H'_k in $\mathcal{H}$ with $x_k \notin H'_k$. The sets H_n for $n \in \mathbb{N}$ and H'_k for $k \in \mathbb{N}$ constitute a countable subfamily of $\mathcal{H}$ with an empty intersection.

For the remainder of the proof we assume that no countable collection $\mathcal{H}'$ in $\mathcal{H}$ satisfies $\cap \mathcal{H}' \subseteq ext(X)$. As the operation of intersecting with X_n is continuous in 2^X we obtain a compact family

$$\{H \cap X_n \mid H \in \mathcal{H}^-\}$$

for each n. Applying Theorem 4.4 for each n separately yields a countable collection

$$\mathcal{H}' = \{H_k \mid k \in \mathbb{N}\} \subseteq \mathcal{H}^-$$

with an empty intersection.

Fix k for a moment. We have countable sets $A \subseteq H_k$, $B \subseteq X \setminus H_k$ with

$$H_k \setminus ext(X) \subseteq co(A); \quad X \setminus (H_k \cup ext(X)) \subseteq co(B).$$

Enumerate the points of A and B and let A_l, resp., B_l be the set of the first l points of A, resp., B. Since H_k is adherent to $\mathcal{H}$, there is a set H_{kl} in $\mathcal{H}$ such that

$$A_l \subseteq H_{kl}; \quad B_l \subseteq X \setminus H_{kl}$$

for each $l \in \mathbb{N}$. By compactness of $\mathcal{H}^-$ the sequence $(H_{kl})_{l=1}^{\infty}$ clusters at some H'_k in $\mathcal{H}^-$. As complementation is a continuous operation in 2^X, we find that the sequence $(X \setminus H_{kl})_{l=1}^{\infty}$ clusters at $X \setminus H'_k$. The increasing sequence $(co(A_l))_{l=1}^{\infty}$ converges to $co(A)$, showing that $co(A) \subseteq H'_k$. Similarly, $(co(B_l))_{l=1}^{\infty}$ is an increasing sequence converging to $co(B)$ and

§4: Infinite Combinatorics   227

$co(B) \subseteq X \setminus H'_k$. This yields

$$H_k \setminus ext(X) = co(A) \setminus ext(X) \subseteq H'_k \setminus ext(X);$$
$$X \setminus (H_k \cup ext(X)) = co(B) \setminus ext(X) \subseteq X \setminus (H'_k \cup ext(X)),$$

showing that $H_k \setminus ext(X) = H'_k \setminus ext(X)$.

All this serves to show that $H_k \setminus ext(X)$ is a cluster point of the sequence $(H_{kl} \setminus ext(X))_{l=1}^{\infty}$. By assumption there is a point $p \in \cap_{k,l} H_{kl} \setminus ext(X)$. Then

$$\{ H_{kl} \setminus ext(X) \mid k, l \in \mathbb{N} \} \subseteq \mathbb{E}(\{p\}, \varnothing),$$

where the right hand set is closed. But then all cluster points $H_k \setminus ext(X)$ for $k \in \mathbb{N}$ contain p, contradicting that $\cap \mathcal{H}' = \varnothing$. ∎

The condition on half-spaces is fulfilled provided each half-space is the hull of a countable set. Note that the 2-cell with the standard convexity does not satisfy this stronger condition, but it does satisfy the hypotheses of the theorem. For an application of the result to vector spaces, see 4.28. Separable compact trees form a different class of examples; see IV§2.34.2. A topological version of the Countable Intersection Theorem is given in Section IV§2. This result involves completely different techniques and is independent of the above theorem.

4.12. Rank and generating degree. Recall (I§2.8) that a subset F of a convex structure X is *convexly independent* provided $a \notin co(F_a^{\wedge})$ for each $a \in F$; it is *convexly dependent* otherwise. As usual, $F_a^{\wedge}$ denotes F with a removed. In contrast with earlier notions of "dependence", no finiteness is required. In fact, as observed in I§2.8,

4.12.1. *A set is convexly independent iff all of its finite subsets are.*

The *rank* of a convex structure X is defined to be the number

$$d(X) \in \{0, 1, ..., \infty\}$$

determined by the following prescription. For each n with $0 \leq n < \infty$,

$d(X) \leq n$ iff each subset of X with more than n points is dependent.

It is clear that rank is a convex invariant. Moreover,

4.12.2. *Any convex structure satisfies $d \geq \max\{h, c, r\}$.* ∎

As for the exchange number, see 4.14.6 below. There is a second, related invariant. Recall that the *width* of a poset is the largest possible cardinality of a set of mutually incomparable points. We need the following combinatorial result on posets.

4.12.3. (*Dilworth Theorem*).[2] *Let X be a poset and let $0 \leq n < \infty$. Then the width of X is at most n iff there exist n totally ordered subsets $X_1, .., X_n \subseteq X$ such that*

---
2.   Dilworth [1950].

$X_1 \cup .. \cup X_n = X$.

Each collection of sets can be seen as a poset under the partial order of inclusion. This leads us to the following definition. The *generating degree* of a convex structure X is the number

$$gen(X) \in \{0, 1, .., \infty\}$$

satisfying $gen(X) \le n$ iff there is a subbase of X of width $\le n$, where $n < \infty$.

We want to compare rank and generating degree of a space. Note that a convex structure is coarse iff the empty family is a subbase and that the empty family has zero width. On the other hand, the rank of a convex structure is zero iff the space is empty.

4.13. Proposition. *If a convex structure X is not coarse, then*

$$d(X) \le gen(X).$$

Proof. If X is not coarse, then $gen(X) \ge 1$ and it suffices to show that

$$gen(X) \le n \implies d(X) \le n$$

for all n with $1 \le n < \infty$. To this end, let $\mathcal{S}$ be a subbase of X of width $\le n$ and let $F \subseteq X$ have more than n points. Without loss of generality F can be taken finite. If F is independent then $a \notin co(F_a^\wedge)$ for all $a \in F$. As each polytope is the intersection of subbasic sets, cf. I§1.7.3, there is a set $S(a) \in \mathcal{S}$ such that

$$F_a^\wedge \subseteq S(a); \quad a \notin S(a)$$

for each $a \in F$. Now F has at least two points, and hence the various sets $S(a)$ are incomparable. But we have more than n of them, a contradiction. ∎

4.14. Examples

4.14.1. Totally ordered sets. A totally ordered set X has a subbase consisting of all lower and upper sets. If X has at least two points, then this subbase is of width 2; hence $gen(X) \le 2$. On the other hand, two distinct points in an S_1 space are obviously independent; hence $d(X) \ge 2$. It follows from Proposition 4.14.1 that for a totally ordered set with more than one point,

$$d(X) = gen(X) = 2.$$

For a tree, the situation is more complicated; see 4.18.3 below.

4.14.2. Vector spaces. Let V be an n-dimensional vector space over a totally ordered field; V is given the standard convexity. It is easily seen that

$$d(V) = \begin{cases} 1, & \text{if } n = 0; \\ 2, & \text{if } n = 1; \\ \infty, & \text{if } n > 1. \end{cases}$$

By Proposition 4.13 and the previous example (in case $n = 1$) it follows that rank and gen of V are equal if the dimension of V is at least one. Both invariants are infinite if the

§4: Infinite Combinatorics   229

dimension is at least two.

4.14.3. H-convexity. Let $\mathbb{R}^n$ be equipped with the H-convexity, symmetrically generated by a family $\mathcal{F}$ of linear functionals. We may and will assume that $t \cdot f \notin \mathcal{F}$ whenever $f \in \mathcal{F}$ and $t \neq 1$. Then

$$d(\mathbb{R}^n) = gen(\mathbb{R}^n) = \begin{cases} 2 \cdot \#\mathcal{F}, & \text{if } \mathcal{F} \text{ is finite;} \\ \infty, & \text{otherwise.} \end{cases}$$

Indeed, consider the usual inner product $x \cdot y$ on $\mathbb{R}^n$, and represent each $f \in \mathcal{F}$ by a unit vector $x(f)$:

$$\forall v \in \mathbb{R}^n: \ f(v) = x(f) \cdot v.$$

The collection of all points $\pm x(f)$ for $f \in \mathcal{F}$ has $2 \cdot \#\mathcal{F}$ members (∞, if $\mathcal{F}$ is infinite), and is obviously independent. This establishes that $d(\mathbb{R}^n) \geq 2 \cdot \#\mathcal{F}$. On the other hand, the canonical subbase of the convexity is a union of $2 \cdot \#\mathcal{F}$ many totally ordered sets, namely,

$$\{ f^{-1}(\leftarrow, t] \mid t \in \mathbb{R} \}; \quad \{ f^{-1}[t, \rightarrow) \mid t \in \mathbb{R} \},$$

where $f \in \mathcal{F}$. Hence $gen(X) \leq 2 \cdot \#\mathcal{F}$. Equality follows from Proposition 4.13.

4.14.4. Solid cubes as meet semilattices. Let the solid n-cube $Q^n = [0,1]^n$ be equipped with the Cartesian order, and consider the coordinate-wise meet operator "$\wedge$". The resulting meet semilattice satisfies

$$d(Q^1, \wedge) = gen(Q^1, \wedge) = 2;$$
$$d(Q^n, \wedge) = gen(Q^n, \wedge) = \infty \ (n > 1).$$

Indeed, the case $n = 1$ is taken care of in the first example. For $n > 1$, the set of all points at unit distance from the origin (standard metric) is infinite and convexly independent.

4.14.5. Solid cubes as lattices. This time, we consider the Cartesian ordered cube Q^n as a distributive lattice by introducing the additional coordinate-wise join operation "$\vee$". Then

$$\forall n \geq 1: \ d(Q^n, \wedge, \vee) = gen(Q^n, \wedge, \vee) = 2n.$$

This can be shown as with H-convexities since by I§4.19 the convexity of the lattice Q^n is the product convexity. Note that the set of all barycenters of the $(n-1)$-faces is independent, which gives an independent set of the largest possible cardinality, namely $2n$.

4.14.6. Graphic cubes. The lattice $\{0,1\}^n$ satisfies the equalities

$$gen(\{0,1\}^n) = 2n;$$
$$d(\{0,1\}^n) = n, \text{ if } n > 1.$$

Indeed, the first formula follows by using the smallest possible subbase of the graphic n-cube, namely the family of all $(n-1)$-faces. On the other hand, the points with exactly one non-zero coordinate form an independent set with n vertices, so $n \leq d$. Suppose F is an independent set with more than n points. Fix $p \in F$ and let H be a half-space with

$F\setminus\{p\} \subseteq H$ and $p \notin H$. Then H is an $(n-1)$–face. If $n = 2$, then F contains three out of four corner points of the square, and hence is dependent. For $n > 2$ a contradiction can be derived by induction. Note that the graphic n–cube satisfies $e = n + 1$ by 2.8.3. This illustrates that the rank need not majorate the exchange number; cf. 4.12.2.

The pairs of examples 2, 3 and 4, 5 illustrate how sensitive rank is with respect to refining a convexity. In fact, each point of an independent set F contributes essentially to the size and shape of the convex set $co(F)$. As polytopes determine a convexity, rank indicates the amount of variation which the shape of a convex set is allowed to have. Topic 4.35.1 and Theorem IV§5.19 are motivating.

4.15. Directional degree. Two subsets S_1, S_2 of X are *compatible* provided they are either comparable (i.e., $S_1 \subseteq S_2$ or $S_2 \subseteq S_1$), or disjoint, or supplementary (i.e., $S_1 \cup S_2 = X$); they are called *incompatible*[3] otherwise. Let $n < \infty$. A family $\mathcal{S}$ of subsets of X has *spread* $\leq n$ provided among every $n+1$ members of $\mathcal{S}$ there always exist two which are compatible. A family consisting of pairwise compatible elements is called a *direction*.

4.15.1. Proposition. *If the family $\mathcal{S} \subseteq 2^X$ is a direction, then so is each of the following.*

(i) $\mathcal{S} \cup \{X \setminus D \mid D \in \mathcal{S}\} \cup \{\varnothing, X\}$.
(ii) *The closure $\overline{\mathcal{S}}$ of $\mathcal{S}$ in the Cantor cube 2^X.*

Proof. The first statement being clear, we just verify the second one. Suppose D_1, D_2 are incompatible sets. We obtain four points as follows.

$$x_{12} \in D_1 \cap D_2; \quad x^{12} \in X \setminus (D_1 \cup D_2); \quad x_1^2 \in D_1 \setminus D_2; \quad x_2^1 \in D_2 \setminus D_1.$$

This yields the following basic neighborhoods in the Cantor cube:

$$U_1 = \mathbb{E}(\{x_{12}, x_1^2\}, \{x^{12}, x_2^1\}), \quad \text{of } D_1;$$
$$U_2 = \mathbb{E}(\{x_{12}, x_2^1\}, \{x^{12}, x_1^2\}), \quad \text{of } D_2.$$

Clearly, each choice of $D_i' \in U_i$ for $i = 1, 2$, yields two incompatible sets. ∎

The *directional degree* of a convex structure X is defined to be the number

$$dir(X) \in \{0, 1, .., \infty\}$$

with the following property for each n with $0 \leq n < \infty$.

$$dir(X) \leq n \Leftrightarrow X \text{ has a subbase of spread } \leq n.$$

As we shall see later, the spread of a family need not be equal to the number of directions required to cover the family (contrasting with the relationship between the chains of a poset and its width as described in 4.12.3). If $\mathcal{S}_i$ for $i = 1, .., n$ are n directions

---

3.   The notion of incompatibility relates with the concept of a "free set" in a Boolean algebra; cf. Birkhoff [1967].

§4: Infinite Combinatorics   231

of sets, such that $\cup_{i=1}^{n} \delta_i$ is a subbase of X, then X is *covered with the directions* $(\delta_i)_{i=1}^{n}$.

Finite spaces have a smallest subbase. Clearly the spread of such a subbase is decisive. In general, it is difficult to keep an eye on every possible subbase. The following result is used to compute *dir* (and *gen*) in more sophisticated cases.

4.16. Theorem. *Let X be a non-empty S_3 convex structure. Then there is a subbase $\mathcal{H}$ of X with the following properties.*

(1) $\mathcal{H}$ *consists entirely of half-spaces.*
(2) *gen (X) equals the width of $\mathcal{H}$ and dir (X) equals the spread of $\mathcal{H}$.*
(3) *Let X have a finite Helly number h. Then each non-trivial half-space $H_0 \subseteq X$ is the intersection of at most $h-1$ members of $\mathcal{H}$. In fact, either H_0 or $X \setminus H_0$ is an intersection of at most $\lfloor h/2 \rfloor$ members of $\mathcal{H}$.*

Proof. Let $\mathcal{H}'$ consist of all convex sets C with the property that there exists a point $p \in X$ with $p \notin C$ and such that C is maximal in this respect (i.e. C is a copoint at p). Obviously, $\mathcal{H}'$ is a subbase of X, and by the property S_3 it consists entirely of half-spaces. Let $\mathcal{H}$ be its closure in 2^X. By Theorem 4.3, this yields another subbase consisting entirely of half-spaces, as required in (1).

Let δ be any subbase of X. Then by 4.3 its closure includes the above constructed family $\mathcal{H}'$. Consequently, $\mathcal{H} \subseteq \bar{\delta}$. By virtue of 4.15.1, the families δ and $\bar{\delta}$ have the same spread. By similar reasoning, these families also have the same width. These numbers do not increase when passing to a smaller family. It follows that width and spread of $\mathcal{H}$ are optimal, establishing (2).

As for (3), let H_0 be a non-trivial half-space, and consider the compact families

$$\mathcal{K} = \{ H \in \mathcal{H} \mid H_0 \subseteq H \}; \quad \mathcal{L} = \{ H \in \mathcal{H} \mid X \setminus H_0 \subseteq H \}.$$

By the property S_3, it follows that

$$\cap \mathcal{K} = H_0 \quad \text{and} \quad \cap \mathcal{L} = X \setminus H_0.$$

Both intersections are non-empty. By the Compact Intersection Theorem 4.4, there are h or less sets in $\mathcal{K} \cup \{X \setminus H_0\}$ with an empty intersection. As $H_0 \neq \emptyset$, we obtain at most $h-1$ members of $\mathcal{K}$ whose intersection is H_0.

The compact family $\mathcal{K} \cup \mathcal{L}$ has an empty intersection. By Theorem 4.4 once more, at most h members of it have an empty intersection. No more than $\lfloor h/2 \rfloor$ of them come from $\mathcal{K}$, or else from $\mathcal{L}$, establishing the result.   ∎

For general spaces a weaker version of (1) and (2) can be obtained; cf. 4.26.2.

4.17. Corollary. *Let X be a non-coarse S_3 convex structure of Helly number $h \leq 3$. Then the spread of **any** subbase of half-spaces equals dir (X). If $h \leq 2$, then the width of **any** subbase of half-spaces equals gen (X).*

Proof. Let $\mathcal{H}$ be a subbase as described in the previous theorem. Note that $\mathcal{H} \neq \emptyset$ since X is not coarse. Then by 4.15.1 the family

$$\mathcal{H}_0 = \mathcal{H} \cup \{ X \setminus H \mid H \in \mathcal{H} \} \cup \{ \emptyset, X \}$$

has the same spread as $\mathcal{H}$. If $h \leq 3$ then by 4.16(3) a non-trivial half-space of X is in $\mathcal{H}$, or else its complement does. Hence $\mathcal{H}_0$ consists of all half-spaces. If $h \leq 2$ then by 4.16(3) the original $\mathcal{H}$ consists of all half-spaces, and consequently $\mathcal{H}_0 = \mathcal{H}$.

In both cases, the family $\mathcal{H}_0$ of *all* half-spaces determines $gen(X)$ and $dir(X)$. Therefore, *any* subbase of half-spaces is optimal. ∎

It is not known whether the last result holds for *gen* if $h = 3$.

4.18. Examples

4.18.1. Euclidean convexity. In $\mathbb{R}^n$, the bounding hyperplane of a standard half-space H is defined to be $Cl(H) \setminus Int(H)$. Two half-spaces of $\mathbb{R}^n$ are compatible iff their bounding hyperplanes are parallel. If $n > 1$, this equivalence relation yields a partitioning of the set of all non-trivial half-spaces into an infinite number of directions (for $n = 1$, there is only one direction). Each direction is the union of two totally ordered families, each one consisting of the complements of the other one. Also, any family of half-spaces forming a direction must be part of an equivalence class.

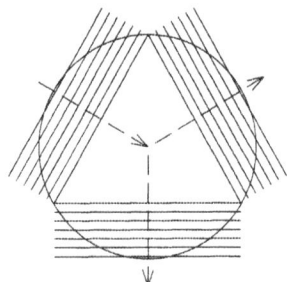

Fig. 1: A direction in a Euclidean 2–cell

In more general spaces the situation will rarely be that simple. As a more representative example, consider the unit disk of $\mathbb{R}^2$. There is no partitioning of the (relative) half-spaces into "maximal" directions, nor need a direction be built up with two totally ordered families of complementary half-spaces (Fig. 1).

4.18.2. H-convexity. Let $\mathbb{R}^n$ be equipped with an H-convexity $\mathcal{C}$, symmetrically generated by a collection $\mathcal{F}$ of linear functionals. Then

$$gen(\mathbb{R}^n, \mathcal{C}) = 2 \cdot dir(\mathbb{R}^n, \mathcal{C}).$$

Indeed, as in 4.14.3 we assume that $\mathcal{F}$ contains no proper multiples of its members. Suppose first that $\mathcal{F}$ is infinite. Then *gen* is infinite by Example 4.14.3. If *dir* is finite, then according to Theorem 4.16, some subbase of half-spaces has a finite spread. But its members are standard half-spaces, and by the first example,, compatibility partitions the subbase into finitely many directions. By the previous example again, each direction gives

§4: Infinite Combinatorics   233

rise to two chains, and then *gen* would be finite.
   The above argument also shows that for each subbase $\mathcal{H}$ of half-spaces
(1) $width(\mathcal{H}) \leq 2 \cdot spread(\mathcal{H})$.
Assume $\mathcal{F}$ is finite. It follows from I§3.24.4 that $\mathcal{H}$ is part of the canonical subbase, consisting of inverse images of half-spaces in $\mathbb{R}$ by members of $\mathcal{F}$. By construction, the spread of this subbase equals $\#\mathcal{F}$, and hence
(2) $spread(\mathcal{H}) \leq \#\mathcal{F}$.
Take $\mathcal{H}$ as the subbase in Theorem 4.16. Then by 4.14.3 and by (1), (2),
$$2 \cdot \#\mathcal{F} = gen(\mathbb{R}^n) = width(\mathcal{H}) \leq 2 \cdot spread(\mathcal{H}) \leq 2 \cdot \#\mathcal{F},$$
establishing the desired result.
   We next consider some examples with a combinatorial flavor.

4.18.3. Proposition. *Let T be a tree with more than one point. Then*
(1) $dir(T) = 1$.
(2) $d(T) = gen(T)$.
(3) *If T is absolutely complete, then $d(T)$ and $gen(T)$ are equal to the number of end points of T.*

Proof of (1). A tree is a semilattice by definition. Hence, as observed in I§1.7, it has a subbase $\mathcal{H}$, consisting of all sets of type $\uparrow(p)$ and their complements. Let $H_1, H_2 \in \mathcal{H}$. Passing to the complement if necessary, we may assume that $H_i = \uparrow(x_i)$ for $i = 1, 2$. If the corresponding points x_1, x_2 are comparable, say $x_1 \leq x_2$, then $H_2 \subseteq H_1$. If these points are incomparable, then they cannot have a common upper bound, and we find that
$$H_1 \cap H_2 = \uparrow(x_1) \cap \uparrow(x_2) = \emptyset.$$
Hence the entire subbase is a direction. Since T has more than one point, the empty family is not a subbase, and the above subbase is optimal.

Proof of (2). A tree is a poset; we use its width as an intermediate parameter. Two auxiliary results are needed to prove (2).

(4) *Let $n \geq m \geq \lceil n/2 \rceil$, let $width(T) < n$, and let*
   $H_k = \uparrow(x_k)$,   *for $k = 1,..,m$;*
   $H_l = T \setminus \uparrow(x_l)$,   *for $l = m+1,..,n$.*
If these half-spaces are pairwise incomparable, then $n = m+1$; $width(T) = m$, the points $x_1,..,x_m$ are pairwise incomparable, and x_{m+1} is comparable with each of them.

   Clearly, $x_1,..,x_m$ are incomparable and $x_{m+1},..,x_n$ are incomparable. This yields $m \leq width(T) < n$, and hence there is at least one point x_l for $l > m$. Each x_l is comparable with each x_k for $k \leq m$. If not, then x_k, x_l have no common upper bound. This means that $\uparrow(x_k) \cap \uparrow(x_l) = \emptyset$, whence $H_k \subseteq H_l$, contradiction. If there are two or more x_l's, then -- as these points are incomparable -- they must be larger than x_k for $k \leq m$. As the latter are mutually incomparable, there can be only one of them. But this yields $m = 1 \geq n - m \geq 2$, contradiction.

We conclude that there is only one point x_l (so $n = m+1$) and, since there are $m = n-1 \geq width(T)$ many mutually incomparable points, that m equals the width.

(5) *If T has a smallest element $\mathbf{0}$, and if there exist two incomparable elements in T whose infimum is $\mathbf{0}$, then*

$$d(T) = width(T) = gen(T).$$

First, let $n \leq width(T)$, and consider n incomparable points $x_1,..,x_n$ in T. For a fixed i and for each $j \neq i$ we have $x_j \in T \setminus \uparrow(x_i)$. As the latter is a half-space missing x_i, convex independence follows at once. Hence $n \leq d(T)$, establishing that $width(T) \leq d(T)$.

By Proposition 4.13, it suffices to show that $gen(T) \leq width(T)$. To this end, let $width(T) < n$ and suppose that $H_1,..,H_n$ are mutually incomparable members of $\mathcal{H}$. Without loss of generality,

$$H_k = \uparrow(x_k), \quad \text{for } k = 1,..,m;$$
$$H_l = T \setminus \uparrow(x_l), \quad \text{for } l = m+1,..,n,$$

and $m \geq n - m$ (if necessary, replace each set by its complement). By (4) we conclude that $width(T) = m$, that $n = m+1$, and that x_{m+1} is comparable with each of the mutually incomparable points x_k for $k \leq m$. As T is a tree, the inequality $x_k \leq x_{m+1}$ can occur at most once for $k \leq m$. If it does occur, then $m = 1$ by the incomparability of the points x_k ($k \leq m$). On the other hand, there exist two incomparable points $u, v \in T$ with $u \wedge v = \mathbf{0}$, showing that the width m is at least two. We conclude that the inequality $x_k \leq x_{m+1}$ does not occur, whence

$$x_{m+1} \leq x_1 \wedge .. \wedge x_m.$$

As the points x_k for $k \leq m$ are incomparable, none of them is a common lower bound of u, v. Similarly, no x_k is a common upper bound of u, v either. But each of u, v is comparable with a member of the maximal antichain $\{x_k \mid k = 1,..,m\}$. Hence there exist $i \neq j \leq m$ such that x_i is comparable with u and x_j is comparable with v. All four possibilities yield $x_i \wedge x_j = \mathbf{0}$, and hence $x_{m+1} = \mathbf{0}$. This implies $H_{m+1} = \emptyset$, a contradiction.

We conclude that there are no more than $width(T)$ many mutually incomparable members of $\mathcal{H}$, and that $gen(T) \leq width(T)$.

The proof of (2) is completed as follows. By I§5.3.3, we may replace the tree order by any base-point order. If T has at least three points, then one of them can be taken as a base-point with the property required in (5). The result is clear if $\#T = 2$.

Proof of (3). Recall (cf. I§5.26 and I§5.3.3) that each end point of a tree is an extreme point and a maximal element of some base-point order. Moreover, if a point p is maximal in one order, then it is maximal in any other base-point order except $\leq_p$. If the base-point order is as in (5), then the base-point is not an extreme point and the end points are the maximal elements. Clearly, the width of a poset is *at least* the number of its maxima. On the other hand, if $width(T) \geq n$, then there is an antichain with n points. Since T

§4: Infinite Combinatorics   235

is an absolutely complete tree, we can enlarge each point of the antichain to obtain n maximal points. So the width is *at most* the number of end points. ∎

4.18.4. Free spaces. *In a free space X with more than one point,*

(1) $dir(X) = 1$; $gen(X) = \#X$.

If X has n points, then

(2) $width(2^X) = \binom{n}{\lfloor n/2 \rfloor}$; $spread(2^X) = \binom{n-1}{\lceil n/2 \rceil}$.

Proof. If x is a real number, then $\lfloor x \rfloor$ resp., $\lceil x \rceil$ denotes the largest integer below x, resp., the least integer above x. The subbase $\mathcal{H}$ indicated in the proof of Theorem 4.16 consists of all subsets $H \subseteq X$ with $X \setminus H$ a singleton, and the values of gen and dir in (1) follow. We next consider the family 2^X of all subsets of X. The relation of inclusion turns 2^X into a poset, subject to some interesting combinatorial results.[4]

Sperner Theorem. *Let $\mathcal{A}$ be an antichain of subsets of a p-point set. Then*

$$\#\mathcal{A} \leq \binom{p}{\lfloor \tfrac{1}{2}p \rfloor}.$$

The maximum is attained for even p on sets of size $\tfrac{1}{2}p$; for odd p, it is attained on sets of size $\tfrac{1}{2}(p-1)$ or $\tfrac{1}{2}(p+1)$.

Brace-Daykin Theorem. *Let $\mathcal{A}$ be a family of pairwise incompatible subsets of a p-point set. Then*

$$\#\mathcal{A} \leq \binom{p-1}{\lceil \tfrac{1}{2}p \rceil}.$$

Application of these theorems yields the desired formulae (2). ∎

The result in 4.18.4 shows that in an S_3 convex structure X the subbase of all half-spaces may have a width larger than $gen(X)$ or a spread larger than $dir(X)$; compare with Corollary 4.17.

4.19. Theorem. *The following are valid for a convex structure X.*
(1) $dir(X) \geq e(X) - 1$ *and if* $c(X) \geq 3$, *then* $dir(X) \geq c(X)$.
(2) *If X is a non-empty median space, then* $dir(X) = e(X) - 1$.
(3) *If X is a non-empty median space and* $e(X) < \infty$, *then there is a segment of X with the same exchange number.*
(4) *If X is a median graph and* $e(X) < \infty$, *then X includes a graphic cube of dimension* $e(X) - 1$.

Proof of (1). Note that a coarse space satisfies $e \leq 1$ and $dir = 0$. We henceforth assume that X is not coarse, so that $e > 1$ and $dir > 0$. Let $0 < n < \infty$ and $e \geq n+1$. Then there is an E-independent set $F = \{x_i \mid i = 0,..,m\}$ with $m + 1 > n$ points. For $i = 0,..,m$,

---
4.   See Anderson [1987, Thm. 1.2.1; Cor. 5.2.4].

let $F_i = F \setminus \{x_i\}$, and consider a point $x \in co(F_0) \setminus \cup_{i=1}^{m} co(F_i)$. If $\mathcal{S}$ is a subbase of X then there exist $S_i \in \mathcal{S}$ such that

(*) $\qquad F_i \subseteq S_i; \quad x \notin S_i \quad (i = 1,..,m)$

since each non-empty polytope is the intersection of subbasic convex sets. For each $i, j \in \{1,..,m\}$ with $i \neq j$, this yields

$$x \notin S_i \cup S_j; \quad x_0 \in S_i \cap S_j; \quad x_j \in S_i \setminus S_j.$$

Hence the spread of $\mathcal{S}$ is at least $m \geq n$, showing that $dir \geq n$.

Let $n \geq 3$, and suppose next that $c \geq n$. Consider a C-independent set $F = \{x_i \mid i = 1,..,n\}$ and a point $x \in co(F) \setminus \cup_{i=1}^{n} co(F_i)$. A similar argument, applied to an arbitrary subbase, provides us with n subbasic sets $S_1,..,S_n$ as in (*). If $i \neq j$, then $S_i \cap S_j$ is non-empty since it contains a point x_k for some $k \neq i, j$. Hence the sets $S_1,..,S_n$ are mutually incompatible and the spread of the subbase is at least n.

Proof of (2). Let X be a median convex structure with more than one point, and assume $dir \geq n \geq 1$. By Theorem 4.16(2) there exist n pairwise incompatible half-spaces $H_1,..,H_n$. For a fixed i the convex sets $X \setminus H_j$ ($j \neq i$) and H_i meet two by two, yielding a point $x_i \in H_i \setminus \cup_{j \neq i} H_j$ for each $i = 1,..,n$. As the sets $X \setminus H_1,..,X \setminus H_n$ meet pairwise, there is a point $x_0 \in X \setminus \cup_{i=1}^{n} H_i$ (if $n = 1$, this requires that $\#X > 1$). By construction, the convex sets H_i ($i = 1,..,n$) and $co\{x_1,..,x_n\}$ meet two by two, yielding a point

$$x \in H_1 \cap .. \cap H_n \cap co\{x_1,..,x_n\}.$$

For each $i \neq 0$, we obtain

$$\{x_j \mid j \neq i\} \subseteq X \setminus H_i; \quad x \in H_i.$$

Hence $co\{x_1,..,x_n\}$ is not covered by the polytopes $co\{x_j \mid j \neq i\}$ for $i \neq 0$. This shows that the set $\{x_0,x_1,..,x_n\}$ is E-independent and that $e \geq n+1$.

Proof of (3). Since $dir = e-1$ there exist $e-1$ mutually incompatible half-spaces $H_1,..,H_{e-1}$ of X. We write $H'_i = X \setminus H_i$ for convenience. Take $a \in \cap_{i=1}^{e-1} H_i$ and $b \in \cap_{i=1}^{e-1} H'_i$. It appears that the relative half-spaces $H_i \cap ab$ of ab are mutually incompatible, and the equality $e(ab) = e$ follows via Theorem 1.11(1).

Proof of (4). Let $n = e-1$, and let H_i, H'_i for $i = 1,..,n$ be as in (3). There are 2^n choice functions assigning to $i \in \{1,..,n\}$ one of the sets H_i, H'_i. Each choice function F yields a set of type

$$C(F) = F(1) \cap .. \cap F(n).$$

By virtue of the fact that X is discrete, there is a smallest convex set D meeting each $C(F)$. For each pair F, F' of complementary choice functions we take a pair of mutual nearest points $x(F) \in C(F)$ and $x(F') \in C(F')$ which are in D. Then the segment $x(F)x(F')$ meets both sets $C(F), C(F')$ in one point. Therefore, it meets each individual set H_i, H'_i and hence it meets each set of type $C(G)$, where G is a choice function as described above. Taking into account that the sets $C(G)$ cover the whole of X, this shows that, whatever complementary pair of choice functions we take, the resulting segment

§4: Infinite Combinatorics   237

equals D and the points $x(G)$ chosen above constitute all members of it.

We conclude that the segment D consist of 2^n points, falling into 2^{n-1} pairwise disjoint spanning pairs. Therefore, each non-trivial relative half-space of it contains exactly 2^{n-1} points and the characterization of graphic cubes in I§6.35.3 is applicable. ∎

4.20. Examples

4.20.1. Trees and their products. Certain subtleties involving low values of c can be illustrated as follows. A tree with more than two points satisfies $dir = 1$, $e = 2$, and $c = 2$. A product of two such trees has $c = 2$ and $e = 3$; cf. Example 2.8.1. This time, $dir = e - 1 = c$. The directional degree of a product can be obtained directly by 4.34.

4.20.2. Superextensions. Consider the superextension $\lambda(n)$ of an n-point free space $F(n)$. The family of all sets of type S^+ for $S \subseteq F(n)$ is a subbase of $\lambda(n)$, consisting of half-spaces. By I§1.8.5, it has the same width and the same spread as the original family $2^{F(n)}$. By Corollary 4.17, Example 4.18.4, and the previous theorem we obtain the following equalities.

(1) $\quad e\,\lambda(n) - 1 = dir\,\lambda(n) = \binom{n-1}{\lceil n/2 \rceil}$.

(2) $\quad gen\,\lambda(n) = \binom{n}{\lceil n/2 \rceil}$.

The inequality $e - 1 \leq c$, combined with (1) and Theorem 4.19(1), yields $c = e - 1$ if $n \geq 4$.

The result on embedding, described in 1.22.5, is a key to the evaluation of the classical convex invariants in general median spaces. If $r < \infty$ is the Radon number of a median algebra M, then $\lambda(r)$ embeds in M. Consequently, $c(M) \geq c(\lambda(r))$, showing that the Eckhoff-Jamison Inequality (reducing to $r \leq c + 1$) is *very* unsharp on the class of median spaces. Here are some values.

r	5	6	7	8	9	10
c	4	10	15	35	56	126

4.21. Theorem. *Let X be an S_3 convex structure of Helly number ≤ 3 and with more than one point. If the Radon number r of X is finite, then*

$$\binom{r-1}{\lceil r/2 \rceil} \leq dir\,X.$$

Proof. Let $F \subseteq X$ be an R-independent set with n points, $n > 1$, and consider a family $\mathcal{F}$ of mutually incompatible subsets of F. As X is even S_4, Theorem 4.9, for each $A \subseteq F$ there is a half-space $H(A)$ such that

$$A \subseteq H(A); \quad F \setminus A \subseteq X \setminus H(A).$$

The selected half-spaces $\{H(A) \mid A \in \mathcal{F}\}$ are pairwise incompatible, showing that the spread of the family of all half-spaces is at least $\#\mathcal{F}$. By 4.18.4, the maximal cardinality attained by $\mathcal{F}$ is $\begin{bmatrix} n-1 \\ \lceil n/2 \rceil \end{bmatrix}$. The result follows from Corollary 4.17. ∎

4.22. Examples

4.22.1. Superextensions. Consider the superextension $\lambda(n)$ of an n-point free space. The formulae in I§1.8.5 illustrate that the (embedded) original set is Radon independent. Hence the Radon number r of $\lambda(n)$ is at least n. If $r > n$ then[5] by 4.20 and the previous theorem,

$$dir(\lambda(n)) = \begin{bmatrix} n-1 \\ \lceil n/2 \rceil \end{bmatrix} < \begin{bmatrix} r-1 \\ \lceil r/2 \rceil \end{bmatrix} \leq dir(\lambda(n)),$$

a contradiction. This shows that Theorem 4.21 is best possible for median spaces.

4.22.2. Cubical complexes. At the end of Section 3, we observed that the Radon number of a median graph need not be equal to the Radon number of its realization as a cubical complex. However, for each $n \geq 3$, we have $r|\lambda(n)| = n$. Indeed, a median graph and its realization have the same exchange number by Theorem 3.17. Hence they have the same directional degree by Theorem 4.19. If $r(|\lambda(n)|) > n$ then $\lambda(n+1)$ embeds, whence its directional degree does not exceed the one of $|\lambda(n)|$. This contradicts with

$$\begin{bmatrix} n-1 \\ \lceil n/2 \rceil \end{bmatrix} < \begin{bmatrix} n \\ \lceil (n+1)/2 \rceil \end{bmatrix} \quad (n \geq 3).$$

We next consider an example of Helly number 3.

4.22.3. H-convexity. Let $d = d_1 + .. + d_k$, where $1 \leq d_i \leq 2$ for $i = 1,..,k$. For each i with $d_i = 2$, let $\mathbb{R}^{d_i}$ be equipped with the symmetric H-convexity, produced by the two coordinate projections and their sum. The corresponding product convexity $\mathcal{C}$ on $\mathbb{R}^d$ has Helly number ≤ 3, it is S_3, and it is coarser than the product convexity $\mathcal{C}'$ obtained by giving each factor the natural convexity. Hence $r(\mathbb{R}^d, \mathcal{C}) \leq r(\mathbb{R}^d, \mathcal{C}')$, and the estimates of Table 2.2 are of use. On the other hand, Theorem 4.21 allows to estimate the Radon number directly. The results are compared in Table 4.1, where $d = \sum_{i=1}^{k} d_i$ and $1 \leq d_i \leq 2$. Each factor of dimension 2 is equipped with the H-convexity, generated symmetrically by the coordinate projections and their sum. If $d_i = 1$ for all i, then Table 2.3 should be consulted. Partitions of d into more than five parts have not been considered in Table 2.2.

4.23. Theorem. *Let X be an S_3 convex structure and let $0 \leq k < \infty$. If each polytope of X is included in the hull of k points of X, then*

$$gen(X) \leq k \cdot dir(X).$$

Proof. Without loss of generality, $dir(X) \leq n < \infty$, and there is a subbase $\mathcal{H}$ of half-spaces which determines dir (Theorem 4.16). Assume first that $X = co\{x_1,..,x_k\}$, let $q = kn$, and consider $q + 1$ non-trivial members H_j for $j = 0,..,q$ of $\mathcal{H}$. Regarding the

---
5.   An alternative argument that $r(\lambda(n)) = n$ is suggested in Topic 1.22.5.

§4: Infinite Combinatorics

Table 4.1:
Estimated Radon number of $\mathbb{R}^d = \mathbb{R}^{d_1} \times .. \times \mathbb{R}^{d_k}$

d	partition	Table 2.2	current	d	partition	Table 2.2	current
3	1 + 2	4	5	8	6×1 + 2	6	5
4	2×1 + 2	5	5		4×1 + 2×2	6	6
	2×2	5	5		2×1 + 3×2	7	6
5	3×1 + 2	5	5		4×2	7	6
	1 + 2×2	6	5	9	7×1 + 2	6	6
6	4×1 + 2	5	5		5×1 + 2×2	7	6
	2×1 + 2×2	6	5		3×1 + 3×2	7	6
	3×2	6	5		1 + 4×2	7	6
7	5×1 + 2	6	5	10	8×1 + 2	6	6
	3×1 + 2×2	6	5		6×1 + 2×2	7	6
	1 + 3×2	7	6		4×1 + 3×2	7	6
					2×1 + 4×2	7	6
					5×2	8	7

indices i as "mod k" integers, we define

$$\mathcal{H}_i = \{H_j \mid x_i \in H_j; x_{i+1} \notin H_j\}$$

for $i = 1,..,k$. Now $\emptyset \neq H_j \neq X$ whence H_j contains at least one x_i and misses at least one x_i. Hence for some i (mod k) it must be true that $x_i \in H_j$ and $x_{i+1} \notin H_j$, showing that

$$\{H_1,..,H_{q+1}\} = \bigcup_{i=1}^{k} \mathcal{H}_i.$$

Consequently, there is an index i with $\#\mathcal{H}_i \geq n+1$. Then there exist two members of $\mathcal{H}_i$ which are compatible. By construction, these sets cannot be disjoint or supplementary and hence they must be comparable. It follows that the width of $\mathcal{H}$ is at most kn; therefore

$$gen(X) \leq width(\mathcal{H}) \leq kn.$$

To establish the general case, assume that $gen > kn$ and consider $kn + 1$ pairwise incomparable half-spaces $H_1,..,H_{kn+1} \in \mathcal{H}$. For each $i \neq j$ among $1,..,kn + 1$ take a point in $H_i \setminus H_j$; the polytope spanned by these points is included in a subspace of type $Y = co\{x_1,..,x_k\}$. Now $gen(Y)$ is at least $kn + 1$ whereas $dir(Y) \leq n$ since Y is a subspace of X, contradicting with the previously established fact. ∎

4.24. Examples

4.24.1. Trees. Let T be an absolutely complete tree. Then T is the hull of its end point set; cf. I§5.26. Assume there are $k > 0$ end points. By 4.18.3,

$$gen(T) = k; \quad dir(T) = 1.$$

Let $X = T^n$. Then X is spanned by k points and the product formulae of 4.34 yield

$$gen(X) = kn; \quad dir(X) = n$$

(the second equality can also be obtained with Theorem 4.19 and Example 2.8.1). This shows that Theorem 4.23 is sharp.

4.24.2. Superextensions. We computed *gen* and *dir* of the superextension $\lambda(r)$ in 4.20. The parameter k (the number of points needed to span the space) can be determined as follows. The r original points span $\lambda(r)$, and each of them is an extreme point; cf. I§1.23.5. Therefore, each spanning set must include the original points. Consequently, $k = r$ is best possible. Some invariants of $\lambda(r)$ in the range $3 \leq r \leq 10$ are given in Table 4.2. The rank of $\lambda(r)$ is determined in Topic 4.37.4.

Table 4.2:
Finite superextensions $\lambda(r)$

r	d	gen	dir	#$\lambda(r)$†
3	3	3	1	4
4	4	6	3	12
5	10	10	4	81
6	15	20	10	2,646
7	35	35	15	1,422,564
8	56	70	35	229,809,982,112
9	126	126	56	$\pm 10^{21}$
10	210	252	126	unknown

† See notes on this section.

Further Topics

4.25. The compact space 2^X

4.25.1. Show that each of the following functions $2^X \times 2^X \to 2^X$ are continuous.
$$(A,B) \mapsto A \cup B; \quad (A,B) \mapsto A \cap B; \quad (A,B) \mapsto A \setminus B.$$

4.25.2. Recall (I§1.30.1) that a convexity can be seen as an algebraic lattice. Show that the Lawson topology corresponds with the inclusion-exclusion topology.

4.25.3. Let $(C_i)_{i \in I}$ be an up-directed family of convexities on a set X and let C be the closure of $\cup_{i \in I} C_i$ in 2^X. We let co and co_i denote the hull operator of C, resp., C_i. Show that C is a convexity on X and that $co(F) = \cap_{i \in I} co_i(F)$ for each finite set $F \subseteq X$.

4.25.4. In circumstances as above, note that if $C_i \supseteq C_j$, we have a CP identity map $(X, C_i) \to (X, C_j)$, leading to an inverse system. Show that (X, C) is the inverse limit.

4.25.5. Let V be a vector space over a totally ordered field, and let $C(\mathcal{F})$ denote the H-convexity generated by a collection $\mathcal{F}$ of linear functionals. Deduce that $(V, C(\mathcal{F}))$ is the inverse limit of spaces of type $(V, C(\mathcal{F}_i))$, with $\mathcal{F}_i$ ranging over all finite subsets of $\mathcal{F}$.

§4: Infinite Combinatorics   241

4.25.6. Let X be a convex structure such that for each finite subset there is a partial order inducing the relative convexity. Then there is a partial order on X inducing the original convexity. Hint. The collection of all partial orders on X is compact in $2^{X \times X}$.

In the terminology of Jamison [1982], the class of spaces which are induced by a poset is a variety.

4.25.7. (cf. Lemma I§6.20(1)). Let M be a median algebra. Show that the collection of all sets of type $H_1 \cup H_2$, where $H_1, H_2 \subseteq M$ are half-spaces, form a compact subbase of the convexity of median stable subsets of M. Deduce that if $Y \subseteq M$ then p is in the median stabilization $med(Y)$ of Y iff there exist half-spaces $H_1, H_2 \subseteq M$ of the (usual) convexity such that

$$p \in H_1 \cap H_2; \quad H_1 \cap H_2 \cap Y = \emptyset.$$

4.26. Invariants and subbases

4.26.1. (cf. Corollary 4.5) If $\mathcal{S}$ is *any* subbase of a space, and if *any* subcollection of sets in $\mathcal{S}$ meeting n by n has a non-empty intersection, then the Helly number of the space is at most n.

4.26.2. Show that for each convex structure X the width, resp., the spread, of the subbase of all copoints equals $gen\ X$, resp., $dir\ X$.

4.26.3. (Hanner [1956]) Let $K \subseteq \mathbb{R}^n$ be a compact convex set with non-empty interior. The K-*convexity*, generated by K is the convexity with, as a subbase $\mathcal{X}$, all homothetic copies of K. Show that if each pairwise intersecting *four*-tuple in $\mathcal{X}$ has a non-empty intersection, then each linked system in $\mathcal{X}$ has a non-empty intersection. Consequently, the resulting K-convexity is of Helly number 2.

If $K \subseteq \mathbb{R}^3$ is an octahedron, then each *triple* of pairwise intersecting sets in $\mathcal{X}$ has a non-empty intersection, whereas the resulting K-convexity has Helly number 4. Hint. Up to an affine isomorphism, the octahedron is the unit disk of the Manhattan norm of $\mathbb{R}^3$.

4.26.4. Let X be S_4. Give a direct proof that $h(X) \leq n$ iff each finite collection of half-spaces meeting n by n has a non-empty intersection (compare Corollary 4.5; consult I§3.20).

4.27. Convexity in Euclidean spaces

4.27.1. Problem. Can Theorem 4.7 be extended to an infinitely symmetrically generated H-convexity in a vector space over an arbitrary totally ordered field?

4.27.2. Problem. Give a description of the Radon number of a (symmetrically generated) H-convexity in the spirit of Theorem 4.7.

4.28. Countable intersection property in vector spaces (Jamison [1977]; compare Klee [1956]). Let V be a vector space over a totally ordered field $\mathbb{K}$. Prove that the following are equivalent:

(i) V has the countable intersection property for convex sets.
(ii) Each half-space of V is the hull of a countable set.
(iii) The field $\mathbb{K}$ is completely sequential and V is countable-dimensional.

A *completely sequential poset* is a partially ordered set in which each non-empty subset has a countable coinitial and cofinal subset.

4.29. Simplex graphs (Bandelt and van de Vel [1989]). Let F be an arbitrary graph. By a *simplex* of F is meant a set of vertices which are pairwise connected with an edge. The collection of all simplices of F is denoted by $\sigma(F)$. Define $S_1, S_2 \in \sigma(F)$ to form an edge provided their symmetric difference consists of at most one point. The resulting graph $\sigma(F)$ is called the *simplex graph* of F. For instance, $\sigma(K_n)$ is a graphic n-cube; the simplex graph of the 5-cycle and the 6-cycle are depicted in Fig. 2.

4.29.1. Show that $\sigma(F)$ is a median graph (the original graph is usually not an induced subgraph), and that the non-trivial half-spaces of $\sigma(F)$ are of the following type.
$$H(v) = \{S \in \sigma(F) \mid v \in S\}, \quad \text{or,} \quad \sigma(F) \setminus H(v) \quad (v \in F).$$

4.29.2. Show that the following are equivalent for an arbitrary graph F.

(i) The chromatic number of F is at most n.
(ii) The simplex graph $\sigma(F)$ can be covered with at most n directions.

4.30. Embedding in products of trees (Bandelt and van de Vel [1989])

4.30.1. Let X be a convex structure and let $p \in X$. Furthermore, let $\mathcal{H}$ be a direction of half-spaces of X. We regard $\mathcal{H}$ as a poset under the partial order of inclusion and, as such, $\mathcal{H}$ is assumed complete. For instance, $\mathcal{H}$ may be finite, or it may be a maximal direction in X. Define
$$T = T(\mathcal{H}; p) = \{H \in \mathcal{H} \mid p \notin H, H \neq \varnothing\} \cup \{X\}.$$
Show that under the partial order of containment "$\supseteq$", the set T is a complete tree.

4.30.2. Let X be a bipartite S_3 graph. Show that, in circumstances as above, T has finite segments. (use I§3.26). Hence T is graphic. In particular, T can be seen as a median graph (a graphic tree).

4.30.3. Let $\mathcal{H}$ be as in (1) and let $\pi(x)$ be the smallest element $H \in T$ with $x \in H$ ("smallest" refers to inclusion). This yields a well-defined CP function
$$\pi = \pi(\mathcal{H}; p): X \to T.$$
If X is a bipartite S_3 graph, then π is EP.

§4: Infinite Combinatorics   243

4.30.4. Let X be a bipartite S_3 graph, and let $\mathcal{H}_i$ for $i = 1,..,n$ be complete directions of half-spaces. For each i we choose $p_i \in X$. We have a corresponding function

$$\pi_i: X \to T_i = T(\mathcal{H}_i; p_i),$$

If the families $\mathcal{H}_i$ are pairwise disjoint, then the function

$$\pi = (\pi_1,..,\pi_n): X \to \prod_{i=1}^{n} T_i$$

is edge preserving relative to the Cartesian product of graphs on the right (cf. I§4.20).

4.30.5. Show that a point-convex S_3 space can be embedded in a product of n trees iff it is generated with n directions. For a bipartite S_3 graph, it can be arranged that this embedding is an embedding of graphs as well. Verify that the median graph suggested in the left Figure 2 cannot be embedded in a product of two trees. On the other hand, the seemingly more complex median graph suggested by the right hand figure does embed in a product of two trees.

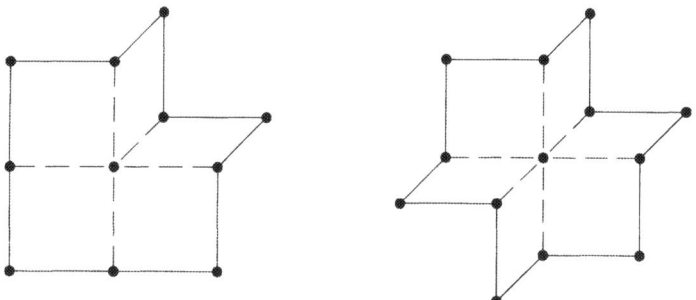

Fig. 2: Embedding in a product of two trees ..
.. not possible!   .. possible!

4.30.6. Use the fact[6] that there exist triangle-free finite graphs of any chromatic number to show that for any $n < \infty$ there exists a finite median graph which is 2–dimensional in the sense that no cubes occur of dimension ≥ 3, and which cannot be embedded in a product of n trees.

4.31. Embedding in products of chains

4.31.1. Modify the above constructions to show that a point-convex S_3 segment can be embedded in an n-cube iff $gen \leq 2n$. If the space is a bipartite S_3 graph, the embedding can be made edge preserving moreover.

Problem. Can the segment restriction be dropped?

---

6.   See Walther and Voss [1974].

4.31.2. Verify that a bipartite S_3 graph G can be embedded in $\{0,1\}^n$ (simultaneously as a graph and as a convex structure) iff the number of non-trivial half-spaces of G is at most $2n$.

4.31.3. (compare Duchet [1988]) Conclude that the Radon number of $\{0,1\}^n$ equals $\lfloor \log_2(2n+2) \rfloor$, thereby establishing the sharpness of the estimate given in 2.23. Hint: use finite superextensions as in 1.22.5.

4.32. Bipartite graphs

4.32.1. Let G and H be connected bipartite graphs, of which G is S_3. Show that an isometric embedding of H in G corresponds with a CP embedding of H as an induced subgraph of G. In either case, H is S_3 and, if H is even S_4, then $r(H) \leq r(G)$.

4.32.2. Show that the Helly number h of an isometric subgraph of $\{0,1\}^n$ satisfies $h \leq n$. For each n, give an example illustrating the sharpness of this inequality.

Hint. The case $n \leq 2$ is simple. Operate by induction on the dimension n of the cube.

4.32.3. Deduce that all isometric subgraphs of $\{0,1\}^3$ are S_4. In contrast, show that an isometric S_4 subgraph of $\{0,1\}^n$ has a Helly number $\leq \log_2(2n+2)$. Fig. 3 displays five non-S_4 isometric subgraphs of $\{0,1\}^4$ which, according to Chepoi [1991], are critical for a characterization of bipartite S_4 graphs.

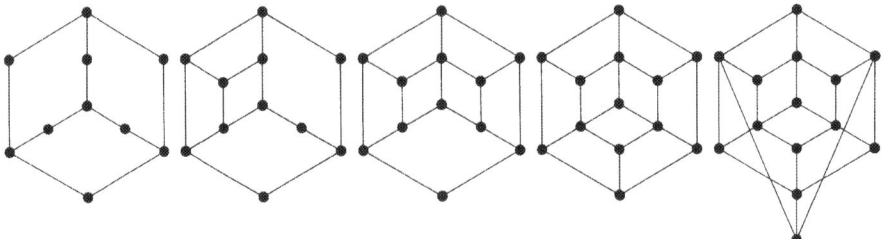

Fig. 3: Five critical isometric subgraphs of the 4–cube

4.33. Symmetric H-convexity

4.33.1. If a symmetric H-convexity is generated by finitely many linear functionals, then each of its polytopes is included in a segment. Note that, in combination with Theorem 4.23, this gives back the formula in 4.14.3.

4.33.2. Let $\mathbb{K}$ be a totally ordered field, and let $f_1,..,f_n$ be the coordinate projections of $\mathbb{K}^n$. Verify that the collection
$$\mathcal{F} = \{f_1,..,f_n, f_1 +..+ f_n\}.$$
minimally linearly dependent. Conclude that the H-convexity, symmetrically generated

by $\mathcal{F}$, has a Helly number equal to $n+1$.

4.33.3. (compare van de Vel [1983g]) Consider an H-convexity, symmetrically generated by a collection $\mathcal{F}$ of linear functionals and of Helly number $h < \infty$. Show that if this convexity has the separation property S_4, then

$$\#\mathcal{F} \geq C(h-1, \lceil h/2 \rceil).$$

This result can be used to check that certain H-convexities fail to be S_4; for instance, none of the following families symmetrically generates an S_4 convexity on $\mathbb{R}^5$.

(i) $\{f_1,...,f_5, \sum_{i=1}^{5} f_i\};$ (ii) $\{f_1,...,f_5, f_1+f_2+f_3, f_2+f_3+f_4, f_3+f_4+f_5\}.$

In either case, f_i denotes the ith coordinate projection of $\mathbb{R}^5$.

4.34. Images, subspaces and products

4.34.1. Show that $d(X) \geq d(Y)$ provided there is a CP surjection $X \to Y$.

4.34.2. Let X be a subspace of Y. Show that if f denotes one of d, gen, dir, then $f(X) \leq f(Y)$.

4.34.3. In a product with two non-empty factors, the following equalities hold.

$$gen = gen_1 + gen_2; \quad dir = dir_1 + dir_2.$$

4.34.4. (Lassak [1983]) Consider the following property of a convex structure X.

(*) If $F \subseteq X$ is independent and if $p \in X$ is such that $F \cup \{p\}$ is dependent, then there is an $x \in F$ such that $x \in co(\{p\} \cup F_x^{\wedge})$.

Note that each matroid has this property. Each non-free convex geometry and each non-free geometric interval space fails to have the property. If X_i is a non-coarse space of finite rank d_i, $i = 1,..,n$, then the rank d of the product X is determined by

$$d = (\sum_{i=1}^{n} d_i) - m,$$

where m is the number of factors with the property (*).

Hints. Concentrate on two factors first; let π_i denote the ith projection of the product. If $D \subseteq X$ is independent, then define D_i to be the set of all $d \in D$ such that $\pi_i(d) \notin \pi_i(D_d^{\wedge})$. Then π_i is injective on D_i, and this set is independent in X_i. Next, show that a product of two non-coarse factors never satisfies (*).

4.35. Rank and breadth. Let X be a convex structure. The family $\mathcal{C}(X)$ of all convex sets can be seen as a lattice under the operations

$$C \wedge D = C \cap D; \quad C \vee D = co(C \cup D) \quad (C, D \in \mathcal{C}(X)).$$

Note that the subcollection $\mathcal{C}_*(X) = \mathcal{C}(X) \setminus \{\emptyset\}$ is a join subsemilattice.

4.35.1. (Jamison [1982]; van de Vel [1983]) Show that if X is not coarse then
$$d(X) = b(\mathcal{C}(X)) = b(\mathcal{C}_*(X)).$$

4.35.2. Compact Intersection Theorem for Rank. If $d < \infty$ and if $\mathcal{S} \subseteq \mathcal{C}(X)$ is a compact set such that $\cap \mathcal{S} = D_0$, then there exist $D_1,..,D_d \in \mathcal{S}$ such that $\cap_{i=1}^{d} D_i = D_0$. Compare with "The Lemma" in Gierz et al [1980, V§1.1]; cf. Theorem 4.4.

4.35.3. Countable Intersection Theorem for Rank. A convex structure X has *sigma-finite rank* provided there is a sequence X_n, $n \in \mathbb{N}$, of subspaces of X of finite rank, such that $X = \cup_{n \in \mathbb{N}} X_n$. Verify that the standard convexity of $\mathbb{R}^n$ for $n > 1$ is not of sigma-finite rank. Show that if X is an S_3 convex structure of sigma-finite rank and if for each half-space $H \subseteq X$ there is a countable set $A \subseteq H$ with $H \setminus ext(X) \subseteq co(A)$, then each family $\mathcal{S}$ of convex sets has a countable subcollection $\mathcal{S}'$ such that $\cap \mathcal{S} = \cap \mathcal{S}'$; cf. Theorem 4.11.

4.36. Segments of median spaces. This topic is inspired by Theorem 4.19(3).

4.36.1. Verify the Erdös-Kleitman [1970] estimate concerning the number of points in a superextension (check with Table 4.2 for $3 \leq r \leq 8$):
$$\#\lambda(r) > 2^n, \quad \text{where } n = \binom{r-1}{\lceil r/2 \rceil} \text{ and } r \geq 3.$$

4.36.2. (van de Vel [1983c]) Problem. Let X be a convex structure of Helly number $h < \infty$ and of exchange number $e < \infty$. Is there a set $F \subseteq X$ such that $\#F = h$ and $e(co(F)) = e$?

4.37. Breadth of median spaces

4.37.1. (compare van de Vel [1984c]; cf. Proposition 1.14) Let L be a distributive lattice. Show that
$$b(L) = e(L) - 1; \quad \binom{r-1}{\lceil r/2 \rceil} \leq 2b.$$

Equality holds at the right if L is order dense.

4.37.2. Show that if L is a distributive lattice and if $b(L) < \infty$, then L embeds as a sublattice of a $b(L)$-dimensional cube and some $b(L)$-dimensional cube embeds in L as a convex subspace.

4.37.3. Let M be a median space with a base-point 0. Then $(M, \leq_0)$ is a median semilattice (cf. Topic I§6.26). Show that the breadth of this semilattice equals the Carathéodory number c of the median space M provided c is at least 2.

4.37.4. Deduce that for $n \geq 2$, the rank of $\lambda(n)$ equals the Carathéodory number of

§4: Infinite Combinatorics   247

$\lambda(n+1)$ (for $n \geq 3$, the latter equals the directional degree of $\lambda(n+1)$). Hint: I§6.39.4.

4.38. Relation between rank, gen, and dir

4.38.1. Show that if *dir* is finite then finiteness of *d* and finiteness of *gen* are equivalent. In a tree, $dir = 1$, but both *d* and *gen* can be infinite.

4.38.2. (compare van de Vel [1986b]) Let X be a median convex structure. Then $gen(X) \leq d(X)^2$ and $dir(X) \leq d(X)$. If each polytope of X is included in the hull of a k-point set, then $gen(X) \leq k \cdot d(X)$.

4.38.3. Problems. Let d_n denote the smallest possible rank of a subspace X of the graphic n-cube such that $med(X) = \{0,1\}^n$. Is it true that $d_n \mapsto \infty$ if $n \mapsto \infty$? Is the result in part (1) still valid if $dir = \infty$? The answer is affirmative for S_3 spaces iff the answer to the previous problem is.

4.39. Rank, generating degree, and convex hyperspaces

4.39.1. (for topological median spaces, van de Vel [1983f]) Let X be a S_4 space and let $\mathcal{H}$ be the collection of all half-spaces of X. Show that

$$d(\mathcal{C}_*(X)) \leq d(X) + width(\mathcal{H}).$$

In particular, $d(\mathcal{C}_*(X)) \leq d(X) + gen(X)$ for median spaces. Note that if X is a (graphic) tree then equality holds. In general, the inequality may be strict: consider the graphic n-cube and its convex hyperspace.

4.39.2. (cf. van de Vel [1983f]) Let T be a graphic tree with $n \geq 2$ end points, let $m > 2n$, and let $T_1,...,T_m$ be non-empty induced subtrees of T. Then there is an index $i \in \{1,..,m\}$ such that each subtree of T including each T_j for $j \neq i$ also includes T_i, and each subtree of T meeting each T_j for $j \neq i$ also meets T_i. Moreover, the lower bound $2n$ is sharp.

4.39.3. Let X be a free n-point space. Show that

$$d(\mathcal{C}_*(X)) = \binom{n}{\lfloor n/2 \rfloor}.$$

This shows that the inequality $d(\mathcal{C}_*(X)) \leq d(X) + gen(X)$ is not valid in general.

Hint. The convex hyperspace of X is isomorphic to the join semilattice $\{0,1\}^n \setminus \{0\}$ (usual convexity of order convex subsemilattices).

4.40. Radon number and products of trees

4.40.1. Let X be an S_4 convex structure of which the half-spaces form n directions. Show that $r \leq r_n$ (where the latter is the Radon number of the solid cube of dimension n; cf. 2.17). If $h \leq 3$, it suffices to assume that X is S_3 and is covered with n directions.

4.40.2. (compare Jamison [1981b]) Deduce that the Radon number of a product of n trees is at most r_n, and that equality holds if each factor tree is dense.

4.40.3. (compare van de Vel [1984c]) Deduce that $\lambda(5)$ does not embed in a product of four trees; $\lambda(6)$ does not embed in a product of nine trees; $\lambda(7)$ does not embed in a product of seventeen trees. These spaces do embed in, respectively, a 5–cube; a 10–cube; an 18–cube.

4.41. Arity of a convex structure (compare Soltan [1984] or Jamison [1974] who uses the term "degree"). Recall (I§1.4) that the *arity* of a space X is the number $a(X) \in \{0, 1, .., \infty\}$, defined by the following rule.

$a(X) \leq n$ (where $n < \infty$), iff each subset of X which includes the hull of its subsets of cardinality $\leq n$ is convex.

Note that $a \leq c$. By definition, interval spaces correspond with the spaces of arity ≤ 2.

4.41.1. Let X be a convex structure and let $0 < a < \infty$. For $A \subseteq X$ we recursively define a sequence $(h^n(A))_{n=0}^{\infty}$ of sets as follows.

$$h^0(A) = A; \quad h^{n+1}(A) = \cup \{co(F) \mid F \subseteq h^n(A); \#F \leq a\}.$$

Show that the operator $co_a: 2^X \to 2^X$, defined by

$$co_a(A) = \bigcup_{n=0}^{\infty} h^n(A),$$

yields a convexity of arity $\leq a$ on X, finer than the original one.

4.41.2. Construct a space of infinite arity.

4.41.3. For a product with two factors, verify that $a = \max\{a_1, a_2\}$. Give an example showing that the arity of a subspace can be larger than the arity of the superspace.

4.41.4. (compare I§1.12) Let $X \to Y$ be a quotient map. Show that $a(X) \geq a(Y)$.

4.41.5. Problem. Determine the arity of a (symmetrically) generated H-convexity (cf. I§4.16.4). What is the arity of a convex hyperspace expressed in terms of the arity of the basic space?

4.42. Depth of median graphs (Bandelt and van de Vel [1991]). The *depth* of an S_4 convex structure is the supreme length of a chain of non-trivial half-spaces. Note that among S_1 spaces, the S_4 spaces of depth 0 are exactly the singletons.

4.42.1. (compare Isbell [1980] and van de Vel [1983e]; cf. I§6.35.3) Show that the finite median graphs of depth one are exactly the graphic n–cubes, $n > 0$.

4.42.2. Show that the following are equivalent for a finite median graph G.

§4: Infinite Combinatorics 249

(i) G is isomorphic to a simplex graph (cf. 4.29).
(ii) The depth of G is ≤ 2.
(iii) There exists a vertex $p \in G$ such that each maximal cube contains p.

Hint (iii) $\Rightarrow$ (i). Take F equal to the set of all elements adjacent to p; two such elements are defined to be adjacent in F if they occur together in a square. For an isomorphism $\sigma(F) \to G$, assign to a simplex of F its p-supremum. Consult I§6.26 and I§6.37.3.

4.42.3. Let $3 \leq r < \infty$. Let $\lambda^*(r)$ denote the superextension $\lambda(r)$ minus its r extreme points; cf. I§1.23. Verify that the depth of $\lambda(r)$ is $r-1$ and that the depth of $\lambda^*(r)$ equals $r-3$. Note that $\lambda^*(5)$ must be a simplex graph; find a "central" vertex as in (iii) above. The corresponding graph F is the intersection graph of the collection of all 2-point sets in a 5-point set (F is the complement of the so-called *Petersen graph*; cf. Fig. 4).

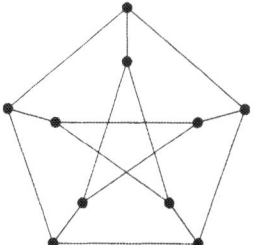

Fig. 4: Petersen graph

4.42.4. Show that if X is a finite S_4 convex structure, then the depth of $\mathcal{C}_*(X)$ equals the depth of X plus one.

Notes on Section 4

The Compactness Theorem 4.2 and the Compact Intersection Theorem 4.4 are due to Jamison [1977]. A particular case of Corollary 4.5 was obtained by Mulder and Schrijver [1979]. We have used this Corollary to determine the Helly number of H-convexities (Theorem 4.7). For real vector spaces, this was done by Boltyanskii [1976], who gave a more complicated proof under slightly different assumptions in terms of positive linear dependence. It differs from the present notion by the fact that linear combinations are considered with positive coefficients only. The application in 4.8 relates with a result of Szökefalvi-Nagy, stating that a K-convexity (cf. 4.26) in Euclidean space has Helly number two iff it is generated by a parallelepiped; see Hanner [1956] or Boltyanskii and Soltan [1978]. Corollary 4.9 is taken from van de Vel [1984]. The proof of the second part is new. The Countable Intersection Theorem, 4.11, is a mild strengthening of a result of Jamison [1977] who used it to generalize a result of Klee [1956] from real vector spaces to vector spaces over totally ordered fields; cf. 4.28.

Rank and independence are fundamental notions in matroid theory; cf. Section I§2.

A (universal-) algebraic approach to independence is presented in the survey paper of Marczewski [1966]. Convex independence of points in general convex structures was introduced by Calder [1971]. The notion of rank comes from Jamison [1981b], who used it to bound the Tverberg numbers from above (cf. 5.16).

The generating degree of a convex structure was introduced by van de Vel [1984f] as a tool for computing the rank. This paper contains the inequality of Proposition 4.14.1, and the part of Theorem 4.16 concerning the generating degree. The directional degree of a convex structure was introduced by the author in [1986b], where the remaining part of Theorem 4.16 and a restricted version of Corollary 4.17 (Helly number 2 case) were obtained. The part of Proposition 4.18.3, on equality of rank and generating degree with the number of end points of an absolutely complete tree, was obtained by the author in [1984f] under topological restrictions. The author's paper [1986b] also contains Theorem 4.19, on equality of the directional degree with the exchange number minus one on the class of median spaces, and Theorem 4.23, estimating *gen* with an upper bound which is twice as large as the present one. Theorem 4.21 is a reformulation of a result in the author's paper [1983g]. In contrast with the techniques in Section 2, this method of estimating the Radon number is based on counting ***non-Radon*** partitions; the result is adequate only when the Helly number is low. See Table 4.1.

The values of $\#\lambda(r)$ in Table 4.2 were obtained for $r \leq 7$ by Verbeek and Brouwer, and were published in Verbeek's dissertation [1972]. Computer assistance was necessary for $r = 7$ as well as for $r = 8$, in which case the counting has been carried out by Mills and Mills [1980]. The values of $\#\lambda(r)$ have been obtained independently for $r \leq 6$ in the context of free distributive lattices; see Monjardet [1974] for a survey.

5. Tverberg Numbers

> The list of convex invariants is closed with the most famous ones: the partition numbers. These are modeled after the Radon number by considering partitions of vertex sets in more than two parts, and they occur in Tverberg's Theorem on $\mathbb{R}^d$. Finiteness of the Radon number implies finiteness of all partition numbers, and in spaces of low Carathéodory or Radon number, the relationship between partition invariants is about the same as in $\mathbb{R}^d$. That this might always be the case is the subject of the Eckhoff conjecture.

5.1. Partition numbers. Let X be a convex structure, and let $F \subseteq X$ be a non-empty indexed set. A (weighted) partition $\{F_1,..,F_k\}$ of F is called a *Tverberg k-partition* provided
$$\bigcap_{i=1}^{k} co(F_i) \ne \emptyset.$$
The kth *partition number* (kth *Tverberg number*, k-*Radon number*) of X is defined to be the number
$$p_k(X) \in \{0,1,..,\infty\}$$
satisfying the following rule. If $n < \infty$ then $p_k(X) \le n$ iff each finite indexed set with more than n points has a Tverberg partition in $k+1$ parts. Note that the first partition number is just the Radon number.

The definition of this new invariant is stated in terms of weighted sets. Unlike the definition of the classical invariants, this makes an essential difference with a definition in terms of ordinary sets (Topic 5.19).

To complete our terminology, a set is *Tverberg (k) dependent* (T_k-*dependent, divisable*) if it admits a Tverberg partition into $(k+1)$ parts, and it is *Tverberg (k) independent* (T_k-*independent*) otherwise. Note that a subset of a T_k-independent set is T_k-independent (heredity).

We first consider the question whether finiteness of the Radon number implies finiteness of all the partition numbers. The answer is affirmative:

5.2. Theorem. *The partition numbers of a convex structure satisfy the following inequalities ($l, m \ge 1$).*
(1) $p_{(l+1)m} \le (1+p_l)p_m.$
(2) $p_{(l+1)(m+1)-1} \le (1+p_l)p_m + p_l.$
Hence if the Radon number is finite then so is p_k for each $k \ge 2$.

Proof of (1). Let $F = \{a_0,..,a_n\}$ be an indexed set with $n \ge (1+p_l)p_m$, and consider a weighted partition of $\{a_1,..,a_n\}$ into $1+p_l$ sets $F_0,..,F_{p_l}$ each of weight $\ge p_m$. Then $F_i \cup \{a_0\}$ has weight $> p_m$ and consequently admits a weighted partition of sets F_i^j,

$j = 0,..,m$, together with a point $b_i \in \cap_{j=0}^{m} co(F_i^j)$. We assume that $a_0 \in F_i^0$ for all i. The resulting set

$$G = \{b_i \mid i = 0,..,p_l\}$$

admits a weighted partition of sets G^k, $k = 0,..,l$, together with a point $b \in \cap_{k=0}^{l} co(G^k)$. Define $F^0 = \cup_{i=0}^{p_l} F_i^0$, and note that $G \subseteq co(F^0)$. For $j > 0$ we let

$$F^{jk} = \cup \{F_i^j \mid b_i \in G^k\}.$$

Note that $G^k \subseteq co(F^{jk})$ for all $j = 1,..,m$. We find that the sets F^{jk} for $j = 1,..,m$ and $k = 0,..,l$, together with F^0, constitute a partition of F and that the hull of each set of the partition contains b. Therefore, F is $T_{(l+1)m}$–dependent.

Proof of (2). This time we consider an indexed set F with more than $(1+p_l)p_m + p_l$ points, i.e. F has weight $n \geq (1+p_l)(1+p_m)$. Divide this set into p_l+1 subsets, each of weight at least $1+p_m$. Proceeding as above (but without considering a "special" point a_0), we arrive at a Tverberg partition of F with $(l+1)(m+1)$ sets, thus establishing (ii).

To verify the final part of the theorem, note that for $l = 1$ we have $p_l = r$ and our formulas reduce to the following.

$$p_{2m} \leq (1+r)p_m;$$
$$p_{2m+1} \leq (1+r)p_m + r.$$

∎

Table 5.1:
Estimates of partition numbers in terms of Radon numbers

p_1	r
p_2	$r^2 + r$
p_3	$r^2 + 2r$
p_4	$r^3 + 2r^2 + r$
p_5	$r^3 + 2r^2 + 2r$
p_6	$r^3 + 3r^2 + 2r$
p_7	$r^3 + 3r^2 + 3r$
p_8	$r^4 + 3r^3 + 3r^2 + r$
p_9	$r^4 + 3r^3 + 3r^2 + 2r$
It is expected that $p_k \leq kr$ (Eckhoff conjecture, see 5.17)	

The upper bound for the Tverberg numbers, obtained with the above formulas, is a polynomial expression in the Radon number. Some of them are given in Table 5.1.

5.3. Full independence. The proof of the main theorem is based on the use of the following concept. Let M be a matroid of finite rank $d(M)$. A non-empty finite set $F \subseteq M$ is *fully independent* provided for each family $\{F_0,..,F_k\}$ of disjoint subsets of F,

(F) $\qquad d(\cap_{i=0}^{k} co(F_i)) = \max\{0; d(M) - \sum_{i=0}^{k}(d(M) - d(F_i))\}.$

The following series of results may be of help to become familiar with the concept

§5: Tverberg Numbers   253

of full independence. Let us first compare with the concept of a set being in general position, defined for matroids in Section 2.

5.4. Proposition. *In a matroid of finite rank, each fully independent subset is in general position.*

Proof. Let M be a matroid of rank $d < \infty$, let $F \subseteq M$ be a fully independent set, and let $G \subseteq F$ consist of m elements, where $m \leq d$. Then $d(G) = m$ by (F) (with $k = 0$). By definition F is in general position. ∎

From now on, we concentrate on projective matroids. In these circumstances, the (projective) dimension has been defined as the rank minus one, cf. I§2.18. Henceforth, as in I§2.7.3, the hull operator of a projective space is denoted by pr. Full independence of a set F in a projective space $\mathbb{P}$ of dimension $d < \infty$ can be expressed alternatively as follows. If $F_0, ..., F_k$ are pairwise disjoint subsets of F, and if F_i has $d_i + 1 \leq d + 1$ elements for each i, then

(*) $\quad dim(\cap_{i=0}^{k} pr(F_i)) = \max \{-1; \sum_{i=0}^{k} d_i - kd \}$.

5.5. Proposition. *Let $\mathbb{P}$ be a projective space of finite dimension d, let F be a fully independent set, and let $P_0, .., P_k$ be projective sets spanned by pairwise disjoint subsets of F. Then $\cap_{i=0}^{k} P_i = \emptyset$ iff the sum of the codimensions of P_i for $i = 0, .., n$ exceeds d.*

Proof. By the previous result we find that each P_i is projectively spanned by a subset of F with $d_i + 1 \leq d + 1$ points, where $dim\, P_i = d_i$. By assumption, $\cap_{i=0}^{k} P_i = \emptyset$ iff $\sum_{i=0}^{k} d_i - kd \leq -1$, which is in turn equivalent to $\sum_{i=0}^{k} (d - d_i) \geq d + 1$. ∎

5.6. Proposition. *Let $\mathbb{P}$ be a projective space of dimension d, let F be a fully independent set, and for $k \geq 1$, let $F_0, .., F_k$ be pairwise disjoint subsets with $\#F_i \leq d + 1$ and $\cap_{i=0}^{k} pr(F_i) \neq \emptyset$. Then*

$$pr\big(pr(F_0) \cup \cap_{i=1}^{k} pr(F_i)\big) = \mathbb{P}.$$

Proof. Let $\#F_i = d_i + 1$ for $i = 0, .., k$. Then by I§2.19 and by (*),

$$dim\, pr\big(pr(F_0) \cup \cap_{i=1}^{k} pr(F_i)\big) =$$

$$= dim\, pr(F_0) + dim\, \big(\cap_{i=1}^{k} pr(F_i)\big) - dim\, \big(\cap_{i=0}^{k} pr(F_i)\big)$$

$$= d_0 + \sum_{i=1}^{k} d_i - (k-1)d - \big(\sum_{i=0}^{k} d_i - kd\big).$$
∎

The results obtained so far indicate that full independence is a rather heavy condition. Do these sets occur frequently enough? The first step in showing this to be the case is the following.

5.7. Proposition. *Let $\mathbb{P}$ be a projective space of dimension d, let F be fully independent, and let $p \notin F$ be such that for $k \geq 0$ and for each disjoint family of sets $F_0,..,F_k \subseteq F$ the following holds.*

(**) *If the set $P = pr\left(\cap_{i=1}^{k} pr(F_i) \cup pr(F_0)\right)$ is not equal to $\mathbb{P}$, then $p \notin P$.*

In these circumstances, $F \cup \{p\}$ is fully independent.

Proof. Let $F_0,..,F_k$ be disjoint subsets of $F \cup \{p\}$ with
$$d_i = \#F_i - 1 \leq d \, (i = 0,..,k).$$
We may assume that $p \in F_0$ and that $F'_0 = F_0 \setminus \{p\}$ is not empty. Let
$$P' = pr\left(\bigcap_{i=1}^{k} pr(F_i) \cup pr(F'_0)\right).$$
Note that $P' \subseteq P$ and that $p \notin pr(F'_0)$ (take $k = 0$ in (**)). We have to show that the dimension $\bar{d}$ of the projective set $\cap_{i=0}^{k} pr(F_i)$ satisfies
$$\bar{d} = \max\left\{-1; \sum_{i=0}^{k} d_i - kd\right\}.$$

First case: The set $\cap_{i=0}^{k} pr(F_i)$ is not empty (in particular, $\bar{d} \geq 0$). Take a point a in this set. If $a \in pr(F'_0)$, then $P' = \mathbb{P}$ by 5.6. If $a \notin pr(F'_0)$, then by the Exchange Law for matroids, we find that $p \in pr(F'_0 \cup \{a\})$, whence $p \in P' \subseteq P$, and $P = \mathbb{P}$ by (**). In either situation we can use the projective dimension formula I§2.19, yielding
$$d = \dim P = \dim\left(\bigcap_{i=1}^{k} pr(F_i)\right) + \dim pr(F_0) - \bar{d},$$
$$= \sum_{i=1}^{k} d_i - (k-1)d + d_0 - \bar{d}.$$
It follows that $0 \leq \bar{d} = \sum_{i=0}^{k} d_i - kd$, which is as required.

Second case: $\cap_{i=0}^{k} pr(F_i) = \emptyset$, i.e., $\bar{d} = -1$. By Proposition I§2.19,
$$d \geq \dim P = \dim\left(\bigcap_{i=1}^{k} pr(F_i)\right) + \dim pr(F_0) - \bar{d}.$$
It follows that
$$-1 = \bar{d} \geq \sum_{i=1}^{k} d_i - (k-1)d + d_0 - d,$$
which yields the correct formula again. ∎

Throughout, we consider Euclidean d-space $\mathbb{R}^d$ to be embedded in *real* projective space $\mathbb{P}^d$ of dimension d.

5.8. Proposition. *Let G be a fully independent set in $\mathbb{P}^d$, let $F = \{p_1,..,p_m\} \subseteq \mathbb{R}^d$ and let U_i be a neighborhood of p_i for $i = 1,..,m$. Then there is a set $F' = \{p'_1,..,p'_m\}$ such that $F' \cup G$ is fully independent and $p'_i \in U_i$ for all i.*

§5: Tverberg Numbers   255

Proof. We show that for each $m > 0$ the set of all m-tuples of $\mathbb{R}^d$ forming a fully independent set with the points of G, constitute a dense subset of $(\mathbb{R}^d)^m$. The argument goes by induction on the number $m \geq 0$. The requirements being "vacuously" fulfilled for $m = 0$, suppose the result has been established for m-tuples, let $p, p_1,..., p_m \in \mathbb{R}^d$, and let U_i and U be neighborhoods of p_i and of p, respectively. For each $i = 1,..,m$, there is a point $q_i \in U_i$, such that the set $F' = \{q_1,..,q_m\}$, together with G, is fully independent. Each point $q \in \mathbb{R}^d$, not in one of finitely many proper projective subsets of $\mathbb{P}^d$ of a type, described in (**), can be added to $F' \cup G$ to form a larger fully independent set. Note that these points q form a dense subset of $\mathbb{R}^d$. In particular, we can replace p with a point $q \in U$ as desired. ∎

Let $\mathbb{P}(\infty)$ denote the $(d-1)$-dimensional space of all points at infinity in $\mathbb{P}^d$. Then $\mathbb{P}(\infty)$ is spanned by a set G with d points. Hence, if F is such that $F \cup G$ is fully independent and if $P_1,..,P_k$ are projective spaces spanned by pairwise disjoint subsets of F, such that the set $P = \cap_{i=1}^{k} P_i$ is of dimension $\bar{d} \geq 0$, then the dimension of $P \cap \mathbb{P}(\infty)$ is given by $\bar{d} + (d-1) - d = \bar{d} - 1$. It follows that $\cap_{i=1}^{k} P_i \not\subseteq \mathbb{P}(\infty)$. This observation is of use in the proof of the main result.

5.9. Theorem (*Tverberg Theorem*). *For each $k \geq 1$ the kth partition number p_k of $\mathbb{R}^d$ satisfies*

$$p_k = k \cdot (d+1).$$

Proof. We first show that a fully independent set $F \subseteq \mathbb{R}^d$ with $m = k(d+1)$ points is T_k-independent, establishing that $p_k \geq k(d+1)$. To this end, let $\{F_0,..,F_k\}$ be a partition of F in $k+1$ parts. By Proposition 5.4, each F_i can be reduced to a set with $d_i + 1 \leq d + 1$ points without affecting the projective hull. As

$$\sum_{i=0}^{k} d_i + (k+1) = \sum_{i=0}^{k} (d_i + 1) \leq m = k(d+1),$$

we have $\sum_{i=0}^{k} d_i \leq kd - 1$ and hence that $\sum_{i=0}^{k} (d - d_i) \geq d + 1$. By Proposition 5.5 it follows that $\cap_{i=0}^{k} pr(F_i) = \varnothing$. In particular, $\{F_0,..,F_k\}$ is not a Tverberg partition.

To obtain the opposite inequality, first note that if an m-set admits no Tverberg partition in $k+1$ parts, then the same is true if its points are moved in a small neighborhood. Therefore, by Proposition 5.8, it suffices to consider fully independent sets F. We consider an additional set of d "spare" points at infinity which, together with F, form a fully independent set. As a consequence, $\cap_{i=0}^{k} pr(F_i)$ cannot consist exclusively of points at infinity whenever F_i for $i = 0, 1,..,k$ is a disjoint family in F.

We have to show that if $\#F > k(d+1)$ then F admits a Tverberg partition in $k+1$ sets. For $k = 1$ the desired result corresponds with Radon's Theorem, 1.8(3). We assume henceforth that $k > 1$ and that the result holds for values $< k$. We consider all families of pairwise disjoint sets of type

(*) $\mathcal{P} = \{F_0,..,F_k\}$ (where $F_i \subseteq F$; $\#F_i = d_i + 1 \leq d + 1$)

with the property that $\cap_{i=1}^{k} co(F_i) \neq \varnothing$. Such families exist by the inductive hypothesis, combined with Carathéodory's Theorem, 1.8(2), and are called "admissible" in the sequel. Assume that none of them is a Tverberg k-partition.

Consider an admissible partition $\mathcal{P}$ as in (*) with the smallest possible distance between $co(F_0)$ and $\cap_{i=1}^{k} co(F_i)$. Take $u \in co(F_0)$ and $v \in \cap_{i=1}^{k} co(F_i)$ realizing this minimal distance. Consider the affine function $f : \mathbb{R}^d \to \mathbb{R}$ defined by

$$f(x) = (v-u) \cdot (v-x).$$

Let $H = f^{-1}(0)$ and $A_i = aff(F_i)$ for each i. Note that (Fig. 1)

$$f(co(F_0)) \geq f(u) > 0; \quad f(\bigcap_{i=1}^{k} co(F_i)) \leq f(v) = 0.$$

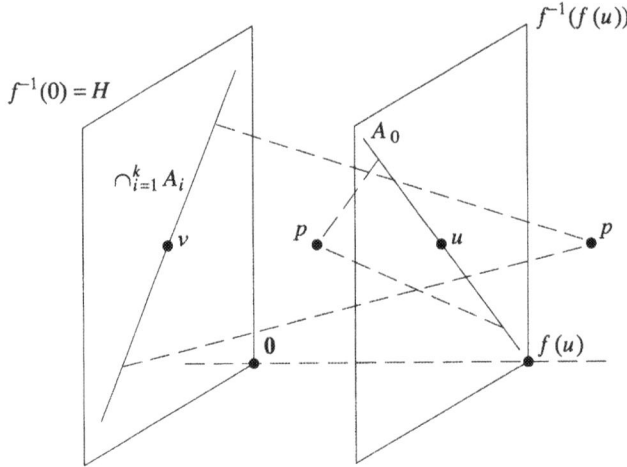

Fig. 1: Assigning p

We may assume that each F_i has been minimized with respect to the property of containing u, or v. This implies that v is in the interior of $co(F_i)$ relative to A_i and hence that $\cap_{i=1}^{k} A_i$ is included in the supporting hyperplane $f^{-1}(0)$. Similarly, $A_0 \subseteq f^{-1}(f(u))$. Hence $\cap_{i=0}^{k} pr(F_i)$ has no points in $\mathbb{R}^d$, and by an earlier remark we conclude that this intersection must be empty. By full independence of F, we have $\sum_{i=0}^{k} d_i - kd \leq -1$, whence the total number of points covered by the disjoint family $\mathcal{P}$ satisfies

$$\#\cup_{i=0}^{k} F_i = \sum_{i=0}^{k} d_i + k + 1 \leq k(d+1).$$

Consequently, at least one point $p \in F$ is not involved. We will add it to one of the sets F_i for $i = 0, 1, .., k$, in such a way that the metric "shift" of the new partition is strictly smaller than the assumed minimal one.

The easiest case is $f(p) < f(u)$: then p is added to F_0 and a contradiction is obtained. So assume that $f(p) \geq f(u) > 0$. The affine sets A_i for $i = 1, ..,k$ having a point in common, it follows from I§7.18 that

§5: Tverberg Numbers

$$pr\bigl(\cap\{A_j \mid j=1,..,k; j \neq i\}\bigr) = \cap\{pr(A_j) \mid j=1,..,k; j \neq i\}.$$

By virtue of Proposition 5.6,

$$\mathit{aff}\bigl(\cap\{A_j \mid j=1,..,k; j \neq i\} \cup A_i\bigr) = \mathbb{R}^d.$$

By Proposition I§7.11(3) there exist two points q_i, r_i such that

$$q_i \in \cap\{A_j \mid j=1,..,k; j \neq i\}; \quad r_i \in A_i; \quad p \in q_i/r_i.$$

We claim that one of the segments pq_i, for $i = 1,..,k$, is disjoint with H. Indeed, assume the contrary. First, note that $r_i \notin H$, for otherwise $p \in q_i/r_i \subseteq H$. By Theorem I§3.19, there is a point

$$x \in \bigcap_{i=1}^{k} co\bigl(\{q_j \mid j=1,..,k; j \neq i\} \cup \{r_i\}\bigr) \subseteq \bigcap_{i=1}^{k} A_i \subseteq H.$$

Take y_i in $co\{q_j \mid j=1,..,k; j \neq i\}$ such that $x \in y_i r_i$. By Theorem I§4.12, the segment py_i meets H as well. Consider the triangle (p, y_i, r_i). Each side meets H and the corner points p and r_i are not in H. The only way out is that two intersection points with H coincide in a corner point -- which must be y_i -- and hence that $x = y_i$ (for else $r_i \in x/y_i \subseteq H$). This yields

$$x \in \bigcap_{i=1}^{k} co\{q_j \mid j=1,..,k; j \neq i\},$$

and by 1.4, the points q_i are affinely dependent, say:

$$q_i \in \mathit{aff}\{q_j \mid j=1,..,k; j \neq i\}.$$

However, the last set is included in A_i and hence $p \in q_i/r_i \subseteq A_i$, contradicting that F is fully independent.

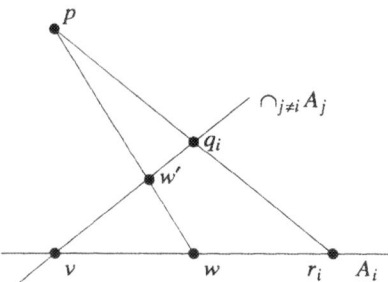

Fig. 2: Finding a point closer to u

So far we have shown that some segment pq_i has positive f-values. The set

$$\cap\{co(F_j) \mid j=1,..,k; j \neq i\}$$

has a non-empty interior relative to

$$\cap\{A_j \mid j=1,..,k; j \neq i\},$$

and $co(F_i)$ has a non-empty interior relative to A_i. By using the Pasch and Peano Properties, we can find a point $w \in (v \circ r_i) \cap co(F_i)$ such that $p \circ w$ meets $v \circ q_i$ in a point

$w' \in \cap_{j \neq i} co(F_j)$ (see Fig. 2). Adding p to F_i, we find that $w' \in \cap_{j=1}^{k} co(F_j)$, and some point of $v \circ w'$ is closer to u than v is. ∎

In regard to the classical Radon Theorem, 1.8(3), the last result can be reformulated this way: in Euclidean space of any dimension, $p_k = k \cdot p_1$. We will henceforth use the familiar "r" again to denote the Radon number.

We now work towards a second result involving a similar upper bound of the Tverberg numbers. The following "local" version of directional degree is of use.

5.10. Pointwise directional degree. Let X be a convex structure, let $\mathcal{S}$ be a subbase for X, and let $p \in X$. Then the *spread of* $\mathcal{S}$ *at* p is the spread of the family

$$\mathcal{S}_p = \{ S \in \mathcal{S} \mid p \notin S \}.$$

The *directional degree of* X *at* p is the minimal spread at p of a subbase. It will be denoted by $dir_p(X)$.

That dir and dir_p can be different is illustrated as follows. In a poset X, it is easily seen that $dir\, X$ is infinite whenever X has an infinite antichain, the elements of which are pairwise bounded from above. In contrast, $dir_p(X) \leq 2$ for each $p \in X$: use the subbase described in I§4.19.

5.11. Proposition. *Let X be a non-empty convex structure.*

(1) *For each $p \in X$,*

   $dir_p(X) \leq dir(X).$

(2) *If $e(X) < \infty$, then there is a point $p \in X$ such that $dir_p(X) \geq e(X) - 1$. If $c(X) \geq 3$, then there is a point $p \in X$ with $dir_p(X) \geq c(X)$.*

(3) *If X is median, then*

   $dir(X) = \sup \{ dir_p(X) \mid p \in X \}.$

Proof. Part (1) is a direct consequence of the definition. Parts (2) and (3) follow from an inspection of the arguments in 4.19 concerning dir, c, and $e - 1$. ∎

Let F be a weighted subset of X, and for $k \geq 1$ let

$$\mathcal{F}_k = \{ co(G) \mid G \subseteq F; \#(F \setminus G) = k \}.$$

The character "#" indicates weight, not cardinality, and complementation has to be interpreted accordingly. A point common to all members of the set $\mathcal{F}_k$ is a k-*core point* of F. The observations in (1) and (2), together with Theorem 4.19, lead to the following result, contrasting with the situation in posets.

5.12. Lemma. *Let the space X be of pointwise directional degree ≤ 2 at a k-core point p of a set F with at least $2k + 2$ points. Then p is in the hull of $k + 1$ pairwise disjoint subsets of F.*

Proof. By Proposition 5.11(2), the Carathéodory number c of X is at most 2. Let p be a k-core point of F. We look for two distinct points $a, b \in F$ with the following

§5: Tverberg Numbers   259

properties.

(1) $p \in ab$.
(2)   If $G \subset F$ is such that $\#F \setminus G = k + 1$ and $p \notin co(G)$, then G contains a or b.

Let $\mathfrak{b}$ be the family of all sets G satisfying the hypothesis of (2). Note that if $G \in \mathfrak{b}$, then for each point $a \notin G$ there is a point $b \in G$ with $p \in ab$. This follows from the fact that p is in the hull of each subset of F missing k points, and from the fact that $c \leq 2$. Hence we are done if $\#\mathfrak{b} \leq 1$. If $\mathfrak{b} = \{G, G'\}$, then take $a \in G \setminus G'$; the desired pair of points obtains by taking a point $b \in G'$ with $p \in ab$. So assume $\#\mathfrak{b} \geq 3$. Let $\mathfrak{s}$ be a subbase of spread ≤ 2 at p. Note that if $p \notin S \in \mathfrak{s}$, then S cannot include two different members of $\mathfrak{b}$ since p is a k-core point. For each $G \in \mathfrak{b}$ there is a set $S_G \in \mathfrak{s}$ including G and missing p. We conclude that there exist $G, G' \in \mathfrak{b}$ with $S_G \cap S_{G'} = \emptyset$. As $\#F \geq 2(k+1)$, we find that $G \cup G' = F$. If G'' is the third and last element of $\mathfrak{b}$, then take $a \in G \cap G''$ and since $a \notin G'$ a point $b \in G'$ can be obtained as above. If $\mathfrak{b}$ has a fourth element G''', then $G'' \cap G''' = \emptyset$ and the point b constructed above is once again a solution. Each point of F is already in two members of $\mathfrak{b}$; there cannot be another set in $\mathfrak{b}$.

Having completed the construction of a, b as in (1) and (2), note that p is a $(k-1)$-core point of $F \setminus \{a,b\}$. Proceeding recursively, we obtain $k+1$ disjoint parts of F the hull of which contains p. ∎

5.13. Proposition. *Let X be a convex structure satisfying one of the following properties.*

(1)   *X has a Radon number ≤ 2;*
(2)   *X is everywhere of pointwise directional degree ≤ 2.*

Then $p_k \leq kr$ for all $k \geq 1$.

Proof. Suppose first[1] that X satisfies the assumption (1). Throughout, a member x' of $co\{x\}$ is called a *child* of x. Conversely, x is a *parent* of all members of $co\{x\}$. Let $F \subseteq X$ be a finite set with at least three points. For each pair $a \neq b$ in F let $F(a,b)$ be the set of points $x \in F$ with the following property.

(†) $\exists a', b' \, \forall a'', b'' : co\{x\} \cap co\{a'',b''\} \neq \emptyset$.

Among all admissible pairs a, b, we choose one with $\#F(a,b)$ maximal. Suppose $c \in F \setminus F(a,b)$. Then there exist a', b' such that $co\{c\} \cap co\{a',b'\} = \emptyset$. As $r \leq 2$, we find that either $co\{a'\} \cap co\{b',c\} \neq \emptyset$, or $co\{b'\} \cap co\{a',c\} \neq \emptyset$. It is easily seen that for some triple a'', b'', c', one of these formulas remains valid for each child of a'', b'', and c'. We suppose that the first formula does. Consider any pair of children b''' of b'' and c'' of c'. Let $x \in F(a,b)$ and take a point a''' in $co\{a''\} \cap co\{b''',c''\}$. Then

$$\emptyset \neq co\{x\} \cap co\{a''',b'''\} \subseteq co\{x\} \cap co\{b''',c''\}$$

Since b'' is a child of b, this is as required in (†). It follows that $F(a,b) \subseteq F(b,c)$ and the

---
1.   This part of the proof can be considerably simplified by assuming X to be point-convex.

inclusion is proper since c is in the second but not in the first set. We conclude that $F(a,b) = F$ for some $a \neq b$ in F.

The first result is established by induction as follows. The case $k = 1$ being obvious, we consider $k \geq 2$. Let $F \subseteq X$ have more than $2k$ points. There exist $a \neq b \in F$ such that $F(a,b) = F$. Then for each $x \neq a$, b in F there is a point x' common to the sets $co\{x\}$ and $co\{a,b\}$. Let F^* be the collection of all selected children x'. By the inductive hypothesis, there is a Tverberg partition in k parts of F^*. Note that the hull of each part is included in $co\{a,b\}$. After replacing all children by their parents and adding the set $\{a,b\}$, we obtain a Tverberg partition of F in $k+1$ parts.

As for the second half of the proposition, let X be a convex structure as in (2). Then the Carathéodory number c of X is at most 2 by 5.11(2). If $h < \infty$ is the Helly number of X, and if $\#F > kh$, then the sets $co(G)$, for $G \subseteq F$ and $\#(F \setminus G) = k$, meet h by h, whence there exist k–core points of F.

By Lemma 5.12 and Levi's inequality, each set with at least $\max\{kr+1, 2k+2\}$ points is T_k–dependent. For $k \geq 2$ we have $2k + 2 \leq 3k$. We conclude that $p_k \leq kr$. If $r = 2$, then part (1) of the proposition applies. ∎

For additional information, see 5.17. We note that two points a, b as in (1) and (2) of Lemma 5.12 exist in spaces satisfying the following property:

(*) If $p \in ab \setminus \{a,b\}$ and if C is a copoint at p, then a or b is in C.

This property obviously holds in posets and in vector spaces; moreover, it is inherited by subspaces. Consequently, every space of Carathéodory number 2 and satisfying (*) will also satisfy the conclusion of Proposition 5.13.

Further Topics

5.14. Images, subspaces and products

5.14.1. (compare Kołodziejczyk [1991]) Show that $p_k(X) \geq p_k(Y)$ for each $k \geq 1$ provided there is a CP surjection $X \to Y$.

5.14.2. Show that the kth Tverberg number of a subspace may be larger than the one of the superspace.

5.14.3. Problem. Compute (or at least estimate) the Tverberg numbers of the set of Gaussian integers, $\mathbb{Z}^n$, equipped with the relative standard convexity. Compute or estimate the Tverberg numbers of the superextension $\lambda(r)$ and of its subspaces arising from the subtraction of one or more of its extreme points.

5.15. Higher-order invariants.
Let X be a convex structure, and let $k \geq 1$. Define $h_k(X)$, $c_k(X)$ $e_k(X) \in \{0, 1, .., \infty\}$ as follows. For $n < \infty$,

(i) $h_k(X) \leq n$ iff each finite non-empty weighted set $F \subseteq X$ with more than n points has a non-empty k–core.

§5: Tverberg Numbers   261

(ii) $c_k(X) \le n$ iff each finite non-empty weighted set $F \subseteq X$ with more than n points satisfies

$$co(F) = \bigcup \{ co(G) \mid G \subseteq F;\ \#(F \setminus G) = k \}.$$

(iii) $e_k(X) \le n$ iff for each finite weighted set $F \subseteq X$ with more than n points and for each $p \in F$ the following holds.

$$co(F \setminus \{p\}) \subseteq \bigcup \{ co(G) \mid p \in G \subseteq F;\ \#(F \setminus G) = k \}.$$

Note that for $k = 1$ these numbers reduce to the original Helly-, Carathéodory-, and exchange number. Complements are taken in the sense of weighted sets.

5.15.1. The next formulae are valid for all convex structures.

$h_k = kh \le p_k$ (Jamison [1981b]);

$p_k \le c(kh - 1) + 1$ (Doignon, Reay and Sierksma [1981]);

$\left. \begin{array}{l} c_k \le c + k - 1; \\ e_k \le c_k + 1 \end{array} \right\}$ (Degreef [1981]).

Note that the finiteness of c implies finiteness of each c_k and of each e_k.

5.15.2. (Degreef [1981]) Show that the example in 2.19.2 satisfies $e_k = \infty$ for all $k \ge 2$ (in spite of the fact that $e = 2$).

5.15.3. (compare Kołodziejczyk [1991]) Show that $h_k(X) \ge h_k(Y)$ for each $k \ge 1$ provided there is a CP surjection $X \to Y$.

5.16. Rank and Tverberg numbers

5.16.1. Show that $h = r = d$ for all matroids. Give an example of a matroid with $c, e \ne d$.

5.16.2. (Jamison [1981b]) Verify the following equalities for the poset $\{0,1\}^n$, equipped with the Cartesian order.

$$h = r = d = \binom{\lfloor \tfrac{1}{2}n \rfloor}{n}.$$

5.16.3. (Jamison [1981b]) Show that $p_k \le kd$ for all convex structures. Deduce that $p_k = kr$ for convex structures with $h = d$. Kołodziejczyk and Sierksma [1990] have obtained the same result in terms of so-called semirank. In general, this invariant is smaller than the rank; both are equal on S_1 spaces.

5.17. Eckhoff conjecture

5.17.1. (compare Jamison [1981b]) Let X be a poset, or a median space of Carathéodory number ≤ 2. Show that $p_k \le kr$; in fact, equality holds for posets.

5.17.2. (compare Hare and Thompson [1975]) Verify that $p_k = kr$ for products of type $D_m(X) \times D_n(X)$. Consult 2.24.

5.17.3. (Lindström [1972], Tverberg [1971]; compare 1.24) Let $\mathcal{A}$ be a collection of more than kn subsets of $<n>$. Show that there exist $k+1$ mutually disjoint subfamilies $\mathcal{A}_0,..,\mathcal{A}_k$ of $\mathcal{A}$ with the property $\cup \mathcal{A}_0 = .. = \cup \mathcal{A}_k$.
Deduce that the kth Tverberg number of the semilattice $S = \{0,1\}^n$ (equipped with the usual convexity or with the convexity of subsemilattices) equals kr.

5.17.4. Problem. (Eckhoff [1979]; compare Calder [1971]) Does the inequality $p_k \leq kr$ hold for all $k \geq 1$ and for all convex structures? This problem is known in the literature as the *Eckhoff conjecture*, although it was formulated by Calder eight years before.

5.18. Partition numbers and products

5.18.1. (compare Hare and Thompson [1975]) If the kth Tverberg numbers of the factors of a product are, respectively, p_k^1, p_k^2, then the corresponding Tverberg number of the product space satisfies

$$\max\{p_k^1, p_k^2\} \leq p_k \leq p_k^1 + p_k^2.$$

The first inequality is sharp; cf. 5.17.2, but the second inequality is unsharp for $k = 1$; cf. Theorem 2.13.

5.18.2. (Lindquist and Sierksma [1981]; compare with the Extension Lemma 2.10) Let $1 \leq n \leq m$. For each n-set $F \subseteq <m>$ a disjoint family $\Phi(F)$ of k subsets of F is given. Let $N_k(\Phi; m,n)$ be the number of distinct families of k disjoint subsets of $<m>$ extending one of the given families $\Phi(F)$, and let $N_k(m,n)$ be the minimum taken over all functions Φ of the above kind. Show that

$$N_k(m,n) \geq k N_k(m-1,n) + N_k(m-1,n-1).$$

Problem (compare 2.25). Is the value $N_k(m,n)$ achieved by the "alternating" function Φ, which is obtained as follows: if $F = \{i_1,..,i_n\}$ with $i_1 < .. < i_n$, then

$$\Phi(F) = \left\{ \; \{ i_{jk+l} \mid 1 \leq j \leq \left\lceil \frac{n}{k+1} \right\rceil \} \mid l = 1,..,k \; \right\}?$$

5.18.3. (Sierksma [1982c]) For $k \leq m$, let $S_k(m)$ denote the total number of k-partitions of $<m>$. This is a *Stirling number of the second kind*, and it is given by the formula

$$S_k(m) = \frac{1}{k!} \sum_{i=1}^{k} (-1)^{k-i} i^m \binom{k}{i}.$$

For $i = 1,..,l$, let X_i be a convex structure of kth Tverberg number p_k^i. Show that the kth partition number p_k of the product is at most

§5: Tverberg Numbers   263

$$\max \left\{ m \mid \sum_{i=1}^{l} N_k(m,n) \le (l-1)S_{(k+1)}(m) \right\}.$$

Illustration. For the triple product $[0,1]^3$ and $k = 2$, the last result gives $p_2 \le 11$; estimating via the Hare-Thompson result gives $p_2 \le 12$, whereas Table 5.1 gives $p_2 \le 20$; Eckhoff's conjecture would yield $p_2 \le 8$. This bound can indeed be obtained by regarding the given cube as a CP image of the cube with standard convexity, and by using Tverberg's Theorem.

5.18.4. Problem. Find a sharp upper bound for the kth Tverberg number of a product of $n \ge 2$ factors. For a study of the second-order exchange number in products, see Degreef [1981].

5.19. Restricted partition numbers. The *restricted k^{th} partition number* $\bar{p}_k$ is defined as the original partition number, but weighted sets are replaced by ordinary ones.

5.19.1. The numbers p_k and $\bar{p}_k$ satisfy $\bar{p}_k \le p_k$, and the inequality may be strict for $k > 1$. This can happen for trivial reasons: if X is finite and $k > \#X$, then $\bar{p}_k = \#X$, whereas p_k tends to ∞ if k does.

If $\mathbb{N}$ is given the coarse convexity, then

$$p_k(X) = \bar{p}_k(X \times \mathbb{N}).$$

5.19.2. (Jamison [1981b]) Show that a poset with an odd Helly number h satisfies

$$\bar{p}_k \le k(h-1)+1.$$

Compare with the unrestricted values in 5.17.1. Hint. The following are equivalent: (i) p is a k-core point of F; (ii) $L(p) \cap F$ and $U(p) \cap F$ both have at least $k+1$ points; (iii) There is a partition of F in $k+1$ parts, each containing p in its convex hull.

5.19.3. Problem. Is $\bar{p}_k \le kr$? The answer is affirmative provided the answer to 5.17.4 is.

5.20. Copoint Intersection Property (compare Jamison [1981b]). Let $m \ge n$. A convex structure X has the property *CIP(m,n)* (*Copoint Intersection Property*) provided that for each $p \in X$ it is true that some n out of each m copoints at p have an empty intersection. Show that a space X has the CIP$(m+1, 2)$ iff $dir_p(X) \le m$ for all $p \in X$.

Notes on Section 5

Partition numbers of abstract convex structures and divisibility of finite sets were first considered by Calder [1971], Bean [1974], and by Hare and Thompson [1975]. Many authors use the "restricted" numbers as defined in 5.19; we follow the definition of Jamison [1981b] (we adapted the definition to the style used for Radon numbers). The polynomial bounds of partition numbers in terms of the Radon number (Theorem 5.2) are also taken from this paper. The inequality for even-indexed partition numbers is due independently to Doignon, Reay, and Sierksma [1981]. Jamison's paper [1981b] also includes Proposition 5.13 and its application to several classes of examples such as posets (improving an earlier result of Bean [1974]) and products of two trees. See 5.16 and 5.17. These results of Jamison are formulated in terms of "copoint intersection properties"; this viewpoint is explained in 5.20. Theorem 5.9 was obtained by Tverberg [1966]; an extension to vector spaces over a totally ordered division ring (skew-field) was given by Doignon and Valette [1977]. The notion of full independence (see 5.3) is taken from this paper. A simplified proof was given by Tverberg in [1981], who used Reay's [1968] notion of strong independence. The argument given here is largely based on Tverberg's second proof. For a comparison of full and strong independence, see Doignon and Valette [1977].

The conjecture in 5.17.3 is attributed to Eckhoff [1979] (but see Calder [1971]) and has lead to much subsequent activity in the area. Jamison's results of [1981b] (on spaces of low Carathéodory or Radon number), and the Boland-Sierksma results of [1983] (concerning spaces with a small number of points) illustrate that it is still a long way to settling the problem. On account of the difficulties in determining the Radon number -- even in quite common situations, like H-convexities, semilattices, or products -- the conjecture may seem somewhat premature. What is actually wanted is a purely combinatorial proof of Tverberg's Theorem.

A by-product of the conjecture has been the study of "higher-order" Helly-, Carathéodory-, and Exchange numbers by several authors. See 5.15. There are many other, sometimes very subtle variations on this theme. See Degreef [1981] or Sierksma [1982b] for a survey of some results and problems in the area. For information on refinements of Radon's and Tverberg's Theorem, see Reay [1982].

CHAPTER **III**

TOPOLOGICAL CONVEX STRUCTURES

1. Topology and Convexity on the same Set

Topological convex structures (tcs's), subspaces and products, convex closure, closure stability and interior stability, local convexity, weak topology.

2. Continuity of the Hull Operator

Lower and upper semicontinuity (LSC and USC) of convex closure (hull) versus local convexity, pure subspaces and relative continuity, non-existence of non-trivial convex closures on the circle.

3. Uniform Convex Structures

Compatible uniformity, relation with uniformly continuous hull operator, Uniform Separation Theorem, Uniform Completion Theorem, Uniform Quotient Theorem, Uniform Factorization Theorem, non-existence of a uniform compactification of Euclidean space.

4. Topo-convex Separation

Lower- and upper CP functionals, neighborhood separation (NS_i), functional separation (FS_i), Urysohn Theorem, compactification via superextensions.

5. Intrinsic Topology

Continuous posets, core topology, uniqueness of a topology on Bryant-Webster spaces of finite dimension, intrinsic weak (IW) topology, characterizing FS_4 in median algebras by the IW topology.

1. Topology and Convexity on the same Set

> In this section we set the terminology and notation that will be used in the remaining part of this monograph. A topology and a convexity on the same set are compatible if all polytopes are closed. The resulting combined structure is called a topological convex structure.
>
> Some largely self-explaining concepts are considered: closure- and interior stability, local convexity, and weak topology (dual compatibility). It turns out that all properties under consideration are preserved by products and that a majority of them is not preserved by passing to subspaces.
>
> In this framework a first study is made of the basic properties of ordinary compact convex sets, pospaces, Lawson semilattices, median algebras, cones, etc..

Throughout this and the next chapter, $Cl(A)$, $Int(A)$, and $Bd(A)$ respectively denote the closure, interior and boundary of a set A.

1.1. Compatibility. Let X be a set equipped with a topology $\mathcal{T}$ and with a convexity $\mathcal{C}$. We say that $\mathcal{T}$ is *compatible* with the convex structure $(X, \mathcal{C})$ (or, with the convexity $\mathcal{C}$) provided all polytopes of $\mathcal{C}$ are closed in $\mathcal{T}$. It follows from I§1.7.3 that

1.1.1. *A topology is compatible with a convexity on the same underlying set, iff the convexity is generated by closed sets.*

A triple $(X, \mathcal{T}, \mathcal{C})$ consisting of a set X, a topology $\mathcal{T}$ on X, and a convexity $\mathcal{C}$ on X, is called a *topological convex structure* (briefly, a *tcs,* or, if no confusion is possible, a *space*) provided $\mathcal{T}$ is compatible with $(X, \mathcal{C})$.

Let X be a T_1 topological space. As a matter of permanent notation and terminology, we propose the following.

$\mathcal{T}_*(X)$: The set of all non-empty *closed* subsets.
$\mathcal{T}_{fin}(X)$: The collection of all non-empty *finite* (closed) sets.
$\mathcal{T}_{comp}(X)$: The collection of all non-empty *compact* (closed) sets.

The space $\mathcal{T}_*(X)$ is called the *(topological) hyperspace* of X. For an S_1 (hence T_1) topological convex structure X, we extend our notation with the following.

$\mathcal{TC}_*(X)$: The collection of all non-empty *closed convex* sets.
$\mathcal{TC}_{comp}(X)$: The collection of all non-empty *compact convex* closed sets.

In phrases like "a compact S_4 space" where, apparently, a topology and a convexity are involved on the same set, it is understood that we have a topological convex structure. In those (rare) cases where this is not what is meant, we will use a rather explicit construction like "a compact space with an S_4 convexity", where topological and convexical ingredients are presented separately. In most situations we will omit explicit reference to the topology and to the convexity, and use only one symbol X, Y,... to denote a topological convex structure. When explicit names are needed, $\mathcal{T}(X)$ is used for the topology of X and $\mathcal{C}(X)$ for the convexity of X.

1.2. Examples. The subsequent list describes most of the examples that will be dealt with in this section. A few other types (cones, spaces of arcs, superextensions) will be considered later on.

1.2.1. Topological vector spaces. Let V be a *topological vector space,* that is, a (real) vector space with a Hausdorff topology, such that the operations

$$+: V \times V \to V, \quad (v,w) \mapsto v+w, \quad \text{and}$$
$$\cdot: \mathbb{R} \times V \to V, \quad (t,v) \mapsto t \cdot v,$$

of vector addition and scalar multiplication are continuous. It can easily be derived from the hull formula I§1.5.1 that the polytopes of the standard convexity are compact in V. Therefore, V is a tcs.

On a related topic, a standard result[1] asserts that a finite-dimensional linear subspace of a topological vector space is closed. Hence the linear and affine matroid of a vector space are topological convex structures.

1.2.2. H-convexity. Let V be a topological vector space and let $\mathcal{F}$ be a collection of continuous linear functionals $V \to \mathbb{R}$. The H-convexity generated by these functionals has a subbase of closed sets. According to 1.1.1, we obtain a topological convex structure. In particular, the Euclidean topology is compatible with any H-convexity on $\mathbb{R}^n$.

1.2.3. Pospaces. Let X be a *partially ordered space (pospace)* that is, a topological space with an additional partial order, the graph of which is closed in $X \times X$. Note that the diagonal in $X \times X$ is the intersection of the closed sets "$\leq$" and "$\geq$". Hence X is Hausdorff. Furthermore, the lower- and upper set of a point $p \in X$ are closed since these sets are precisely the intersection of "$\leq$" with $X \times \{p\}$ and $\{p\} \times X$, respectively. Consequently, all segments ab of X are closed, being of type $\{a,b\}$ or $[a,b]$. A poset has Carathéodory number ≤ 2 (cf. II§1.22) and hence

$$co(F) = \bigcup \{ab \mid a, b \in F\},$$

which is a closed set if F is finite. See 1.6.1 for a stronger statement.

1.2.4. Topological semilattices. Let S be a *topological (meet) semilattice,* that is, a Hausdorff space with a continuous meet operator

$$\wedge: S \times S \to S.$$

In particular, the corresponding partial order has a closed graph, and hence for a finite set $F \subseteq X$ the polytope

---
1. Köthe [1960, p. 256].

§1: Topology and Convexity on the same Set

$$co(F) = \{x \mid \exists a \in F: \wedge F \le x \le a\}$$

is closed. The case where S is a tree (a *topological tree*) will receive considerable attention.

1.2.5. Topological lattices. Let L be a *topological lattice*. Here, L is a Hausdorff space with continuous meet and join operators

$$\wedge, \vee: L \times L \to L.$$

As each polytope of L is in fact a segment, it follows easily that the given topology is compatible with the lattice convexity.

1.2.6. Topological median algebras. Let X be a *topological median algebra*, that is, a Hausdorff space with a continuous median operator

$$m: X \times X \times X \to X.$$

For $n \ge 2$ let $F = \{a_1,...,a_n\} \subseteq X$ and consider the following recursively defined functions:

$$f_2(x) = m(a_1, a_2, x); \quad f_{k+1}(x) = m(a_{k+1}, f_k(x), x).$$

It is easily seen that all f_k are continuous. Now $x \in co(F)$ iff x equals its gate in $co(F)$, which by Theorem I§6.10 is the infimum of F in the poset $(X, \le_x)$. Note that $f_2(x)$ is the infimum of a_1, a_2, and that f_{k+1} takes the infimum of $f_k(x)$ and a_{k+1}. Hence

$$co\{a_1,...,a_n\} = \{x \mid f_n(x) = x\},$$

which is a closed set since X is T_2 and f_n is continuous.

By Lemma I§6.15, a median metric space is a topological median algebra. In a different direction, a distributive topological lattice and a topological tree are examples of topological median algebras. For the first, this follows directly from the defining formula of the median, which is in terms of the join and meet operation of the lattice. For the second assertion, use the construction of a median as devised in I§1.22, or consult I§6.

1.2.7. Convex hyperspaces. In I§1.8.2 we introduced the hyperspace of a convex structure relative to a generating "protopology". If X is a topological convex structure, then a natural choice for this protopology is the intersection of the topology with the convexity. The collection $\mathcal{TC}_*(X)$ of all non-empty convex closed subsets of X is equipped with the Vietoris convexity and with the relative Vietoris topology. This convexity is generated by sets of type

$$<D> \cap \mathcal{TC}_*(X), \quad <D,X> \cap \mathcal{TC}_*(X),$$

where $D \subseteq X$ is convex closed. The Vietoris topology of $\mathcal{TC}_*(X)$ is generated by the sets of type $<D>$, $<D,X>$, where D is closed in X. By 1.1.1, $\mathcal{TC}_*(X)$ is a topological convex structure. It is called the *convex hyperspace* of X. For a non-compact Hausdorff tcs X, we will occasionally consider the subspace $\mathcal{TC}_{comp}(X)$ of all *compact* convex sets.

If $\mathcal{A} \subseteq 2^X$ is a collection of sets, then $T \subseteq X$ is called a *transversal set* of $\mathcal{A}$ provided $T \cap A \ne \emptyset$ for all $A \in \mathcal{A}$. The collection of all closed transversals of $\mathcal{A}$ is denoted by $\bot \mathcal{A}$. Such sets are of use to describe the convex closure of certain families. For instance, if $\mathcal{A} = \{C_i \mid i = 1,...,n\}$ is a collection of non-empty convex closed sets, then (cf. I§1.8.3)

$co(\mathcal{A}) = co^*(\cup \mathcal{A}) \cap \bigcap_{T \in \underline{\mathcal{A}}} <T,X> \cap \mathcal{IC}_*(X)$.

1.3. Subspaces and products. Let X be a topological convex structure and let the subset Y of X be equipped with the relative topology $\mathcal{T} \mid Y$ and the relative convexity $\mathcal{C} \mid Y$, where

$$\mathcal{T} \mid Y = \{ C \cap Y \mid C \in \mathcal{T} \}; \quad \mathcal{C} \mid Y = \{ C \cap Y \mid C \in \mathcal{C} \}.$$

By the relative hull formula I§1.9.1, $\mathcal{T} \mid Y$ is compatible with $\mathcal{C} \mid Y$. The resulting tcs is called a *subspace* of X. The term "subspace" has an established meaning in topology; we used subspaces in the general theory of convexity as well. It is understood that in case of a topological convex structure this term is used with both meanings simultaneously, unless stated explicitly to the contrary.

Let $(X_i)_{i \in I}$ be a family of topological convex structures. It follows from the product formula in I§1.10.2 that the polytopes of $\prod_{i \in I} X_i$ are closed in the product topology, which is therefore compatible with the product convexity. This gives rise to a topological convex structure, called the *product* of the family $(X_i)_{i \in I}$.

1.4. Theorem. *Let X_i for $i \in I$ be a collection of topological convex structures. Then the following are true.*

(1) *Each convex subset of X with a non-empty interior is a product set. In particular, the convex open sets of X are exactly the sets of type $\prod_i O_i$, where each factor O_i is convex open set of X_i and $O_i = X_i$ for all but finitely many $i \in I$.*

(2) *Each convex set C of X is a dense subset of $\prod_i \pi_i(C)$, where $\pi_i \colon X \to X_i$ is the ith projection.*

(3) *The convex closed sets of X are exactly the sets of type $\prod_i C_i$, where C_i is a convex closed subset of X_i.*

Proof. As to (1), let $C \subseteq X$ be convex such that $Int(C) \neq \emptyset$. Fix a point $p \in Int(C)$ and let $B = \prod_i B_i$ be a basic neighborhood of p, included in C. Then each B_i is a neighborhood of p_i and all but finitely many factors B_i are equal to the corresponding X_i. Due to the latter fact, the family of all product sets P with $B \subseteq P \subseteq C$ is inductively ordered. So we may assume that B is maximal in C. Let $q = (q_i)_{i \in I} \in C \setminus B$, and consider $x \in \prod_i (B_i \cup \{q_i\})$. Define an auxiliary point y as follows.

$$y_i = \begin{cases} p_i, & \text{if } x_i \notin B_i; \\ x_i, & \text{if } x_i \in B_i. \end{cases}$$

Then $y \in B$ and the polytope product formula gives $x \in co\{q,y\} \subseteq C$. This shows that $\prod_i (B_i \cup \{q_i\}) \subseteq C$, contradicting the maximality of B.

It is clear that sets of type $\prod_i O_i$, with O_i convex open and equal to X_i for almost all i, are convex and open in the product space. The converse follows from the above result.

As to (2), let $C \subseteq X$ be convex and non-empty, let $p = (p_i)_{i \in I}$ be a point of $\prod_i \pi_i(C)$, and take a basic neighborhood $N = \prod_i N_i$ of p. Then $N_i = X_i$ for all i except

§1: Topology and Convexity on the same Set

$i = i_1,..,i_k$ and $p_i \in \pi_i(C)$ for all i. Fix $q \in C$ and construct a point x with i^{th} coordinate equal to q_i for all i except $i \in \{i_1,..,i_k\}$; in the latter case we take the ith coordinate equal to p_i. For such i we fix a point $p(i) \in C$ with $\pi_i(p(i)) = p_i$. We find that $x \in N$ and

$$x \in \prod_{i \in I} co\{p_i,q_i\} \subseteq co\{p(i_1),..,p(i_k),q\} \subseteq C.$$

Part (3) follows directly from (2). ∎

1.5. Closure and hull. In I§1.2 we introduced *closure operators* as a common generalization of the convex hull operator and the Kuratowski operator in topology. Actually, a topological convex structure X yields two additional closure operators:

$$\overline{co}(A) = Cl(co(A));$$
$$co^*(A) = \bigcap \{C \mid A \subseteq C \in \mathcal{T}(X) \cap \mathcal{C}(X)\}.$$

The second one is called the *convex closure operator* of X. Formally, with the abbreviations $\mathcal{T} = \mathcal{T}(X)$ and $\mathcal{C} = \mathcal{C}(X)$,

$$\overline{co} = cl_{\mathcal{T}} \circ co_{\mathcal{C}};$$
$$co^* = cl_{\mathcal{T} \cap \mathcal{C}}.$$

If $C = co^*(A)$, then C is said to be *spanned by* A. Evidently, $\overline{co}(A) \subseteq co^*(A)$ for each $A \subseteq X$. In general, the two can be distinct.

1.6. Examples. In pospaces and in semilattices, some special formulas are valid for convex closure and hull.

1.6.1. Proposition. *If X is a pospace and if $K \subseteq X$ is compact, then*

$$co^*(K) = \bigcup \{[u,v] \mid u \leq v;\ u, v \in K\} = co(K).$$

Proof. The second equality is valid since[2] $c \leq 2$. We just show that $co(K)$ is closed. Let $p \notin co(K)$. The set $Y = (K \times K) \cap (\text{graph of } \leq)$ is compact. For each pair $(u,v) \in Y$, either $u \not\leq p$ or $p \not\leq v$. The first relation remains valid within neighborhoods N_u of u and W_p^{uv} of p; the second relation remains valid within neighborhoods N_v of v and W_p^{uv} of p. In the first case the following defines a Y-neighborhood of (u,v):

$$N_{uv} = (N_u \times K) \cap (\text{graph of } \leq).$$

In the second case we consider the Y-neighborhood

$$N_{uv} = (K \times N_v) \cap (\text{graph of } \leq)$$

of (u,v). By the compactness of Y there is a finite subset F of $Y \times Y$ such that the neighborhoods N_{uv} for $(u,v) \in F$ cover $Y \times Y$. This yields a neighborhood $W = \bigcap_{(u,v) \in F} W_p^{uv}$ of p with the following property. If $u \leq v$ are in K and if $x \in W$ then $(u,v) \in N = N_{u'v'}$ for some $(u',v') \in F$. Then either $N = (N_{u'} \times K) \cap (\text{graph of } \leq)$ -- hence $u \not\leq x$ -- or $N = (K \times N_{v'}) \cap (\text{graph of } \leq)$ -- hence $x \not\leq v$, showing that $W \cap co(K) = \emptyset$. ∎

---

2. c is the Carathéodory number of the pospace; see II§1.20.

1.6.2. Proposition. *If K is a compact subset of a compact semilattice, then*

(*) $\quad co^*(K) = \bigcup \{\, [\inf(K), k] \mid k \in K \,\}$.

Proof. If C is a convex set including K then C contains $\wedge F$ for each finite set $F \subseteq K$. These points form a net converging to $\inf(K)$. Hence if C is also closed, then $\inf(K) \in C$. Consequently, the right hand set in equality (*) is included in $co^*(K)$. The set $\bigcup_{k \in K} [\inf(K), k]$ is obviously convex; it is closed by virtue of Proposition 1.6.1. ∎

1.7. Stability. A topological convex structure X is *closure stable* provided the closure of each convex subset is convex. Dually, X is called *interior stable* provided the interior of each convex subset is convex.

1.8. Proposition. *Let X be a tcs.*
(1) $\quad X$ is closure stable iff $\overline{co} = co^*$.
(2) $\quad X$ is interior stable iff the convex hull of each open set in X is open.

Proof. Part (1) is an easy consequence of the definitions. As to (2), let X be interior stable and let $O \subseteq X$ be open. Then $Int(co(O))$ includes O by definition of interior, and is convex by assumption. By the definition of convex hull, it includes $co(O)$ and hence both sets are equal. Conversely, if the hull of each convex set is open, then for each convex set C we find that $co(Int(C)) \subseteq Int(C)$ by definition of interior, and equality follows from the extensiveness of the convex hull operator. ∎

1.9. Examples

1.9.1. Proposition. *A (real) topological vector space with its standard convexity is both closure- and interior stable.*

Proof. Let V be a topological vector space, let $C \subseteq V$ be convex, and let $x, y \in C^-$. Then there exist two nets $(x_i)_{i \in I}$ and $(y_i)_{i \in I}$ converging to x, resp., to y (by the homogeneity of V, there is no loss of generality in assuming that both nets involve the same index set). By the continuity of the algebraic operations we find that the net

$$(t \cdot x_i + (1-t) \cdot y_i)_{i \in I}$$

converges to $t \cdot x + (1-t) \cdot y$, which is therefore in the closure of C. Consequently, $co\{a, b\} \subseteq C^-$, showing that C is convex.

On the other hand let $O \subseteq V$ be an open set, and let $p \in co(O)$. Then there exist points $u_i \in O$ and real numbers $t_i > 0$ such that

$$p = \sum_{i=1}^{n} t_i \cdot u_i; \quad \sum_{i=1}^{n} t_i = 1.$$

Multiplication with a scalar $\neq 0$ and addition with a fixed vector both yield a homeomorphism of V. It follows that the set $\sum_{i=1}^{n} t_i \cdot O$ is a neighborhood of p included in $co(O)$, whence $co(O)$ is open. ∎

§1: Topology and Convexity on the same Set   273

The proof of closure stability extends to convex subspaces of V, but the proof of interior stability does not. A convex subspace of V need not be interior stable; consult 1.22.

1.9.2. Proposition. *An H-convexity on Euclidean space is closure - and interior stable.*

Proof. Let C be convex. By invariance under translations (cf. I§1.13.7) it can be assumed that the origin $\mathbf{0}$ is in the *intrinsic core* of C, that is: the relative interior of C in its affine hull (see Section 5 for a general definition). Then

$$C^- = \bigcap \{t \cdot C \mid t > 1\}$$

and closure stability follows from the invariance under homotheties. On the other hand, the intrinsic core of C is given by

$$\bigcup \{t \cdot C \mid 0 \leq t < 1\},$$

which is a union of a chain of H-convex sets. The interior of C is either empty, or equals the intrinsic core of C. ∎

1.9.3. Proposition. *A topological lattice is closure stable.*

Proof. Let C be convex. Then C is a sublattice and by the continuity of the lattice operations it follows easily that C^- is a sublattice. Let $a \leq b$ be in C^- and let $a \leq x \leq b$. Consider a net of points $b_i \in C$ converging to b. Then $a \wedge b_i \leq x \wedge b_i \leq b_i$ and $x \wedge b_i$ converges to $x \wedge b = x$. It suffices to show that $x \wedge b_i \in C^-$; in other words, we may assume that $b \in C$. The result then follows by operating on a in the same way. ∎

Interior stability can fail to hold in a topological lattice, as is shown by the sublattice of $[0,1]^2$, described in Fig. 1: the interior of the rectangle C is not convex.

1.9.4. Proposition. *A topological median algebra is closure stable.*

Proof. Let C be convex in a topological median algebra X. By the continuity of the median operator $m: X^3 \to X$, we find that

$$m(\bar{C} \times \bar{C} \times X) \subseteq \bar{C},$$

expressing that $\bar{C}$ is convex. ∎

In Fig. 1, we described a distributive lattice which is not interior stable. Such a lattice is a median algebra as well; cf. Corollary I§6.12.

1.9.5. Order convexity. A (compact) pospace need not be closure stable, nor interior stable. Consider the following subspace of $\mathbb{R}^2$ with the Cartesian order (Fig. 3).

$$X = A \cup B \cup C.$$

Here, A is the line segment joining $(-2,1)$ with $(-1,0)$, B is the line segment joining $(-2,0)$ to $(2,0)$, and C joins $(1,0)$ to $(2,-1)$. Note that X is compact. The set $A \cup C \setminus \{(-1,0),(1,0)\}$ is an antichain and hence it is order convex. Its closure contains

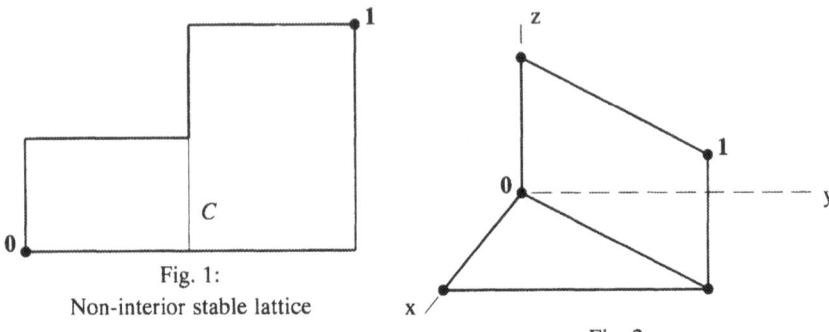

Fig. 1:
Non-interior stable lattice

Fig. 2:
Non-closure/interior stable semilattice

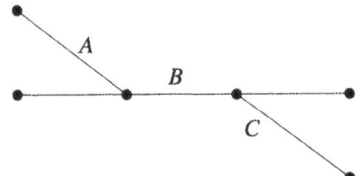

Fig. 3: Non-closure/interior stable pospace

$(-1,0)$ and $(1,0)$, but not the points in between, and hence it is not a convex set. It is easily seen that the convex set B has a non-convex interior.

1.9.6. Semilattices. A topological semilattice need not be closure stable or interior stable. Indeed, consider the following semilattice (cf. Fig. 2)

$$S = \{(x,y,0) \mid 0 \le y \le x\} \cup \{(x,y,z) \mid 0 \le x = y \le 1; 0 \le z \le 1\}.$$

This is a compact subsemilattice of the 3–cube, regarded as a meet semilattice (coordinate-wise meet). The second term of the union defining S is not order convex, although it is the closure of its S–interior, which is convex. On the other hand, the first term of the union is a convex set with a non-convex interior. See Corollary 1.15 for conditions leading to closure stable semilattices.

The behavior of stability in products is investigated in the next result.

1.10. Theorem. *Let X_i for $i \in I$ be a collection of non-empty topological convex structures with product X.*
(1) *X is closure stable iff all factors X_i are.*
(2) *X is interior stable iff all factors X_i are.*

Proof. The statement (1) is a direct consequence of 1.4(2). As to (2), let $C \subseteq X$ be convex. If $Int(C) = \emptyset$ then we are done. If $Int(C) \ne \emptyset$, then C is a product of convex sets by (1), and the result easily follows. ∎

§1: Topology and Convexity on the same Set

Passing to a subspace usually destroys both closure - and interior stability. For instance, $[0,1]^n$ is closure stable as a meet semilattice by Corollary 1.15 below. An example is given in 1.9.5 of a non-closure stable subsemilattice of the 3-cube. An example involving (non-) interior stability is given in topic 1.22. Alternatively, consult 1.9.3.

1.11. Local and proper local convexity. A tcs X is *(properly) locally convex at a point* $p \in X$ provided p has a neighborhood base of convex (open) sets. If X is (properly) locally convex at each point then X is *(properly) locally convex* for short. No example is known yet of a space which is locally convex but not properly locally convex.

1.11.1. *A subspace of a (properly) locally convex tcs is (properly) locally convex.* ∎

1.11.2. *A product of (properly) locally convex spaces is (properly) locally convex.* ∎

The first result is evident; the second is a consequence of Theorem 1.4. In order to verify proper local convexity of a space, it is often useful to consider a stronger condition as in the next result.

1.12. Proposition. *If X is a point-convex space such that for each convex closed set $C \subseteq X$ and for each open set $O \supseteq C$ there is a convex closed set D with $C \subseteq Int(D) \subseteq D \subseteq O$, then X is properly locally convex.*

Proof. Let $p \in X$ and let U be a neighborhood of X. Construct a sequence of convex closed sets $(D_n)_{n \in \mathbb{N}}$ such that $D_n \subseteq Int\, D_{n+1}$ by using the assumption first on $\{p\} \subseteq Int\, U$, and (inductively) on $D_n \subseteq Int\, U$. Then $\cup_{n \in \mathbb{N}} D_n$ is a convex open neighborhood of p. ∎

The condition concerning convex closed neighborhoods will be reconsidered in Sections 2 and 4. We note that this condition is somewhat unrealistic for non-compact sets. For instance, it is not fulfilled in Euclidean space of dimension ≥ 2. See 2.2 for details. In Section 3 we will consider a modification of the condition involving uniform spaces.

1.13. Examples. We investigate local convexity in pospaces, compact semilattices, trees, and compact distributive lattices. The following fact is needed.

1.13.1. *If X is a compact pospace and $O \subseteq X$ is open, then the sets*
$$O_L = \{x \mid L(x) \subseteq O\}; \quad O_U = \{x \mid U(x) \subseteq O\}$$
are convex and open.

Proof. Both sets are clearly convex. We just show that O_L is open. Let $L(p) \subseteq O$. For each $x \in X$ we determine a pair of neighborhoods U_{xp} of x and V_{px} of p as follows. If $x \leq p$, then both are taken equal to O. If $x \not\leq p$, then the neighborhoods are chosen such that $x' \not\leq p'$ whenever $x' \in U_{xp}$ and $p' \in V_{px}$. By the compactness of X, there is a neighborhood V_p of p such that $X \times V_p$ is covered by the sets of type $U_{xp} \times V_{px}$. If $p' \in V$ and $x' \leq p'$, then the pair (x',p') is covered by $O \times O$ only. Hence $L(p') \subseteq O$. ∎

1.13.2. Proposition. *Let X be a compact pospace. Then each convex closed subset of X has a basis of convex closed neighborhoods in X. In particular, X is properly locally convex.*

Proof. Let $C \subseteq O$ where C is non-empty convex closed and O is open. Then $C = L(C) \cap U(C)$, where $L(C)$ and $U(C)$ are closed. By normality we obtain open sets $O_- \supseteq L(C)$ and $O_+ \supseteq U(C)$ with $O_- \cap O_+ \subseteq O$. We have

$$C \subseteq \{x \mid L(x) \subseteq O_-\} \subseteq O_-; \quad C \subseteq \{x \mid U(x) \subseteq O_+\} \subseteq O_+.$$

The previous result shows that the sets in the middle are open. Intersecting these sets gives a convex open neighborhood P of C included in O. Take any closed neighborhood $D \subseteq P$ of C. By Proposition 1.6.1, $co(D)$ is a convex closed neighborhood of C with the desired property. The second part follows from 1.12. ∎

By a *Lawson semilattice* is meant a compact semilattice in which each point has a neighborhood base consisting of subsemilattices.

1.13.3. Proposition. *A compact semilattice is locally convex iff it is a Lawson semilattice. In fact, each convex closed set of a Lawson semilattice has a basis of convex closed neighborhoods. In particular, a Lawson semilattice is properly locally convex.*

Proof. Let S be a Lawson (meet) semilattice and let $p \in S$. If N is a neighborhood of p then by Proposition 1.13.2 we obtain an order convex neighborhood $U \subseteq N$ of p. In turn we find a subsemilattice neighborhood $V \subseteq U$ of p. Then the order convex hull of V is included in N and is easily seen to be a subsemilattice as well. This establishes the local convexity of Lawson semilattices. The converse is trivial.

We verify that S is even properly locally convex. Let $C \subseteq O$, where C is convex closed and O is open. By Proposition 1.13.2, there is an order convex neighborhood P of C with $P \subseteq O$. As S is a Lawson semilattice, the operator

$$\wedge \colon \mathcal{F}_*(S) \to S$$

is continuous. It maps the family $\langle C \rangle$ into $C \subseteq P$ and hence there is an open set $\mathcal{P}$ around $\langle C \rangle$ which is mapped into P. Regard a hyperspace as a compact pospace and regard $\langle C \rangle$ as the lower set of C. The statement in 1.13.2 yields is a basic neighborhood $\mathcal{U} = \langle P_1,...,P_n \rangle$ of C with $\langle C' \rangle \subseteq \mathcal{P}$ for each $C' \in \mathcal{U}$. Take a closed set D with $C \subseteq Int(D) \subseteq D \subseteq \cup_{i=1}^n P_i$. Then $D \in \mathcal{U}$ and hence $\langle D \rangle \subseteq \mathcal{P}$. The collection Q of all infima $\wedge D'$, for $D' \subseteq D$ closed and non-empty, is a compact semilattice. This gives

$$C \subseteq Int(D) \subseteq D \subseteq Q \subseteq P.$$

The last set being order convex, we use Proposition 1.13.2 once more to conclude that the order convex hull of Q is a closed convex neighborhood of C included in O. ∎

1.13.4. Corollary. *A compact tree is locally convex.*

Proof. According to a result of Lawson[3], a compact semilattice of finite breadth is

---
3. Lawson [1980]

§1: Topology and Convexity on the same Set   277

a Lawson semilattice. Almost by definition, a compact tree has breadth ≤ 2. Consequently it is locally convex. ∎

1.13.5. Proposition. *A compact distributive lattice is locally convex iff it is completely distributive.*

Proof. We use the fact[4] that a compact distributive lattice L is completely distributive iff the corresponding meet - and join semilattices are Lawson semilattices. This is the case if L is locally convex. Conversely, let $p \in L$ and let N be a neighborhood of p. Take a neighborhood $V_- \subseteq N$ of p which is an order convex meet-subsemilattice. Then choose a neighborhood $V_+ \subseteq V_-$ which is a join subsemilattice. It follows from distributivity that the meet semilattice V generated by V_+ is actually a sublattice of V_-. The order convex hull of V is included in N and (as in 1.13.3) gives a convex neighborhood of p. ∎

For a description of local convexity in the more general class of median spaces, see 2.8.1, where much stronger properties are considered.

1.13.6. Spaces which are not locally convex. There exists a compact, convex subspace X of a topological vector space such that X is not locally convex[5]. One can proceed as follows to find an example of a compact semilattice which is not locally convex[6]. Let S be the convex hyperspace of X. The semilattice operation is

$$C \vee D = co^*(C \cup D) = \bigcup \{t \cdot C + (1-t) \cdot D \mid 0 \leq t \leq 1\},$$

where the second equality holds by virtue of the hull formula in Proposition I§2.14. To see that S is not locally convex, just observe that if $T \subseteq S$ is a subsemilattice, then the elements of T are convex subsets of X, the union of which is convex.

1.14. Proposition. *Let X be locally convex and S_4. Then X is closure stable iff the closure of each half-space is convex.*

Proof. Suppose that the closure of each half-space is convex. Let $C \subseteq X$ be convex and let $p \notin C^-$. By local convexity there is a convex neighborhood N of p with $N \cap C = \emptyset$. Then take a half-space H extending C but missing N. We find that H^- is convex and $p \notin H^- \supseteq C^-$. This shows that C^- is the intersection of convex sets. ∎

The corresponding statement about interior stable convexity is *not* valid: see IV§2.40 for a procedure leading to a natural counterexample.

In a complete meet semilattice S, each non-empty set which is bounded from below, has an infimum. This implies that each pair of points $a, b \in S$ with an upper bound

---

4. Gierz et al [1980, Prop. VII§2.9]
5. The first example of such a space has been given by Roberts [1977]. The details of its construction would lead us too far.
6. For a second example of such a semilattice, see Gierz et al [1980, VI§4]. As with Roberts' example, the details are rather involved.

admits a supremum, $a \vee b$. The (partial) operator "$\vee$" is called a *join*. A complete meet semilattice S is *join-continuous* provided for each down-directed set $D \subseteq S$ and for each $x \in S$ with $\{x\} \cup D$ bounded from above,

$$x \vee \inf(D) = \inf\{(x \vee d) \mid d \in D\}.$$

For instance, a complete tree is join-continuous. On the other hand, a complete and completely distributive lattice by definition is a join-continuous semilattice.

1.15. Corollary. *Let S be a compact meet semilattice. Then:*

(1) *The closure of each lower set in S is convex.*
(2) *If S is a join-continuous Lawson semilattice, then S is closure stable.*

Proof of (1). Let $H \subseteq S$ be a lower set. As H is a subsemilattice and as "$\wedge$" is continuous, it follows easily that H^- is a subsemilattice. We verify that it is order convex. Let $a \leq b$ in H^-, and consider a net of points $b_i \in H$ converging to b. If $a \leq x \leq b$ then $\lim_i (x \wedge b_i) = x \wedge b = x$, where $x \wedge b_i \in H$ since H is a lower set.

Proof of (2). By Proposition 1.14, it suffices to show that the closure of a halfspace $H \subseteq S$ is convex. By I§3.2.2, H is a lower set or an upper set. The first case has been handled in (1). Let H be an upper set. As before, it suffices to show that H^- is order convex. Let $a \leq x \leq b$ with $a, b \in H^-$. Note that H is a down-directed set, and by the compactness of S it follows that $\inf(H)$ exists. Now $\inf(H) \in H^-$, and $a = \inf(H)$ without loss of generality. Consider a down-directed net of points $a_i \in H$ converging to a (for instance, the set H is such a net) and consider a net of points $b_j \in H$ converging to b. We fix the index j for a while, and we put $x_j = x \wedge b_j$. Note that $a \leq x, b_j$. By join-continuity,

$$x_j = x_j \vee a = x_j \vee (\bigwedge_i a_i) = \bigwedge_i (x_j \vee a_i),$$

and hence $\lim_i (x_j \vee a_i) = x_j$. As H is an upper set, it follows that $x_j \vee a_i \in H$ and hence that $x_j \in H^-$. We finally release j, yielding

$$\lim_j (x_j) = \lim_j (x \wedge b_j) = x.$$

This shows that $x \in H^-$. ∎

1.16. Weak topology. A topology $\mathcal{T}$ of (closed sets of) the underlying set of a convex structure X is a *weak topology* provided it has a subbase of $\mathcal{C}$–convex sets. Observe the duality with the concept of compatibility (cf. 1.1). It appears that the term *subbase* can be used with two different meanings: a subbase (of closed sets) for a topology will be referred to as a *closed subbase*; a subbase for a convexity is called a *convex subbase*. Weak topologies are usually obtained from a topological convex structure $(X, \mathcal{T}, \mathcal{C})$ by considering the coarser topology $\mathcal{T}_w$, generated by $\mathcal{T} \cap \mathcal{C}$. It is called the *weak topology* of $(X, \mathcal{T}, \mathcal{C})$ (the term "weak topology" will be justified in 4.15).

If the topology of tcs X is changed into the weak topology, then the resulting topological convex structure is denoted with X_w. A function from X to a topological space Y is

§1: Topology and Convexity on the same Set

called *weakly continuous* provided it is continuous as a function $X_w \to Y$. Here is a short list of useful results.

First, observe that (with the above notation) $\mathcal{T}_w \subseteq \mathcal{T}$ and consequently that

$$\mathcal{T}_w \cap \mathcal{C} \subseteq \mathcal{T} \cap \mathcal{C} \subseteq \mathcal{T}_w.$$

This shows that $\mathcal{T}_w \cap \mathcal{C} = \mathcal{T} \cap \mathcal{C}$. In words:

1.17.1. *In a topological convex structure a convex set is closed iff it is weakly closed (that is, closed in the corresponding weak topology).* ∎

1.17.2. *If $(X, \mathcal{T}, \mathcal{C})$ is a topological convex structure, then so is $(X, \mathcal{T}_w, \mathcal{C})$.* ∎

As a consequence of Theorem 1.4(3) we have

1.17.3. *The weak topology of a product equals the product of the weak topologies.* ∎

1.17.4. *If $(X, \mathcal{T}, \mathcal{C})$ is closure stable, then so is $(X, \mathcal{T}_w, \mathcal{C})$.* ∎

1.18. Superextensions. Let $\mathcal{S}$ be a collection of subsets of X. Recall (I§1.8.4) that $\lambda(X, \mathcal{S})$ denotes the superextension of X relative to a collection of subsets $\mathcal{S}$. Its convexity is generated by the family $\mathcal{S}^+$, consisting of all sets of type S^+, which in turn consist of all mls's in $\mathcal{S}$ having S as a member. The topology on $\lambda(X, \mathcal{S})$, generated by the closed subbase $\mathcal{S}^+$, is regarded to be standard. We collect some facts for later use.

1.18.1. Proposition. *Let $\mathcal{S}$ be a family of subsets of a set X.*
(1) $\lambda(X, \mathcal{S})$ *is a compact convex structure of Helly number ≤ 2. Its topology is a weak topology.*
(2) *If $\mathcal{S}$ is a T_1 family, then the canonical injection $l: X \to \lambda(X, \mathcal{S})$ is an embedding of the topological convex structures, generated by the closed and convex subbases $\mathcal{S}$, resp., $\mathcal{S}^+$.*
(3) *If $\mathcal{S}$ is a T_1 family, and if Y is a subspace of $\lambda(X, \mathcal{S})$ of Helly number ≤ 2, such that $l(X) \subseteq Y$, then Y is dense in $\lambda(X, \mathcal{S})$.*
(4) *If $\mathcal{S}$ is a normal T_1 family, then $\lambda(X, \mathcal{S})$ is locally convex.*

Proof of (1), Let $\mathcal{A}$ be a collection of subbasic sets meeting with finitely many at the same time. Then the sets $S \in \mathcal{S}$ with $S^+ \in \mathcal{A}$ form a linked system extending to a maximal one in $\mathcal{S}$, say m. By construction, $m \in \cap \mathcal{A}$, and compactness of the superextension follows from the Alexander Subbase Theorem in topology.[7] A superextension is a topological convex structure by 1.1.1 and its topology is a weak topology by construction.

Proof of (2). This follows from the fact that $l^{-1}(S^+) = S$ for each $S \in \mathcal{S}$.

Proof of (3). Let Y be a subspace of Helly number ≤ 2 including $l(X)$. To see that

---
7.   Engelking [1977, p. 280]

Y is dense in $\lambda(X, \mathcal{S})$ we proceed as follows. Suppose $p \notin \overline{Y}$. As $\mathcal{S}^+$ is a closed subbase, we find $S_i \in \mathcal{S}$ for $i = 1,..,n$ such that

$$p \notin \bigcup_{i=1}^{n} S_i^+ \supseteq Y.$$

For each i we obtain $T_i \in p$ with $T_i \cap S_i = \emptyset$. This yields

$$p \in \bigcap_{i=1}^{n} T_i^+ \subseteq \lambda(X, \mathcal{S}) \setminus Y.$$

As p is a linked system, the sets T_i^+ meet pairwise on $l(X) \subseteq Y$. As Y has Helly number ≤ 2, it follows that these sets have a common point in Y, contradiction.

Proof of (4): According to I§3.21.2, each pair of distinct points can be screened with subbasic sets. In a compact space, this yields that each point has a neighborhood base, consisting of intersections of (finitely many) subbasic sets. ∎

Henceforth, if X is a tcs with a convexity $\mathcal{C}$ and a topology $\mathcal{T}$ (of closed sets), then $\lambda(X)$ will be used as shorthand to denote the superextension of X with respect to its convex closed sets, $\lambda(X, \mathcal{T} \cap \mathcal{C})$.

1.19. Proposition. *Let X be a Hausdorff, locally convex and closure stable tcs.*

(1) *If X is compact, then $X = X_w$, i.e. X has the weak topology.*
(2) *If X has connected convex sets and is locally compact and S_4, then $X = X_w$.*

Proof of (1). Let A be closed in X, and let $p \in X \setminus A$. For each $a \in A$ there is a neighborhood N_a of a with $p \notin \overline{N}_a$ and a convex neighborhood $V_a \subseteq N_a$ of a. The compact set A admits a finite cover of type $\{V_{a_i} \mid i = 1,..,n\}$. Then $\cup_{i=1}^{n} \overline{V}_{a_i}$ is a weakly closed set including A and not containing p.

Proof of (2). The assumptions imply that each point of X has a compact convex neighborhood. By (1), the weak topology of X induces the original topology on each compact convex set. We only have to show, therefore, that a compact convex neighborhood U of $p \in X$ is at the same time a weak neighborhood of p. To this end, note that $Bd(U)$ is a closed subset of U which can be covered with finitely many convex closed sets C_i, $i = 1,..,n$, such that $p \notin \cup_{i=1}^{n} C_i$. Fix another convex neighborhood V of p with $V \subseteq U \setminus \cup_{i=1}^{n} C_i$. As X is S_4, by Theorem I§3.8 each of the sets

$$D_i = \{x \mid co\{p,x\} \cap C_i \neq \emptyset\} \quad (i = 1,..,n)$$

is convex. Now $V \cap D_i = \emptyset$ since V is convex and $p \in V$, showing that $p \notin \overline{D}_i$. Finally, for each $x \in X \setminus U$ the connected segment $co\{p,x\}$ meets some C_i in a point of $Bd(U)$, showing that $X \setminus U$ is covered with finitely many convex closed sets $\overline{D}_i$, $i = 1,..,n$. It follows that U includes a weak neighborhood of p. ∎

Further Topics

1.20. Quotients

1.20.1. Let X, Y be compact Hausdorff convex structures and let $q: X \to Y$ be a convexity preserving and convex-to-convex map of X onto Y. Show that Y is closure stable provided X is.

1.20.2. (van de Vel [1993]) Show that the compact tree which looks like a character "T" is a CP and CC image of the closed 2–disk. Deduce that interior stability in not inherited by CP and CC images.

1.21. Cones (compare van de Vel [1993]). Let X be a tcs (topological convex structure). The cone $\Delta(X)$ is equipped with the quotient topology relative to the natural quotient function $X \times [0,1] \to \Delta(X)$ and with the convexity described in I§2.4.1.

1.21.1. Verify that $\Delta(X)$ is a topological convex structure.

1.21.2. Show that $\Delta(X)$ inherits each of the following properties from X. Closure- resp., interior stability; (proper) local convexity.

1.21.3. Let $\Delta(X)$ now be convexified as the quotient of the product tcs $X \times [0,1]$. Show that if X has more than one point, then $\Delta(X)$ is not closure stable.

1.22. Interior stability

1.22.1. (O'Brien [1976]) Show that a compact convex subspace X of a locally convex vector space is interior stable iff the following midpoint function is open.
$$m: X \times X \to X, \quad (u,v) \mapsto \tfrac{1}{2}(u+v).$$
The convex hull in $\mathbb{R}^3$ of a circle and two additional points on a straight line passing through -- but not coplanar with -- the circle, is not interior stable. Hint: if the midpoint function is open, then the set of extreme points is closed.

1.22.2. Verify that if X is an interior stable space and if Y is an open subspace of X, then Y is interior stable.

1.22.3. (Kay [1977]) Let X be a tcs. For each point $p \in X$ and for each open set $O \subseteq X$, consider the collection
$$pO = \bigcup \{px \mid x \in O\}.$$
Show that if X is join-hull commutative and if $pO \setminus \{p\}$ is open in X for each $p \in X$ and for each open set $O \subseteq X$, then X is interior stable. Note that the condition is not necessary for X to be interior stable.

1.22.4. A *topological Boolean algebra* is a topological lattice which is Boolean, such that the operation of taking complements is continuous. Cantor cubes are a relevant type of example; cf. II§4.25.1. Show that a topological Boolean algebra is interior stable (being a topological lattice, it is closure stable as well). Deduce that each half-space of a topological Boolean algebra is (i) open or without interior, and (ii) closed or dense.

1.22.5. Problem. Is interior stability inherited by the weak topology of a space?

1.23. Locally convex spaces

1.23.1. (Jamison [1974]; compare Proposition 1.19) Let X be a compact tcs with a metrizable topology. Show that if X has the weak topology, then the points at which X is locally convex constitute a dense G_δ–subset.

1.23.2. Show that the geodesic convexity of a Ptolemaic metric space is properly locally convex.

1.23.3. Problems. Is there an example of a space which is locally convex, but not properly locally convex? Can the restriction to spaces with connected convex sets be removed from Proposition 1.19? Is it true that a compact, locally convex, and JHC space admits a neighborhood base of convex sets at each of its convex closed sets?

1.24. On a theorem of Tietze. A subset C of a tcs X is *locally a convex set* provided for each $p \in C$ there is a neighborhood N such that $C \cap N$ is convex. The tcs X has the *Tietze Property* provided each compact and connected set is convex whenever it is locally a convex set.

1.24.1. (Klee [1951]) Show that a topological vector space has the Tietze Property. This fact, applied to $\mathbb{R}^n$, is the original result of Tietze [1928]. Deduce that each H-convexity on $\mathbb{R}^n$ has the Tietze Property.

1.24.2. (Jamison [1978]). Let S be an arc-wise connected topological meet semilattice. Prove that S has the Tietze property.

Hints. Let T be a continuum which is locally a convex set. Note that T cannot have infinitely many minima. Consider a minimal subcontinuum $T_0 \subseteq T$ which is a lower set of T. Verify that T_0 is the union of finitely many order intervals. If T has more than one minimum, then a contradiction is waiting at some maximum of T_0. Consider the unique minimum of T as the lower bound of S, and show that no element of T can be approximated from below by elements not in T.

1.24.3. Show that a connected median space with compact segments has the Tietze Property. Hint. A median space is a topological semilattice in each base-point order.

1.24.4. Problem. Which classes of topological convex structures posess the Tietze Property? Concerning connected graphs, Chepoi [1991] has considered a counterpart of

§1: Topology and Convexity on the same Set

the Tietze Property; cf. I§6.22.3.

1.25. Vector convexity (van de Vel and Verheul [1990]; cf. Lassak [1984]). Let V be a vector space. A convexity $\mathcal{C}$ on V is called a *vector convexity* provided

(V-1) $\mathcal{C}$ consists of standard convex sets.
(V-2) $\mathcal{C}$ is stable under homotheties with positive coefficients.

It easily follows that a vector convexity is stable under translation. A vector convexity $\mathcal{C}$ is *symmetric* provided it is stable under homotheties with negative coefficients as well. Examples of vector convexities are: H-convexity, K-convexity (cf. Topic II§4.26.3) and affine convexity. The geodesic convexity derived from a norm is a symmetric vector convexity. The join of (symmetric) vector convexities is a (symmetric) vector convexity. Throughout, V is a topological vector space with a vector convexity $\mathcal{C}$.

1.25.1. A vector convexity $\mathcal{C}$ is both closure- and interior stable. If, in addition, $\mathcal{C}$ is symmetric, and if $H \in \mathcal{C}$ is a standard open half-space, then H is a half-space of $\mathcal{C}$ and its bounding hyperplane is in $\mathcal{C}$. In general, polytopes of $\mathcal{C}$ need not be closed.

1.25.2. Let $O \in \mathcal{C}$ be an open set, and let $C \subseteq Bd(O)$ be a standard convex set. Then there is a standard open half-space $H \in \mathcal{C}$ disjoint from C and including O. Hint: consider a maximal open extension $H \in \mathcal{C}$ of O, disjoint from C and show that its boundary is a standard convex set.

1.25.3. In addition, let $\mathcal{C}$ be symmetric. Show that an open set $O \in \mathcal{C}$ can be separated from a standard convex subset of its boundary with a continuous CP linear functional $V \to \mathbb{R}$, and that O can be separated from each point $p \notin O$ with a continuous CP linear functional $V \to \mathbb{R}$. Give an example in $\mathbb{R}^2$ showing that in the last result, "point" cannot be replaced by "member of $\mathcal{C}$".

Conclude that a symmetric vector convexity on a topological vector space admits a non-trivial convex body iff it admits a non-trivial, continuous, CP, and linear functional.

1.26. Continuity of real functionals (for a restricted notion of locally convex "convexity topological space", Guay and Naimpally [1988]; compare 5.6 and 5.14). An interval space with an additional topology is *locally star-shaped* provided each point has a basis of neighborhoods which are star-shaped. Note that a tcs with a continuous segment operator and a metric space with the geodesic interval operator are locally star-shaped. Let X be a locally star-shaped tcs and let $f : X \to \mathbb{R}$ be a CP and CC function. Show that f is continuous iff its fibers $f^{-1}(t)$ for $t \in \mathbb{R}$ are closed.

Hint. Let $f : X \to Y$ be a function between topological spaces, let $p \in X$, let $\mathcal{U}(p)$ be a neighborhood base of X at p, and let $\mathcal{V}(f(p))$ be a neighborhood base of Y at $f(p)$. Assume that $f(U)$ is connected for each $U \in \mathcal{U}(p)$ and that $f^{-1}(Bd(V))$ is closed for each $V \in \mathcal{V}(p)$. Then f is continuous at p.

1.27. Weak topology in median spaces (van de Vel and Verheul [1991])

1.27.1. Let X be a median metric space, let $\mathbf{0} \in X$, and let $C \subseteq X$ be star-shaped at $\mathbf{0}$. Then C is weakly compact iff it is metrically complete and each $\mathbf{0}$–increasing sequence in C is metrically bounded. For an example of a weakly compact star-shaped set which is not metrically bounded, consider the following subset of $\ell_1(\mathbb{N})$.

$$C = \{(x_n)_{n=1}^{\infty} \mid \sum_{n=1}^{\infty} \frac{x_n}{n} \leq 1\}.$$

1.27.2. Conclude that a median metric space is metrically complete iff all closed metric disks are weakly compact.

1.27.3. Let X be a locally star-shaped median space, and let $Y \subseteq X$ be a median stable subset. Show that the (intrinsic) weak topology of Y equals the relative weak topology, derived from X.

1.27.4. Show that in a locally convex median space the closed half-spaces yield a subbase for the weak topology. This subbase is normal T_1 if the space is complete.

1.28. Superextensions (van Mill [1977]). Let X be a compact Hausdorff tcs of Helly number ≤ 2, and let $\mathcal{S}$ be a convex closed subbase of X. If $Y \subseteq X$ is such that

$$\forall S_0, S_1 \in \mathcal{S}: S_0 \cap S_1 \neq \emptyset \Rightarrow S_0 \cap S_1 \cap Y \neq \emptyset,$$

then X is isomorphic to $\lambda(Y, \mathcal{S} \mid Y)$.

1.29. Mixers (van Mill and van de Vel [1979]). Let X be a compact Hausdorff space. By a *mixer* on X is meant a mapping

$$m: X^3 \to X$$

satisfying the Absorption Law (M-1); cf. I§1.5.6.

1.29.1. Show that each absolute retract (AR) admits a mixer (the converse is a notorious open problem).

1.29.2. Let $f: S^n \to X$ be a map of the n–sphere ($n > 0$) into X. Prove that f can be extended to the $(n+1)$–disk. If U is a neighborhood of $x \in X$ then there is a neighborhood $V \subseteq U$ of x such that each map of an n–sphere into V extends to a map of the $(n+1)$-disk into U. In technical terms, X is C^n and LC^n for all $n > 0$.

Hint. Represent a point x of the $(n+1)$–disk by three points of the bounding n–sphere as follows (Fig. 4). The first and second are the projections of x along the $(n+1)^{th}$ axis to the sphere. The third is the projection of x from $b = (-1, 0, .., 0)$.

1.29.3. Show that a continuum (that is, a compact Hausdorff connected space) with a mixer is locally connected and that a metric continuum with a mixer is path connected.

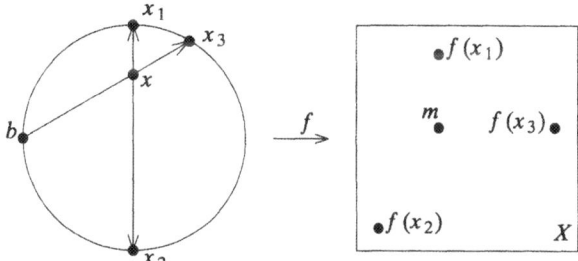

Fig. 4: Representing a point x on a sphere and mapping it to X

1.29.4. Let m be a mixer on X and let a, $b \in X$ be such that $m(a,b,x) = x$ for all $x \in X$. Show that if X is path connected then X is *contractible*, that is, there is a map

$$F: X \times [0,1] \to X$$

such that $F(x, 0) = x$ and $F(-, 1)$ is a constant function.

1.29.5. Let X now be a median continuum. Conclude that all convex sets of X are connected and that all compact convex sets are locally connected. If, in addition, X is metrizable, then all polytopes are contractible. Conclude that a compact connected tree is locally connected, and even an absolute retract if it is metrizable.

Notes on Section 1

The notion of a topological convex structure (tcs) has been introduced by Jamison in [1974]. Restricted or deviating notions were formulated by Deak [1966], Fuchssteiner [1970], Bryant [1975], Kay [1977], Guay [1978], and by van Mill and van de Vel [1978]. Concentrating on compact spaces, Jamison made a study of closure stability, local convexity, and continuity of the convex closure operator. Many of these results have been reconsidered here, sometimes in slightly more general circumstances.

The terms "closure stability" and "interior stability" were suggested by Jamison (communication). Almost invariably, compatibility of a topology with a mathematical structure leads to compatibility of this topology with the corresponding convexity. All examples of topological convex structures, given in 1.2, have been considered before (in an a-topological setting) in Chapter I. For topological vector spaces we refer to Köthe [1960]. For topological (semi)lattices we refer to Gierz et al [1980] or to Hofmann and Mostert [1966]. A standard reference for pospaces is Nachbin [1965]. Convex hyperspaces were studied by van Mill and van de Vel [1979b] and by van de Vel [1983], [1984d].

The fact, that the polytopes of a topological median algebra are closed, is new, as is Theorem 1.4 (on product spaces). However, it was shown by Jamison [1974] that in a product of compact convex structures with a continuous convex closure operator, each convex closed set is a product set. The relationship between local convexity and Lawson semilattices or completely distributive lattices belongs to the folklore of (semi-) lattice theory. See Gierz et al [1980] for an account of these matters. Local convexity of compact pospaces is a well-known fact; cf. Nachbin [1965]. The stronger statements that are established here (involving neighborhoods of convex closed sets) seem to be new. It is still an open problem (cf. 1.23.3) to show that a compact, locally convex, and JHC space admits a neighborhood base of convex sets around each of its convex closed sets. Such a result on compact convex subsets of topological vector spaces would largely avoid the need for re-embedding these sets into locally convex vector spaces (this is made possible by a rather sophisticated result which was first obtained by Jamison, O'Brien and Taylor [1976]). Corollary 1.15, on closure stability in semilattices, is given by the author in [1984e]. The notion of weak topology was introduced by the author in [1983d]. Parts (1), (2) and (4) of Proposition 1.18.1 on superextensions belong to the folklore of the subject. Part (3) seems new. Proposition 1.19, on weak topology in (locally) compact spaces, is new. Weak compactness in median spaces has been investigated in depth by Verheul and the author [1991]. See 1.27. A theory of vector convexity (cf. 1.25) has been developed by Lassak [1984]; it was refined and exploited by van de Vel and Verheul [1990].

2. Continuity of the Hull Operator

> The convex hull operator and the convex closure operator of a topological convex structure (tcs) can be seen as mappings between certain parts of topological hyperspaces. Continuity of the second implies continuity of the first; there is an example showing that the converse is not always valid. Both conditions are shown to be equivalent for compact median algebras and for compact semilattices. There is an intimate connection with local convexity. Continuity in products and subspaces is considered. We end with a remarkable result, stating that the circle admits no continuous hull operator except for the identity.
> Many results in this section are restricted to compact spaces; for non-compact spaces the use of uniform structures seems appropriate; see next section.

2.1. Set-valued functions. By a *multifunction* (*set-valued function*) of a set X into a set Y is meant a binary relation $F \subseteq X \times Y$ such that for each $x \in X$ the set (*point-value, value set*)

$$F(x) = \{y \in Y \mid (x,y) \in F\}$$

is non-empty. Note that F can be seen as a function $X \to 2^Y$, or even $X \to \mathcal{J}_*(Y)$, provided Y is a topological space and the value sets are closed. We occasionally use the notation

$$F: X \multimap Y.$$

Let X, Y be topological spaces and let $F: X \multimap Y$ be a multifunction. We say that F is *lower semi-continuous* (*LSC*) provided for each open set $O \subseteq Y$ the set

$$F^{-1}(O) = \{x \mid F(x) \cap O \neq \emptyset\}$$

is open in X. The multifunction F is *upper semi-continuous* (*USC*) provided for each open set $O \subseteq Y$ the set

$$\overleftarrow{F}(O) = \{x \mid F(x) \subseteq O\}$$

is open in X. A multivalued function which is both lower- and upper semi-continuous is said to be *continuous* for short. Note that if all point-values of F are closed, then continuity of F amounts to continuity as a function into the hyperspace of Y.

2.2. Continuity of the hull operator. If X is an S_1 topological convex structure, then two functions into the hyperspace of X are of interest, namely,

$$co : \mathcal{J}_{fin}(X) \to \mathcal{J}_*(X) \qquad \text{(hull operation on finite sets);}$$
$$co^* : \mathcal{J}_{comp}(X) \to \mathcal{J}_*(X) \qquad \text{(convex closure operation on compact sets).}$$

(cf. 1.7). Note that an S_1 tcs is T_1, and hence that finite sets are closed. In this section we will investigate the condition that one of these operators is continuous. More specifically, we will look at two constituents of continuity, namely lower semicontinuity

(LSC) and upper semicontinuity (USC).

The following example illustrates that it is unreasonable to expect the convex closure operation to be continuous (specifically, USC) at a noncompact closed set. Let the plane be provided with the standard convexity, and consider the following open neighborhood of the horizontal axis A:

$$O = \{(x,y) \mid |y| < e^{-|x|}\}.$$

For any hyperspace neighborhood $\mathcal{U}$ of A there exists a closed set $A' \in \mathcal{U}$ (even a finite one) such that $co^*(A') \not\subseteq O$.

For this reason, we define a tcs X to be *continuous* provided it is S_1 and its convex closure operation is continuous on $\mathcal{J}_{comp}(X)$. In this section we largely restrict to compact spaces, where the situation is somewhat simpler. On non-compact spaces, good results can be obtained by using uniform structure; see next section. We begin with a study of LSC and USC operators.

2.3. Proposition. *Let X be a T_3 and S_1 space. Then $\mathcal{JC}_*(X)$ is closed in $\mathcal{J}_*(X)$ in each of the following situations.*

(1) *The operator*

$$co : \mathcal{J}_n(X) \to \mathcal{J}_*(X)$$

is LSC for each $n \in \mathbb{N}$.

(2) *The space X is of arity m, and co is LSC on $\mathcal{J}_m(X)$.*

In either case, X is closure stable, and co^* coincides with $\overline{co}$.

Proof. Let co be LSC on the hyperspaces $\mathcal{J}_n(X)$, $n \in \mathbb{N}$. If $A \subseteq X$ is closed and non-convex then, by domain-finiteness, there is a finite subset $F = \{a_1,..,a_n\}$ of A with $co(F) \not\subseteq A$. Note that n can be taken equal to m if X is of arity m. Let $p \in co(F) \setminus A$. As X is T_3 we obtain disjoint neighborhoods O of A and P of p. By assumption we find neighborhoods P_i of a_i such that if $x_i \in P_i$ for each $i = 1,..,n$), then the hull $co\{x_1,..,x_n\}$ meets P. No member of the hyperspace neighborhood

$$<O> \cap <P_1,...,P_n,X>$$

of A can be a convex set.

If $\mathcal{JC}_*(X)$ is closed in $\mathcal{J}_*(X)$ and if $C \subseteq X$ is convex then, C^- being the limit of the net $co(F)$ for $F \subseteq C$ finite, we conclude that $C^- \in \mathcal{JC}_*(X)$. ∎

2.4. Proposition. *Let X be a topological convex structure with a USC operator*

$$co : \mathcal{J}_{fin}(X) \to \mathcal{J}_*(X).$$

(1) *If X is S_1 then X is locally convex.*

(2) *If the topology of X is regular, then each base-point quasi-order has a closed graph in $X \times X$.*

(3) *If X has compact polytopes and a finite Carathéodory number, then the convex hull of each compact subset of X is compact.*

§2: Continuity of the Hull Operator

Proof. (1). Let U be an open neighborhood of $p \in X$. Then $\{p\} = co\{p\} \subseteq U$ and co being USC on $\mathcal{J}_{fin}(X)$, there is a neighborhood V of p such that $co(F) \subseteq U$ for each finite set $F \subseteq V$. Briefly: $co(V) \subseteq U$, where $co(V)$ is a convex neighborhood of p.

(2). Let $b \in X$ and suppose $x_1 \not\leq_b x_2$. Then $x_1 \notin bx_2$ and there exist disjoint neighborhoods U_1 of x_1 and V of bx_2. This leads us to a neighborhood U_2 of x_2 such that $bx_2' \subseteq V$ whenever $x_2' \in U_2$. No two points $x_1' \in U_1$ and $x_2' \in U_2$ satisfy $x_1' \leq_b x_2'$.

(3). If $c(X) < \infty$, then for each subset K of X we have

$$co(K) = \bigcup \{co(F) \mid F \subseteq K, \#F \leq c(X)\}.$$

It is a general result[1] that the image of a compact set under a USC, compact valued multifunction is compact. If $K \subseteq X$ is compact, then so is the family of all $F \subseteq K$ with at most c points. Regard co as a compact-valued USC multifunction of $\mathcal{J}_c(X)$ into X. ∎

The second part largely covers a proposition on pospaces obtained in 1.9.5.

2.5. Theorem. *Let X be a compact S_1 space. Then:*
(1) *co^* is LSC, iff co is LSC on the hyperspace of finite sets, iff $\mathcal{JC}_*(X)$ is compact.*
(2) *co^* is USC, iff for each $C \in \mathcal{JC}_*(X)$ and for each open set $O \supseteq C$ there is a convex closed set D with*

$$C \subseteq Int(D) \subseteq D \subseteq O.$$

Proof of (1). If co is LSC on the hyperspace of finite sets, then $\mathcal{JC}_*(X)$ is closed in $\mathcal{J}_*(X)$ by Proposition 2.3. Furthermore, the topological hyperspace $\mathcal{J}_*(X)$ is compact.[2]

Assume next that $\mathcal{JC}_*(X)$ is compact, let $A \in \mathcal{J}_*(X)$, and let $co^*(A)$ meet the open set $O \subseteq X$. Suppose there is a net $(A_i)_{i \in I}$ in $\mathcal{J}_*(X)$, converging to A and such that $co^*(A_i) \cap O = \emptyset$ for all i. By compactness of the hyperspace of X, the sets $co^*(A_i)$ cluster at a set C. This C must be convex since $\mathcal{JC}_*(X)$ is closed, and it is also disjoint from O. Now $A_i \subseteq co^*(A_i)$ for all i implies $A \subseteq C$. But then $co^*(A) \subseteq C$, and hence $C \cap O \neq \emptyset$, contradiction. Hence co^* is LSC.

The remaining implication is trivial.

Proof of (2). Assume first that co^* is USC, and let $C \subseteq O$, where C is convex closed and O is open. Then there exists a basic hyperspace neighborhood $<P_1,..,P_n>$ of C mapping into $<O>$ under the operator co^*. The sets $P_1,..,P_n$ are open in X. By the normality of X, there is a closed set $B \subseteq \bigcup_{i=1}^n P_i$ with $Int(B) \supseteq C$. Then $B \in <P_1,..,P_n>$ and $D = co^*(B)$ is as desired.

As for the converse, let A be closed and let $O \supseteq co^*(A)$ be open. Fix a convex closed set D with $co^*(A) \subseteq Int(D) \subseteq D \subseteq O$. Then $<Int(D)>$ is a hyperspace neighborhood of A and each of its members A' satisfies $co^*(A') \subseteq O$. ∎

---

1. Smithson [1972, Prop. 2.3]
2. Engelking [1977, p. 306].

Propositions 1.13.2 and 1.13.3 imply, respectively, that a compact pospace and a Lawson semilattice have a USC convex closure operator.

2.6. Theorem. *Let X be a compact S_1 space, let $n \in \mathbb{N}$, and let $c(X) \le n$. Then the following assertions are equivalent.*

(1) $co^* : \mathcal{J}_*(X) \to \mathcal{J}_*(X)$ *is continuous.*
(2) $co : \mathcal{J}_{fin}(X) \to \mathcal{J}_*(X)$ *is continuous.*
(3) $co : \mathcal{J}_n(X) \to \mathcal{J}_*(X)$ *is continuous.*

Proof. We only verify that (3) implies (1). Suppose $A \subseteq X$ is a closed non-convex set. Then there exist a finite set $F \subseteq A$ and a point $p \in co(F) \setminus A$. As $c(X) \le n$, we may assume that $\#F \le n$. By (1), there exist open sets $O_1,..,O_k$ of X such that $F \in <O_1,..,O_k>$ and $p \notin co(F')$ whenever $F' \in <O_1,..,O_k>$. Then $<O_1,..,O_k,X>$ is a hyperspace neighborhood of A, no member of which is convex. This shows that $\mathcal{IC}_*(X)$ is compact.

In regard to Theorem 2.5, it suffices to show that each compact convex set C has a neighborhood base of convex closed sets. To this end, let $(N_i)_{i \in I}$ be the down-directed net of all closed neighborhoods of C, and let $x \in \cap_{i \in I} co(N_i)$. For each $i \in I$ we fix a finite set $F_i \subseteq N_i$ with $x \in co(F_i)$. We can take care that $\#F_i \le c$. By compactness of the hyperspace of all sets of cardinality $\le n$, the sets F_i cluster at a set F with at most n points. Evidently, $F \subseteq \cap_{i \in I} N_i = C$. By the continuity of co, we find that $x \in co(F) \subseteq C$. By Proposition 2.4, the hull of a compact set is closed. This shows that $C = \cap_{i \in I} co^*(N_i)$. It easily follows that each neighborhood of C includes one of the sets $co(N_i)$. ∎

2.7. Theorem. *Let X be a JHC and S_1 space.*

(1) *If X is locally convex and if the segment operator is USC and compact-valued, then all polytopes of X are compact and co is USC on $\mathcal{J}_{fin}(X)$.*
(2) *If the segment operator is LSC, then co is LSC on $\mathcal{J}_{fin}(X)$.*
(3) *If the segment operator is continuous and compact-valued, then X is locally convex iff the operator*

$$co : \mathcal{J}_{fin}(X) \to \mathcal{J}_*(X)$$

is continuous.

Proof of (1). We show, by induction on $n \ge 1$, that if F is an n-point set, then $co(F)$ is compact and $co : \mathcal{J}_{fin}(X) \to \mathcal{J}_*(X)$ is USC at F. The statement being evident for $n = 1$ (by S_1, resp., by local convexity), consider $n \ge 2$. If $q \in F$, then by join-hull commutativity,

$$co(F) = \cup \{qx \mid x \in co(F_q^\wedge)\}.$$

This allows to regard $co(F)$ as the image of the compact set $co(F_q^\wedge)$ under a USC compact-valued function, namely the segment operator. Topological considerations yield that $co(F)$ is compact. Let $co(F) \subseteq O$. Fix $p \in F$ and let $G = F \setminus \{p\}$. The segment operator being USC, for each $x \in co(G)$ there is a neighborhood W_x of x and a neighborhood $N_{p,x}$ of p such that $x'p' \subseteq O$ whenever $x' \in W_x$ and $p' \in N_{p,x}$. As $co(G)$ is compact,

§2: Continuity of the Hull Operator   291

it can be covered with finitely many sets of type $Int(W_x)$. The union W of the selected sets W_x yields a hyperspace neighborhood $<W>$ of G, and the intersection of the corresponding sets $N_{p,x}$ yields a neighborhood N_p of p. The induction hypothesis yields an open set $N \supseteq G$ such that $co(N) \subseteq W$. By join-hull commutativity, for each $p' \in N_p$ and for each finite set $G' \subseteq N$ we find that

$$co(\{p'\} \cup G') = \bigcup \{p'x \mid x \in co(G')\} \subseteq O.$$

Consequently, $F \subseteq N \cup N_p$, and for each finite set $F' \subseteq N_p \cup N$ we have $co(F') \subseteq O$ by virtue of the hull formula I§2.14 for a JHC convexity.

Proof of (2). The space X is of arity 2 and Proposition 2.3 applies.

Proof of (3). Continuity of co implies local convexity by Proposition 2.4(1). The converse is a combination of (1) and (2). ∎

2.8. Some consequences. In regard to Theorem 2.5, the Propositions 1.13.2 and 1.13.3 imply that compact pospaces and Lawson semilattices always have a USC convex closure operator. On a compact pospace, the convex closure operator will rarely be LSC, whereas in Lawson semilattices this can be achieved under reasonable additional assumptions (cf. 2.8.2 below). The results are at best for compact median algebras.

2.8.1. Proposition. *In a compact median algebra, continuity of co^*, continuity of co, and local convexity are equivalent. In particular, a compact tree and a compact completely distributive lattice are continuous tcs's.*

Proof. We only have to verify that local convexity implies continuity of convex closure operator. As to LSC, in regard to Theorem 2.7(2), it suffices to verify that the segment operator is LSC. This follows easily from the fact that for $a, b \in X$,

$$co\{a,b\} = \{m(a,b,x) \mid x \in X\}$$

(cf. I§4.2 and I§6.10(5)).

We concentrate on the proof that co^* is USC. Let $C \subseteq X$ be non-empty convex closed, and let $p \in X \setminus C$. The convex sets C, and $co\{p,x\}$ for $x \in C$, meet two by two. Since X is compact and of Helly number ≤ 2, we find that $\cap_{x \in C} px$ meets C. A point c in this set is the $\leq_p$-minimum of C, and hence is the nearest point of p in C. Fix disjoint convex neighborhoods N_c of c and N_p of p, and consider a half-space H with

$$N_p \subseteq H; \quad N_c \cap H = \emptyset.$$

By the closure stability of X (Proposition 1.9.4), $\overline{H}$ and $\overline{X \setminus H}$ constitute a convex screening of p, c and hence of p, C (Screening Lemma, I§5.11). In particular,

$$C \subseteq Int(\overline{X \setminus H}) \subseteq \overline{X \setminus H} \not\ni p.$$

Compactness of X implies that C has a neighborhood base of closed convex sets. Upper semi-continuity now follows from Theorem 2.5.

As to the last part of the theorem, a compact tree is both a Lawson semilattice (see the remarks at the end of 1.13.3) and a topological median algebra (cf. 1.2). The corresponding convexities are identical. Regarded as a Lawson semilattice, X is locally convex, and regarded as a compact median algebra, X is continuous.

A compact completely distributive lattice is locally convex by Proposition 1.13.5, and hence it is continuous. ∎

2.8.2. Proposition. *The following statements are equivalent for a compact meet semilattice S*

(1) $co^* : \mathcal{J}_*(S) \to \mathcal{J}_*(S)$ *is continuous.*
(2) $co : \mathcal{J}_{fin}(S) \to \mathcal{J}_*(S)$ *is continuous.*
(3) S *is locally convex and the following multifunction is LSC.*

$$S \supseteq \{(x,y) \mid x \leq y\} \to \mathcal{J}_*(S); \quad (x,y) \mapsto [x,y].$$

These statements are all valid if for each open set $O \subseteq S$ the lower set $L(O)$ is also open.

Proof. (1) $\Rightarrow$ (2) is evident. As to (2) $\Rightarrow$ (3), just note that $[x,y] = co\{x,y\}$ for $x \leq y$. We verify that (3) $\Rightarrow$ (1). It follows from Proposition 2.5 and Theorem 1.13.3(2) that co^* is USC. As S is an interval convex structure, it suffices by 2.7 to verify that the segment operator of S is LSC. But this follows directly from the formula

$$co\{a,b\} = [a \wedge b, a] \cup [a \wedge b, b],$$

together with the assumption of (3).

To establish the remaining part of the Proposition, let $a \leq b$ in S and let $O \subseteq S$ be an open set meeting $[a,b]$ in x, say. Then $b \wedge x = x \in O$ and hence there exist neighborhoods P_b of b and O_x of x with $P_b \wedge O_x \subseteq O$. Take $P_a = L(O_x)$. If $a' \in P_a$ and $b' \in P_b$ are such that $a' \leq b'$, then consider a point $x' \in O_x$ with $a' \leq x'$ and put $y = x' \wedge b$. It follows that $y \in O \cap [a',b']$. ∎

Note that the last condition of the proposition is equivalent with lower semicontinuity of the multifunction $S \multimap S;\ x \mapsto U(x)$. That the statement is not equivalent to continuity of the hull operator can be seen as follows. Let S be a compact tree shaped like a character "Y". One of the end points is taken as a base-point. Then S has a continuous hull operator by Proposition 2.8.1, but the operator U fails to be LSC at the ramification point.

The previous results illustrate that continuity of the convex hull operator on the hyperspace of all finite sets, or of the convex closure operator on the hyperspace of all compact sets, often go together. The properties are not equivalent, however. For a counterexample, see Topic 2.20.

2.8.3. Non-continuous H-convexity. The following describes a symmetric, point-convex H-convexity with a discontinuous segment function. Let $V \subseteq S^2$ be the set of all unit vectors $v = (v_1, v_2, v_3)$ such that $|v_1| \leq |v_2|$. We consider the H-convexity $\mathcal{C}$ on $\mathbb{R}^3$ generated by the family $\mathcal{F}$ of functionals $x \mapsto v \cdot x$ for $v \in V$ (the dot refers to the inner product of $\mathbb{R}^3$). Note that $\mathcal{F}$ is symmetric and point-separating. Throughout, we let e_i denote the standard i^{th} unit vector.

§2: Continuity of the Hull Operator

Let $v \in V$ and let $H = \{x \mid v \cdot x \le t\} \in \mathcal{C}$. If H contains both e_2 and $-e_2$ then
$$|v \cdot e_1| = |v_1| \le |v_2| = |v \cdot e_2| \le t.$$
It follows that $e_1 \in co\{e_2, -e_2\}$. On the other hand, consider the following sequences of points in V:
$$v^+(n) = (n, n+1, n^2+n); \quad v^-(n) = (-n, n+1, n^2+n),$$
where $n \in \mathbb{N}$. We compute the convex set
$$C_n = \{x \mid v_n^+ \cdot x = 0; \; v_n^- \cdot x = 0; \; -1 \le e_2 \cdot x \le +1 \}.$$
The condition that x is orthogonal to both v_n^+ and v_n^- implies that x is a multiple of $(0, 1, \frac{-1}{n})$. The third condition on x yield that the resulting scalar factor stays within the interval $[-1, +1]$. Hence C_n is a line segment joining the points
$$e_2 - \frac{e_3}{n}; \quad -e_2 + \frac{e_3}{n}.$$
These points tend to e_2 and $-e_2$, respectively, as $n \mapsto \infty$, from which it follows that the sequence of $\mathcal{C}$-segments C_n converges to the standard line segment joining $e_2, -e_2$. Hence $\mathcal{C}$ does not have a continuous segment operator.

Note that this example has many other properties, such as closure - and interior stability (cf. 1.9.2), as well as local convexity. For a result on (uniform) continuity of the hull operator for H-convexities, see 3.10.2 below.

2.9. Theorem. *Let X, Y be compact Hausdorff tcs's and let $f : X \to Y$ be a CP and CC map of X onto Y. If X is continuous, then so is Y.*

Proof. The argument involves the diagram below. The map $\mathcal{J}_*(f)$ on top is defined by $A \mapsto f(A)$, and it is continuous and surjective. The map $\mathcal{JC}_*(f)$ at the bottom is the restriction of $\mathcal{J}_*(f)$ to the hyperspace of compact convex sets. It is well defined and surjective since f is CP and CC. Now $\mathcal{J}_*(X)$ and $\mathcal{J}_*(Y)$ are compact Hausdorff spaces, and hence $\mathcal{JC}_*(X)$ and $\mathcal{JC}_*(Y)$ are compact as well. The diagram is commutative since f is CP and CC. Elementary considerations on (topological) quotients yield the desired result. ∎

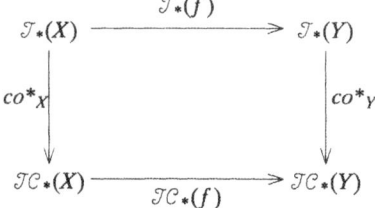

2.10. Theorem. *The product of Hausdorff continuous tcs's is continuous.*

Proof. Let X_i be Hausdorff and continuous for each $i \in I$. The hull operator of X_i is co_i, and π_i denotes the ith projection. Construct a closure operator on $X = \prod_{i \in I} X_i$ as follows. For $A \subseteq X$ closed,

$$cl(A) = \prod_{i \in I} co^*_i(\pi_i(A)).$$

It is easy to verify that $cl : \mathcal{T}_{comp}(X) \to \mathcal{T}_*(X)$ is continuous. By virtue of Theorem 1.4(3) we find that $cl = co^*$ and hence that the product convexity is continuous. ∎

2.11. Pure subspaces. A particular difficulty in studying subspaces relates with the fact that a relatively closed convex set is the trace of a closed set and of a convex set, but not necessarily of a *closed convex* set. For instance, let $Q = [0,1]^\infty$ be the Hilbert cube and consider the compact subspace $X = T \cup \{e\}$, where T is the set of all points $x = (x_n)_{n=1}^\infty \in Q$ such that $x_n \neq 0$ for at most one n and $e = (e_n)_{n=1}^\infty$ satisfies $e_n = 1$ for all n. Then T is a relatively convex and closed subset of X. Its convex hull in Q consists of all points x such that $x_n \neq 0$ for finitely many n only. The closure of this hull contains the point e.

The next definition anticipates on this phenomenon. A subspace Y of a tcs X is called *pure* provided each relatively convex closed set in Y is the trace of a convex closed set of X. A convex closed subspace is evidently pure, and in a closure stable space every non-closed convex subspace is pure as well.

2.12. Examples

2.12.1. *A compact subspace of a pospace is pure.* ∎

2.12.2. *A compact subsemilattice of a compact semilattice is pure.* ∎

The first result follows directly from the fact that (in the given circumstances) the convex hull of a compact set is closed; cf. Proposition 1.6.1. The second result follows by virtue of the convex closure formula, given in Proposition 1.6.2.

2.13. Lemma. *Let X be a compact continuous convex structure and let Y be a closed subspace. If $\mathcal{TC}_*(X) \mid Y$ is closed in the hyperspace of Y, then Y is a pure subspace of X, and Y is continuous.*

Proof. The family $\mathcal{TC}_*(X) \mid Y$ contains all polytopes of Y. Let $C \subseteq Y$ be relatively convex and (relatively) closed. The net of all Y–polytopes included in C converges in the hyperspace of Y to the closure of its union, which equals C. Therefore, $C \in \mathcal{TC}_*(X) \mid Y$, and Y is pure in X.

Let $C \subseteq Y$ be relatively convex and closed and let $O \supseteq C$ be relatively open. As Y is a pure subspace of X, there is a convex closed set $D \subseteq X$ such that $C = D \cap Y$. Then $P = O \cup (X \setminus Y)$ is an open set in X including D, and there is a convex closed neighborhood $D' \subseteq P$ of D in X. Then the relativization of D' to Y is a convex closed neighborhood of C. By Theorem 2.5, Y is a continuous convex structure. ∎

To illustrate the complexity of the continuous subspace problem, we present two results. Recall (II§1.21) that a subset of a convex structure is *free* provided it is convex

§2: Continuity of the Hull Operator   295

and independent. A convex structure is a *convex geometry* provided it satisfies the
anti-Exchange Law: For each pair of distinct points p, q such that $p, q \notin co(A)$,
if $p \in co(\{q\} \cup A)$ then $q \notin co(\{p\} \cup A)$.
See Topic I§2.24 for examples.

2.14. Proposition. *Let X be a compact, Hausdorff, continuous and JHC convex geometry with decomposable segments and a finite Carathéodory number c. Then a closed subspace Y of X is continuous iff the collection of all free sets in Y with at most c points is compact.*

Proof. It is easy to see that in any compact Hausdorff, continuous convex structure the collection of all free sets with at most n points must be compact. After specifying to $n = c$, this establishes the necessity of the condition. We require the following intermediate result (throughout, co denotes X's hull operator).

2.14.1. *Let $C \subseteq Y$ be relatively convex and closed. Then*
$$co(C) = \bigcup \{ co(F) \mid F \subseteq C; F \text{ is } Y\text{-free}; \#F \leq c \}.$$

Indeed, let $x \in co(C)$, let $\mathcal{P}$ be the set of all X-polytopes containing x, which can be spanned with a finite subset of C, and let $n = n(x)$ be the smallest possible cardinality of such a spanning set. Note that by definition of the Carathéodory number c, there exists a set $F \subseteq C$ with $\#F \leq c$ and $x \in co(F)$, cf. II§1.7(2). Therefore, $n \leq c$. A second observation is that $\mathcal{P}$ has minimal elements. To this end, consider a decreasing chain $co(F_i)$, $i \in I$, in $\mathcal{P}$. Note that the collection of all sets in C with at most n points is compact. Hence the net $(F_i)_{i \in I}$ clusters at a set $F \subseteq C$ with at most n points. The net $(co(F_i))_{i \in I}$ clusters at $co(F)$ by the continuity of co. As the sets F_i decrease with i, it follows that $co(F) = \cap_{i \in I} co(F_i)$. Evidently, $x \in co(F)$. Hence, $\#F = n$. We are now in a position to apply Zorn's lemma, yielding that $\mathcal{P}$ has a minimal element $co(F)$.

Let $F \subseteq C$ be such that $\#F \leq n$ and $co(F) \in \mathcal{P}$ is minimal. Note that F is convexly independent. Assume to the contrary that there is a point $p \in co_Y(F) \setminus F$. Then $p \in C$ since C is relatively convex. As X is JHC and has decomposable segments, we can apply the Subdivision Property II§1.2:
$$co(F) = \bigcup \{ co((F \setminus \{a\}) \cup \{p\}) \mid a \in F \}.$$
Hence there is $a \in F$ such that $x \in co((F \setminus \{a\}) \cup \{p\}) \subseteq co(F)$. By minimality, the inclusion is an equality. As F is independent and X is a convex geometry, $p \in F$.

This completes the proof of the auxiliary statement 2.14.1 and we proceed as follows. Assume that the family
$$\mathcal{F} = \{F \subseteq Y \mid F \text{ free}, \#F \leq c\}$$
is compact. Consider a net of compact relatively convex subsets K_i of Y. The hyperspace of Y being compact, there is a finer net K_i converging to a compact set $K \subseteq Y$ (for simplicity, we re-use the original index). As co^* is continuous we find that $co^*(K_i) \mapsto co^*(K)$. As $c < \infty$ it follows that the hull of a compact set in X is closed (cf. 2.7) and hence we

may drop the stars. To see that K is convex, take $x \in co(K)$. Now $co(K_i) \mapsto co(K)$, and we may pass to a finer net if necessary to obtain a net $x_i \mapsto x$ with $x_i \in co(K_i)$. For each i we apply 2.14.1 to find a free set $F_i \in K_i$ with at most c points, such that $x_i \in co(F_i)$. Again by passing to a finer net, we find that F_i converges to a set F. This F is free and has at most c points by the compactness of $\mathcal{F}$, whereas $F_i \subseteq K_i$ for all i implies $F \subseteq K$. Also, $x_i \in co(F_i)$ implies $x \in co(F)$ by the continuity of the hull operator. But then $x \in co(F) \cap Y = co_Y(F)$ and hence $x \in F \subseteq K$ since F is free. This shows that K is relatively convex, and completes the proof that $\mathcal{IC}_*(X) \mid Y = \mathcal{IC}_*(Y)$ is compact. The result follows from Lemma 2.13. ∎

The previous result describes all continuous compacta in Euclidean space. A minor improvement is presented in 2.23. For a different example, let T be a compact tree. Then T is continuous (Proposition 2.8.1), JHC (Theorem I§6.10(3)) and a convex geometry with decomposable segments (left to the reader). Moreover, $c(T) \leq 2$. It is clear that any two-point set in an S_1 space is independent. It can be verified that in any compact subset of T the collection of all relatively convex sets with no more than two points is compact. Consequently,

2.15. Corollary. *All compact subspaces of a compact tree are continuous.* ∎

In the next result we present a different solution of the continuous subspace problem in terms of a "rotundity" condition.

2.16. Proposition. *Let X be a compact, Hausdorff and continuous convex structure which is S_2 and has connected convex sets. Let Y be a compact subspace such that each relatively convex set of Y included in $Bd(Y)$ has at most one point. Then Y is a continuous tcs and a pure subspace of X.*

Proof. Throughout, co^* denotes the convex closure operator of X. By Lemma 2.13, it suffices to show that $\mathcal{IC}_*(X) \mid Y$ is compact. To this end, consider a net $(A_i)_{i \in I}$ in $\mathcal{IC}_*(X) \mid Y$ and let $A \in \mathcal{I}_*(Y)$ be a cluster point. Passing to a finer net if necessary, we may assume that $\lim_i A_i = A$. Take a point $p \in co^*(A) \cap Y$. Now $co^*(A_i)$ converges to $co^*(A)$ by the continuity of co^*, and passing to a finer net if necessary, we can pick a point $p_i \in co^*(A_i)$ for each i, such that p_i converges to p. Observe that if $p_i \in Y$ then $p_i \in A_i$ since $A_i \in \mathcal{IC}_*(X) \mid Y$. It follows that if p_i is eventually in Y then

$p = \lim_i p_i \in \lim_i A_i = A$.

We henceforth assume that $p_i \notin Y$ for all (sufficiently large) indices i; in particular, $p \in Bd(Y)$. By modifying the choice of the points p_i we reduce this situation to the previous one as follows.

Take any point $a \in A$. As before, we obtain a net a_i converging to a with $a_i \in A_i$. Since X is S_2, by Proposition 2.4(2), the base-point quasi-order $\leq_i$ of p_i is a partial order with a closed graph. Hence for each index i there exists a ($\leq_i$)-minimal element q_i in Y such that $q_i \leq_i a_i$. Note that

§2: Continuity of the Hull Operator   297

$$q_i \in co\{a_i,p_i\} \cap Y \subseteq co^*(A_i) \cap Y = A_i.$$

As $p_i \notin Y$ and as all convex sets are connected, it follows from the minimality of q_i that $q_i \in Int(Y)$. Passing to a finer net if necessary, we may assume that q_i converges to q. Then $q \in Y$, and $q \in co\{a,p\}$ by continuity. Suppose $q \neq p$. Then by assumption on $Bd(Y)$, there exists a point $r \in co\{q,p\} \cap Int(Y)$. Once more, we obtain a net of points $r_i \in co\{q_i,p_i\}$ converging to the point r. This net is eventually in $Int(Y)$ and consequently $r_i <_i q_i$ for i large enough. This contradicts the minimality of q_i.

Having produced a net of points $q_i \in A_i$ converging to p we conclude that $p \in A$ and hence that A is convex.   ∎

Note that the previous result is effective in case $Int(Y) \neq \varnothing$ only. Instead of S$_2$, one may assume that the segment operator is geometric. See Proposition I§5.2. If X is a compact subspace of Euclidean space with a continuous convexity coarser than the standard one (certain H-convexities, for instance), and if $Y \subseteq X$ is such that each point of $Bd(Y)$ is extreme in the standard convexity, then 2.16 applies.

A well-known result in topology asserts that the image of a connected set under a continuous function is connected. This result does not apply simply in case of a continuous hull operator and a connected subset of a convex structure, because a hull operator acts on sets, not on points. The "expected" conclusion is nevertheless correct:

2.17. Theorem. *Let X be a point-convex tcs and let the operator*

$$co : \mathcal{F}_{fin}(X) \to \mathcal{F}_*(X)$$

be continuous. If $A \subseteq X$ is connected, then so is $co(A)$.

Proof. Without loss of generality we have $co(A) = X$. Suppose X decomposes as $P \cup Q$, where P, Q are non-empty, disjoint, and open in X. Define

$$\mathcal{F}_P = \{F \mid \varnothing \neq F \subseteq A, \#F < \infty, co(F) \subseteq P\};$$
$$\mathcal{F}_Q = \{F \mid \varnothing \neq F \subseteq A, \#F < \infty, co(F) \cap Q \neq \varnothing\}.$$

Note that $\mathcal{F}_P$ contains the singletons corresponding with the points of P. Hence $\mathcal{F}_P \neq \varnothing$; similarly, $\mathcal{F}_Q \neq \varnothing$. The first set is open since co is USC and the second one is open since co is LSC. These sets are obviously disjoint and cover $\mathcal{F}_{fin}(A)$. This leads to a contradiction since $\mathcal{F}_{fin}(A)$ is connected when A is a connected T$_1$ space.   ∎

This result has a remarkable consequence.

2.18. Corollary. *The only continuous* S$_1$ *convexity on the circle is free.*

Proof. As is well-known, the hyperspace of all subcontinua of the circle can be represented as

$$E^2 = \{x \in \mathbb{R} \mid ||x|| \leq 1\},$$

such that a point p of the boundary circle corresponds with the continuum $\{p\}$. Consider a topological S$_1$ convexity on the circle such that the convex closure operator co^* is continuous. Then co^* maps singletons to singletons, and it maps the hyperspace of subcontinua into itself. With the above representation, $co^*: E^2 \to E^2$ maps the bounding circle identically onto itself. As a consequence of the well-known Brouwer Fixed Point Theorem[3], it follows that the model of co^* is surjective. In direct terms, all subcontinua of the circle are convex. As any finite subset of the circle is the intersection of subcontinua, the desired result follows easily. ∎

It is not known whether this result holds for higher dimensional spheres as well. Some additional information is given in Topics 2.24 and 2.25.

2.19. Topological convex systems. We introduced convex systems in Section I§1 as sets with a partial convexity (the universal set need not be convex). It is easy to design a natural and non-trivial partial convexity on the circle (or on spheres of any dimension) such that the partial hull operator is continuous. The (expected) impossibility to extend this to a full, continuous convexity -- see Corollary 2.18 -- provides additional motivation to study convex systems. At this moment, we shall be satisfied with introducing the following concept.

A *topological convex system* is a triple $(X, \mathcal{T}, \mathcal{C})$, consisting of a set X, a topology $\mathcal{T}$ on X, and a partial convexity $\mathcal{C}$ on X with closed polytopes.

Further Topics

2.20. Continuity of co and of co^* (Jamison [1974]). Consider the following points in the separable Hilbert space ℓ_2. For $n > 0$, let $e(n)$ be the sequence with only one non-zero, namely $1/n$ at the nth place, and let $e(0)$ denote the null sequence. Furthermore, $p(n) = \sum_{k \geq n} 2^{n-k-1} \cdot e(k)$ for $n > 0$. Let X be the (compact) subspace of ℓ_2 consisting of all points defined so far.

2.20.1. Verify that for each $k \geq n$,
$$p(n) = \frac{e(n)}{2} + \frac{e(n+1)}{4} + .. + \frac{e(k-1)}{2^{k-n}} + \frac{p(k)}{2^{k-n}}.$$

Let $F \subseteq X$ be finite. Deduce that if $x \in co_X(F) \setminus F$ then $x = p(n)$ and F contains $e(n),..,e(k-1),p(k)$ for some pair $k > n$. Conclude that $co: \mathcal{T}_{fin}(X) \to \mathcal{T}_*(X)$ is continuous.

2.20.2. Verify that $A = \{e(n) \mid n \geq 0\}$ is a compact relatively convex set and that X is the only closed convex neighborhood of A. Deduce that $co^*: \mathcal{T}_*(X) \to \mathcal{T}_*(X)$ is not continuous.

Note. The tcs X is a subspace of ℓ_2 and hence it is locally convex. This example disproves a conjecture of Kay [1977].

---

3. See van Mill [1988] or IV§6.13 for a full proof.

§2: Continuity of the Hull Operator 299

2.21. Convex closure versus local convexity. Let X be a compact Hausdorff, JHC and S_4 space such that the segment operator is continuous. Theorem 2.7 asserts that $co: \mathcal{J}_{fin}(X) \to \mathcal{J}_*(X)$ is continuous iff X is locally convex.

Problem. Under the given assumptions, are these statements equivalent to the continuity of $co^*: \mathcal{J}_*(X) \to \mathcal{J}_*(X)$? All known examples (e.g., compact convex sets in vector spaces, compact median algebras) confirm this.

2.22. Semilattices. Problem. Characterize closure stability resp., continuity in topological semilattices. Are these properties equivalent for Lawson semilattices?

2.23. Continuous subspaces of $\mathbb{R}^d$ (Jamison [1974]; compare Proposition 2.14). For a convex structure X we let $\mathcal{F}_n(X)$ denote the collection of all free subsets of X with at most n points. Show that if X is a compact subspace of $\mathbb{R}^d$, then X is continuous iff $\mathcal{F}_d(X)$ is compact.

Hint: show that compactness of $\mathcal{F}_{d+1}(X)$ implies compactness of $\mathcal{F}_d(X)$. A relevant fact is, that if a $(d+1)$-set in $\mathbb{R}^d$ is Carathéodory independent, then its convex hull has a non-empty interior.

2.24. Continuous convexity on a sphere (Jamison [1974])

2.24.1. (compare Corollary 2.18). Show that for each $n > 0$ the n-sphere has no continuous convexity which is stable under rotations, except for the free convexity.

2.24.2. Problem. Is there a continuous convexity on the n-sphere other than the free convexity?

2.25. Continuous convexity on special sets (Jamison [1974])

2.25.1. Construct a non-free continuous convexity on a torus and on the figure "eight". This may illustrate that Corollary 2.18 does not extend to spaces "with a hole".

2.25.2. Show that the standard convexity of $[0,1]$ is the coarsest topological (S_1) convexity with a continuous hull operator.

2.26. H-Convexity

2.26.1. (van de Vel [1983g]; compare 3.10.2). Let $\mathbb{R}^n$ be equipped with an H-convexity generated by a finite number of linear functionals, and let X be a compact convex subspace of $\mathbb{R}^n$. Show that X is continuous.

2.26.2. Extend this result to compact subspaces X such that $X \setminus Int(X)$ consists of (ordinary) extreme points of X.

2.26.3. Problem. (Compare 2.8.3 and part (1) above) Characterize continuity of the hull or convex closure operator for a (symmetric) H-convexity in $\mathbb{R}^n$.

2.27. Pure subspaces

2.27.1. Let the Euclidean plane be equipped with the standard convexity. Consider the following two subspaces.

$\{x \in \mathbb{R}^2 \mid 1 \le ||x|| \le 2\}$ (circular annulus);

$[-2,+2]^2 \setminus (-1,+1)^2$ (rectangular annulus).

Show that the former is pure and continuous, and that the latter is pure but not continuous.

2.27.2. Let X be a locally convex median space and let Y be a median stable subspace of X. Show that Y is pure.

2.27.3. Let $\mathcal{S}$ be a closed, convex, and normal T_1 subbase for X. Show that X is a pure subspace of $\lambda(X, \mathcal{S})$ iff each convex closed set in X is a subbasic intersection.

2.27.4. Let $f : X \to Y$ be a closed, continuous, CP and CC surjection and let $A \subseteq Y$. If $f^{-1}(A)$ is a pure subspace of X, then A is a pure subspace of Y.

2.27.5. Let X be a Hausdorff S_1 tcs with compact polytopes and of finite Carathéodory number, such that the convex hull operator is upper semi-continuous on the hyperspace of finite sets. Then every compact subspace of X is pure.

2.28. Hyperspaces

2.28.1. Let X be a compact Hausdorff space with a USC convex closure operator. Show that the following are equivalent.

(i) X is continuous.
(ii) The sets of type
(*) $<D> \cap \mathcal{IC}_*(X)$ and $<D,X> \cap \mathcal{IC}_*(X)$ ($D \subseteq X$ convex closed),
      form a closed subbase for the relative Vietoris topology of $\mathcal{IC}_*(X)$.
(iii) $\mathcal{IC}_*(X)$ has the weak topology.

Hint (iii) $\Rightarrow$ (i). Let $B \in \overline{\mathcal{IC}_*(X)} \setminus \mathcal{IC}_*(X)$. Show first that there exist convex closed families $(\mathcal{S}_i)_{i=1}^n$ in $\mathcal{IC}_*(X)$, such that $\cup_{i=1}^n \mathcal{S}_i$ is a weak neighborhood of B not containing $co^*(B)$. Conclude that for some i there is a cofinal collection of finite subsets $F \subseteq B$ with $co(F) \in \mathcal{S}_i$, and derive a contradiction.

2.28.2. The topological hyperspace $\mathcal{I}_*(X)$ of X is the convex hyperspace corresponding with the free convexity of X. Show that the Vietoris convexity of $\mathcal{I}_*(X)$ is continuous. Note that this convexity also obtains from the natural join semilattice structure of the hyperspace.

2.28.3. (compare van Mill and van de Vel [1978], [1979b]) Recall (1.2.7) that $\perp \mathcal{A}$ denotes the transversal of a family $\mathcal{A} \subseteq \mathcal{I}_*(X)$. It is easily seen that $\perp(\mathcal{A})$ is always closed. Show that the transversality operator of $\mathcal{I}_*(X)$ is continuous. Hint. First, $\perp(\mathcal{A})$ is a

§2: Continuity of the Hull Operator   301

convex set of $\mathcal{J}_*(X)$ for each family $\mathcal{A} \subseteq \mathcal{J}_*(X)$. Therefore, $\bot$ factors through $\mathcal{JC}_*(\mathcal{J}_*(X))$. Second, it follows from Part (1) that the canonical convex subbase of $\mathcal{JC}_*(\mathcal{J}_*(X))$ is at the same time a closed subbase. Hence it suffices to verify that the inverse images of subbasic convex sets of type

$$<\mathcal{A}>, \quad <\mathcal{A}, \mathcal{J}_*(X)>$$

(with $\mathcal{A} \subseteq \mathcal{J}_*(X)$ convex and closed) are closed.

2.28.4. (compare van de Vel [1983e]) Show that the closed half-spaces of $\mathcal{J}_*(X)$ (convexified as in the Part (2)) are precisely the sets of type (*) above.

2.29. Gates (compare van Mill and van de Vel [1978]). Let X be a compact median algebra.

2.29.1. Show that the convex hyperspace $\mathcal{JC}_*(X)$ is a compact median algebra with a median operator defined by

$$m(A,B,C) = co(A \cup B) \cap co(B \cup C) \cap co(C \cup A).$$

2.29.2. Note that each non-empty compact convex set of X is gated. Show that the resulting function

$$X \times \mathcal{JC}_*(X) \to X,$$

assigning to a pair (x,C) the gate of x in C, is continuous.

2.29.3. Koch's Arc Theorem[4] states that if an open set O of a compact pospace has no local minimum, then each point of O is on an "arc" (a totally ordered subcontinuum) meeting the boundary of O. Use this result on a metrizable space X to produce a homeomorph of $[0,1]$ in $\mathcal{JC}_*(X)$, connecting some singleton with X. Conclude that the underlying topological space of a median metrizable continuum is contractible (sharpening the conclusion in 1.29).

Notes on Section 2

Continuity of the convex closure operator on certain parts of its domain (all compact sets, all n-point sets, all finite sets, ..) has been studied by Jamison [1974], Kay [1977], and van Mill and van de Vel [1981]. Proposition 2.4 is essentially due to Jamison [1974]. Theorem 2.5, characterizing lower- and upper semi-continuity of the convex closure operator in compacta, is a result of Jamison [1974]. Theorem 2.7, on the equivalence of local convexity and continuity of the hull operator, is new; a restricted version is given by Kay [1977]. The characterization of continuity of the hull operator for median spaces (Proposition 2.8.1) is new; the corresponding result for Lawson semilattices, 2.8.2, is

---
4.   Gierz et al [1980, Thm. VI,5.9 p. 299].

given by van de Vel [1984e]. Example 2.8.3 (that symmetric H-convexity in $\mathbb{R}^n$ need not be continuous) is taken from Boltyanskii and Soltan [1978].

Our study of continuity in subspaces is based on results of Jamison [1974], where somewhat restricted versions of Propositions 2.14 and 2.16 have been obtained. The concept of purity for subspaces is also due to Jamison, who gives Theorems 2.17 (on connectedness of the hull of a connected subset) and 2.18 (on the nonexistence on the circle of a continuous convexity other than the discrete one); in fact, Jamison attributes this result to Nadler.

The example given in 2.20 (that continuity of the hull operator on the hyperspace of finite sets need not imply continuity of the convex closure operator) is again due to Jamison [1974]. Continuity of the transversality operator in the hyperspace of a compact space (cf. 2.28) is an application of continuous convexity and is due to van Mill and van de Vel [1979b]. No direct proof of this fact is known.

3. Uniform Convex Structures

A uniformity on a set is compatible with a convexity on the same set provided the hull operator is uniformly continuous. For compact Hausdorff spaces, this leads back to the theory developed in the previous section. In a topological vector space with the standard convexity, and in compact median algebras, we are lead to local convexity. For compact semilattices, a minor additional condition is required. In addition, we study cones, finitely generated H-convexities, convex hyperspaces, and spaces of (continuous) arcs.

The main results are on uniform separation, on completion (in particular, compactification), and on quotients. The last two results have been derived for uniform convex systems. The problem remains unsolved whether uniform convex systems can be extended to uniform convex structures. A solution is given for spaces of arcs. A somewhat unexpected consequence of the theory is that the standard Euclidean convexity in dimensions $n > 1$ cannot be compactified.

3.1. Uniformity. By a *uniform structure* is meant a set X together with a filter μ of subsets of $X \times X$ such that[1]

(U1) $\forall U \in \mu \; \forall x \in X : (x,x) \in U$.

(U2) $\forall U \in \mu : U^{-1} \in \mu$.

(U3) $\forall U \in \mu \; \exists V \in \mu : V \circ V \subseteq U$.

The members of μ are called *entourages* and μ is called a *uniformity on X*. In each point $x \in X$ a neighborhood filter is obtained by the collection

$$\{ U(x) \mid U \in \mu \}.$$

This leads to the *uniform topology of* (X, μ). Alternatively, a uniformity can be described as a non-empty collection μ of covers of X with the following properties.

(U'1) Each cover refined by a member of μ is in μ.

(U'2) Each pair of covers in μ has a common refinement in μ.

(U'3) Each member of μ has a star refinement in μ.

With this presentation, μ is called a *covering uniformity* and its members are called *uniform covers*.

The transformation from one type of structure to the other goes as follows. If U is a uniform entourage, then $\{U(x) \mid x \in X\}$ is a uniform cover. If $\mathcal{U}$ is a uniform cover, then $\cup_{U \in \mathcal{U}} (U \times U)$ is a uniform entourage. We do not a priori impose a separation property.

---

1.   We refer to Engelking [1977] for information on the theory of uniform structures. We use the standard notation U^{-1} to describe the inverse relation of $U \subseteq X \times X$, and $U \circ V$ to describe a composite relation. Also, $U(x)$ denotes the image of x under the relation U.

In either setting it is possible to define uniform neighborhoods of a set A. If U is a uniform entourage, then

$$U(A) = \{x \mid \exists a \in A : (a,x) \in U\}.$$

If $\mathcal{U}$ is a uniform cover, then

$$star(A, \mathcal{U}) = \bigcup \{U \in \mathcal{U} \mid U \cap A \neq \emptyset\}.$$

We usually omit reference to the uniformity of a uniform structure, and we use notations like $\mu(X)$ when explicit names are needed.

3.2. Compatible uniformity. Let $(X, \mathcal{C})$ be a convex structure and let μ be a covering uniformity on X. We say that μ is *compatible with* $(X, \mathcal{C})$ provided for each uniform cover $\mathcal{U} \in \mu$ there is a uniform cover $\mathcal{V} \in \mu$ such that

(*) $\quad \forall C \in \mathcal{C}: co(star(C, \mathcal{V})) \subseteq star(C, \mathcal{U}).$

A cover $\mathcal{V}$ with the property (*) is referred to as a *corresponding* or an *associated* refinement of $\mathcal{U}$. If μ is given in terms of entourages, there is an equivalent description as follows. Let $U, V \in \mu$ be entourages. Then $V \in \mu$ is associated to $U \in \mu$ provided

$$\forall C \in \mathcal{C}: co(V(C)) \subseteq U(C).$$

The uniformity μ is compatible with $(X, \mathcal{C})$ (or, with $\mathcal{C}$) if for each $U \in \mu$ there is an associated $V \in \mu$. The triple $(X, \mu, \mathcal{C})$ is called a *uniform convex structure* if μ is compatible with $\mathcal{C}$ and if all polytopes are closed in the uniform topology. In particular, a uniform convex structure induces a topological convex structure.

A uniform convex structure X is *metric* provided its uniformity is generated by a metric d; in these circumstances, the topological convex structure X is said to be *metrizable* and the metric d is *compatible* with the convex structure.

A topological convex structure is *uniformizable* (*metrizable*) provided there is a (metric) uniformity which generates the topology and is compatible with the convexity. We begin with an equivalent description of compatibility. Let $\mathcal{U}$ be a cover of X and let $U \subseteq X \times X$ be an entourage. Two sets $A, B \in \mathcal{F}_*(X)$ are $\mathcal{U}$–*close* provided $A \subseteq star(B, \mathcal{U})$ and $B \subseteq star(A, \mathcal{U})$, and they are U–close if $B \subseteq U(A)$ and $A \subseteq U(B)$.

3.3. Lemma. *Let X be a convex structure and let U, V be two uniform entourages of X. Then the following assertions are equivalent.*

(1) $co(V(C)) \subseteq U(C)$ *for all convex sets* $C \subseteq X$.
(2) $co(V(P)) \subseteq U(P)$ *for all polytopes* $P \subseteq X$.
(3) *If $F, G \subseteq X$ are finite and V–close, then $co(F), co(G)$ are U–close.*

Proof. Evidently, (1) $\Rightarrow$ (2). Assume (2) and let $F, G \subseteq X$ be V–close. Then $G \subseteq V(F)$ and consequently

$$co(G) \subseteq co(V(co(F))) \subseteq U(co(F)).$$

Changing the role of F and G then yields (3).

Assume (3) and let $C \subseteq X$ be convex. If $x \in co(V(C))$ then there is a finite set the standard convexity of a locally convex metrizable vector space is metrizable. ∎

Note that the compatible uniformity need not be the canonical one. For $X = \mathbb{R}$, the uniformity based on all finite convex open covers and the uniformities described in 3.10.3 below are compatible as well. It is not known whether there exist other compatible uniformities on locally convex linear spaces of higher dimension. A partial (negative) answer is given in 3.16 below.

If X is a locally compact subset which is intrinsically locally convex, then[4] X can be affinely embedded into a locally convex linear space. Consequently, such a space X is uniformizable as a convex structure. A corresponding result for non-(locally) compact convex subsets is not known.

3.10.2. Proposition. *Let X be $\mathbb{R}^n$ equipped with a finitely generated H-convexity and with the standard uniformity. Then X is a uniform convex structure.*

Proof. Let $K \subseteq \mathbb{R}^n$ be a closed wedge at 0, that is, K is a non-empty set satisfying

$$K + K \subseteq K; \quad \forall t \geq 0: tK \subseteq K.$$

(compare I§1.19.1). We let $x \cdot y$ denote the inner product of $x, y \in \mathbb{R}^n$. Then the set

$$K^* = \{x \mid \forall y \in K : x \cdot y \leq 0\}$$

is also a closed wedge and $K^* \cap K = \{\mathbf{0}\}$.

Let $\mathcal{F}$ be a finite collection of linear functionals generating an H-convexity $\mathcal{C}$ on $\mathbb{R}^n$. We assume that each f_i is a norm one functional. Let $K \in \mathcal{C}$ be a wedge, $K \neq \mathbb{R}^n$. Then (cf. I§1.10.1) K can be written as the intersection of subbasic sets, say

$$K = \bigcap_{i=1}^{k} f_i^{-1}(L_i) \quad (f_1,...,f_n \in \mathcal{F}),$$

where $L_i \subseteq \mathbb{R}$ is a left end of type $(-\infty, a_i)$ or $(-\infty, a_i]$. As $\mathbf{0} \in K$, we have $a_i \geq 0$, and as $tK \subseteq K$ for t positive, $a_i > 0$ is impossible. So $a_i = 0$ and $L_i = (-\infty, 0]$ since $0 \in K$. Consider the (standard) convex closed set

$$D = D(K) = \{x \in K^* \mid f_i(x) \leq 1\}.$$

To see that D is compact, it suffices to verify that each ray R through $\mathbf{0} \in D$ meets D in a bounded set. To this end, let $x \in D \cap R \setminus \{\mathbf{0}\}$. If $f_i(x) \leq 0$ for all $i = 1,..,k$, then by using the above representation of K, we see that $x \in K$. But then $x \in K^* \cap K = \{\mathbf{0}\}$. So $f_i(x) > 0$ for at least one i and $f_i(tx) > 1$ for large enough t.

Having shown that $D = D(K)$ is compact, we fix a number $v(K) > 0$ such that all points of D have norm $< v(K)$. As $\mathcal{C}$ is finitely generated, there are but finitely many distinct closed wedges in $\mathcal{C}$, and we consider $v = \max_K v(K)$.

This allows us to prove that the convexity is uniform. Let $\varepsilon > 0$, and take $\delta = \varepsilon/v$. We verify that

---
4.   Lawson [1976]; Roberts [1978].

§3: Uniform Convex Structures   307

$$\bigcup_{n=0}^{\infty} C_n \subseteq \bigcup_{n=0}^{\infty} U_n(C_n) = U_0(C_0).$$

The set on the left is convex open and is in between $V_0(C_0)$ and $U_0(C_0)$. ∎

The previous result implies that a uniformizable convex structure is properly locally convex. If X is a topological convex structure with a metric d inducing the topology of X, then d is compatible iff for each $r > 0$ there is an $s > 0$ with $co\, B(C,s) \subseteq B(C,r)$ for each non-empty convex set C. According to Theorem 3.8, it can be taken care of that some convex open set is in between $B(C,s)$ and $B(C,r)$. In general, the open disks themselves need not be convex (not even the disks centered at one point).

3.9. Theorem. *Let X be a uniform convex structure with compact polytopes. Then the convex hull of a totally bounded set is totally bounded. If X is uniformly complete, then the closed convex hull of a compact set is compact.*

Proof. Let $A \subseteq X$ be totally bounded and let $\mathcal{U}$ be a uniform cover of X. Consider a sequence of uniform covers

$$\mathcal{V} \leq \mathcal{U}' \leq \mathcal{U},$$

where $\mathcal{U}'$ is a star refinement of $\mathcal{U}$ and $\mathcal{V}$ is associated to $\mathcal{U}'$. Fix a finite set $F \subseteq A$ with $A \subseteq star(F, \mathcal{V})$. Then,

$$co(A) \subseteq co(star(F, \mathcal{V})) \subseteq star(co(F), \mathcal{U}').$$

Polytopes being compact, a finite subcover $\{U'_1,..,U'_n\}$ of $\mathcal{U}'$ can be found on $co(F)$. After choosing $U_i \in \mathcal{U}$ such that $star(U'_i, \mathcal{U}') \subseteq U_i$ for $i = 1,..,n$, we find that

$$co(A) \subseteq star(co(F), \mathcal{U}') \subseteq \bigcup_{i=1}^{n} U_i.$$

This shows that $co(A)$ is totally bounded; the second half of the theorem follows easily. ∎

3.10. Examples. The next list is a continuation of 3.7. This time, we consider non-compact spaces as well.

3.10.1. Proposition. *Let X be a topological vector space, equipped with the standard convexity and with the canonical translation-invariant uniformity. Then X is uniform iff it is locally convex. If, in addition, X is (topologically) metrizable, then it is metrizable as a convex structure.*

Proof. Let $\mathcal{U}$ be the cover of all translates of a neighborhood U of the origin **0**. If V, W are neighborhoods of **0** such that $V \subseteq U$ is convex and $W - W \subseteq V$, then

$$star(C, \mathcal{W}) \subseteq C + V \subseteq star(C, \mathcal{U}),$$

where $\mathcal{W}$ denotes the cover of all W-translates and the set $C + V$ is convex. Conversely, if some uniformity on X is compatible with the standard convexity, then X has a basis of convex open sets by Theorem 3.8, whence X is locally convex.

Finally, a metrizable linear space admits a translation-invariant metric[3] and hence

---
3.   Köthe [1960, p.166].

We refer to Example 2.20, exhibiting a compact tcs X (a subset of the separable Hilbert space ℓ_2, actually), such that the hull operator is continuous on $\mathcal{J}_{fin}(X)$, whereas the convex closure operator fails to be continuous on the whole of $\mathcal{J}_*(X)$.

3.7. Examples. The last corollary can be used to handle median spaces, semilattices, and lattices with the convexity of order convex meet semilattices.

3.7.1. Proposition. *A compact median space is uniform iff it is locally convex.*

Proof. Use 2.8.1 and 3.6. ∎

In particular, a compact distributive lattice is uniform iff it has small lattices. The situation in semilattices and in non-distributive lattices is discussed in the next results. Concerning semilattices, by 2.8.2 and 3.6 we have:

3.7.2. Proposition. *A compact semilattice is a uniform convex structure iff (i) it is locally convex, and (ii) the intervals $[a,b]$ depend continuously on the pair of end points $a \leq b$.* ∎

3.7.3. Corollary. *A compact lattice with the convexity of all order-convex meet subsemilattices is a uniform convex structure iff it has small meet semilattices.*

Proof. The interval operator is USC by virtue of the fact that the graph of the order is closed. Let $a \leq b$ and let O be an open set meeting $[a,b]$ in x. By the continuity of the join operator, there exist neighborhoods P of a and O_1 of x such that $a' \vee x' \in O$ for $a' \in P$ and $x' \in O_1$. By the continuity of the meet operator, there exist neighborhoods Q of b and O_2 of x such that $b' \wedge x' \in O_1$ for $b' \in P$ and $x' \in O_2$. Hence, if $a' \in P$, $b' \in Q$ and $a' \leq b'$, then the element $x' = a' \vee (b' \wedge x)$ is in $[a',b'] \cap O$, establishing LSC. The result follows from the previous proposition. ∎

3.8. Theorem. *Let X be a uniform convex structure. Then for each $U \in \mu(X)$ there is a $V \in \mu(X)$ with the following property. If $C \subseteq X$ is convex, then there is a convex open set $O \subseteq X$ such that $V(C) \subseteq O \subseteq U(C)$.*

Proof. We fix two sequences of entourages $(U_n)_{n=0}^{\infty}$ and $(V_n)_{n=0}^{\infty}$, where $U_0 = U$, $U_{n+1} \circ U_{n+1} \subseteq U_n$, and V_n is associated to U_{n+1} for each n. We verify that $V = V_0$ is as required. To this end, let $C = C_0$ be convex and (recursively)

$$C_{n+1} = co(V_n(C_n)).$$

Then

$$C_n \subseteq Int(C_{n+1}); \quad C_{n+1} \subseteq U_{n+1}(C_n).$$

The second formula yields that

$$U_{n+1}(C_{n+1}) \subseteq U_{n+1}(U_{n+1}(C_n)) \subseteq U_n(C_n).$$

Therefore,

§3: Uniform Convex Structures

$F \subseteq V(C)$ with $x \in co(F)$. For each $a \in F$ we obtain a point $b \in C$ which is V-close to a. This yields a finite set $G \subseteq C$ which is V-close to F. Consequently, $co(G)$ is U-close to $co(F)$ and in particular there is a point $u \in co(G)$ such that u and x are U-close. Hence $x \in U(C)$. ∎

A uniformity μ on a space X can be "lifted" to the hyperspace $\mathcal{J}_*(X)$ in the following way.[2] The sets of type

$$[A, \mathcal{U}] = \{ B \in \mathcal{J}_*(X) \mid A, B \text{ are } \mathcal{U}\text{-close}\}, A \in \mathcal{J}_*(X),$$

form a covering of $\mathcal{J}_*(X)$, and these covers constitute a uniformity base for the so-called **Hausdorff uniformity** of $\mathcal{J}_*(X)$. Note that on the subspace $\mathcal{J}_{comp}(X)$ of all *compact* sets, the uniform topology equals the usual Vietoris topology. The constructions can be adapted in case the uniformity is described in terms of entourages, where the following notation is used.

$$[U] = \{ (A, B) \in \mathcal{J}_*(X) \times \mathcal{J}_*(X) \mid A, B \text{ are } U\text{-close}\}.$$

3.4. Corollary. *Let X be a topological convex structure and let μ be a uniformity on X generating the topology of X. Then μ is compatible with X iff the operator*

$$co: \mathcal{J}_{fin}(X) \to \mathcal{J}_*(X)$$

is uniformly continuous. ∎

3.5. Corollary. *A uniformizable convex structure is closure stable.*

Proof. Continuity of the convex hull operator on the hyperspace of finite sets implies closure stability by 2.3. ∎

We next draw attention to compact Hausdorff topologies. As it is well-known, such a topology is induced by a unique uniformity on the underlying set.

3.6. Corollary. *The following are equivalent for a compact Hausdorff topological convex structure X.*

(1) *X is a uniform convex structure*
(2) *The convex closure operator $co^*: \mathcal{J}_*(X) \to \mathcal{J}_*(X)$ is continuous.*

Proof. (1) $\Rightarrow$ (2). The continuity of the operator co on $\mathcal{J}_{fin}$ and the third axiom of convexity imply that the non-empty closed convex sets of X form a closed subset of $\mathcal{J}_*(X)$. By the definition of uniform convexity, and by virtue of closure stability, each compact convex set has a neighborhood base (in X), consisting of convex closed sets. The first assertion implies that co^* is LSC, the second one implies that co^* is USC; cf. 2.5.

(2) $\Rightarrow$ (1). The operator co^* is uniformly continuous by the compactness of $\mathcal{J}_*(X)$ and its restriction to $\mathcal{J}_{fin}(X)$ equals co by the fact that polytopes are closed. ∎

---

2. Engelking [1977, p. 572]

(*) $co(B(C,\delta)) \subseteq B(C,\varepsilon)$

for all polytopes C (cf. Lemma 3.3). We only use that $C \in \mathcal{C}$ is compact. Take $p \in co(B(C,\delta)) \setminus C$, and let $q \in C$ be the (unique) metric nearest point of p in C. Without loss of generality, $q = 0$. Then $x \cdot p \leq 0$ holds for all $x \in C$, and hence for all x in the closed $\mathcal{C}$-cone $K = \cup_{t>0} tC$. So $p \in K^*$. Let K be represented as above. Then

$$p \in co(B(C,\delta)) \subseteq co(B(K,\delta)) \subseteq \bigcap_{i=1}^{k} f_i^{-1}(-\infty,\delta),$$

where the last inclusion obtains from the fact that each f_i is a norm one functional. So we have $f_i(p) < \delta$ for all i. Then $p/\delta \in D(K)$ and we conclude that $\|p\| < \delta\nu = \varepsilon$. This means that p is at a distance $< \varepsilon$ of C, establishing (*). ∎

We refer to 2.8.3 for an example of an infinitely generated symmetric H-convexity with a discontinuous hull operator.

3.10.3. Proposition. *If a connected and locally connected tree X is completely regular, then the corresponding convex structure is uniformizable, and even metrizable if the topological space X is.*

Proof. Let μ be any covering uniformity on X (there is one by complete regularity). For $\mathcal{U} \in \mu$, let $\mathcal{U}'$ denote the cover, consisting of all components of members of $\mathcal{U}$. Note that the members of $\mathcal{U}'$ are open by local connectedness. The coverings so obtained constitute a basis for another uniformity μ', which is easily seen to be compatible with the tree convexity. Note that μ' is complete if μ is. If d is a metric for μ, then μ has a countable base of coverings $\mathcal{U}$, and the corresponding coverings $\mathcal{U}'$ then yield a countable base for μ'. It follows that μ' is metric too.[5] ∎

3.10.4. Proposition. *Let X be a uniform convex structure such that the convex closure of the union of two compact convex sets is compact. Then the Hausdorff uniformity on $\mathcal{IC}_*(X)$ is compatible with the Vietoris convexity of $\mathcal{IC}_*(X)$. In particular, $\mathcal{IC}_*(X)$ is metrizable provided X is.*

Proof. Let U be a uniform entourage on X and consider the following sequence:

$$W \subseteq V \subseteq U' \subseteq \overline{U}' \subseteq U,$$

where W and V are associated to, respectively, V and U'. We will show that for each polytope $\mathcal{S} \subseteq \mathcal{IC}_{comp}(X)$,

$co([W](\mathcal{S})) \subseteq [U](\mathcal{S})$,

that is, $[W]$ is associated to $[U]$ (cf. Lemma 3.3). Let $C_1,..,C_n \in \mathcal{IC}_{comp}(X)$ and

$D \in co[W](co\{C_1,..,C_n\})$.

We use the sets

---
5.   Engelking [1977, p.533].

310   Chap. III: Topological Convex Structures

$$E_1 = co^*(V(D)); \quad E_2 = co^*(\bigcup_{i=1}^{n} C_i); \quad E = E_1 \cap E_2.$$

Note that E is compact and convex. We first verify that D and E are U-close. There exist compact convex sets $D_1,..,D_p$ in $[W](co\{C_1,..,C_n\})$ such that $D \in co\{D_1,..,D_p\}$. For each $j = 1,..,p$, there is a compact convex set $A_j \in co\{C_1,..,C_n\}$ which is W-close to D_j. This yields

$$\bigcup_{j=1}^{p} D_j \subseteq W(\bigcup_{j=1}^{p} A_j) \subseteq W(co^*(\bigcup_{j=1}^{p} A_j)) \subseteq W(E_2),$$

whence

$$D \subseteq co^*(\bigcup_{j=1}^{p} D_j) \subseteq co^*W(E_2) \subseteq Cl(V(E_2)).$$

Hence, if $x \in D$ then there is a point in E_2 which is $\bar{V}$-close to x. It follows that

$$\emptyset \neq \bar{V}(\{x\}) \cap E_2 = \bar{V}(\{x\}) \cap E,$$

since $\bar{V}(\{x\}) \subseteq E_1$. Consequently, $D \subseteq U(E)$. On the other hand,

$$E \subseteq E_1 \subseteq Cl(U'(D)) \subseteq U(D),$$

showing that D and E are U-close.

We next verify that $E \in co\{C_1,..,C_n\}$. Note that $E \subseteq E_2 = co^*(\bigcup_{i=1}^{n} C_i)$; it remains to be shown that E meets $co\{c_1,..,c_n\}$ whenever $c_1 \in C_1,..,c_n \in C_n$. As $A_j \in co\{C_1,..,C_n\}$, we obtain a point $a_j \in A_j \cap co\{c_1,..,c_n\}$ for each $j = 1,..,p$. Now $co\{a_1,..,a_p\} \subseteq E_2$, and it suffices to show that E_1 meets $co\{a_1,..,a_p\}$. As A_j is W-close to D_j, we obtain a point $d_j \in D_j$ which is W-close to a_j. Therefore,

$$\{d_1,..,d_n\} \subseteq W(co\{a_1,..,a_p\}),$$

and consequently

(*) $co\{d_1,..,d_n\} \subseteq V(co\{a_1,..,a_p\}).$

Since D is in $co\{D_1,..,D_n\}$, it meets the left hand set of (*), whence

$$D \cap V(co\{a_1,..,a_p\}) \neq \emptyset.$$

We conclude that

$$\emptyset \neq V(D) \cap co\{a_1,..,a_n\} \subseteq E_1 \cap co\{a_1,..,a_p\},$$

as desired.   ∎

3.10.5. Let us finally illustrate that there is no relationship between convexity induced by a metric (geodesic convexity) and metrizability as considered in this section. For a simple and natural example, consider $\mathbb{R}^3$ with the Cartesian norm:

$$||x|| = \max_{i=1,2,3} |x_i| \quad (x = (x_1, x_2, x_3)).$$

This space has no non-trivial geodesically convex open sets (cf. I§6.32.3), and hence its hull operator cannot be uniformly continuous.

§3: Uniform Convex Structures   311

3.11. Theorem (*Uniform Separation Theorem*). *Let X be a uniform S_4 convex structure. Then each two non-proximate convex sets can be separated by a uniformly continuous CP function $X \to [0,1]$.*

Proof. Let C, D be convex sets which are non-proximate, that is:
$$\exists U \in \mu: U(C) \cap U(D) = \varnothing.$$
Let $V \in \mu(X)$ be associated to U as in 3.8. This gives us two convex open sets O, P of X with
$$V(C) \subseteq O \subseteq U(C); \quad V(D) \subseteq P \subseteq U(D).$$
We find that $O \cap P = \varnothing$ and by the axiom S_4 there is a half-space H with the properties
$$O \subseteq H; \quad P \cap H = \varnothing.$$
For convenience, we label H as $H_{1/2}$. Note that both pairs of convex sets
$$(C, X \setminus H_{1/2}); \quad (H_{1/2}, D)$$
are non-proximate (use an entourage W with $W \circ W \subseteq V$), and hence that the above "screening" process can be iterated. In a routine way, this leads to a collection of half-spaces H_t, for $0 < t < 1$ dyadic, such that the following hold for all dyadic numbers s, t.
(1) H_t is increasing with t.
(2)   Both pairs $(C, X \setminus H_t)$ and (H_t, D) are non-proximate.
(3)   If $s < t$ then the pair $(H_s, X \setminus H_t)$ is non-proximate.

Define a function $f: X \to [0,1]$ by the prescription
$$f(x) = \inf \{t \mid x \in H_t\}.$$
It is understood that in the right hand expression, t is a dyadic variable and the infimum of the empty set is 1. In particular, f maps C to 0 and D to 1 by (2). By (1), we find that
$$f(x) = \sup \{t \mid x \notin H_t\},$$
(with sup $(\varnothing) = 0$), and it is not difficult to conclude[6] that the sets $f^{-1}[0,r], f^{-1}[r,1]$ are convex for each $r \in [0,1]$. This shows f to be CP. On the other hand, a uniform cover of $[0,1]$ admits a Lebesgue number $t = 2^{-n}$. To see that f is uniformly continuous, it suffices to verify that the sets $f^{-1}[kt, (k+2)t]$ for $k = 0,1,..,2^n-2$ constitute a uniform cover of X. This is immediate from (2) and (3). ∎

The non-proximity condition reflects the well-known criterion in vector spaces that two convex sets can be separated by a continuous linear functional iff the origin is not adherent to the algebraic difference of the sets.

3.12. Uniform convex systems. We proceed with some results which (with some extra efforts) are established for convex *systems*. We refer to Section I§1.17ff for an introduction to the subject. We adapt the relevant definitions as follows. Let $(X, \mu, \mathcal{C})$ be a

---
6.   This technique is discussed in some detail in Section 4.

triple consisting of a uniformity µ and a partial convexity C on the set X. Inspired by Corollary 3.4, we say that µ is *compatible* with (X, C) if the (partial) hull operator is uniformly continuous. Explicitly, for each entourage $U \in \mu$ there is an entourage $V \in \mu$ such that if F, G are V-close finite admissible sets, then $co(F)$ and $co(G)$ are U-close. In these circumstances, we say that V is *associated* to U. If µ is compatible with (X,C) and if all polytopes are closed in the µ-topology, then the triple (X, μ, C) is a *uniform convex system*. It is clear what should be understood by a *uniformizable convex system* and a *metric* or *metrizable convex system*.

3.13. Examples

3.13.1. Subsets. Let X be a uniform convex structure and let $Y \subseteq X$. A convex system can be defined on Y by considering the subsets of Y which are convex in X. This system is evidently uniform. It is an open problem whether -- conversely -- each uniform convex system extends to a uniform convex structure such that each convex set of the system is a convex subspace of the extension. A partial solution for spaces of arcs is given at the end of this section.

3.13.2. Compact convex surfaces. Let S be the boundary of a compact convex body B of $\mathbb{R}^n$, and let $p \in Int(B)$. We consider the convex system on S, consisting of all traces on S of wedges at p; cf. I§1.19.2. This system turns out to be non-uniform (see the discussion after 3.14 below); therefore we have to build in some restrictions. Consider $n+1$ closed standard half-spaces $H_0,..,H_n$ of $\mathbb{R}^n$ such that $p \notin \cup_{i=0}^n H_i$ and $S \subseteq \cup_{i=0}^n H_i$. As $p \notin H_i$, we see that $H_i \cap S$ is a closed convex set of the surface. We restrict the given partial convexity to those convex sets which are a subset of $H_i \cap S$ for some $i = 0,..,n$. By using stereographic projection with center p onto the bounding hyperplane of H_i, we see that $H_i \cap S$ is uniformly isomorphic to a compact convex subspace of $\mathbb{R}^n$. Now S is the union of $n+1$ sets of this kind. Therefore, the (relative) natural uniformity on S makes the partial hull operator uniformly continuous.

Some other examples will be discussed below.

3.14. Theorem (*Uniform Completion Theorem*). *Let X be a uniform convex system with compact polytopes. Then there is a partial convexity on the completion X^* turning X^* into a uniform convex system, such that each convex set of X is a convex subspace of X^* and the closure of each convex set in X^* is convex. If X is a genuine convex structure, then so is X^*. In these circumstances, X is a convex subspace, and the extended convexity on X^* is unique with these properties.*

Proof. We rely on the following results from the theory of uniform spaces.[7]

---

7. As to (1), see Engelking [1977, Thm. 8.3.10]. The result (2) for metric spaces is in Engelking [1977, p. 371]. The general (uniform) case can be derived from the metric case.

§3: Uniform Convex Structures

(1) Let X be a uniform space with a dense subspace X', and let $f: X' \to Y$ be a uniformly continuous function into a complete uniform space Y. Then there is one and only one uniformly continuous extension of f over X.

(2) The hyperspace of compacta of a complete uniform space is complete.

Let $\mathcal{A}$ denote the subspace of $\mathcal{T}_{fin}(X)$, consisting of all admissible sets. As usual, $\mathcal{TC}_{comp}(X)$ denotes the hyperspace of all compact convex sets, and $\mathcal{T}_{comp}(X)$ denotes the hyperspace of all compacta. It is clear that $\mathcal{T}_{comp}(X)$ is a uniform subspace of $\mathcal{T}_{comp}(X^*)$ and that $\mathcal{A}$ is a uniform subspace of $\mathcal{T}_{fin}(X^*)$. We regard the convex hull operator of X as a uniformly continuous function

$$h = co_X : \mathcal{A} \to \mathcal{T}_{comp}(X^*),$$

where the range space is complete by (2). Let $\mathcal{A}^* = Cl(\mathcal{A}) \cap \mathcal{T}_{fin}(X^*)$. Application of (1) yields a uniformly continuous extension

$$h^* : \mathcal{A}^* \to \mathcal{T}_{comp}(X^*).$$

Note that if $\mathcal{A} = \mathcal{T}_{fin}(X)$ (that is, if X is a genuine convex structure), then $\mathcal{A}^* = \mathcal{T}_{fin}(X^*)$. Let $F \in \mathcal{A}^*$. We verify the following two statements.

(i) $F \subseteq h^*(F)$. Indeed, there is a Cauchy net $(F_i)_{i \in I}$ converging to F, such that $F_i \in \mathcal{A}$ for all $i \in I$. As $F_i \subseteq h(F_i)$ for all i, the result follows at once.

(ii) If $G \subseteq X^*$ is a finite set with $G \subseteq h^*(F)$, then $G \in \mathcal{A}^*$ and $h^*(G) \subseteq h^*(F)$. Indeed, for each uniform entourage U of X^* consider a uniform entourage V such that V–close members of $\mathcal{A}^*$ map to U–close sets under h^*. Let $F_V \in \mathcal{A}$ be V–close to F. As $G \subseteq h^*(F)$, there is a finite set $G_V \subseteq h(F_V)$ which is U–close to G. Note that $G_V \in \mathcal{A}$. It follows that the net $(G_V)_V$ converges to G, whence $G \in \mathcal{A}^*$. As $G_V \subseteq h(F_V)$ for all V, it follows that $h_V(G_V) \subseteq h_V(F_V)$. Passing to the limit, we find $h^*(G) \subseteq h^*(F)$.

Operators with the properties (i) and (ii) determine a convex system on X^*, the convex sets C of which are characterized by the property that each non-empty finite set $F \subseteq C$ is in $\mathcal{A}^*$ and $h^*(F) \subseteq C$. The partial hull operator coincides with h^* and $\mathcal{A}^*$ is the collection of all non-empty admissible finite sets. See I§2.21; for genuine convex structures the result is given in I§2.2. So we arrive at a uniform convex system on X^*. As h^* extends $h = co_X$, it is clear that each convex set in X is a convex subspace of X^*. Let $C \subseteq X^*$ be convex, A routine argument shows that $h^*(F) \subseteq \overline{C}$ for each $F \in \mathcal{A}^*$ included in $\overline{C}$ and that each non-empty finite subset of $\overline{C}$ is in $\mathcal{A}^*$. Hence $\overline{C}$ is convex.

If X is a genuine convex structure, then all finite sets of X^* are admissible, whence X^* is a convex structure. Its hull operator is the unique extension of X's hull operator. ■

This result illustrates why the convex sets of certain convex systems have to be chosen small enough. For instance, consider a sphere with the convex system of all wedges at the origin (cf. 3.13.2). If this system were uniform, then the uniform extension of the hull operator would lead us to convex sets which are closed half-spheres. Intersections of such sets produce pairs of diametrical points. As nearby pairs span connected sets, the hull operator is clearly discontinuous. A proper restriction is described in 3.13.2.

On a related subject, if a uniform space X is totally bounded, then its completion X^* is a compactification of X. It would be of intrinsic interest to have such compactifications available for Euclidean convexity. For $n = 1$, we see that $\mathbb{R}$ is isomorphic (as a tcs) with its open subspace $(0,1)$. Therefore, $[0,1]$ is a compactification as desired. The situation is different in dimensions > 1.

3.15. Lemma. *Let V be a locally convex vector space and let μ be a compatible uniformity on V. Then μ includes all covers consisting of two open half-spaces.*

Proof. Let $\{O_1, O_2\}$ be a cover as announced. We assume that none of the half-spaces involved are trivial. Let $H_i = V \setminus O_i$ for $i = 1, 2$ and take $x \in O_1 \cap O_2$. There is a cover $\mathcal{U} \in \mu$ with

$$x \notin star(H_1, \mathcal{U}) \cup star(H_2, \mathcal{U}).$$

Let $\mathcal{V} \in \mu$ be an associated refinement. As H_1 and O_2 have parallel bounding hyperplanes, and $x \notin co(star(H_1, \mathcal{V}))$ it follows that $co(star(H_1, \mathcal{V})) \subseteq O_2$. The same goes with 1, 2 interchanged. If $V \in \mathcal{V}$ meets H_1 then $V \subseteq O_2$, and if V meets none of H_1, H_2, then $V \subseteq O_1 \cap O_2$. So $\mathcal{V}$ refines $\{O_1, O_2\}$ and the latter is a uniform cover. ∎

3.16. Proposition. *There is no compatible totally bounded uniformity on $\mathbb{R}^n$, $n > 1$*

Proof. As $\mathbb{R}^2$ embeds as a convex subspace of $\mathbb{R}^n$ for $n > 1$, it suffices to consider the case $n = 2$. Suppose μ is a totally bounded, compatible uniformity on $\mathbb{R}^2$. By the previous result, we obtain two covers $\{O_1, O_2\}, \{P_1, P_2\} \in \mu$, where

$$O_1 = \{(x,y) \mid x < 1\}, \quad O_2 = \{(x,y) \mid x > -1\},$$
$$P_1 = \{(x,y) \mid y < 1\}, \quad P_2 = \{(x,y) \mid y > -1\}.$$

Consider the four-element cover $\mathcal{U}$, consisting of all sets of type $O_i \cap P_j$. Then $\mathcal{U} \in \mu$

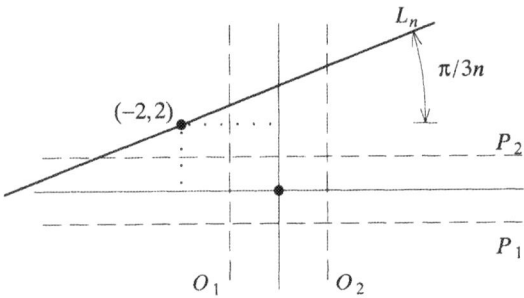

Fig. 1: Disproving the existence of a compactification

and there is an associated refinement $\mathcal{V}$ of $\mathcal{U}$. As μ is totally bounded, $\mathcal{V}$ may be taken finite. By virtue of Theorem 3.8, $\mathcal{V}$ may be taken as a convex (open) cover. Let $(L_n)_{n=1}^{\infty}$ be a sequence of lines through $(1,1)$, where L_n has a slope $\pi/3n$ (Fig. 1). Then

$$star(L_n, \mathcal{U}) = O_1 \cup P_2.$$

§3: Uniform Convex Structures

For each $n \in N$ there is a $V_n \in \mathcal{V}$ covering a right upper end of L_n. Consequently, there exist $m > n$ and $V \in \mathcal{V}$ covering a right upper end of both L_m and L_n. But then

$$co\,(star(L_n,\mathcal{V})) \supseteq O_2 \cap P_1 \not\subseteq star(L_n,\mathcal{U}).$$ ∎

In regard to the fact that completions of totally bounded uniform spaces correspond with compactifications of the original space, we can rephrase the last result as follows. For $n > 1$, there is no compact uniform convex structure including $\mathbb{R}^n$ with Euclidean convexity as a dense convex subspace. We shall see in the next section, however, that there exist compact uniform spaces with $\mathbb{R}^n$ as a (non-convex) subspace.

We now concentrate on a procedure to construct quotient spaces.

3.17. Theorem (*Uniform Quotient Theorem*). *Let X be a uniform convex system with compact polytopes, and let $\mu' \subseteq \mu(X)$ be a compatible subuniformity. Then the relation*

$$R = \cap \{U \mid U \in \mu'\}$$

is an equivalence relation on X. Let $q: X \to X/R$ be the quotient function, where X/R is equipped with the quotient partial convexity and let the uniformity μ/R on X/R consist of all entourages U with $(q \times q)^{-1}(U) \in \mu'$. Then $(X/R, \mu/R)$ is a uniform convex system and q is a uniformly continuous, CP and CC function.

Proof. For a subset V of $X \times X$, we denote $(q \times q)(V)$ more briefly by $\tilde{V}$. Note that if $V \in \mu'$ then $\tilde{V} \in \mu/R$. We first verify that μ/R is a uniformity. Let $U \in \mu/R$ and let $V \in \mu'$ be such that

$$V \circ V \circ V \subseteq (q \times q)^{-1}(U).$$

Then $\tilde{V} \circ \tilde{V} \subseteq U$. Indeed, take $(\tilde{v}_1, \tilde{v}_2) \in \tilde{V} \circ \tilde{V}$, say: $(\tilde{v}_1, \tilde{v}), (\tilde{v}, \tilde{v}_2) \in \tilde{V}$. Then there exist $v_1, v_2, v, v' \in X$ such that

$$q(v_1) = \tilde{v}_1;\quad q(v_2) = \tilde{v}_2;\quad q(v) = \tilde{v} = q(v');\quad (v_1,v), (v',v_2) \in V.$$

As $R \subseteq V$, we have $(v,v') \in V$ and hence $(v_1,v_2) \in V \circ V \circ V \subseteq (q \times q)^{-1}U$.

We next verify that if $C \subseteq X$ is a convex set, then its saturation with respect to R is an induced convex set. Let $F = \{x_1,..,x_n\}$ be admissible and let x_i be R-equivalent to $c_i \in C$ for each i. As μ' is compatible with the partial convexity of X, for each $U \in \mu'$ there is a $V \in \mu'$ such that V–close admissible sets have U–close hulls. Hence, if $x \in co(F)$, then (as F is V–close to the set $G = \{c_1,...,c_n\}$) we see that $U(x)$ meets $co(G)$ for each $U \in \mu'$. As μ' has a basis of entourages which are closed in the μ–topology, and as $co(G)$ is compact, x is R–equivalent to some point of $co(G)$.

The previous argument shows that q is CC (it is CP by construction). We finally show that μ/R is compatible with the quotient convexity. Let $U \in \mu/R$, let $V \in \mu'$ be such that $V \circ V$ is associated to $(q \times q)^{-1}(U)$, and let $W \in \mu'$ be such that $W \circ W \subseteq V$. If $D \subseteq X/R$ then

(∗) $q^{-1}(\tilde{W}(D)) \subseteq V(q^{-1}(D)).$

Indeed, if $q(x) \in \tilde{W}(D)$ then $(q(x), d) \in \tilde{W}$ for some $d \in D$. By the construction of $\tilde{W}$, there is a pair $(x',v) \in W$ mapped to $(q(x),d)$ by $q \times q$. So $v \in q^{-1}(D)$, $(x,v) \in W \circ W$, and consequently $x \in V(q^{-1}(D))$.

Suppose F, G are W–close admissible sets in X/R. As q is both CP and CC, there is an admissible set $G' \subseteq X$ such that $G = q(G')$; cf. I§1.18.1. By (*), we have $G' \subseteq V(q^{-1}(co_R(F)))$. Hence each element of G' is V–close to some point mapping into $co_R(F)$. Let $H \subseteq co_R(F)$ be the image of the selected points. As before, we obtain a finite admissible set $H' \subseteq X$ such that $q(H') = H$. As $R \subseteq V$, G' is $V \circ V$–close to H'. Hence $co(G')$ and $co(H')$ are $(q \times q)^{-1}(U)$–close. As $co(H') \subseteq q^{-1}co_R(F)$, we conclude that

$$co(G') \subseteq (q \times q)^{-1}(U)(q^{-1}(co_R(F))) = q^{-1}(U(co_R(F))).$$

So $co_R(G) = q(co(G')) \subseteq U(co_R(F))$. After reversing the role of F and G, we find that $co_R(F)$ and $co_R(G)$ are U–close. ∎

We will refer to X/R as a **uniform CP quotient of X**. An equivalence relation R as in Theorem 3.17 can be characterized on compact convex *structures* by two conditions:

(1) The graph of R is closed in $X \times X$.
(2) The R–saturation of a convex set is convex.

Indeed, the fact that the relation R is compact yields for each $U \in \mu(X)$ a $V \in \mu(X)$ such that $R \circ V \subseteq U \circ R$. Due to the compactness of X, each $\mu(X)$–entourage including R also includes an entourage of type $U \circ R$, where $U \in \mu(X)$. These sets form a basis for a subuniformity μ'. Indeed, if U, V, $W \in \mu(X)$ are such that $R \circ W \subseteq V \circ R$, $W \subseteq V$, and $V \circ V \subseteq U$, then

$$(W \circ R) \circ (W \circ R) = W \circ (R \circ W) \circ R \subseteq W \circ (V \circ R) \circ R \subseteq U \circ R.$$

To see that μ' is compatible, consider $U \in \mu'$, let $U' \in \mu'$ satisfy $U' \circ U' \subseteq U$, and let $V \in \mu(X)$ be associated to U'. Then for each convex set $C \subseteq X$, we have $co(V(R(C))) \subseteq U'(R(C))$ since $R(C)$ is convex. As $U' \circ R \subseteq U' \circ U' \subseteq U$, we see that $V \circ R \in \mu'$ is associated to U.

This illustrates that the Uniform Quotient Theorem is essentially the same as Theorem 2.9 in case of compact convex structures. We do not know if the conditions (1) and (2) (formulated in terms of induced convex sets) characterize uniform CP quotients in case of compact convex *systems*.

3.18. Examples: cones and projective spaces.

3.18.1. Proposition. *The cone of a compact continuous tcs is a continuous tcs.*

Proof. Recall that a cone $\Delta(X)$ is a quotient of type $X \times [0,1] / X \times \{1\}$, where the convexity of $X \times [0,1]$ is defined as follows. A pair (x,t) is in between the points $(x_1, t_1), \ldots, (x_n, t_n)$ provided

$$x \in co\{x_i \mid t_i \leq t\}; \quad \min_i t_i \leq t \leq \max_i t_i.$$

See I§2.4.1. The space $X \times [0,1]$ is compact. Its canonical uniformity is based on entourages of type

$$U_r = \{((x,s),(y,t)) \mid (x,y) \in U; \ |s-t| < r\},$$

§3: Uniform Convex Structures   317

where U is a uniform entourage of X and $r > 0$. Let $\varepsilon > 0$ and U as above. Assume that the sets
$$\{(x_i,s_i) \mid i = 1,..,m\}; \quad \{(y_j,t_j) \mid j = 1,..,n\}$$
are V_ε-close, where V is a uniform entourage of X such that V-close sets have U-close hulls. Let (x,s) be in the hull of the first set, and put
$$F = \{i \mid s_i \leq s\}.$$
By construction, $x \in co\{x_i \mid i \in F\}$. Let G be the collection of all j such that (y_j,t_j) is V_ε-close to some point (x_i,s_i) with $i \in F$. The sets $\{x_i \mid i \in F\}$ and $\{y_j \mid j \in G\}$ are V-close and hence there is a point $y \in co\{y_j \mid j \in G\}$ which is U-close to x. Let t be the maximum of t_j for $j \in G$ and of s. As the sets
$$\{s_i \mid i \in F\} \quad \text{and} \quad \{t_j \mid j \in G\}$$
are ε-close, and as $s_i \leq s$ for all $i \in F$, we find that $0 \leq t - s < \varepsilon$. Hence (y,t) is U_ε-close to (x,s). Note that the set of all j such that $t_j \leq t$ includes G. We conclude that
$$y \in co\{y_j \mid t_j \leq t\}$$
and hence that
$$(y,t) \in co\{(y_j,t_j) \mid j = 1,..,n\}.$$
The result follows from an application of the Uniform Quotient Theorem. ∎

3.18.2. A convex system in real projective space. Consider the uniform convex system on the n-sphere S^n as described earlier. After identifying all antipodal pairs, we arrive at the real projective space $\mathbb{P}^n$ of dimension n. The equivalence relation involved can be obtained with a subuniformity based on uniform covers $\mathcal{U}_r$ ($r > 0$) of the following type. For $p \in S^n$ let p' denote its antipode. Then $\mathcal{U}_r$ consists of all sets $B(p,r) \cup B(p',r)$ for $p \in S^n$, where $B(p,r)$ is the set of all points $x \in S^n$ at angular distance $< r$ radials from p.

In order that the subuniformity be compatible with the convex system on S^n, we have to restrict the diameter of the spherical convex sets to less than $d < \pi/2$ radials. The original uniformity of the sphere is based on coverings $\mathcal{B}_r$, $r > 0$, which consist of all metric disks of radius $< r$. Let $r > 0$ and take $s > 0$ such that s-close admissible sets on the sphere have r-close hulls. We may assume that $d + s < \pi/2$. Let $F_1, F_2 \subseteq S^n$ be finite admissible sets which are $\mathcal{U}_s$-close. Note that each point of F_2 is s-close to some point of F_1 or to its antipode. Elementary metric considerations involving the inequality $d + s < \pi/2$ yield that F_2 is s-close to either F_1 or its antipode. Hence $co(F_2)$ is r-close to either $co(F_1)$ or the antipodal hull.

We conclude that the subuniformity is compatible, that $\mathbb{P}^n$ is a uniform quotient, and (in particular) that the quotient map $S^n \to \mathbb{P}^n$ is CP and CC.

3.19. Corollary (*Uniform Factorization Theorem*). *Let X be a compact, continuous convex system, let Y be a compact topological space, and let $f : X \to Y$ be continuous. Then there is a uniform quotient $\tilde{X}$ of X with a quotient map $q : X \to \tilde{X}$, such that*

318   Chap. III: Topological Convex Structures

(1)   *The (topological) weight of $\tilde{X}$ is at most that of Y.*
(2)   *There is a continuous function $\tilde{f}: \tilde{X} \to Y$ with $\tilde{f} \circ q = f$.*

Proof. Let w denote the weight of Y. Then there is a base ν of cardinality w for the uniformity of Y, based on all finite open covers. Consider the uniformity

$$\mu_0 = \{ g^{-1}(\mathcal{V}) \mid V \in \nu \}$$

on X. We recursively construct μ_{n+1} from μ_n by adding to each pair of covers in μ_n a common, associated, open star refinement. Then $\cup_{n=0}^{\infty} \mu_n$ can be taken as a basis of cardinality w for a compatible subuniformity μ' on X. Let $q: X \to \tilde{X}$ be the uniform quotient map, derived from μ' (cf. Proposition 3.17). As $\mu_0 \subseteq \mu'$, two points identified by q are also identified by f. The result follows. ∎

3.20. Spaces of arcs. The definitions given in I§2.4.2 are now adapted to the situation where a topology is involved on a poset. An (**order**) **arc** A joining two points $a \leq b$ of a pospace Y is a connected and totally ordered subset of the order interval $[a,b]$ containing both a and b. Note that A is a maximal chain of $[a,b]$. A pospace Y is **arc-wise connected** provided each pair of points $a \leq b$ in Y can be joined with an arc. The graph of the order relation being closed in $Y \times Y$, all arcs are closed in Y. The set $\Gamma(Y)$ of all arcs in Y can be topologized as a (closed) subspace of the hyperspace of compacta, $\mathcal{T}_{comp}(Y)$. We let $\Lambda(Y)$ denote the subspace of $\Gamma(Y)$, consisting of all maximal arcs. The space Y is usually taken compact. This allows us to consider the canonical (relative) hyperspace uniformity on $\Gamma(Y)$ and on $\Lambda(Y)$.

Let S be a compact join semilattice with a top element **1** and with a lower bound **0**. A convex structure can be defined on $\Lambda(S)$ as in the discrete case (cf. I§3.3): If $A_1,..,A_n$ are maximal arcs in S, then the hull of $A_1,..,A_n$ is defined to be the set of all maximal arcs A which are **in between** $A_1,..,A_n$, that is: A is built up with points of type

(1) $a_1 \vee .. \vee a_n$,   with $a_1 \in A_1,..,a_n \in A_n$.

In general, a compact join semilattice S only has a top element **1**, and maximal arcs need not join the same pair of points. We define a collection of arcs in S to be admissible provided all of its members join the same pair of points. The hull of a finite admissible collection is defined as above, and a convex system obtains. The **straightening** of an n-tuple $(a_1,..,a_n)$ in a join semilattice is the n-tuple $(b_1,..,b_n)$ with $b_i = \vee_{j \leq i} a_j$.

3.21. Lemma. *Let S be a Lawson join semilattice, and let $U \in \mu(S)$.*

(1) *There is an entourage $V \in \mu(S)$ such that for all $n \in \mathbb{N}$ the straightening of two V-close n-tuples are U-close.*
(2) *There is an entourage $V \in \mu(S)$ such that the following is true. If $A, B \subseteq S$ are chains and if*

$$a_0 = \inf(A) \leq a_1 \leq .. \leq a_n = \sup(A);$$
$$b_0 = \inf(B) \leq b_1 \leq .. \leq b_n = \sup(B),$$

are subsets of A, B respectively, such that a_i, b_i are V-close and b_i, b_{i+1} are

§3: Uniform Convex Structures

V–close for all i, then A, B are U–close.

Proof of (1). As S is compact, each non-empty subset has a supremum. The operation $\sup : H(S) \to S$ is continuous if S is a Lawson semilattice.[8] Hence there is an entourage $V \in \mu(S)$ such that $\sup A$ and $\sup B$ are U–close for each pair A, B of V–close sets.

Proof of (2). Consider the following sequence of entourages in $\mu(S)$.
$$V \subseteq 2V \subseteq V_2 \subseteq V_1 \subseteq 2V_1 \subseteq U.$$
The entourage V_2 is chosen such that for each $x \in S$ the order-convex hull of $V_2(x)$ is included in $V_1(x)$ (local order-convexity; cf. 1.13.2). If $b \in B$, say: $b_i \leq b \leq b_{i+1}$, then $b_i, b_{i+1} \in V(b_i) \subseteq V_2(b_i)$, whence $b \in V_1(b_i)$. As $b_i \in V_1(a_1)$, we find that $b \in U(a_i)$. Conversely, let $a \in A$, say: $a_i \leq a \leq a_{i+1}$, Then $a_i \in V(b_i)$ and $a_{i+1} \in 2V(b_i)$. Hence $a_i, a_{i+1} \in V_2(b_i)$ and it follows that $a \in V_1(b_i) \subseteq U(b_i)$. ∎

3.22. Proposition. *Let S be an arc-wise connected Lawson join semilattice. Then $\Gamma(S)$ is a uniform convex system which is metrizable if the semilattice is metrizable as a topological space.*

Proof. Recall that $\Gamma(S)$ is a compact subspace of $\mathcal{T}_*(S)$. The (subspace) topology of $\Gamma(S)$ is generated by a uniformity, based on entourages of type
$$\{(A,B) \mid A, B \in \Gamma(S) \text{ are } V\text{–close}\} \quad (V \in \mu(S))$$
Note that this is a metric uniformity if $\mu(S)$ is. A polytope $co\{A_1,..,A_n\}$ in $\Gamma(S)$ can be seen as the intersection of $\Gamma(S)$ with the closed subset
$$< \bigvee_{i=1}^{n} A_i > = \{A \in \mathcal{T}_*(S) \mid A \subseteq \bigvee_{i=1}^{n} A_i\}$$
of the hyperspace, which is a compact set.

Let $U \in \mu(S)$ and consider the following refining sequence of uniform entourages.

(1) $V_1 \subseteq U$ is chosen as in Lemma 3.21(2);
(2) $V_2 \subseteq V_1$ is chosen as in Lemma 3.21(1);
(3) V–close sets in S have V_2–close suprema (in particular, $V \subseteq V_2$).

Let $\{A_1,.., A_m\}$ and $\{B_1,.., B_n\}$ be V–close admissible sets of arcs in $\Gamma(S)$. Assume that all arcs of the first set join $\mathbf{0}_A$ with $\mathbf{1}_A$, and that all arcs of the second set join $\mathbf{0}_B$ with $\mathbf{1}_B$. We verify that the respective hulls are U–close. To this end, let $A \in co\{A_1,.., A_m\}$ and let
$$s_0 = \mathbf{0}_A \leq s_1 \leq .. \leq s_k = \mathbf{1}_A$$
be an increasing sequence in A with V–close successive points. Fix $l \in \{0,.., k\}$ for a moment. The point s_l can be represented as
$$s_l = a_{l,1} \vee .. \vee a_{l,m} \quad (a_{l,i} \in A_i).$$

---
8.   cf. Gierz et al [1980, p. 289].

Choose a point $b_{l,j} \in B_j$ for each $j = 1,..,n$ in such a way that the sets

$$\{a_{l,i} \mid i = 1,..,m\}; \{b_{l,j} \mid j = 1,..,n\}$$

are V–close. By (3), s_l is V_2–close to the point

$$t'_l = b_{l,1} \vee .. \vee b_{l,n} \in \bigvee_{j=1}^{n} B_j.$$

We may assume that $t'_0 = \mathbf{0}_B$ and that $t'_1 = \mathbf{1}_B$. By (2) and Lemma 3.21(1), the straightening $(t_l)_{l=0}^{k}$ of $(t'_l)_{l=0}^{k}$ is V_1–close to $(s_l)_{l=0}^{k}$. Note that t_l, being a supremum of points in $\vee_{j=1}^{n} B_j$, is an element of this (sub-)semilattice. We extend the chain of all points t_l to a maximal chain $B \subseteq \vee_{j=1}^{n} B_j$. The latter being an arc-wise connected semilattice, we find that B is an arc joining $\mathbf{0}_B, \mathbf{1}_B$. By construction, we have $B \in co\{B_1,..,B_n\}$. Finally, by (1) and Lemma 3.21(2), B is U–close to A. Reversing the role of the arcs A_i and B_j, we conclude that the resulting polytopes are U–close, as announced. ∎

It is an open problem to extend uniform convex systems to uniform convex structures in such a way that convex sets of the former are convex sets of the latter. However, we have a solution for spaces of arcs.

3.23. Lemma. *Let S be a Lawson join semilattice. Then there is a continuous function $e: \Gamma(S) \to S$, mapping an arc to its minimum.*

Proof. Let $a = e(A)$ and let $N \subseteq S$ be a neighborhood of a. There exist points $u, v_1,..,v_n \in S$ such that the set[9]

$$\{x \in S \mid x \leq u\} \setminus (\cup_i \{x \mid x \not\leq v_i\})$$

is a neighborhood of a included in N. Consider the following set.

$$\mathcal{U} = \bigcap_{i=1}^{n} <S \setminus \downarrow(v_i)> \cap <\uparrow(u), S> \cap \Gamma(S).$$

Clearly, $\mathcal{U}$ is a neighborhood of A and each arc $B \in \mathcal{U}$ ends in U. ∎

3.24. Theorem. *Let S be a connected Lawson join semilattice with a continuous interval operator. Then there is a connected Lawson join semilattice T with a continuous interval operator and with a lower bound, together with a topological embedding $\Gamma(S) \to \Lambda(T)$, such that each convex subspace of $\Gamma(S)$ is a convex subspace of $\Lambda(T)$. Moreover, T is metrizable if S is.*

Proof. By 3.7.2, S is a uniform convex structure (usual convexity). The convex hyperspace $\mathcal{IC}_*(S)$ of S is a connected Lawson semilattice under the join operation

$$C_1 \vee C_2 = Cl\,co\,(C_1 \cup C_2) \quad (C_1, C_2 \in \mathcal{IC}_*(S)).$$

We construct an intermediate embedding

---

9.   cf. Gierz et al [1980, pp. 144, 282].

§3: Uniform Convex Structures

$I: \Gamma(S) \to \Gamma(\mathcal{JC}_*(S))$

as follows. If A is an arc in S with end point $e(A)$, then

$I(A) = \{[e(A), a] \mid a \in A\}$.

Then I is continuous since the end point map e is continuous (Lemma 3.23), and since the order-intervals of S depend continuously on the end points. Clearly, the operator I is injective, and hence it is an embedding of compact topological spaces. For each admissible collection $(A_i)_{i=1}^n$ of arcs in S we have

$I(co\{A_1, .., A_n\}) = co\{I(A_1), .., I(A_n)\}$

(first hull in $\Gamma(S)$, second in $\Gamma(\mathcal{JC}_*(S))$). As a consequence, I embeds convex subspaces of $\Gamma(S)$ as convex subspaces of $\Gamma(\mathcal{JC}_*(S))$.

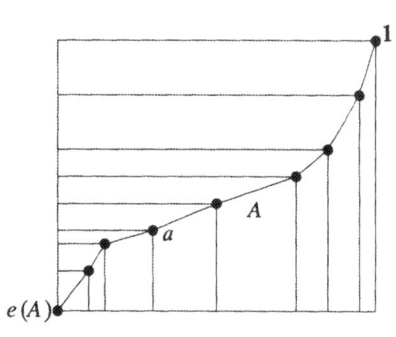

Fig. 2: The embedding I Fig. 3: The embedding J

We next consider the cone $\Delta(S)$, which is another uniform S_4 convex structure by Proposition 3.18.1. We let p denote its apex. Now $\mathcal{JC}_*(\Delta(S))$ is a connected Lawson semilattice (Topic 3.27). The desired semilattice is a convex subspace of it, namely,

$T = \mathcal{JC}_p(\Delta(S))$,

where the subscript "p" indicates that we only consider sets containing p. The lower bound of this new semilattice is $\{p\}$. Moreover, T is connected, and T is metrizable if S is. To complete the proof, we construct a second embedding

$J: \Gamma(\mathcal{JC}_*(S)) \to \Lambda(T)$

as follows. Let $\mathcal{A} \subseteq \mathcal{JC}_*(S)$ be an arc joining the two convex sets $A_0 \subseteq A_1$, and put

$J(\mathcal{A}) = \mathcal{A}_0 \cup \mathcal{A}_1 \cup \mathcal{A}_2$,

where $\mathcal{A}_0$ consists of all subcones of type

$\{[(x,t)] \mid x \in A_0; t \geq t_0\}$ $(t_0 \in [0,1])$,

$\mathcal{A}_1$ consist of the cones $\Delta(A)$ over the elements $A \in \mathcal{A}$, and $\mathcal{A}_2$ consists of all sets of type

$$\Delta(A_1) \cup \{[(x,t)] \mid x \in S; t \geq t_2\} \quad (t_2 \in [0,1]).$$

See Fig. 3; the brackets [.] refer to an equivalence class. All members of $J(\mathcal{A})$ are convex sets of the cone convexity of $\Delta(S)$, the operation J is easily seen to be continuous (and hence it is a topological embedding), and for each admissible collection $(\mathcal{A}_i)_{i=1}^n$ of arcs in $\mathcal{TC}_*(S)$ we have

$$J(co\{\mathcal{A}_1,..,\mathcal{A}_n\}) = co\{J(\mathcal{A}_1),..,I(\mathcal{A}_n)\}$$

(first hull in $\Gamma(\mathcal{TC}_*(S))$, second in $\Lambda(T)$). The desired embedding $\Gamma(S) \to \Lambda(T)$ is the composed map $J \circ I$. ∎

Further topics

3.25. Compatible uniformity in vector spaces. Let X be a convex subset of a locally convex vector space V. By Proposition 3.10.1, the hull operator of X is uniformly continuous with respect to the standard (translation invariant) uniformity of V. A uniformity on X is *exotic* provided it is different from the relative standard uniformity and it makes the hull operator uniformly continuous. Observe that exotic uniformities exist on $\mathbb{R}$. Show that $[0,\infty)^n$ admits an exotic (in fact, a totally bounded) uniformity for each $n \geq 1$. Problem: does there exist an exotic uniformity in a locally convex vector space of dimension > 1? Verify that the answer is affirmative iff it is for $X = \mathbb{R}^2$.

3.26. Compact metric trees (cf. van de Vel [1993]). Prove that each compact, connected, metric tree is a uniform CP quotient of the 2–cell (standard convexity). Hint. Try first to obtain the trees with finitely many ramifications. This can be done recursively by creating a ramification point via the introduction of a polygon (with edges on the circle) in the partition. This result was suggested in I§1.16.4 and illustrated for a tree with one ramification point.

3.27. Convex hyperspaces: the semilattice viewpoint (compare van de Vel [1993]). Let X be a compact uniform convex structure. Then $\mathcal{TC}_{comp}(X)$ is a compact semilattice under the join operator $(C,D) \mapsto co^*(C \cup D)$. In addition, it has small $\cup$–semilattices, and its intervals depend continuously on their end points. Conclude that the semilattice $\mathcal{TC}_*(X)$ is a uniform convex structure.

3.28. Completion of convex systems. As is well-known, a uniformly continuous function extends in a unique way to a uniformly continuous function of the completion. Show that if the original function is a CP and CC map of uniform convex systems, then the extension is CP and CC.

3.29. Special metrics (van Mill and van de Vel [1978]; continued 4.33). Let X be a topological convex structure. A metric d on X is *adapted* provided

(i)   it generates the topology of X, and
(ii) for each convex set $C \subseteq X$ and for each $r > 0$ the following set is convex.

$$D(C,r) = \{x \mid d(x,C) \le r\}.$$

It is possible to give an equivalent definition in terms of open disks. Observe that an adapted metric is compatible with the convexity.

3.29.1. If T is a compact tree and if d is a convex metric generating the topology of T, then d is adapted. Give an example of an adapted metric on $[0,1]$ (natural convexity) which is not a convex metric.

3.29.2. (Franchetti [1971]) Let X be a convex structure and let d be a metric on X such that all open disks

$$B(p,r) = \{x \in X \mid d(p,x) < r\} \quad (p \in X; r > 0)$$

are convex. Show that $diam(A) = diam(co(A))$ for all non-empty sets $A \subseteq X$.

3.29.3. Let (X,d) be a compact metric space, and let I be a non-expansive interval operator on X, that is, for all $u_1, u_2, v_1, v_2 \in X$ and for all $\varepsilon > 0$,

$$d(u_1,u_2) < \varepsilon \text{ and } d(v_1,v_2) < \varepsilon \quad \text{implies} \quad d_H(I(u_1,v_1),I(u_2,v_2)) < \varepsilon,$$

where d_H denotes the Hausdorff metric on the hyperspace of X. Show that d is an adapted metric.

3.29.4. The study of topo-convex separation properties in the next section provides a method to construct adapted metrics in particular situations. The question remains unsolved whether a metrizable convex structure admits an adapted metric generating the compatible uniformity. The same question, with "hull operator" replaced by "segment operator", has a negative answer. Consider a compact convex subspace X of a (topologically) metrizable vector space, such that X is not locally convex. The segment operator of X is uniformly continuous, but there can be no metric on X relative to which the segment operator is non-expansive: cf. part (3) of this topic.

3.30. The torus. Show that the convex system, defined on the n-dimensional torus in I§1.31.3, is uniform provided the size of the convex sets is slightly restricted. Compare with the argument on projective spaces in 3.18.2.

3.31. On the Uniform Factorization Theorem (cf. van Mill and van de Vel [1986]). Let X be a compact uniform convex system and let $f: X \to Y$ be a map. If $dim(Y) \le n$ (Lebesgue dimension) then there is a factorization $f = g \circ q$ of f with a uniform quotient map $q: X \to \tilde{X}$ as in 3.19, and such that $dim(\tilde{X}) \le n$.

3.32. Spaces of arcs (van de Vel [1991]). Let S be a Lawson semilattice with universal bounds **0** and **1**. Show that $\Lambda(S)$ is JHC, S_4, and that its convex sets are connected. Hints. As to JHC and S_4, the methods presented in the discrete case (cf. I§2.17.1 and I§4.14.3) can be adapted. Use Koch's Arc Theorem.[10]

10.   Full citation in 2.29.3.

3.33. Some problems. Let X be a uniform convex structure. Is there a basis of entourages $U \in \mu(X)$ such that $U(C)$ is convex for each convex set C? For metric uniformities, this is essentially the problem on adapted metrics, mentioned in 3.29.4 above.

If X is a uniform convex structure with a metrizable underlying space, must X be metrizable as a convex structure?

If X is a continuous convex structure with a uniformizable underlying space, must X be uniformizable as a convex structure?

Is it possible to extend a uniform convex system X to a uniform convex structure X^*, such that each convex set of X is a convex subspace of X^*? Can properties like S_4, connectedness of convex sets, and compactness of polytopes be preserved? For a solution in case of spaces of arcs, see Theorem 3.24.

Notes on Section 3

The notion of a uniform or metric convex structure has been introduced by the author in [1993] (first preprint dated 1980) primarily with the intention of formulating and proving a general selection theorem (cf. IV§3.5 below). Wieczorek [1983] considers a concept of "quasi-uniformly convex" space to describe topological convex structures with a uniformity base of entourages V such that $V(C)$ is convex for each convex set C. In our terminology, such entourages are self-associated. See Topics 3.29 and 3.33.

Notions comparable with uniform convex structures have also appeared in a context of "convexity with parameters". Here, the initial emphasis is on convex combinations of points, and a set C is considered convex if all convex combinations of its points are in C. The efforts have been directed almost exclusively to selection theorems and, more recently, to minimax and equilibrium theorems. The results require either a restrictive set of "algebraic" conditions (e.g., Michael [1959], Curtis [1985]), or an additional condition stating that some hull-like operator is uniformly continuous (e.g., Pasicki [1987] and [1990], Bielawski [1987]). Because of the rather specialized goals, the general (geometric) theory of such structures is fairly poor.

Our paper [1993] contains a.o. the equivalence of the existence of a compatible uniformity to uniform continuity of the hull operator (Corollary 3.4), and the Uniform Separation Theorem 3.11. Uniform convexity in convex hyperspaces was studied in our paper [1984d]. The Uniform Quotient Theorem, 3.17, and the Factorization Theorem, 3.19, are taken from van Mill and van de Vel [1986], where information is included concerning the dimension of the quotient. The extension to convex systems is new.

The impossibility to "compactify" Euclidean convexity in dimensions > 1 (Proposition 3.16) was proved by the author in [1983b]. A different viewpoint on compactification will be developed in the next section. The uniformity of a convex system in spaces of arcs, Theorem 3.22, was obtained in our paper [1991], together with the extension to a uniform convex structure of arcs, Theorem 3.24.

4. Topo-convex Separation

A topological convex structure allows for separation properties involving convex closed sets. Among many possible combinations, we consider a sequence of weak axioms NS_2–NS_4 involving neighborhood separation and a sequence of strong axioms FS_2–FS_4 involving separation with CP functionals. The conditions can be compared with the Hausdorff property (T_2), regularity (T_3), and normality (T_4) in topology. All valid implications are mentioned and many standard examples are investigated. The NS_3 property relates with upper semi-continuity of the convex closure operator and with closedness of base-point orders. In an FS_3 space, the weak topology is generated by the continuous CP functionals, thus providing some justification of our terminology.

Among the main results are: a characterization of the "regularity axiom" FS_3 in terms of pure embedding in compact locally convex median spaces, a characterization of the existence of normal T_1 convex subbases in terms of (not necessarily pure) embeddings, and the existence of a functorial pure extension of FS_3 spaces to compact locally convex median spaces.

In this section, **all convex structures are** S_1 unless stated explicitly to the contrary. For topological convex structures, this implies the topological axiom T_1.

4.1. Functionals. A function of type $X \to L$, where L is a complete totally ordered set, will henceforth be called a *functional of X*. A functional f gives rise to a chain of subsets of X consisting of all sets of type $f^{-1}(\leftarrow, t]$. Formally, this chain can be indexed with the elements of L. Conversely, let $(A_t)_{t \in L}$ be an *indexed chain of sets in X*, that is: L is a totally ordered set and $s < t$ in L implies $A_s \subseteq A_t$. We do not assume that L has universal bounds (which when present are denoted by $\mathbf{0}$ and $\mathbf{1}$). Consider the *Dedekind completion* L^* of $L \cup \{\mathbf{0}, \mathbf{1}\}$. A functional $f : X \to L^*$ can be defined by

$$f(x) = \inf \{t \in L \mid x \in A_t\}.$$

Two functionals with a common domain are *equivalent* provided they induce the same chain of sets. In other words, the functionals f_i for $i = 1, 2$ are equivalent iff there is an OP (= order preserving) isomorphism $l : f_1(X) \to f_2(X)$ with $l \circ f_1 = f_2$. If l is only required to be OP and surjective, then f_1 is a *refinement* of f_2. Refinement yields a quasi-order on the class of functionals. Two functionals f and g are mutually refining iff they are equivalent. Although the functionals of a non-empty set form a ***proper class***, it is always possible to have a *set* of representatives by passing to functionals into completed chains of subsets. This allows to construct "maximal refinements" within certain classes of functionals. See 4.21.1 and IV§2.5.3.

By convention, a totally ordered set is equipped with the order convexity. This convexity is the join of two coarser ones, namely, the lower and upper convexity; cf. I§1.5.2. If X is a convex structure, then a functional $f : X \to L$ is *lower CP* provided it inverts lower sets of L into convex sets of X, and f is *upper CP* provided it inverts upper sets of L into convex sets of X. Note that $f : X \to L$ is CP iff it is both lower and upper CP.

If X is a topological space then a functional $f: X \to L$ is *lower semi-continuous*[1] (*lsc*) provided sets of type $f^{-1}(\leftarrow, t]$ are closed; f is *upper semi-continuous* (*usc*) provided all sets of type $f^{-1}[t, \to)$ are closed. Note that f is continuous with respect to the order topology precisely if it is lsc and usc.

In most cases, L is simply $\mathbb{R}$, but even in this case, it may be profitable to operate with chains of sets. We need two auxiliary results.

4.2. Lemma. *Let $(A_t)_{t \in L}$ be an indexed chain of subsets of a set X with a densely ordered index set L, and let $f: X \to L^*$ be the corresponding functional into the Dedekind completion L^* of $L \cup \{0, 1\}$. Then for each $t \in L$,*

$$f^{-1}(\leftarrow, t] = \cap \{A_s \mid t < s\};$$
$$f^{-1}[t, \to) = X \setminus \cup \{A_s \mid s < t\}. \qquad \blacksquare$$

The segments of type $(\leftarrow, t]$ (resp., of type $[t, \to)$) for $t \in L$ constitute a subbase of the lower topology and of the lower convexity (resp., of the upper topology and of the upper convexity) in L^*. Hence, if X is a tcs, then

(i) f is lsc (resp., lower CP) provided A_t is closed (resp., convex) for each $t \in L$.
(ii) f is usc (resp., upper CP) iff for each $t \in L$ the set

$$\cup \{A_s \mid s < t; s \in L\}$$

is open (resp., concave) in X.

By way of example, let V be a topological vector space and let $O \ne V$ be a non-empty convex open subset containing the origin. Consider the chain of all sets of type $t \cdot O$ for $t > 0$. The completed index set corresponds with the extended real half-line $[0, \infty]$. The resulting functional f is lower CP and usc by (i) and (ii). Note that $V = \cup_{t > 0} t \cdot O$, showing that the value "∞" is not taken. Since the set $\cap_{t > r} t \cdot O$ is closed in V for each $r \ge 0$, the first formula in Lemma 4.2 implies that f is lsc as well. The functional f is known as the *Minkowski functional of O*.

To investigate continuous CP functionals, the following additional result is convenient.

4.3. Lemma. *Let $(A_t)_{t \in L}$ and $(B_s)_{s \in M}$ be indexed chains of convex closed sets in a tcs X with densely ordered index sets L and M, and let*

$$\iota: L \to M$$

be an order-reversing bijection. Suppose that for all $t, t' \in L$ the following hold.

(i) $A_t \cup B_{\iota(t)} = X$.
(ii) *If $t < t'$ then $A_t \cap B_{\iota(t')} = \varnothing$.*

Then the functionals determined by L and M are equivalent, continuous and CP.

---

1. The addition "semi" is superfluous but traditional. It seems to reflect depreciation of the related "half" topology. We do not extend this attitude to convexity.

§4: Topo-convex Separation   327

Proof. Let $\iota^*: L^* \to M^*$ be the canonical anti-isomorphism extending ι. Note that the conditions imposed on L and on M are symmetric. We have associated functions
$$f: X \to L^*; \quad g: X \to M^*$$
which are both lsc and lower CP by the previous result. Let $x \in X$, $t \in L$, and $s = \iota(t)$. Suppose $f(x) \geq t$. Then $x \notin A_{t'}$ for all $t' < t$ in L. If $s' > s$ in M, then $s' = \iota(t')$ for some $t' < t$, yielding $x \in B_{\iota(t')}$ by (i). Hence $g(x) \leq s$. Conversely, if $g(x) \leq s$, then $x \in B_{s'}$ for all $s' > s$ in M. Let $t' < t$ in L. As L is order dense, there is $t'' \in L$ with $t' < t'' < t$. As $x \in B_{\iota(t'')}$, condition (ii) yields $x \notin A_{t'}$, and we conclude that $f(x) \geq t$.

As L and M are dense in, respectively, L^* and M^*, this shows that $g = \iota^* \circ f$. Therefore, both functions are continuous and CP. ∎

4.4. Neighborhood separation. A space X is said to be:

NS$_2$, if for each pair of distinct points p, $q \in X$ there exists a convex closed neighborhood N of p with $q \notin N$.

NS$_3$, if for each convex closed set C and for each point $p \notin C$ there is a convex closed neighborhood N of C with $p \notin N$.

NS$_{3+}$, if for each convex closed set C and for each polytope P disjoint from C there is a convex closed neighborhood N of C with $N \cap P = \emptyset$.

NS$_4$, if for each pair C, D of disjoint convex closed sets there is a convex closed neighborhood N of C with $N \cap D = \emptyset$.

The label "NS" stands for **Neighborhood Separation**. We list some elementary observations. Remember that a tcs is assumed point-convex unless stated explicitly to the contrary.

4.4.1. NS$_4$ ⇒ NS$_{3+}$ ⇒ NS$_3$ ⇒ NS$_2$ ⇒ *Hausdorff.*

4.4.2. *A product of* NS$_i$ *spaces is* NS$_i$ *for* $i = 2, 3, 3^+, 4$.

4.4.3. *A subspace of an* NS$_2$ *space is* NS$_2$. *A pure subspace of an* NS$_i$ *space is* NS$_i$ *for* $i = 3, 3+$.

4.4.4. *A closed, continuous, CP + CC image of an* NS$_4$ *space is* NS$_4$.

The second of these statements is a direct consequence of Theorem 1.4 on products.

4.5. Proposition. *Let X be a topological convex structure.*

(1) *If X is compact and* NS$_2$, *then X is locally convex.*
(2) *If X has compact polytopes and is* NS$_3$, *then X is* NS$_{3+}$.
(3) *If X is compact, then the properties* NS$_3$, NS$_{3+}$, NS$_4$ *are all equivalent to upper semi-continuity of the convex closure operator.*

Proof. (1). Suppose X is compact, let $p \in X$, and let O be an open neighborhood of p. For each $q \in X \setminus O$ we take a convex closed neighborhood $N(q)$ of p with $q \notin N(q)$. After reducing to a finite cover $(X \setminus N(q_i))_{i=1}^n$ of $X \setminus O$, we obtain a convex closed neighborhood $N = \cap_{i=1}^{n} N(q_i)$ of p with $N \subseteq O$.

(2). Let C and P be disjoint convex sets of which C is closed and P is a polytope. For each $p \in P$ we have a convex closed neighborhood N_p of C with $p \notin N_p$. There is a finite collection $F \subseteq P$ such that the sets $X \setminus N_p$ for $p \in F$ cover the compact set P. Then $\cap_{p \in F} N_p$ is a convex closed neighborhood of C missing P.

(3). It is clear that upper semi-continuity of co^* implies NS$_4$. It is therefore sufficient to show that NS$_3$ implies upper semicontinuity of co^*. By Theorem 2.5(2) we only have to verify that for any neighborhood U of a convex closed set C there is a convex closed neighborhood N of C with $N \subseteq U$. This can be done using NS$_3$ with an argument as in the proof of (1). ∎

4.6. Examples

4.6.1. Lawson semilattices and pospaces. A compact pospace and a Lawson semilattice have a USC convex closure operator by 1.13.2 and 1.13.3. Hence by 4.5(3) these spaces are NS$_4$.

4.6.2. Standard convexity. A topological vector space of dimension ≥ 2 is never NS$_4$. Indeed, for $\mathbb{R}^2$ this follows by considering the disjoint convex closed sets

$$\{(x_1, x_2) \mid x_1 \geq 0; x_2 \geq 0; x_1 \cdot x_2 \geq 1\}; \quad \{(x_1, x_2) \mid x_1 = 0\}.$$

For a general vector space of dimension ≥ 2, the result follows from the fact that $\mathbb{R}^2$ embeds as a convex closed subspace. However, all locally convex vector spaces are NS$_{3+}$; see 4.13.1 below, where an even stronger property is considered. As a consequence, a compact convex subspace of a locally convex vector space is NS$_4$.

4.6.3. Proposition. *Let X be a compact Hausdorff tcs which is* NS$_4$. *Then the cone of X is also* NS$_4$.

Proof. Recall (cf. I§2.4.1) that the cone ΔX of a convex structure X is a CP and CC image of a convex structure, defined on $X \times [0,1]$ by the following hull formula.

$$co\{(x_1, t_1), ..., (x_n, t_n)\} = \{(x, t) \mid x \in co\{x_i \mid t_i \leq t\}; \min_i t_i \leq t \leq \max_i t_i\}.$$

The topology of ΔX is the quotient of the product topology on $X \times [0,1]$; cf. 1.21. By 4.4, it suffices to verify that the auxiliary space is NS$_4$. This is easy, considering the fact that the "horizontal" sections of a convex set increase with the height of the section (up to a certain limit level). ∎

More examples will be given after introducing stronger separation properties in 4.9 below.

4.7. Proposition. *In an* NS$_3$ *space X the following are true.*

(1) *The weak topology of X is regular.*
(2) *If $C \subseteq X$ is spanned by a compact set, then there is a minimal compact set spanning C.*

§4: Topo-convex Separation

(3) *If all polytopes of X are compact, then CUP and JHC are equivalent for X.*

Proof of (1). In the weak topology, each point p has a neighborhood base, consisting of sets of type
$$U = X \setminus (\bigcup_{i=1}^{n} C_i),$$
where the sets C_i are convex closed. Under assumption of NS_3, we obtain convex closed sets D_i with $C_i \subseteq Int(D_i)$ and $p \notin D_i$ for all i. Then $V = X \setminus (\bigcup_{i=1}^{n} D_i)$ is a weak neighborhood of p with $\overline{V} \subseteq U$.

Proof of (2). Let $(K_i)_{i \in I}$ be a chain of compacta with $co^*(K_i) = C$ and let $K = \cap_{i \in I} K_i$. If $p \in C \setminus co^*(K)$ then fix a convex closed neighborhood N of $co^*(K)$ with $p \notin N$. Now $\cap_{i \in I} K_i \subseteq Int(N)$ and hence $K_i \subseteq Int(N)$ for some i. This yields $C = co^*(K_i) \subseteq N$, a contradiction.

Proof of (3). JHC implies CUP as we saw in I§2.13(2). For the reverse implication, let P be a polytope and $p \notin P$. If $a \notin \cup_{x \in P} px$ then for each $x \in P$ we can fix a convex neighborhood $C(x)$ of px with $a \notin C(x)$. As P is compact, a finite number of sets $C(x_1), .., C(x_n)$, suffices to cover P. By CUP,
$$a \notin \bigcup_{i=1}^{n} C(x_i) = \bigcup_{i=1}^{n} co(\{p\} \cup C(x_i)) \supseteq co(\{p\} \cup P).$$
The result follows from I§2.13(1). ∎

We now switch to a description of separation in terms of functionals. A functional $f: X \to L$ **separates two sets** A and B in X provided there exist $s < t$ in L with
$$A \subseteq f^{-1}(\leftarrow, s]; \quad B \subseteq f^{-1}[t, \rightarrow).$$
If $\mathcal{F}$ is a collection of functionals of X and if $\mathcal{P}$ is a collection of pairs of sets in X, then $\mathcal{F}$ **separates (the pairs of)** $\mathcal{P}$ provided for each pair $(A,B) \in \mathcal{P}$ there is a functional $f \in \mathcal{F}$ separating (A,B). Certain informal usage of the above expressions is more or less self-explaining. The family $\mathcal{F}$ **separates points** in case it separates all pairs of distinct singletons in X, and $\mathcal{F}$ **separates convex closed sets from points** if it separates all pairs, consisting of a convex closed set of X and a point of X outside the convex set.

We are lead to the following results on separation by lower CP functionals.

4.8. Theorem

(1) *If a tcs X is NS_3, and if $Y \subseteq X$ is a compact subspace, then for each pair of closed sets A, $B \subseteq X$ of which A is convex, and which do not intersect on Y, there is a lsc and lower CP functional f of X such that $f \mid Y$ is continuous and f separates $A \cap Y$ and $B \cap Y$.*

(2) *The collection of all continuous lower CP functionals on a tcs X separates convex closed sets from general closed sets, iff each convex closed subset has a neighborhood base in X, consisting of convex closed sets.*

Proof of (1). Let A, $B \subseteq X$ be as announced and define $A(0) = A$; $A(1) = X$. Let $D_0 = \{0,1\}$, and for $n > 0$, let D_n be the set of dyadic numbers in $[0,1]$ which can be expressed with a denominator 2^m for some $m \leq n$. Suppose a convex closed set $A(d)$ of X has been constructed for each $d \in D_n$, such that the following hold (Fig. 1).

(i) If $d, d' \in D_n$ and $d < d'$, then $A(d) \cap Y \subseteq \text{Int } A(d')$.
(ii) If $d \in D_n$ and $d < 1$, then $A(d) \cap Y \cap B = \emptyset$.

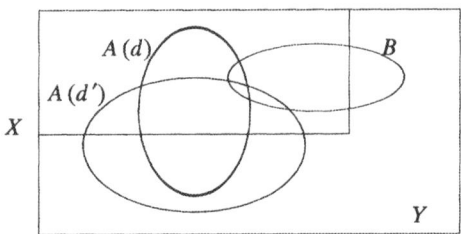

Fig. 1: Construction of a functional

Consider two consecutive elements d_1 and d_2 of D_n, and let $d = \frac{1}{2}(d_1 + d_2)$. As $A(d_1)$ is disjoint from the compact set $Z = (Y \cap B) \cup (Y \setminus \text{Int}(A(d_2)))$, we can operate as in 4.5(2) to find a convex closed neighborhood $A(d)$ of $A(d_1)$ in X disjoint from Z.

This completes the induction, and we arrive at an indexed chain of relatively convex closed sets $A(d) \cap Y$, for $d \in [0,1]$ dyadic, satisfying (i) and (ii). By Lemma 4.2, the corresponding functional

$$f: Y \to [0,1], \quad f(y) = \inf \{d \mid y \in A(d)\}$$

is continuous and lower CP.

There is a second indexed chain of convex closed sets $co^*(A(d) \cap Y)$, and the corresponding functional -- which extends f to X -- is lsc and lower CP by Lemma 4.2. Clearly, f is 0 on $A \cap Y$ and 1 on $Y \cap B$.

Proof of (2). Suppose first that for each convex closed set C and for each open set $O \supseteq C$ there is a convex closed neighborhood of C included in O. The argument to construct a continuous lower CP functional on X, separating C from $X \setminus O$, is similar to (but simpler than) the argument above. We leave the details to the reader.

Conversely, let $f: X \to L$ be a continuous lower CP functional of X such that $\sup f(C) < \inf f(D)$. If these elements aren't neighbors in L, then choose t strictly between them; take $t = \sup f(C)$ otherwise. In either case, a convex closed neighborhood of C disjoint from D is given by $f^{-1}(\leftarrow, t]$. ∎

Let X be a topological space, formally equipped with the free convexity. Then X is NS_4 iff the topological space X is normal. Therefore, part (2) of the above result yields the classical **Urysohn Theorem:** A space X is normal iff each pair of disjoint closed subsets of X can be separated with a continuous function into $\mathbb{R}$.

§4: Topo-convex Separation

The separation properties considered so far are heavily asymmetric: separation of a pair of sets is carried out with a closed convex set and an open concave set. Moreover, the properties do not relate with the axioms S_2, S_3, S_4 considered before. We now present a fairly strong set of axioms meeting with these requirements. Conditions of intermediate strength are described in Topic 4.28.

4.9. Functional separation. A tcs X is said to be:

FS_2, if for each pair of distinct points p, $q \in X$ there exists a continuous CP functional of X separating p and q.

FS_3, if for each convex closed set C and for each point $q \notin C$ there exists a continuous CP functional of X separating C and q.

FS_{3+}, if for each pair, consisting of a convex closed set C and a polytope P disjoint from C there exists a continuous CP functional of X separating C and P.

FS_4, if for each pair of disjoint and non-empty convex closed sets C, D there exists a continuous CP functional of X separating C and D.

Each of these separation properties states that the family of all continuous CP functionals separates a suitable collection of pairs. The label "FS" stands for *Functional Separation*. At particular occasions one may wish to see real-valued functionals. They can easily be produced as follows. Suppose $f : X \to L$ is a continuous CP functional separating two non-empty convex closed sets C, D:

$$\sup f(C) = c < \inf f(D) = d.$$

There is a natural retraction $L \to [c,d]$ which is continuous and CP. Then use the argument in 4.8 to separate the points c, d with a continuous, CP, and real-valued functional. The resulting composition $X \to L \to [c,d] \to \mathbb{R}$ is as desired.

The following are clear:

4.9.1. $FS_4 \Rightarrow FS_{3+} \Rightarrow FS_3 \Rightarrow FS_2$.

4.9.2. $FS_i \Rightarrow S_i + NS_i$ *for* $i = 2, 3, 4$ *whereas* $FS_{3+} \Rightarrow S_4 + NS_{3+}$.

As to the last implication, S_4 follows from the Polytope Screening Characterization I§3.8.

4.9.3. *A subspace of an* FS_2 *space is* FS_2. *A pure subspace of an* FS_3 *space is* FS_3.

4.9.4. *The product of a family of* FS_i *spaces is* FS_i *for each* $i = 2, 3, 3^+, 4$.

4.9.5. *A space* X *is* FS_i *iff* X_w *is* $(i = 2, 3, 3+, 4)$.

The fourth result involves the fact that convex closed sets in a product space must be product sets, cf. 1.4(3).

4.10. Theorem. *Let X be a tcs such that the collection of convex closed sets is normal. If A and B are disjoint, convex and closed, then there is a continuous CP functional* on X separating A and B.

Proof. As in the proof of Theorem 4.8, let $D_0 = \{0,1\}$, and for $n > 0$, let D_n be the set of dyadic numbers in $[0,1]$ which can be expressed with a denominator 2^m for some $m \le n$. We construct two families

$$(A(t))_{t \in D}; \quad (B(t))_{t \in D}$$

of convex closed sets in X with the following properties.

(i) $A(t)$ increases with t and $B(t)$ decreases with t.
(ii) $\forall t \in D: A(t) \cup B(t) = X$.
(iii) $\forall s < t$ in $D: A(s) \cap B(t) = \emptyset$.

At the level D_0 we define $A(0) = A$, $A(1) = X$, $B(0) = X$ and $B(1) = \emptyset$. At the level D_1, consider a convex closed screening $A(\frac{1}{2})$, $B(\frac{1}{2})$ of A, B. Let $n \ge 1$ and suppose $A(t)$ and $B(t)$ have been constructed as in (i)-(iii) for all $t \in D_n$. Let $t \in D_{n+1} \setminus D_n$, say: $t = 2k+1/2^{n+1}$ for some k with $0 \le k < 2^n$. By (iii), the sets $A(k/2^n)$ and $B(k+1/2^n)$ are disjoint. Hence there is a convex closed screening of this pair; the resulting convex closed sets are taken as $A(t)$, $B(t)$.

This completes the inductive construction of the sequences. The desired functional is obtained via Lemma 4.3. ∎

Note that the functional constructed in the above proof is actually equivalent to a real-valued one.

4.11. Corollary. *Let $X \to Y$ be a closed surjective mapping which is CP and CC. If X is FS_4, then so is Y.*

Proof. The collection of all pairs of disjoint convex closed sets in Y is easily seen to be a normal separation structure. ∎

Compare with the situation in I§3.11; there is no direct way to turn CP functionals of X into CP functionals of Y.

4.12. Corollary. *In a compact Hausdorff and closure stable tcs,*

(1) FS_4 *is equivalent to* $S_4 + NS_4$.
(2) FS_{3+} *implies* FS_4.

In particular, a continuous, compact Hausdorff and S_4 tcs is FS_4.

Proof of (1). It has already been observed in 4.9.2 that FS_4 implies S_4 and NS_4. Conversely, let $C_1, C_2 \subseteq X$ be convex closed and disjoint subsets of an S_4 and NS_4 compact Hausdorff space X. As X is topologically normal we obtain open sets $O_i \supseteq C_i$ with $O_1 \cap O_2 = \emptyset$. By Proposition 4.5(3), co^* is USC. Hence there exist convex sets P_i with

$$C_i \subseteq Int P_i \subseteq P_i \subseteq O_i \quad (i = 1, 2).$$

Fix a half-space H with $P_1 \subseteq H$ and $H \cap P_2 = \emptyset$. Then $\overline{H}, \overline{X \setminus H}$ is a convex closed screening of C_1, C_2. The result follows from Theorem 4.10(3).

§4: Topo-convex Separation   333

Note that a continuous convexity is closure stable, establishing the final part of the corollary.

Proof of (2). Let X be FS_{3+}. As observed in 4.9.2, X is S_4 whereas X is NS_4 by Proposition 4.5(3). The result follows from the first part. ∎

It is usually a lot easier to verify S_4 than to verify FS_4. This makes (1) into a particularly useful result.

4.13. Examples. The Uniform Separation Theorem 3.11 implies that a uniform S_4 convex structure is FS_3, and even FS_{3+} if all polytopes are compact. This leads to the first class of examples.

4.13.1. Proposition. *Let V be a locally convex vector space. Then V is* FS_{3+}. ∎

Note that a topological vector space V of dimension ≥ 2 is not NS_4 and hence it is not FS_4. If X is a compact convex subspace of V, and if X is **intrinsically** locally convex, then the hull operator of X is continuous on the hyperspace of finite subsets (cf. 2.7) and X is S_4. It does not follow at once, however, that the entire convex closure operator is continuous. So, unfortunately, we cannot use Corollary 4.12 to show that X is FS_4.

An extension of Proposition 4.13.1 to join spaces with an "intrinsic" topology will be given in the next section.

4.13.2. Proposition. *A Lawson semilattice is* FS_4.

Proof. Note that a Lawson semilattice S is NS_4 by 4.6.1. Unfortunately, S need not be closure stable (cf. 1.9.6) and Corollary 4.12 is not applicable. We give a direct proof based on Theorem 4.10.

Let $C, D \subseteq S$ be disjoint convex closed sets. By 1.13.3, we obtain convex closed disjoint neighborhoods C' of C and D' of D. As in I§3.12.2 it follows that e.g., $\downarrow(C') \cap D' = \varnothing$ and hence that $C' \cap \uparrow(D') = \varnothing$. Now $\uparrow(D') = \uparrow(d)$ for $d = \inf(D')$, and we put

$$H = \uparrow(d); \quad K = CL(S \setminus \uparrow(d)).$$

The last set is convex by Corollary 1.15(1). Then H and K are closed by definition, and it is not difficult to verify that both are convex. Now $D \cap K = \varnothing$ since $D \subseteq Int\, D' \subseteq D' \subseteq H$, and it easily follows that K, H is a convex closed screening of C, D. Then apply Theorem 4.10. ∎

4.13.3. Proposition. *A locally convex median space with compact segments is* FS_4.

Proof. Consider a pair of disjoint and non-empty convex closed sets. The compactness of segments can be used to see that both sets are gated. Then by I§5.11(4) there is a pair of "mutual" gates. As the space in consideration is Hausdorff and locally convex, these points admit disjoint convex neighborhoods. Separate them with a half-space H. By closure stability, the pair built with the closures of H and its complement, provides us with a convex closed screening of the mutual nearest point pair, and hence of the original sets (Screening Lemma, I§5.11(2)). The result follows from Theorem 4.10. ∎

Distributive lattices have been characterized among lattices by the property S_4; cf. I§3.12.3. We conclude from 1.13.5 and the above proposition that a compact distributive lattice is FS_4 iff it is completely distributive. The combination of FS_4 and non-compactness of the underlying space appears to be quite exceptional. Note that a compact locally convex median algebra is a uniform tcs, and FS_{3+} follows at once. In fact, Corollary 4.12 yields the property FS_4.

4.13.4. Proposition. *A compact pospace is FS_3. In fact, every pair of disjoint convex closed sets that can be screened with convex (not necessarily closed) sets, can also be separated with a continuous CP functional.*

Proof. Let X be a compact pospace. We may assume that X has a smallest element $\mathbf{0}$ and a largest element $\mathbf{1}$ (if necessary, add an isolated maximum or minimum and observe that the relative convexity on a subset of a poset is the order convexity, cf. I§4.19). Let $C_i \subseteq X$ for $i = 1, 2$ be non-empty convex closed sets and let D_1, D_2 be a convex screening:

$$C_1 \subseteq D_1 \setminus D_2; \quad C_2 \subseteq D_2 \setminus D_1; \quad D_1 \cup D_2 = X.$$

We assume $\mathbf{0} \in D_1$, whence $\mathbf{1} \in D_2$. Then

$$\downarrow(C_1) = co(C_1 \cup \{\mathbf{0}\}) \subseteq D_1,$$

where the lower set $\downarrow(C_1)$ is closed. Similarly, the upper set $\uparrow(C_2)$ is a closed convex set included in D_2. This allows us to assume without loss of generality that C_1 (resp., C_2) is a lower (resp., upper) set. By 4.6.1 and Theorem 4.8(2), there is a continuous lower CP function $f : X \to [0,1]$ with $C_1 \subseteq f^{-1}(0)$ and $C_2 \subseteq f^{-1}(1)$. For each $t \geq 0$ we conclude that $f^{-1}[0,t]$ is a convex set containing $\mathbf{0}$. Hence this set is a lower set and, in particular, it is a half-space. It follows that f is a CP map.

As every pospace is at least S_3, cf. I§4.14.2, the first part of the proposition follows at once. ∎

We note that a CP functional f, defined on a poset with universal bounds $\mathbf{0}$ and $\mathbf{1}$, is order preserving or order reversing according to whether $f(\mathbf{0}) \leq f(\mathbf{1})$ or $f(\mathbf{1}) \leq f(\mathbf{0})$.

4.13.5. Proposition. *The convex hyperspace of a compact continuous S_4 space is FS_4.*

Proof. Let X be a continuous S_4 space. In regard to Corollary 4.12, convex closed sets can be screened from polytopes by convex closed sets. By Corollary I§3.13, the convex hyperspace is S_4. By Proposition 3.10.4, the convex hyperspace is continuous. Another application of Corollary 4.12 gives the desired result. ∎

If all convex closed sets of the convex hyperspace are assumed to be subbasic intersections, then the above argument on disjoint polytopes can be used for a proof that

§4: Topo-convex Separation

disjoint convex closed sets of the convex hyperspace can be screened with convex closed sets. Then FS$_4$ follows at once (cf. Theorem 4.10). However, this assumption is not known to be valid if X fails to be continuous.

4.13.6. Proposition. *The cone of a compact* FS$_4$ *space is* FS$_4$.

Proof. Let X be a compact convex structure. Its cone ΔX is a closed, continuous, CP and CC image of the "auxiliary" space $X \times [0,1]$. By Corollary 4.11, it suffices to show that the latter is FS$_4$. If $C, D \subseteq X \times [0,1]$ are disjoint convex closed sets, then consider

$$t_C = \sup \pi_2(C); \quad t_D = \sup \pi_2(D).$$

Suppose e.g., $t = t_C \leq t_D$. Then the top t–level C_t of C is disjoint from the t–level D_t of D. Fix a convex closed screening C', D' of these sets in X. Then $C' \times \{t\}$ is disjoint from the closed set D and we obtain $\varepsilon > 0$ with

$$C' \times [t-\varepsilon, t+\varepsilon] \cap D = \emptyset.$$

The following sets form a convex closed screening of $C, D \subseteq X \times [0,1]$:

$$\hat{C} = C_t \times [0, \min\{1, t+\varepsilon\}];$$

$$\hat{D} = D_t \times [0, \min\{1, t+\varepsilon\}] \cup X \times [\min\{1, t+\varepsilon\}, 1]. \quad \blacksquare$$

4.14. Functional generating. Let X be a convex structure and let $\mathcal{F}$ be a collection of functionals on X. The range spaces of the various functionals may be distinct. We say that X (rather, its convexity) is **(functionally) generated by** $\mathcal{F}$ provided the collection

$$\{f^{-1}(\leftarrow, t] \mid t \in f(X); f \in \mathcal{F}\}$$

is a convex subbase of X. For instance, an H-convexity is functionally generated by a set of real-valued linear functionals. To be consistent with our permanent assumption of S$_1$, the family $\mathcal{F}$ should separate the points of the convex structure. If X is functionally generated by $\mathcal{F}$, then each $f \in \mathcal{F}$ is a lower CP function. If the convexity of X is generated by the subbase

$$\{f^{-1}(\leftarrow, t] \mid t \in f(X); f \in \mathcal{F}\} \cup \{f^{-1}[t, \rightarrow) \mid t \in f(X); f \in \mathcal{F}\},$$

then X (rather, its convexity) is **symmetrically generated by** $\mathcal{F}$. This extends the terminology introduced in I§1.8.1.

4.15. Theorem. *If X is* FS$_3$, *then the weak topology of X is symmetrically generated by the collection of all continuous CP functionals $X \to \mathbb{R}$. In particular, X_w is properly locally convex.*

Proof. Each convex closed set of X is the intersection of sets of type $f^{-1}(S)$, where $f: X \to \mathbb{R}$ is a continuous CP functional and $S \subseteq \mathbb{R}$ is a closed lower or upper set. The second part is clear. $\blacksquare$

In the theory of topological vector spaces, the term "weak topology" is used to describe the topology generated by the set of all continuous linear functionals. This agrees with our terminology in case of FS_3 spaces. Let V be a topological vector space which is FS_3. Then V_w is locally convex and hence (cf. 4.13.1) it is FS_{3+}. As observed in 4.9.5, this yields that the original space V is FS_{3+}. We note[2] that an FS_3 vector space need not be locally convex.

We now turn to a study of separation properties of median convexity.

4.16. Theorem. *Let X be a Hausdorff topological convex structure with compact segments and of Helly number ≤ 2. Then the following are equivalent.*

(1) *Each pair of distinct points in X can be screened with convex closed sets.*

(2) *Each pair of disjoint convex closed subsets of X can be screened with convex closed sets.*

(3) *The space X is FS_4 (and hence S_4).*

In either case, each non-empty convex closed subset of X is gated, X is closure stable, the median operator of X is weakly continuous, and X_w is a properly locally convex median space.

Proof. We first show that (1) implies (2). Let $C \subseteq X$ be a non-empty convex closed set, and let $b \in X$. Then the sets bc for $c \in C$ and C meet pairwise. As X is of Helly number ≤ 2, these sets are finitely intersecting. Segments being compact, we find that $\bigcap_{c \in C} bc$ meets C. A point in this intersection is the gate of b in C. Having shown that each non-empty convex closed set is gated in X, it follows from I§5.11(4) that each pair of non-empty convex closed sets has a pair of mutual gates. The result follows directly from (1) and the Screening Lemma, I§5.11(2). The implication (2) $\Rightarrow$ (3) follows from Theorem 4.10, and (3) $\Rightarrow$ (1) is trivial.

As X_w is topologically generated by the convex closed sets of X, each convex closed set is the intersection of closed half-spaces. Consequently, the closed half-spaces of X constitute a subbase for the topology of X_w. If m denotes the median operator of X and if H is a half-space of X, then -- as the reader can easily verify --

$$m^{-1}(H) = (H \times H \times X) \cup (H \times X \times H) \cup (X \times H \times H).$$

Weak continuity of m follows directly, whereas proper weak local convexity follows from Theorem 4.15.

We finally show that X is closure stable. To this end, note that the median algebra X_w is closure stable, cf. 1.9.4. Hence it suffices to show that $Cl(C) \supseteq Cl_w(C)$ (weak closure at the right) for each convex set $C \subseteq X$. Let $x \in Cl_w(C)$ and fix a point $p \in C$. Then for each convex weak neighborhood N of x the sets C, N and xp meet two by two, and hence they have a point in common. By weak local convexity, $x \in Cl_w(C \cap xp)$. Now xp is compact by assumption and is weakly Hausdorff by (1). Hence the original topology and the weak topology agree on xp. We conclude that

---

2. See Klee [1951] for a counterexample.

§4: Topo-convex Separation   337

$$x \in Cl(C \cap xp) \subseteq Cl(C).$$ ∎

If X is a compact locally convex median algebra, then $X_w = X$ by virtue of Proposition 1.19, showing that compact locally convex median algebras agree with compact spaces of Helly number ≤ 2 in which every two points can be screened with convex closed sets. This is used in the next result.

4.17. Corollary. *Let $\mathcal{S}$ be a normal T_1 subbase[3] of the tcs X. Then the superextension $\lambda(X,\mathcal{S})$ is a compact locally convex median space and the median stabilization of the (embedded) subspace X is dense. If $\mathcal{T}$ is any T_1 subbase of a tcs Y and if $f: Y \to X$ is a function such that $f^{-1}(S) \in \mathcal{T}$ for each $S \in \mathcal{S}$, then there is one and only one continuous CP extension $\lambda(Y,\mathcal{T}) \to \lambda(X,\mathcal{S})$ of f.*

Proof. By Proposition I§3.21.2, each pair of distinct points can be screened with subbasic members. The first part of the result now follows from the previous theorem and from Proposition 1.18.1(3).

As for the second part, we recall from Proposition 1.18.1 that there exist natural embeddings $l_X: X \to \lambda(X,\mathcal{S})$ and $l_Y: Y \to \lambda(Y,\mathcal{T})$. Let $m \in \lambda(Y,\mathcal{T})$ and consider the linked system

(*) $\tilde{p} = \{S \in \mathcal{S} \mid f^{-1}(S) \in m\}$.

We first show that $\tilde{p}$ extends to one and only one maximal linked system. It suffices to verify that there is no pair $S_1, S_2 \in \mathcal{S}$ of disjoint sets both meeting all members of $\tilde{p}$. Assume the contrary. Consider a screening with members of $\mathcal{S}$ as follows.

$$S_1 \cap S'_2 = \emptyset = S_2 \cap S'_1; \quad S'_1 \cup S'_2 = X.$$

By assumption, S'_1 and S'_2 are not in $\tilde{p}$, and hence there exist $T_1, T_2 \in m$ with

$$f^{-1}(S'_i) \cap T_i = \emptyset \quad (i = 1, 2).$$

As $f^{-1}(S'_1) \cup f^{-1}(S'_2) = Y$, it follows that T_1 and T_2 are disjoint, a contradiction.

We let $\lambda(f)(m)$ be the unique mls extending $\tilde{p}$. Note that if $m = l_Y(y)$ then $\tilde{p}$ equals $l_X(f(y))$, which is an mls. This shows that the resulting function

$$\lambda(f): \lambda(Y,\mathcal{T}) \to \lambda(X,\mathcal{S})$$

is an extension of f with respect to the indicated embeddings.

To see that $\lambda(f)$ is CP, let m and m_i for $i = 1,..,n$ be mls's in $\mathcal{T}$ and let $p = \lambda(f)(m); p_i = \lambda(f)(m_i)$. We show that if $m \in co\{m_1,..,m_n\}$ then $p \in co\{p_1,...,p_n\}$. Let $\tilde{p}$ denote the linked system constructed as in (*). By Proposition I§1.7.3, the polytope spanned by the points p_i is a subbasic intersection; there is a family $\mathcal{S}_0 \subseteq \mathcal{S}$ with

$$co\{p_1,..,p_n\} = \cap \{S^+ \mid S \in \mathcal{S}_0\}.$$

This reduces the problem to showing that each $S_0 \in \mathcal{S}_0$ meets each $S \in \tilde{p}$. To this end,

---
3.   As agreed in 1.16, and unless stated to the contrary, "subbase" refers to both the topology and the convexity.

assume $S_0 \cap S = \emptyset$ and consider a screening $S'_0, S' \in \mathcal{S}$ of S_0, S:

$$S \cap S'_0 = \emptyset = S_0 \cap S'; \quad S' \cup S'_0 = X.$$

As $S_0 \in p_i$ we find that $f^{-1}(S') \notin m_i$ for all i. Now $f^{-1}(S') \cup f^{-1}(S'_0) = Y$, and hence $f^{-1}(S'_0) \in m_i$ for all i. Consequently $f^{-1}(S'_0) \in m$ and $S'_0 \in \tilde{p}$. However, S'_0 is disjoint from $S \in \tilde{p}$.

We next verify that $\lambda(f)$ is continuous. To this end, let $S_0 \in \mathcal{S}$ and suppose that $l \notin \lambda(f)^{-1}(S_0^+)$. Then the system

$$\{S \in \mathcal{S} \mid f^{-1}(S) \in l\} \cup \{S_0\}$$

is not linked. Take $S \in \mathcal{S}$ such that $f^{-1}(S) \in l$ and $S \cap S_0 = \emptyset$. By normality, there exist $S', S'_0 \in \mathcal{S}$ screening S, S_0:

$$S \cap S'_0 = \emptyset = S_0 \cap S'; \quad S' \cup S'_0 = X.$$

Then $f^{-1}(S'_0) \notin l$ and hence $\lambda(Y, \mathcal{T}) \setminus (f^{-1}(S'_0))^+$ is a neighborhood of l disjoint from $\lambda(f)^{-1}(S_0^+)$.

Our proof that a continuous CP extension of f is unique involves a little trick. Consider the "universal" S_2 quotient

$$q: \lambda(Y, \mathcal{T}) \to Q$$

(cf. I§3.35; there is no need to topologize Q). Let $\hat{Y}$ be the median stabilization of

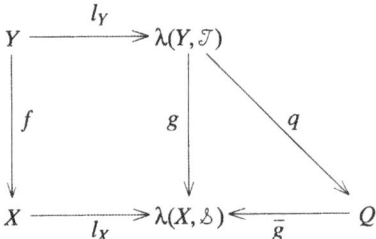

$q \circ l_Y(Y)$. Then $q^{-1}(\hat{Y})$ is a subspace of $\lambda(X, \mathcal{T})$ with Helly number 2. By Proposition 1.18.1, part (3), we find that $q^{-1}(\hat{Y})$ is dense in $\lambda(Y, \mathcal{T})$. Suppose g is a continuous CP extension over $\lambda(Y, \mathcal{T})$ of f. Then g factors through a CP function $\bar{g}: Q \to \lambda(X, \mathcal{S})$. Now $\bar{g}$ is median preserving, and hence its values on the median stabilization $\hat{Y}$ are completely determined by its values on $q l_Y(Y)$. So if g extends f, then it agrees with $\lambda(f)$ on the dense subspace $q^{-1}(\hat{Y})$ and consequently the functions agree on the whole of $\lambda(Y, \mathcal{T})$. ∎

We note that $\lambda(f)^{-1}(S_0^+)$ is **not** simply equal to $(f^{-1}(S_0))^+$, unless f is surjective. The argument on the uniqueness of the extension of f can be simplified if the subbase $\mathcal{T}$ of the domain is normal too. We leave the details to the reader.

4.18. Corollary. *For a topological convex structure X the following are equivalent.*

(1)   *X has the separation property* FS_3.

§4: Topo-convex Separation   339

(2) X_w embeds as a pure subspace of a compact locally convex median space.

Proof. Let X be FS_3. Then there is a collection $\mathcal{F}$ of continuous CP functions $X \to [0,1]$, separating convex closed sets and points. This leads to a CP embedding of convex structures, $X \to [0,1]^{\mathcal{F}}$, mapping $x \in X$ to $(f(x))_{f \in \mathcal{F}}$. By Theorem 4.15, this is also an embedding with respect to the weak topology of X. By construction, the embedding is pure.

Conversely, let X_w be embedded as a pure subspace of a compact locally convex median algebra M. If $C \subseteq X$ is relatively convex closed and if $p \in X \setminus C$, then there is a convex closed set D of M with $D \cap X = C$. In particular, $p \notin D$, whence by Theorem 4.16(3), there is a continuous CP functional separating between D and p. ∎

4.19. Corollary. *There is a functor μ on the category of FS_3 spaces and CP maps into the category of compact locally convex median spaces, together with a natural transformation ν of the identity functor to μ, such that*

(i)   *The function ν_X embeds X as a pure weak subspace of $\mu(X)$.*
(ii) *The median stabilization of $\nu_X(X)$ in $\mu(X)$ is dense.*
(iii) *If $f: X \to Y$ is a continuous CP function of FS_3 spaces X and Y, then $\mu(f): \mu(X) \to \mu(Y)$ is the only continuous CP extension of f.*

Proof. Consider the following equivalence relation "≡" in $\lambda(X) = \lambda(X, \mathcal{IC}_*(X))$.

$a \equiv b$ iff no CP map $\lambda(X) \to [0,1]$ separates a from b.

Let $q = q_X: \lambda(X) \to \mu(X)$ be the corresponding quotient map. The Helly number of $\mu(X)$ is ≤ 2 since it is a CP image of $\lambda(X)$. By construction, each CP map $\lambda(X) \to [0,1]$ factors through $\mu(X)$, and it follows that two distinct points of $\mu(X)$ can be separated by continuous CP functionals. By Theorem 4.16, $\mu(X)$ is a compact locally convex median space. We have a composed CP map

$$\nu_X: X \xrightarrow{l_X} \lambda(X) \xrightarrow{q_X} \mu(X),$$

where l_X is the canonical embedding. That ν_X embeds X as a pure weak subspace of $\mu(X)$ can be seen as follows. A pair of type C, p with $C \subseteq X$ convex closed and $p \in X \setminus C$ can be separated by a continuous CP function $\alpha: X \to [0,1]$ such that, for instance, $\alpha(p) = 0$; $C \subseteq \alpha^{-1}(1)$. By Corollary 4.17, there is a continuous CP extension

$$\lambda(\alpha): \lambda(X) \to \lambda[0,1] \approx [0,1]$$

(the last isomorphism is simply the canonical embedding of $[0,1]$ in its superextension, which is surjective since the interval itself is median). Up to embedding X in $\lambda(X)$, we find that p maps to 0 and that C is included in a convex closed set $\lambda(\alpha)^{-1}(1)$.

As for the density of the median stabilization $med(X)$, note that $q^{-1}(med(X))$ is a Helly number 2 subspace of $\lambda(X)$ including X, and as such it must be dense; cf. Proposition 1.18.1. Hence its q-image is also dense.

As for the final part of the result, let f be as announced. Any CP extension to $\mu(X)$ is median preserving by I§6.11. Hence, building the median stabilization of X step by

340   Chap. III: Topological Convex Structures

step[4], we find at each stage that the values of the extension are canonical. Then f can have at most one CP extension to the median stabilization of X, and hence it has at most one *continuous* CP extension to $\mu(X)$.

It remains to be verified that there exists an extension (the fact that it is unique

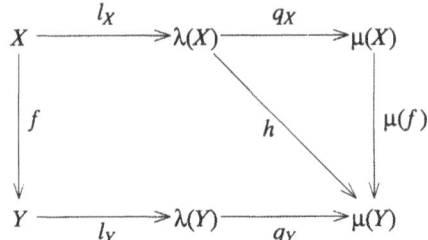

implies that μ is functorial). We have quotient maps

$$q_X: \lambda(X) \to \mu(X); \quad q_Y: \lambda(Y) \to \mu(Y).$$

By Corollary 4.17, the map

$$q_Y \circ l_Y \circ f : X \to \mu(Y)$$

extends to a CP map

$$h : \lambda(X) \to \lambda(\mu(Y)) \approx \mu(Y)$$

making the trapezium part of the diagram commutative. Two points of $\lambda(X)$ with a different h-image can be separated by a continuous CP functional into $[0,1]$ (compose with a functional of $\mu(Y)$), and hence h factors through a map $\mu(X) \to \mu(Y)$ which is as desired. ∎

The functor μ can be seen as a compactification procedure for FS_3 spaces. The result is sharply contrasting with Proposition 3.16, stating that Euclidean space of dimension > 1 cannot be embedded as a *convex* dense subspace of a compact continuous space. For an alternative approach to the construction of μ, consult Topic 4.30.

The existence of a normal T_1 subbase can be regarded as another separation property; cf. Proposition I§3.21.1. It can be characterized in terms of embedding into median spaces. Comparing with Corollary 4.18, note the absence of "subspace purity" in the next result.

4.20. Proposition. *For a compact tcs X the following are equivalent.*

(1) X has a normal, T_1, convex subbase consisting of closed sets.
(2) X can be embedded in a compact locally convex median algebra.

Proof of (1) ⇒ (2). Let $\mathcal{S}$ be a normal, T_1, convex subbase of closed sets. Compactness of X and normality of $\mathcal{S}$ imply that $\mathcal{S}$ is a closed subbase for X. By Proposition 1.18.1, X embeds in the superextension $\lambda(X, \mathcal{S})$ which, by Corollary 4.17, is a compact

---
4.   See I§6.34 for a detailed description.

§4: Topo-convex Separation   341

locally convex median algebra.

Proof of (2) ⇒ (1). Suppose X embeds in a compact locally convex median algebra M. Since the closure of the median stabilization $med(X)$ is median stable, we may and will assume that $M = Cl(med(X))$. Let $\mathcal{S}$ consist of all sets $S \subseteq X$ such that there is a CP map $f : M \to \mathbb{R}$ and a closed half-space $K \subseteq \mathbb{R}$ with $S = f^{-1}(K) \cap X$. By construction, the involved functionals separate polytopes and points of X, and hence $\mathcal{S}$ is a convex subbase of X. Let $S_1, S_2 \in \mathcal{S}$ be disjoint. Without loss of generality, there are CP maps $f_i : M \to \mathbb{R}$ with
$$S_i = f_i^{-1}(-\infty, 0] \cap X \quad (i = 1, 2).$$
The corresponding subsets $H_i = f_i^{-1}(-\infty, 0]$ of M are disjoint. Indeed, suppose $H_1 \cap H_2 \neq \emptyset$. As M is a normal topological space, there exist open sets $U_i \supseteq H_i$ with $U_1 \cap U_2 \subseteq M \setminus X$. As f_i is a closed map, there exists a $t_i > 0$ with $O_i = f_i^{-1}(-\infty, t_i) \subseteq U_i$. The open set $O_1 \cap O_2$ meets the dense set $med(X)$. However, by II§4.25.7 two halfspaces of $med(X)$ with a non-empty intersection must also intersect on X, a contradiction.

Having shown that H_1, H_2 are disjoint convex closed sets of M, we use the property FS$_4$ to obtain a continuous CP functional $f : M \to \mathbb{R}$ separating these sets, say
$$H_1 \subseteq f^{-1}(-\infty, 0); \quad H_2 \subseteq f^{-1}(0, \infty).$$
The resulting subbasic pair of sets
$$S'_1 = f^{-1}(-\infty, 0] \cap X; \quad S'_2 = f^{-1}[0, \infty) \cap X,$$
is a screening in X of S_1 and S_2. ∎

See 4.27 for additional information on (not necessarily pure) embedding in compact locally convex median algebras.

Further topics

4.21. Functionals

4.21.1. Show that each continuous (lower) CP functional has a maximal refinement which is continuous and (lower) CP.

4.21.2. Show that each continuous CP functional of a topological vector space can be refined by a continuous (real-valued) linear functional.

4.21.3. (compare Michael [1951] for the topological data) Let X be a compact Hausdorff space, let L be a complete totally ordered space, and let $f : X \to L$ be continuous and CP. Show that the following induced functions are continuous and CP.
$$f_+ : \mathcal{IC}_*(X) \to L, \ C \mapsto \sup f(C); \quad f_- : \mathcal{IC}_*(X) \to L, \ C \mapsto \inf f(C).$$

4.21.4. (cf. 4.13.5) Use the previous result and Proposition 3.10.4 for a direct proof that if X is continuous and FS$_4$ then $\mathcal{IC}_*(X)$ is FS$_4$.

4.21.5. Let X be compact, let M be a compact locally convex median algebra including X as a pure subspace with $Cl\,(med\,(X)) = M$. If $f: X \to L$ is an lsc and lower CP functional into a complete chain L, then there is an extension $M \to L$ of f which is lsc and lower CP. Hint: use Lemma 4.2.

4.22. Functional generating and embedding. Throughout, Let X be symmetrically generated by a collection $\mathcal{F}$ of functionals separating the points of X.

4.22.1. Show that the function
$$F: X \to \mathbb{R}^{\mathcal{F}}: x \mapsto (f(x))_{f \in \mathcal{F}}$$
is a CP embedding of X into the product space $\mathbb{R}^{\mathcal{F}}$.

4.22.2. Suppose that for each $f_1, f_2 \in \mathcal{F}$, the sum $f_1 + f_2$ is a CP function. Show that each linear combination of members of $\mathcal{F}$ with positive coefficients is a CP functional.

4.22.3. (Choquet and Meyer [1963]) With the additional assumption of part (2), Show that the function F, defined in part (1), is an embedding of X into the *vector space* $\mathbb{R}^{\mathcal{F}}$. Hint: by (2), each function of type $(f_1,...,f_n): X \to \mathbb{R}^n$, where $f_i \in \mathcal{F}$ and $\mathbb{R}^n$ carries the standard convexity, is CP.

4.23. Strong convexity (compare van de Vel [1983d]). A subset C of a tcs X is called *strongly convex* provided $co^*(K) \subseteq C$ for each compact set $K \subseteq C$.

4.23.1. Show that in each of the following spaces, all convex open sets are strongly convex.
(i) Locally convex vector spaces.
(ii) Locally convex median spaces with compact segments.
(iii) Lawson semilattices.
(iv) Partially ordered spaces.

4.23.2. If each factor of a product has the property that all convex open subsets are strongly convex, then the same property holds for the product space.

4.23.3. If a compact Hausdorff space X has the property that its convex open sets are strongly convex, then so does each CP + CC continuous image of X.

4.23.4. Let X be a JHC and NS_3 space, such that the convex closure of a compact set is compact. Show that every convex open subset of X is strongly convex.

4.23.5. Problem. Find general conditions under which each convex open set of a space is strongly convex. For instance, can the condition of join-hull commutativity be removed from (4)? Or, can the condition concerning the convex closure of compact sets be removed?

§4: Topo-convex Separation

4.24. Functionally convex open sets. Let X be a topological convex structure. A set $C \subseteq X$ is *functionally convex* provided there is a continuous lower CP functional $f : X \to L$ such that $C = f^{-1}(D)$ for some convex subset of L.

4.24.1. Show that a functionally convex open set is strongly convex (cf. 4.23).

4.24.2. Let X be a compact NS_4 space and let $O \subseteq X$ be a strongly convex open F_σ-set. Show that O is functionally convex open.

4.24.3. Show that all convex open sets in a topological vector space are functionally convex open.

4.24.4. Problem. Is it true that all convex open sets of a Lawson semilattice are functionally convex open? And in a compact locally convex median algebra?

4.24.5. Let X be a compact FS_4 space, and let $H \subseteq X$ be a half-space such that H is functionally convex open and $X \setminus H$ is functionally convex closed. Prove that there is a totally ordered set L, a CP map $f : X \to L$, and an open half-space $P \subseteq L$ with $H = f^{-1}(P)$.

4.25. Compact convex sets. Let X be a JHC topological convex structure with compact segments.

4.25.1. If X is NS_3 and compact, and if $C_1, \ldots, C_n$ are compact convex subsets, then $co \left(\cup_{i=1}^n C_i \right)$ is compact.

4.25.2. If X is NS_{3+}, and if $C_1, \ldots, C_n$ are compact convex, then $co \left(\cup_{i=1}^n C_i \right)$ is weakly compact (note that this hull is closed if the weak topology is T_2). In particular, X has weakly compact polytopes.

4.26. Dense median hull. Let M be a locally convex median algebra with compact segments, and let $X \subseteq M$ be such that $med(X)$ is dense in M.

4.26.1. Show that for each open half-space $O \subseteq M$ the set $co(O \cap X)$ is dense in O.

4.26.2. Deduce that if the subspace X is pure and closure stable, then $\overline{O} \cap X = Cl_X(O \cap X)$ for each open half-space $O \subseteq M$. Compare with the concept of "continuous position" in IV§2.21 below.

4.27. Embedding in compact median algebras

4.27.1. Show that a topological convex structure embeds weakly in a compact locally convex median algebra iff each disjoint pair, consisting of a polytope and a point, can be separated with a continuous (real) CP functional. This property could be described by the label FS_2+.

4.27.2. Note that $FS_3 \Rightarrow FS_{2+} \Rightarrow FS_2$ and that FS_{2+} is equivalent to the existence of a normal T_1 convex subbase of closed sets. Give an example of a compact space which is FS_{2+} but not FS_3.

Problem. Is there an example of an FS_2 space which is not FS_{2+}?

4.27.3. Let $\mathcal{C}$ be a symmetric H-convexity. Almost by definition, $\mathcal{C}$ is FS_{2+}. Show that the canonical (symmetric) subbase of $\mathcal{C}$ is normal and, that $\mathcal{C}$ is FS_3 provided it is finitely generated.

Problem. Is every symmetrically generated H-convexity FS_3?

4.28. Separation and screening. The following are separation properties of intermediate strength. A space is said to be:

CCS_2, if distinct points can be screened with convex closed sets.

CCS_{2+}, if each pair, consisting of a polytope and a point in its complement, can be screened with convex closed sets.

CCS_3, if each pair, consisting of a convex closed set and a point in its complement, can be screened with convex closed sets.

CCS_{3+}, if each pair, consisting of a convex closed set and a polytope disjoint from it, can be screened with convex closed sets.

CCS_4, if each pair of disjoint convex closed sets can be screened with convex closed sets.

The label "CCS" refers to the term *Convex Closed Screening*.

4.28.1. Note that

$$CCS_4 \Rightarrow CCS_{3+} \Rightarrow CCS_3 \Rightarrow CCS_{2+} \Rightarrow CCS_2;$$
$$CCS_i \Rightarrow S_i \ (i = 2, 3); \quad CCS_{2+} \Rightarrow S_3; \quad CCS_{3+} \Rightarrow S_4.$$

4.28.2. A CCS_3 space is weakly locally convex.

4.28.3. $FS_i \Rightarrow CCS_i \Rightarrow NS_i$ for $i = 2, 2+, 3, 3+, 4$. Moreover, $CCS_4 \Leftrightarrow FS_4$.

4.28.4. Problem. Is there an example of a space which is CCS_i but not FS_i for $i = 2, 2+, 3, 3+$?

4.29. Superextensions of topological spaces. Let X be a compact Hausdorff space and let $\lambda(X) = \lambda(X, \mathcal{F}_*(X))$ be the superextension of X with respect to the collection of all closed sets.

4.29.1. (van de Vel [1983d]) Show that the closed half-spaces of $\lambda(X)$ are exactly the sets of type A^+, for $A \subseteq X$ closed. Note that (up to embedding) the singleton half-spaces are exactly the extreme points of X. Deduce that two compact Hausdorff spaces are homeomorphic iff their superextensions are isomorphic (as topological convex structures), and that the superextension of a compact space is the convex closure of its extreme points.

§4: Topo-convex Separation 345

Remark. It follows from certain results in the next section that a set-theoretic CP isomorphism between superextensions of compact Hausdorff spaces is also a topological isomorphism (a homeomorphism).

4.29.2. (van Mill and van de Vel [1979b]; cf. 2.28) Show that $\lambda(X)$ is the fixed point set of the transversality operator

$$\bot: \mathcal{T}_*(\mathcal{T}_*(X)) \to \mathcal{T}_*(\mathcal{T}_*(X))$$

(compare Monjardet's [1975] notion of "ipsoduality"; cf. I§6.34.3). In addition, show that $\lambda(X)$ is a subspace of the iterated hyperspace $\mathcal{T}_*(\mathcal{T}_*(X))$. Conclude that $\lambda(X)$ is metrizable iff X is (compare Verbeek [1972]).

4.30. Functorial median extensions

4.30.1. Let μ' be a functor on the category of FS_3 spaces into the category of compact locally convex median algebras and let ν' be a natural transformation of the identity functor to μ'. Assume the following properties.

(i) if X is an FS_3 space then ν'_X is an embedding of X in $\mu'(X)$ with a dense median stabilization.
(ii) If $f: X \to Y$ is a CP map, then $\mu'(f): \mu'(X) \to \mu'(Y)$ is a continuous CP extension of f with respect to the canonical embeddings of X and Y.

Show that there is a natural isomorphism $\mu \approx \mu'$ commuting with the natural embeddings. Note that the purity of $\nu'_X(X)$ in $\mu'(X)$ follows from this.

4.30.2. For an FS_3 space X let $\mathcal{F}(X)$ denote the collection of all $[0,1]$-valued continuous CP functionals of X. Consider the canonical embedding of X into $[0,1]^{\mathcal{F}(X)}$, and let $\tau(X)$ be the closure of its median stabilization. Verify that each CP map $X \to X'$ extends uniquely to a CP map $\tau(X) \to \tau(X')$. Show that $\tau(X)$ is (naturally) isomorphic with $\mu(X)$ by an isomorphism which preserves the embeddings $X \to \tau(X)$ and $X \to \mu(X)$.

4.30.3. If $f: X \to Y$ is a surjective CP map between FS_3 spaces, then $\mu(f): \mu(X) \to \mu(Y)$ is surjective. Give counterexamples to the above statement in case "surjective" is replaced throughout by "injective", or by "bijective".

4.30.4. Let X be a compact FS_3 space. Show that if $H \subseteq \mu(X)$ is a closed half-space then $co^*(H \cap X) = H$.

4.31. Inverse limits.
Let $\pi_{ij}: X_i \to X_j$ for $i \leq j$ in I be an inverse system of compact Hausdorff convex structures, and let X be the inverse limit (both in a convex and a topological sense). All bonding maps π_{ij} are assumed to be continuous, CP, CC and surjective.

4.31.1. It is a topological result that X is compact and Hausdorff, and that the limit projections $\pi_i: X \to X_i$ are continuous and surjective (cf. Engelking [1977]). Show that

π_i is CC (it is CP by construction). Conclude that X is S_4 provided all spaces X_i are.

4.31.2. Show that each of the properties NS_i and FS_i for $i = 2, 3, 3+, 4$, as well as closure stability and continuity of the convex closure operator, is inherited by X.

4.31.3. Show that $\mu(X) \approx \lim_{\leftarrow} \mu(X_i)$, i.e., μ is a *continuous functor*. Hint: the canonical map $\mu(X) \to \lim_{\leftarrow} \mu(X_i)$ is an isomorphism provided each continuous CP functional of X extends to a continuous CP functional of $\lim_{\leftarrow} \mu(X_i)$. Model the functional with chains of closed sets and use Lemma 4.3.

4.31.4. Let X be a compact locally convex median space. Show that X is isomorphic to the inverse limit of the system of all polytopes in X with, as bonding maps, the restricted gate projections.

4.32. Topological properties

4.32.1. Let the tcs X be FS_3. Show that the topological space $\mu(X)$ is connected / compact metrizable / separable provided X is.

4.32.2. Give an example of a compact totally disconnected FS_3 space X with a connected median extension $\mu(X)$.

4.32.3. Show that

(*) $$\mu(\prod_{i \in I} X_i) \approx \prod_{i \in I} \mu(X_i)$$

Hint: let $X = \prod X_i$. It suffices to show that if a half-space $H \subseteq X$ extends to a closed half-space of the left-hand set of (*), then it extends to a closed half-space of the right-hand set of (*). Use Theorem 1.4.

4.33. Special metrics (with reference to parts (1), (2) and (3), van Mill and van de Vel [1978]). The concept of an adapted metric for a topological convex structure has been defined in 3.29.

4.33.1. If d is any metric on a compact space X, then the metric inherited by $\lambda(X) = \lambda(X, \mathcal{I}_*(X))$ from the iterated hyperspace of X (cf. 4.29) is adapted.

4.33.2. Let X be a compact median space and let the metric d be adapted to X. Note that X is locally convex. Let C be a non-empty convex closed set. Show that the gate map $p: X \to C$ has the following properties.

(i) p is a *metric nearest point function*, that is,
$$\forall x \in X: d(x,C) = d(x,p(x))$$
(in general, there may be points of C other than $p(x)$ which are nearest to x).

(ii) p is a *metric contraction*, that is,

§4: Topo-convex Separation

$$\forall x_1, x_2 \in X: d(p(x_1),p(x_2)) \leq d(x_1,x_2).$$

4.33.3. Show that if a compact locally convex median algebra has a metrizable underlying space, then it admits an adapted metric. Also, if X is a compact FS_3 space, then there is a metric d generating the topology of X and such that all open or closed cells (with singleton centers) are convex.

It appears from the results in the previous section that the existence of an adapted metric is a rather heavy condition, which cannot be met by many FS_3 spaces. One should not expect the last result to be improved to cells with non-singleton centers.

4.33.4. (compare Carruth [1968]) If X is a compact pospace with a metrizable topology, then there is a *radially convex metric* d on X, i.e.,

$$\forall x_1, x_2, x \in X: x_1 \leq x \leq x_2 \Rightarrow d(x_1,x_2) = d(x_1,x) + d(x,x_2)$$

(*Urysohn-Carruth Metrization Theorem*). Hint: embed the FS_3 space X into $\mu(X)$ and use 4.32.1 in combination with part (3) above.

4.33.5. (compare Hofmann [1970]) Let X be a compact locally convex median algebra and let d be an adapted metric on X. Show that for each quintuple $a, b, c, u, v \in X$ the following inequality is valid.

$$\max\{d(a,u),d(b,v)\} \geq d(acu,bcv).$$

Hint: use part (2) on $X \times X$. Deduce that each Lawson semilattice S admits an *ultrametric d*, that is,

$$\forall a, b, x, y \in S:\ d(a \wedge x, b \wedge y) \leq \max\{d(a,x),d(b,y)\}.$$

4.33.6. By a famous result of Bing [1949], each compact, metrizable, connected, and locally connected space X admits a convex metric. Assuming this fact, show that $\lambda(X) = \lambda(X, \mathcal{T}(X))$ admits a metric which is both convex and adapted.

Problem. Let X be a compact, connected and locally convex median algebra. By 1.29, X is locally connected. Does X admit a metric which is simultaneously adapted and convex?

Notes on Section 4

In the literature on the subject, lower (upper) CP functionals are met under various names; the terms "quasi convex" and "quasi concave" are the most frequent ones. This terminology refers to the well-known concepts of convex and concave functionals in convex analysis. The general approach to functionals with the aid of indexed chains of sets (Lemmas 4.2 and 4.3) seems to be new. The method has been inspired by the traditional proof of Urysohn's Theorem in topology (cf. Engelking [1977, p. 63]) and by the construction of Minkowski functionals in a locally convex vector space (cf. the remarks following Lemma 4.2).

There are many possibilities to describe topo-convex separation properties; those treated here are not the only ones. The neighborhood separation property NS_3 is implicit in Jamison's [1974] study of continuity of the convex closure operator, where a result comparable with Theorem 4.8 was obtained. The property is also considered by Keimel and Wieczorek [1988]. Property NS_4 in compact pospaces (cf. 4.6.1) is due to Nachbin [1965]. The functional separation axioms FS_i and the related separation structures were introduced by van de Vel [1983d], where Theorem 4.10 and part of Corollary 4.12 have been obtained. For vector spaces, the separation property FS_3 has implicitly been considered by Bourgin [1943]. In vector spaces, the property is implied by -- but not equivalent to -- local convexity, as Klee [1951] has shown. The FS_3 property of compact pospaces, Proposition 4.13.4, corresponds with a result on separation in terms of order preserving functionals. Compare with Nachbin's [1965] results on "normally ordered" pospaces.

The conclusion of Theorem 4.15 (on functional generating of convexities) has often been involved in the definition of convex structures. This is the case, for instances, in papers on synthetic Choquet theory (see Section IV§2 for this subject). Theorem 4.16, on the equivalence of various topo-convex separation properties in spaces of Helly number ≤ 2, is a result of the author in [1983d]. Verbeek [1972] describes the procedure in Theorem 4.17 to extend CP maps to superextensions. Uniqueness of such extensions has been obtained by van Mill and van de Vel [1978] under somewhat restricted circumstances. Corollary 4.19, on (functorial) pure extension of FS_3 spaces to median spaces, has been suggested to the author by E. Wattel (communication). Together, these results present the viewpoint of compactification theory in a context of topological convex structures. The application of median compactification to special metrics (cf. Topic 4.33, parts 4 and 5) seems to be new. Proposition 4.20, on the equivalence between the existence of normal subbases and the existence of embeddings in compact locally convex median algebras, is new.

5. Intrinsic Topology

One may wonder how flexible a topology is in the framework of a topological convex structure. Partial answers are provided in the present section, where it is shown that in median spaces, or in finite-dimensional Bryant-Webster (BW) spaces, or in unconditionally complete continuous semilattices, the convexity component of the structure gives an accurate idea of how the accompanying topology looks like.

We will consider two procedures to construct a topology from convexity data: the first involves the core of a convex set, and the second involves completeness of sets relative to the base-point orders. This leads to a "core" topology based on convex open sets and to an "intrinsic weak" topology with a subbase of relatively complete convex sets.

The core topology of a complete BW space without boundary is Hausdorff, locally convex and FS_{3+}; its polytopes are compact, and its hull operator is continuous. A finite-dimensional topology with these (or even weaker) properties is unique and the standard convexity of $\mathbb{R}^n$ gives back the Euclidean topology.

The intrinsic weak (IW) topology leads to several standard topologies as well: the Lawson topology of a continuous semilattice, the interval topology of a complete and distributive lattice and, again, the Euclidean topology of $\mathbb{R}^n$. For median algebras, the weak topology of an FS_4 space is invariably equal to the intrinsic weak topology. Hence, for *compact* locally convex median algebras there is simply no choice left of another compact topology. To round up the picture, the core- and intrinsic weak topologies are identical for finite-dimensional, complete BW spaces without boundary, and for polytopes in complete median algebras.

Unless stated to the contrary, *all spaces in consideration are* S_1.

A theme common to all topics of this section is the one of continuous posets, which is a generalization of continuous semilattices.

5.1. Continuous posets. In a poset X there is a so-called *way-below relation* $\ll$ defined as follows.

$x \ll y$ iff for each up-directed set $D \subseteq X$ with $y \leq \sup D$,
there exists a point $d \in D$ with $x \leq d$.

Note that $x \ll y$ implies $x \leq y$: just take $D = \{y\}$. One should not confuse $\ll$ with a strict order: it is well possible that $x \ll x$ for some $x \in X$, e.g., if x is a (local) minimum. On the other hand, if X happens to be totally ordered, then $x < y$ obviously implies that $x \ll y$. In fact, this is already true when X is a tree. The relation $x \ll y$ is interpreted as *robust* approximation of y by x. The larger x, the *sharper* the approximation. The following are easy to verify:

5.1.1. $x' \leq x \ll y \leq y'$ *implies* $x' \ll y'$.

5.1.2. *If* $x_i \ll y$ *for* $i = 1, 2$, *and if* $x = x_1 \vee x_2$ *then* $x \ll y$.

The poset X is *continuous* provided it is (conditionally!) complete and for each point

$a \in X$, the set $\{x \mid x \ll a\}$ is up-directed with supremum a. Thus, in a continuous poset it is possible to have arbitrarily sharp and robust approximations of each point.

5.1.3. *If $a \ll b$ in a continuous poset, then there exists a point x with $a \ll x \ll b$.*

Proof. Let D be the set of all points u with $u \ll v \ll b$ for some v. By the above formulae and the definition of continuity, the set D is up-directed. We show that the supremum p of D must be b. If not, then $b \not\leq p$ and hence there is a point v with $v \ll y$ and $v \not\leq p$. Similarly, there is a point u with $u \ll v$ and $u \not\leq p$. But $u \in D$, which is a contradiction. As $a \ll b$ and sup $(D) = y$, we find $x \leq u \ll v \ll b$ for some u, v. ∎

Let X be an S_2 convex structure. Each point $b \in X$ gives rise to a partial order $\leq_b$ on X, and hence to a way-below relation denoted by $\ll_b$. It follows from an observation above that every complete tree order is continuous. Considering the exposition in I§7.1, the following is an immediate consequence.

5.1.4. *In a complete and S_4 space with decomposable segments, all base-point orders are continuous.* ∎

Continuity of base-point orders in median spaces is characterized in Theorem 5.18 below.

5.2. Core of a set. Let X be an interval space, and let $N \subseteq X$. A point $p \in X$ is called a *core point* of N provided $p \in N$ and for each non-empty collection $A \subseteq X$, if $\bigcap_{a \in A} ap = \{p\}$ then there is a finite set $F \subseteq A$ with $\bigcap_{a \in F} ap \subseteq N$. The *core of a set* $N \subseteq X$ is defined to be the collection of all core points of N. We denote it by a straightforward *core* (N). As usual, the definition extends to general convex structures by considering the segment operator. In general, the core of a convex set need not be convex. It is easy to produce a counterexample in a tree with three branches emanating at one point. See 5.31 for other details about the core operator.

Here are a few general facts and examples.

5.2.1. Proposition. *Let $N_i \subseteq X_i$ for each $i \in I$. The following holds in the product space $\prod_{i \in I} X_i$.*

$$\text{core}\left(\prod_{i \in I} N_i\right) = \begin{cases} \prod_{i \in I} \text{core}(N_i), & \text{if } N_i = X_i \text{ for all but finitely many } i \in I; \\ \varnothing, & \text{otherwise.} \end{cases}$$

Proof. The first formula can be seen as follows. If A is a subset of the product space then by the Product Polytope Formula, I§1.10.3, we have

(*) $\qquad \bigcap \{ap \mid a \in A\} = \prod_{i \in I} \bigcap_{a \in A} a_i p_i,$

where $a = (a_i)_{i \in I}$ and $p = (p_i)_{i \in I}$. If the left hand intersection equals $\{p\}$, then each factor at the right is a singleton, and hence for each i with $N_i \neq X_i$ there is a finite set F_i with $\bigcap_{a \in F_i} a_i p_i \subseteq N_i$. Then $F = \cup_i F_i$ (union taken over those i with $N_i \neq X_i$) is a finite set with $\bigcap_{a \in F} ap \subseteq N$. This establishes the inclusion "⊇". The other one is easy.

§5: Intrinsic Topology   351

As for the second formula, suppose $a_j \in X_j \setminus N_j$ for each j in an infinite index set $J \subseteq I$. Let $p \in \prod_{i \in I} N_i$ and let $F \subseteq J$ be non-empty and finite. We define a point $a(F)$ as follows.

$$a(F)_i = \begin{cases} p_i, & \text{if } i \in (I \setminus J) \cup F; \\ a_j, & \text{if } i \in J \setminus F. \end{cases}$$

All factor spaces being S_1, it follows from the formula (*) above that $\cap_F co\{p, a(F)\} = \{p\}$. The intersection of a finite number of segments $co\{p, a(F_k)\}$ for $k = 1,..,m$ equals $co\{p, a(F)\}$, where $F = \cup_{k=1}^m F_k$. Hence it is not included in $\prod_{i \in I} N_i$. ∎

5.2.2. Proposition. $core\ (\cap_{i=1}^n N_i) = \cap_{i=1}^n core\ N_i$.   ∎

The argument is straightforward. The restriction to finitely many intersecting sets is essential, as one can see from the previous proposition.

In a BW space, owing to the Ramification Property, two segments ending in p are either included in a single line or meet in p only. The following is a direct consequence.

5.2.3. Proposition. *Let X be a BW space, and let $N \subseteq X$. Then $p \in N$ is a core point of N iff for each $a \in X$ there is a point $y \in a \circ p$ with $yp \subseteq N$.*   ∎

This corresponds with the usual notion of "core" in real vector spaces. For finite-dimensional BW spaces the situation is as follows.

5.2.4. Proposition. *Let X be a BW space and let $F = \{a_1,..,a_n\}$ be an affine basis of X. Then the core of $co(F)$ includes $a_1 \circ .. \circ a_n$, and equality holds if X has no boundary.*

Proof. Let $p \in a_1 \circ .. \circ a_n$ and let $a \in X = aff(F)$. We show that $yp \subseteq co(F)$ for some $y \in a \circ p$ (cf. Fig. 1). If $a \in co(F)$ then we are done. In the remaining case, by the Affine Hull Formula I§7.11, we have $a \in v/u$ for some $u \neq v$ in $co(F)$. If $p = u$ then $y = v$

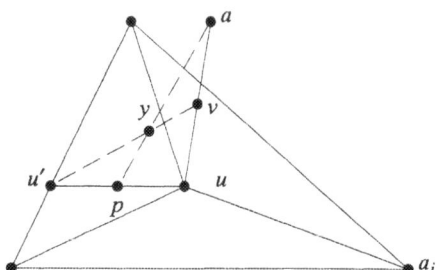

Fig. 1: Core formula for polytopes

will do. Assume $p \neq u$. By the Subdivision Property II§1.2(2) we have

$$co(F) = \bigcup_{i=1}^n co(\{u\} \cup F_i),$$

where F_i denotes F minus a_i. Then, for instance, $p \in co(\{u\} \cup F_i)$. By join-hull commutativity there is a point $u' \in co(F_i)$ with $p \in uu'$. As $p \notin co(F_i)$ we even have $p \in u \circ u'$. By the Pasch Property, there is a point $y \in u' \circ v \cap a \circ p$. In particular $y \circ p \subseteq co(F)$, showing that $p \in core(F)$.

We finally show that $core(F) \subseteq a_1 \circ .. \circ a_n$ provided X has no boundary. To this end, take $p \in co(F) \setminus a_1 \circ .. \circ a_n$. Then $p \in co(F_i)$ for some i and $F_i = \{a_1,..,a_i^\wedge,..,a_n\}$. Let $c \in a_1 \circ .. \circ a_n$ and let $d \in p/c$ be distinct from p. If $d \circ p$ contains a point $v \in co(F)$, then by the hull formula of I§7.7 we find that

$$p \in c \circ v \subseteq a_1 \circ .. \circ a_n,$$

a contradiction. We conclude that $p \notin core(F)$. ∎

5.3. Theorem. *Let X be a complete and S_4 convex structure, let $p \in X$ and let $N \subseteq X$. Then the following are equivalent.*

(1) *p is a core point of N.*
(2) *For each $b \in X$, if $D \subseteq X$ is up-directed in $(X, \leq_b)$ and if $\sup_b D = p$ then there is a $d \in D$ such that $dp \subseteq N$.*
(3) *If $D \subseteq X$ is down-directed in $(X, \leq_p)$ and if $\inf_p D = p$, then there is a $d \in D$ such that $dp \subseteq N$.*

If, in addition, all base-point orders are continuous, then these statements are equivalent to

(4) *For each $b \in X$ there is a point c such that $c \ll_b p$ and $pc \subseteq N$.*

Proof of (1) ⇒ (2) and (2) ⇒ (3). Let $p \in core\ N$, and let $D \subseteq X$ be b-up-directed with $b - \sup D = p$. By Theorem I§4.7 and Proposition I§5.2(2), the orders $\leq_b$ and $\leq_p$ are mutually inverse on bp, whence $p = p - \inf_p D$, that is: $\cap_{d \in D} bp = \{p\}$. As p is a core point of N, we obtain a finite set $F \subseteq D$ with $\cap_{d \in F} dp \subseteq N$. Now D is up-directed with respect to b and hence it is down-directed with respect to p, so that some $d \in D$ is a lower bound of F in the poset $(X, \leq_p)$. Hence $dp \subseteq N$. The second implication obtains from a similar argument on b-suprema and p-infima.

Proof of (3) ⇒ (1). By Theorem I§5.20, each base-point order $\leq_p$ is a semilattice order. Hence each non-empty set $A \subseteq X$ can be turned into a down-directed set by adding all points of type $\inf F$ for $F \subseteq A$ finite.

For the remainder of the proof, we assume that $\leq_b$ is a continuous partial order. By Theorem I§5.20 and by 5.1.2 the set

$$D = \{x \mid x \ll_b p\}$$

is up-directed from the viewpoint of b, and we have $b - \sup D = p$ by continuity of $\leq_b$. Hence, assuming (2), we conclude that $dp \subseteq N$ for some $d \in D$.

Conversely, let N satisfy (4). Take $b \in X$, and let D be up-directed in $(X, \leq_b)$, such that $p = b - \sup D \in N$. By assumption there is a point c with $c \ll_b p$ and $pc \subseteq N$. By the definition of "way-below", there is a point $d \in D$ such that

§5: Intrinsic Topology   353

$c \leq_b d \leq_b p$.

By Proposition I§5.2(2), we find that $d \in cp$, whence $dp \subseteq pc \subseteq N$. ∎

5.4. Core topology. The *core topology* of a convex structure is the one generated by the open base of all convex sets O with $O = core(O)$. This gives a properly locally convex topology, but in the absence of topological separation properties, this need not be significant. As a consequence of Theorem 5.3(2) we have:

5.4.1. Proposition. *In a complete and S_4 space X a convex set O equals its core iff $X \setminus O$ is complete in X. Consequently the core topology of X is generated by the closed base of all concave sets which are complete in X.* ∎

That the family of convex sets, each equal to their own core, acts as an open base for a topology, follows from 5.2.2. Clearly, the union of a family of sets, each equal to its own core, again has this property. Consequently,

5.4.2. Proposition. *A convex set C is core-open iff $C = core(C)$.* ∎

If $\mathbb{R}^2$ is given the standard convexity, then the complement of the set of all points (t,t^2) with $t \in \mathbb{Q}$ equals its own core. However, this non-convex set fails to be open in the core topology, as our subsequent study of BW spaces will show.

The position of the core topology among "reasonable" topologies for a convex structure is described in the next result.

5.4.3. Proposition. *Let X be a properly locally convex Hausdorff space with compact segments. Then the topology of X is coarser than the core topology.*

Proof. Since the topology of X is generated by convex open sets, it suffices to show that $core(O) = O$ for $O \subseteq X$ convex open. Let $p \in O$ and let $A \subseteq X$ be non-empty such that $\cap_{a \in A} ap = \{p\}$. If $\cap_{a \in F} ap \not\subseteq O$ for each finite $F \subseteq A$ then by the compactness of segments it follows that the sets pa for $a \in A$ meet on $X \setminus O$. ∎

The internal and the relative core topologies of a convex subspace need not be the same. For instance, consider the closed unit 2–cell C in Euclidean plane. The bounding circle of C is a discrete subspace of C with respect to the core topology, whereas the relative topology of C, inherited from the core topology of the plane, is the standard topology. Here, the bounding circle is a connected set. The following result (with a straightforward argument) describes what is best possible.

5.4.4. Proposition. *If C is convex in X, then the core topology of C is finer than the relative core topology of C derived from X.* ∎

5.5. Theorem. *The core topology of a product space is the product of the factor core topologies.*

Proof. Let $X = \prod_{i \in I} X_i$. By Proposition 5.2.1, the sets of type $\prod_{i \in I} C_i$ with

$C_i \subseteq X$ convex and core-open, and such that $C_i = X_i$ for all but finitely many i, are open in the core topology of X. Conversely, let $C \subseteq X$ be non-empty, convex and core-open. We show that C is a product of convex sets; by 5.2.1 again, all factor sets will then be core-open and all but finitely many of them equal the corresponding full factor space.

To this end, let $p \in C$. Henceforth, $\pi_i : X \to X_i$ denotes the ith projection. If C is a non-product set then there is a point

$$q \in \prod_{i \in I} \pi_i(C) \setminus C.$$

As C is convex, there is a point $q(F) \in C$ for each finite set $F \subseteq I$, such that if $j \in F$ then the jth coordinate of $q(F)$ and of q are equal. Consequently, if a point $a(F)$ satisfies $\pi_i(a(F)) = \pi_i(q)$ for all $i \notin F$, then $q \in co\{a(F), q(F)\}$ and hence $a(F) \notin C$. We completely determine $a(F)$ by requiring moreover that $\pi_i(a(F)) = \pi_i(p)$ for each $i \in F$. As each factor is S_1, we find that $\cap_F co\{p, a(F)\} = \{p\}$, and as the segments $co\{p, a(F)\}$ constitute a down-directed collection, one of them must be included in C, contradiction. ∎

5.6. Theorem. *Let X be complete and S_4, let Y have core-closed singletons and let $f : X \to Y$ be a CP + CC surjection. Then f is continuous with respect to the core topologies of X and Y iff all fibers of f are closed in X.*

Proof. If f is continuous then its fibers are evidently closed. Conversely, assume the latter to be true. Let $C \subseteq Y$ be a core-open convex set and take $p \in f^{-1}(C)$. Consider a down-directed family of segments ap for $a \in A$ such that $\cap_{a \in A} ap = \{p\}$. We show that

$$\bigcap_{a \in A} f(ap) = \{f(p)\}.$$

That $f(p)$ is in the left-hand set is clear. Let $q \neq f(p)$. By assumption there is a convex core-open set $O \subseteq X$ with $p \in O$ and $f^{-1}(q) \cap O = \emptyset$. Then $ap \subseteq O$ for some $a \in A$ and hence $q \notin f(ap)$, establishing the formula. As f is CP and CC we have $f(ap) = f(a)f(p)$. Consequently,

$$\bigcap_{a \in A} f(a)f(p) = \{f(p)\}$$

and as $C \subseteq Y$ is core-open, we find that $f(a)f(p) \subseteq C$ for some $a \in A$. For this a we have $ap \subseteq f^{-1}(C)$. By Theorem 5.3(2), it follows that $p \in core\, f^{-1}(C)$. ∎

We note that no general conditions are known under which a core topology is T_1. We now prepare for the first main result on core topology in BW spaces.

5.7. Lemma. *Let X be a JHC and S_3 convex structure with decomposable segments and let X be equipped with a locally convex and Hausdorff topology in which all segments are compact and connected. Then the relative topology of a segment is exactly the order topology. If, moreover, the segment operator is LSC, then all polytopes are compact and the hull operator is continuous on the hyperspace of all finite sets.*

Proof. Under the assumption of S_2, a decomposable segment is totally ordered. Let $a, b \in X$. Considering that X is locally convex, by Proposition I§5.2(3) each point of ab

§5: Intrinsic Topology   355

has a relative neighborhood base of order-convex sets. As X is Hausdorff, it follows that the relative topology is finer than the order topology of ab. The former is compact and the latter is Hausdorff. Therefore, they are equal.

Let $O \supseteq ab$ be an open set. As the relative topology of ab is just the (compact connected) order topology, there is a finite sequence of convex sets $U_1,..,U_n$ such that

(i)   the sets *Int* (U_i) cover ab;
(ii) *Int* $(U_i) \cap$ *Int* $(U_{i+1}) \cap ab \neq \emptyset$ for $i = 1,..,n-1$;
(iii) $a \in$ *Int* (U_1); $b \in$ *Int* (U_n).

As the segment operator is LSC, there exist neighborhoods $V_a \subseteq U_1$ of a and $V_b \subseteq U_n$ of b such that for each $a' \in V_a$ and $b' \in V_b$ the segment $a'b'$ meets each of the sets *Int* $U_i \cap$ *Int* U_{i+1} for $i = 1,..,n-1$. The segment $a'b'$ being decomposable and U_i being convex, it follows that

$$a'b' \subseteq \bigcup_{i=1}^{n} U_i \subseteq O.$$

This establishes the continuity of the segment operator, and the result follows from Proposition 2.7(3).   ∎

5.8. Theorem. *Let X be a complete BW space without boundary. Then the core topology turns X into a topological convex structure which is locally convex, Hausdorff, and both closure - and interior stable. Moreover, its polytopes are compact and its hull operator is continuous on the hyperspace of finite sets.*

Proof. The argument is subdivided into statements of independent interest.

5.8.1. Proposition. *Let X be a BW space, let $H \subseteq X$ be a hyperflat and let $a, b \in X \setminus H$ such that $a \circ b \cap H \neq \emptyset$. Then there is a partition of X into three convex sets,*

$$X = H \cup H/a \cup H/b,$$

such that core$(H/a) = H/a$ and core$(H/b) = H/b$. In particular, H is closed in the core topology of X.

Proof. A basic but obvious fact used below is that a line (i.e. a one-dimensional affine set) not included in an affine subspace meets this subspace in at most one point. Hence $H \cap H/a = \emptyset = H \cap H/b$. Now suppose that $x \in (H/a) \cap (H/b)$. There exist points $x_a \in H \cap x \circ a$ and $x_b \in H \cap x \circ b$. Take a point $c \in H \cap a \circ b$. Two applications of the Pasch Property yield (Fig. 2)

$$a, b \in \text{aff}\{x_a, x_b, c\} \subseteq H,$$

a contradiction.

We next show that the sets described in 5.8.1 cover X. To this end we use the affine hull formula of I§7.11:

$$X = \text{aff}(\{a\} \cup H) = H \cup (a \circ H)/H \cup (H/a).$$

We only have to show that $(a \circ H)/H \subseteq H/b$. Let $x \in (a \circ H)/H$. Then there exist $u, v \in H$

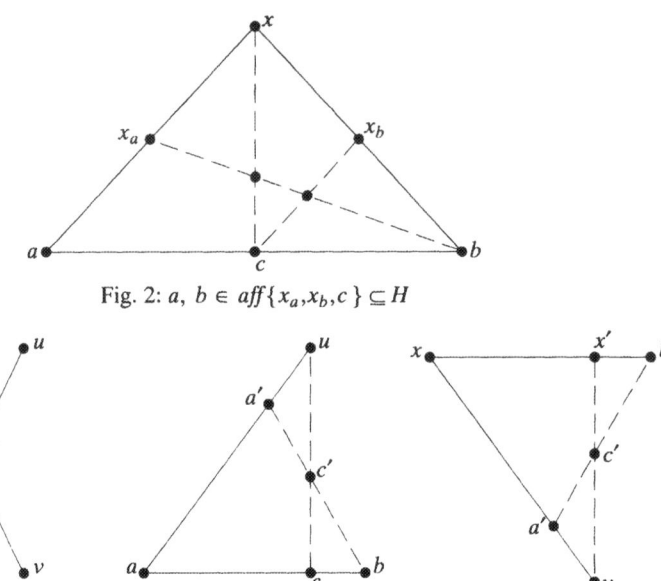

Fig. 2: $a, b \in aff\{x_a, x_b, c\} \subseteq H$

Fig. 3: Proof of $X = H \cup H/a \cup H/b$

together with a point $a' \in a \circ u \cap x \circ v$ (cf. Fig. 3). Now $a' \circ b$ and $u \circ c$ meet in some point $c' \in H$ by the Pasch axiom (J-4). By (J-3) the ray c'/v meets $x \circ b$ in a point x'. Note that $x' \in c'/v \subseteq H$, whence $x \in H/b$.

As each of the sets H, H/a, H/b is convex, it remains to be verified that the second and third of them are equal to their core. We do this for H/a. Consider $x \in H/a$ and take

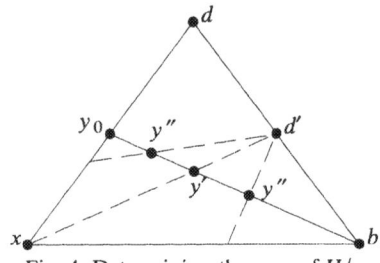

Fig. 4: Determining the core of H/a

$d \in X$. We have to check that $yx \subseteq H/a$ for some $y \in d \circ x$. If $d \in H/a \cup H$ then $y = d$ will do. In the remaining case, we have $d \in H/b$ and it suffices to find a point $y \in H \cup H/a$ of $d \circ x$. Take any point $y_0 \in d \circ x$ and consider the elements

$$d' \in H \cap d \circ b; \quad y' \in y_0 \circ b \cap d' \circ x$$

(cf. Fig. 4). If $y_0 \in H/a \cup H$ then we are done. So assume $y_0 \in H/b$ and let

§5: Intrinsic Topology

$y'' \in H \cap y_0 \circ b$. Note that $y'' \neq y'$ for otherwise $x \in H$. Looking from b, the point y'' is either before or after y'. By applying (J-3), in the first case we find a point of H in $x \circ b$ (a contradiction) and in the second case we find a point y of H in $y_0 \circ x$. This y is the desired point. ∎

5.8.2. Proposition. *Let a, b be distinct points of a BW space X. Then there is a hyperflat H with $a \circ b \cap H \neq \varnothing$. Consequently, the core topology of X is Hausdorff.*

Proof. Let $a \neq b$ and fix a point $c \in a \circ b$. Extend $\{c\}$ to a collection E maximal with the properties of being affinely independent and $a \notin \mathit{aff}(E)$. Then $\mathit{aff}(E)$ is a hyperflat met by $a \circ b$. By 5.8.1, we obtain open convex sets $O_a = \mathit{aff}(E)/b$, $O_b = \mathit{aff}(E)/a$, with

$$a \in O_a; \quad b \in O_b; \quad O_a \cap O_b = \varnothing.$$
∎

5.8.3. Proposition. *In a complete BW space without boundary, all segments are compact relative to the core topology, and the segment operator is LSC.*

Proof. Each segment is an unconditionally complete totally ordered set and hence its core topology (which equals the order topology) is compact. The relative core topology of a segment is Hausdorff by 5.8.2. As the former is finer than the latter (Proposition 5.4.4), both topologies are equal.

Suppose that the segment ab meets the convex open set O. Disregarding the trivial case where a or b are in O, assume $c \in a \circ b \cap O$. Consider two points $a_0 \in a/c$ and $b_0 \in b/c$, and let

$$V_a = a_0 \circ O; \quad V_b = b_0 \circ O.$$

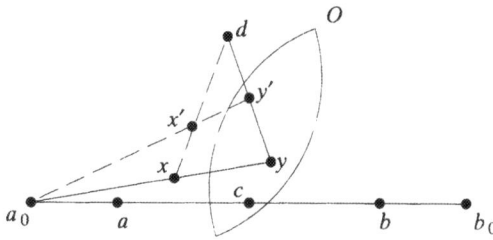

Fig. 5: LSC operator $(a,b) \mapsto ab$

See Fig. 5. Note that $a \in V_a$ and $b \in V_b$. We show that the convex set V_a is open. Take $x \in V_a$, say, $x \in a_0 \circ y$ with $y \in O$, and consider $d \in X$. There is a point $y' \in d \circ y$ with $y' \circ y \subseteq O$ and by the Pasch property (J-4) there is a point $x' \in y' \circ a_0 \cap d \circ x$. It easily follows from (J-3) that $x' \circ x \subseteq V_a$. A similar argument shows that V_b is open.

Now consider $a' \in V_a$, $b' \in V_b$. We have $O \cap a_0 \circ b_0 \neq \varnothing$, whereas by the Arithmetic Laws of I§7.5.1) and by the convexness of O,

$$a_0 \circ b_0 \subseteq (a'/O) \circ (b'/O) \subseteq (a' \circ b')/(O \circ O) = (a' \circ b')/O.$$

Thus $(a' \circ b')/O$ meets the convex set O, showing that $a' \circ b'$ meets O. ∎

We have proved so far that the core topology is locally convex and Hausdorff, and that there is an LSC and compact-valued segment operator. By Lemma 5.7, all polytopes are compact and the hull operator is continuous on the hyperspace of finite subsets. In particular, X is closure stable. To complete the proof of the theorem, we just establish the following fact.

5.8.4. Proposition. *The core of a convex set C in a BW space is convex and Int (C) = core (C).*

Proof. Let $x \in$ *Int* (C). Then there is a convex open set $O \subseteq$ *Int* (C) containing x, and if $d \in X$ then a final subset of $d \circ x$ remains in O and hence in C. So $x \in$ *core* (C). The converse follows from the fact that for C convex, *core* (*core* C) = *core* C. The proof is based on the Pasch axiom (J-4) and is left to the reader. ∎

5.9. Corollary. *Let X be a complete BW space without boundary, and let $\mathcal{T}$ be a locally convex and Hausdorff topology on X, relative to which all segments of X are compact. Then the core topology of X is finer than $\mathcal{T}$, and all polytopes are compact in $(X, \mathcal{T})$.*

Proof. By the first part of Lemma 5.7, the relative topology on each segment equals the order topology. Let U be a convex neighborhood of a point $a \in X$. Then $a \in$ *core* (U), for otherwise there is a segment ba with $b \neq a$ and $ba \cap U = \{a\}$. But then a is an isolated point of the relative topology on ba, which is certainly not equal to the order topology. By 5.8.4, U is a neighborhood of a in the core topology. Consequently, as $\mathcal{T}$ is locally convex, it is coarser than the core topology. The latter has compact polytopes by Theorem 5.8. But then the same goes relative to $\mathcal{T}$. ∎

It is not known whether a topology with properties as $\mathcal{T}$ above leads to a continuous segment operator. If so, then the second part of the above result would follow from Lemma 5.7. If the condition of local convexity is changed into **proper** local convexity, then the the conclusions of 5.9 follow directly by Proposition 5.4.3.

5.10. Corollary. *Let X be a complete BW space without boundary and of finite dimension. Then there is one and only one topology on X which is locally convex and Hausdorff, and such that all segments are compact.*

Proof. The core topology satisfies the above requirements. Suppose $\mathcal{T}$ is a second topology on X of this type. By the previous corollary the core topology is finer than $\mathcal{T}$. To obtain equality, assume first that the dimension of X is at most one. Then X is simply a totally ordered set with the order convexity, and it is easy to see that a Hausdorff locally convex topology for which all segments are compact equals the order topology (this is actually a particular case of the results below involving median convexity).

We next establish the following fact by induction on the dimension n of X.

5.10.1. *For each point a and for each convex core-open neighborhood U of a in an n-dimensional BW space without boundary there is an $(n+1)$-set $F \subseteq U$ such that*

§5: Intrinsic Topology   359

$a \in core\ co(F)$.

Proof. The result is obvious if $n \le 1$. We proceed by induction. Let $b_0 \ne a$ in U, let $a' \in a/b_0$ in U, and let H be a hyperplane through a' (use a construction as in 5.8.2) not passing through a. Take $b_1,..,b_n \in H \cap U$ such that

$$a' \in core_H\ co\{b_1,..,b_n\}.$$

It appears from the core formula in 5.2.4 that the collection of all b_i, for $i = 0,..,n$, is as required.

Continuing with the original argument, assume X is n-dimensional, where $n > 1$. Let $a \in X$, let U be a convex neighborhood of a in the core topology, and let $F = \{b_0,..,b_n\} \subseteq U$ be the vertices of a polytope $P = co(F)$ as in 5.10.1. The corresponding boundary $R = P \setminus core\ P$ is the union of all polytopes

$$co\{b_0,..,\hat{b_i},..,b_n\}$$

for $i = 0,..,n$. By Corollary 5.9, these sets are $\mathcal{T}$-compact. Hence there is a convex $\mathcal{T}$-neighborhood V of a disjoint from R. If V is not included in P, then some segment $ax \subseteq V$ joins a with a point $x \notin P$. Being a core-connected set, ax crosses the core-boundary R, a contradiction. We conclude that $V \subseteq U$ and hence that U is also a $\mathcal{T}$-neighborhood. Thus, $\mathcal{T}$ is finer than the core topology. ∎

5.11. Corollary. *A complete BW space without boundary is* FS_{3+} *relative to its core topology.*

Proof. Let X be a complete BW space without boundary. We rely on the fact (Topic 5.26.2) that the hyperflats of X correspond with the half-space boundaries. We proceed in three steps.

Step I. If H_0 is a hyperplane of X and if $p \notin H_0$, or if H_1, H_2 are disjoint hyperplanes of X with $p \in H_1 \circ H_2$, then there is a hyperplane through p disjoint from the given one(s).

We concentrate on the second situation. A routine argument shows that $p \circ H_1 \cap p \circ H_2 = \emptyset$. As X is S_4, there is a half-space H with

$$p \circ H_1 \subseteq H;\quad p \circ H_2 \cap H = \emptyset.$$

The common boundary $Bd(H)$ of H and $X \setminus H$ is a hyperplane. The fact that $p \in H_1 \circ H_2$ directly implies that this hyperplane passes through p. Moreover, it is not difficult to see that $p \circ H_i$ is core-open, from which disjointness of $Bd(H)$ with H_i follows for $i = 1, 2$.

The argument for the first situation is essentially the same, and involves S_4 for the disjoint pair $p, p \circ H$.

Step II. If H_0 is a hyperplane of X then there is a partition of X into hyperplanes, one of which equals H_0.

Indeed, consider a maximal family $\mathcal{H}$ of pairwise disjoint hyperplanes and with

$H_0 \in \mathcal{H}$. Let $p \in X$ not be covered by these hyperplanes. For each $H \in \mathcal{H}$, consider the open half-space bounded by H and missing p, and let $\mathcal{O}$ denote the resulting family. Suppose first that p is not between (in the sense of Step I) two of the selected hyperplanes. Then $\mathcal{O}$ is a totally ordered collection. If it has a last element O then $p \notin \overline{O}$ and Step I yields a new hyperplane disjoint from all members of $\mathcal{H}$, contradiction. Hence $\mathcal{O}$ has no largest element. Consequently, the boundary of $\cup \mathcal{O}$ is a hyperplane disjoint from all the given ones, a contradiction.

Next, assume p is between two members of $\mathcal{H}$. This time, $\mathcal{O}$ is built with two chains. If p is adherent to the union of one of them, then the boundary H of the union is a hyperplane through p. Since X has no boundary, two intersecting hyperplanes actually *cross*: the two open half-spaces induced by one hyperplane meet the two open half-spaces induced by the other hyperplane. Therefore, H is disjoint from all members of $\mathcal{H}$, which is a contradiction.

We may assume that both chains in $\mathcal{O}$ have a largest element. Now p is between the corresponding hyperplanes. By Step I, we obtain a hyperplane through p and disjoint from the two neighboring ones. Hence this new hyperplane is disjoint from all members of $\mathcal{H}$.

Step III. Proof that X is FS_{3+}. Let C be convex closed, let $n > 0$, and let $P = co(\{p_1,..,p_n\})$ be disjoint from C. As co is continuous on the hyperspace of finite sets, cf. 5.8, there exist convex neighborhoods W_i of p_i ($i = 1,..,n$) such that the set $P' = co(\{p'_1,..,p'_n\})$ is disjoint from C whenever $p'_i \in W_i$ for all i. As X is JHC and interior-stable, it follows that the union of all polytopes of type P' yields a convex neighborhood W of P disjoint from C. Consider a half-space H including C and missing W. Then $\overline{H}$ is a half-space (by closure - and interior stability) which is disjoint from P.

The corresponding hyperflat belongs to a partition of X by hyperflats (Step II). Replace each hyperplane by the two closed half-spaces bounded by it. The resulting family of half-spaces decomposes in two indexed chains of sets as in Lemma 4.3. This yields a continuous CP functional X separating C and P. Its fibers are exactly the hyperplanes of $\mathcal{H}$, as it can be deduced from the formulae in Lemma 4.2. ∎

The remainder of this section is devoted to a study of the intrinsic weak topology.

5.12. The intrinsic weak topology. Let X be an S_2 convex structure. The *intrinsic weak topology* of X (briefly, the *IW topology*) is generated by the closed subbase of all convex sets which are complete in X. Note that this is indeed a weak topology. Here are some general observations.

5.12.1. *An IW-closed subset of an up-complete S_3 convex structure X is complete in X. In particular, a convex set is IW-closed in X iff it is complete in X.*

Proof. Let $A \subseteq X$ be closed and $p \notin A$. Suppose $b \in X$ and $D \subseteq A$ is a b-down-directed collection. As X is even complete, cf. I§5.19, D has an infimum p. Suppose $p \notin A$. Then there exist convex sets $C_1,..,C_n$, each of which is complete in X, such that

§5: Intrinsic Topology   361

$$A \subseteq \bigcup_{i=1}^{n} C_i; \quad p \notin \bigcup_{i=1}^{n} C_i.$$

Then D is eventually included in some set C_i. The latter being complete in X, the b-infimum of D exists and is in C_i, contradicting that $p \notin C_i$. By Proposition I§5.19, the set A is also up-complete in X. ∎

A second observation is that the intersection of subsets, which are complete in X, is complete in X. If $D \subseteq C \subseteq X$ are such that D is complete in C and C is complete in X, then D is complete in X. As a direct consequence of these facts, we have

5.12.2. *If C is a convex subset which is complete in X, then the IW topology of C equals the relative IW topology of X.* ∎

As an application of completeness in product sets, cf. I§5.17(2), we obtain the following result.

5.12.3. *In a product of convex structures, the IW topology is the product of the factor IW topologies.* ∎

A subset C of a convex structure X is **strongly complete in X** provided each collection of segments which is finitely intersecting on C has a non-empty intersection with C; cf. I§5.30.

5.12.4. Proposition. *Let X be a convex structure with a geometric segment operator. Then any subset of X, which is strongly complete in X, is complete in X.*

Proof. Let $C \subseteq X$ be strongly complete in X, let $c \in X$ with a corresponding basepoint order $\leq_c$, and let $A \subseteq C$ be a up-directed set with an upper bound $d \in X$. Consider the set $B = \bigcap_{a \in A} da$ of all upper bounds of A below d in $(X, \leq_c)$. The family of all segments ab with $a \in A$ and $b \in B$ is finitely intersecting. Indeed, let $a_1 b_1, ..., a_n b_n$ be of the previously described type. The set A being up-directed, there is an element $a \in A$ such that $a_i \leq_c a$ for all $i = 1, ..., n$. By I§5.2(3), we have $a \in a_i b_i$ for all i. It follows that the set

$$\bigcap \{ ab \cap C \mid a \in A, b \in B \}$$

is non-empty. Any member of the intersection is an upper bound of A smaller than each $b \in B$ -- hence it is the supremum of A. This shows that C is up-complete in X. To see that it is down-complete in $(X, \leq_c)$, use Proposition I§5.19. ∎

Our last observation largely motivates the definition of the IW-topology.

5.12.5. Proposition. *Let X be a convex structure with a topology relative to which all segments are closed and compact. Then each closed subset of X is strongly complete in X. If, in addition, the segment operator of X is geometric, then all closed sets are complete in X.*

Proof. As for the first part, it is an elementary topological fact that a finitely intersecting family of closed sets, of which at least one is also compact, has a non-empty intersection. As for the second part, apply the previous proposition. ∎

We note that an S_3 convex structure has a geometric segment operator; cf. Theorem I§4.7. If a topological space is Hausdorff, then compact subsets are closed and the formulation of the last proposition can be simplified.

5.13. Examples. We compute the IW topology for Lawson semilattices, for distributive lattices, and for complete Bryant-Webster spaces.

5.13.1. Proposition. *Let S be an unconditionally complete continuous semilattice. Then the IW topology of S is exactly the Lawson topology.*

Proof. Let S be equipped with the Lawson topology. Then S is a compact Hausdorff semilattice and by Proposition 5.12.5, it follows that each closed subset of S is complete in S. Conversely, let $C \subseteq S$ be *convex* and complete in S. As usual, **0** denotes the smallest element of S (which exists by compactness). The base-point order $\leq_0$ equals the semilattice order. An up-directed subset of C has an upper bound in $(S, \leq)$, and (by the completeness of C in S) it has a supremum in C. A similar argument shows that the infimum of a non-empty subset of C is in C (C is a semilattice, so one can add all infima of finite subsets). It follows[1] that the subsemilattice C is closed relative to the Lawson topology of S.

To complete the proof, note that each point of S has a neighborhood base consisting of closed convex sets (cf. Proposition 1.13.3), from which it easily follows that the convex closed sets in S form a subbase for the Lawson topology. ∎

5.13.2. Proposition. *Let L be an unconditionally complete and distributive lattice. Then the IW topology of L is the interval topology.*

Proof. By I§5.18.2, completeness of a distributive lattice L is equivalent to completeness of L as a convex structure. Moreover, L has a smallest element **0** and a largest element **1**. If a non-empty convex set C is complete in L, then it is gated. Let $p : L \to C$ denote its gate map. By I§5.12(4), p is CP and CC, whence

$$C = p(L) = p(\mathbf{01}) = p(\mathbf{0})p(\mathbf{1}).$$

Consequently, C is an order interval. Conversely, every order interval is a polytope, and hence it is complete in L because it is a gated set. ∎

5.13.3. Proposition. *The IW topology of a Bryant-Webster space X is Hausdorff. If X is complete and has no boundary then the IW topology equals the weak core topology. The core- and IW topologies are equal if, in addition, X is (affinely) finite-dimensional.*

Proof. By virtue of Proposition 5.8.2, for each pair of distinct points $a, b \in X$ there exists a pair of core-closed half-spaces H_a, H_b screening a, b. By definition of the core

---
1.   Lawson [1973b].

§5: Intrinsic Topology 363

topology, both half-spaces are complete in X and hence they are closed in the IW topology. This establishes the Hausdorff property.

Assume next that X is complete and has no boundary. All core-closed convex sets are IW-closed by Proposition 5.12.5. We verify the converse. Suppose C is convex and complete in X, and let $p \notin C$. If $p \notin \mathit{aff}(C)$ then we are done since affine subsets are core-closed; cf. 5.26. So assume $p \in \mathit{aff}(C)$ and fix a basis $B \subseteq C$ of $\mathit{aff}(C)$. As the internal core topology of an affine subset equals the relative core topology (cf. 5.26), we may and will assume that $X = \mathit{aff}(C)$. There is a finite set $B_0 \subseteq B$ with $p \in A_0 = \mathit{aff}(B_0)$. Take a point $q \notin \mathit{co}(B_0)$ in the (open) join of p with a core point of $\mathit{co}(B_0)$ relative to A_0, and let D_0 be a relative half-space of A_0 containing q and disjoint

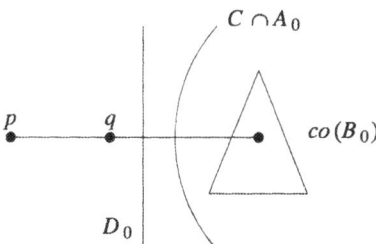

Fig. 6: Initial B_0 and D_0

from C (cf. Fig. 6). One easily verifies that p is a core point of D_0 relative to A_0. Well-order $B \setminus B_0$ by indexing it with ordinal numbers, say:

$$B \setminus B_0 = \{b_\alpha \mid \alpha < \kappa\}.$$

For each α we let L_α be a line through b_α meeting $\mathit{co}(B_0)$ in a core point. We construct the following sets by transfinite induction.

$$A_\alpha = \begin{cases} \mathit{aff}(A_{\alpha-1} \cup \{b_\alpha\}) \\ \bigcup_{\beta<\alpha} \mathit{aff}(A_\beta) \end{cases} \quad D_\alpha = \begin{cases} D_{\alpha-1}/L_{\alpha-1} & \text{(non-limit } \alpha\text{)}; \\ \bigcup_{\beta<\alpha} \mathit{aff}(D_\beta) & \text{(limit } \alpha\text{)}. \end{cases}$$

At each stage one can verify that p is a core point of D_α relative to A_α. This process leads us to a convex core neighborhood $D = \bigcup_{\alpha<\kappa} D_\alpha$ of p which is disjoint from C.

We have shown so far that the IW topology is the weak core topology. In particular, X is locally convex and its segments are compact in the IW topology. The last part of the proposition now follows from Corollary 5.10. ∎

5.14. Theorem. *Let X, Y be convex structures of which X is complete and S_4, and let $f: X \to Y$ be a CP and CC surjection, the fibers of which are strongly complete in X. Then f preserves infima of down-directed sets in any base-point order and f is continuous in the respective IW-topologies of X and Y.*

Proof. We first show that if $b \in X$ and $q \in Y$, then the upper set $C = U_b(f^{-1}(q))$ is complete in $(X, \leq_b)$. To this end, let $D \subseteq C$ be down-directed in the base-point order of b. Then the collection of segments $\{bd \mid d \in D\}$ is down-directed and its members all meet the set $f^{-1}(q)$. The latter being strongly complete in X, we conclude that $\cap_{d \in D} bd$ meets $f^{-1}(q)$. Any point c in this intersection is a lower bound of D which is in $f^{-1}(q)$. As X is complete, the b-infimum of D exists, and is at least c. Therefore, this infimum is in $U_b(f^{-1}(q))$. By Proposition I§5.19, it follows that C is complete in X.

We next show that f preserves infima of down-directed sets. Let $D \subseteq X$ be a b-down-directed set with infimum p. Then, as f preserves base-point orders, we find that $f(p)$ is a lower bound of $f(D)$ in $(X, \leq_{f(b)})$. Suppose that $q \geq_{f(b)} f(p)$ is another lower bound of the set $f(D)$. Then $q \in \cap_{d \in D} f(p)f(d)$ and hence $D \subseteq U_b(f^{-1}(q))$. By the first part of the proof, we conclude that $p \in U_b(f^{-1}(q))$. It follows that $q \leq_{f(b)} f(p)$ and hence that $f(p)$ is the $f(b)$-infimum of $f(D)$.

Let $C \subseteq Y$ be convex and complete in Y, and let $D \subseteq f^{-1}(C)$ be a down-directed set with a b-infimum x. Then $f(x)$ is the $f(b)$-infimum of $f(D)$ and it follows that $f(x) \in C$ because C is complete in Y. Therefore, $x \in f^{-1}(C)$ and we conclude that $f^{-1}(C)$ is complete in X. ∎

5.15. Examples. The following may illustrate the sharpness of certain assumptions in 5.14.

5.15.1. Totally ordered spaces. Consider the totally ordered spaces

$$X = [0,1) \cup \{2\}; \quad Y = [0,2].$$

Both are complete and the IW topology is the order topology (= interval topology). The inclusion function $X \to Y$ has closed (singleton) fibers, it is CP, and it is CC with respect to the image space (not with respect to the whole Y). However, this function is not continuous in the order topologies.

5.15.2. Standard open disk. Let B denote the open unit disk in $\mathbb{R}^2$, equipped with the standard convexity. This gives a complete BW space without boundary and of finite dimension. Consequently, the core and IW topologies of B coincide. Throughout, $\pi_i: \mathbb{R}^2 \to \mathbb{R}$ denotes the ith projection ($i = 1, 2$). We construct a function $f: B \to [-1,1]$ as follows.

$$f(x) = \begin{cases} \pi_2(x), & \text{if } \pi_1(x) = 0; \\ -1, & \text{if } \pi_1(x) < 0; \\ +1, & \text{if } \pi_1(x) > 0. \end{cases}$$

Then f is a discontinuous surjective CP functional which is not CC.

It is not known whether the assumption on strong completeness of fibers in the superspace can be replaced by the usual (relative) completeness property (compare with Theorem 5.6).

5.16. Theorem. *Let X be a complete S_3 convex structure.*

(1)   *If X is median, then all polytopes of X are closed and compact in the intrinsic weak*

§5: Intrinsic Topology   365

   topology. *In particular, X is a topological convex structure.*
(2) *If X has median segments, then a convex set is complete in X iff it is strongly complete in X.*
(3) *If X has median segments, then a surjective CP + CC function of X into a convex structure is continuous in the respective IW topologies iff its fibers are closed.*

Proof of (1). All non-empty polytopes are gated and hence they are complete in X. As the intersection of complete convex sets is complete and convex, it follows that a polytope P is compact iff every chain of non-empty complete convex subsets of P has a non-empty intersection. We verify this "intersecting chain" property for a polytope $P = co\{a_1,..,a_n\}$.

Let $\leq_k$ be the base-point order with base-point a_k. The above statement being trivial for $n = 1$, we proceed by induction on n. Let $(C_i)_{i \in I}$ be a chain of convex sets which are complete in P. The index set is totally ordered by the prescription

$$i \leq j \Leftrightarrow C_i \supseteq C_j.$$

For each $i \in I$ we let $x_i \in C_i$ be the gate of a_n. For $i \leq j$ we find $x_i \leq_n x_j$. For $k = 1,..,n$ let u_k be the a_k-infimum of the points x_i, and assume that for some $i_0 \in I$,

$$co\{u_1,..,u_{n-1}\} \cap C_{i_0} = \emptyset.$$

Then there is a half-space H such that

$$u_1,..,u_{n-1} \notin H; \quad C_{i_0} \subseteq H.$$

If $a_k \in H$ for some $k \leq n-1$ then $u_k \in a_k x_{i_0} \subseteq H$, a contradiction. As x_{i_0} is in the complete set P, it follows that the remaining vertex a_n is in H. If $i \leq i_0$ then $x_i \in a_n x_{i_0}$, and hence the a_1-infimum u' of the points x_i for $i \leq i_0$ satisfies $u' \in a_n x_{i_0}$. For $i \geq i_0$ we have $x_i \in C_i \subseteq C_{i_0}$. Hence the a_1-infimum u'' of x_i for $i \geq i_0$ is in C_{i_0} by completeness of the latter. Consequently, $u', u'' \in H$. However, u_1 is the a_1-infimum of u' and u'', whence $u_1 \in H$, a contradiction.

We conclude that the sets

$$C_i \cap co\{u_1,..,u_{n-1}\} \quad (i \in I)$$

form a chain of non-empty complete convex subsets of $co\{u_1,..u_{n-1}\}$. By the induction hypothesis, the intersection of this chain is non-empty.

Proof of (2). X being complete, we find that all segments are complete in X which is why they are closed. By 5.12.2, the relative IW topology of a segment equals its intrinsic IW topology, which is compact by the first part of the Theorem. The result follows from Propositions 5.12.4 and 5.12.5.

Proof of (3). This follows from (2) in combination with Theorem 5.14. ∎

As a particular application of (1), we obtain the well-known fact that the interval topology of an unconditionally complete lattice is compact. A second application obtains from (3): each linear functional of a real vector space is IW-continuous. Note that such a functional is either surjective or constant. The previous result fails to hold for surjective CP functionals; cf. 5.15.2.

We now investigate under which conditions a median space has a Hausdorff IW-topology turning the median operator continuous. The **Infinite Distributive Law** of a lattice L states that for each $a \in L$ and for each non-empty set $B \subseteq S$,

$$(\vee B) \wedge a = \vee \{ b \wedge a \mid b \in B \}; \quad (\wedge B) \vee a = \wedge \{ b \vee a \mid b \in B \}.$$

An unconditionally complete lattice which is infinitely distributive is known as a **complete Heyting algebra**.

5.17. Lemma. *For a complete median space the following are equivalent.*
(1) *All gate functions are continuous in the IW topology.*
(2) *Every segment is infinitely distributive.*

Proof. Throughout, X denotes a complete median convex structure.

Proof of (1) $\Rightarrow$ (2). Let $b, c \in X$ and let $A \subseteq bc$ be non-empty. We regard bc as a lattice with $b = \mathbf{0}$ and $c = \mathbf{1}$. For $x \in bc$, the restricted gate function $p: bc \to xc$ satisfies $p(a) = a \vee x$. It is surjective and continuous by assumption. By Theorem 5.16(3), we find that

$$(\wedge A) \vee x = p(\inf {}_b A) = \inf {}_{p(b)} p(A) = \inf {}_x p(A) = \inf {}_b p(A) = \wedge \{ a \vee x \mid a \in A \}.$$

A similar argument works for the dual formula.

Proof of (2) $\Rightarrow$ (1). Let C be a non-empty convex set with a gate function

$$p: X \to C.$$

Note that p is a surjective CP + CC function; cf. Proposition I§5.12(4). By Theorem 5.16(3) it suffices to show that the fibers of p are complete in X. To this end, take $c \in C$ and let $A \subseteq p^{-1}(c)$ be down-directed in the base-point order of b. By I§6.37.4, the convex set $D = co(\{b\} \cup C)$ is gated; let

$$q: X \to D$$

be the gate function. The Transitivity Lemma I§5.11(3) implies that $p \circ q = p$, whence $q(A) \subseteq p^{-1}(c)$. As q is a CP function, it preserves base-point orders. Hence $q(A)$ is down-directed in the order of $q(b) = b$. Moreover, for each $a \in A$ the convex set ba meets D whence $q(a) \in ba$, that is: $q(a) \leq_b a$. By the completeness of X, the b–infima of A and of $q(A)$ exist, and we find for each $a \in A$ that

$$\inf {}_b q(A) \leq \inf {}_b A \leq_b a.$$

This implies that $\inf {}_b A$ is in the segment $co\{\inf {}_b q(A), a\}$. Therefore, in order to show that $\inf {}_b A$ is in $p^{-1}(c)$, it suffices to show that $\inf {}_b q(A)$ is in $p^{-1}(c)$.

If $a \in A$ then $q(a) \in bc$. Assuming the contrary, we obtain a half-space H with $q(a) \in H$ and $b, c \notin H$. By join-hull commutativity, there is a point $x \in C$ with $q(a) \in bx$. Hence H meets C in x and, as $p(a) = c \notin H$, we conclude that $a \notin H$. However, $X \setminus H$ meets D (e.g., in b) and hence $q(a) \in X \setminus H$, a contradiction.

§5: Intrinsic Topology 367

We have shown so far that $q(A) \subseteq bc$. The gate function $p: X \to C$ being CP and CC, we have
$$p(bc) = p(b)p(c) = p(b)c \subseteq bc.$$
Therefore, p restricts to a gate function
$$bc \to p(b)c \subseteq bc.$$
In the lattice bc (with $c = 0$ and $b = 1$) the above function operates as $y \mapsto y \wedge p(b)$. Intervals being infinitely distributive, we conclude that $\vee q(A)$ is mapped to the point
$$\vee \{ q(a) \wedge p(b) \mid a \in A \},$$
which equals c since $q(a) \wedge p(b) = p(q(a)) = p(a) = c$ for all $a \in A$. Therefore, $b - \inf q(A) \in p^{-1}(c)$. ∎

We arrive at the following major result on IW topology in median spaces.

5.18. Theorem. *For a complete median space X the following are equivalent.*

(1) X is FS_4 with respect to the IW topology (in particular, each base-point order is closed).
(2) X is a topological median algebra with respect to the IW topology (in particular, X is Hausdorff).
(3) Each segment of X is a completely distributive lattice.
(4) Each base point order of X is continuous.

Proof of (1) ⇒ (2). As singletons are convex closed, the property FS_4 implies that X is Hausdorff. All segments are compact by Theorem 5.16. Hence by Theorem 4.16, X is a topological median algebra with respect to the IW topology.

Proof of (2) ⇒ (3). All segments of X are distributive lattices by I§6.10. The IW topology of a segment bc equals the relative IW topology of X and hence is Hausdorff. The operators $m(b, -, -)$, resp., $m(c, -, -)$, correspond with the inf- and sup-operation relative to $\leq_b$, and hence bc is a compact topological lattice. By Proposition 5.13.2, the IW topology equals the interval topology of bc. Then bc is completely distributive.[2]

Proof of (3) ⇒ (4). Consider $b \in X$ as a base-point, and let $c \in X$. The lattice $(bc, \leq_b)$ is completely distributive and hence it is a continuous lattice. To see that $(X, \leq_b)$ is a continuous semilattice, it suffices to show that the way-below relation of bc is part of the way-below of $(X, \leq_b)$. To this end, let $x \ll_b c$ in bc, and let $D \subseteq X$ be an up-directed set such that $c \leq \sup D$. Consider the gate map $p: X \to bc$. By Lemma 5.17, the up-directed set $p(D)$ satisfies
$$c = p(\sup D) = \sup p(D).$$

---

2. According to Raney [1953], an unconditionally complete and distributive lattice is completely distributive if (and only if) it is a topological lattice with respect to the interval topology.

Hence there is a point $d \in D$ with $x \leq_b p(d)$. Now $p(d) \in bd$ since the latter is a convex set containing d and meeting bc. Hence $x \in bp(d) \subseteq bd$, that is, $x \leq_b d$.

Proof of (4) $\Rightarrow$ (1) via (3). We use the fact[3] that an unconditionally complete and distributive lattice which is continuous in both directions is completely distributive. In this situation, moreover, each pair of points can be separated with a lattice homomorphism into [0,1] which is continuous in the interval topology. In a continuous poset, a lower set of type $L(c)$ is evidently continuous. It follows that each segment bc is continuous with respect to the base point orders $\leq_b$ and $\leq_c$. By the above cited results, the segments of X are completely distributive lattices and every pair of distinct points of bc can be separated with a lattice homomorphism into [0,1] which is continuous with respect to the interval topology of bc. Such functions are CP, and the interval topology is the relative IW topology. By Lemma 5.17, the composition of such maps with the projection onto bc is continuous. This shows that distinct points of X can be separated with CP maps, and the property FS_4 follows from Theorem 4.16. ∎

Some of the applications of Theorem 5.18 involve the following result, which provides additional motivation to study intrinsic topology of median algebras.

5.19. Lemma. *Let X, X' be FS_2 spaces with compact segments and of Helly number ≤ 2. If $f : X \to X'$ is an isomorphism of convex structures, then f is a homeomorphism of the corresponding weak topologies.*

Proof. For reasons of symmetry, it suffices to show that f is a closed function with respect to the corresponding weak topologies (i.e., f maps weakly closed sets to weakly closed sets). To this end, let $A \subseteq X$ be weakly closed and let $f(p) \notin f(A)$. Then $p \notin A$ and hence there exist convex closed sets C_i for $i = 1,..,n$, such that

$$A \subseteq \bigcup_{i=1}^{n} C_i; \quad p \notin \bigcup_{i=1}^{n} C_i.$$

Suppose $f(p)$ is adherent to $\cup_{i=1}^{n} f(C_i)$ (weak closure). Then $f(p)$ is adherent to one of the summands, say, to $f(C_i)$. However, C_i is gated by virtue of Proposition 5.12.3. By Proposition I§5.12(3), $f(C_i)$ is gated. Let $q \in f(C_i)$ be $f(p)$'s gate. Note that $f(p) \notin f(C_i)$ since f is injective, and hence that $q \neq f(p)$. By the property FS_2 there exists an open half-space $O \subseteq X'$ such that $f(p) \in O$ but $q \notin O$. Note that an open half-space is weakly open. But then O meets $f(C_i)$ and by I§5.11(1) we conclude that q should also be in O. This shows that $f(p)$ is not adherent to $\cup_{i=1}^{n} f(C_i)$ and, a fortiori, not to $f(A)$. ∎

5.20. Corollary. *Let X be an FS_2 space with compact segments and of Helly number ≤ 2. Then the weak topology of X is Hausdorff and equals the IW topology of X. In particular, all gate functions of X are weakly continuous.*

---
3. Gierz et al [1980, Props. VII§2.8-9].

§5: Intrinsic Topology 369

Proof. All base point orders in X are closed by Proposition 4.7. Hence by 5.12.3, each convex closed subset of X is complete in X and hence is IW-closed. It follows that the weak topology $\mathcal{T}_w$, derived from the given topology, is coarser then the IW topology $\mathcal{T}_{IW}$. By the property FS_2, every two distinct points of X can be screened with convex closed sets and hence with intrinsically closed convex sets. By Theorem 5.18, X is FS_4 relative to $\mathcal{T}_{IW}$. Note that $(X, \mathcal{T}_w)$ inherits the property FS_4 from the original space. Finally, the compactness of all segments in the original space implies that X is complete. Then by Theorem 5.14, all segments are intrinsically compact.

We are now in a position to apply Lemma 5.19 to the identity map $(X, \mathcal{T}_{IW}) \to (X, \mathcal{T}_w)$, to the effect that $\mathcal{T}_{IW} = \mathcal{T}_w$. The last part follows from Lemma 5.17. ∎

For instance, X may be taken as a locally convex median algebra with compact segments (cf. 4.13.3). Corollary 5.20 implies that for such spaces there can be no misunderstanding about the corresponding weak topology.

5.21. Corollary. *Let X, Y be FS_2 spaces with compact segments and of Helly number ≤ 2, both equipped with the weak topology. If $f : X \to Y$ is injective, CP, and continuous, then f is an embedding.*

Proof. Note that the subspace $f(X)$ of Y is S_3 and has Helly number ≤ 2 (cf. I§3.15; II§1.10). Consequently, f induces an isomorphism $X \approx f(X)$ of convex structures, cf. I§6.9. This shows that f is an embedding of convex structures. By Lemma 5.19, f is an isomorphism with respect to the weak topology of the spaces X and $f(X)$. It remains to be verified that the weak topology $\mathcal{T}_w$ of $f(X)$ equals the relative topology $\mathcal{T}_r$ of $f(Y)$ inherited from Y. Note that the former topology is the IW topology of $f(X)$ by Corollary 5.20.

As f is continuous in the given topologies, we find $\mathcal{T}_r \subseteq \mathcal{T}_w$. The Screening Lemma I§5.11(2) implies that distinct points of $f(X)$ can be screened with relatively convex closed sets. The intervals of $f(X)$ are compact in $\mathcal{T}_r$ because they are so in the finer topology $\mathcal{T}_w$. Now Theorem 4.16 can be applied, showing that $f(X)$ is FS_4 in the topology $\mathcal{T}_r$. Finally, we apply Corollary 5.20 to conclude that $\mathcal{T}_w$ is the weak topology corresponding to $\mathcal{T}_r$. In particular, $\mathcal{T}_w \subseteq \mathcal{T}_r$, and equality follows. ∎

As a direct consequence of Theorem 5.16 and Corollary 5.20, we have:

5.22. Corollary. *Let X, Y be FS_2 spaces with compact segments and of Helly number ≤ 2. Then a surjective CP function $f : X \to Y$ is weakly continuous iff its fibers are closed.* ∎

We now compare the two intrinsic topologies on median algebras.

5.23. Theorem. *Let X be a complete median space with a compatible IW topology. Then the following are true.*

(1) *The core topology of X is finer than the IW topology.*

(2) *If a convex subset C is complete in X, then the core topology and the relative core topology of C are equal and the gate map onto C is core-continuous.*

(3) *The IW topology and the core topology of X coincide on each polytope of X.*

Proof of (1). The IW topology makes X FS$_4$ by Theorem 5.18. Hence X is properly locally convex by Theorem 4.15. Its segments are compact by Theorem 5.16. The result now follows from 5.4.3.

Proof of (2). Let $p: X \to C$ be the gate map. By Theorem 5.18, each segment is completely distributive -- hence infinitely distributive. By Lemma 5.17, the fibers of p are complete in X. By (1), all fibers of p are core-closed. It follows from Theorem 5.6 that p is core-continuous. As p restricts to the identity of C we find that the core topology of C is coarser than the relative core topology. The converse is always true; cf. 5.4.4.

Proof of (3). By (2) we may assume that X is a polytope, say: $X = co\{a_1,..,a_n\}$. Let $O \subseteq X$ be convex and core open and let $b \in O$. For each i we find that $ba_i \cap O$ is core-open in ba_i. We regard this segment as a distributive lattice with $a_i = 0$ and $b = 1$. By virtue of Topic 5.29.1 we conclude from the continuity of base-point orders that the sets of type

$$U(x) \quad (x \ll b \text{ in } ba_i),$$

constitute a neighborhood base of b in the core topology of this lattice. The (well-known) fact that these sets are also neighborhoods of b for the interval (= IW) topology can be seen as follows. Let $x \ll b$. For each closed half-space H of the lattice with a non-empty IW-interior and with $1 = b \in H$, let $x(H)$ be the gate of 0 in H. As the gate projection onto H is CP + CC, we conclude that $H = bx(H)$. Also, $\sup_H x(H) = b$ (think of a suitable closed half-space separating between the supremum and b). Consequently there exist closed half-spaces $H_1,..,H_m$ with

$$\bigcap_{i=1}^{n} H_i = \bigcap_{i=1}^{m} bx(H_i) \subseteq U(x).$$

The former set is a IW-neighborhood. To complete the proof of (3), let

$$p_i: X \to ba_i$$

be the gate map, and observe that for each convex set C containing b,

$$\bigcap_{i=1}^{n} p_i^{-1}(C \cap ba_i) = C$$

(see I§6.37.2). Applying this formula to the convex set O and using the continuity of p_i with respect to the IW topologies, we arrive at the desired result. ∎

That the core and IW topology can be different for complete median spaces (even for "finite-dimensional" ones) is illustrated in 5.28 below. Let us finally consider some topological aspects of the amalgamation process, studied in I§5.

5.24. Theorem. *Let X_1, X_2 be complete and topological median algebras with respect to the IW topology. If $X_1 \cap X_2$ is a non-empty convex subspace of both X_1, X_2 which is complete in either space, then the amalgam of X_1 and X_2 is a complete topological median algebra with respect to the IW topology, and X_1, X_2 are closed subspaces.*

second variable. Prove that if $co^*(f(\{z\} \times X)) = Y$ for each $z \in Z$, then f is continuous.

Hint. To begin, settle the problem if X, Y are polytopes. Two ingredients are needed for this. First, a compact median polytope is locally polytopal in the sense that each point has a neighborhood base consisting of polytopes. Second, if X is locally polytopal and if Y is locally convex, then on the set of all continuous CP functions $X \to Y$, the topology of pointwise convergence equals the compact–open topology. To handle the general case, operate carefully on the inverse systems consisting of polytopes of X resp., Y, and of the corresponding nearest point projections. Consult 4.31 and I§6.37.7.

5.31. Core of a convex set. Let X be a convex structure and let $C \subseteq X$ be convex. For each ordinal number α, define $core^\alpha(C)$ by transfinite recursion as follows.

$$core^0(C) = C;$$
$$core^\alpha(C) = \begin{cases} core(core^{\alpha-1}(C)) & \text{(non-limit } \alpha\text{);} \\ \bigcap_{\beta<\alpha} core^\beta(C) & \text{(limit } \alpha\text{).} \end{cases}$$

Whatever α is, give an example of a convex set (e.g., in a tree) such that the transfinite sequence $(core^\beta(C))_{\beta < \alpha}$ is strictly decreasing.

Problem. Let $C \subseteq X$ be convex. Under which conditions is $core(C)$ convex, resp., open in the core topology of X?

5.32. Separation properties of intrinsic topologies

Problem (compare Theorem 5.8). Under which circumstances is the core topology T_1? Or T_2?

Problem (compare Theorems 5.18 and 5.23). Under which circumstances is the IW topology T_2?

5.33. Compactness and intrinsic topologies

5.33.1. (compare Proposition 5.12.5) Let X be an S_3 convex structure with a topology in which all segments are closed and compact. Let $b \in X$ and let $A \subseteq X$ be a down- (resp., up-) directed set with infimum (resp., supremum) p relative to the base-point order $\leq_b$. Note that A can be regarded as a net in X. Show that p is a limit of A. Conclude that all closed subsets are complete in X.

5.33.2. Problem. Find necessary and/or sufficient conditions on a convex structure in order that all intervals (resp., polytopes) are compact with respect to the core topology or the IW topology.

5.34. An example. The standard square $[-1,1]^2$ is convexified as a (completely distributive) lattice. Observe that the IW and core topology coincide with the Euclidean topology. Consider the following sublattice (Fig. 7):

$$X = [-1,0] \times [-1,1] \cup [0,1] \times (0,1].$$

In the relative topology, X is locally convex and Hausdorff and the median operator is

§5: Intrinsic Topology 373

5.28.3. Let κ be an infinite cardinal number and let $T(\kappa)$ be the tree, consisting of κ copies of $[0,1]$ emanating from one point. Show that the core topology and the IW topology of this "one-dimensional" object are distinct.

5.29. Intrinsic topology in (semi)lattices

5.29.1. Let X be a continuous poset and let $p \in X$. Assume that X is either a complete meet semilattice or a complete and distributive lattice. Show that if $b \ll p$, then the upper set $U(b)$ is a neighborhood of p in the core topology of X. Conversely, if an upper set U is a neighborhood of p in the core topology of X and if b is a lower bound of U, then $b \ll p$.

5.29.2. Show that the set of all maximal elements in a semilattice S is closed and discrete relative to the core topology of S.

5.29.3. Deduce that the core topology of an unconditionally complete and continuous semilattice S equals the Lawson topology iff the set of maximal elements of S is finite.

5.29.4. Show that in a topological semilattice S, all base-point orders are closed.

5.29.5. Let S, T be Lawson meet semilattices and let $f: S \to T$ be a surjective homomorphism preserving the (partial) join operation. Show that f is CC (in addition to being CP; cf. I§1.13.3) and that f is continuous iff all fibers of f are closed. Hint: use Theorem 5.14.

5.30. Intrinsic topology in median spaces (except for (1), van de Vel [1984])

5.30.1. Let X be a complete median algebra. Show that all segments of X are infinitely distributive iff the upper sets $U_b(a)$ for $a, b \in X$ are IW-closed in X.

5.30.2. Show that the following are equivalent for a distributive lattice L.
(i) For each $x \leq y$ in L, the sublattice xy is completely distributive.
(ii) L is an order-convex sublattice of a completely distributive lattice.

5.30.3. Show that a median algebra is IW-compact iff it is absolutely complete.

5.30.4. Let (X, m) be a complete median algebra. Note that $m(a,b,c)$ is the gate of a in bc, and hence that m is IW-continuous in each variable *separately* under the conditions described in Lemma 5.17.
Problem. Find necessary and sufficient conditions on X under which $m: X^3 \to X$ is IW-continuous.

5.30.5. Let X, Y be locally convex median algebras with compact segments and suppose that the respective topologies are weak topologies. Let Z be any compact Hausdorff space and let $f: Z \times X \to Y$ be continuous in each variable separately and CP in the

5.26.4. Let $C \subseteq X$ be convex. Show that the set $ext(C)$ of standard extreme points of C is a closed discrete subspace of C in the internal core topology of C. Conclude that the relative and internal core topologies of a convex subspace can be different, and that a convex set is internally core-compact iff it is a polytope.

5.26.5. Show that the relative core topology, the internal core topology and the IW topology are identical on each polytope.

5.26.6. (for real vector spaces, Meyer and Kay [1973]; continued IV§1.21) Let $f : X \to \mathbb{R}$ be an injective CP function. Prove that X is at most 1–dimensional (affine dimension).

Hint: Argue by contradiction. For each $t \in f(\mathbb{R})$ there is a half-space $f^{-1}(\infty, t) = H_t$ of X. Verify that the corresponding hyperflats $Bd(H_t)$ (see (2) above) for $t \in f(\mathbb{R})$ either coincide or are disjoint. If $f(x) = t$ then choose $x' \in Bd(H_t) \cap H_t$ and verify that the real segments $[f(x'), f(x)]$ are pairwise disjoint.

5.27. Vector spaces

5.27.1. (compare Kakutani and Klee [1963]) Let V be a real vector space. The *finite topology* of V consists of all sets whose intersection with each finite-dimensional linear subspace F of V is intrinsically closed in F. Note that the core topology is coarser than the finite topology. Show that if V is at most countable-dimensional, then V is properly locally convex in the finite topology. Conclude that in these circumstances, the finite topology equals the core topology.

5.27.2. Let the vector space V be equipped with the core topology. Prove that scalar multiplication and addition of vectors,

$$\mathbb{R} \times V \to V, \quad (t,v) \mapsto t \cdot v;$$
$$V \times V \to V, \quad (v_1, v_2) \mapsto v_1 + v_2;$$

are continuous. Hint: as for addition, concentrate on continuity in each variable separately. Then use local convexity.

Klee and Kakutani op. cit. have shown that, relative to the finite topology, addition is continuous in each variable separately, and is discontinuous as a function in two variables, unless the vector space is countable-dimensional.

5.28. Intrinsic topology in trees

5.28.1. Show that the core topology of a tree T consists of all sets $O \subseteq T$ with the following property. If $C \subseteq T$ is a convex totally ordered subset, then $O \cap C$ is open in the order topology of C.

5.28.2. Let T be a complete tree. Show that the IW topology makes T into a locally convex median algebra with compact segments. If T is absolutely complete, then this median algebra is compact (cf. 5.30.3 below).

§5: Intrinsic Topology

Proof. Let $X = X_1 \cup X_2$ be the amalgam. Clearly, a convex set $C \subseteq X$ is complete in X iff $C \cap X_i$ is complete in X_i for each i. It follows that each X_i is a closed subspace of X. The gate map $p_1 : X \to X_1$ is the identity on X_1, whereas on X_2 it equals the gate map onto $X_1 \cap X_2$. Hence p_1 is continuous with respect to the IW topologies of X and of X_1. In a similar way, it follows that the gate map $p_2 : X \to X_2$ is IW-continuous.

Each X_i is FS_2 by Theorem 5.18. We establish this fact for X: two distinct points a, b of X can be separated with a continuous CP functional $X \to \mathbb{R}$. For a, b in the same X_i, this follows by composing a separating functional $X_i \to \mathbb{R}$ with the projection p_i. For $a \in X_1 \setminus X_2$ and $b \in X_2$, just separate a and $p_1(b)$ with a functional $X_1 \to \mathbb{R}$ and compose with the map p_1. ∎

According to Theorem II§3.16, the realization $|G|$ of a finite median graph G admits only one median convexity such that G is a subspace, and each solid cube (with its standard median convexity) is a convex subspace. This convexity can be obtained with a sequence of amalgamations involving the realized cubes. By appealing to the previous result at each stage of this process, we arrive at the following conclusion.

5.25. Corollary. *Let G be a finite median graph and let $|G|$ be equipped with the canonical topology (in which all realized cubes are compact subspaces with the natural topology) and with the canonical median convexity. Then the resulting space is a compact, connected and locally convex median algebra.* ∎

Taking into account Theorem II§3.13, it follows that if a compact connected cubical polyhedron P carries a median convexity with convex standard cubes, then the median operator of P is continuous and P is locally convex.

Further Topics

5.26. Intrinsic topology in BW spaces. Let X be a complete BW space without boundary, equipped with the core topology.

5.26.1. (Cantwell [1978]) Let $Y \subseteq X$ be an affine subset (flat). Show that the core topology of Y equals the relative topology and that Y is closed in X.

5.26.2. Let $H \subseteq X$ be a non-trivial half-space. Show that $\overline{H} \setminus Int\, H$ is a hyperflat (i.e., a maximal, proper affine subspace). Note that completeness is essential; the intended result is not valid in the vector space $\mathbb{Q}^2$.

5.26.3. (Cantwell [1978]; continued IV§1.24) If $K \subseteq X$ is compact, then $aff(K)$ is of finite (affine) dimension. Conclude that if X is locally compact, then $dim(X) < \infty$.

Hint. Suppose $B \subseteq K$ is an infinite affinely independent set spanning $aff(K)$. In $aff(B)$ each flat of type $aff(B \setminus \{b\})$ with $b \in B$ is a hyperplane, and the intersection of these hyperplanes is empty.

§5: Intrinsic Topology   375

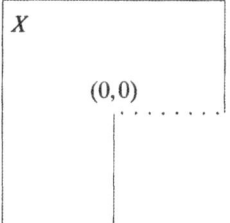

Fig. 7: A compact Hausdorff, non-compatible IW topology

continuous. Show that in the *internal* IW topology, X is compact and Hausdorff, but its median operator is discontinuous. In fact, the join operator "$\vee$" of the lattice X is discontinuous. Strauss [1968] claims that the interval topology of a complete and distributive lattice is Hausdorff iff the lattice is completely distributive. However, as the above example illustrates, this result works only if the interval topology is compatible. The correct result is given by Gierz et al [1980]. The example also shows that the relative IW topology on a median stable subspace need not be the internal IW topology.

5.35. Comparing the core and IW topology
Problem (compare Proposition 5.13.3 and Theorem 5.23). Under which conditions is the IW topology equal to the weak core topology? Do these topologies coincide on polytopes of complete (S_4) spaces?

5.36. Topo-convex properties. There are quite a bit of unsolved problems involving separation and intrinsic topology. The following is a sample list.
Problem. Let X be complete and S_4 and let $f : X \to Y$ be CP. Is it true that f is IW-continuous iff its fibers are closed (compare 5.14 and 5.29)?
Problem. Characterize closure- or interior stability in the core or IW topology. Even for median spaces, no criterion for closure stability is known.
Problem. Characterize the properties NS_i ($i = 2,3,3+,4$) for a convex structure equipped with the core or with the IW topology.
Problem. Idem, for the properties FS_i ($i = 2,3,3+,4$).

5.37. Polyhedra. Let P be a compact polyhedron and let $\Delta(P)$ be its (topological) cone. Show that $\Delta(P)$ can be given the structure of a locally convex median algebra such that all realized simplices of some triangulation of P are relatively convex subsets in P. Hint: consider the simplex graph associated with the vertex set of a suitable triangulation of P; cf. II§4.29.

Notes on Section 5

Continuous posets have been designed as a generalization of the very fruitful notion of continuous (semi-)lattice. See Lawson [1979] for an extensive treatment of the subject. In real vector spaces, the term "core" was first used by Klee [1951]. In [1972], Bryant and Webster extended it to join spaces. For general convex sets, a (still restricted) definition was given by Kay [1977]. The definition presented here is new.

Intrinsic topologies have been considered for linear spaces (see Klee [1951], who uses the term "convex core topology", or Kakutani and Klee [1963]), as well as for semi-lattices (cf. Lawson [1973b]). The core topology has been constructed for join spaces by Bryant [1975]. The author [1984] constructed the IW topology for general convex structures and discussed a "fine" topology for median spaces which (by Theorem 5.23) equals the core topology. The core and IW topologies are implicitly considered by Cantwell [1978] for line spaces. Bryant [1975] has shown that the core topology of a join space satisfies $a \in Cl(a \circ b)$, and that the segment- and ray operator $(a,b) \mapsto a \circ b$ resp., a/b, are both lower semi-continuous. Parts of the proof of Theorem 5.8 are taken from this paper. Proposition 5.8.1 is also given by Kay [1977]. Bryant op. cit. actually discusses a variant of Pasch-Peano spaces (cf. I§4), which is difficult to classify in usual terms. The unicity of topologies on finite-dimensional join spaces, Corollary 5.10, is new; it corresponds with a classical result of Tychonov [1935] for Euclidean spaces. Theorem 5.11 has been obtained for finite-dimensional spaces by Cantwell [1978]. The relationship between partitions of a join space into hyperflats and CP functionals is briefly mentioned in Kay [1977].

The characterization in Theorem 5.6 of core-continuous CP functions in terms of fibers being closed, is new. A corresponding result for IW topologies, 5.14, has been obtained for median spaces by van de Vel [1984]. The examples presented in 5.13 and the results 5.14-22 are taken from this paper too. A restricted version of Lemma 5.19 was obtained by the author in [1983e]: there is at most one compatible compact Hausdorff topology on a median space. This result (which has a counterpart in semilattice theory, cf. Lawson [1973b]) motivated much of the subsequent theory of IW topology for median spaces.

The intrinsic topology of amalgams of median spaces (cf. 5.25) provides a more general approach to the results of the author [1983e] on cubical polyhedra.

CHAPTER **IV**

MISCELLANEOUS

1. Embedding Bryant-Webster Spaces into Vector Spaces

Perspective correspondence, Desargues Property and its dual, Quadrangle Lemma, midpoints, the Embedding Theorem.

2. Extremality, Pseudo-boundary and Pseudo-interior

Synthetic extremality, functional betweenness, Fan-Krein-Milman Theorem, Choquet-Krein-Milman Theorem, Pseudo-boundary and pseudo-interior, δ–extremality, Bauer Maximum Principle, Continuous position, Countable Intersection Theorems.

3. Continuous Selection

Contractibility of convex open covers, selection over a finite-dimensional or infinite-dimensional domain, Selection Theorems of Michael, Nadler and Curtis, Approximation Theorem of Beer, Absolute retracts, Hilbert cube manifolds.

4. Dimension Theory

Convex small inductive dimension *cind*, characterization via hyperplanes or via CP mappings onto cubes, equality theorems for various dimension functions.

5. Dimension and Convex Invariants

Additional inequality $h \leq c$, the inequality $h \leq cind + 1$, Radon numbers of median spaces versus Radon numbers of cubes, low values of c, e versus join-hull commutativity, rank versus dimension of convex hyperspaces.

6. Fixed Points

Invariant Cube Theorem, nonexpansive mappings, labeled crystals: a generalization of Sperner labeling, Kakutani Fixed Point Property, KKM Function Principle, Minimax Theorem, Invariant Arc Theorem.

1. Embedding Bryant-Webster Spaces into Vector Spaces

In view of the similarities between the geometry of Bryant-Webster (BW) spaces and of (convex subspaces of) vector spaces, it is natural to ask how far these classes really are apart. One indication of a close agreement is the result that in complete BW spaces without boundary, all lines are isomorphic with the real line.

In this section, a counterpart is given of the classical and deep result in synthetic geometry, that spaces satisfying Desargues' property can be described with coordinates taken from a division ring (skew field). All extensible BW spaces of dimension ≥ 3 satisfy the former condition; the Moulton plane is a counterexample in dimension two. For complete, extensible spaces, Desargues' property is necessary and sufficient for the existence of an embedding as core-open sets in real vector spaces. The argument involves a continuous "midpoint" operation on rays and an intermediate matroid embedding in a projective space which somehow preserves the segments of the BW space.

The main result can also be interpreted as an intrinsic characterization of a standard class of examples in convexity theory, consisting of all (core-open) convex subspaces of real vector spaces. The efforts needed to establish this result may also illustrate how complicated the recovery[1] of algebraic data from geometric data really is.

1.1. Perspectivity. Let X be a BW space, let $L, M \subseteq X$ be two distinct coplanar lines and let c be a point of the plane spanned by these lines. We assume moreover that $c \notin L \cup M$. The *perspective correspondence (perspectivity)* between L and M with c as a *center of perspectivity* is defined to be the set of all pairs of points (l,m) with $l \in L$, $m \in M$ and such that l, m, c are collinear (compare Topic I§3.29). We denote this correspondence by

$$c : L \overset{c}{\overline{\overline{\wedge}}} M \quad \text{or} \quad L \overset{c}{\overline{\overline{\wedge}}} M.$$

Clearly, a perspectivity between L and M is at least a partial function[2] $L \to M$. If $U \subseteq L$ and $V \subseteq M$ are subsets such that the given perspectivity maps U into V, then we also write $c : U \overline{\overline{\wedge}} V$.

Recall (cf. I§7.22) that an (open) ray at a point p is a set of type

$$p \circ a \cup \{a\} \cup a/p \qquad\qquad p \quad \underline{\quad p \circ a \quad} \bullet a \quad\ a/p$$

where $a \neq p$. In this section, we adopt the viewpoint that a ray R at p *ends* at p. In particular, we will consider on R the inverse of the base-point order based at p. To illustrate this viewpoint, the name "∞" will be used frequently to describe the end point of a ray. Each

---

1. Solutions may exist which do not correspond under an affine isomorphism of the related vector spaces. See I§3.29.1 and 1.22.3 for related details.
2. By a *partial function* $f : X \to Y$ is meant a function into Y with domain included in X.

line through p can be thought of as being composed of two rays at p. The following is not difficult to verify.

1.1.1. *Let R, S be rays ending at p and let $c \in \text{aff}(R \cup S) \setminus (\text{aff } R \cup \text{aff } S)$. Then the domain and the range of $c: R \underset{\wedge}{=} S$ are convex subsets of R and S, respectively, and the perspectivity is a CP isomorphism between them. In an extensible BW space, the domain and range are relatively open subsets of the respective rays.*

We mentioned in I§7.22 that all lines in a complete extensible BW space of dimension > 1 are mutually isomorphic. The following result claims more.

1.2. Proposition. *Let X be a complete BW space without boundary and of dimension ≥ 2. Then each line of X is isomorphic with the real line.*

Proof. All lines of X are isomorphic with an open segment of a line, cf. I§7.22.2. Since X is at least two-dimensional, there exist three non-collinear points a, b, c. Consider a sequence $(a_m)_{m=1}^{\infty}$ of distinct points of the open segment $a \circ b$, such that $a_{m+1} \in a \circ a_m$. By the completeness of X, this sequence has an infimum a_∞ as seen from the base-point a. Replacing a with this infimum if necessary, we may assume that a is the infimum of a countable subset of $a \circ b$.

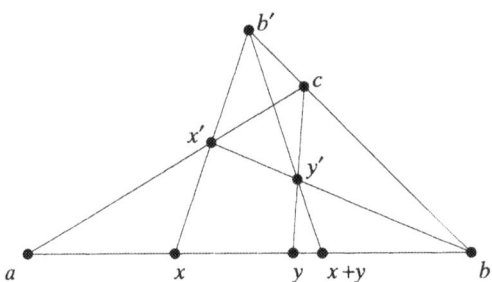

Fig. 1: Addition in a segment

We construct an operator
$$+: (a \circ b)^2 \to a \circ b$$
as follows (see Fig. 1). Fix $x, y \in a \circ b$. Since X has an empty boundary, there is a point $b' \in b/c$. Note that $b \in x/a \cap c/b'$. Hence, by the Pasch Property (J-4), there is a point $x' \in a \circ c \cap x/b'$. Similarly, $a \in x'/c \cap y/b$, providing a point $y' \in x' \circ b \cap y \circ c$. Finally, the open segment $y' \circ b'$ meets $y \circ b$ in a point which we define to be $x + y$. The passage from x to $x + y$ can be carried out as a composition of three perspective projections:

(i) $b': a \circ b \underset{\wedge}{=} a \circ c; \quad x \mapsto x'$;
(ii) $b: a \circ c \underset{\wedge}{=} y \circ c; \quad x' \mapsto y'$;
(iii) $b': y \circ c \underset{\wedge}{=} y \circ b; \quad y' \mapsto x + y$.

It follows that for each y, the operation $x \mapsto x + y$ is the restriction of an isomorphism $ab \approx yb$ with $a \mapsto y$ and $b \mapsto b$. This allows to conclude that adding y at the right preserves

§1: Embedding Bryant-Webster Spaces into Vector Spaces

the base-point order of a.

With this operation at hand, we define $n \cdot x$ recursively for each $n \in \mathbb{N}$ as follows.

$$1 \cdot x = x; \quad (n+1) \cdot x = n \cdot x + x.$$

We verify the following

Archimedian Property. *If $y < x$ then $n \cdot y > x$ for some $n \in \mathbb{N}$.*

Indeed, assume the contrary, and let $y_\infty \leq x$ be the supremum (as seen from a) of $(n \cdot y)_{n=1}^\infty$. Then $y_\infty \in y \circ b$ and hence there is a point $z \in a \circ b$ with $z + y = y_\infty$. Now $z < y_\infty$ and hence $n \cdot y > z$ for some $n \in \mathbb{N}$. But then $(n+1) \cdot y > z + y = y_\infty$, a contradiction.

Let a_m for $m \in \mathbb{N}$ be a countable decreasing set in $a \circ b$ with $a = \inf_m a_m$. We claim that the collection

$$\{ n \cdot a_m \mid m, n \in \mathbb{N} \}$$

is dense in $a \circ b$. To this end, let $x < y$ and fix $z \in a \circ b$ with $x + z = y$. Then $a_m < z$ for some $m \in \mathbb{N}$. Consider the largest number n with $n \cdot a_m \leq x$. We find that

$$x < (n+1) \cdot a_m \leq x + a_m < x + z = y,$$

as desired.

Now $a \circ b$ is a complete, densely and totally ordered set with a countable dense subset and without extremalities. By a standard result,[3] $a \circ b$ is isomorphic with $\mathbb{R}$. ∎

The condition on the dimension of X is essential: see 1.18 below.

1.3. Perspective triangles. Let X be a BW space. A *triangle* in X is a triple of non-collinear (i.e., independent) points and a *dual triangle* in X is a triple of non-concurrent lines. Note that a triple involves a listing order. If (a, b, c) is a triangle then the *side opposite to* a (resp., to b; c) is the line determined by the remaining points b, c (resp., c, a and a, b). Denoting these lines by A, B, C (in order of appearance), we obtain a dual triangle (A, B, C) associated to the given one.

Two triangles (a, b, c) and (a', b', c') are *centrally perspective* from a point p provided $a \neq a'$, $b \neq b'$, $c \neq c'$ and p is collinear with each pair of points a, a'; b, b'; c, c'. Two dual triangles (A, B, C); (A', B', C') are *axially perspective* from a line (axis) P provided $A \neq A'$; $B \neq B'$; $C \neq C'$ and P is concurrent with each pair of sides A, A'; B, B'; C, C'. In particular, these three pairs of sides each have a non-empty intersection (which consists of a single point). Brackets for denoting triangles will often be omitted. The following is a fundamental condition in affine or projective geometry.

Desargues Property. *If two triangles are centrally perspective, and if the corresponding pairs of opposite sides each consist of two distinct, intersecting lines, then the dual triangles are axially perspective.*

---
3.   Kamke [1965, p. 108].

There are a few "degenerate" situations where this property holds, regardless of the space in consideration. For instance, if p equals one of the vertices of the given triangles -- say: $a = p$ -- then $c' \in B \cap B'$ and $b' \in C \cap C'$, whereas $A \cap A'$ by definition is on the line joining b', c'.

1.4. Examples. We check the Desargues Property in vector spaces and in the Moulton plane.

1.4.1. Proposition. *A vector space of dimension ≥ 2 over a totally ordered field has the Desargues Property.*

Proof. Let the triangles (a, b, c) and (a', b', c') be centrally perspective from p. Without loss of generality, $p = \mathbf{0}$. Furthermore, $a \neq a'$; $b \neq b'$; $c \neq c'$. The corresponding pairs of opposite sides

$$A, A'; \quad B, B'; \quad C, C'$$

are distinct and are assumed to intersect in, respectively, l, m, n. The result now follows from a translation of these data into affine algebra:

We have $a = \alpha a'$; $b = \beta b'$; $c = \gamma c'$ for some coefficients α, β, $\gamma \neq 1$. To find l, we must solve the equation

$$tb + (1-t)c = sb' + (1-s)c'$$

for s and t, giving

$$l = \frac{1-\gamma}{\beta-\gamma} \cdot b + \frac{1-\beta}{\gamma-\beta} \cdot c.$$

Note that $\beta \neq \gamma$, since the lines through b, c and b', c' are not parallel. Similarly,

$$m = \frac{1-\alpha}{\gamma-\alpha} \cdot c + \frac{1-\gamma}{\alpha-\gamma} \cdot a; \quad n = \frac{1-\beta}{\alpha-\beta} \cdot a + \frac{1-\alpha}{\beta-\alpha} \cdot b.$$

We obtain that

$$(1-\alpha)(\beta-\gamma)l + (1-\beta)(\gamma-\alpha)m + (1-\gamma)(\alpha-\beta)n = \mathbf{0},$$

and the sum of the coefficients is zero. Hence l, m, n are collinear. ∎

1.4.2. Proposition. *The Moulton plane does not satisfy the Desargues Property.*

Proof. We refer to I§7.9.3 for a description of the Moulton plane. Consider the following two triangles, perspective from the origin (Fig. 2).

(i) (a,b,c) with $a = (\frac{4}{3}, -2)$; $b = (7,0)$; $c = (0,6)$.

(ii) (a',b',c') with $a' = (-4,3)$; $b' = (-3,0)$; $c' = (0,3)$.

The points of intersection of corresponding opposite edges are l, m, n. Rather than relying on a ruler or verifying by computation that they are not collinear, just argue as follows. Among all lines involved, only the one through a, $\mathbf{0}$, a' is not Euclidean. The Euclidean line through a', $\mathbf{0}$ meets the line C in a point a! different from a. If, in the

§1: Embedding Bryant-Webster Spaces into Vector Spaces   383

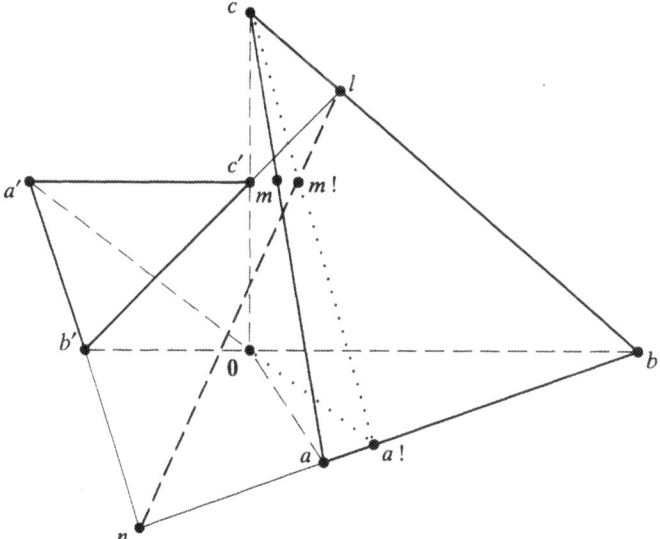

Fig. 2: Failure of the Desargues Property in the Moulton plane

given configuration, a is replaced with this point, we obtain the usual Desargues configuration in **Euclidean** plane. Reconsidering l, m, n in these circumstances gives a different result m! for m only. In *this* position, m will be on the line through l, n. ∎

For our next result we consider the

Dual Desargues Property: *If two dual triangles are axially perspective and if the corresponding pairs of opposite vertices yield distinct, intersecting lines, then these triangles are centrally perspective.*

1.5. Proposition. *A BW space satisfying the Desargues Property also satisfies the dual Desargues Property.*

Proof. Let the given triangles be (a, b, c) and (a', b', c'), with duals (A, B, C) and (A', B', C'), respectively. By assumption, the intersection points l (of A, A'), m (of B, B'), and n (of C, C') are on a line P. We show that the lines L, M, N formed by the pairs of vertices a, a'; b, b' and c, c' pass through one point. To this end, consider the following triangles (the corresponding duals are listed right below them):

(m, c, c'); (n, b, b');
(N, B', B); (M, C', C).

Note that m is not collinear with c, c' for otherwise the lines B and B' coincide with the line through c, c'. Similarly, n is not collinear with b, b'. Furthermore, $B' \neq C'$ since (a', b', c') is a triangle. Similarly, $B \neq C$. Finally, $N \neq M$ by assumption.

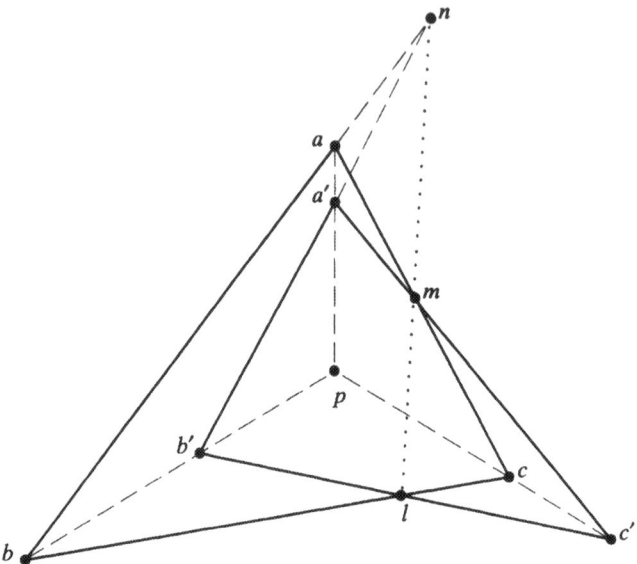

Fig. 3: Dual Desargues Property

The two triangles in consideration are centrally perspective from the point l. Application of the Desargues Property yields that the intersection points of the pairs of corresponding opposite sides,

a of B, C; a' of B', C'; p of N, M,

are collinear.   ∎

The following result is a counterpart for BW spaces of a classical result in affine or projective geometry. We are now exclusively concerned with the matroid convexity of a BW space; consequently, the usual notation ab for a segment will refer to the *affine segment* between the two points, that is, to the line spanned by these points.

1.6. Theorem. *A BW space of dimension ≥ 3 and without boundary has the Desargues Property.*

Proof. Consider two triangles (a, b, c) and (a', b', c') which are centrally perspective from p. We adopt all previous notation; e.g., A is the side opposite to a, .. , l is the intersection point of A, A', etc.. Assume first that p is not coplanar with a, b, c. The intersection points of the corresponding pairs of sides all lie within the intersection of the two planes spanned by a, b, c and a', b', c'. The affine space spanned by all six points is 3–dimensional by the assumptions involved in the Desargues Property. By the Affine Dimension Formula, I§7.14, we find that the intersection of the two triangle planes is a line, as desired.

§1: Embedding Bryant-Webster Spaces into Vector Spaces 385

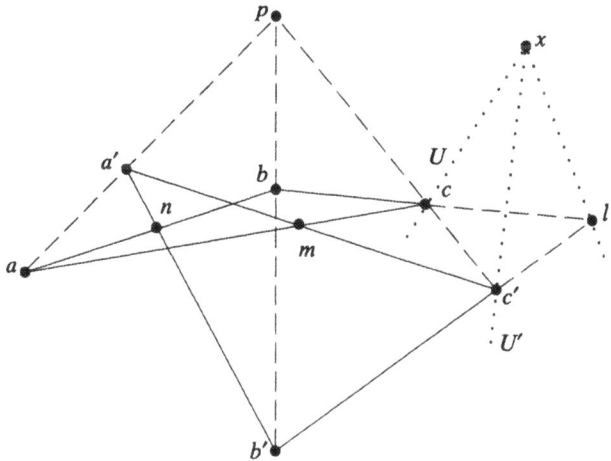

Fig. 4: Moving c, c' off the plane a, b, p

We consider the remaining possibility that p (and hence a', b', and c') are coplanar with a, b, c. The strategy is to move c, c' to suitable positions $\bar{c}$, $\bar{c}'$ outside the plane under consideration. To this end, fix a point $x \notin aff\{a,b,c\}$. We have perspective projections

$$b: cx \underset{\sim}{\overline{\sim}} lx; \quad b': lx \underset{\sim}{\overline{\sim}} c'x$$

such that c, l correspond under b and c', l correspond under b'. By 1.1.1 we obtain a composed CP isomorphism ϕ between a convex neighborhood U of c in cx and a convex neighborhood U' of c' in $c'x$. See Fig. 4. If $\phi(\bar{c}) = \bar{c}'$, then the points $\bar{c}$, $\bar{c}'$, p are common to the planes spanned by $\bar{l}$, b, b' and by x, c, c'. Hence these points are collinear. As a consequence, $\bar{c}'$ is the intersection of the lines $p\bar{c}$; $c'x$.

We now operate the same way with a, a' instead of b, b', and with the intersection point m instead of l. This leads to line neighborhoods V, V' of c, c' in cx, $c'x$, respectively, consisting of points corresponding under the composed projection

$$\psi: cx \overset{a}{\underset{\sim}{\overline{\sim}}} mx \overset{a'}{\underset{\sim}{\overline{\sim}}} c'x.$$

If $\bar{c} \in V$ then -- exactly as before -- $\psi(\bar{c})$ is the intersection of the lines $p\bar{c}$; $c'x$, showing that ϕ and ψ agree on the common part of their domain. After fixing $\bar{c} \in U \cap V \setminus \{c\}$, we obtain two collinear triples, one consisting of $\bar{c}$, b and of a point $\bar{l}$ on the line lx, the other one consisting of $\bar{c}$, a and of a point $\bar{m}$ on the line mx. The point $\bar{c}' = \phi(\bar{c}) = \psi(\bar{c})$ is collinear with both b', $\bar{l}$ and with a', $\bar{m}$.

The triangles a, b, $\bar{c}$ and a', b', $\bar{c}'$ are perspective from p, and p is not coplanar with any one of them. By the first part of the proof, we find that the intersection points

n, of the lines ab; $a'b'$; $\quad \bar{l}$, of the lines $b\bar{c}$; $b'\bar{c}'$; $\quad \bar{m}$, of the lines $\bar{c}a$; $\bar{c}'a'$

are collinear. By using the collinear triples x, $\bar{l}$, l and x, $\bar{m}$, m, we see that the original intersection points l, m, n are coplanar with a, b, c, a', b', c' and with $\bar{l}$, $\bar{m}$, x. The intersection of these planes is one-dimensional, as desired. ∎

We now prepare for a sequence of auxiliary results. The following general principle in topology will frequently be called upon.

1.7. Lemma (*Singleton Intersection Principle*). *Let X be a normal topological space and let $\mathcal{P}$ be a collection of pairs (A,B) with A, $B \subseteq X$ closed, such that $A \cap B$ has precisely one point. If $\mathcal{P}$ is given the subspace topology of $\mathcal{T}_*(X) \times \mathcal{T}_*(X)$, then the function*

$$\mathcal{P} \to X; \ (A,B) \mapsto \text{unique point of } A \cap B$$

is continuous.

Proof. Let O be an open neighborhood of $A \cap B$. Then $A \setminus O$ and $B \setminus O$ are disjoint closed sets and by normality they have disjoint X-neighborhoods O_A and O_B, respectively. Now $<O,O_A>$ is a hyperspace neighborhood of A and $<O,O_B>$ is a hyperspace neighborhood of B. If $A' \in <O,O_A>$, $B' \in <O,O_B>$ then

$$A' \cap B' \subseteq (O \cup O_A) \cap (O \cup O_B) = O.$$ ∎

This result can be applied in case X is a polytope in a complete BW space without boundary and $\mathcal{P}$ consists of pairs of intersecting segments which are not on one line. Implicit use will be made of the fact that segments depend continuously on their end points and that polytopes are core-compact and Hausdorff (hence normal); cf. III§5.8.

By a *quadrangle* in a BW space X is meant an ordered set of four coplanar points, no three of which are collinear, together with the six lines ("sides") joining two out of four points. If two quadrangles are given, then two vertices (one from each of the quadrangles) "correspond" provided they occur at the same place with respect to the involved listing. By a corresponding pair of lines we mean two lines (one from each quadrangle) joining corresponding points.

1.8. Lemma (*Quadrangle Lemma*). *Let X be a Desarguesian BW space without boundary, let $P \subseteq X$ be a line and let (a, b, c, d) and $(\bar{a}, \bar{b}, \bar{c}, \bar{d})$ be two quadrangles such that five pairs of corresponding lines each meet on P. If the sixth side of (a, b, c, d) and of $(\bar{a}, \bar{b}, \bar{c}, \bar{d})$ both meet P, then this pair intersects on P as well.*

Proof. Three intermediate results are needed.

1.8.1. *Let X be a two-dimensional BW space without boundary, let $p \in X$, and let $a_1,..,a_n \in X$ be points, no two of which are collinear with p. Then there exists a neighborhood N of p such that for **any** choice of n points $b_1,..,b_n$ in N, the lines a_ib_i for $i = 1,..,n$ meet two by two.*

Indeed, for each couple a_i, a_j of distinct points, we use extensibility to find a point p_{ij} with

$$p \in p_{ij} \circ a_i \circ a_j.$$

§1: Embedding Bryant-Webster Spaces into Vector Spaces

Note that the three vertices involved at the right are affinely independent. Hence, by virtue of Proposition III§5.2.4, the set $N_{ij} = p_{ij} \circ a_i \circ a_j$ is a neighborhood of p. By the associativity of join, (J-3), any point $x \in N_{ij}$ extends away from a_i to a point of $p_{ij} \circ a_j$. The same goes with i, j interchanged. It follows from the Pasch Property (J-4) that two lines, each joining a point of N_{ij} with a_i, resp., a_j, must intersect. The desired neighborhood of p is $\cap_{i \neq j} N_{ij}$.

For the remaining part of the proof, only the matroid structure of X is relevant; henceforth, the segment notation ab refers to the line passing through a and b. Next comes a result of independent interest.

1.8.2. *Let X be as above, consider a point p and an axis (line) P, and let (a, b, c) be a triangle such that $p, a, b, c \notin P$ and p is not on one of the sides ab, bc, ca of the triangle. Suppose each of these sides meets the axis P and let N be any neighborhood of p. Then there is a convex neighborhood U of p in the line ap and, for each $a' \in U$, a triangle (a', b', c') with vertices in N, which is both centrally perspective (from p) and axially perspective (from P) with (a, b, c).*

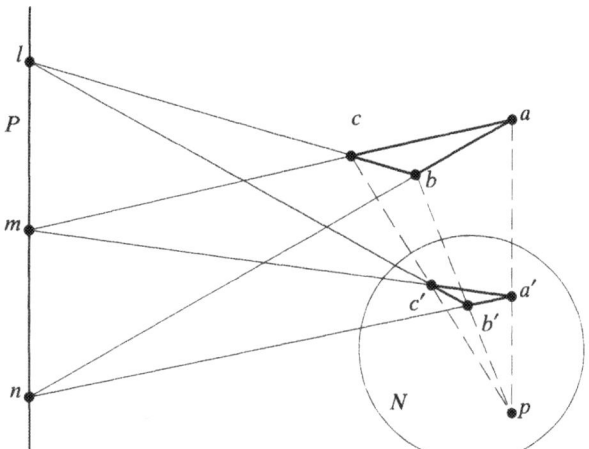

Fig. 5: Axially perspective triangles with a given axis P

The proof goes as follows (Fig. 5). Let $l, m, n \in P$ be the intersection of, respectively, bc, ca and ab with P. Passing to a smaller neighborhood N if needed, by 1.8.1, we may assume that two lines, each joining a point of the finite set $\{a, b, c, l, m, n\}$ with a point of N, intersect. Take any point $a' \in ap \cap N$. Then the pairs of lines bp, na' and cp, ma' intersect in, say, b' and c', respectively. By the Singleton Intersection Principle, applied to a sufficiently large, compact, Hausdorff polytope of X, we can move a' a bit closer to p to achieve that $b', c' \in N$. Such a position of a' can be determined at both sides of p on ap, to delimit a line neighborhood of p as required. We assume that the present choice of a' lies within the resulting line neighborhood of p, and we proceed with the above obtained points. The triangles $(a', b', p); (m, l, c)$ are axially perspective: the intersection points of corresponding pairs of sides are n, b, a. The lines through corresponding vertices meet two by two, and the Desargues Property yields that these triangles are centrally perspective. As $a'm$ and pc both pass through c', the latter must be the point of concurrence. In particular, the points l, b', c' are collinear.

This leads to the conclusion that the triangles (a, b, c) and (a', b', c') are axially perspective from P whereas by construction, they are centrally perspective from p.

Remark 1. The construction of b' above depends only on the knowledge of a, b, a', p, P, *not* on c. Indeed, b' has been obtained as the meet of bp with na', where n is the meet of P with ab.

Remark 2. In a different viewpoint, the construction of b', c' involves the intersection points m, n of the two sides ac, ab with the axis P (the intersection l of the third side with P is used at a later stage to legitimate the conclusion of axial perspectivity). This directly leads to the following by-product, finishing the preliminary considerations.

1.8.3. Modified Desargues Property. *Let (a, b, c) and (a', b', c') be centrally perspective triangles such that both pairs of sides ab, $a'b'$ and ac, $a'c'$ intersect on a line P, which is met by the individual remaining sides bc, $b'c'$. Then these triangles are axially perspective from P.*

We are now ready for a proof of the lemma. The intersection of a particular side -- e.g., of ab -- with the axis P is denoted by "s" with the side name as a label; currently, s_{ab}. We assume that $s_L = s_{\bar{L}}$ for each pair L, $\bar{L}$ of corresponding sides, except for the pair cd, $\bar{c}\bar{d}$.

Let $p \in X$ be chosen outside P and outside any of the lines joining two of the points $a,..,d,\bar{a},..,\bar{d},s_{ab},..,s_{\bar{c}\bar{d}}$. By (1), there is a convex neighborhood N of p such that any two lines joining one of the listed points with a point of N have a non-empty intersection. We apply 1.8.2 to each of the triangles (a, b, c), (a, b, d) and (a, c, d). Starting with a point a' in a sufficiently small ap-neighborhood of p, let (a', b', c') be perspective with (a, b, c) from both p and P. If we operate on (a, b, d) instead, then a perspective triangle (a', b', d') results, where by Remark 1, the point b' is the same as before. Finally, starting with (a, c, d), and using the same observation twice, we arrive at a perspective triangle (a', c', d').

Now consider the triangle $(\bar{a}, \bar{b}, \bar{c})$: its sides intersect P in the same points as the corresponding sides of (a, b, c) do. Hence (a', b', c') is axially perspective with it and, as its vertices are in the neighborhood N, we conclude that the lines $a'\bar{a}$, $b'\bar{b}$, $c'\bar{c}$ meet two by two. The (usual) Desargues Property implies that these lines concur, say at $\bar{p}$. Dealing with $(\bar{a}, \bar{b}, \bar{d})$ in a similar way, we conclude that this triangle is centrally perspective with (a', b', d'). The center of perspectivity is, of necessity, equal to $\bar{p}$.

This shows, in particular, that the triangle $(\bar{a}, \bar{c}, \bar{d})$ is centrally perspective with (a', c', d'). The modified Desargues Property then yields that $s_{cd} = s_{\bar{c}\bar{d}}$. ∎

1.9. Midpoints. Let X be a BW space of dimension ≥ 2, let R be a ray ending at ∞, and let $a \neq b$ in R. By a *midpoint* of a, b is meant a point $c \in a \circ b$ with the following property (Fig. 6).

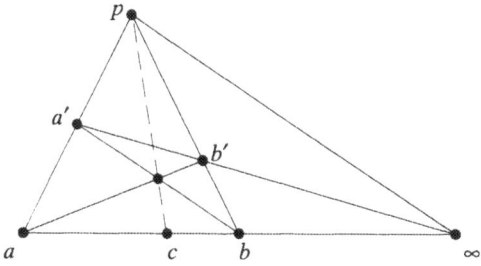

Fig. 6: Midpoint construction

There exists a ray $R' \neq R$ at ∞, a point $p \notin R \cup R' \cup \{\infty\}$, such that if a', b' are perspective images of a, b under $p: R \barwedge R'$, then p, c and the intersection point of $a \circ b'$, $a' \circ b$ are collinear[4]. If $a = b$, then the midpoint is defined to be this point. Given $a, b \in R$, the configuration described in the above definition can be effectively obtained in any BW space of dimension ≥ 2. The elementary geometric argument is left to the reader.

As a direct consequence of the Quadrangle Lemma, we have:

1.9.1. *In a Desarguesian BW space without boundary, each pair of points on a ray has only one midpoint.*

The unique midpoint of a, b on a fixed ray R will be denoted by $a \blacksquare_R b$, or simply $a \blacksquare b$ if there can be no misunderstanding about the ray R. Another consequence of the Quadrangle Lemma is the following.

1.9.2. *In a Desarguesian BW space without boundary, perspectivity preserves midpoints: Let $R \neq R'$ be rays at ∞ and $p \notin R \cup R' \cup \{\infty\}$. If the perspectivity $p: R \barwedge R'$ maps a to a' and b to b', then it maps $a \blacksquare_R b$ to $a' \blacksquare_{R'} b'$.*

We arrive at a sequence of four lemmas involving midpoints. Throughout, X is a complete extensible BW space with Desargues' property and of dimension ≥ 2.

1.10. A sequence of lemmas

1.10.1. Lemma. *Let R be a ray with end point ∞. Then:*
(1) *The midpoint function $R^2 \to R$ is continuous.*
(2) *If $a \in R$, then the function $x \mapsto a \blacksquare_R x$ is a homeomorphism of $\{a\} \cup a \circ \infty$ onto itself.*

---
4. The reader acquainted with projective geometry will recognize the construction of the *harmonic fourth* to a, b, ∞.

Proof of (1). Continuity in a point (x,y) with $x \neq y$ is a direct consequence of the continuity of the segment operator and of the Singleton intersection principle. Continuity at pairs of type (x,x) follows from local convexity and from the fact that the midpoint of two elements lies in their join.

Proof of (2). Let $a \leq b < \infty$ on R. We construct a "doubled" point $2b$ by the method of "addition" (cf. Fig. 7; compare 1.2).

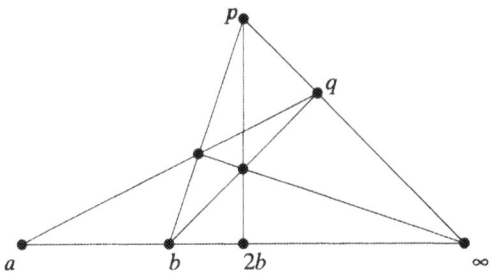

Fig. 7: Doubling a point

Let $T \neq R$ be a ray through ∞ and fix two auxiliary points $p, q \in T$ such that $q \in p \circ \infty$. Project b to aq from p, project this point onto bq from ∞ and finally project this image to R from p. The resulting point $2b$ depends continuously on b by the Singleton intersection principle. We consider the intersection c of the segments aq and $p2b$ as a center of perspectivity,

$$c : T \overset{\sim}{\underset{\sim}{}} R,$$

for which $p \mapsto 2b$, $q \mapsto a$ and for which ∞ is kept fixed. Hence $p \blacksquare q$ goes to $a \blacksquare 2b$. Observe that the line through c and b meets T by definition in $p \blacksquare q$. We conclude that $b = a \blacksquare 2b$.

This shows that the operation of doubling a point of $\{a\} \cup a \circ \infty$ is a continuous right inverse to the midpoint operation. As $b < 2b$ and $a = 2a$, it follows that the former operation is even bijective and hence that both are homeomorphisms. ∎

Two segments ab and cd on the ray R are *equi-perspective* provided there exist distinct rays S, $T \neq R$ ending at ∞ and such that S cuts the angle[5] RT, together with two points $u, v \in S$ and two points $p, q \in T$ such that the perspectivities

$$p : S \overset{\sim}{\underset{\sim}{}} R; \quad q : S \overset{\sim}{\underset{\sim}{}} R$$

map the segment uv onto ab and cd, respectively.

1.10.2. Lemma. *Let ab and cd be equi-perspective segments on R. Then each half of ab is equi-perspective with each half of cd.*

---
5.   I.e., S meets each line segment joining a point of R with a point of T; cf. I§7.22.

§1: Embedding Bryant-Webster Spaces into Vector Spaces 391

Proof. First, consider distinct rays R, S, T ending at ∞ and such that S cuts the angle RT. Suppose $p \in T$ and the perspectivity

$$p : S \overline{\wedge} R$$

maps the segment $uv \subseteq S$ onto ab. As perspectivities preserve midpoints, we conclude that projection from p maps the first (resp., second) half of uv to the first (resp., second) half of ab.

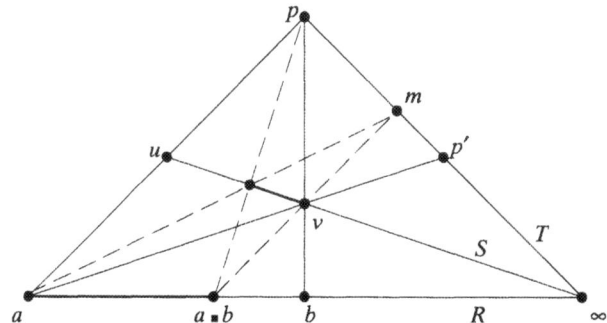

Fig. 8: Equi-perspectivity of half-segments

We next consider the two perspectivities (cf. Fig. 8)

$$v : R \overline{\wedge} T ; \quad a : T \overline{\wedge} S.$$

The first gives $b \mapsto p$ and $a \mapsto p'$ (defining p'), whence v maps $a \blacksquare b$ to $m = p \blacksquare p'$. The second perspectivity gives $p \mapsto u$ and $p' \mapsto v$, whence $m \mapsto u \blacksquare v$. It follows that the second half of uv corresponds with the first half of ab under the perspectivity $m : S \overline{\wedge} R$.

Assume for a moment that the line through b, u meets T. Then, in a similar way, it is possible to find a center of perspectivity on T for which the first half of uv corresponds with the second half of ab. Let ab be equi-perspective with cd. This gives a segment uv on S (as before) and two centers of perspectivity on a third ray T ending at ∞, for which uv corresponds with ab and with cd respectively. We also assume that all lines through one of a, b, c, d and one of u, v meet T. Then, by using suitable halves of uv, the previous argument at once yields the desired result.

Suppose that the intersection points as described above do not all exist. In this case we use Lemma 1.10.1 to find w such that $v = u \blacksquare w$. After projecting S to R from p, we obtain a "doubled segment" of ab and of cd. To these doubled segments we apply the first part of the proof, yielding perspective centers on T for which vw corresponds with the first half of each doubled segment, namely, ab and cd. As $v < w$ on S, the extension of each of a, b, c, d away from each of v, w contains a point of T. ∎

1.10.3. Lemma. *Let aa' and bb' be equi-perspective segments on R such that $a < a' \leq b' < b$. Then*

$a \blacksquare b = a' \blacksquare b'$.

Proof. Let $S \neq T$ be rays ending at ∞ such that S cuts the angle RT, let u, v on S, and let p, $q \in T$ be centers of perspectivity on T such that $p: S \overset{\wedge}{=} R$ maps uv onto aa' and

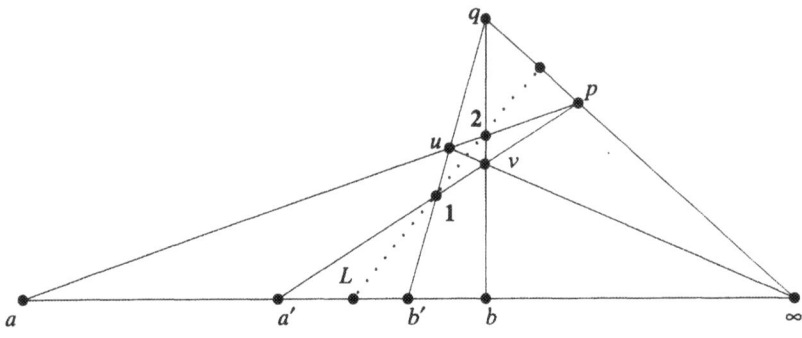

Fig. 9: Midpoint formula

$q: S \overset{\wedge}{=} R$ maps uv onto bb'. Join-hull commutativity implies that $q < p$ in the (reversed base-point) order of T; cf. Fig. 9. The Pasch Property yields two points $\mathbf{1} \in a'p \cap b'q$ and $\mathbf{2} \in ap \cap bq$. Now use the perspectivities

$\mathbf{1}: R \overset{\wedge}{=} T$ and $\mathbf{2}: T \overset{\wedge}{=} R$.

Under **1**, we have $a' \mapsto p$, $b' \mapsto q$ and hence $a' \blacksquare b' \mapsto q \blacksquare p$. Under **2**, we find that $p \mapsto a$, $q \mapsto b$ and hence that $p \blacksquare q \mapsto a \blacksquare b$. The first perspectivity yields that the line L through $a' \blacksquare b'$ and $p \blacksquare q$, which passes through **1**, must also go through **2** (cf. the quadrangle construction of a midpoint). The second perspectivity then yields that L goes through $a \blacksquare b$. We conclude that $a \blacksquare b = a' \blacksquare b'$. ∎

We are lead to the following construction. Fix two points $o < e$ on the open ray R. A sequence of finite subsets of oe is constructed by induction as follows. Let $D_1 = \{o, e\}$. Given D_n, we let D_{n+1} consist of all points of D_n, together with the midpoints of successive pairs of D_n. To facilitate the computations below, the following notation will be adopted: $d_1^0 = o$; $d_1^1 = e$, and if the sequence

$$d_n^0 < d_n^1 < .. < d_n^{2^n},$$

lists the elements of D_n in order, then we put

$$d_{n+1}^{2k} = d_n^k; \quad d_{n+1}^{2k+1} = d_{n+1}^{2k} \blacksquare d_{n+1}^{2k+2}.$$

It follows by induction from Lemma 1.10.2 that all segments spanning successive elements of D_n are equi-perspective.

Finally, we let $D = \cup_{n=1}^{\infty} D_n$. We refer to D as the set of **dyadic points** of oe. As an application of Lemma 1.10.3, we show that the midpoint of any two dyadic elements is again dyadic. Specifically:

1.10.4. Dyadic Midpoint Formula.

$$d_n^k \bullet d_n^l = d_{n+1}^{k+l} \qquad (0 \le k \le l \le 2^n).$$

Proof. The result is valid for $n = 1$ by construction. We proceed by induction on n and on $l - k$. If $l - k \le 1$, then the result is valid by construction. If $l - k \ge 2$, then consider the points d_n^{k+1} and d_n^{l-1}. Lemma 1.10.3 and the induction hypothesis yield that the desired midpoint equals the midpoint of the last named points. ∎

1.10.5. *D is a dense subset of the segment oe.*

Proof. Assume to the contrary that for some $p \in oe$,

$$\sup L(p) \cap D = a < b = \inf U(p) \cap D.$$

Now $L(p) \cap D$ is a (totally ordered) net converging to its supremum, and $U(p) \cap D$ is a net converging to its infimum. Hence the net $d' \bullet d''$ for $d' \in L(p) \cap D$ and $d'' \in U(p) \cap D$ converges to $a \bullet b$ by continuity. But $d' \bullet d'' \in D$ and $a \bullet b \in a \circ b$, a contradiction. ∎

These efforts lead to the following result of independent interest.

1.11. Proposition. *Let X, $\tilde{X}$ be complete Desarguesian BW spaces without boundary, let R, $\tilde{R}$ be rays in X, resp., $\tilde{X}$ ending at ∞, resp., $\tilde{\infty}$ and let $o < e$ on R and $\tilde{o} < \tilde{e}$ on $\tilde{R}$. Then there is a midpoint-preserving isomorphism of segments*

$$oe \approx \tilde{o}\tilde{e}.$$

Proof. In each of the segments, we construct the set of dyadic elements as devised above. An order-isomorphism f' between these sets is defined by assigning to the point with label d_n^k (where $0 \le k \le 2^n$) in oe the dyadic element of $\tilde{o}\tilde{e}$ bearing the same label. As both segments in consideration are absolutely complete and totally ordered sets, and as the dyadic sets are dense, we obtain a unique continuous extension of f' to a map $f: oe \to \tilde{o}\tilde{e}$. By construction, this extension is an isomorphism. It preserves the midpoint operation on a dense set of pairs. As f and the midpoint function are both continuous, it follows that f preserves midpoints throughout. ∎

We are now ready to handle the isomorphism of two-dimensional "simplices".

1.12. Proposition. *Let X, $\tilde{X}$ be complete extensible BW spaces with Desargues' property and let $a, b, c \in X$, $\tilde{a}, \tilde{b}, \tilde{c} \in \tilde{X}$ be independent points. Then the triangles $co\{a,b,c\}$ and $co\{\tilde{a},\tilde{b},\tilde{c}\}$ are isomorphic.*

Proof. Let $p \in b/a$ and $q \in c/a$ and consider p, q as the end points of a ray through a and, respectively, b, c. We operate the same way in $\tilde{X}$, yielding rays at $\tilde{p}$, $\tilde{q}$ through $\tilde{a}$ and, respectively, $\tilde{b}$, $\tilde{c}$. On each ray this leads to a continuous midpoint function, and Proposition 1.11 provides us with two midpoint preserving maps

$$f_b: ab \to \tilde{a}\tilde{b}; \quad f_c: ac \to \tilde{a}\tilde{c}.$$

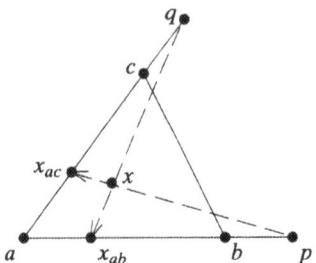

Fig. 10: Isomorphism of triangles

The desired isomorphism ϕ is constructed as follows (Fig. 10). Let $x \in co\{a,b,c\}$. By perspective projection of x from q, resp., p, we obtain two elements $x_{ab} \in ab$ and $x_{ac} \in ac$, which we map into $\tilde{X}$ by f, g. The join of $\tilde{x}_{ab} = f(x_{ab})$ with $\tilde{p}$ and of $\tilde{x}_{ac} = f(x_{ac})$ with $\tilde{q}$ meets in a point $\tilde{x}$ (Pasch Property), which we define to be $\phi(x)$. Continuity of ϕ follows from the Singleton Intersection Principle. We next show that f is a CP function. As the process is reversible, this yields that ϕ is a CP isomorphism.

Consider $x \neq y$ in $co\{a,b,c\}$. Define the midpoint of xy to be the point z whose coordinates are $x_{ab} \cdot y_{ab}$ and $x_{ac} \cdot y_{ac}$. By considering the appropriate quadrangles, it is easily seen that $z \in xy$. By construction, $\phi(z)$ is the midpoint of $\tilde{x}\tilde{y}$. By virtue of 1.10.5, we obtain a dense subcollection of xy mapped by ϕ to a dense subcollection of $\tilde{x}\tilde{y}$. As ϕ is continuous, we conclude that $f(xy) \subseteq \tilde{x}\tilde{y}$. Consequently, ϕ is CP. ∎

We now extend the previous result to higher dimensions. For technical reasons, we prefer to work with an "open hull" operator $co°$ as described below. Let B be a subset of a BW space X. Then we define $co°(B)$ to be the core of $co(B)$ relative to $aff(B)$. It is not difficult to see that $co°(B)$ can be recursively constructed as follows. Let

$$B = \{b_\alpha \mid \alpha < \kappa\},$$

where α and κ are two ordinals. Then $co°(\{b_0\}) = \{b_0\}$ and for each $\alpha < \kappa$,

(*) $\qquad co°\{b_\beta \mid \beta \leq \alpha\} = b_\alpha \circ \left(\bigcup \{ co°\{b_\gamma \mid \gamma \leq \beta\} \mid \beta < \alpha \} \right).$

1.13. Lemma. *Let X, $\tilde{X}$ be BW spaces without boundary and let $A_0 \subseteq X$, $\tilde{A}_0 \subseteq \tilde{X}$ be affinely independent sets of at least three points and suppose there is an isomorphism*

$$f_0 : co°(A_0) \to co°(\tilde{A}_0); \quad x \mapsto \tilde{x}.$$

If $a_0 \in X \setminus A_0$ and $\tilde{a}_0 \in \tilde{X} \setminus \tilde{A}_0$ are affinely independent of A_0 and $\tilde{A}_0$, respectively, then there is an isomorphism between $co°(A_0 \cup \{a_0\})$ and $co°(\tilde{A}_0 \cup \{\tilde{a}_0\})$, extending f_0.

Proof. Take $a_1 \in A_0$. We put

$A = A_0 \cup \{a_0\}; \quad A_1 = A \setminus \{a_1\};$

$\tilde{A} = \tilde{A}_0 \cup \{\tilde{a}_0\}; \quad \tilde{A}_1 = \tilde{A} \setminus \{\tilde{a}_1\}.$

Fix $b \in a_0/a_1$ and $\tilde{b} \in \tilde{a}_0/\tilde{a}_1$ and let

§1: Embedding Bryant-Webster Spaces into Vector Spaces

$$\pi_b: co(A_1) \to co(A_0); \quad \pi_{\tilde{b}}: co(\tilde{A}_1) \to co(\tilde{A}_0)$$

be the perspective projections from b, resp., $\tilde{b}$. Then π_b and $\pi_{\tilde{b}}$ are CP isomorphisms and we consider the isomorphism

$$f_1 = \pi_{\tilde{b}}^{-1} \circ f_0 \circ \pi_b : co(A_1) \to co(\tilde{A}_1).$$

We construct a function

$$g: co^\circ(A) \to co^\circ(\tilde{A})$$

as follows. By appealing to the formula in (*) above, we see that each point x of $co^\circ(A)$ has two "coordinates" $x_0 \in co^\circ(A_0)$ and $x_1 \in co^\circ(A_1)$, obtained by perspective projection from a_1 resp., a_0 into the opposite face. These coordinates are mapped to $\tilde{x}_0$ resp., $\tilde{x}_1$ by f_0 and f_1. Two applications of the Pasch Property yield that the segments $\tilde{a}_0\tilde{x}_0$, $\tilde{a}_1\tilde{x}_1$ meet (Fig. 11). The resulting point $\tilde{x}$ is defined to be $g(x)$.

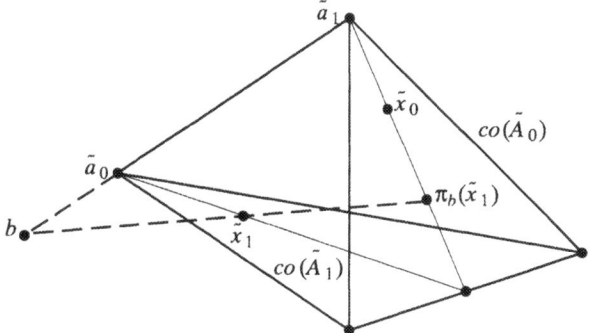

Fig. 11: Isomorphism of simplices

To see that g is CP we use the following representation (which works only because $co(A_0)$ and $co(A_1)$ are at least two-dimensional): $co^\circ(A)$ is isomorphic with the subspace of the product $co^\circ(A_0) \times co^\circ(A_1)$, consisting of all pairs (x_0, x_1) such that

$$\dim \mathrm{aff}\{a_0, a_1, x_0, x_1\} = 2.$$

The isomorphism is given by the pair of perspective projections from a_i onto $co(A_i)$ for $i = 0, 1$. A similar representation works with the corresponding sets in $\tilde{X}$. Then g is represented by the pair of CP mappings f_0, f_1 and hence is CP itself.

By considering a similar process in the opposite direction, $\tilde{X} \to X$, we can conclude that g is a CP isomorphism. ∎

1.14. Proposition. *Let B be a basis for a complete and extensible BW space with the Desargues Property. If $\#B \geq 3$ then $co^\circ(B)$ is CP isomorphic with the open convex hull of an affine basis of the same cardinality in a real vector space.*

Proof. Let $\tilde{B}$ be an affine basis for a real vector space such that $\#\tilde{B} = \#B$. If $\#B = 3$ then Proposition 1.12 gives the desired result. Note that the result works for $\#B = 2$ as well; cf. Proposition 1.2. We assume $\#B > 3$. Fix a bijection $f : B \to \tilde{B}$; $b \mapsto \tilde{b}$ and well-order B, $\tilde{B}$ such that f is an order-isomorphism. The collection of all isomorphisms $co^\circ(S) \approx co^\circ(\tilde{S})$, where $S \subseteq B$ and $\tilde{S} \subseteq \tilde{B}$ are initial subsets corresponding under f, is clearly inductively ordered by the relation of extension. We consider a maximal element of it, $co^\circ(S) \approx co^\circ(\tilde{S})$. Then S has at least three elements. If $S \neq B$, then there is a point $b \in B$ such that

$$S = \{s \mid s < b\}; \quad \tilde{S} = \{\tilde{s} \mid \tilde{s} < \tilde{b}\}.$$

Lemma 1.13 then allows to extend the CP isomorphism to an isomorphism

$$co^\circ(\{s \mid s \leq b\}) \approx co^\circ(\{\tilde{s} \mid \tilde{s} \leq \tilde{b}\})$$

with a properly larger domain, a contradiction. ∎

1.15. Theorem (*Embedding Theorem*). *Let X be a complete BW space without boundary. If X is at least three-dimensional, or two-dimensional and Desarguesian, then X embeds as a core-open subspace of a real vector space.*

Proof. Let B be a basis for the matroid convexity of X and let V be the vector space which is freely generated by B. By Proposition 1.14, we obtain a CP isomorphism

$$co_X^\circ(B) \approx co_V^\circ(B).$$

For convenience, we will identify the two sets along such an isomorphism. Let $\mathbb{P}$ be the projective extension of V (cf. I§2.7.3). A *matroid* embedding $g : X \to \mathbb{P}$ is constructed as follows.

Let $x \in X$ and fix two points p_1, p_2 of $co^\circ(B)$. Now $co^\circ(B) = core\, co(B)$, hence we obtain two points $q_i \in x \circ p_i$ for $i = 1, 2$. Define $g(x)$ to be the intersection in $\mathbb{P}$ of the lines $p_i q_i$ for $i = 1, 2$.

We first show that $g(x)$ does not depend on the choice of the auxiliary points p_i, q_i. To this end, consider one extra pair of points p_3, q_3 with $q_3 \in x \circ p_3$. Note that the points q_i can be moved freely along $x \circ p_i$ (but within $co^\circ(B)$). Move q_1 so close to p_1 that p_1/p_2 meets q_1/q_2 within $co^\circ(B)$. Similarly, we move q_3 towards p_3 such that

$$p_3/p_2 \cap q_3/q_2 \cap co^\circ(B) \neq \emptyset \neq p_3/p_1 \cap q_3/q_1 \cap co^\circ(B).$$

By the Desargues Property of X, the triangles (p_1, p_2, p_3) and (q_1, q_2, q_3) are axially perspective within $co^\circ(B)$. Reconsidering the situation in $\mathbb{P}$, we find that these triangles are centrally perspective, and hence that all three projective lines $p_i q_i$ meet in $g(x)$.

We next show that the function g is an embedding of matroids. To distinguish between the standard and matroid convexities of the various spaces, we reserve the term "convex set" exclusively with respect to BW spaces. In all other situations, we use typical matroids terms (line, flat, affine or projective set, etc.). All matroids in consideration are interval spaces by (weak) join-hull commutativity (cf. I§2.15 and I§7.12(2)), and hence it suffices to verify that lines of X map onto lines of $g(X)$ (cf. I§1.12). First, consider three distinct collinear points $l, m, n \in X$, say: $m \in l \circ n$. We aim at constructing two centrally perspective triangles of $co^\circ(B)$, which are axially perspective with respect to the line

§1: Embedding Bryant-Webster Spaces into Vector Spaces

through l, m, n. Let $a \in co°(B)$ be coplanar with l, m, n and take any point $b \in a \circ n \cap co°(B)$. Now $b \circ l$ meets $a \circ m$ in a point c by the Pasch Property. Taking b closer to a if necessary, we may assume that $c \in co°(B)$. Next, fix $p \in co°(B)$ not coincident with one of the lines passing through two of the points a, b, c, l, m, n. Application of 1.8 then yields a triangle (a',b',c') in $co°(B)$ which is centrally perspective with (a,b,c) from p. Reconsidering this situation in the projective space $\mathbb{P}$, we conclude that these triangles are axially perspective. By construction, the intersection points of the corresponding pairs of sides are $g(l)$, $g(m)$, $g(n)$, which are therefore collinear.

We have shown so far that g is a well-defined CP injection. Suppose next that $g(l)$, $g(m)$, $g(n)$ are distinct collinear points. Choose a point p in $co°(B)$ outside the corresponding $\mathbb{P}$-line. Going back to X, we find that $p \notin aff\{l,m,n\}$ since g is CP. Join each of l, m, n to p and pick a point l', m', n' in $co°(B)$ on the way. If l, m, n are not collinear, then $aff\{l',m',n',p\}$ is three-dimensional. Now g is the identity on $co°(B)$, but in $\mathbb{P}$, the points l', m', n', p span a plane and hence are dependent.

At the next stage of proof, some elementary topological considerations are of use. Recall that $\mathbb{P}$ is a matroid quotient of the (punctured) vector space $V \times \mathbb{R}$. Let the latter be given the core topology (with reference to the *standard* convexity). The quotient map q can now be used to introduce a quotient topology on $\mathbb{P}$ as well. Note that the relative topology on a closed subset C of $\mathbb{P}$ equals the quotient topology associated with the restriction $q^{-1}(C) \to C$. Hence each projective subset of finite projective dimension is topologically a quotient of a punctured Euclidean space partitioned with the remainders of all lines through the origin. It follows that each projective set of finite (projective) dimension is compact and Hausdorff, and that the projective lines are topological copies of a circle.

As a consequence, a line joining two distinct points is built with two segments meeting in the original pair of points. A second consequence is that the function g is continuous on each finite-dimensional affine subspace of X: use the Singleton Intersection Principle. As g is injective, the last two observations lead to the following.

1.15.1. *Under the matroid embedding g, a segment joining two points of X maps to one of the two $\mathbb{P}$-segments joining the image points.*

A projective hyperflat P of $\mathbb{P}$ inverts to a linear hyperflat of the punctured vector space, and hence P is closed. Its complement is an open set, in which each pair of points can be joined with precisely one (circle) segment. It is easily seen to be a complete BW space without boundary, and its relative topology is the core topology. The final intermediate result is the following. We do no longer distinguish between X and $g(X)$.

1.15.2. *There is a projective hyperflat $P \subsetneq \mathbb{P}$ disjoint from X.*

Indeed, let $P_0 \subseteq \mathbb{P}$ be any projective hyperflat and suppose it meets X. Put $H_0 = P_0 \cap X$. Observe that X and $\mathbb{P}$ share a matroid basis, so $X \not\subseteq P_0$. If $x \in X \setminus H_0$ then by the Relative Hull Formula I§1.9.1,

$$X = X \cap pr(\{x\} \cup P_0) = \mathit{aff}_X(\{x\} \cup H_0).$$

Hence $H_0 = P_0 \cap X$ is a hyperflat of X. According to III§5.8.1, it divides X into two convex relatively open sets C_1 and C_2. Let A denote the affine space $\mathbb{P}\setminus P_0$. If $a, b \in C_i$, then among the two $\mathbb{P}$–segments joining a, b there is only one meeting P_0; hence the other segment corresponds with the X–segment by 1.15.1 and is included in C_i. It follows that both sets C_i are convex and open in A. By virtue of Theorem III§5.8, there is a hyperflat $H \subseteq A$ separating between these sets. Let $P = pr(H)$. If X meets P then (as $X \subseteq C_1 \cup C_2 \cup P_0$) it also meets $P \cap P_0$, say, in p. Let $q \in X\setminus(P \cup P_0)$ and take $r \in p/q$. Note that one of q, r must be in C_1, the other one in C_2. In the affine space A, we see that the line through q, r is parallel with H, although it connects a point of C_1 with a point of C_2.

By 1.15.2, X embeds as a convex subspace of a space of type $\mathbb{P}\setminus P$, where P is a projective hyperflat in $\mathbb{P}$. This space is easily copied onto the original vector space V. Note that the embedded copy of X is core-open since $\mathit{aff}(X) = \mathbb{P}\setminus P$ and X has no boundary points. ∎

The following examples may illustrate the sharpness of the main theorem.

1.16.1. A spherical BW space. Consider the complete BW space X defined in I§7.9.2. This space is the intersection of the n-sphere S^n with a half-space of $\mathbb{R}^{n+1}$ which is maximal with the property of missing the origin. The convexity of X consists of the intersections of X with (proper) wedges at the origin. Note that X has boundary points. Two relevant properties of this space have been observed in I§7.9.2.

(i) Each polytope of X is isomorphic with a convex subspace of Euclidean n-space.

Therefore, if two triangles are centrally perspective from a seventh point, and if the corresponding pairs of sides meet in three more points, then the convex hull of this ten-point configuration can be copied into Euclidean n-space to conclude that the last mentioned three points are collinear. Hence X has Desargues' property.

(ii) Regarded as an affine matroid, X is isomorphic with the n-dimensional real projective matroid of dimension n.

In particular, two coplanar lines of X always intersect, making it clear that if $n \geq 2$, then X can not be embedded in a real vector space.

1.16.2. The Moulton plane. This is a non-Desarguesian BW space of dimension 2. As Desargues' property is obviously inherited by convex subspaces, it is clear that this space can not meet the embedding requirements of Theorem 1.15.

Further Topics

1.17. Addition on a line (compare Sperner [1938]; cf. Stevenson [1972]). Show that in a BW space with the Desargues property, the addition operation defined in 1.2 depends only on the end points of the segment, not on the auxiliary triangle considered in 1.2. Moreover, addition is commutative, associative and has a neutral element. Finally, the cancellation law is fulfilled:

$$a + b_1 = a + b_2 \Rightarrow b_1 = b_2.$$

1.18. Long lines. Let X be a well-ordered set. The *long line based on X* is the set $X \times [0,1)$, equipped with the *lexicographic order*,

$$(x,s) \leq (y,t) \Leftrightarrow \text{either } x < y, \text{ or } x = y \text{ and } s \leq t.$$

Show that a long line is a complete BW space and that it cannot be embedded in $\mathbb{R}$ if the basic set X is uncountable.

1.19. Bundles of lines. Throughout, let X be a complete, extensible BW space of dimension ≥ 3.

1.19.1. (in dimension three, Coxeter [1947]) By a *trihedron* in X is meant a triple of concurrent, non-coplanar lines. The point of concurrence will be referred to as the apex of the trihedron. If A,B,C is a trihedron, then by the affine hull formula I§7.11 the plane spanned by A, B equals A/B or B/A. Such a plane is called a *trihedron face* Let A,B,C and A',B',C' (with $A \neq A'$; $B \neq B'$; $C \neq C'$) be two trihedra at the same apex, such that each pair of corresponding faces intersects in a line. Show that if the resulting three lines are coplanar, then the planes spanned by corresponding lines of the trihedra are coaxial, that is: they pass through one line.

Hint: consider a hyperflat not passing through the apex and cutting all six lines $A,...,C'$ as well as the three intersection lines described above. Apply the dual Desargues theorem on the intersection points.

1.19.2. *Lemma on Roofs* (compare Rubinstein [1970], Cantwell and Kay [1978]; in three dimensions, Coxeter [1947]). Let L_1, L_2, M, M' be lines in X. If each of the pairs L_1, L_2; L_1, M; L_2, M; L_1, M'; L_2, M' are coplanar, then the same is true of the pair M, M'. Hint: if the dimension of the space is at least four, then this is an easy consequence of the affine dimension formula, I§7.14. In dimension three, use the above result.

1.19.3. Let A, B be two distinct coplanar lines in X. We consider the set $\mathcal{B}(A,B)$ of all lines C which are either outside the plane spanned by A, B but coplanar with each of A, B, or are included in this plane and coplanar with some line of the previous type. The collection $\mathcal{B}(A,B)$ is called the *bundle of lines* determined by A, B. Note that if A, B meet, then $\mathcal{B}(A,B)$ consists of all lines through the point of intersection.

Show that for each point $p \in X$ there is exactly one line in the bundle $\mathcal{B}(A,B)$ passing through p (except, of course, if p happens to be the intersection of the original lines A, B).

1.20. Equi-perspectivity. Show that the relation of segments on a fixed ray being equi-perspective is an equivalence relation provided the BW space is complete, extensible, and Desarguesian.

1.21. Injective CC functions (compare Meyer and Kay [1973])

1.21.1. Let X and Y be complete BW spaces without boundary, where X is of affine dimension > 1, and let $f : X \to Y$ be an injective CC function. Show that f is CP; in fact, it is a CP embedding of X into Y.

Hint. The inverse function $g : f(X) \to X$ is CP and the BW space $f(X)$ has no boundary points. Let L be a line of X. Its f-image is convex; assume it is of dimension > 1 and apply III§5.26.6.

Give an example showing that the dual statement, on surjective CP functions between complete BW spaces without boundary being CC, is false.

1.21.2. Let V and W be real vector spaces such that V is at least two-dimensional and let $f : V \to W$ be injective and CC. Conclude that f is an affine isomorphism of V with an affine subspace of W.

1.21.3. Deduce that an isometry between two rotund Banach spaces is an affine isomorphism (this is a particular case of the Mazur-Ulam Theorem; cf. Day [1962, p. 110]).

1.22. Linearization (Mah, Naimpally and Whitfield [1976]). Let X be a convex structure and let $a \neq b \in X$. Recall that the line spanning two points a, b is the collection

$$L(a,b) = \{x \in X \mid x \in ab \text{ or } a \in xb \text{ or } b \in ax\};$$

cf. I§7.19. By a *linearization family* of X is meant a separating collection $\mathcal{F}$ of CC functionals $X \to \mathbb{R}$ with the following properties.

(LF-1) There is a distinguished point $\mathbf{0} \in X$ with $f(\mathbf{0}) = 0$ for all $f \in \mathcal{F}$.
(LF-2) If $L \subseteq X$ is a line and $f \in \mathcal{F}$ then the restriction $L \to \mathbb{R}$ of f is either constant or bijective.
(LF-3) If f, $g \in \mathcal{F}$ both separate the points a, $b \in X$ then there exist λ, $\mu \in \mathbb{R}$ such that g equals $\lambda f + \mu$ on the line spanned by a, b.

1.22.1. If X is a convex structure with a linearization family $\mathcal{F}$, then the following prescriptions of scalar multiplication and addition are well-defined, and yield a vector space structure on the set X.

(i) $\lambda \cdot a = b$ iff $b \in L(a, \mathbf{0})$ and for each $f \in \mathcal{F}$ we have $f(b) = \lambda f(a)$.
(ii) $a + b = c$ iff $f(a) + f(b) = f(c)$ for all $f \in \mathcal{F}$.

Moreover, each $f \in \mathcal{F}$ is a linear function. Hint. Establish (i) first. As for (ii), if $a \neq b$

§1: Embedding Bryant-Webster Spaces into Vector Spaces

then consider a functional f separating a, b, take the point x on $L(a,b)$ with $f(x) = \frac{1}{2}(f(a)+f(b))$ and define $c = 2x$. Then show that this is independent of the functional $f \in \mathcal{F}$ in consideration.

1.22.2. Show that if, in addition, the convex structure X is JHC and its base-point quasi orders are actually partial orders, then X is isomorphic with the standard convex structure defined by the vector space $(X, +, \cdot)$.

Remarks.

(i) The field of reals can be replaced here with any totally ordered field. Szafron and Weston [1976] have developed an "internal" variant of the notion of linearization family in terms of families of flats partitioning the space. These families are indexed by a totally ordered field. An additional feature is that conditions are given to the effect that some given topology of X is compatible with the resulting linear operations.

(ii) Guay and Naimpally [1975] have obtained a characterization of convex subspaces of a vector space by the following modification of (LF-2)

(LF-2′) If $L \subseteq X$ is a line and $f \in \mathcal{F}$, then the restriction $L \to \mathbb{R}$ of f is either constant or injective.

1.22.3. A *linearization* of a convex structure $(X, \mathcal{C})$ is a (real) vector space structure $(X, +, \cdot)$ on the set X such that $\mathcal{C}$ is exactly the collection of all sets convex in $(X, +, \cdot)$. Note that parts (1) and (2) provide conditions under which a space has a linearization. Show that any two linearizations of X are isomorphic under an affine homomorphism of the corresponding vector spaces.

1.23. Line spaces (cf. I§7.20). A *line space* in the sense of Cantwell [1974] is a Bryant-Webster space with the additional requirement that the totally ordered lines are order-isomorphic with the real line.

1.23.1. Note that line spaces other than a single line correspond exactly with complete and extensible BW spaces of dimension ≥ 2.

1.23.2. (Cantwell [1978]) Let X be a line space of finite affine dimension $n \geq 2$. Prove directly that X is (topologically) homeomorphic with $\mathbb{R}^n$.

Hint. Operate by induction on $n \geq 1$. If $H \subseteq X$ is an n-dimensional hyperflat, then fix two points p_1, p_2, one at each side of H. Let K_i be the closed half-space bounded by H and containing p_i, and consider the mapping f of X onto H defined as follows. If $x \in K_1$ (say) then $f(x)$ is the only point in H collinear with x, p_2. On the other hand, H determines a CP functional $g: X \to \mathbb{R}$ as devised in III§5.11. If h is a homeomorphism of H with $\mathbb{R}^n$, then the mapping pair $(h \circ f, g): X \to \mathbb{R}^{n+1}$ yields a homeomorphism of X with an open subspace of $\mathbb{R}^{n+1}$.

1.24. Bounded sets. Let X be a complete and extensible BW space of finite (affine) dimension. If a convex set $C \subseteq X$ meets each line of X in a closed interval, then C is compact in the core topology of X.

Hint. First, reduce the problem to a situation where $aff(C) = X$. Take $p \in Int(C)$ and use the Singleton intersection principle, combined with a result in III§5.26, to show that the boundary of C is locally homeomorphic with the boundary of a compact convex neighborhood of p.

It is easy to construct a non-compact closed convex set in ℓ_1 meeting each line segment in a closed interval.

1.25. Distance geometry (cf. I§7.23). Let X be a complete, convex and externally convex metric space with the weak Euclidean Four-point Property.

1.25.1. Show that if $p \in X$ and if C is a convex closed subset of a finite-dimensional affine subspace of X, then p has a unique (metric) nearest point in C.

Hint. Existence via 1.24; unicity via the weak Euclidean Four-point Property.

1.25.2. Let p, q, $r \in X$ be distinct points. Then the following are equivalent:

(i) The nearest point of p in the (metric) line qr is q.
(ii) $d(p,q)^2 + d(q,r)^2 = d(p,r)^2$ (where d is the given metric).
(iii) If $r' \neq r$ is a point on the line qr such that $d(r,q) = d(r',q)$, then $d(p,r) = d(p,r')$.

If either of these conditions is fulfilled, then we express this by the phrase "pq is orthogonal to rq". This suggests that orthogonality is a property of lines rather than of points. Indeed, if p satisfies (i), then so does each element of the line pq.

1.25.3. Let (X_1, d_1) and (X_2, d_2) be metric spaces and let the product set X_0 be equipped with the metric d, defined by

$$d((a_1,a_2),(b_1,b_2)) = \sqrt{d_1(a_1,b_1)^2 + d_2(a_2,b_2)^2}.$$

Show that if both factors (or factor metrics) are convex / are externally convex / have the weak Euclidean Four-point Property, then so does (X_0, d).

Conclude that all convex, externally convex metric spaces with the weak Euclidean Four-point Property have the Desargues Property.

1.25.4. (Compare Blumenthal [1953]) Prove that if X is locally compact, then X is isometric with Euclidean space of some dimension.

Hint: X is of finite dimension (say: n) by III§5.26.3. Construct an affine basis $\mathbf{0}, e_1, .., e_n$ of X such that $\mathbf{0}e_i$ for $i = 1, .., n$ are pairwise orthogonal and the distance of e_i to $\mathbf{0}$ equals the corresponding Euclidean distance. Appealing to Theorem 4.15, consider a convex (CP) embedding of X into $\mathbb{R}^n$ and compose it with an affine isomorphism, mapping the basis of X to an affine basis of $\mathbb{R}^n$ including the origin, such that the non-zero basic vectors are mutually orthogonal in the standard sense. By induction on the dimension n, show that the resulting embedding is an *isometry* onto a convex subspace of $\mathbb{R}^n$.

§1: Embedding Bryant-Webster Spaces into Vector Spaces

Finally, observe that each line of X is isometric with the standard metric space $\mathbb{R}$; cf. I§7.23.3.

Notes on Section 1

The isomorphism of lines in complete extensible BW spaces with the real line, Theorem 1.2, is taken from Doignon [1976]. The "addition" used in its proof is a standard operation in projective geometry (a corresponding "multiplication" exists as well). For a synthetic theory of affine and projective geometries, we refer to Stevenson [1972]. The formulation of the Desargues Property and of its dual are similar to the classical formulation in affine or projective spaces. That extensible BW spaces of dimension ≥ 3 have the Desargues Property (Theorem 1.6) is a modification of a standard result in synthetic affine geometry. The argument given here is due to Doignon [1976].

Results on the isomorphism of BW spaces (or the like) with convex subspaces of linear spaces were first obtained in three-dimensional spaces by adding "ideal" points (viz., bundles of lines, 1.19); see Coxeter [1947] for a detailed account. For two-dimensional spaces, a method involving the Desargues Property was elaborated by Sperner [1938], who essentially obtained Proposition 1.12 on the isomorphism of 2–simplices. Later on, both Blumenthal [1953] and Busemann [1955] developed a technique to handle particular classes of metric spaces of finite (geometric) dimension. Rubinstein [1970] outlined a proof of an embedding theorem for a class of at least four-dimensional spaces. His method involves ideal points as described in Coxeter op. cit., combined with ideas on parallelism. Cantwell and Kay [1978] obtained the result in dimensions ≥ 3 by a similar method. The most general result of all is formulated in Theorem 1.15. It has been obtained by Doignon [1976], who based the first part of his proof on Sperner's results in the two-dimensional case, and obtained the extension to higher dimensions by adapting Busemann's method of embedding into projective spaces from finite-dimensional metric spaces to BW spaces of any dimension.

The argument presented here starts with Sperner's Quadrangle Lemma, 1.8, and goes via Sperner's result on isomorphism of simplices, Proposition 1.12. The approach to this result with midpoints is new and relatively fast. A somewhat comparable method has been outlined by Busemann in metric spaces; however, Busemann's argument neglects some serious problems, as is signaled by Soetens [1973]. The remaining part of the procedure is taken from Doignon's paper with a few modifications (we are more involved with topological considerations). The application of the Main Theorem to distance geometry in 1.25 is a new approach to results in Blumenthal [1953].

2. Extremality, Pseudo-boundary and Pseudo-interior

Extremality and the theorem of Krein-Milman are classics in the theory of convex structures. We first consider general closure spaces with a synthetic notion of extremality, and derive a general Krein-Milman type theorem. Examples involve extremality defined by functional betweenness, by representing probability measures, and by proper support points (or, pseudo-boundary points, as we prefer to call them). This results into several sets of "extreme points" spanning a given compact convex set. In general, these sets are difficult to compare. Among the main results is a version of Bauer's Maximum Principle, stating that maxima of lower CP functionals can be found at an extreme point.

The study of the pseudo-boundary and of the complementary pseudo-interior of a set is continued outside the framework of extremality. A basic assumption in this study is that convex sets are in continuous position. A characterization is given of the existence of pseudo-interior points for all convex closed sets in terms of a countable intersection property for convex open sets.

Throughout, *all spaces are assumed* S_1.

2.1. Extremality: a synthetic approach. An *extremality structure* is a triple $(X, \mathcal{C}, \mathcal{E})$ built with a closure space $(X, \mathcal{C})$ and a family $\mathcal{E}$ of non-empty subsets of X with the following properties.

(ES-1) If $E \in \mathcal{E}$ and if $C \subseteq X$ is closed such that $E \not\subseteq C$, then there is a set $E' \in \mathcal{E}$ with $E' \subseteq E$ and $E' \cap C = \varnothing$.

(ES-2) If $E \in \mathcal{E}$ has more than one point, then there is a closed set C with $E \cap C \neq \varnothing \neq E \setminus C$.

(ES-3) $X \in \mathcal{E}$ and $\mathcal{E}$ is downward inductive, that is: each chain has a lower bound in $\mathcal{E}$.

The members of $\mathcal{E}$ are interpreted as (*synthetic*) *extreme sets* and points $x \in X$ with $\{x\}$ extreme are called (*synthetic*) *extreme points*. The collection of all extreme points is denoted by $Ext(X)$ or by $Ext_{\mathcal{E}}(X)$; the capital "E" is used to distinguish between the present notion and one in terms of concave singletons (cf. I§1.23, II§4.10). We usually omit explicit reference to the protopology or to the extremality from our notation.

2.2. Proposition. *If X is a extremality structure, then:*

(1) *Each extreme set contains an extreme point.*
(2) $X = cl(Ext(X))$.

Proof. Let E_0 be an extreme set. By (ES-3) and Zorn's Lemma, there is a minimal extreme set E with $E \subseteq E_0$. If E has more than one point, then by (ES-2) we can fix a closed set C with both C, $X \setminus C$ meeting E. By (ES-1), E can be decreased to an extreme set disjoint from C, contradicting minimality.

This establishes (1). As for (2), suppose $X \neq cl(Ext(X))$. Now, X is extreme by (ES-3) and $X \not\subseteq cl(Ext(X))$, whence by (ES-1) there is an extreme set E disjoint from

$cl(Ext(X))$. However, E meets $Ext(X)$ by (1), contradiction. ∎

We will take time out for describing various ways in which extremality in the above sense occurs, and for giving some examples in each case.

2.3. Functional generating. In addition to generating a convexity on a set X, a family $\mathcal{F}$ of functionals on X can also be used to produce a closure space, with a protopology consisting of all intersections of sets in the family

$$\mathcal{S} = \{ f^{-1}(\leftarrow, t] \mid t \in f(X); f \in \mathcal{F} \}$$

(compare III§4.14). We call this the *protopology generated by* $\mathcal{F}$, or briefly, the $\mathcal{F}$-*protopology*. Its members are referred to as $\mathcal{F}$-*closed sets* and the corresponding closure operator is denoted by $cl_\mathcal{F}$. If Y is any subset of X, then the intersections of $\mathcal{F}$-closed sets with Y constitute a protopology generated by the family $\mathcal{F}_Y$, consisting of all restrictions to Y of functionals in $\mathcal{F}$. This protopology is a basis for the relative convexity of Y and its closure operator $cl_{\mathcal{F}_Y}$ satisfies

$$cl_{\mathcal{F}_Y}(A) = cl_\mathcal{F}(A) \cap Y \quad (A \subseteq Y).$$

2.4. Functional betweenness and extremality. Let X be a convex structure generated by a point-separating collection $\mathcal{F}$ of functionals. Let $F \subseteq X$ be a finite set, say: $F = \{p_1, \ldots, p_n\}$. A point p is *Fan $\mathcal{F}$-between* (briefly, *Fan between*) *the set F* provided

$$\forall f \in \mathcal{F}: f(p) \geq \max \{ f(p_i) \mid i = 1, \ldots, n \} \text{ implies } f(p) = f(p_1) = \ldots = f(p_n).$$

Observe that p is $\mathcal{F}$-between F iff $p \in co(F)$ and $f(p) < \max(f(F))$ for each $f \in \mathcal{F}$ which is not constant on F. The collection $\beta(F; \mathcal{F})$ of all points which are $\mathcal{F}$-between F can be described alternatively as follows.

2.4.1. *For each $f \in \mathcal{F}$ let $m_f = \max f(F)$. Then $\beta(F; \mathcal{F})$ is the intersection of $co(F)$ with all sets of type $f^{-1}(\leftarrow, m_f)$, where $f \in \mathcal{F}$ is not constant on F.* ∎

Let Y be a subset of X. A set $E \subseteq Y$ is *Fan $\mathcal{F}$-extreme* in Y with respect to a set of functionals $\mathcal{F}$ (less accurately, a *Fan extreme set*) provided for each $p \in E$, if p is $\mathcal{F}$-between a finite set $F \subseteq Y$ then $F \subseteq E$. The resulting set of extreme points is denoted by $Ext_\mathcal{F}(Y)$. Obviously, $ext(Y) \subseteq Ext_\mathcal{F}(Y)$. The following is a straightforward application of the definitions.

2.4.2. *Let $Y \subseteq X$ be closed, let $f \in \mathcal{F}$ and let $m = \max f(Y)$. Then the set $f^{-1}(m) \cap Y$ is $\mathcal{F}$-extreme in Y.* ∎

Proposition 2.6 below presents conditions under which the axioms of extremality are fulfilled in the collection of closed $\mathcal{F}$-extreme sets relative to the $\mathcal{F}$-protopology.

2.5. Examples

2.5.1. Standard convexity. Let V be a locally convex vector space and let $\mathcal{F}$ consist of all linear functionals $V \to \mathbb{R}$. If $F \subseteq V$ is finite then p is Fan $\mathcal{F}$-between F iff p is a

§2: Extremality, Pseudo-boundary and Pseudo-interior

core point of $co(F)$ relative to the affine hull of F (an *intrinsic core point* of $co(F)$).

This can be seen with the aid of 2.4.1 as follows. First, note that a linear functional is constant on F iff it is constant on the affine hull of F. If p is an intrinsic core point of $co(F)$ and if f is non-constant on F, then take $x \in aff(F)$ with $f(p) < f(x)$. The segment px is mapped isomorphically onto a real segment by f, whereas a non-trivial end at p is within $co(F)$. Therefore, $f(p)$ is not maximal in $f(co(F))$.

Conversely, if p is not an intrinsic core point of $co(F)$, then consider a point $x \in aff(F)$ such that

$$px \cap co(F) = \{p\}.$$

The convex set $xp \setminus \{p\}$ is disjoint from $co(F)$. Hence it extends to an open half-space O of V, disjoint from $co(F)$. This determines a continuous linear functional f on V with the bounding hyperplane of O as a fiber. Without loss of generality, $f(x) > f(p)$. Then $f(p) = \max f(F)$ and f is not constant on F, for otherwise $x \in aff(F) \subseteq \overline{O} \setminus O$.

It follows directly that the Fan extreme points of a set $Y \subseteq V$ are precisely the points $y \in Y$ such that $\{y\}$ is a relative half-space of Y.

2.5.2. H-convexity. If $\mathcal{F} \subseteq \mathcal{E}$ are two collections of functionals on a set X, then, evidently, each subspace Y of X satisfies

$$Ext_{\mathcal{F}}(Y) \subseteq Ext_{\mathcal{E}}(Y).$$

Consider $\mathbb{R}^n$ with an H-convexity, generated by a (point-separating) collection $\mathcal{F}$. By 2.5.1 and by the previous observation we have $Ext_{\mathcal{F}}(Y) \subseteq ext(Y)$, where "ext" on the right refers to the set of standard extreme points. The inclusion can be proper. For a simple example, consider $\mathbb{R}^2$ with the H-convexity which is symmetrically generated by the two coordinate projections, and let Y be the unit disk. Then $Ext(Y) = \{\pm(1,0), \pm(0,1)\}$, whereas the usual extreme points cover the unit circle.

2.5.3. Chains of closed convex sets. Let X be a topological convex structure. We consider the collection $\mathcal{F}$ of all continuous lower CP functionals on X which cannot effectively be refined (cf. Topic III§4.21.1). To illustrate that even such a canonical choice of $\mathcal{F}$ can lead to a trivial $\mathcal{F}$-betweenness relation, let $X = \mathbb{R}$ and consider the indexed chain $(C_t)_{0 \leq t}$, defined by

$$C(t) = \begin{cases} [-t/2, t], & \text{if } 0 \leq t < \tfrac{1}{2}; \\ [-t, t], & \text{if } \tfrac{1}{2} \leq t. \end{cases}$$

By Lemma III§4.2, the corresponding functional $f : X \to \mathbb{R}$ is continuous and lower CP, and one can verify that it has no continuous, lower CP, proper refinement. This functional can be described as follows (see Fig. 1).

$$f(t) = \begin{cases} |t|, & \text{if } 0 \leq t \text{ or } t \leq -\tfrac{1}{2}; \\ -2t, & \text{if } -\tfrac{1}{4} \leq t \leq 0; \\ \tfrac{1}{2}, & \text{if } -\tfrac{1}{2} \leq t \leq -\tfrac{1}{4}. \end{cases}$$

Now $-\tfrac{1}{4}$ is not $\mathcal{F}$-between $\{-\tfrac{1}{2}, 0\}$ since $f(-\tfrac{1}{4}) = f(-\tfrac{1}{2}) = \tfrac{1}{2} > f(0) = 0$.

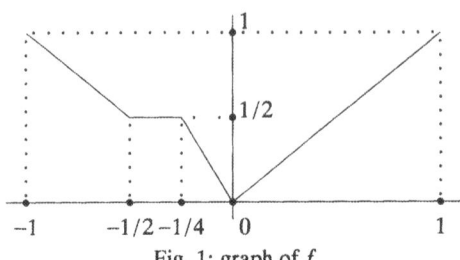

Fig. 1: graph of f

By modifying the numerical data, it can be seen that no point is $\mathcal{F}$–between $-\frac{1}{2}$ and 0 (or between any other pair of distinct points of $\mathbb{R}$). Alternatively, one can use the fact that, given $a_1 < a_2 < a_3$ and $b_1 < b_2 < b_3$, there is a CP isomorphism $\mathbb{R} \approx \mathbb{R}$ mapping a_i to b_i for $i = 1, 2, 3$.

The next results involve some terminology on the separation of sets by functionals, introduced and studied in Section III§4.

2.6. Theorem (*Fan-Krein-Milman Theorem*). *Let X be a tcs, let $Y \subseteq X$ be compact, and let $\mathcal{F}$ be a collection of continuous lower CP functionals of X separating convex closed sets of X from points of Y. Then the collection of all closed Fan $\mathcal{F}$-extreme sets in Y is an extremality with respect to the $\mathcal{F}$-protopology of Y, and*

$$co^*(Y) = co^*(Ext_{\mathcal{F}}(Y)).$$

Proof. Let $\mathcal{E}$ be the collection of all closed Fan extreme sets in Y. We first verify the axioms of extremality with respect to the $\mathcal{F}$-protopology of Y. If $(E_i)_{i \in I}$ is a chain in $\mathcal{E}$, then $\cap_{i \in I} E_i \neq \emptyset$ by compactness. If $p \in \cap_{i \in I} E_i$ is Fan between $a_1,..,a_n$, then for each $i \in I$ we have $E_i \supseteq \{a_1,..,a_n\}$ and it follows that $\cap_{i \in I} E_i$ is $\mathcal{F}$-extreme. This establishes (ES-3).

As to (ES-1), let $C \subseteq X$ be $\mathcal{F}$-closed, let $E \in \mathcal{E}$, and suppose $E \not\subseteq C$. Take a point $x \in E \setminus C$ and consider a functional $f \in \mathcal{F}$ with $f(x) > \sup f(C)$. Let $t = \sup f(E)$, and note that the supremum is attained for at least one point of the compact set E. The set $E' = E \cap f^{-1}[t, \infty)$ is closed and disjoint from C; we show that it is $\mathcal{F}$-extreme. To this end, let $p \in E'$ be $\mathcal{F}$-between $a_1,..,a_n \in Y$. As E is $\mathcal{F}$-extreme, we find that these points are all in E. But then $f(p) \geq t \geq \max_{i=1}^{n} f(a_i)$, and hence $f(p) = f(a_1) = .. = f(a_n)$. It follows that all points a_i are in E'.

As to axiom (ES-2), note that X is S_1 by (permanent) assumption. In particular, each pair of distinct points of Y is separated by a member of $\mathcal{F}$. Let $E \in \mathcal{E}$. If $p \neq q$ are in E, then fix an $f \in \mathcal{F}$ with, for instance, $f(p) < f(q)$. The set $C = f^{-1}(-\infty, f(p)]$ is $\mathcal{F}$-closed and $E \cap C \neq \emptyset \neq E \setminus C$.

Having verified the axioms, we find that by Proposition 2.2,

$$Y = cl_{\mathcal{F}}(Ext_{\mathcal{F}}(Y)) \cap Y.$$

As the members of $\mathcal{F}$ are continuous and lower CP, all $\mathcal{F}$-closed sets of X are convex

§2: Extremality, Pseudo-boundary and Pseudo-interior   409

closed. As $\mathcal{F}$ separates convex closed sets of X from points of Y, the protopology of Y consists exactly of all traces of convex closed sets of X. Consequently,

$$Y \subseteq co^*(Ext_{\mathcal{F}}(Y)) \subseteq co^*(Y),$$

and another application of co^* yields the desired result. ∎

Note that the existence of a family of continuous and lower CP functionals separating convex closed sets of X from points of X implies the separation property NS_3 of X.

2.7. Radon probability and support. Let X be a compact space. The smallest family of sets including all open sets and which is stable for complements and for countable unions is called the *Borel algebra* of X. Among its members are the familiar open sets, closed sets, G_δ-sets and F_σ-sets.

A *Radon probability measure* on X is a (positive) regular[1] measure, defined on the Borel algebra of X, such that the measure of X equals 1. Let $\mathcal{M}(X)$ be the collection of all such measures. Note that each continuous function $X \to \mathbb{R}$ is μ-integrable for each $\mu \in \mathcal{M}(X)$. If $A \subseteq X$ is closed, then $\mathcal{M}(A)$ can be regarded as the set of all $\mu \in \mathcal{M}(X)$ such that $\mu(A) = 1$.

The *support* of $\mu \in \mathcal{M}(X)$ is the collection of all $x \in X$ such that each open neighborhood N of x satisfies $\mu(N) > 0$. The resulting set is denoted by $Supp(\mu)$. Note that this is a closed set and, by the regularity of μ, that $\mu(Supp(\mu)) = 1$.

2.8. Choquet extremality. Let $\mathcal{F}$ be a point-separating collection of continuous functions $X \to \mathbb{R}$, and let Y be a compact subspace of X. For $p \in Y$ we let $\mathcal{M}(p;\mathcal{F})$ denote the collection of all $\mu \in \mathcal{M}(Y)$ with the property

$$\forall f \in \mathcal{F}: f(p) \leq \int_Y f \, d\mu.$$

Such measures are said to *represent* the point p. Note that if the family $\mathcal{F}$ is *symmetric* (i.e., if $-f \in \mathcal{F}$ whenever $f \in \mathcal{F}$), then the above inequality may be replaced by an equality. Due to the fact that $\int_Y f \, d\mu \leq \max f(\overline{Supp(\mu)})$, if μ is a measure representing p, then $p \in cl_{\mathcal{F}}(Supp(\mu))$.

A set $E \subseteq Y$ is said to be a *Choquet $\mathcal{F}$-extreme set* (less accurately, a *Choquet extreme set*) provided for each $p \in E$ and for each Radon measure μ representing p, it is true that $Supp \, \mu \subseteq E$. The corresponding collection of *Choquet extreme points* will be denoted by $Ext_{Ch}(Y)$. Note that a point p is Choquet extreme iff the trivial probability measure, assigning measure 1 to $\{p\}$, is the only measure representing p. The following result provides a standard example.

2.9. Proposition. *Let Y be a non-empty compact convex subspace of a locally convex vector space V. Then, relative to the collection $\mathcal{F}$ of all (restrictions of) continuous linear functionals on Y,*

---

1. Regularity of a Borel measure μ means that for each closed set A and for each $\varepsilon > 0$ there exists an open set $O \supseteq A$ with $\mu(O \setminus A) < \varepsilon$.

$Ext_{Ch}(Y) = ext(Y)$.

Proof. First, observe that if $p = \sum_{i=1}^{n} t_i p_i$, with $0 \le t_1,..,t_n$ and $\sum_{i=1}^{n} t_i = 1$, then there is a Radon probability measure μ on Y determined by the prescriptions

$$\mu(\{p_i\}) = t_i \quad (i = 1,..,n).$$

This measure clearly represents p. Hence, if $p \in Ext_{Ch}(Y)$, then p is an ordinary extreme point. The converse implication requires the following auxiliary result.

Lemma. *If $\mu \in \mathcal{M}(Y)$, then there is one and only one point in Y represented by μ.*

Proof. For each $f \in \mathcal{F}$ we consider the hyperflat $H(f)$ of all points y with $f(y) = \int_Y f d\mu$. If $f_1,..,f_n \in \mathcal{F}$ then a continuous linear function is defined by

$$F: V \to \mathbb{R}^n; \quad F(y) = (f_1(y),..,f_n(y)).$$

Suppose that the point $q = (\int f_1 d\mu,..,\int f_n d\mu)$ is not in the compact convex set $F(Y)$. Then there is a linear functional h on $\mathbb{R}^n$ such that $h(q) > \sup h(F(Y))$. We represent h by means of a vector a in the sense that $h(v) = a \cdot v$ (inner product) for all $v \in \mathbb{R}^n$. If $a = (a_1,..,a_n)$, then define

$$g = \sum_{i=1}^{n} a_i f_i.$$

Note that $g = h \circ F$. We find that

$$\int g d\mu = \sum_{i=1}^{n} a_i \int f_i d\mu = a \cdot q > \sup g(Y),$$

which is impossible.

We conclude that the hyperflats $H(f)$ for $f \in \mathcal{F}$ have the finite intersection property. Hence these sets meet in a point y, which by construction is represented by μ. As $\mathcal{F}$ separates the points of Y, we find that y is uniquely determined by this requirement. ∎

To complete the proof of the proposition, suppose that $p \notin Ext_{Ch}(Y)$. Then there is a Radon probability measure μ representing p, such that $Supp(\mu) \ne \{p\}$. Let $q \ne p$ in $Supp(\mu)$ and let C be a convex closed neighborhood of q missing p. If $\mu(C) = 1$, then consider a continuous linear functional f with $f(p) < \inf(f(C))$. We find that

$$f(p) = \int_Y f d\mu = \int_C f d\mu \ge \inf(f(C)),$$

a contradiction. So $0 < \mu(C) = t < 1$. Define two new measures μ_1, μ_2 on Y as follows. If $B \subseteq Y$ is a Borel set, then so are $B \cap C$ and $B \setminus C$, and we put

$$\mu_1(B) = \frac{\mu(B \cap C)}{t}; \quad \mu_2(B) = \frac{\mu(B \setminus C)}{(1-t)}.$$

Now $\mu_1, \mu_2 \in \mathcal{M}(Y)$ and we consider the respective representations $p_1, p_2 \in Y$. As $\mu(B) = t \mu_1(B) + (1-t)\mu_2(B)$ for each Borel set B, we find that $p = tp_1 + (1-t)p_2$, showing that p is not an ordinary extreme point. ∎

§2: Extremality, Pseudo-boundary and Pseudo-interior 411

2.10. Theorem (*Choquet-Krein-Milman Theorem*). *Let X be a tcs, let $Y \subseteq X$ be compact and let $\mathcal{F}$ be a collection of continuous lower CP functionals $X \to \mathbb{R}$ separating convex closed sets of X from points of Y. Then the family of all closed Choquet $\mathcal{F}$-extreme sets in Y is an extremality with respect to the $\mathcal{F}$-protopology of Y and*
$$co^*(Y) = co^*(Ext_{Ch}(Y)).$$

Proof. The axioms (ES-2) and (ES-3) are obvious. As for (ES-1), let $E \subseteq Y$ be a Choquet extreme set, let $C \subseteq X$ be $\mathcal{F}$-closed and suppose $E \not\subseteq C$; say: $p \in E \setminus C$. Then there is a functional $f \in \mathcal{F}$ such that $\sup f(C) < f(p)$. Note that f, being continuous, attains a maximal value on E. Without loss of generality, this maximum t is taken at p. The set $E' = f^{-1}[t, \to) \cap E$ is closed and disjoint from C. We verify that it is extreme. Let $\mu \in \mathcal{M}(X)$ be a measure representing a point $x \in E'$. Then $Supp\,\mu \subseteq E$. If $Supp\,\mu \not\subseteq f^{-1}(t)$, then there is a closed set $B \subseteq E$ of measure $\mu(B) > 0$, such that $B \cap f^{-1}(t) = \emptyset$. If t_0 is the maximum taken by f on B, then $t_0 < t$ and we obtain a contradiction by the elementary rules of integration:
$$f(x) \leq \int_X f d\mu = \int_E f d\mu = \int_{E \cap B} f d\mu + \int_{E \setminus B} f d\mu \leq t_0 \cdot \mu(E \cap B) + t \cdot \mu(E \setminus B) < t.$$

Having verified the axioms of synthetic extremality, we apply Proposition 2.2:
$$Y = cl_{\mathcal{F}} Ext_{Ch}(Y) \cap Y.$$

Proceeding as in the proof of 2.6 yields the desired result. ∎

The above proof of (ES-1) is based on an argument which actually shows that if t is a maximum of f on Y, then $f^{-1}(t) \cap Y$ is Choquet extreme in Y.

2.11. Pseudo-boundary and pseudo-interior. Let X be a topological convex structure and let $A \subseteq X$. A point p is a *pseudo-boundary point* of A relative to X provided there is a convex open set $O \subseteq X$ with the following properties.
$$O \cap A \neq \emptyset; \quad A \subseteq co^*(O); \quad p \in A \setminus O.$$

The collection of all such points is called the *pseudo-boundary* of A relative to X and is denoted by $\partial_X(A)$. The complementary set $\iota_X(A) = A \setminus \partial_X(A)$ is called the *pseudo-interior* of A relative to X. If $A = X$ then the subscript "X" will be dropped and the sets $\partial(X)$, $\iota(X)$ are called the *absolute pseudo-boundary* resp., the *absolute pseudo-interior* of X. Some formulae involving product spaces are presented in Topic 2.32. The following ones will probably look familiar. Let $X = X_1 \times X_2$. We use subscripts X, 1, 2 to refer to X, X_1, X_2 respectively.
$$\iota_X(C_1 \times C_2) = \iota_1(C_1) \times \iota_2(C_2);$$
$$\partial_X(C_1 \times C_2) = (\partial_1 C_1 \times C_2) \cup (C_1 \times \partial_2 C_2).$$
The next result gives some additional information on pseudo-interiority.

2.12. Proposition. *Let X be closure stable and S_4.*
(1) *If $A \subseteq X$ and if $p \in A$, then $p \in \iota_X(A)$ iff $Int(H) \cap A \neq \emptyset$ for each closed half-*

space $H \subseteq X$ containing p and not including A.

(2) If $A \subseteq Y \subseteq X$ then
$$\iota_X(A) \subseteq \iota_X(co^*(A) \cap Y).$$

Proof of (1). Let $O \subseteq X$ be a convex open set and let $p \notin O$. Let $H \subseteq X$ be a maximal convex set with the properties $p \in H$ and $H \cap O = \emptyset$. Then H is a half-space by the axiom S_4 and is closed by virtue of closure stability. Observe that if $P = X \setminus H$, then
$$Int(H) = X \setminus \overline{P} = X \setminus co^*(P).$$
The result follows directly from these observations.

Proof of (2). Let $p \notin \iota_X(co^*(A) \cap Y)$. By (1), there is a closed half-space $H \subseteq X$ containing p and not including $co^*(A) \cap Y$. Hence $A \not\subseteq H$, and $p \notin \iota_X(A)$ by (1). ∎

By a *proper support point* of a set A in a vector space V is meant a point $p \in A$ such that there exists a continuous linear functional f of V with $f(p) = \sup f(A)$. The addition "proper" refers to the additional assumption that f is not constant on A. Application of 2.12(1) to a topological vector space yields that the pseudo-boundary points of a subset are exactly its proper support points.

It is not true that the operator ι_X is monotonic in general. For instance, if $A, B \subseteq \mathbb{R}^2$ are taken as $[0,1] \times \{0\}$ and $[0,1] \times [0,1]$, respectively, then $A \subseteq B$ but the non-empty sets $\iota_{\mathbb{R}^2}(A)$ and $\iota_{\mathbb{R}^2}(B)$ are disjoint.

2.13. Examples: Trees, Tychonov cubes, and some spaces with no pseudo-interior points.

2.13.1. Proposition. *In a connected and locally connected tree T, all polytopes are compact, the convex sets are exactly the connected sets, and ∂T is the set of all end points of T.*

Proof. We first verify that all segments of T are compact. The median $m(a,b,c)$ of the points $a, b, c \in T$ is the maximum of $a \wedge b$, $b \wedge c$, $c \wedge a$ (cf. I§1.22). This gives a continuous operator $m: T^3 \to T$. Now $m(a,b,c)$ is the b-infimum of a and c, whence $(T, \leq_b)$ is a topological tree for any base-point $b \in T$. A segment of type ab corresponds with the lower set $L(a)$ in the base point order of b. In addition to being totally ordered, $L(a)$ is connected and locally connected, being a retract of T under the mapping $x \mapsto x \wedge a$. The topology of a totally ordered pospace is at least as fine as the order topology, whence each of its connected sets is order convex. If, in addition, a totally ordered space is connected, then each connected neighborhood of a point must include an order-neighborhood. It follows that $L(a)$ carries the order topology. As is well-known, a connected totally ordered space with a maximum and a minimum is compact. By virtue of Theorem III§2.7, all polytopes are compact.

Connectedness of all segments yields that all convex sets are connected. To see that, conversely, all connected sets are convex, we first show that the segment ab consists exactly of a, b and of those points $p \in T$ which cut between a and b, that is: the points

§2: Extremality, Pseudo-boundary and Pseudo-interior 413

a, b are in different components of $T \setminus \{p\}$ (then p is a *cut point of T between* a *and* b). Segments being connected, a point outside of ab cannot cut between a, b. Suppose next that $p \in ab$; $p \neq a, b$. We consider the base-point order of b as the given tree order. Note that $p < a$. Consider the sets

$$O = \{x \mid x \not\geq p\}; \quad P = \{x \mid x > p\}.$$

Clearly, $a \in P$, $b \in O$, and

$$T \setminus \{p\} = O \cup P; \quad O \cap P = \emptyset.$$

The set O is open, as one can deduce from the properties of general pospaces. To obtain a decomposition of $T \setminus \{p\}$ as desired, it suffices to show that P is open as well. Let $c \in P$ and consider a connected neighborhood N of c such that $c \wedge x \neq p$ for all $x \in N$. The subset $c \wedge N$ of the totally ordered space $L(c)$ is connected and hence order convex. Furthermore, $c \wedge x > p$ holds in case $x = c$. If the inequality $c \wedge x < p$ occurs for some $x \in N$, then $c \wedge N$ contains the point p, a contradiction. This shows that $N \subseteq P$ and completes the proof that O, P constitute a decomposition of T. In particular, p is a cut point of T between a, b. Now assume that C is a connected set. If $a, b \in C$, then any cut point between a and b is in C and it follows that C is a convex set of the interval space T.

To establish the final part of the proposition, suppose first that $p \in T$ is an end point. The set $O = T \setminus \{p\}$ is convex open, and since T is connected we find that $co^*(O)$ properly includes O. Hence $p \in \partial T$. If p is not an end point and if $O \subseteq T$ is a non-empty convex open set with $p \notin O$, then fix $b \in O$ and choose $p' >_b p$. By local convexity, there is a convex neighborhood U of p' disjoint from bp. If $U \cap O \neq \emptyset$ then (as the Helly number of T is 2) we have $U \cap O \cap bp' \neq \emptyset$. But $bp' = bp \cup pp'$, where the first summand is disjoint from U and the second is disjoint from O, a contradiction. Hence $co^*(O) \neq T$, as required for p to be a pseudo-interior point. ∎

The pseudo-boundary and pseudo-interior of a set are usually considered relative to the given superspace. The resulting sets may well be distinct from the absolute ones, as the next example illustrates. A less natural, but more spectacular example is presented in Topic 2.35.3.

2.13.2. Proposition. *Let* α *be an ordinal number and let the vector space* $V = \mathbb{R}^\alpha$ *be given the product topology and the standard convexity. Then the convex subspace* $Q = [0,1]^\alpha$ *of V has the following relative and absolute pseudo-interior.*

$$\iota_V(Q) \stackrel{(1)}{=} (0,1)^\alpha; \quad \iota(Q) \stackrel{(2)}{=} \bigcup_{n=2}^\infty [\frac{1}{n}, 1-\frac{1}{n}]^\alpha.$$

These sets are distinct if α *is infinite.*

Proof. It is clear that $\iota_V(Q) \subseteq (0,1)^\alpha$. Suppose $p \in (0,1)^\alpha$. Evidently, each CP isomorphism of V mapping Q onto itself, will map the relative pseudo-interior of Q into itself. Considering the CP isomorphism $f : V \to V$, defined by

$$f(x)_i = \begin{cases} x_i, & \text{if } p_i \le \tfrac{1}{2}; \\ 1-x_i, & \text{if } p_i \ge \tfrac{1}{2}, \end{cases}$$

we can achieve that all coordinates of p are in $(0, \tfrac{1}{2}]$. Let $H \subseteq V$ be a closed half-space not including Q, and such that $p \in H$. We aim at an application of 2.12(1). For each $i < \alpha$ we consider the point $a(i)$ of which all coordinates are zero, except for the ith one which equals 1. These points constitute a linear base for a dense linear subspace of V (the whole of V, if V is finite-dimensional), and hence the hyperflat $H \setminus Int(H)$ cannot contain them all. If some $a(i)$ is in $Int(H)$ then the latter meets Q as required. We are left with the possibility that some $a(i)$ is not in H. By assumption, the point

$$p(i) = tp+(1-t)a(i)$$

is in Q for some $t > 1$. As $a(i) \notin H$, we find that $p(i) \in Int(H) \cap Q$. We conclude that $p \in \iota_V(Q)$, which proves the first equality.

A computation of the absolute pseudo-interior is more complicated. Throughout, we let x_i denote the ith coordinate of a point $x \in \mathbb{R}^\alpha$. We have to show that a point $p \in Q$ is in $\iota(Q)$ iff no sequence of coordinates of p tends to 0 or to 1. We use the evident principle, that an isomorphism of a space into itself permutes the absolute pseudo-interior points.

Suppose first that p_{i_n} converges to 0. Without loss of generality, $p_{i_n} \le 1/n$. It is a standard fact that the series $\Sigma 1/n^2$ converges to $\pi^2/6$. The following defines a relative closed half-space of Q containing p and without interior points:

$$H = \{x \mid \sum_{i=1}^{\infty} \frac{1}{n} x_{i_n} \le \frac{\pi^2}{6}\}.$$

By 2.12(1), we conclude that $p \in \partial(Q)$. The possibility that some sequence of p-coordinates tends to 1 can be reduced to the previous situation by considering the CP isomorphism

$$f: Q \to Q: x \mapsto 1-x.$$

This establishes the inclusion from left to right in the equality (2). To prove the opposite inclusion, let $b \in Q$, have all its coordinates equal to $1/2$. The above introduced isomorphism f has the property that $b \in xf(x)$ for each $x \in Q$. Consequently, if $H \subseteq Q$ is a closed relative half-space with $b \in H$, then f maps the relatively open set $Q \setminus H$ homeomorphically into H and hence into $Int(H)$. It follows easily that $b \in \iota(Q)$. Next, let all coordinates of $b(n)$ be equal to $1/n$ ($n \ge 2$). An isomorphism $Q \approx Q$ mapping $b = b(2)$ to $b(n)$ is constructed as follows (Fig. 2). For $n > 2$ consider the point c_n in the standard plane, common to the lines through $(0,1),(1,0)$ and through $(0,1/n),(1/2,0)$. Then c_n can be seen as a center of perspective projection of $[0,1] \times \{0\}$ onto $\{0\} \times [0,1]$ mapping $1/2$ to $1/n$. This leads to a CP isomorphism $Q \approx Q$ under which (copies of) b and $b(n)$ correspond. It follows at once that $b(n) \in \iota(Q)$.

For each subset $I \subseteq \alpha$, we have a CP isomorphism $g = g_I$ of Q such that

§2: Extremality, Pseudo-boundary and Pseudo-interior 415

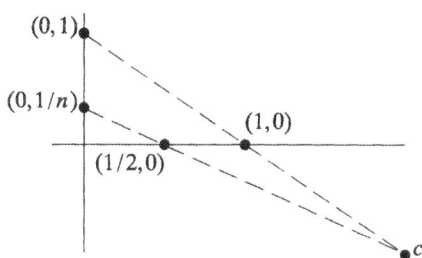

Fig. 2. Relative pseudo-interior of a cube: planar construction.

$$g(x)_i = \begin{cases} 1-x_i, & \text{if } i \in I; \\ x_i, & \text{otherwise.} \end{cases}$$

By using such isomorphisms, it follows that the compact set $A(n)$, consisting of all points with coordinates among $1/n$, $1-1/n$, is included in $\iota(Q)$. Recall (cf. III§4.23) that a subset of a tcs is **strongly convex** provided it includes the convex closure of each of its compact subsets. By III§4.23, each convex relatively open set in Q is strongly convex. Hence $\iota(Q)$ is strongly convex (being an intersection of such sets), leading us to the conclusion that $co^*(A(n)) \subseteq \iota(Q)$. But

$$co^*(A(n)) = [1/n, 1-1/n]^\alpha,$$

and the result follows. ∎

For a comparison of absolute and relative pseudo-boundary in median spaces, see 2.35.

2.13.3. Empty pseudo-interior. Let X be a topological space without isolated points. X is equipped with the discrete convexity, which is continuous and S_4. Then $\partial X = X$ since for each $p \in X$ the set $X \setminus \{p\}$ is convex open and its closure is X.

Here is a less trivial example with empty pseudo-interior. Let X in addition be compact. By Proposition III§3.10.4, the Vietoris convexity on the hyperspace $\mathcal{T}_*(X)$ is continuous and S_4. If $A \in \mathcal{T}_*(X)$ and $a \in A$, then the closed half-space $<\{a\},X>$ of $\mathcal{T}_*(X)$ contains A and has no interior points. By 2.12, it follows that $A \in \partial(\mathcal{T}_*(X))$.

For an example in Hilbert space, see 2.29.2.

2.14. δ–Extremality. Let X be a tcs and let $Y \subseteq X$. We use pseudo-interiority as a kind of "strict betweenness" and define extremality with a procedure as before. A set $E \subseteq Y$ is δ_X-*extreme* in Y provided for all $p \in E$, if $A \subseteq Y$ and $p \in \iota_X(A)$, then $A \subseteq E$. We note that δ–extremality depends on the superspace X. If $Y = X$, then we occasionally drop the subscript. The following is a simple application of the definitions.

2.14.1. *If $O \subseteq X$ is a convex open set meeting Y and if $Y \subseteq co^*(O)$, then $Y \setminus O$ is a δ_X-extreme set of Y.* ∎

The resulting set of all δ_X-extreme points of a set Y is denoted by $Ext_{\delta_X}(Y)$. Application of 2.12(2) yields that the set A, occurring in the definition of extremality, may be assumed to be the trace of a convex closed set on Y. Explicitly:

2.14.2. *Let X be S_4 and closure stable. If $E \subseteq Y \subseteq X$, then E is δ_X-extreme in Y iff $C \cap Y \subseteq E$ for each $p \in E$ and for each convex closed set $C \subseteq X$ with $p \in \iota_X(C \cap Y)$.* ∎

Under the assumptions of the last result, a point $p \in Y$ is a δ_X-extreme point of Y iff $p \in \delta_X(C)$ whenever C has more than one point and is the trace on Y of a closed convex set of X which contains p.

We first give conditions in order that δ-extreme points are ordinary extreme points. A topological convex structure X has **Fuchssteiner's Property** provided for each convex open set $O \subseteq X$ and for each $a \in O$ and $b \in \overline{O} \setminus O$, the segment ab meets $\overline{O} \setminus O$ in b only. The condition clearly holds in complete BW spaces and in connected trees.

2.15. Proposition. *Let X be a closure stable, JHC and S_4 space with Fuchssteiner's Property. If X is connected, then each δ_X-extreme point of a compact convex subspace of X is an ordinary extreme point.*

Proof. Let $P \subseteq X$ be a polytope, say: $P = co(F)$ where F is a finite set. We first show that if $p \in Ext_{\delta_X}(P)$ then $p \in F$. To this end, suppose $G \subseteq F$ is minimal with the property that $p \in co(G)$. If G is a one-point set then we are done. In the opposite case, p is in $\partial_X co(G)$ and by definition of extremality there is a convex open set $O \subseteq X$ such that

$$co(G) \cap O \neq \emptyset; \quad p \notin O; \quad co(G) \subseteq \overline{O}.$$

Let H be a convex set, maximal with the properties that $p \in H$ and $H \cap O = \emptyset$. Then H is a half-space by the assumption of S_4 and is closed by the assumption of closure stability. If we replace O by the open half-space $X \setminus H$, then the above listed properties of O persist. The half-space O meets G, whereas $G \not\subseteq O$ since $p \notin O$. Hence by join-hull commutativity, there exist $u \in co(G \setminus O)$ and $v \in co(G \cap O)$ such that $p \in uv$. By Fuchssteiner's Property, we find that $p = u$, showing that p is in the hull of $G \setminus O$, which is properly smaller than G.

If $C \subseteq X$ is a compact convex set, then $Ext_{\delta_X}(C) \subseteq ext(C)$. Indeed, if $p \in Ext_{\delta_X}(C)$ then $p \in Ext_{\delta_X}(P)$ for each polytope $P \subseteq C$ containing p and having more than one point. We conclude that $p \notin co(F)$ for each finite set $F \subseteq C \setminus \{p\}$, in other words, that $p \in ext(C)$. ∎

This result applies to several classes of examples. First, a complete BW space with the core topology and a locally convex topological vector space are easily seen to satisfy the properties listed in the result above. We do not know (even for locally convex vector spaces) whether each ordinary extreme point of a compact convex set is δ-extreme. However, if X is a compact convex subspace of a vector space, then the set of ordinary extreme points of X is evidently equal to the set of δ_X-extreme points of X.

§2: Extremality, Pseudo-boundary and Pseudo-interior   417

On the other hand, each compact convex subset of a connected and locally connected tree satisfies all hypotheses of the previous proposition. We just verify Fuchssteiner's Property. Let O be convex open, let $a \in O$ and $b \in \overline{O} \setminus O$. Then the segment ab is totally ordered as seen from a and O cuts off a relatively open front end of it. If $c <_a b$ is in $\overline{O} \setminus O$ and if U is a convex neighborhood of b missing c, then O, U, ab are pairwise intersecting convex sets without a common point, contradiction.

2.16. Theorem. *Let X be a connected NS_3 space and let $Y \subseteq X$ be compact. Then the set of closed δ_X–extreme subsets of Y is an extremality with respect to the protopology on Y consisting of all traces of convex closed sets and*
$$co^*(Y) = co^*(Ext_{\delta_X}(Y)).$$

Proof. As for (ES-1), let $C \subseteq X$ be convex closed, let E be δ_X–extreme and closed in Y and suppose $E \not\subseteq C$. Take a point $p \in E \setminus C$. As X is NS_3, there is a sequence $(C_n)_{n=0}^{\infty}$ of sets not containing p such that $C_0 = C$ and C_{n+1} is a convex closed neighborhood of C_n for all n. Then $\cup_{n=0}^{\infty} C_n$ is a convex open set including C and missing p. Owing to the compactness of E, there is a maximal convex open set O with the properties
$$C \subseteq O; \quad E \not\subseteq O.$$
Note that $E \subseteq co^*(O)$, for otherwise another application of NS_3 would yield a convex open set $O' \supseteq co^*(O)$ which does not include E. But then $O \subset O'$ by the connectedness of X, contradiction. The set $E' = E \setminus O$ is closed and is disjoint from C. If $q \in E'$ is a pseudo-interior point of a closed set $B \subseteq Y$ then $B \subseteq E$ since E is an extreme set. We find that $B \subseteq co^*(O)$ and since q is pseudo-interior to B, it follows that $B \cap O = \emptyset$. Hence $B \subseteq E'$, and E' is extreme.

The condition (ES-2) follows from the (permanent) assumption of S_1. As for (ES-3), chains of extreme sets have a non-empty intersection by the compactness of Y. This intersection is an extreme set in Y.

The desired formula follows from Proposition 2.2 with the observation that the involved closure operator on Y is given by the assignment $B \mapsto Y \cap co^*(B)$. ∎

The following theorem extends a well-known result in traditional convexity on so-called "convex" functionals.

2.17. Theorem (*Bauer Maximum Principle*). *Let Y be a non-empty compact subspace of a connected NS_3 space X, and let $f : X \to \mathbb{R}$ be a usc and lower CP functional. Then f attains its maximum over Y at a δ_X–extreme point of Y.*

Proof. There is a totally ordered collection of non-empty closed sets $f^{-1}[t, \infty) \cap Y$, for t ranging over $f(Y)$. By the compactness of Y, these sets have a common point. Hence f has a supreme value s at this point. We consider the convex open set $O = f^{-1}(\infty, s)$ of X. Note that O meets Y in a proper subset (unless f is constant, in which case the result is obvious). Let $P \supseteq O$ be a convex open set, maximal with the property that $P \cap Y$ is a proper subset of Y. If $co^*(P)$ does not include Y, then take $p \in Y \setminus co^*(P)$

and use Proposition III§4.8(1) to produce a convex open set P' of X with $p \notin P' \supseteq co^*(P)$. The connectedness of X implies that $P' \neq P$, contradiction.

As observed in 2.14.1, the closed set $Y \setminus P$ is δ_X-extreme in Y. Such extreme sets constitute a synthetic extremality by Theorem 2.16. By Proposition 2.2, $Y \setminus P$ contains a δ_X-extreme point. Since

$$Y \setminus P \subseteq f^{-1}(s) \cap Y,$$

we conclude to the desired result. ∎

For a deeper study of pseudo-boundaries and pseudo-interiors, the following concept is indispensable.

2.18. Continuous position. Let X be a topological convex structure. A set $A \subseteq X$ is said to be *in continuous position* (within X) provided for each convex open set $O \subseteq X$ meeting A,

$$A \cap \overline{O} = Cl_A(A \cap O).$$

The bar on the left refers to closure in X. Note that the inclusion from right to left is valid for purely topological reasons. The phrase "in continuous position" refers to the resemblance of the formulae

$$A \cap Cl(O) \subseteq Cl(A \cap O),$$
$$f(Cl(A)) \subseteq Cl f(A),$$

the second one being a well-known criterion for continuity of a function f. Here are some elementary results, valid in any tcs.

2.18.1. *Being in continuous position is a transitive property.*

Explicitly, if $Y \subseteq X$ is in continuous position relative to X and if $Z \subseteq Y$ is in continuous position relative to the subspace Y, then Z is in continuous position in X.

2.18.2. *A convex set C is in continuous position provided for each $a, b \in C$ the segment ab is in continuous position.*

Proof. Suppose C is not in continuous position. Then there is a convex open set $O \subseteq X$ and a point $a \in X$ with

$$O \cap C \neq \emptyset; \quad a \in \overline{O} \cap C; \quad a \notin Cl_C(O \cap C).$$

Take $b \in O \cap C$. We find that

$$O \cap ab \neq \emptyset; \quad a \in \overline{O} \cap ab; \quad a \notin Cl_{ab}(O \cap ab),$$

showing that ab is not in continuous position. ∎

2.19. Proposition. *Let X be a closure stable, S_4 and weakly locally convex*[2] *space, and let $C \subseteq X$ be a closed convex set. Then the following are equivalent:*

2. that is: locally convex in the corresponding weak topology.

§2: Extremality, Pseudo-boundary and Pseudo-interior   419

(1) C is in continuous position.
(2)   For each open half-space $O \subseteq X$,
$$O \cap C \neq \emptyset \Rightarrow \overline{O \cap C} = \overline{O} \cap C.$$
(3)   If $H \subseteq X$ is a closed half-space not including C then $Int_C(H \cap C) = (Int\, H) \cap C$.

Proof. The implications (1) $\Rightarrow$ (2) $\Rightarrow$ (3) are elementary; we concentrate on (3) $\Rightarrow$ (1). Let $O \subseteq X$ be a convex open set meeting C and let $x \in C \setminus \overline{O \cap C}$. By closure stability, the set $\overline{O \cap C}$ is convex and hence weakly closed. As X is weakly locally convex, there is a convex (weak) C–neighborhood $P \subseteq C$ of x disjoint from $O \cap C$ and hence with O. Regarding the third axiom of convexity, consider a convex set $H \subseteq X$, maximal with the properties
$$P \subseteq H;\quad H \cap O = \emptyset.$$
Then H is a half-space by S_4 and is closed by virtue of closure stability. We find that $C \not\subseteq H$ since O meets C. Hence by (3),
$$x \in Int_C(H \cap C) = Int\, H \cap C.$$
It appears that $Int\, H$ is an X–neighborhood of x, disjoint from O, showing that $x \notin \overline{O} \cap C$. Consequently,
$$\overline{O \cap C} \supseteq \overline{O} \cap C.$$
The opposite inclusion is trivial. ∎

2.20. Proposition. *Let X be a closure stable, point-convex, NS_3 space, and let K be a compact set in continuous position. If K is included in a connected subset of X, then K is connected.*

Proof. Suppose $K = K_1 \cup K_2$, where the sets K_i are closed, non-empty and disjoint. Let the collection $\mathcal{O}$ consist of all convex open sets meeting K_1 and disjoint from K_2. To see that $\mathcal{O}$ is non-empty, take a point p of K_1 and apply Theorem III§4.8(1) with reference to the compact subspace K and the convex set $\{p\}$. This yields a continuous lower CP functional f on X and a point $t \in f(X)$ such that $f(p) < t$ and $f(K_2) \subseteq [t, \infty)$. The set $f^{-1}(\infty, t)$ is in $\mathcal{O}$. It is clear that $\mathcal{O}$ is inductively ordered by inclusion. Let O be a maximal member. The set K being in continuous position, we find that
$$\overline{O} \cap K = \overline{O \cap K} \subseteq K_1.$$
Another application of Theorem III§4.8(1) (this time with reference to the convex closed set $C = \overline{O}$) yields a convex open set $P \supseteq \overline{O}$ missing K_2. However, as K is included in a connected set, we find that P is strictly larger than O, which is a contradiction. ∎

Although results like the last proposition may suggest a closer connection between continuous positions and continuity of the hull or the convex closure operator, no results are known in this direction.

2.21. The continuous positions property. A topological convex structure X is said to have the *Continuous Positions Property* (abbreviated: *CPP*) provided each convex set is in continuous position in X. By Proposition 2.18.2, it suffices that all segments are in continuous position within X. As the closure of a convex set is the same in the original topology as in the corresponding weak topology, it follows that X_w has the CPP whenever X has. The Product Polytope Formula I§1.10.3 and Theorem III§1.4 on products lead to the following result.

2.22. Proposition. *The product of a family of spaces with the CPP has the CPP.* ∎

Note that if all factors are non-empty, then the converse is true as well. The issue is that each factor can be seen as a convex subspace of which each relatively open convex set extends to a convex open set of the product.

CPP is, in general, not inherited by compact convex subspaces. For instance, let X be the unit square of the standard plane $\mathbb{R}^2$, and consider the following two convex subsets of X.

$$O = (0,1]^2 \cup \{0\} \times (0,\tfrac{1}{2}); \quad C = \{0\} \times [0,1].$$

Then O meets C and $C \subseteq \overline{O}$, but $\overline{O \cap C}$ is the set $\{0\} \times [0,\tfrac{1}{2}]$. Hence, C is not in continuous position within X.

Let us describe some situations in which the CPP occurs.

2.23. Proposition. *In a tcs with connected convex sets, Fuchssteiner's Property implies the Continuous Positions Property.*

Proof. Let C be convex and let O be a convex open set meeting C. Suppose $x \in C \cap \overline{O}$ and $x \notin O$. Fix $y \in O \cap C$. Fuchssteiner's Property (cf. 2.15) implies that

$$x \in xy \setminus O \subseteq \{x\}.$$

Now xy is a connected subset of C, and hence

$$x \in Cl_C(xy \setminus \{x\}) \subseteq Cl_C(xy \cap O) \subseteq Cl_C(C \cap O).$$

This shows that $C \cap \overline{O} \subseteq Cl_C(C \cap O)$, and the result follows. ∎

Particular examples are: topological vector spaces, complete join spaces without boundary, and topological trees. The latter also occur via the next result.

2.24. Proposition. *A weakly locally convex space of Helly number* ≤ 2 *has the CPP. In fact, if C, D are intersecting convex sets, then*

$$\overline{C} \cap D = Cl_D(C \cap D).$$

Proof. We only derive the inclusion from left to right. Let $x \in \overline{C} \cap D$ and let N be a convex neighborhood of x. Then the sets N, C, D meet two by two. As the Helly number is ≤ 2, it follows that $N \cap C \cap D \neq \emptyset$. Local convexity then yields that $x \in Cl_C(C \cap D)$. ∎

Then O and C are convex, whereas O is relatively open. Now

§2: Extremality, Pseudo-boundary and Pseudo-interior 421

$$\overline{O} \cap C = C; \quad \overline{O \cap C} = \{0\} \times [0, \tfrac{1}{2}].$$

2.25. Proposition. *Let X be a uniform S_4 convex structure such that the convex closure of the union of two compact convex sets is compact. If X has the CPP, then so does the convex hyperspace of X.*

Note that the condition concerning the convex closure is fulfilled if either X is complete (as a uniform space; cf. Proposition III§3.9), or X is join-hull commutative (use I§2.14). We will state and prove two auxiliary results first.

2.25.1. Lemma. *Let X be a uniform convex structure with compact polytopes, and let $O \subseteq X$ be a convex open set. Then the following closure formulas are valid.*

$$Cl(<O> \cap \mathcal{IC}_{comp}(X)) = <Cl(O)> \cap \mathcal{IC}_{comp}(X);$$
$$Cl(<O,X> \cap \mathcal{IC}_{comp}(X)) = <Cl(O),X> \cap \mathcal{IC}_{comp}(X).$$

Proof. The inclusions from left to right are elementary. Let $D \subseteq Cl(O)$ be a compact convex set, and consider a basic hyperspace neighborhood $<O_1,..,O_n>$ of D, where the sets O_i are open in X. By Theorem III§3.8, there is a convex open set P with $D \subseteq P \subseteq \cup_{i=1}^{n} O_i$. For each $i = 1,..,n$ there is a point $x_i \in O_i \cap P \cap O$. The polytope $D' = co\{x_1,..,x_n\}$ is compact and

$$D' \in <O_1,..,O_n>; \quad D' \in <O>.$$

This shows that D is in the closure of $<O> \cap \mathcal{IC}_{comp}(X)$.

Suppose next that $D \cap Cl(O) \neq \emptyset$, and let $<O_1,..,O_n>$ be a basic hyperspace neighborhood of D, where each O_i is an open set of X. Let P be a convex open set such that $D \subseteq P \subseteq \cup_{i=1}^{n} O_i$. As D meets $Cl(O)$ there is a point $x \in P \cap O$. Take $x_i \in D \cap O_i$ for $i = 1,..,n$. The polytope $D' = co\{x,x_1,..,x_n\}$ is included in P and it meets O as well as each O_i, whence

$$D' \in <O_1,..,O_n>; \quad D' \in <O,X>.$$

This shows that D is in the closure of $<O,X> \cap \mathcal{IC}_{comp}(X)$. ∎

Note that the proof of the previous lemma can be adapted to the situation where each compact convex set of X has a neighborhood base of convex sets. This is the case, for instance, if X is compact and NS_3.

2.25.2. Lemma. *Let X be a closure stable space such that*

(i) *Each pair of distinct points can be screened with convex closed sets.*
(ii) *the convex closure of the union of two compact convex sets is compact.*

If $\mathcal{H}$ is a closed half-space of $\mathcal{IC}_{comp}(X)$, then there is a closed half-space $H \subseteq X$ such that

$$\mathcal{H} = <H> \cap \mathcal{IC}_{comp}(X) \quad \text{or} \quad \mathcal{H} = <H,X> \cap \mathcal{IC}_{comp}(X).$$

Proof. First, observe that each minimal element C of $\mathcal{H}$ is a singleton. Indeed, if C_1, C_2 are a convex closed screening in C of two supposedly distinct points, then $C \in co\{C_1,C_2\}$ and one of C_1, C_2 is a member of $\mathcal{H}$. Let H be the set of all singletons

422   Chap. IV: Miscellaneous

in $\mathcal{H}$. The assignment $x \mapsto \{x\}$ being an embedding (cf. I§3.30.1), H is a closed half-space of X. If $C \subseteq H$ is a compact convex set, and if $F \subseteq C$ is finite, then

$$co(F) \in co\{\{x\} \mid x \in F\} \subseteq \mathcal{H}$$

(first hull in X, second hull in $\mathcal{IC}_{comp}(X)$). As the net of polytopes included in C converges to C, we conclude that $C \in \mathcal{H}$. Conversely, if $C \in \mathcal{H}$ then (as we observed above) a minimal member of $\mathcal{H}$ included in C is a singleton. Consequently, $C \in <H,X>$.

So far, this shows that

$$<H> \subseteq \mathcal{H} \subseteq <H,X>.$$

We verify that one of the two inclusions is, in fact, an equality.

(i) If $\bigcup \mathcal{H} = X$, then $\mathcal{H} = <H,X>$.

Indeed, let $D \in <H,X>$. Then D includes a member of $\mathcal{H}$ (at least, some singleton will do). Clearly, there is a maximal convex closed set $D_0 \subseteq D$ which is a member of $\mathcal{H}$. If $x \in D \setminus D_0$, then $x \in C$ for some $C \in \mathcal{H}$. Now

$$D \cap co^*(D_0 \cup C) \in co\{C,D_0\} \subseteq \mathcal{H},$$

and the set on the left is properly larger than D_0, a contradiction. Thus, $\mathcal{H} = <H,X>$.

(ii) If $\bigcup \mathcal{H} \neq X$, then $\mathcal{H} = <H>$.

Indeed, let $x \in X \setminus (\bigcup \mathcal{H})$. We may assume that $\mathcal{H} \neq \emptyset$ and hence that $H \neq \emptyset$. Let $u \in H$. If $v \in \bigcup \mathcal{H} \setminus H$, then observe that $\{v\} \notin \mathcal{H}$ and $xu \notin \mathcal{H}$ (since $xu \not\subseteq \bigcup \mathcal{H}$). Therefore, uv being in the hyperspace hull of $\{v\}$ and xu, we conclude that $uv \notin \mathcal{H}$. On the other hand, some $C \in \mathcal{H}$ contains v and

$$uv \in co\{\{u\}, co^*(C \cup \{u\})\},$$

which shows that $uv \in \mathcal{H}$, a contradiction. This shows that $\bigcup \mathcal{H} \subseteq H$, and the equality $\mathcal{H} = <H> \cap \mathcal{IC}_{comp}(X)$ follows easily. ∎

We now turn to the actual proof of Proposition 2.25. Assume that X has the CPP. The convex hyperspace $\mathcal{IC}_{comp}(X)$ is uniform by Proposition III§3.10.4. By 2.18.2 and Proposition 2.24, it suffices to show that for each open half-space $\mathcal{O}$ meeting a segment $C_1 C_2$ of $\mathcal{IC}_{comp}(X)$,

$$Cl(\mathcal{O}) \cap C_1 C_2 \subseteq Cl(\mathcal{O} \cap C_1 C_2).$$

For the remainder of the proof, let $C \in Cl(\mathcal{O}) \cap C_1 C_2$, let $<O_1,..,O_n>$ be a hyperspace neighborhood of C, and let $P \subseteq X$ be a convex open set such that

$$C \subseteq P \subseteq \overline{P} \subseteq \bigcup_{i=1}^{n} O_i.$$

In regard to Lemma 2.25.2, $\mathcal{O}$ is of one of the following types.

Type 1: $\mathcal{O} = <O,X> \cap \mathcal{IC}_{comp}(X)$ for some open half-space $O \subseteq X$. We have $C \subseteq co^*(C_1 \cup C_2)$ and one of C_1, C_2 is in $\mathcal{O}$. Therefore, O meets $co^*(C_1 \cup C_2)$. By CPP,

§2: Extremality, Pseudo-boundary and Pseudo-interior   423

$$\overline{O} \cap co^*(C_1 \cup C_2) = Cl(O \cap co^*(C_1 \cup C_2)).$$

As C meets $\overline{O}$ (Lemma 2.25.1), we conclude that $P \cap O \cap co^*(C_1 \cup C_2) \neq \emptyset$. Let x be a member of this set and consider the compact set $C' = co^*(C \cup \{x\})$. Then $C \subseteq C' \subseteq co^*(C_1 \cup C_2) \cap \overline{P}$, from which it follows that $C' \in C_1C_2 \cap <O_1,..,O_n>$. As $C' \in <O,X> = \mathcal{O}$, this shows that $C \in Cl(\mathcal{O} \cap <O_1,..,O_n>)$.

Type 2: $\mathcal{O} = <O> \cap \mathcal{IC}_{comp}(X)$ for some open half-space $O \subseteq X$. This time, one of C_1, C_2 is included in O. If $c_j \in C_j$ ($j = 1, 2$), then there is a point $u \in C \cap c_1c_2$ and $u \in P \cap O_i$ for some i. As $C \subseteq \overline{O}$, we have

$$u \in \overline{O} \cap c_1c_2 = Cl(O \cap c_1c_2).$$

This yields a point

$$x \in P \cap O_i \cap O \cap c_1c_2.$$

Let $P(c_1,c_2)$ be a convex open set of X with

$$x \in P(c_1,c_2) \subseteq \overline{P(c_1,c_2)} \subseteq P \cap O_i \cap O.$$

By the continuity of the segment operator, there exist neighborhoods $U(c_1,c_2)$ of c_1 and $V(c_1,c_2)$ of c_2 such that $c'_1c'_2$ meets $P(c_1,c_2)$ for all $c'_1 \in U(c_1,c_2)$ and $c'_2 \in V(c_1,c_2)$. The set $C_1 \times C_2$ being compact, there is a finite collection of sets of type $U(c_{1k})$ (where $k = 1,..,p$) and $V(c_{2l})$ (where $l = 1,..,q$) such that

$$U(c_{1k}) \times V(c_{2l}) \subseteq U(c_{1k},c_{2l}) \times V(c_{1k},c_{2l})$$

for all k, l involved, and such that the left hand product sets cover $C_1 \times C_2$. By CPP,

$$C \subseteq \overline{O} \cap co^*(C_1 \cup C_2) \subseteq Cl(O \cap co^*(C_1 \cup C_2)).$$

As C is covered by the sets $P \cap O_i$ for $i = 1,..,n$, there exist points

$$x_i \in P \cap O_i \cap O \cap co^*(C_1 \cup C_2).$$

Consider the convex closed set

$$C' = co^*\Big(\bigcup_{k=1}^{p} \bigcup_{l=1}^{q} \big(P(c_{1k},c_{2l}) \cap co^*(C_1 \cup C_2) \big) \cup \{x_1,..,x_n\} \Big).$$

Note that $C' \subseteq \overline{P}$ and C' meets each O_i, whence

$$C' \in <O_1,..,O_n>.$$

Also, C' is the convex closure of the union of finitely many members of $\mathcal{O}$, whence $C' \in \mathcal{O}$. Finally, if $c_j \in C_j$ ($j = 1, 2$) then there exist $k \in \{1,..,p\}$ and $l \in \{1,..,q\}$ with $c_1 \in U(c_{1k})$ and $c_2 \in V(c_{2l})$. Consequently, c_1c_2 meets $P(c_{1k},c_{2l})$. As c_1c_2 is included in $co^*(C_1 \cup C_2)$, it follows that C' meets c_1c_2. Clearly, $C' \subseteq co^*(C_1 \cup C_2)$, and we conclude that C' is in the hyperspace segment C_1C_2. Summarized, C is in the closure of $\mathcal{O} \cap C_1C_2$, completing the proof of Proposition 2.25. ∎

We now prepare for some criteria on the existence of pseudo-interior points.

2.26. Lemma. *Let X be a closure stable S_4 space such that $\iota_X(C) \neq \varnothing$ for all $C \in \mathcal{IC}_*(X)$. If $D \subseteq X$ is convex closed and if $\mathcal{H}$ is a covering of D with closed half-spaces of X, then there exists $H_0 \in \mathcal{H}$ with*

$$D \subseteq \bigcup \{ Int(H) \mid H \in \mathcal{H} \} \cup H_0.$$

Proof. Let the convex closed set $E \subseteq D$ be defined as follows.

$$E = \bigcap \{ \overline{X \setminus H} \mid H \in \mathcal{H} \} \cap D.$$

If $E = \varnothing$, then D is covered with the interiors of the members of $\mathcal{H}$ and the choice of a half-space $H_0 \in \mathcal{H}$ (as required above) is irrelevant. So assume $E \neq \varnothing$. Then there is a point $x \in \iota_X(E)$. Let $H_0 \in \mathcal{H}$ be such that $x \in H_0$. By (5.11.2), either $E \subseteq H_0$ as desired, or $E \cap Int(H_0) \neq \varnothing$. The second possibility conflicts with the construction of E. ∎

We are lead to the following results. An a-topological theorem on countable intersections has been obtained in II§4.11.

2.27. Theorem. (*Countable Intersection Theorem, IIA*) *Let X be a closure stable, S_4 and properly locally convex space with the continuous positions property. If the underlying topological space is separable and completely metrizable, then the following assertions are equivalent.*

(1) *Each non-empty convex closed set has a non-empty pseudo-interior relative to X.*
(2) *Each covering of X with closed half-spaces has a countable subcovering.*
(3) *If C is convex closed and if $\mathcal{O}$ is a family of convex open sets such that $\bigcap \mathcal{O} = C$, then there is a countable subfamily $\mathcal{O}' \subseteq \mathcal{O}$ with $\bigcap \mathcal{O}' = C$.*

Proof of (1) $\Rightarrow$ (2). Let $\mathcal{H}$ be a covering of X with closed half-spaces. By Lemma 2.26, there exists $H_0 \in \mathcal{H}$ such that

$$X = \bigcup \{ Int(H) \mid H \in \mathcal{H} \} \cup H_0.$$

The topological space X being separable and metrizable, there is a countable collection of sets $H_n \in \mathcal{H}$, $n \in \mathbb{N}$, such that $\{ Int(H_n) \mid n \in \mathbb{N} \}$ covers $X \setminus H_0$. The desired covering of X consists of the sets H_n for $n = 0, 1, 2, ...$

Proof of (2) $\Rightarrow$ (3). Let C be a convex closed set and let $\mathcal{O}$ be a family of convex open sets such that $\bigcap \mathcal{O} = C$. If $x \notin C$ then $x \notin O$ for some $O \in \mathcal{O}$. If D is a convex set maximal with the properties $x \in D$ and $D \cap O = \varnothing$, then D is a closed half-space by S_4 and by closure stability. This shows that, without loss of generality, the family $\mathcal{O}$ consists of open half-spaces. Assume first that $C = \varnothing$. Then the complements of members of $\mathcal{O}$ yield a covering of X with closed half-spaces. Some countable closed subcovering of it leads to a countable subfamily of $\mathcal{O}$ with empty intersection.

We now settle the general case. For each $x \notin C$, consider a convex open neighborhood $P(x)$ disjoint from C. Then $X \setminus C$ can be covered with countably many convex open sets P_n for $n \in \mathbb{N}$, taken among the sets $P(x)$. For each n there is a countable subfamily $\mathcal{O}_n$ of $\mathcal{O}$ such that $\bigcap \mathcal{O}_n \cap P_n = \varnothing$. Then $\bigcup_{n \in \mathbb{N}} \mathcal{O}_n$ is a countable subfamily of $\mathcal{O}$ and its

§2: Extremality, Pseudo-boundary and Pseudo-interior   425

intersection equals C.

Proof of (3) $\Rightarrow$ (1). Let C be convex closed. By definition, $\iota_X(C)$ is the intersection of the family $\mathcal{O}$ consisting of all convex open sets O meeting C and such that $C \subseteq \overline{O}$. By CPP, the latter is equivalent to the statement that $O \cap C$ is relatively dense in C. Being a closed subset of a completely metrizable space, C is a Baire space. Hence the family $\mathcal{O}$ is countably intersecting. By (3), it follows that $\cap \mathcal{O}$ is non-empty. ∎

Observe that the implications (1) $\Rightarrow$ (2) $\Rightarrow$ (3) hold without the assumption of completeness and without CPP. The following is an adapting of the previous result to median spaces.

2.28. Theorem. (*Countable Intersection Theorem, IIB*) *Let X be a compact, locally convex, median algebra. Then the following assertions are equivalent.*

(1) *Each non-empty convex closed set has a non-empty pseudo-interior relative to X.*
(2) *Let $\mathcal{O}$ be a family of convex open sets in X such that for each $O \in \mathcal{O}$ there is a closed set $A \subseteq O$ meeting all members of $\mathcal{O}$. Then $\cap \mathcal{O} \neq \emptyset$.*

Observe that a family $\mathcal{O}$ as in (2) is "countably intersecting". Indeed, suppose $O_n \in \mathcal{O}$ for $n \in \mathbb{N}$. Take a closed set $A_1 \subseteq O_1$ meeting all $O \in \mathcal{O}$. Since O_1 is strongly convex (cf. III§4.23), we have $co^*(A_1) \subseteq O_1$, whence A_1 may be considered to be convex closed. Take a closed set $A_2 \subseteq O_2$ meeting all $O \in \mathcal{O}$. Without loss of generality, A_2 contains a point of A_1 and is convex. Proceeding by induction, we arrive at a sequence of closed convex sets $A_n \subseteq O_n$ for $n \in \mathbb{N}$ which is finitely intersecting. Then $\emptyset \neq \cap_n A_n \subseteq \cap_n O_n$ by the compactness of X.

Proof of (1) $\Rightarrow$ (2). Let $\mathcal{O}$ be a family of convex open sets as in (2). As in the proof of 2.27, it may be assumed that the family $\mathcal{O}$ consists of open half-spaces. If $\cap \mathcal{O} = \emptyset$, then the family
$$\mathcal{H} = \{X \setminus O \mid O \in \mathcal{O}\}$$
of closed half-spaces covers X. Lemma 2.26 provides $H_0 \in \mathcal{H}$ such that
$$X = \cup \{Int(H) \mid H \in \mathcal{H}\} \cup H_0.$$
We verify that
(*) $\quad \mathcal{I}_*(X) = \bigcup_{H \in \mathcal{H}} \langle H \rangle \cup \langle H_0, X \rangle.$

Let $A \subseteq X$ be a non-empty compact set disjoint from H_0. Then $co^*(A) \cap H_0 = \emptyset$ by the strong convexity of convex open sets in X, cf. III§4.23. Hence $co^*(A)$ is covered by the sets $Int(H)$ for $H \in \mathcal{H}$. As $co^*(A)$ is compact, there exist $H_i \in \mathcal{H}$ for $i = 1,..,n$ with $co^*(A) \subseteq \cup_{i=1}^n H_i$. As the Helly number of X is ≤ 2, some H_i includes $co^*(A)$.

Formulating the equality (*) in terms of the original family $\mathcal{O}$, we conclude that no closed subset of $O_0 = X \setminus H_0$ meets every $O \in \mathcal{O}$.

Proof of (2) $\Rightarrow$ (1). Let C be convex closed and consider the family $\mathcal{O}$, consisting of all convex open sets O meeting C and such that $C \subseteq \overline{O}$. Note that X has the CPP by 2.24. By definition, $\iota_X(C) = \bigcap \mathcal{O}$. For each $O \in \mathcal{O}$ consider a closed set $A \subseteq C \cap O$ with a non-empty C-interior. As each member of $\mathcal{O}$ meets C in a relatively dense subset, we find that A meets all other members of $\mathcal{O}$. By (2), it follows that $\mathcal{O}$ has a non-empty intersection. ∎

2.29. Examples. Recall that a *Fréchet space* is a locally convex, completely metrizable vector space. We examine the (non-)existence of pseudo-interior points in Fréchet spaces, nonseparable Hilbert spaces, and locally compact median algebras.

2.29.1. Proposition. *In a separable Fréchet vector space V, each non-empty convex closed set has a non-empty relative pseudo-interior. Consequently, each countably intersecting family of convex open sets in V has a non-empty intersection.*

Proof. Let $C \subseteq V$ be a non-empty convex closed set, and consider a countable dense subset $\{p_n \mid n \in \mathbb{N}\}$ of C. We first assume that C is compact. Then the point

$$p = \sum_{n=1}^{\infty} 2^{-n} p_n$$

is well-defined. If f is a continuous linear functional on V, then

$$f(p) = \sum_{n=1}^{\infty} 2^{-n} f(p_n).$$

Hence if $f(p_n) \geq f(p)$ for all n then none of these inequalities can be strict. In this situation, f is constant on the set of all points p_n -- hence on the whole of C. Similarly, $f(p_n) \leq f(p)$ for all $n \in \mathbb{N}$ implies that f is constant on C. The only possibility left is that

$$f(p_k) < f(p) < f(p_l)$$

for some $k, l \in \mathbb{N}$. This shows that p is not a proper support point of C and by a remark following 2.12 we arrive at the desired result.

We now treat the general case. Take a point $q \in C$. For each n we consider a point $q_n \in p_n \circ q$ at a distance less than $1/n$ from q. In a completely metrizable locally convex vector space, the convex closure of a compact set is compact (cf. III§3.9). Hence the convex closure D of

$$\{q_n \mid n \in \mathbb{N}\} \cup \{q\}$$

is a compact set. By the previous argument, there is a point $p \in \iota_V(D)$. We verify that $p \in \iota_V(C)$. To this end, let H be a closed half-space of V not including C (say: $p_n \notin H$) and such that $p \in H$. If $D \nsubseteq H$ then $Int\, H$ is non-empty by assumption on p. If $D \subseteq H$ then by CPP,

$$q_n \in Int_{p_n q}(H \cap p_n q) = Int\, H \cap p_n q.$$

In particular, $Int\, H \neq \emptyset$. By 2.12, we conclude that $p \in \iota_V(C)$.

The last part of the proposition follows from Theorem 2.27. ∎

§2: Extremality, Pseudo-boundary and Pseudo-interior 427

2.29.2. The following is an example of a non-separable Hilbert space and a non-empty convex closed set with an empty pseudo-interior. Let S be an uncountable set and consider the Hilbert space $\ell_2(S)$ of all square summable sequences over S, that is, the set of all functions $x: S \to \mathbb{R}$ such that $x(s) = 0$ for all but countably many $s \in S$ -- say: s_n for $n \in \mathbb{N}$ -- and such that $\sum_{n=1}^{\infty} x(s_n)^2$ converges. In these circumstances, the series $\sum x(s_n)^2$ converges absolutely and hence the order of summation is irrelevant. We use the simpler notation $\sum x(s)^2$ (and the like) in case the corresponding series is absolutely convergent. This yields a well-known example of a Hilbert space, with an inner product defined by

$$x \cdot y = \sum x(s) y(s).$$

To obtain a closed convex set with an empty pseudo-interior, we consider the set C of all $x \in \ell_2(S)$ taking non-negative values only. This C is obviously convex. If $p \in C$ then $p(s) = 0$ for some $s \in S$ since S is uncountable. The evaluation map

$$f: \ell_2(S) \to \mathbb{R}, \quad x \mapsto x(s),$$

is a continuous linear functional such that $f(x) \geq 0$ for all $x \in C$, whereas $f(p) = 0$ and $f(x) > 0$ for some $x \in C$. So C is properly supported at p. ∎

2.29.3. Proposition. *Let X be a compact, locally convex median algebra on a separable and first countable topological space. Then each non-empty convex closed subset of X has a non-empty relative pseudo-interior.*

Proof. The topological assumptions on X lead to a sequence $(O_n)_{n \in \mathbb{N}}$ of non-empty open sets such that each dense open set of X includes O_n for some n. For each $n \in \mathbb{N}$ we apply the Urysohn Theorem (cf. III§4.8) to obtain a non-zero continuous functional $f_n: X \to [0,1]$ which is zero outside of O_n. Consider the function

$$w: \mathcal{J}_*(X) \to \mathbb{R}, \quad w(A) = \sum_{n=1}^{\infty} 2^{-n} \sup f_n(A).$$

It is topological routine to verify that w is well-defined and continuous. Here are two additional properties of w.

(1) $A \subseteq B$ implies $w(A) \leq w(B)$.
(2) If $A \in \mathcal{J}_*(X)$ and $\mathrm{Int}\, A = \emptyset$, then $w(A) < w(X)$.

The first statement is obvious. As for the second, if $\mathrm{Int}(A) = \emptyset$, then $A \cap O_n = \emptyset$ for some n and hence f_n is zero on A. Since all functions f_k in consideration are nonnegative and since f_n is non-zero, we conclude that $w(A) < w(X)$.

Consider the following subsets of $\mathcal{J}_*(X)$.

$$\mathcal{M} = \{ M \mid w(M) > w(\overline{X \setminus M}) \}; \quad \mathcal{N} = \{ N \mid w(N) \geq w(\overline{X \setminus N}) \}.$$

We first show that $\mathcal{M}$ is a linked system. If $M_1, M_2 \in \mathcal{M}$ are disjoint, then $M_1 \subseteq X \setminus M_2$ and $M_2 \subseteq X \setminus M_1$, whence by (1),

$w(M_1) \leq w(\overline{X \setminus M_2}) < w(M_2)$.

The same holds if the indices 1, 2 are interchanged. The combined inequalities yield a contradiction. We next show that $\mathcal{n}$ is closed in $\mathcal{J}_*(X)$. To this end, consider the following functions:

$$f : \mathcal{J}_*(X) \times \mathcal{J}_*(X) \to \mathcal{J}_*(X); \quad f(A,B) = A \cup B;$$
$$g : \mathcal{J}_*(X) \times \mathcal{J}_*(X) \to [0,1] \times [0,1]; \quad g(A,B) = (w(A), w(B));$$
$$h : \mathcal{J}_*(X) \times \mathcal{J}_*(X) \to \mathcal{J}_*(X); \quad h(A,B) = A.$$

Note that

$$\mathcal{n} = h\big(f^{-1}\{X\} \cap g^{-1}\{(t_1, t_2) \mid t_1 \geq t_2\}\big).$$

Since f and g are continuous and since h is a closed function (by the compactness of the hyperspace), the set $\mathcal{n}$ is closed.

Each closed set $A \subseteq X$ such that $Int\, A = \varnothing$ is disjoint from some member of $\mathcal{m}$. Indeed, we have $\overline{X \setminus A} = X$ and (2) implies that $w(A) < w(\overline{X \setminus A})$. Hence $A \notin \mathcal{n}$ and as the latter is closed, we obtain a hyperspace neighborhood $\mathcal{U}$ of A disjoint from $\mathcal{n}$. Choose an open set $P \subseteq X$ such that $A \subseteq P$ and $\overline{P} \in \mathcal{U}$. Now $Cl(X \setminus \overline{P}) \subseteq X \setminus P$ and hence $w(\overline{P}) < w(X \setminus P)$. The set $X \setminus P$ is in $\mathcal{m}$ and is disjoint from A.

The space X being compact and of Helly number 2, there exists a point

$$p \in \cap \{co^*(M) \mid M \in \mathcal{m}\}.$$

We claim that $p \in \iota(X)$. Suppose $O \subseteq X$ is convex open and $p \notin O$. If $X \setminus O$ has an empty interior, then it is disjoint from some $M \in \mathcal{m}$. However, convex open sets of X are strongly convex (cf. Topic III§4.23). Consequently $p \in co^*(M) \subseteq O$. We conclude that $\overline{O} \neq X$.

To complete the proof, let $C \subseteq X$ be a non-empty convex closed set. The gate function $p: X \to C$ is continuous by Corollary III§5.20, whence C is separable and first countable. Therefore, $\iota(C) \neq \varnothing$ by the previous proof, and it remains to be observed that $\iota(C) = \iota_X(C)$. This is true because a convex relatively open subset O of C extends to a convex open set $p^{-1}(O)$ of X. ∎

Further Topics

2.30. On the classical Krein-Milman Theorem (Wieczorek [1989]). Let X be a convex structure and let $A \subseteq X$. A point $p \in X$ is **strictly between** a finite set F provided $p \in co(F)$ and $p \notin co(G)$ if G is a proper subset of F. A set $E \subseteq A$ is **strictly extreme in** A provided for each $p \in E$, if p is strictly between a finite set $F \subseteq A$ then $F \subseteq E$. Note that strictly extreme singletons correspond with standard extreme points. A function $f: A \to \mathbb{R}$ is **strictly convex** provided for each finite set $F \subseteq A$ and for each point $p \in A$ which is strictly between F, either f is constant on F or $f(p) < \max f(F)$. Note that a strictly extreme functional is lower CP.

§2: Extremality, Pseudo-boundary and Pseudo-interior   429

2.30.1. If f is strictly convex and if $m = \max f(A)$, then $f^{-1}(m)$ is a strictly extreme set of A.

2.30.2. Let X be a Hausdorff S_1 tcs such that the family of usc strictly convex functionals of X separates between convex closed sets and points. If $K \subseteq X$ is compact, then
$$co^*(K) = co^*(ext(K)).$$

2.30.3. (compare Soltan [1987]) Let (X, ρ) be a metric space. A function $f: X \to \mathbb{R}$ is ρ–*convex* provided it has the following property. For each $u \neq v$ and for each $x \in X$ which is geodesically between u, v,
$$f(x) \leq \frac{\rho(x,v)}{\rho(u,v)} \cdot f(u) + \frac{\rho(u,x)}{\rho(u,v)} \cdot f(v).$$

Show that ρ–convex functionals are strictly convex. Conclude that if the family of continuous ρ–convex functionals separates convex closed sets of X from points, then each compact set is included in the convex closure of its (standard) extreme points. Prove that if ρ is derived from a norm and if the above mentioned separation property holds, then the norm is rotund and the geodesic convexity is standard.

Hint. Use the theory of vector convexity (cf. Topic III§1.25) to show that each pair of points can be separated with a CP functional of the geodesic convexity.

2.31. Shilov boundary. Let X be a topological convex structure, and let $K \subseteq X$ be a compact subset. A closed set $A \subseteq K$ is a *Shilov subset of K* provided $co^*(A) = co^*(K)$. The results in 2.6, 2.10, and 2.16 provide examples of Shilov sets.

2.31.1. Let X be NS_3 and let $K \subseteq X$ be compact. Show that the following are equivalent for a closed set $A \subseteq K$.

(i) A is a Shilov subset of K.
(ii) If $f: X \to \mathbb{R}$ is an lsc and lower CP functional, and if $f|K$ is continuous, then $\sup f(A) = \sup f(K)$.

The notion of a Shilov set is usually phrased in terms of functionals as in (ii).

2.31.2. Verify that each compact set of an NS_3 space X has a *minimal* Shilov set. A *smallest* Shilov set is called a *Shilov boundary*. Verify that a space in which each compact subspace has a Shilov boundary is a convex geometry.

2.31.3. Let X be a closure stable, JHC and S_4 space with Fuchssteiner's property. If X is connected, then each compact convex subspace of X has a Shilov boundary, namely, the closure of the set of all (ordinary) extreme points.

2.32. Product spaces. Let X_j for $j \in J$ be topological convex structures with product X, and let π_j be the jth projection $X \to X_j$.

2.32.1. For each $j \in J$ let ι_j denote the pseudo-interiority operator of X_j. Show that if $Y \subseteq X$ be non-empty.

$$\iota_X(Y) = Y \cap \left(\prod_{j \in J} \iota_j(\pi_j Y) \right).$$

2.32.2. Let all factor spaces X_j be S_4 and closure stable. Let Ext and Ext_j refer to extremality with respect to the boundary operators δ_X and δ_{X_j}, respectively. Show that if $C \subseteq X$ is convex, then

$$Ext(C) = C \cap \left(\prod_{j \in J} Ext_j(\pi_j C) \right).$$

2.33. Spanning sets (compare van de Vel [1983d]). Let X be a connected, closure stable $NS_3 + S_4$ space, and let A be a subset of X with $\overline{co}(A) = X$.

2.33.1. If A is compact then $\overline{co}(A \cap \partial(X)) = X$.

2.33.2. If A_0 is a compact subset of $\iota_X(A)$, then $\overline{co}(A \setminus A_0) = X$.

2.33.3. Assume moreover that X is compact. Observe that X admits maximal elements with respect to each base-point order. Let $B \subseteq X$ consist of all pseudo-boundary points of X which are at the same time maximal with respect to a given point $b \in \iota(X)$. Show that $co^*(B) = X$. Give examples showing that the condition $b \in \iota(X)$ is necessary, that maximal points need not be pseudo-boundary points, and that pseudo-boundary points need not be maximal points.

2.34. Trees in topology

2.34.1. (compare Whyburn [1968] and Proposition 2.13.1). Let T be a connected and locally connected Hausdorff space in which any two points can be separated by removal of a third one. The *cut point order* $\leq_b$ of $b \in T$ is given by

$$x \leq_b y \Leftrightarrow x \in \{b, y\} \text{ or } x \text{ is a cut point between } b, y.$$

Show that $(T, \leq_b)$ is a topological tree with compact segments satisfying

$$xy = \{x, y\} \cup \{z \mid z \text{ separates between } x, y\},$$

and that the cut point order $\leq_b$ equals the base-point order of b. A *compact* connected space, in which two distinct points can be separated by a third one, is locally connected.

2.34.2. Let T be a separable tree. Show that if each countable subfamily of a family $\mathcal{H}$ of half-spaces has a non-empty intersection, then $\cap \mathcal{H} \neq \emptyset$.

2.35. Absolute and relative pseudo-interior

2.35.1. (van de Vel [1983d]) Let X be an FS_2 topological convex structure with compact segments and of Helly number ≤ 2, and let $C \subseteq X$ be a convex closed set. Show that $\iota(C) = \iota_X(C)$ and that $\iota(C)$ is a dense subset of C provided it is non-empty.

§2: Extremality, Pseudo-boundary and Pseudo-interior   431

Problem. Does the previous result extend to spaces of finite Helly number?

2.35.2. Extend Proposition 2.29.3 from compact to locally compact spaces.

2.35.3. Let X be a properly locally convex tcs and let $\Delta(X)$ be its cone. Note that X reappears as a convex subspace of $\Delta(X)$ (the "zero level"). Show that for each convex set $C \subseteq X$ with more than one point, $\iota_{\Delta(X)}(C) = \varnothing$.

2.36. Existence of pseudo interior points (van de Vel [1983d])

2.36.1. Let X, Y be closure-stable, S_4 and properly locally convex tcs's with a separable, metrizable underlying space. Suppose there is a continuous CP surjection $X \to Y$. Show that if each non-empty convex closed subset of X has a non-empty relative pseudo-interior, then the same goes for Y.

Problem. Can the conditions of separability or metrizability be removed?

2.36.2. Formulate and prove a similar result for separable and first countable median spaces.

2.36.3. Let X be an uncountable tcs which is topologically discrete and convexly free. Show that $\iota(\lambda(X)) \neq \varnothing$ and that $\iota(C) = \varnothing$ for some non-empty compact convex set in $\lambda(X)$. Note that there is a continuous CP surjection $\lambda(X) \to C$.

2.36.4. A convex structure X is *convexly homogeneous* provided for each pair of points $x_1, x_2 \in X$ there is a CP isomorphism $X \approx X$ carrying x_1 to x_2. Note that a Cantor cube is convexly homogeneous. Show that a non-trivial locally convex median algebra on a separable, first countable continuum is not convexly homogeneous.

Problem. Is there an example of a non-trivial, locally convex, median continuum which is convexly homogeneous?

2.37. Superextensions

2.37.1. (van de Vel [1983d]) Let X be a compact topological space and let $\lambda(X) = \lambda(X, \mathcal{T}_*(X))$. Show that the following are equivalent:
(i) Each compact convex set in $\lambda(X)$ has a pseudo-interior point, and
(ii) The space X has the *Shrinking Property*, that is: if $\mathcal{O}$ is a family of open sets in X and if each $O \in \mathcal{O}$ can be "shrunk" to a closed set $A \subseteq O$ meeting all members of $\mathcal{O}$, then the members O of $\mathcal{O}$ can be shrunk simultaneously to closed sets $A(O) \subseteq O$ such that the family $\{A(O) \mid O \in \mathcal{O}\}$ is linked.

2.37.2. (C.F. Mills; communicated by the author [1983d]) Show that a compact space with the Shrinking Property has countable cellularity, that is, there is no uncountable family of pairwise disjoint non-empty open sets. Deduce that a compact metrizable space has the Shrinking Property.[3]

---
3.   K.P. Hart informed me that a direct proof of this fact is possible.

Problem. Is the Shrinking Property of compact spaces equivalent to one of the properties: countable cellularity / separability / metrizability?

2.38. Continuous position (van de Vel [1988b])

2.38.1. Show that an open set and a dense set of a tcs are in continuous position.

2.38.2. Let H be a Hilbert space. Show that a closed subset of H is in continuous position iff it is a convex set. Hint. First, prove the result for compact sets in Euclidean space. Then deduce the result for compact sets in locally convex linear spaces. Finally, observe that the unit ball of H is weakly compact.

2.39. Topological Hilbert cubes

2.39.1. (van de Vel [1986]) Let X be a compact locally convex median algebra on a metrizable underlying space, and let $H \subseteq X$ be a dense half-space with a dense complement. Show that X is homeomorphic to the Hilbert cube $Q = [0,1]^\omega$.

Hint. Consider an adapted metric (cf. III§4.33) on X. If $P \subseteq X$ is a polytope at Hausdorff distance to X less than $\varepsilon > 0$, then the gate map $p : X \to P$ is ε–close to identity. Apply Torunczyk's Theorem.[4]

2.39.2. (compare van Mill [1980]; van de Vel [1983d]). Let Y be a metric continuum with more than one point. Show that the pseudo-boundary of $\lambda(Y)$ is a convex (!) subset. Conclude that $\lambda(Y)$ is homeomorphic with the Hilbert cube Q. By the way: *no* homeomorphism between the compact median algebras $\lambda(Y)$ and Q can be median preserving.

Remark. van Mill obtained his result before Torunczyk's Hilbert cube characterization was available. Its proof is more complicated than the one outlined here.

2.40. Interior stability

2.40.1. It is an elementary fact that if C is a convex set in a finite-dimensional topological vector space, then $Int(C) = Int(Cl(C))$. Use this fact, together with the CPP to show that the *relative* interior of a *relative* half-space of C is convex.

2.40.2. (compare Lemma III§1.14) Give an example of a compact locally convex S_4 space X, in which the interior of each half-space is convex, and such that X is not interior stable.

---

4.   Full citation in 3.18 below.

§2: Extremality, Pseudo-boundary and Pseudo-interior   433

Notes on Section 2

The concepts of (synthetic) extremality and of functional betweenness have been introduced by Fan [1963], who derived the general Proposition 2.2 and its application to $\mathcal{F}$-extremality, Proposition 2.6. Fan considered betweenness with respect to segments.

Generating functionals have been a popular approach to abstract convexity during the sixties. Except for Fan's paper, it has also been employed in the prominent work of Bauer [1961], presenting a general approach to Choquet extremality. Proposition 2.9, on compact convex sets in locally convex vector spaces, is standard; the argument is taken from Phelps [1966]. Proposition 2.10 extends a classical result of Bauer [1961], who assumes addition stability of the generating functionals, and combines the result with information on the existence of a Shilov boundary (cf. 2.31). Stability of a set of functionals under addition is a standard assumption in generalized Choquet theory. As we already stated in III§4.22, it implies that the functionally generated convex structure can be embedded in a vector space. The approach to extremality via pseudo-boundaries, Theorem 2.16, is due to the author [1983d]. The examples in 2.13 are taken from this paper.

The three viewpoints on extremality, developed in this section, have rather similar looks. Considerable efforts have been spent to find other approaches, aimed at giving a unified treatment. We mention the work of Fuchssteiner [1971], Lassak [1986], and Wieczorek [1989]. See, for instance, 2.30. These abstractions are essentially reducible to Fan's synthetic approach. The original Bauer's maximum principle is given in Bauer [1960]. The present version, Theorem 2.17, is new. It is possible to attach such a principle to each of the notions of extremality, developed in 2.4 and 2.8 as well. This requires somewhat technical conditions (usually called concavity or convexity) to the effect that a functional can be added to the generating set of functionals without altering Fan betweenness in finite sets or without altering a point's representing measures. See, for instance, Davies [1967].

The concept of a set being in continuous position is taken from van de Vel [1983d], where Propositions 2.23 and 2.24 can be found. This paper further contains the Countable Intersection Theorems 2.27 and 2.28, together with the examples in 2.29.1 and 2.29.3. Example 2.29.2 is a remake of an example given by Klee [1956]. The Continuous Positions Property of convex hyperspaces, Proposition 2.25, is taken from van de Vel [1983g].

The concept of a Shilov boundary (cf. 2.31) appears in almost any treatment of abstract Choquet theory. The existence of such a boundary is usually obtained via Choquet extremality by assuming (among other things) stability of functionals under addition.

3. Continuous Selection

The main result of this section is that a lower semi-continuous (LSC) multi-function with convex closed value sets, mapping a topological space into a uniform S_4 convex structure, admits a continuous selection under suitable additional conditions. The domain is at least a normal topological space. As to the range space, all polytopes are compact, all convex sets are connected, and the uniformity is metric.

Application of this result to Fréchet vector spaces, to connected metric trees, and to spaces of arcs leads to some well-known selection theorems. In combination with results on convex hyperspaces or on spaces of arcs, it also leads to the approximation of USC convex valued functions by continuous single-valued functions and to a selection theorem in spaces of arcs. Other applications involve the topology and geometry of uniform convex structures. Applications concerning fixed points are postponed to Section 6.

In this section, *all spaces are assumed* S_1.

3.1. Simplicial complexes (cf. I§1.19.5). We briefly recall some conventions, notation, and terminology. We let $|S|$ be the realization of a simplicial complex S and $|S|^q$ its q-skeleton, that is, the union of all realised simplices of dimension $\leq q$. A family $\mathcal{O}$ of subsets of a set X induces a simplicial complex with vertex set $\mathcal{O}$; its simplices are the finite subsets of $\mathcal{O}$ with a non-empty intersection. The geometric realization $|\mathcal{O}|$ of $\mathcal{O}$ is called the *nerve* of $\mathcal{O}$. A simplex $\{O_1,..,O_n\}$ of $\mathcal{O}$ is *on a subset* A of X provided $\cap_{i=1}^{n} O_i \cap A \neq \emptyset$. The simplices on A constitute a subcomplex $\mathcal{O}(A)$ of $\mathcal{O}$.

If $\mathcal{P}$ is a second family of subsets of X, then a *refinement function* $\alpha: \mathcal{P} \to \mathcal{O}$ is a simplicial map satisfying $P \subseteq \alpha(P)$ for each $P \in \mathcal{P}$. In these circumstances, $\mathcal{P}$ is called a *refinement* of $\mathcal{O}$, in symbols: $\mathcal{P} \leq \mathcal{O}$.

3.2. Theorem. *Let X be a tcs with connected convex sets and with compact polytopes, and let $C \subseteq X$ be non-empty and convex.*

(1) *If X is closure stable and S_4 and if $\mathcal{O}$ is a family of convex relatively open subsets of C covering C, then the polyhedron $|\mathcal{O}(C)|$ is contractible.*

(2) *If each polytope of X is FS_4 and if $\mathcal{S}$ is a family of convex relatively closed subsets of C covering C, then the polyhedron $|\mathcal{S}(C)|$ is contractible.*

Proof of (1). We first verify that the theorem holds for a finite convex open cover $\mathcal{O}$ of a convex set C. Let s be the total number of simplices occurring in $\mathcal{O}(C)$. If $s = 1$ we are done. We proceed by induction as follows. Each simplex of $\mathcal{O}(C)$ extends to a maximal one. If only one maximal simplex occurs, then the result is evident. Suppose σ_1 and σ_2 are distinct maximal simplices. Then $\sigma_1 \cap C$ and $\sigma_2 \cap C$ are disjoint convex sets, whence by the Kakutani Separation Property there is a half-space $H \subseteq X$ with

$$\cap \sigma_1 \cap C \subseteq H; \quad \cap \sigma_2 \cap C \subseteq X \setminus H.$$

Let $C_1 = Cl(C \cap H)$ and $C_2 = Cl(C \setminus H)$. Then

$$|\mathcal{O}(C_1)| \cup |\mathcal{O}(C_2)| = \mathcal{O}_C.$$

On the other hand, if a simplex σ is in $\mathcal{O}(C_i)$ for $i = 1, 2$ then the non-empty connected set $\cap \sigma$ meets C_1, C_2 and hence it meets $C_1 \cap C_2$. This shows that

$$\mathcal{O}(C_1) \cap \mathcal{O}(C_2) = \mathcal{O}(C_1 \cap C_2).$$

Note that $\cap \sigma_1$ is disjoint from C_2 and that $\cap \sigma_2$ is disjoint from C_1 since $\cap \sigma_1$, $\cap \sigma_2$ are open sets. Consequently, the simplicial complexes $\mathcal{O}(C_1)$, $\mathcal{O}(C_2)$, $\mathcal{O}(C_1 \cap C_2)$ each have less than s simplices and by inductive assumption, the nerves are contractible. Hence[1] the nerve of

$$\mathcal{O}(C) = \mathcal{O}(C_1) \cup \mathcal{O}(C_2)$$

is contractible.

We next verify that the result holds for an arbitrary convex open cover $\mathcal{O}$ of a convex set C. Let $K \subseteq |\mathcal{O}(C)|$ be compact. Then[2] there is a compact polyhedron $P \subseteq |\mathcal{O}(C)|$ including K. Let $\mathcal{O}'$ be the set of vertices of P. For each simplex σ of P we fix a point in $\cap \sigma \cap C$. The convex hull of the selected points is a compact subset C_0 of C and it is covered by a finite subcollection $\mathcal{O}''$ of $\mathcal{O}$. Let $\mathcal{P} = \mathcal{O}' \cup \mathcal{O}''$. Then

$$K \subseteq P \subseteq \mathcal{P}(C_0),$$

and $\mathcal{P}(C_0)$ has a contractible nerve by the first part of the proof. This shows that $|\mathcal{O}(C)|$ is weakly homotopy trivial and hence[3] that this polyhedron is contractible.

Proof of (2). The argument is rather similar to the previous one. Consider a finite cover $\mathcal{B}$ first. For each maximal simplex σ on C take a point in $\cap \sigma \cap C$. This leads to a polytope in C on which $\mathcal{B}$ has the same nerve. Operating by induction on the number of maximal simplexes, the axiom FS_4 can be used to produce smaller complexes. The general case is proved exactly as in the first part. ∎

We now proceed with a proof of a selection theorem. We first concentrate on multifunctions with a finite-dimensional domain.

3.3. Proposition. *Let X be an S_4 topological convex structure with compact polytopes and with connected convex sets, and let d be a compatible metric. If Y is a finite-dimensional paracompact space and if $F: Y \multimap X$ is an LSC multivalued function with convex and d-complete value sets, then F admits a continuous selection.*

Proof. By Theorem III§3.8, the uniformity of X has a base of convex open covers. Let $\mathcal{O}_0$ be such a cover and consider the multifunction

---

1. Borsuk [1967, p. 90]
2. Spanier [1966, p. 113]
3. Spanier [1966, p. 405]. Weak homotopy triviality refers to the fact that each mapping of a sphere into a space extends to the enclosed disk. In this context, the term has nothing to do with "weak topology".

§3: Continuous Selection 437

$$S_0: Y \multimap |\mathcal{O}_0|, \quad S_0(y) = |\mathcal{O}_0(F(y))|.$$

In words: $S_0(y)$ is the realised complex of all simplices which are on $F(y)$. The greater part of our efforts are spent to a proof of the following result.

Lemma. *There is a continuous selection s of S_0 with the following properties.*

(i) *If $q = dim(Y)$, then $s(Y) \subseteq |\mathcal{O}_0|^q$.*

(ii) *Each $y \in Y$ has a neighborhood W_y such that $s(W_y)$ is relatively compact.*

Let $\mathcal{O}'_0$ and $\mathcal{O}_1$ be uniform convex open covers of which the first one star refines $\mathcal{O}_0$ and the second one is associated to $\mathcal{O}'_0$ as in III§3.2. We consider a composed refinement map

$$\alpha_1: \mathcal{O}_1 \xrightarrow{\gamma} \mathcal{O}'_0 \xrightarrow{\beta} \mathcal{O},$$

where β is chosen such that $star(O', \mathcal{O}'_0) \subseteq \beta(O')$ for all $O' \in \mathcal{O}'_0$. The multifunction $S_1: Y \multimap |\mathcal{O}_1|$ is defined by $S_1(y) = |\mathcal{O}_1(F(y))|$. Note that $|\alpha_1|$ maps $S_1(y)$ into $S_0(y)$. We first verify the following.

(1) *If $A \subseteq Y$ is non-empty and $\cap_{a \in A} S_1(a) \neq \emptyset$, then the restriction*

$$\bigcap_{a \in A} S_1(a) \to \bigcap_{a \in A} S_0(a)$$

of $|\alpha_1|$ is null homotopic.

Indeed, for each simplex σ of $\cap_{a \in A} S_1(a)$ we select a point in $\cap \sigma \cap F(a)$, and we let $C(a)$ denote the hull of these points. We also fix a point $p \in A$ for reference. The sets $C(a), C(p)$ are $\mathcal{O}'_0$-close, being the hull of $\mathcal{O}_1$-close sets. Hence γ maps $\cap_{a \in A} S_1(a)$ into $\mathcal{O}'_0(C(p))$. This complex is contractible by Theorem 3.2(1).

So it suffices to show that β maps the last named complex into $\cap_{a \in A} S_0(a)$. To this end, let $\sigma \subseteq \mathcal{O}'_0$ be a simplex on $C(p)$ and fix a point $x \in \cap \sigma \cap C(p)$. For each $a \in A$ the set $C(a)$ is $\mathcal{O}'_0$-close to $C(p)$. Consequently, there is a point $x_a \in C(a)$ such that x, x_a occur together in some $O'_a \in \mathcal{O}'_0$. Hence for each $O' \in \sigma$ we have

$$\beta(O') \supseteq star(O', \mathcal{O}'_0) \supseteq O' \cup O'_a,$$

showing that $x_a \in \cap \beta(\sigma)$ and that the image simplex is in $\cap_{a \in A} S_0(a)$.

Repeating the above constructions yields a sequence of uniform convex open covers and refinement functions

$$\mathcal{O}_q \xrightarrow{\alpha_q} \mathcal{O}_{q-1} \to .. \xrightarrow{\alpha_1} \mathcal{O}_0,$$

together with a sequence of multifunctions

$$S_j: Y \multimap |\mathcal{O}_j|, \quad S_j(y) = |\mathcal{O}_j(F(y))| \quad (j = 0,..,q),$$

with the following property:

(2) *If $A \subseteq Y$ is non-empty and if $\cap_{a \in A} S_{j+1}(a) \neq \emptyset$ then the restriction*

$$\bigcap_{a \in A} S_{j+1}(a) \to \bigcap_{a \in A} S_j(a)$$

of $|\alpha_{j+1}|$ is homotopy trivial.

As F is LSC, there is an open cover
$$F^{-1}(\mathcal{O}_q) = \{F^{-1}(O) \mid O \in \mathcal{O}_q\}$$

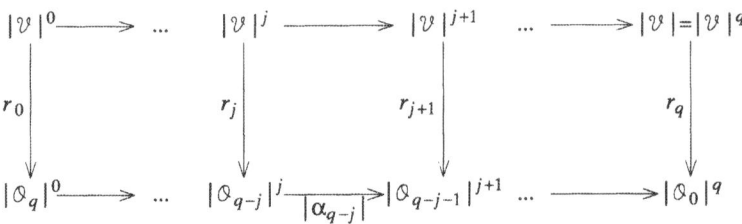

of Y, which admits a q-dimensional open refinement $\mathcal{V}$. For each simplex $\sigma \subseteq \mathcal{V}$ let $A(\sigma) = \cap \sigma$. Note that $A(\sigma)$ is a non-empty subset of Y. For each $V \in \mathcal{V}$, there is a set $r_0(V) \in \mathcal{O}_q$ such that $V \subseteq F^{-1}(r_0(V))$. By construction,

(3;0) If $\sigma \subseteq \mathcal{V}$ is a simplex, then r_0 maps $|\sigma|^0$ to the 0-skeleton of $\cap_{a \in A(\sigma)} S_q(a)$.

We proceed by induction to construct mappings r_j for $j \leq q$ as suggested in the diagram and such that

(3;j) For each simplex $\sigma \subseteq \mathcal{V}$, the function r_j maps $|\sigma|^j$ into the j-skeleton of
$$\cap_{a \in A(\sigma)} S_{q-j}(a).$$

For each $j < q$, the step from j to $j+1$ is made as follows. Let $\mu \subseteq \mathcal{V}$ be a $(j+1)$-simplex. The restriction of r_j,
$$|\mu|^j \to (\cap_{a \in A(\mu)} S_{q-j}(a))^j,$$
composed with α_{q-j}, is null homotopic in the complex
$$\cap_{a \in A(\mu)} S_{q-j-1}(a),$$
and hence within its $(j+1)$-skeleton[4]. This homotopy determines r_{j+1} on μ, and fitting these pieces of maps together yields a mapping r_{j+1} as suggested. Statement (3;j+1) is clear for simplices σ of dimension $\leq j+1$. Suppose σ is of dimension $> j+1$. If μ is a $(j+1)$-face of σ we have $A(\mu) \supseteq A(\sigma)$, whence
$$\cap_{a \in A(\mu)} S_{q-j-1}(a) \subseteq \cap_{a \in A(\sigma)} S_{q-j-1}(a).$$
Hence $|\sigma|^{j+1}$ is mapped into the $(j+1)$-skeleton of the last complex.

This completes the construction of the diagram, producing a mapping $r = r_q$. As Y is paracompact, there is a partition of unity[5] $(g_V)_{V \in \mathcal{V}}$ subordinate to $\mathcal{V}$, These functions are used as the affine coordinates of a map
$$g: Y \to |\mathcal{V}|.$$

---

4.   This is the Cellular Approximation Theorem, cf. Spanier [1966, p. 404].
5.   Engelking [1977, p. 374].

§3: Continuous Selection 439

For each $y \in Y$ we fix a neighborhood W_y on which only finitely many mappings g_V do not vanish. Then g has the following property.

(4) For each $y \in Y$, the set $g(W_y)$ is included in a compact full subcomplex of $|\mathcal{V}|$ built with the vertices V satisfying $supp\,(g_V) \cap W_y \neq \emptyset$.

The desired selection of S_0 is defined as the composition $r \circ g$. We check its properties. Let $y \in Y$ and let
$$\sigma = \{V_0,..,V_p\}$$
be the smallest simplex of V with $g(y) \in |\sigma|$. Then $p \leq q$ and $y_i \in \cap_{i=0}^p V_i = A(\sigma)$. By (3;q), we find that
$$rg(y) \in \Big(\bigcap_{a \in A(\sigma)} S_0(a)\Big)^q \subseteq S_0(y)^q,$$
showing that rg is a selection of S_0 ranging into the q-skeleton of $|\mathcal{O}_0|$. By (4), we also conclude that $s(W_y)$ is relatively compact in $|\mathcal{O}_0|$, establishing the lemma.

We now start constructing a selection from F. Let $\varepsilon > 0$ and let $\mathcal{O} = \{O_i \mid i \in I\}$ be a uniform convex open cover of X with sets of diameter $< \varepsilon$. Consider a multifunction
$$S: Y \multimap |\mathcal{O}|$$
as above and let s be a selection of S satisfying properties (1) and (2) of the Lemma. Then s can be decomposed into its affine coordinate functions $s_i: Y \to [0,1]$, which are all continuous. For each $y \in Y$ we put
$$C_i(y) = \begin{cases} O_i, & \text{if } s_i(y) \geq 1/q+1; \\ X, & \text{otherwise.} \end{cases}$$
Consider a new multifunction defined as follows.
$$G: Y \multimap X, \quad G(y) = Cl\Big(\bigcap_{i \in I} C_i(y) \cap F(y)\Big).$$
Note that $G(y)$ is non-empty since s selects from S. In addition, the point $s(y)$ is in a simplex of $\mathcal{O}^q$, so one of its coordinates satisfies $s_i(y) \geq 1/q+1$. This gives $C_i(y) = O_i$; in particular $G(y) \subseteq Cl(O_i)$ is of diameter $< \varepsilon$. We verify that G is LSC at each point $y \in Y$. Let $P \subseteq X$ be an open set meeting $G(y)$, and consider a point
$$x \in \bigcap_{i \in I} C_i(y) \cap F(y) \cap P.$$
Each set $C_i(y)$ (see (ii) of the lemma) is open and all but finitely many are equal to X. Hence there is a $\delta > 0$ such that

(5) $B(x, \delta) = \{x \mid d(x,x') < \delta\} \subseteq \bigcap_{i \in I} C_i(y) \cap P.$

Consider a neighborhood W_1 of y with the property

(6) $\forall y' \in W_1: F(y') \cap B(x, \delta) \neq \emptyset.$

The relatively compact set $s(W_y)$ is included in a finite subcomplex of $|\mathcal{O}|$.[6] This means
-----
6.   Spanier [1966, p. 113].

that all but finitely many s_i are identically zero on W_y, and consequently there is a neighborhood $W_2 \subseteq W_y$ of y such that if $i \in I$ and $s_i(y) < 1/q+1$ then $s_i(y') < 1/q+1$ for all $y' \in W_2$. This readily implies that

(7) $\quad \forall y' \in W_2, \forall i \in I : C_i(y) \subseteq C_i(y')$.

If $y' \in W_1 \cap W_2$, then by (6) there is a point $x \in F(y') \cap B(x, \delta)$. By (5) and (7), we find that $x \in G(y')$.

A proof that G is LSC is now complete. Repeat the previous procedure to obtain a sequence $(F_n)_{n=0}^{\infty}$ of LSC multifunctions with convex closed values, such that

$$F_0 = F; \quad F_n(y) \subseteq F_{n-1}(y); \quad \text{diam } F_n(y) < 1/n.$$

As $F(y)$ is complete, the set $\cap_{n=0}^{\infty} F_n(y)$ consists of one point $f(y)$ only. The resulting function $f : Y \to X$ selects from F and is easily seen to be continuous. ∎

The following observation is of use in the proof of the general Selection Theorem.

3.4. Proposition. *Let $\mathcal{X}$ be a family of subsets of a space X, including all singletons. Assume that every LSC multifunction $F : Y \multimap X$ with values in $\mathcal{X}$ admits a continuous selection. If $A \subseteq Y$ is closed, then each partial selection $f : A \to X$ of F extends to a full selection of F.*

Proof. This follows from the fact that the multifunction F', defined by $F'(y) = F(y)$ for $y \notin A$ and $F'(y) = \{f(y)\}$ for $y \in A$, is LSC. ∎

3.5. Theorem (*Selection Theorem*). *Let X be a topological S_4 convex structure with compact polytopes, with connected convex sets, and with a compatible metric d.*

(1) *If X is separable and if Y is a T_4 space, then each LSC multifunction $Y \multimap X$ with compact convex values admits a continuous selection.*

(2) *If Y is a paracompact space, then each LSC multifunction $Y \multimap X$ with convex and d-complete values admits a continuous selection.*

Proof. Let $\varepsilon > 0$ and let $\delta > 0$ be associated to ε (as in III§3.8). Let F denote the given multivalued function. Take a locally finite open cover $\mathcal{P}$ refining the covering $\mathcal{B}_{\delta/2}$ of all $\delta/2$–disks in X, and let $\mathcal{V} \leq F^{-1}(\mathcal{P})$ be an open cover of Y which will be specified later. Define a map

$$r_0 : |\mathcal{V}|^0 \to X$$

as follows. For each $V \in \mathcal{V}$ select $P_V \in \mathcal{P}$ such that $V \subseteq F^{-1}(P_V)$, and choose a point $r_0(V) \in P_V$. Next, define a multivalued function

$$R : |\mathcal{V}| \multimap X,$$

as follows. Let σ be the smallest simplex of $\mathcal{V}$ with $u \in |\sigma|$. Then

$$R(u) = co\{r_0(V) \mid V \in \sigma\}.$$

This multifunction is easily seen to be LSC. If $r_n : |\mathcal{V}|^n \to X$ is a partial selection of R over $|\mathcal{V}|^n$, and if σ is an $(n+1)$–simplex of $|\mathcal{V}|$, then Proposition 3.3 and the observation

§3: Continuous Selection   441

in 3.4 lead to a partial selection of R over $|\sigma|$ extending r_n. This yields a continuous extension r_{n+1} of r_n selecting from R over $|\mathcal{V}|^{n+1}$. Proceeding by induction, we finally obtain a continuous selection

$$r: |\mathcal{V}| \to X$$

of R. In each of the cases (1), (2), we now specify the open cover $\mathcal{V}$.

Case (1): X being separable, we may assume that $\mathcal{P}$ is countable moreover. As F is compact-valued, it follows that $F^{-1}(\mathcal{P})$ is countable and point-finite. Hence[7] there exists a (countable) star-finite open refinement $\mathcal{V}$ of $F^{-1}(\mathcal{P})$.

Case (2): As Y is paracompact, $F^{-1}(\mathcal{P})$ admits a locally finite open refinement $\mathcal{V}$.

In both cases we infer that there exists a partition of unity $(g_V)_{V \in \mathcal{V}}$ subordinate to $\mathcal{V}$. This yields a map

$$g: Y \to |\mathcal{V}|,$$

and we show that

(*) $\forall y \in Y: d(rg(y), F(y)) < \varepsilon.$

Let $\sigma = \{V_0,..,V_k\}$ be the smallest simplex of $\mathcal{V}$ with $g(y) \in |\sigma|$. Then $g_{V_i}(y) > 0$, so $y \in V_i$. Now $r_0(V_i)$ was chosen to be a point of some $P_i \in \mathcal{P}$ with $V_i \subseteq F^{-1}(P_i)$. Hence P_i meets $F(y)$ for each i. As $\mathcal{P}$ refines $\mathcal{B}_{\delta/2}$ and as δ is associated to ε, we find that

$$rg(y) \in co\{r(V_0),..,r(V_k)\} \subseteq co\, B(F(y),\delta) \subseteq B(F(y),\varepsilon),$$

establishing (*).

Having shown that in each of the cases (a), (b), there is a "uniform approximation" of F by maps $Y \to X$, we proceed as follows to obtain a selection. Let $n \in \mathbb{N}$, let $\varepsilon > 0$ be associated to $1/2n$, and let $g: Y \to X$ be ε-close to F. For each $y \in Y$, define

$$G(y) = Cl\left(co\, F(y) \cap B(g(y), \varepsilon) \right).$$

Note that $G(y) \neq \emptyset$ and that $diam\, G(y) \leq 1/n$ since

$$co\, B(g(y), \varepsilon) \subseteq B(g(y), 1/2n).$$

Let $O \subseteq X$ be an open set meeting $G(y)$ and take a point

$$x \in O \cap co(F(y) \cap B(g(y), \varepsilon)).$$

By the third axiom of convexity there is a finite subcollection $\{x_1,..,x_k\}$ of the argument at the right, such that $x \in co\{x_1,..,x_k\}$. The operator co being LSC on $\mathcal{F}_{fin}(X)$, there is a $\delta > 0$ with $co\{x'_1,..,x'_k\} \cap O \neq \emptyset$ for each k-tuple of points $x'_i \in B(x_i, \delta)$. By taking a smaller δ if necessary, we may assume in addition that $d(g(y),x_i) + 2\delta < \varepsilon$ for all i. Consider a neighborhood V of y such that for all $y' \in V$ and for each $i = 1,..,k$,

(i) $d(g(y'), x_i) < d(g(y), x_i) + \delta$;
(ii) $F(y')$ meets the set $B(x_i, \delta)$.

If $y' \in V$ then take $x'_i \in F(y') \cap B(x_i, \delta)$ for each $i = 1,..,k$. We find that

---
7.   Engelking [1977, p.394]

$$d(g(y'),x'_i) \le d(g(y'),x_i) + d(x_i,x'_i) < d(g(y),x_i) + 2\delta < \varepsilon.$$

This yields

$$\{x'_1,..,x'_k\} \subseteq F(y') \cap B(g(y'),\varepsilon);$$
$$co\{x'_1,..,x'_k\} \subseteq G(y').$$

As $x'_i \in B(x_i,\delta)$ for each i, we find that O meets $co\{x'_1,..,x'_k\}$. Consequently, $G(y')$ meets O for each $y' \in V$.

By inductive application of this procedure, we arrive at a sequence $(F_n)_{n=0}^{\infty}$ of LSC multifunctions $Y \multimap X$ with $F_0 = F$ and such that $F_{n+1}(y)$ is a closed subset of $F_n(y)$ of diameter $< 1/n$ for each $n > 0$ and $y \in Y$. There is only one point $f(y)$ in the set $\cap_{n=0}^{\infty} F_n(y)$, and the resulting function $f: Y \to X$ is clearly a continuous selection of F. ∎

3.6. Fréchet vector spaces. We showed in Proposition III§3.10.1 that a locally convex vector space which is (topologically) metrizable admits a compatible metric. The Kakutani Separation Property (Proposition I§4.14.1) and the (evident) fact that polytopes are compact and convex sets are connected lead us to the following application.

3.7. Theorem (*Michael's Selection Theorem*). *Let V be a Fréchet vector space, let Y be a paracompact space, and let $F: Y \multimap V$ be a LSC multifunction with convex closed values. Then F has a continuous selection.* ∎

3.8. Trees. It is shown in Proposition III§3.10.3 that a connected, locally connected and (topologically) metrizable tree admits a compatible metric. On the other hand, Proposition 2.13.1 guarantees that all polytopes of such a tree are compact and that its convex sets are connected. Finally, a tree is a semilattice and hence it is S_4. In conclusion,

3.9. Theorem (*Nadler's Selection Theorem*). *Let T be a connected and locally connected, metrizable tree. Then there is a continuous function $f: \mathcal{T}_{comp}(T) \to T$ such that $f(C) \in C$ for all $C \in \mathcal{T}_{comp}(T)$.* ∎

3.10. Convex hyperspaces. Let X be a uniform S_4 convex structure. By Proposition III§3.10.4, the Hausdorff uniformity of $\mathcal{TC}_{comp}(X)$ is compatible with the Vietoris convexity. In addition, if the original uniformity on X is derived from a metric, then the Hausdorff uniformity is derived from the corresponding Hausdorff metric. This metric is complete on the hyperspace of compacta, $\mathcal{T}_{comp}(X)$, provided the original metric is, which implies that the Hausdorff metric is complete on the closed subspace $\mathcal{TC}_{comp}(X, \mathcal{C})$. The Kakutani Separation Property S_4 appears from Corollary I§3.13.

If X has connected convex sets and if the convex closure of the union of two compact convex sets is compact, then convex sets are connected and polytopes are compact in $\mathcal{TC}_{comp}(X)$ as one can easily prove.

These data are needed to obtain a theorem on approximation of USC multifunctions by single-valued ones. If X, Y are metric spaces, then a USC closed-valued function

§3: Continuous Selection   443

$Y \multimap X$ has a closed graph in $Y \times X$, and it makes sense to consider the Hausdorff distance between such functions ($Y \times X$ with Cartesian metric). Aside of the previously discussed results, we need two more facts. The first is a special application of selection to convex hyperspaces. Recall that an *enlargement* of a multifunction is one with larger value sets.

3.11. Lemma. *Let X be a complete, metric, S_4 convex structure with compact polytopes and with connected convex sets, and let $F: Y \multimap X$ be a USC function with compact convex values. Then for each $\varepsilon > 0$ there is a continuous enlargement of F with compact convex value sets, and which is ε–close to F.*

Proof. Let $\mathcal{C}$ denote the convexity of X and let d be a complete compatible metric. As we noted above, the corresponding Hausdorff metric is complete on $\mathcal{IC}_{comp}(X, \mathcal{C})$.

Consider the following multifunction.

$$\mathcal{F}: Y \multimap \mathcal{IC}_{comp}(X, \mathcal{C}), \quad \mathcal{F}(y) = \{C \mid F(y) \subseteq C \in \mathcal{IC}_{comp}(X, \mathcal{C})\}.$$

Clearly, the collection $\mathcal{F}(y)$ is convex closed. To see that $\mathcal{F}$ is LSC at $y \in Y$, consider a basic open set $<O_1,..,O_n>$ of the hyperspace, and let

$$C \in <O_1,..,O_n> \cap \mathcal{F}(y).$$

Then $C \subseteq \cup_{i=1}^{n} O_i$ and by Theorem III§3.8, there is a convex open set $O \subseteq X$ with

$$F(y) \subseteq C \subseteq O \subseteq \overline{O} \subseteq \bigcup_{i=1}^{n} O_i.$$

As $\mathcal{F}$ is USC, there is a neighborhood W of y such that $F(y') \subseteq O$ whenever $y' \in W$. Let $x_i \in C \cap O_i$ for $i = 1,..,n$ and let $y' \in W$. The set

$$C' = Cl\, co\bigl(F(y') \cup \{x_1,..,x_n\}\bigr)$$

is compact by Theorem III§3.9. As $\overline{O}$ is convex, we find that $C' \subseteq \overline{O}$, whence

$$C' \in \mathcal{F}(y') \cap <O_1,...,O_n>.$$

Note that a continuous selection of $\mathcal{F}$ is just a continuous enlargement of F with compact convex value sets. To complete the proof of the lemma, we restrict the "choice margins" of $\mathcal{F}$ in such a way that any continuous selection of the shrunken map is as required. To this end, let $\delta > 0$ be such that δ–close subsets of X have $\varepsilon/2$–close hulls. For each $y \in Y$ pick a neighborhood U_y with the following properties.

(1) $diam\, U_y < \varepsilon$.
(2) $\forall y' \in U_y: F(y') \subseteq B(F(y), \delta)$.

Choose a locally finite open refinement $\mathcal{V}$ of $\{U_y \mid y \in Y\}$ and a closed shrinking[8] $\mathcal{D} = \{D_V \mid V \in \mathcal{V}\}$ of $\mathcal{V}$, that is: D_V is a closed set included in V for all $V \in \mathcal{V}$. Each $y \in Y$ has a neighborhood of type

(3) $W(y) = \cap \{V \mid y \in V \in \mathcal{V}\} \setminus \cup \{D_V \mid y \notin D_V\}$.

Finally, for each $V \in \mathcal{V}$ we fix a point $y(V)$ with $V \subseteq U_{y(V)}$. Consider the set

---
8.   Dugundji [1966, VII.6.1].

(4) $\quad C_V(y) = \begin{cases} B(F(y(V)), \delta), & \text{if } y \in D_V; \\ X, & \text{otherwise.} \end{cases}$

By (3), we find that $C_V(y) \subseteq C_V(y')$ for $y' \in W(y)$. This readily implies lower semi-continuity of the multifunction $\mathcal{b}: Y \multimap \mathcal{JC}_{comp}(X, \mathcal{C})$, defined by the prescription

$$\mathcal{b}(y) = Cl\, co\big(\cap_{V \in \mathcal{V}} <C_V(y)> \cap \mathcal{F}(y)\big)$$
$$= Cl\, co\, \{\, C \mid F(y) \subseteq C \in \mathcal{JC}_{comp}(X, \mathcal{C});\quad \forall V \in \mathcal{V}: C \subseteq C_V(y)\,\}$$

(convex hull taken in the convex hyperspace). The set $\mathcal{b}(y)$ is evidently closed and convex. By virtue of (2), it contains the "element" $F(y)$. Extract a continuous selection G from $\mathcal{b}$. We verify that if $x \in G(y)$, then (y,x) is ε-close to some point of the graph of F. Let $V \in \mathcal{V}$ be such that $y \in D_V$, and consider the corresponding point $y(V)$ with $V \subseteq U_{y(V)}$. Then $d(y, y(V)) < \varepsilon$ by (1), whereas by (4) and the definition of $\mathcal{b}(y)$,

$$G(y) \subseteq Cl\, co\, C_{V(y)}$$
$$= Cl\, co\, B(F(y(V)), \delta)$$
$$\subseteq Cl\, B(F(y(V)), \varepsilon/2)$$
$$\subseteq B(F(y(V)), \varepsilon).$$

Hence, x is ε-close to some point of $F(y(V))$, establishing the result. ∎

We now present the second step of the approximation procedure.

3.12. Lemma. *Let X be a metric S_4 convex structure with connected convex sets and with compact polytopes, and let Y be a metric space without isolated points. If $G: Y \multimap X$ is continuous and has compact convex values, then for each $\varepsilon > 0$ there is a continuous (single-valued) selection of G, the graph of which is ε-close to G.*

Proof. Let $\mathcal{U}$ be a locally finite open cover of X with sets of diameter $< \varepsilon$. For each $U \in \mathcal{U}$ let

$$V(U) = \{y \mid G(y) \cap U \neq \varnothing\}.$$

This set is open since G is LSC. For each $y \in Y$ there is an open set $O \subseteq X$ such that $G(y) \subseteq O$ and O meets only finitely many $U \in \mathcal{U}$. As G is USC, there is a neighborhood $W(y)$ of y such that $G(y') \subseteq O$ whenever $y' \in W(y)$. In particular, $W(y)$ meets only finitely many of the sets $V(U)$, $U \in \mathcal{U}$, showing that this cover of Y is locally finite.

For each $U \in \mathcal{U}$ we take a closed discrete set $S(U) \subseteq V(U)$ which is ε-close to $V(U)$. As Y has no isolated points and as the open sets $V(U)$ constitute a locally finite collection, care can be taken that $S(U) \cap S(U') = \varnothing$ for $U \neq U'$. The set $A = \cup_{U \in \mathcal{U}} S(U)$ yields a closed and discrete subspace of Y, and a continuous partial selection

$$g: A \to X$$

of G is obtained as follows. For $a \in A$, say: $a \in S(U)$, choose $g(a) \in G(a) \cap U$. Any extension of g to a full continuous selection of G is ε-close to the graph of G. ∎

We finally arrive at the following generalization of **Beer's Approximation Theorem**.

§3: Continuous Selection

3.13. Theorem. *Let X be a metric S_4 convex structure with connected convex sets and with compact polytopes. Let Y be a metric space without isolated points, and let $F: Y \multimap X$ be a USC, compact convex-valued function. Then for each $\varepsilon > 0$ there is a map $Y \to X$ the graph of which is ε–close to F.*

Proof. According to III§3.14, X extends to a complete metric convex structure X^*. The discussion in 3.10 shows that the completion inherits all properties of X currently considered. Moreover, X is a dense convex subspace of X^*. Enlarge F to a continuous multifunction $G: Y \multimap X^*$ which is $\varepsilon/3$–close to F (Lemma 3.11). Next, select a map $g: Y \to X$ from G having its graph at a distance less than $\varepsilon/3$ away from G (Lemma 3.12). Finally, a simple application of the Selection Theorem yields a map $f: Y \to X \subseteq X^*$ which is $\varepsilon/3$–close to g (even in the sup-metric). The mapping f is the desired approximation to F. ∎

We next concentrate on continuous selection in spaces of arcs. As before, we begin with collecting the prerequisites.

3.14. Spaces of arcs. Let S be a connected Lawson join semilattice. Being a compact space, S has a largest element **1** and it has minimal elements, but there is not necessarily a smallest one. The space of all arcs (totally ordered subcontinua) of S is denoted by $\Gamma(S)$; its subspace of all maximal arcs by $\Lambda(S)$. The first space is compact. If S has a smallest element **0**, then $\Lambda(S)$ consists precisely of all arcs joining **0** with **1**. The resulting space is compact and S_4, and it has connected convex sets (cf. Topic III§3.32).

A uniform convex system on $\Gamma(S)$ has been constructed in III§3.20. Its subspace $\Lambda(S)$ is a genuine convex structure provided S has a lower bound. Theorem III§3.24 states that $\Gamma(S)$ extends to a uniform convex structure of type $\Lambda(T)$ (where T is a bounded and connected Lawson join semilattice) in such a way that convex sets of $\Gamma(S)$ are also convex in $\Lambda(T)$. Its uniformity is metric provided the topology of the original semilattice S is metric. The following is a generalization of the *Curtis Selection Theorem*.

3.15. Theorem. *Let S be a connected Lawson semilattice with a continuous interval operator. Then each LSC multifunction of a paracompact space into $\Gamma(S)$ admits a continuous selection provided its value sets are convex and closed.* ∎

For the remainder of this section, we will consider some applications of continuous selections to the structure of the underlying space. For applications to (semi)lattices we refer to the Topics Section. If $\mathcal{U}$ is a cover of a set X, then two functions $f, g: Y \to X$ are $\mathcal{U}$–*close* provided for each $y \in Y$ there is a $U \in \mathcal{U}$ with $f(y), g(y) \in U$.

3.16. Theorem. *Let X be a metrizable S_4 convex structure with compact polytopes and with connected convex sets, let D be a dense convex subset of X, and let $\mathcal{U}$ be an open cover of X. Then there is a mapping $f: X \to D$ which is $\mathcal{U}$–close to identity and such that f maps each compact set into a polytope.*

Proof. Let $\mathcal{V}$ be a convex open cover refining $\mathcal{U}$, and let $\mathcal{W}$ be a locally finite open cover which is a star refinement of $\mathcal{V}$. For each set $W \in \mathcal{W}$ we select a point $x \in W \cap D$. This yields a polytope-valued function $F: X \multimap X$, defined by the prescription

$$F(x) = co\{x_W \mid x \in W\}.$$

It is easily verified that F is LSC and that F maps each compact set into a polytope included in D. Take a continuous selection f of F. Then

$$f(x) \in F(x) \subseteq co\{star(\{x_W \mid x \in W\}, \mathcal{W})\}$$

and the last set is included in some member of $\mathcal{V}$. Clearly, f is as desired. ∎

We recall that a metric space X is an **absolute retract** provided X is a retract of each metric space in which it is embedded as a closed subspace. Equivalently[9] X is an absolute extensor, that is, for each metric space Y, for each closed subspace $A \subseteq Y$, and for each mapping $f: A \to X$ there is a continuous extension of f over Y.

3.17. Theorem. *Let X be a metrizable S_4 convex structure with compact polytopes and with connected convex sets. Then each non-empty convex set in X is an absolute retract.*

Proof. Observe that the assumptions on X are inherited by a convex subspace of X. We just show that X is an absolute extensor. Let A be a closed subset of a metric space Y and let $g: A \to X$ be a mapping. Let $\rho(y, A)$ denote the distance of y to A. The covering

$$\mathcal{B} = \{B(y, \rho(y, A)/2) \mid y \in Y \setminus A\}$$

of $Y \setminus A$ admits a locally finite open refinement $\mathcal{U}$. For each $U \in \mathcal{U}$ take $p_U \in \mathcal{U}$ and $a_U \in A$ such that

$$\rho(p_U, a_U) < 2 \cdot \rho(p_U, A).$$

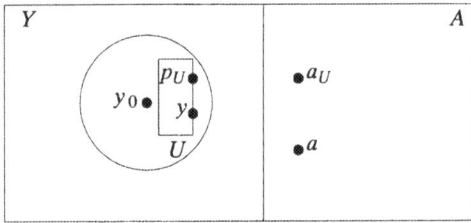

Fig. 1: Extending g

Consider a multifunction $F: Y \multimap X$ defined as follows.

$$F(y) = \begin{cases} \{g(y)\}, & \text{if } y \in A; \\ co\{a_U \mid y \in U \in \mathcal{U}\}, & \text{otherwise.} \end{cases}$$

If $y \in Y \setminus A$, then $V = \cap\{U \mid y \in U \in \mathcal{U}\}$ is a neighborhood of y and $F(y') \supseteq F(y)$ for

---
9. Hanner [1952].

§3: Continuous Selection   447

each $y' \in V$. Consequently, F is LSC. To see that F is LSC (in fact, continuous) at $a \in A$, consider a convex neighborhood N of $g(y)$ in X and let $\delta > 0$ be such that the $B_A(a, 9\delta)$ is included in $g^{-1}(N)$ (the subscript A refers to the metric subspace A). We verify that $F(y) \subseteq N$ for each $y \in D(a, \delta)$. For $y \in A$ this is evident. Let $y \in Y \setminus A$. As $U \subseteq B(y_0, \rho(y_0, A)/2)$ for some point y_0, we can see that $\delta > \rho(y, A) > \rho(y_0, A)/2$. Therefore,

$$\rho(p_U, a_U) < 2 \cdot \rho(p_U, A) \le 2 \cdot (\rho(p_U, y_0) + \rho(y_0, A)) < 2 \cdot (\delta + 2\delta) = 6\delta.$$

On the other hand,

$$\rho(p_U, a) \le \rho(p_U, y) + \rho(y, a) < 2\delta + \delta = 3\delta.$$

We conclude that $a_U \in B_A(a, 9\delta) \subseteq g^{-1}(N)$ for each $U \in \mathcal{U}$ with $y \in U$, and $F(y) \subseteq N$ follows directly.

By Theorem 3.5, F has a continuous selection, which necessarily extends the mapping g over Y.   ■

3.18. Theorem. *Let X be a (locally) compact metrizable S_4 convex structure with connected convex sets and with compact polytopes. If every pair of polytopes of X can be approximated by a pair of disjoint polytopes -- in particular, if there is a pair of disjoint, dense, and convex subsets in X -- then X is a Hilbert cube (manifold).*

Proof. Throughout, Q denotes the Hilbert cube. We rely on the following result.[10]

Torunczyk Theorem. *Let X be a locally compact AR, such that each pair of mappings*

$$f, g : Q \to X$$

can be approximated with mappings having disjoint images. Then X is a Hilbert cube manifold, and even a Hilbert cube if X is compact.

The space X is an AR by Theorem 3.17. Let $\mathcal{O}$ be a convex open cover of X, and consider a barycentric refinement $\mathcal{V}$ of some star refinement of $\mathcal{O}$. Furthermore, let $\mathcal{W}$ be a finite open cover of Q refining both $f^{-1}(\mathcal{V})$ and $g^{-1}(\mathcal{V})$. Choose a point $t_W \in W$ for each $W \in \mathcal{W}$ and consider the following polytopes.

$$P(f) = co\{f(t_W) \mid W \in \mathcal{W}\};$$
$$P(g) = co\{g(t_W) \mid W \in \mathcal{W}\}.$$

By assumption, we obtain a pair of disjoint polytopes $P'(f)$, $P'(g)$ which are $\mathcal{V}$-close to $P(f), P(g)$, respectively. For each $W \in \mathcal{W}$ we take two points as follows: $y_W \in P'(f)$ is $\mathcal{V}$-close to $f(t_W)$, and $z_W \in P'(g)$ is $\mathcal{V}$-close to $g(t_W)$. This leads to the following LSC polytope-valued multifunctions.

$$F : Q \multimap X : F(t) = co\{y_W \mid t \in W\};$$
$$G : Q \multimap X : G(t) = co\{z_W \mid t \in W\}.$$

Take two continuous selections f' and g' of F and G, respectively. If $t \in Q$ then each

---
10.   Torunczyk [1980, p.36].

point y_W with $t \in W$ is in the set
$$star(star(f(t), \mathcal{V})\mathcal{V}),$$
which in turn is included in some $O \in \mathcal{O}$. As O is convex, we find that $F(t) \subseteq O$ and, consequently, that $f(t)$ and $f'(t)$ are $\mathcal{O}$–close. Similarly, $g(t)$ is $\mathcal{O}$–close to $g'(t)$. The images of f', g' are evidently disjoint. ∎

3.19. Examples

3.19.1. Standard compact convex sets. The simplest case where the previous result applies is the following. Let C be a compact infinite-dimensional convex set of a topological vector space. It is clear that every pair of polytopes in C can be approximated with a pair of disjoint polytopes. Hence, if C is metrizable and locally convex, then C is homeomorphic to the Hilbert cube. This result is known as *Keller's Theorem*.

3.19.2. Superextensions. Consider the superextension $\lambda(X)$ of a non-trivial metric continuum X. This topological convex structure is compact by construction, connected, and (topologically) metrizable (cf. III§4.32). Its convex sets are connected (cf. III§1.29.5), and here is a method to produce two disjoint dense convex subsets. Elementary topological considerations yield two disjoint dense subsets A_1, A_2 of the original space X. Then the desired convex sets are
$$C_i = \cup \{F^+ \mid F \subseteq A_i, F \text{ finite}\} \quad (i = 1, 2).$$
The fact that $\lambda(X)$ is homeomorphic to the Hilbert cube is *van Mill's Theorem*.

3.19.3. Proposition. *Let S be an arc-wise connected metric Lawson semilattice with a top element. If S is not reduced to an arc, then $\Lambda(S)$ is homeomorphic to the Hilbert cube.*

Proof. If S is not an arc, then there is a point of S each neighborhood of which has incomparable elements. There clearly is a largest point x_0 with this property. Note that each arc joining the bottom element $\mathbf{0}$ with the top element $\mathbf{1}$ passes through x_0, and that the upper set $\uparrow(x_0)$ is a (possibly degenerate) arc κ_0. See Fig. 2.

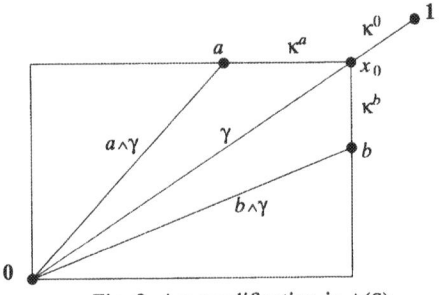

Fig. 2: Arc modification in $\Lambda(S)$

§3: Continuous Selection   449

Let $\delta > 0$ and take $\rho > 0$ such that ρ–close pairs of points have δ–close infima. Consider two incomparable points a, b in some order-convex neighborhood N of x_0 of diameter $< \delta$. Let κ_a, κ_b be two arcs joining x_0 with a, b, respectively. Being included in N, these arcs have diameter $< \delta$.

If γ is an arc joining **0** to **1**, then define

$$\gamma_a = (a \wedge \gamma) \cup \kappa_a \cup \kappa_0.$$

This yields a new arc joining **0** to **1**. As a and x_0 are ρ–close, the sets $a \wedge \gamma$ and $x_0 \wedge \gamma$ are δ–close. Note that the second arc is the part of γ from **0** up to x_0. Furthermore, γ and γ_a share the arc κ_0. As $diam$ $\kappa_a < \delta$, we conclude that γ and γ_a are δ–close.

In a similar way, we can modify γ into an arc

$$\gamma_b = (b \wedge \gamma) \cup \kappa_b \cup \kappa_0$$

joining **0**, **1**, and which is δ–close to γ.

This is used to approximate two polytopes

$$P = co\{\alpha_1,..,\alpha_n\}; \quad Q = co\{\beta_1,..,\beta_m\}$$

in $\Lambda(S)$ with disjoint polytopes as follows. Let $\varepsilon > 0$ and let $\delta > 0$ be such that δ–close sets in $\Lambda(S)$ have ε–close hulls. The above procedure yields two polytopes of type

$$P^a = co\{\alpha_1^a,..,\alpha_n^a\}; \quad Q^b = co\{\beta_1^b,..,\beta_m^b\},$$

which are ε–close to the respective originals. Each member of the first polytope is an arc through a, whereas each member of the second polytope passes through b. Therefore, the approximating polytopes are disjoint. ∎

Further topics

3.20. Usc functions in Lawson semilattices (van de Vel [1993]). According to a result of Lawson [1973b, Cor. 16], a semilattice admits at most one compatible, compact Hausdorff topology. For a continuous join semilattice S, this topology is generated by the open sets of type

(1) $A(x) = \{y \mid y \gg x\}, \quad x \in S;$

(2) $S \setminus L(x) = \{y \mid y \not\leq x\}, \quad x \in S,$

where $\gg$ is the way-above relation of S (this corresponds to the way-below relation for meet semilattices).

If Y is a space and if S is a continuous semilattice, then a function $Y \to S$ is *lsc* (resp. *usc*) provided it inverts the sets of type (1) (resp., (2)) into open sets of Y. It is easily verified that $f : Y \to S$ is lsc (usc) iff its lower graph (upper graph)

$$\{(y,s) \mid s \leq f(y)\} \quad (\text{resp.} \quad \{(y,s) \mid s \geq f(y)\})$$

is closed.

Throughout, S is an arc-wise connected metrizable Lawson join semilattice, the intervals of which depend continuously on their end points. We consider the space $[Y, S]$ of usc functions of a locally compact space Y into S. This is known to be a continuous semilattice with the following way-above relation. Let $f_0, f_1 \in [Y,S]$. Then $f_0 \ll f_1$ iff there exist open sets $U_i, V_i \subseteq Y$ such that $U_i^- \subseteq V_i$ and $\cup_{i=1}^n U_i = Y$, together with points $s_i \ll t_i$ of S, such that for all i,

$$y \in V_i^- \Rightarrow f_0(y) \leq s_i; \quad y \in U_i^- \Rightarrow t_i \leq f_1(y).$$

3.20.1. Let Y be a locally compact space and let $f_0, f_1 : Y \to S$ be usc functions such that $f_0 \ll f_1$. Show that there exists a continuous function $f : Y \to S$ with $f_0 \ll f \ll f_1$. Hint. With the above notation, consider a multifunction $F : Y \multimap S$, defined by

$$F(y) = [\,\inf\{s_i \mid y \in U_i^-\},\ \inf\{t_i \mid y \in V_i\}\,]$$

3.20.2. Show that each usc function $Y \to S$ is the infimum of a down-directed family of continuous functions $Y \to S$.

3.20.3. In addition, let Y be separable and metrizable. Conclude that each usc function $Y \to S$ is the infimum of an decreasing sequence of continuous functions $Y \to S$. Hint ([1980, p. 172]). The topological weight of the space $[Y,S]$ is the maximum of weight(Y) and weight(S).

3.21. Extension in lattices (van de Vel [1993]). Let L be a compact, connected, metrizable topological lattice with small meet-semilattices, let Y be a T_4 space, and let $f_0, f_1 : Y \to L$ be such that f_0 is usc, f_1 is lsc, and $f_0 \leq f_1$. Let $A \subseteq L$ be closed and let $f' : A \to L$ be a continuous function such that $f_0 | A \leq f' \leq f_1 | A$. Prove that there is a continuous function $f : Y \to L$ extending f', and such that $f_0 \leq f \leq f_1$.

Hint. L can be seen as a continuous (join-) semilattice. Hence, its topology must be the "intrinsic" one. This topology also makes the meet-operation continuous.

Note that if $A \subseteq Y$ is closed, then the function $f : Y \to [0,1]$, defined by $f(y) = 0$ for $y \in A$ and $f(y) = 1$ for $y \notin A$, is usc and the function $1-f$ is usc. With these examples in mind, the previous result can be seen to include the classical theorems of Urysohn and Tietze in topology.

3.22. Extending functions that are close. Let X be a metrizable S_4 convex structure with connected convex sets and with compact polytopes. Let Y be a paracompact space. Show that for each open cover $\mathcal{U}$ of X there is an open refinement $\mathcal{V}$ with the following property. If $A \subseteq Y$ is closed, if f, g is a pair of $\mathcal{V}$-close continuous functions on A and if f' is a continuous extension of f over Y, then g has a continuous extension g' over Y such that f' and g' are $\mathcal{U}$-close.

3.23. Another selection theorem. Let X be a metrizable S_4 convex structure with connected convex sets and with compact polytopes. Let Y be a paracompact space and let $F : Y \multimap X$ be a multifunction such that the set $F^{-1}(x) \subseteq Y$ is open for each $x \in X$. Show that the multifunction $y \mapsto coF(y)$ has a continuous selection (this result is sometimes called the *Browder Selection Theorem*).

§3: Continuous Selection

Hint. Consider a locally finite open cover $\mathcal{U}$ refining the cover of sets $F^{-1}(x)$, $x \in X$. To each $U \in \mathcal{U}$ assign a point $x \in X$ such that $U \subseteq F^{-1}(x)$. Use this prescription to define a suitable convex closed-valued and LSC multifunction of Y into X.

3.24. Compressibility (Wieczorek [1991]). Let X be a tcs and let Y be a topological space. Then Y is *compressible in X* provided for each indexed open cover $\{O_1,..,O_n\}$ of Y and for each n-tuple $(x_1,..,x_n)$ of points in X there is a mapping $f: Y \to X$ such that
$$f(y) \in co\{x_i \mid y \in O_i\}.$$

3.24.1. Let X be an S_4 tcs with connected convex sets and with compact polytopes, such that the convex closure operator is continuous on each subspace which is a polytope. Verify that each normal space is compressible into X.

3.24.2. Rederive the Browder Selection Theorem (cf. Topic 3.23) with the following changes: X is a tcs and Y is a compact space compressible in X.

3.25. Possible sharpenings the selection theorem. First, we remind the reader of some problems concerning uniform convex structures, proposed in Topic III§3.33. Solution of some of these may be of help to reformulate the selection theorem in more convenient terms.

Next, note that the axiom S_4 is involved in the selection theorem only with reference to Theorem 3.2 on contractible covers. Is it possible to remove the Kakutani separation property from the hypotheses? In a different direction, one may ask if the result still valid provided uniform continuity of the hull operator is weakened to continuity on the hyperspace of all compact resp., of all finite sets.

In proving a selection theorem for certain parameterized convexities, both Bielawski [1987] and Pasicki [1987] use some sort of "pre-hull" operator, which is required to be uniformly continous. In the present setting, the following may be the appropriate counterpart. By a *pre-hull operator* of a convex structure X is meant a set operator $p: 2^X \to 2^X$ such that (i) $A \subseteq p(A)$; (ii) $A \subseteq B$ implies $p(A) \subseteq p(B)$; and (iii) a set $C \subseteq X$ is convex iff $p(F) \subseteq C$ for each finite set $F \subseteq C$. Compare with Wieczorek's [1989] "spot functions". Note that $co(A) = \cup_{n=1}^{\infty} p^n(A)$ for each $A \subseteq X$ provided p is a pre-hull operator. For instance, if the convexity of X is obtained from an interval operator I, then the prescription
$$p(A) = \cup \{I(u,v) \mid u, v \in A\}$$
determines a pre-hull operator. Can the selection theorem be proved under the modified assumption that some pre-hull operator is uniformly continuous? The result is unlikely to hold without further assumptions since the interval operator of any topological vector space yields a uniformly continuous pre-hull operator with respect to the canonical translation-invariant uniformity.

3.26. Convex systems. Problem. Let X be a metrizable S_4 convex *system* with connected convex sets and with compact polytopes, and let $F: Y \multimap X$ be an LSC multifunction with convex closed value sets, defined on a paracompact space Y. Does F admit a continuous selection? The answer is affirmative if the answer to the last problem in III§3.33 is.

Does the Selection Theorem hold if, in addition, each point of the convex system X has a convex neighborhood (a "local convex system")?

Notes on Section 3

Continuous selection in "abstract" convex structures has first been considered by Michael in [1959]. His result involves a rather different concept of convexity, based on the ability to take convex combinations of points. Modifications of the Michael axioms have been studied more recently by Curtis [1985], Pasicki [1987], and Bielawski [1987]. A somewhat different approach involving a concept of compressibility has been taken by Wieczorek [1987] (cf. Topic 3.24). The Selection Theorem in 3.5 was obtained by the author in [1993] (first manuscript dated 1980). The subsequent corollaries refer to a selection theorem of Michael [1956] in Fréchet spaces (cf. 3.7), one of Nadler [1978] in trees (cf. 3.9), and a result of Beer [1983] on approximation of USC multifunctions (cf. 3.13).

The (dis)similarities of convexity with Michael's notion of [1959] have been discussed in our paper [1993]. Aside of the fact that convex combinations are used by Michael as the primary concept, the resulting structure is some sort of "convex system" in the sense that not every set has a convex hull. Although we spent some efforts trying to obtain a counterpart in the present theory, we only managed to produce such a result on spaces of arcs, leading to Curtis' [1985] result, Theorem 3.15. See the problem description in Topic 3.26.

Theorem 3.16 should be compared with Klee's concepts of admissible vector spaces and approachable mappings, Klee [1960, p. 293]. Theorem 3.18 and Proposition 3.19.3 on Hilbert cube manifolds, as well as the topics 3.20 and 3.21 on approximation in (semi) lattices, are all taken from the author's paper [1993]. Approximation of usc functions into the (extended) real line has been considered by Gierz et al [1980] and by Tong [1952]. Constructing functions in between an lsc and a usc real function has been considered by Tong op. cit. and by Dowker [1951]. The quoted results on IR are at the roots of continuous selection theory. See the introductory section of Michael [1956].

4. Dimension Theory

A topological convex structure (tcs) is subject to a theory of topological dimension, applied to its underlying topological space. In this section, we describe a "convexity" model of dimension, called the convex (small inductive) dimension, $cind$.[1] For a tcs X with connected convex sets, compact polytopes, and with at least the separation property FS_3, $cind$ appears to be well-behaved. For instance, $cind(X) \leq n$ iff $cind(H) \leq n-1$ for all hyperplanes of X, and $cind(X) \geq n$ iff there is a surjective CP mapping of X to the standard median n-cube.

This leads to a "Countable Sum Theorem" for $cind$, and to the fact that $cind$ is determined by the dimension of the polytopes. For separable metrizable spaces, or for compact (not necessarily metrizable) spaces all topological dimension functions coincide with $cind$. In contrast, there is a zero-dimensional topological space with a natural infinite dimensional convexity.

In this section, *all spaces are assumed* S_1.

Throughout, the term *number* refers to a member of the set $\{-1, 0, 1, ..., \infty\}$. It is agreed that $\infty \pm 1 = \infty$.

4.1. Convex closed screening. We recall that a *convex closed screening* of two sets A, B is a pair (C,D) of convex closed sets such that $A \subseteq C \setminus D$, $B \subseteq D \setminus C$, and $C \cup D = X$. If the phrase "of A, B" is omitted, we mean that (C,D) screens at least some pair of singletons. A set $C \subseteq X$ is a *separator* of two non-empty sets $A, B \subseteq X$ provided there exist disjoint open sets $O \supseteq A$ and $P \supseteq B$ such that $X \setminus C = O \cup P$. Note that a separator is closed.

4.2. Convex dimension. The *convex (small inductive) dimension* of a topological convex structure X is the number $cind(X)$ satisfying the following rules.

(CIND-1) $cind(X) = -1$ iff $X = \emptyset$.
(CIND-2) $cind(X) \leq n+1$ (where $n < \infty$) iff each pair, consisting of a convex closed set C and a point $p \in X \setminus C$, has a convex closed screening (A,B) such that

$$cind(A \cap B) \leq n.$$

Observe that (in the terminology of Topic III§4.28) a space of finite convex dimension at least has the separation property CCS_3, and that $cind$ can't see the difference between the original and the weak topology of X. Clearly,

4.2.1. $cind(C) \leq cind(X)$ for each convex subset C of a tcs X. ∎

Since convex subsets of a product with finitely many factors are products of convex sets, Proposition I§1.10.2, the following is evident.

---
1. Other convex models are described in the topics section.

4.2.2. Additive Law of *cind:* If X and Y are tcs's, then

$$cind\,(X \times Y) = cind\,(X) + cind\,(Y).$$ ∎

We begin with comparing *cind* to the small inductive dimension function of the underlying topological space.

4.3. Proposition. *Let X be a tcs of which the weak topology is separable and metrizable. Then* $ind\,(X_w) \le cind(X)$.

Proof. We may assume that $cind\,(X) < \infty$ and that X carries a weak topology. The result is valid if $cind\,(X) = -1$. Assume $cind\,(X) \le n$, where $n < \infty$, and let the proposition be valid if $cind < n$. Consider a closed set $A \subseteq X$ and a point $p \notin A$. As we are dealing with a weak topology of X, there exist convex closed sets $C_1,..,C_m$ such that $A \subseteq \cup_{i=1}^m C_i$ and $p \notin \cup_{i=1}^m C_i$. For each $i = 1,..,m$ there is a convex closed screening D_i, E_i of p, C_i such that $cind(D_i \cap E_i) \le n-1$. Then $D = \cap_{i=1}^m D_i$ is a closed neighborhood of p disjoint from A and

$$Bd(D) \subseteq \bigcup_{i=1}^m (D_i \cap E_i).$$

The induction hypothesis yields that $ind\,(D_i \cap E_i) \le n-1$. Since we are dealing with a separable metric space, the "Sum Theorem" holds for ind.[2] This yields

$$ind(Bd(D)) \le ind\bigl(\bigcup_{i=1}^m (D_i \cap E_i)\bigr) \le n-1.$$

Therefore, $ind(X) \le n$. ∎

4.4. Example. *There exists an infinite-dimensional closure stable* FS_3 *space with a 0–dimensional separable metric weak topology.*

Proof. First, let X be the square of the irrationals, regarded as a subspace of the standard plane. The fact that for each $a, b \in X$ the relative segment ab is dense in the plane segment ab yields that

(1) $Cl_{\mathbb{R}^2}(C)$ is a planar convex set for each relatively convex set $C \subsetneq X$.
(2) X is FS_3 and closure stable.

Consider the relatively convex closed set

$$A = \{(x,y) \mid x \le y\} \cap X$$

with the point $p = (\pi, \pi/2) \in X \setminus A$. Let C_1, C_2 be relatively convex closed sets of X screening A, p. The plane closure D_i of C_i is convex by (1) and $D_i \cap X = C_i$. In particular, D_1 and D_2 are a screening of $Cl(A)$ and p in the plane. As D_1 and D_2 are closed half-spaces of the plane, there exist minimal closed half-spaces $E_i \subseteq D_i$ covering $\mathbb{R}^2$. Then $E_1 \cap E_2$ is a line parallel to the diagonal, whence

---

2.   Engelking [1977, Chapter 7], or van Mill [1988, Thm. 4.3.7].

§4: Dimension Theory

$$\emptyset \neq E_1 \cap E_2 \cap X \subseteq D_1 \cap D_2 = C_1 \cap C_2.$$

It follows that $cind(X) = 1$.

Let Y be the product of countably many copies of X. Then Y is a 0–dimensional separable metric space (in fact, it is homeomorphic to the irrational line), Y is FS$_3$ by (2) and III§4.9.4, and Y is closure stable by (2) and Theorem III§1.10. Finally, the product topology of Y is a weak topology (cf. III§1.17.3). The Additive Law of cind yields that Y has convex sets of arbitrary high cind, and it follows that $cind(Y) = \infty$. ∎

4.5. Lemma. *Let X be a tcs.*

(1) *If (C_1, C_2) be a screening pair of convex closed sets, then there is a minimal convex closed screening pair (D_1, D_2) with $D_i \subseteq C_i$ for $i = 1, 2$.*

(2) *Let X be closure stable, FS$_3$, and let all convex sets be connected. If (C_1, C_2) is a minimal convex closed screening pair and if $C = C_1 \cap C_2$, then for each dense convex set $B \subseteq X$ the set $B \cap C$ is dense in C.*

Proof. The first part is a simple application of Zorn's Lemma. As to (2), suppose $p \in C \setminus Cl(B \cap C)$. By FS$_3$, there is an open half-space $H \subseteq X$ with $p \in H$ and $H \cap B \cap C = \emptyset$. If H meets both the open sets $X \setminus C_1$ and $X \setminus C_2$, then there exist points $b_i \in B \cap H \setminus C_i$ ($i = 1, 2$). As convex sets are connected, there is a point $b \in b_1 b_2 \cap C$. But then

$$b \in b_1 b_2 \cap C \subseteq H \cap B \cap C,$$

a contradiction. So assume $H \subseteq C_2$. We obtain a convex closed set $C'_1 = C_1 \setminus H$ with $p \in C_1 \setminus C'_1$ and $C'_1 \cup C_2 = X$. This contradicts the minimality of the given screening pair. ∎

4.6. Proposition. *Let X be a non-empty, closure stable, and FS$_3$ space with connected convex sets. If $H \subseteq X$ is a half-space, then*

$$cind(\overline{H} \setminus H) \leq cind(X) - 1.$$

Proof. We may assume that $cind(X) < \infty$. The implication

$$cind(X) \leq n+1 \implies cind(\overline{H} \setminus H) \leq n$$

is valid for $n = -1$. Suppose it is valid for $m < n$, where $n \geq 0$. If $cind(X) \leq n+1$ then by 4.2.1, $cind(\overline{H}) \leq n+1$. Let $C \subseteq \overline{H} \setminus H$ be a relatively convex closed set and let $p \in \overline{H} \setminus (H \cup C)$. Then there is a convex closed screening A, B of C and p in $\overline{H}$ such that $cind(A \cap B) \leq n$. By Lemma 4.5, we may assume that (A, B) a minimal screening pair, and by part (2) of this lemma, we conclude that $H \cap A \cap B$ is a dense subset of $A \cap B$. In particular,

$$(\overline{H} \setminus H) \cap A \cap B = (A \cap B) \setminus H = Cl(H \cap A \cap B) \setminus (H \cap A \cap B).$$

As $H \cap A \cap B$ is a relative half-space of $A \cap B$, the induction hypothesis yields that $cind((\overline{H} \setminus H) \cap A \cap B) \leq n-1$. This shows that each relatively convex closed set C of $\overline{H} \setminus H$ and each point $p \notin C$ of $\overline{H} \setminus H$ can be screened with relatively convex closed sets meeting in a set of dimension $\le n-1$. Therefore, $cind(\overline{H} \setminus H) \le n$. ∎

A set of type $\overline{H} \setminus H$, with $H \subseteq X$ an open half-space, will be called a *hyperplane* of the tcs X.

4.7. Corollary. *In a closure stable* FS_3 *space X with connected convex sets the following statements are equivalent for each number n.*

(1) $cind(X) \le n + 1;$
(2) $cind(H) \le n$ for each hyperplane H of X.

Proof. The implication (1) $\Rightarrow$ (2) is given by the previous result. Conversely, suppose each hyperplane of X is at most n–dimensional. By FS_3, a convex closed set C and a point $p \notin C$ can be separated with a continuous functional $f: X \to \mathbb{R}$, say: $f(C) \subseteq (-\infty, 0]$ and $f(p) > 0$. Let $H = f^{-1}(-\infty, f(p)/2)$. Then $Cl(H)$ and $Cl(X \setminus H)$ yield a convex closed screening of C and p. The intersection of the screening sets is equal to the boundary of H, and hence it is of dimension $\le n$. ∎

A direct application of this result is, that $cind(\mathbb{R}^n) = n$ with respect to any point-convex, symmetric H-convexity on $\mathbb{R}^n$.

4.8. Corollary. *In a closure stable* FS_3 *space with connected convex sets, a convex set and its closure have the same convex dimension.*

Proof. Let X be as announced and let $C \subseteq X$ be convex. Without loss of generality, C is dense. We have $cind(C) \le cind(X)$. The implication

$$cind(C) \le n \Rightarrow cind(X) \le n$$

is valid if $n = -1$. Suppose it is valid for $m < n$, where $n \ge 0$. Let $cind(C) \le n$ and consider a convex closed set $D \subseteq X$, together with a point $p \in X \setminus D$. By FS_3, there is an open half-space O with $D \subseteq O$ and $p \notin \overline{O}$. Consider a minimal convex closed screening D_1, D_2 of D, p with $D_1 \subseteq \overline{O}, D_2 \subseteq X \setminus O$. In particular, $D_1 \cap D_2 \subseteq Bd(O)$. As C is dense, topological considerations yield that $\overline{O} \cap C = Cl_C(O \cap C)$. Hence, $Bd(O) \cap C$ is the relative boundary of $O \cap C$ in C. By Proposition 4.6,

$$cind(D_1 \cap D_2 \cap C) \le cind(Bd(O) \cap C) \le n - 1.$$

As C is dense and convex, we infer from 4.5 that $D_1 \cap D_2 \cap C$ is a dense subset of $D_1 \cap D_2$. By induction,

$$cind(D_1 \cap D_2 \cap C) = cind(D_1 \cap D_2).$$

It follows that $cind(X) \le n$. ∎

The following is an application involving some special properties of interval operators (cf. I§7).

4.9. Corollary. *Let X be a closure stable* FS_3 *space with connected convex sets. If X has the Ramification Property, then each segment is 1–dimensional.*

§4: Dimension Theory   457

Proof. Let $a, b \in X$. As the segment operator of an S_3 space is geometric, we obtain a partial order $\leq_b$ by the prescription

$$u \leq_b v \Leftrightarrow bu \subseteq bv$$

(this is the base-point order of b; cf. I§5). If $x \in ab \setminus \{b\}$ and if $u, v >_b x$ then the Ramification Property implies that u and v are comparable in the base-point order of b. Therefore, the convex set

$$C = \cup \{\uparrow_b(x) \mid x \in ab \setminus b\}$$

is a chain containing a. Convex sets being connected, some decreasing net of points $x \in ab \setminus \{b\}$ converges to b. By closure stability, we conclude that $\bar{C} = ab$. Observe that $cind(C) \leq 1$, whence by Corollary 4.8, the segment ab is one-dimensional.   ∎

4.10. Corollary. *In a closure stable FS_3 space with connected convex sets and of finite dimension, each dense half-space has a non-empty interior. In fact, its interior meets every non-empty convex open set of the space.*

Proof. Let X be as announced, let $H \subseteq X$ be a dense half-space, and let $O \neq \emptyset$ be convex open. Then $H \cap O$ is a relatively dense half-space of O, whence

$$cind(O \setminus H) < cind(O)$$

by Proposition 4.6. By Corollary 4.8, the set $O \setminus H$ is not dense in O, showing that

$$\emptyset \neq Int_O(O \cap H) \subseteq Int(H).$$ ∎

For properly locally convex spaces, with further conditions as in Corollary 4.10, each dense half-space has a dense interior. We now switch to a series of results involving CP maps.

4.11. Lemma. *Let X, Y be closure stable FS_3 spaces with connected convex sets, and let $f: X \to Y$ be a closed, continuous and CP function of X onto Y. Then $cind(X) \geq cind(Y)$.*

Proof. We verify that

$$cind(Y) \geq n \Rightarrow cind(X) \geq n$$

for each $n < \infty$. The case $n = -1$ being trivial, assume the statement is valid for $m < n$, where $n \geq 0$. If $cind(Y) \geq n$, then by Corollary 4.7 there is an open half-space $O \subseteq Y$ with $cind(Bd(O)) \geq n-1$. The set $P = f^{-1}(O)$ is an open half-space of X and $f(Bd(P)) = Bd(O)$ since f is closed and surjective. The result follows from an application of the inductive assumption to the restricted mapping $Bd(P) \to Bd(O)$.   ∎

We will extend this result to not necessarily closed CP maps. Some preparatory work on maps into cubes is needed first.

4.12. Lemma. *Let X be a closure stable FS_3 space with connected convex sets, let the n-cube $[-1,1]^n$ (where $0 \leq n < \infty$) be equipped with the standard median convexity, and let d be the Euclidean metric of $[-1,1]^n$. If $f: X \to [-1,1]^n$ is a CP map such that the* *distance of each corner point to $f(X)$ is less than 1, then the image of f includes a closed n-cube. Moreover, if f includes all corner points of $[-1,1]^n$, then it is a surjection.*

Proof. We consider $n > 0$. Let V be the set of corner points of our cube Q, labeled as v_i for $i = 1,..,2^n$. For each i we fix a point $x_i \in X$ with $d(f(x_i), v_i) < 1$, and we let

$$r = \max \{ d(f(x_i), v_i) \mid i = 1,..,2^n \}.$$

Define $W_0 = (1-r) \cdot V$; in other words, W_0 is the set of corner points of the subcube $C = [-1+r, 1-r]^n$. Let $p: Q \to C$ be the gate map; in particular, p is continuous and CP. Observe that if $w_i = (1-r) \cdot v_i$, then $v_i w_i \cap C = \{w_i\}$ and $f(x_i) \in v_i w_i$ (Fig. 1). This shows that $pf(x_i) = w_i$ for all i.

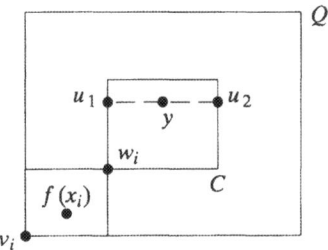

Fig. 1: Mapping into a cube.

For $0 \leq k \leq n$ let W_k be the union of all k-faces of C. We have just shown that $W_0 \subseteq pf(X)$. Assume $W_k \subseteq pf(X)$ and let $y \in W_{k+1}$. Then there exist $u_1, u_2 \in W_k$ such that $y \in u_1 u_2$ and all but one coordinate of u_1, u_2 agree. By assumption, there exist $x_1, x_2 \in X$ such that $pf(x_j) = u_j$ for $j = 1, 2$. As $p \circ f$ is CP, we have $pf(x_1 x_2) \subseteq u_1 u_2$. Now $u_1 u_2$ equals the standard line segment joining u_1, u_2. Therefore, the last inclusion is equality by the connectedness of convex sets in X. We conclude that $y \in pf(X)$, showing that $W_{k+1} \subseteq pf(X)$.

The inductive argument shows that $C \subseteq pf(X)$. It is easily seen that p maps points of $Q \setminus C$ into the boundary of C. Therefore, if $D \subseteq Int(C)$ is another n-cube, then $D \subseteq f(X)$. Note that if $f(X)$ includes all vertices of Q, then C could have been taken equal to Q and $pf(X) = f(X) = Q$ in this case. ∎

This leads us to another major result.

4.13. Theorem. *Let X be a closure stable FS_3 space with connected convex sets, and let $0 \leq n < \infty$. If $C \subseteq X$ is a convex set with $cind(C) \geq n$, then there is a continuous CP function $f: X \to [0,1]^n$ with $f(C) = [0,1]^n$. If all polytopes of X are compact, then the converse is also true.*

Proof. The case $n = 0$ being trivial, let $n > 0$ and assume the result is valid for convex sets of dimension $< n$. By FS_3, the family of all CP maps $X \to [0,1]$ separates relatively convex closed sets of C from points in C. Hence by definition, there is a CP map $h: X \to [0,1]$ such that the boundary of the relative half-space $O = C \cap h^{-1}(0,1]$ is of

§4: Dimension Theory

dimension $\geq n-1$. By inductive assumption, there is a CP map $g: X \to [0,1]^{n-1}$ with $g(Bd_C(O)) = [0,1]^{n-1}$. Consider the following CP map.
$$f = (g,h): X \to [0,1]^n, \quad f(x) = (g(x), h(x)).$$

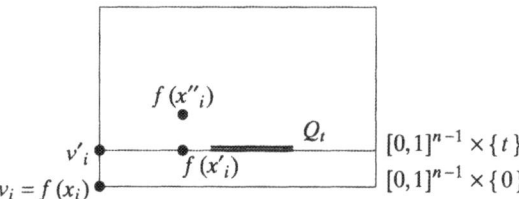

Fig. 2: Finding a cube in the image.

Let $\{v_i \mid i = 1,..,2^{n-1}\}$ be the set of vertices of $[0,1]^{n-1} \times \{0\}$. For each i let $x_i \in X$ be such that $f(x_i) = v_i$, and let $V_i \subseteq B(v_i, \frac{1}{4})$ be a convex neighborhood of v_i in $[0,1]^n$. Then $U_i = f^{-1}(V_i)$ is a convex neighborhood of x_i and there is a point $x''_i \in O \cap U_i$. Let
$$t = \tfrac{1}{2} \cdot \min \{f(x''_i) \mid i = 1,..,2^{n-1}\}.$$
Then $t > 0$ and if $O_t = f^{-1}(t,1]$ we have $x_i \notin O_t, x''_i \in O_t$. Consequently, there is a point
$$x'_i \in x_i x''_i \cap Bd_C(O_t).$$
Observe that $x'_i \in x_i x''_i \subseteq U_i$. Let $v'_i \in [0,1]^{n-1} \times \{t\}$ be the corner point at distance t from v_i. Then
$$d(v'_i, f(x'_i)) \leq d(v'_i, v_i) + d(v_i, f(x'_i)) \leq t + \tfrac{1}{4} \leq \tfrac{1}{2}.$$
By Lemma 4.12, $f(Bd_C(O_t))$ includes an $(n-1)$–cube $Q_t = \prod_{k=1}^{n-1} [a_k, b_k] \times \{t\}$, where $0 \leq a_k < b_k \leq 1$ for all k. Let
$$Q_0 = \prod_{k=1}^{n-1} [a_k, b_k] \times \{0\}; \quad Q = \prod_{k=1}^{n-1} [a_k, b_k] \times [0,t].$$
Now $Q_0 \subseteq f(Bd_C(O)) \subseteq f(C)$. For each point $y \in Q$ there exist $u_0 \in Q_0$ and $u_t \in Q_t$, such that $y \in u_0 u_t$, and u_0, u_t differ in only one coordinate. As in the proof of 4.12, we conclude that $y \in f(C)$. This shows that $f(C)$ includes an n–cube. After composing f with the gate map onto this cube, we obtain a continuous CP function of X mapping C onto an n–cube.

If all polytopes of X are compact, then the converse obtains as follows. Let $f: X \to [0,1]^n$ be a CP map with $g(C) = [0,1]^n$ for some convex set C. For each vertex of the cube we take one pre-image in C, and we let $F \subseteq C$ be the resulting set. By Lemma 4.12, f maps $co(F)$ onto $[0,1]^n$ as well. By Lemma 4.11, $cind(co(F)) \geq n$, yielding the result. ∎

This result has several consequences.

4.14. Corollary. *Let X, Y be closure stable FS_3 spaces with connected convex sets, and such that all polytopes of X are compact. If $f: X \to Y$ is a CP map with a dense* image, then $cind(X) \geq cind(Y)$.

Proof. We verify that
$$cind(Y) \geq n \implies cind(X) \geq n$$
for each $n < \infty$. The case $n = -1$ being trivial, assume the statement is valid for $m < n$, where $n \geq 0$, and let $cind(Y) \geq n$. By Theorem 4.13, there is a CP map $g: Y \to [0,1]^n$ of Y onto the n-cube. The composed map $g \circ f: X \to [0,1]^n$ is CP and has a dense image which, by Lemma 4.12, includes a closed n-cube Q. The composition of $g \circ f$ with the gate map of Q is a CP surjection of X onto a cube, and another application of Theorem 4.13 yields the desired result. ∎

4.15. Corollary. *Let X be a closure stable FS_3 space with connected convex sets and with compact polytopes. Let $n < \infty$. Then $cind(X) \geq n$ iff $cind(P) \geq n$ for for some polytope P of X.*

Proof. If $cind(P) \geq n$ for some polytope $P \subseteq X$ then, of course, $cind(X) \geq n$. Conversely, if $cind(X) \geq n$, then consider a CP map onto an n-cube and take one pre-image of each corner point. The resulting polytope maps to a subset of the cube including the vertex set. Application of Lemma 4.12 yields the result. ∎

4.16. Corollary *(Countable Sum Theorem for cind).* *Let X be a closure stable FS_3 space with connected convex sets and with compact polytopes. Let $n < \infty$. If $(C_k)_{k=1}^{\infty}$ is a sequence of convex subsets with $cind(C_k) \leq n$ for all k, and if C is a convex set with $C \subseteq \cup_{k=1}^{\infty} C_k$, then $cind(C) \leq n$.*

Proof. As $cind(C) \geq n$, there is a polytope $P \subseteq C$ with $cind(P) \geq n$, together with a CP map f of P onto an n-cube Q. For each $k \in \mathbb{N}$ the set $D_k = C_k \cap P$ is compact and convex. Then $f(D_k)$ is a compact subset of Q and $\cup_{k=1}^{\infty} f(D_k) = Q$. By the Baire Category Theorem,[3] some set $f(D_k)$ includes an n-dimensional subcube Q'. In a by now routine manner, this leads to the conclusion that the convex set D_k has at least n dimensions. ∎

The following application relates with the theory of pseudo-interior (cf. Section 2).

4.17. Theorem. *Let X be a closure stable FS_3 space with connected convex sets. If C is a non-empty convex subset of X of finite dimension, then the intersection of all relatively dense convex subsets of C is relatively dense in C.*

Proof. The result is obviously valid for convex sets C of dimension ≤ 0. Proceeding by induction, let $cind(C) \leq n > 0$. Let E be the set of points common to all dense convex subsets of C. Consider two distinct points of C and a minimal convex screening between them with intersection D. We can take care that $cind(D) \leq n-1$. By Lemma 4.5, each dense convex subset of C induces a dense convex subset of D. By the inductive

---
3. Engelking [1977, p. 253]

§4: Dimension Theory   461

assumption $D \cap E \neq \emptyset$.

If $p \in C \setminus \bar{E}$, then consider a minimal convex screening between $\bar{E} \cap C$ and p with intersection D of dimension $\leq n-1$. The previous argument provides us with a point of D which is in E, a contradiction. ∎

We next consider a characterization of *cind* in terms of simultaneous separation. We require the following auxiliary result.

4.18. Lemma. *Let X be a tcs with connected convex sets, let $n \geq 1$, and let $f: X \to [0,1]^n$ be a surjective CP map. For each $i = 1,..,n$, let (A_i, B_i) be the pair of inverse images of the ith pair of opposite faces, and let M_i be a convex closed separator of A_i and B_i. Then $\cap_{i=1}^n M_i \neq \emptyset$.*

Proof by induction on n. The statement holds for $n = 1$ since, by the connectedness of X, each separator between A_i and B_i is non-empty. Suppose the statement to be valid for n, and consider a surjective CP map $f: X \to [0,1]^{n+1}$. With A_i and B_i defined as above, observe that the restrictions

$$A_{n+1} \to [0,1]^n \times \{0\} \quad \text{and} \quad B_{n+1} \to [0,1]^n \times \{1\}$$

are surjective CP maps, and $M_i \cap A_{n+1}$ is a convex separator of $A_i \cap A_{n+1}$ and $B_i \cap A_{n+1}$ for each $i = 1,..,n$. By assumption, $\cap_{i=1}^n (M_i \cap A_{n+1}) \neq \emptyset$. In a similar way, we find that $\cap_{i=1}^n (M_i \cap B_{n+1}) \neq \emptyset$. As $\cap_{i=1}^n M_i$ is a connected set meeting A_{n+1} and B_{n+1}, it meets M_{n+1} too. ∎

4.19. Theorem. *Let X be a closure stable FS_3 space with connected convex sets and with compact polytopes. For each finite $n > 0$ the following assertions are equivalent.*

(1) $cind(X) \leq n$.

(2) *For each collection $(A_i, B_i)_{i=1}^{n+1}$ of pairs of convex closed sets that can be separated by a CP map into $\mathbb{R}$, there is a sequence $(C_i)_{i=1}^{n+1}$ of convex closed sets, such that for each i, the set C_i is a separator of the ith pair (A_i, B_i) and such that $\cap_{i=1}^{n+1} C_i = \emptyset$.*

Proof. (1) $\Rightarrow$ (2). For each $i = 1,..,n+1$, let $f_i: X \to \mathbb{R}$ be a CP map separating the ith pair of convex closed sets (A_i, B_i). We may assume that A_i maps to 0, that B_i maps to 1, and that $f(X) \subseteq [0,1]$. Let $f = (f_1,..,f_{n+1})$. If the open cube $(0,1)^{n+1}$ is included in $f(X)$, then $cind(X) \geq n+1$ by virtue of 4.13. Let $p_i: [0,1]^{n+1} \to [0,1]$ denote the ith projection. Consider a point $y \in (0,1)^{n+1} \setminus f(X)$, and let $P_i = p_i^{-1}[0, p_i(y))$. Then

$$\bigcap_{i=1}^{n+1} Bd(P_i) \cap f(X) = \{y\} \cap f(X) = \emptyset.$$

Now $f^{-1}(P_i)$ is an open half-space of X and f maps its boundary into $Bd(P_i)$. Therefore,

$$\bigcap_{i=1}^{n+1} Bd(f^{-1}(P_i)) = \emptyset.$$

Observe that $Bd(f^{-1}(P_i))$ is a hyperplane separating the pair (A_i, B_i).

We verify (2) $\Rightarrow$ (1) by contraposition. If $cind(X) > n$, then by Theorem 4.13 there is a surjective CP map $f: X \to [0,1]^{n+1}$. With the ith projection p_i defined as above, let

$A_i = f^{-1}p_i^{-1}(0)$ and $B_i = f^{-1}p_i^{-1}(1)$ for $i = 1,..,n+1$. By Lemma 4.18, any set of $n+1$ convex separators, one for each pair (A_i, B_i), has a non-empty intersection. ∎

We have two major results on equality of dimension functions. The first one involves the so-called cohomological dimension[4] cd_G relative to an abelian coefficient group $G \neq 0$.

4.20. Theorem. *Let X be a compact closure stable FS_{3+} space with connected convex sets. Then for each non-trivial abelian group G,*

$$cd_G(X) = dim(X) = ind(X) = Ind(X) = cind(X).$$

Proof. The following (in)equalities are valid.[5]

$$cd_G[0,1]^n = n; \quad cd_G(X) \leq dim(X) \leq ind(X) \leq Ind(X).$$

In addition, we verify that $cind(X)$ is both a lower and an upper bound of the sequence of inequalities.

To see that $cind(X) \leq cd_G(X)$, let $n < \infty$ and $n \leq cind(X)$. By Theorem 4.13, there is a surjective CP map $f: X \to [0,1]^n$ (standard median cube). The fibers of f are convex. If C is a non-empty compact convex set of X, and if $\mathcal{U}$ is an open cover of C then, as X is properly locally convex, there is a convex open cover of C refining $\mathcal{U}$. This refinement has a contractible nerve by Theorem 3.2(1). Therefore, C has trivial Čech cohomology. We are now in a position to apply the Vietoris-Begle Theorem for Čech cohomology with general coefficients[6]. For each closed set $A \subseteq [0,1]^n$, the map f induces an isomorphism of cohomology groups,

$$\check{H}^*([0,1]^n, A; G) \to \check{H}^*(X, f^{-1}(A); G).$$

It follows from the definition of cohomological dimension that $cd_G(X) \geq cd_G[0,1]^n$.

We next verify that $Ind(X) \leq cind(X)$, completing the proof of the theorem. To this end, we consider the statement

Q(n) The union of finitely many convex closed sets having $cind \leq n$ has $Ind \leq n$.

For $n = -1$ this easy. For $n = 0$, all non-empty convex sets in consideration are singletons, and the statement is elementary. For $n > 0$, we proceed by induction. Let $C_1,..,C_p$ be convex closed sets satisfying $cind(C_k) \leq n$ for $k = 1,..,p$. Let $Y = \cup_{k=1}^p C_k$, let $A \subseteq Y$ be closed, and let $U \supseteq A$ be a relatively open set of Y. By FS_2 and compactness of X, the open half-spaces of X constitute an open subbase for X. This yields a finite number of open half-spaces O_{ij} ($i = 1,..,q, j = 1,..,r$), such that

---

4. The reader is referred to Cohen [1954] for further information on cohomological dimension.

5. For the equality and the first inequality, see Cohen [1954]. For the second and third inequality, see Engelking [1977, 7.1.2, 7.2.7].

6. See Lawson [1973, 5.1]. This result is an extension of the classical Vietoris-Begle Mapping Theorem.

§4: Dimension Theory

if $O = \bigcup_{i=1}^{q} \bigcap_{j=1}^{r} O_{ij} \cap Y$ then $A \subseteq O \subseteq Cl(O) \subseteq U$.

We let Bd_k denote the boundary operator of C_k ($k = 1,..,p$). By Proposition 4.6, $cind(Bd_k(O_{ij} \cap C_k)) \leq n-1$. Furthermore, we have

$$Bd_Y(O) \subseteq \bigcup_{i=1}^{q} \bigcap_{j=1}^{r} Bd_Y(O_{ij} \cap Y);$$

$$Bd_Y(O_{ij} \cap Y) \subseteq \bigcup_{k=1}^{p} Bd_k(O_{ij} \cap C_k).$$

This yields

$$Ind(Bd_Y(O)) \leq Ind\left(\bigcup_{k=1}^{p} \bigcup_{i=1}^{q} \bigcap_{j=1}^{r} Bd_k(O_{ij} \cap C_k)\right) \leq n-1,$$

showing that $Ind(Y) \leq n$.

Having completed the inductive proof of the statements Q(n), we conclude at once that $Ind(X) \leq cind(X)$. ∎

4.21. Theorem. *Let X be a separable, metrizable S_4 convex structure with connected convex sets and with compact polytopes. Then*

$$dim(X) = ind(X) = Ind(X) = cind(X).$$

Proof. There is a countable dense subset $\{x_n \mid n \in \mathbb{N}\}$ of X, which we use to construct the following convex sets.

$$D_n = co\{x_1,..,x_n\}; \quad D = co\{x_n \mid n \in \mathbb{N}\}.$$

By the Countable Sum Theorem in topology[7], we have

$$ind(D) = \sup\{ind(D_n) \mid n \in \mathbb{N}\},$$

whereas by the Countable Sum Theorem for *cind*, Corollary 4.16,

$$cind(D) = \sup\{cind(D_n) \mid n \in \mathbb{N}\}.$$

Polytopes being compact and NS$_3$, each D_n has the weak topology. Combining the previous equalities with the inequality of Proposition 4.3 yields

$$ind(D) \leq cind(D).$$

On the other hand, if $cind(D) \geq n$, then $cind(D_k) \geq n$ for some k (Countable Sum Theorem for *cind*). Hence $ind(D) \geq ind(D_n) \geq n$, where the last inequality follows from Theorem 4.20. This shows that $ind(D) = cind(D)$.

We now extend this result to the space X. First, $cind(D) = cind(X)$ by Corollary 4.8. By the (elementary) Subspace Theorem in topology, we have $ind(D) \leq ind(X)$. We complete our proof by showing that $ind(D) \leq n$ implies $ind(X) \leq n$ for all n. We use the result[8] that for a separable metric space X, the inequality $ind(X) \leq n$ holds iff each

7. Engelking [1977, 7.2.1]; van Mill [1988, Thm. 4.3.7].
8. Engelking [1977, 7.4.13]; van Mill [1988, Thm. 4.6.4].

function $f: A \to S^n$, defined on a closed subset A of X and ranging into the unit n-sphere S^n, can be extended over X. As the image space is an absolute neighborhood retract (ANR), there is an extension f' of f to a closed neighborhood A' of A. Let $\mathcal{U}$ be an open cover of X such that $f(U)$ is included in an open hemisphere of S^n for each $U \in \mathcal{U}$, and such that $star(A, \mathcal{U}) \subseteq A'$. Theorem 3.16 provides us with a mapping $g: X \to D \subseteq X$ which is $\mathcal{U}$-close to identity. As $ind(D) \leq n$, the restriction of f' to $A' \cap D$ extends to a map $f'': D \to S^n$. If $x \in A'$ then f'' maps x and $g(x)$ into an open hemisphere. Hence the mappings f and $f'' \circ g | A$ are homotopic. As one of them extends over X, so does the other. ∎

Further Topics

4.22. Dimension functions (van de Vel [1982]). Let A be a hereditary class of topological convex structures ("hereditary" means that the class A contains all convex subspaces of its members). We consider a (class) function $\mathcal{P}$, defined on A and assigning to $X \in A$ a collection $\mathcal{P}(X)$ of subset pairs of X, such that the following conditions are fulfilled.

(1) If $(A, B) \in \mathcal{P}(X)$, then A, B are disjoint convex closed sets which can be separated by a hyperplane of X.
(2) If $a \neq b \in X$ can be separated by a hyperplane, then $(\{a\}, \{b\}) \in \mathcal{P}(X)$.

For instance, consider the class A_0 of all FS_2 spaces, and let $\mathcal{P}_0(X)$ be the set of all pairs of disjoint singletons. Alternatively, consider the class A_1 of all topological convex structures, and let $\mathcal{P}_1(X)$ denote the class of all pairs of convex closed sets in X which can be separated by a hyperplane.

For a class function $\mathcal{P}$ as above, defined on a hereditary class A, the corresponding dimension function $cind_\mathcal{P}$ is defined as follows.

(1) $cind_\mathcal{P} = -1$ iff $X = \emptyset$.
(2) $cind_\mathcal{P} \leq n+1$ iff each pair $(A, B) \in \mathcal{P}(X)$ can be screened with a pair of convex closed sets A', B', such that

$cind_\mathcal{P}(A' \cap B') \leq n$.

Prove that if $\mathcal{P}$ and $\mathcal{Q}$ are class functions of the above type, and if X is a closure stable FS_3 convex structure with connected convex sets, occurring in the domain of $\mathcal{P}$ and $\mathcal{Q}$, then

$cind_\mathcal{P}(X) = cind_\mathcal{Q}(X)$.

Hint. Some of the arguments at the beginning of this section can be adapted to the general situation.

4.23. Trees. Show that the following are equivalent for a locally connected tcs X.

(i) X is a connected tree.
(ii) X is a closure stable FS_3 space with connected convex sets and of dimension ≤ 1.

§4: Dimension Theory 465

4.24. G_δ-half-spaces. Let X be a closure stable finite-dimensional FS_3 space with connected convex sets. If every convex closed set is a G_δ-set in X, then each half-space of X is both a G_δ-set and an F_σ-set.

4.25. Mapping onto cubes (van de Vel [1982]). Consider the following subset of the Hilbert cube: $X = \cup_{k=0}^\infty X_k$, where $X_0 = \{\mathbf{0}\}$ and
$$X_k = \{(x_n)_{n=1}^\infty \mid 2^{-k} \le x_n \le 2^{-k+1} \ (n \le k); \ x_n = 2^{-n+1}(n > k)\}$$
The Hilbert cube is equipped with the standard median convexity. Show that the subspace X is compact, connected, and median stable, and that it is an infinite-dimensional tcs without surjective CP maps onto the the Hilbert cube.

4.26. Weak continuity in median spaces (van de Vel [1984b])

4.26.1. Let X be a weakly locally convex, connected median space with compact segments of finite dimension. Show that each half-space of X is either open or closed. Hint. By Theorem 4.17, the intersection of all dense convex subsets of a segment is a dense subset.

4.26.2. Deduce that if X, Y are as in the first part, then each surjective CP function $X \to Y$ is weakly continuous.

4.27. Dimension of median spaces (van de Vel [1984c]). Almost all results of this section require a tcs with connected convex sets. In this and the next topic, we outline a theory that works for median spaces without connectedness. Throughout, X represents a weakly locally convex median space with compact segments.

4.27.1. Show that $c, d \in X$ are in different components iff there exist convex closed sets $C, D \subseteq X$ with
$$c \in C; \quad d \in D; \quad C \cup D = X; \quad C \cap D = \emptyset.$$
Conclude that $cind(X) \le 0$ iff all components of X are singletons.

4.27.2. Show that a convex set and its closure have the same convex dimension.

4.27.3. Show that a convex set D satisfies $cind(D) \le n+1$ iff for each open half-space $O \subseteq X$, such that some component of X meets both O and $X \setminus O$, we have $cind(Bd(O) \cap D) \le n$.

4.27.4. Use the previous results to show that $cind(X) \le n$ iff some component C of X satisfies $cind(C) \le n$ and that
$$cd_G(X) = ind(X) = Ind(X) = dim(X) = cind(X)$$
(cd_G is the cohomological dimension with coefficient group G). Hint: Adapt the proof of 4.20.

4.28. Embedding of median compacta (van de Vel [1984c]). Let $n < \infty$ and let X be a compact n-dimensional median algebra. Note that X is locally convex.

4.28.1. Show that the prescription

$$u \equiv v \Leftrightarrow u, v \text{ belong to the same component}$$

is a congruence relation on X. Relative to the quotient topology and quotient convexity, the space dX of components of X is a compact median algebra. Let $d: X \to dX$ be the quotient map.

4.28.2. Prove that there is an n-dimensional connected quotient space qX of X with a quotient map $q: X \to qX$, such that $(q,d): X \to qX \times dX$ is an embedding of median algebras.

Hint: Let C_i for $i \in I$ be the family of all components of X. For each $i \in I$ let $p_i: X \to C_i$ be the gate map. Take q equal to the function

$$(p_i)_{i \in I}: X \to \prod_{i \in I} C_i.$$

4.28.3. Let X be a compact distributive lattice of dimension $n < \infty$. Derive the following inequalities for the breadth:

$$n \le b(X) \le n + b(dX).$$

Give examples illustrating the sharpness of these bounds.

4.29. Some problems in dimension theory. The condition that all convex sets be connected is a rather heavy one. Use of the free convexity of a topological space makes it clear, however, that some assumptions are necessary to develop an acceptable theory of convex dimension.

4.29.1. Develop a theory under the assumption of the Continuous Positions Property (CPP). This condition is suggested by the results on median spaces in Topic 4.27.

4.29.2. Can the results of this section be extended to spaces which are not necessarily closure stable? The motivation for this problem is, that Lawson semilattices need not have that property, whereas some excellent results have been developed on the relation between breadth and (cohomological) dimension; cf. Lawson [1970] and [1971].

4.29.3. Develop a theory of convex dimension based on a definition in the style of the Lebesgue covering dimension.

Notes on Section 4

The first attempt to develop a theory of dimension in convex structures goes back to Bryant and Webster [1977]. Here, dimension is simply the affine dimension of the matroid, associated to a BW space (see I§2 and I§7). The concept of convex small

§4: Dimension Theory

inductive dimension, *cind*, was introduced in the author's paper [1983d] (VU report[9] dated 1979) in the context of a theory of support points. A first systematic study of *cind* was undertaken in our paper [1982] (VU report 1980). This paper contains the characterization of *cind* in terms of hyperplanes (Theorem 4.7) and in terms of mappings onto cubes (Theorem 4.13), together with the Corollaries in 4.14 (on images under CP maps), 4.15 (on dimension of polytopes), and 4.16 (Countable Sum Theorem).

Corollary 4.17 (on dense intersections of convex sets) was obtained by the author in [1983d]. Theorem 4.19, on characterizing dimension in terms of sequences of screening pairs, is new. It is close in style to the treatment of dimension in topology.

Equality of topological and convex dimension was first studied in our paper [1982], where Example 4.4 was given (an infinite-dimensional convexity on a zero-dimensional topological space), together with a predecessor of Theorem 4.21. This result was phrased in terms of LC^n and C^n convex sets. The present result is in terms of uniform convex structures, where connectedness of convex sets leads to higher forms of connectedness as cited above. See Section 3. Theorem 4.20 (on equality of some dimension functions on compact tcs's) was obtained by van Mill and van de Vel [1986] (VU report 1981) for uniform convexity; the weakening of the assumptions to the current ones, and the part of the result involving cohomological dimension were achieved later by the author in [1984e] (VU report 1982).

---

9. Internal report series of the *Vrije Universiteit Amsterdam*. We exceptionally give the manuscript data because the original chronology has been somewhat disturbed.

5. Dimension and Convex Invariants

The classical results of Helly, Carathéodory and Radon on $\mathbb{R}^n$ can be interpreted as relations between certain combinatorial invariants and the topological dimension n of $\mathbb{R}^n$. This viewpoint is explored in a wider class of spaces, satisfying conditions as in the previous section. The Carathéodory and Exchange numbers are most tightly connected with dimension, the Radon number tends to infinity if the dimension does, and the Helly number is almost unconnected with the dimension.

The Radon number of an n-dimensional median space equals the Radon number of the standard median n-cube, or is one larger. The latter phenomenon can occur only in restricted dimensions and cubical polyhedra are used to prove the sharpness of this result. Join-hull Commutativity holds in spaces with $c \leq 3$ and $e \leq 3$. The dimension of a convex hyperspace equals the rank of the basic topological space.

In this section, *all spaces are assumed* S_1.

As before, the term *number* refers to a member of the set $\{-1, 0, 1, .., \infty\}$. We begin with an auxiliary result which is independent of dimension theory. It involves conditions that are almost standard throughout this section.

5.1. Lemma. *Let X be a closure stable* FS_{3+} *space with compact polytopes and with connected convex sets. Let $C_1, .., C_n$ ($n \geq 2$) be closed convex sets in X such that $\cup_{i=1}^n C_i$ is convex and $\cap \{C_i \mid i \neq j\} \neq \emptyset$ for each $j = 1, .., n$. Then $\cap_{i=1}^n C_i \neq \emptyset$.*

Proof. For $n = 2$ the statement follows from the connectedness of convex sets. Suppose $n > 2$ and let the result be valid for $m < n$ sets. Consider n sets $C_1, .., C_n$ meeting $n-1$ by $n-1$, such that $\cap_{i=1}^n C_i = \emptyset$. For each $i = 1, .., n$ we take a point $x_i \in \cap_{j \neq i} C_j$ and we let

$$D_i = C_i \cap co\{x_1, .., x_n\}.$$

By Corollary III§4.12, the compact space $co\{x_1, .., x_n\}$ is FS_4. Hence there is a convex closed separator D of $\cap_{i=1}^{n-1} D_i$ and D_n. For each $i = 1, .., n-1$ the set $\cap_{j \neq i,n} D_j$ meets D_i (say, in u_i) as well as D_n (say, in v_i). As $u_i v_i$ is connected, it meets D in some point w_i. Now $w_i \in \cap_{j \neq i,n} D_j \cap D$, and we have shown that the sets

$$D_i \cap D, \quad \text{for} \quad i = 1, .., n-1,$$

meet $n-2$ by $n-2$. The union of these sets is convex. The inductive assumption yields that $\cap_{i=1}^{n-1} D_i \cap D \neq \emptyset$, a contradiction. ∎

This leads to an additional relation between the invariants h and c.

5.2. Theorem. *Let X be a closure stable* FS_{3+} *space with compact polytopes and with connected convex sets. Then $h(X) \leq c(X)$.*

Proof. We verify that $c(X) \leq n$ implies $h(X) \leq n$ for $0 \leq n < \infty$. The case $n = 0$ being trivial, we assume $n \geq 1$. Let $F \subseteq X$ and $\#F = m > n$. The sets $C_a = co(F \setminus \{a\})$ for $a \in F$ meet $m-1$ by $m-1$. As F is C-dependent, we have $\cup_{a \in F} C_a = co(F)$. Lemma 5.1 implies that $\cap_{a \in F} C_a \neq \emptyset$, which shows that F is H-dependent. ∎

We now investigate the relations between the classical invariants and *cind*.

5.3. Theorem. *Let X be a closure stable* FS_{3+} *space with compact polytopes and with connected convex sets. Then $h(X) \leq cind(X) + 1$.*

Proof. We may assume that $cind(X) = n < \infty$. It is easy to see that $h(X) = 0$ iff $X = \emptyset$ (i.e., iff $cind(X) = -1$), and that $h(X) = 1$ iff X is a one-point set (i.e., iff $cind(X) = 0$). We proceed by induction, assuming the result to hold in dimensions $< n$, where $1 \leq n < \infty$. Suppose F is an H-independent set with $m > n + 1$ points, say:

$$F = \{x_1, .., x_m\}.$$

By FS_{3+}, there is a hyperplane[1] H separating between the sets

$$C = \bigcap_{i=1}^{m-1} co(F \setminus \{x_i\}) \quad \text{and} \quad D = co(F \setminus \{x_m\}).$$

For each $i = 1, .., m-1$, the set

$$\cap \{co(F \setminus \{x_j\}) \mid j = 1, .., m-1; j \neq i\}$$

meets C (in x_m, for instance) and D (in x_i, for instance). By the connectedness of convex sets, it meets H. This shows that the sets

(*) $\qquad co(F \setminus \{x_i\}) \cap H \quad (i = 1, .., m-1)$

meet $m-2$ by $m-2$. By Proposition 4.6, we have $cind(H) \leq cind(X) - 1$, whereas $h(H) \leq n$ by inductive assumption. Therefore, the convex sets in (*) have a non-empty intersection, contradicting that $C \cap H = \emptyset$. ∎

In the next results, r_n denotes the largest integer m such that $\binom{m}{\lfloor \frac{1}{2}m \rfloor} \leq 2n$ (notation of II§2).

5.4. Theorem. *Let X be a closure stable* FS_{3+} *space with compact polytopes and with connected convex sets. Let $cind(X) \geq n \geq 0$, where $n < \infty$. Then $r(X) \geq r_n$. In addition, $r(X) = \infty$ iff $cind(X) = \infty$.*

Proof. By Theorem 4.13, there is a surjective CP map $X \to [0,1]^n$ of X to the standard median n-cube. The Radon number of the cube equals r_n; cf. II§2.17. The first part of the result follows from Theorem II§1.10.

As to the second part, let $cind(X) = \infty$. Clearly, $\lim_{n \to \infty} r_n = \infty$, and the previous argument yields $r(X) = \infty$. Conversely, if $cind(X) < \infty$ then $r(X) < \infty$. In fact, for each finite number n,

---
1. Recall that a hyperplane is the boundary of an open half-space.

§5: Dimension and Convex Invariants   471

$$cind(X) \le n \Rightarrow r(X) < 2^n.$$

This is certainly true in dimensions ≤ 0. We proceed by induction. If $cind(X) = n+1$ (where $n \ge 0$) and yet $r(X) \ge 2^{n+1}$, then consider a Radon independent set with 2^{n+1} points and divide it into two sets F_1, F_2, each with 2^n points. Enumerate their elements in an arbitrary order. As $co(F_1) \cap co(F_2) = \emptyset$, there is a convex set C of dimension $\le n$ separating F_1 and F_2. Let $x_i \in C$ be a point of the segment connecting the ith point of F_1 with the ith point of F_2. It is easy to see that the collection $\{x_i \mid i = 1,..,2^n\}$ is Radon independent, a contradiction. ∎

Note that the Helly number of an infinite-dimensional space (with further properties as in the last theorem) can be finite. For example, consider the Hilbert cube with its standard median convexity. Theorem 5.4 can be considerably improved for median spaces, as the next two results may show.

5.5. Theorem. *Let X be a connected median space with compact polytopes, and let $cind(X) = n$, where $0 \le n < \infty$. Then $r(X) = r_n$ or $r(X) = r_n + 1$. The second equality can occur only in those dimensions n where r_n is even and $\binom{r_n}{\frac{1}{2}r_n+1} \le n$.*

Proof. As X_w inherits the assumptions made on X, we may assume that X has the weak topology. First, $cind < \infty$ implies that distinct points of X can be screened with convex closed sets. By Theorem III§4.16, this implies that X is FS$_4$ and (weakly) locally convex. Hence, by Proposition 2.24, X has the Continuous Positions Property (CPP). Finally, all gate maps of X are (weakly) continuous by Corollary III§5.20, whence all convex sets of X are (weakly) connected. We assume $n > 0$ and consider two possibilities for $r = r(X)$.

(i). r is even. Let $F \subseteq X$ be an R-independent set with r points. For each Radon partition $\{F_1, F_2\}$ with $\#F_1 = \#F_2$, we fix a preference "F_1 above F_2" and and we let $\mathcal{P}$ denote the set of all pairs (F_1, F_2) obtained this way. Note that $r \ge r_1 = 2$, so $\mathcal{P} \ne \emptyset$. For each $(F_1, F_2) \in \mathcal{P}$ we consider an open half-space $O = O(F_1, F_2)$ with

$$F_1 \subseteq O; \quad \overline{O} \cap co(F_2) = \emptyset.$$

Let $H(F_1, F_2)$ be the corresponding boundary hyperplane. The convex sets $\overline{O}(F_1, F_2)$ and $X \setminus O(F_1, F_2)$ for $(F_1, F_2) \in \mathcal{P}$ meet two by two. Indeed, the two sets corresponding to a single pair (F_1, F_2) meet in $H(F_1, F_2)$, whereas for distinct pairs (F_1, F_2) and (G_1, G_2) the sets F_i and G_j intersect for all combinations of i, j. As $h(X) \le 2$, we conclude that for each collection $\mathcal{P}' \subseteq \mathcal{P}$ and for each pair $(F_1, F_2) \in \mathcal{P} \setminus \mathcal{P}'$,

$$\cap \{H(G_1, G_2) \mid \{G_1, G_2\} \in \mathcal{P}'\} \cap P(F_1, F_2) \ne \emptyset,$$

where $P(F_1, F_2)$ denotes any one of $\overline{O}(F_1, F_2)$, $X \setminus O(F_1, F_2)$. By CPP, if $O(F_1, F_2)$ meets a convex set C then its relative boundary in C equals $H(F_1, F_2) \cap C$. Repeated use of this fact, in combination with Proposition 4.6, yields that

$$0 \le cind\bigl(\cap\{H(F_1,F_2)\mid (F_1,F_2)\in \mathcal{P}\}\bigr)\le n - \#\mathcal{P}.$$

Consequently,

$$n \ge \#\mathcal{P} = \tfrac{1}{2}\binom{r}{\tfrac{1}{2}r},$$

showing that $r \le r_n$. Combining this with Theorem 5.4, we find that $r = r_n$.

(ii). r is odd, say: $r = 2p + 1$ (where $p \ge 1$). Consider an R-independent set $F = \{x, x_1,\ldots, x_{2p}\}$. This time, $\mathcal{P}$ denotes the collection of all pairs (F_1, F_2), where $\#F_1 = p$, $\#F_2 = p+1$, and $x \in F_1$. For each pair $(F_1, F_2) \in \mathcal{P}$, we also consider an open set $O = O(F_1, F_2)$ with

$$F_1 \subseteq O; \quad O \cap co(F_2) = \emptyset.$$

If (F_1, F_2) and (G_1, G_2) are distinct members of $\mathcal{P}$, then each of the sets F_i and G_j meet for each combination of i, j. Arguing as before, we find

$$0 \le cind\bigl(\cap\{H(F_1,F_2)\mid \{F_1,F_2\}\in \mathcal{P}\}\bigr)\le n - \#\mathcal{P},$$

showing that

$$n \ge \mathcal{P} = \binom{2p}{p-1} \ge \tfrac{1}{2}\binom{2p}{p}.$$

Hence $r-1 \le r_n$, whereas $r_n \le r$ by Theorem 5.4. ∎

A comparable (a-topological) result has been described in Theorem II§4.21. The slight ambiguity in the formula determining r cannot be eliminated. We will construct an example of an n-dimensional median continuum of Radon number $r_n + 1$ in each predicted dimension n.

5.6. Proposition. *Let $n > 0$ be such that r_n is even and $\binom{r_n}{\tfrac{1}{2}r_n+1} \le n$. Then there exists a median continuum of dimension n and of Radon number equal to $r_n + 1$.*

Proof. The superextension $\lambda(r)$ of the free r-point space $\{1,\ldots,r\}$ is a finite median graph of Radon number r. Its realization $|\lambda(r)|$ as a cubical polyhedron is given the unique median convexity extending the graphic convexity, and such that each cube of the complex is a standard median cube. See Theorem II§3.16. As observed in II§4.22.2, the Radon number of the realized complex $|\lambda(r)|$ equals r. Its dimension equals the maximum dimension of a solid cube, which is determined by the maximum (combinatorial) dimension of a graphic cube. This maximum is one less than the exchange number of $\lambda(r)$ by Theorem II§4.19. Hence, by formula (1) in II§4.20.2, we obtain

$$cind(|\lambda(r)|) = \binom{r-1}{\lceil \tfrac{1}{2}r \rceil}.$$

Let $s \ge 2$ be even and

$$n(s) = \binom{s}{\tfrac{1}{2}s+1}; \quad E_s = \{n \mid r_n = s; n(s) \le n\}.$$

The members of E_s are the exceptional dimensions corresponding with the Radon number

§5: Dimension and Convex Invariants   473

Table 5.1: Exceptional dimensions

s	E_s
2	1
4	4
6	15, 16, 17
8	56, 57, .. , 62
10	210, 211, .. , 230
12	792, 793, .. , 857

$r_n = s$ (cf. Table 5.1). Let $n \in E_s$, let $k = n - n(s)$, and consider the space

$$X = |\lambda(s+1)| \times [0,1]^k.$$

The dimension of $|\lambda(s+1)|$ equals

$$\left\lceil \frac{s}{\lceil \frac{1}{2}(s+1)\rceil} \right\rceil = \left\lceil \frac{s}{\frac{1}{2}s+1} \right\rceil = n(s),$$

showing that X is n-dimensional. In regard to the first factor, the Radon number of X is at least $s+1$. By Theorem 5.5, $r(X)$ cannot be strictly larger. ∎

The determination of the invariants of Carathéodory and Sierksma in terms of dimension is not so easy; additional assumptions seem to be necessary. One of them is Join-hull Commutativity; the next result (which is independent of dimension theory) partially justifies the use of it.

5.7. Proposition. *Let X be an S_2 and NS_3 space with connected convex sets and with compact polytopes. If $e(X) \le 3$ and $c(X) \le 3$, then X is JHC.*

Proof. Let $P \subseteq X$ be a polytope and $a \in X \setminus P$. We verify that if $x \in co(\{a\} \cup P)$, then $x \in ap$ for some $p \in P$. The collection of all compact convex sets $D \subseteq P$ with $x \in co(\{a\} \cup D)$ is not empty. If $(D_i)_{i \in I}$ is a chain of such sets with intersection D, and if $x \notin co(\{a\} \cup D)$, then by NS_3 there is a closed convex neighborhood N of $co(\{a\} \cup D)$ with $x \notin N$. However, $D_i \subseteq N$ for some N, whence $x \in co(\{a\} \cup D) \subseteq N$.

With the aid of Zorn's Lemma, this shows that there is a *minimal* compact convex set $D \subseteq P$ with $x \in co(\{a\} \cup D)$. Domain finiteness yields that D is a polytope, say: $D = co(F)$, where $F = \{b_1,..,b_n\}$. We consider F to be minimal as well. As $c(X) \le 3$, there is a 3–point set $G \subseteq F \cup \{a\}$ with $x \in co(G)$. Note that $a \in G$ since $x \notin D$. So $n \le 2$. If $n = 1$ we are done; assume $n = 2$. As $b_1 b_2$ is connected, there is a point $b \in b_1 b_2 \setminus \{b_1, b_2\}$. As $e(X) \le 3$, we find that

$$co\{b_1, b_2, a\} \subseteq co\{b_1, b, a\} \cup co\{b_2, b, a\} \cup co\{b_1, b_2, b\}.$$

The third summand is in D and hence it does not contain x. We have, for instance, $x \in co\{b_1, b, a\}$. By S_2, the segment $b_1 b$ is properly smaller than $b_1 b_2$, a contradiction. ∎

Two auxiliary results are needed to obtain an upper bound for c and e.

5.8. Lemma. *Let X be a closure stable FS_3 space with connected convex sets. If $S \subseteq X$ is a convex closed separator of the sets A, B, then there is a convex closed screening (C,D) of (A,B) with $C \cap D = S$.*

Proof. Let $O, P \subseteq X$ be open sets such that
$$A \subseteq O, \quad B \subseteq P, \quad O \cup P = X \setminus S.$$
If $x \in P$, then there is a half-space H_x with $S \subseteq H_x$ and $x \notin H_x$. If $X \setminus H_x$ meets O, then it meets S by the connectedness of convex sets. It follows that
$$O \cup S = \cap \{ H_x \mid x \in P \}.$$
showing that $O \cup S$ is a convex set. It is also a closed set, being the complement of P. A similar argument works for $P \cup S$. The sets $O \cup S$ and $P \cup S$ yield the desired screening. ∎

5.9. Lemma. *Let X be a tcs with connected convex sets, let $n \geq 2$, and for each $i = 1,..,n$ let (C_i, D_i) be a pair of convex closed sets such that $C_i \cup D_i = X$. If $\cap_{i=1}^n (C_i \cap D_i) = \varnothing$, then there is a choice $E_i \in \{C_i, D_i\}$ for $i = 1,..,n$ such that $\cap_{i=1}^n E_i = \varnothing$.*

Proof. Let $M_i = C_i \cap D_i$ for $i = 1,..,n$. We operate by induction on $n \geq 2$.

If $n = 2$, we have $M_2 \subseteq D_1 \setminus C_1$ or $M_2 \subseteq C_1 \setminus D_1$ since M_2 is connected. In the first case, either $C_2 \subseteq D_1 \setminus C_1$ or $D_2 \subseteq C_1 \setminus D_1$. If not, then D_2 and C_2 would each have a point in C_1. As C_1 is connected, it should contain a point of M_2. The second case is handled similarly.

As to the induction step, we have $(\cap_{i=1}^n M_i) \cap M_{n+1} = \varnothing$, and hence there is a choice of E_i among C_i and D_i for each $i = 1,..,n$, such that $(\cap_{i=1}^n E_i) \cap M_{n+1} = \varnothing$. Hence $\cap_{i=1}^n E_i \subseteq C_{n+1} \setminus D_{n+1}$ or $\cap_{i=1}^n E_i \subseteq D_{n+1} \setminus C_{n+1}$ since $\cap_{i=1}^n E_i$ is connected. In the first case, take $E_{n+1} = D_{n+1}$, and in the second case, take $E_{n+1} = C_{n+1}$. ∎

5.10. Theorem. *Let X be a closure stable FS_3 space with connected convex sets and with compact polytopes. If X is JHC, then*
$$c(X) \leq cind(X) + 1; \quad e(X) \leq cind(X) + 1.$$

Proof. We treat the exchange number first. For all finite numbers $n \geq -1$,
$$cind(X) \leq n \;\Rightarrow\; e(X) \leq n+1.$$
This is evident for $n \leq 0$; we assume $n \geq 1$. Let F be a finite set with more than $n+1$ points, and assume F is E-independent. In particular, any $(n+2)$-point subset $\{x_0, x_1,...,x_{n+1}\}$ of F is E-independent since X has the CUP (cf. II§1.16.1). In the sequel, adding one or more subscripts to F refers to removing the points with exposed labels from F. By E-independence, some "face" of $co(F)$ is not covered by the other ones, say,
$$p \in co(F_0) \setminus \bigcup_{i=1}^{n+1} co(F_i).$$
By Theorem 4.19, there exist $n+1$ convex closed sets M_i ($i = 1,..,n+1$) such that

(1) M_i separates between p and $co(F_i)$ in $co(F)$.

§5: Dimension and Convex Invariants   475

(2) $\bigcap_{i=1}^{n+1} M_i = \emptyset$.

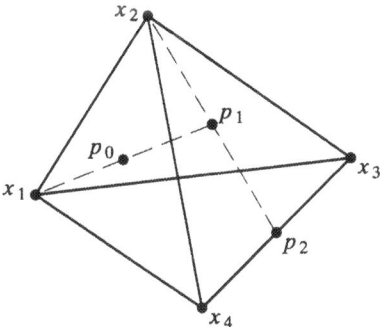

Fig. 1: Construction of the points p_j

By Lemma 5.8, there exist convex closed screenings (C_i, D_i) of $(\{p\}, co(F_i))$ with $C_i \cap D_i = M_i$. By Lemma 5.9, there is a choice

$$E_i \in \{C_i, D_i\} \quad (i = 1, .., n+1)$$

with $\bigcap_{i=1}^{n+1} E_i = \emptyset$. Define

$$I = \{i \mid E_i = C_i\}; \quad J = \{j \mid E_j = D_j\}.$$

As $\bigcap_{i=1}^{n+1} C_i$ and $\bigcap_{i=1}^{n+1} D_i$ are non-empty (consider the respective points p and x_0), we see that I and J are non-empty. Furthermore, $I \cup J = \{1, .., n+1\}$ and $I \cap J = \emptyset$. Let $J = \{1, .., k\}$ for convenience of notation. By JHC, there is a sequence of points $(p_j)_{j=0}^k$ with $p_0 = p$ and with the following properties (Fig. 1).

$$p_j \in x_{j+1} p_{j+1} \quad (j = 0, 1, .., k-1).$$
$$p_j \in co(F_{0,1,2,..,j}) \quad (j = 0, 1, .., k).$$

Observe that $p_k \in co(F_{0,1,..,k}) \subseteq \bigcap_{j \in J} D_j$. On the other hand, we have $p_0 \notin \bigcup_{i \in I} D_i$. Proceeding by induction, assume $j < k$ and $p_j \notin \bigcup_{i \in I} D_i$. Observe that $x_{j+1} \in D_i$ since $j+1 \neq i$. If $p_{j+1} \in D_i$ for some $i \in I$, then $p_j \in x_{j+1} p_{j+1} \subseteq D_i$, contradiction. We conclude that

$$p_k \in co(F) \setminus (\bigcup_{i \in I} D_i) \subseteq \bigcap_{i \in I} C_i.$$

But then p_k is common to all E_i, contradiction.

Only minor modifications of the above argument are required to obtain the result on the Carathéodory number. ∎

It is possible to obtain the conclusion of the previous result without JHC in special circumstances, as the next result illustrates. It involves the Continuous Positions Property (CPP), studied in Section 2.21.

5.11. Theorem. *Let X be a closure stable FS_3 space with compact polytopes and with the Continuous Positions Property (CPP). If $C \subseteq X$ is a connected convex set of dimension ≤ 2, then $c(C) \leq 3$ and $e(C) \leq 3$.*

Proof. By Theorem 2.23, each compact subset of C in continuous position within X is connected. As polytopes are compact, we conclude that all convex subsets of C are connected.

Let $F = \{a, a_1, ..., a_n\}$ be a subset of C with $n + 1 > 3$ elements. We will show that

(1) $\quad co(F) \subseteq \bigcup_{i=1}^{3} co(F \setminus \{a_i\})$,

to the effect that both $c \leq 3$ and $e \leq 3$. To this end, assume (1) is false and consider a point

$$x \in P = co(F) \setminus \bigcup_{i=1}^{3} co(F \setminus \{a_i\}).$$

Consider an open half-space O_1 of X, maximal with the properties

$$F \setminus \{a_1\} \subseteq O_1; \quad x \notin O_1.$$

In particular, $a_1 \notin O_1$. Note that $x \in Bd(O_1)$ by virtue of the fact that X is connected. By CPP, we even have $x' \in Bd(O_1 \cap co(F))$. By Theorem 4.17, the intersection of all dense convex subsets of $Bd(O_1 \cap co(F))$ is a dense subset. As P is a neighborhood of x, we can replace x by a point which, moreover, belongs to each dense convex subset of $Bd(O_1 \cap co(F))$. Next, we operate with a_2 and a_3 in about the same way: there exist open half-spaces O_2, O_3 of X such that

$$F \setminus \{a_i\} \subseteq O_i; \quad x \in Bd(O_i) \quad a_i \notin O_i.$$

for $i = 2, 3$ (and for $i = 1$). We define

$$E_i = Bd(O_i) \cap co(F) \quad (i = 1, 2, 3).$$

Note that $\#E_1 > 1$. Indeed, as $a_2 \in O_1$ and $a_1 \notin O_1$ there is a point u in $a_1 a_2 \cap Bd(O_1)$. Then $u \in E_1$ and u is distinct from x since u is in $co(F \setminus \{a_3\})$ and x is not. Now E_i is a relative hyperplane of $co(F)$ and as such,

$$cind(E_i) < cind(co(F)) \leq 2.$$

Hence $h(E_i) \leq 2$ by Theorem 5.3, and E_i is JHC by Proposition II§1.15. Its exchange number is is at most $cind(E_i) + 1$ by Theorem 5.10. Consequently, E_i is a tree (see the characterization in Topic II§1.27.1). We conclude from Proposition 2.13.1 that the convex subsets of E_i are precisely the connected subsets and that x (being in each dense convex subset of E_1) is not an end point of E_1.

We have two relative hyperplanes $Bd(O_2) \cap E_1$ and $Bd(O_3) \cap E_1$ of E_1. Hence both consist of one point -- necessarily x. On the other hand, as $a \in O_1$ and $a_1 \notin O_1$, we have

$$\emptyset \neq aa_1 \cap Bd(O_1) \cap co(F) \subseteq O_2 \cap O_3 \cap E_1.$$

So, the connected sets $O_2 \cap E_1$ and $O_3 \cap E_1$ intersect and have x as their common

§5: Dimension and Convex Invariants   477

boundary. This yields
$$O_2 \cap E_1 = O_3 \cap E_1.$$
By CPP,
$$E_1 \setminus Cl(O_2) = E_1 \setminus Cl(O_2 \cap E_1) = E_1 \setminus Cl(O_3 \cap E_1) = E_1 \setminus Cl(O_3),$$
and this set is non-empty since x is not an end point. We conclude that
$$Int(X \setminus O_2) \cap Int(X \setminus O_3) \cap E_1 \neq \emptyset.$$
As E_1 is the boundary of $O_1 \cap co(F)$, there is a point (Fig. 2)
$$y \in Int(X \setminus O_2) \cap Int(X \setminus O_3) \cap O_1 \cap co(F).$$

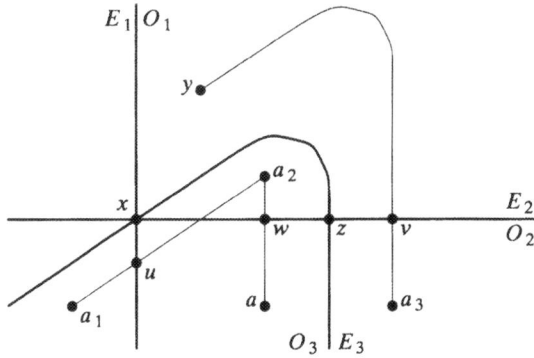

Fig. 2. Location of the various points

We have
$$ya_3 \subseteq (X \setminus O_3) \cap O_1 \cap co(F); \quad y \notin O_2; \quad a_3 \in O_2;$$
$$a_2 a \subseteq O_3 \cap O_1 \cap co(F); \quad a_2 \notin O_2; \quad a \in O_2.$$
Hence there exist points
$$v \in (X \setminus O_3) \cap O_1 \cap Bd(O_2) \cap co(F);$$
$$w \in O_3 \cap O_1 \cap Bd(O_2) \cap co(F).$$
Consequently, the segment vw is included in $O_1 \cap E_2$ and it meets E_3. We conclude that there is a point $z \in O_1 \cap E_2 \cap E_3$. Note that $z \neq x$ since $x \notin O_1$.

This leads to a contradiction as follows. As $a \in O_3 \cap co(F)$ and $a_3 \notin O_3 \cap co(F)$, the segment aa_3 meets E_3. As $aa_3 \subseteq O_2$, we see that O_2 meets E_3. Then the relative hyperplane
$$Bd(O_2) \cap E_3 = Bd(O_2) \cap co(F) \cap E_3 = E_2 \cap E_3$$
of E_3 is zero-dimensional -- hence a singleton -- whereas x, z are both in $E_2 \cap E_3$.

This forces us to conclude that formula (1) is correct. ∎

5.12. Examples

5.12.1. Consider a symmetric H-convexity in the plane. Its Carathéodory and exchange number are at most three by Theorem 5.11. Therefore, a planar symmetric H-convexity is JHC by Theorem 5.7.

5.12.2. Theorems 5.10, resp., 5.11, are not valid without the assumption of JHC resp., CPP, as the following example shows. Let the 3–cube $[0,1]^3$ be equipped with the standard median convexity, and consider the subspace X of all points on a proper face through the origin. Then X is a compact, two-dimensional FS_3 space with connected convex sets, and $c \leq 3$ as a subspace of the standard median 3–cube. On the other hand, $e = 4$ (see Topic II§3.19.3). Apparently, X satisfies neither JHC nor CPP.

5.13. Corollary. *Let X be a closure stable, JHC and FS_{3+} space with connected convex sets and with compact polytopes. If $cind(X) < \infty$, then*

$$h(X) = cind(X) + 1 \iff c(X) = cind(X) + 1.$$

Proof. Let $n = cind(X)$. Then $h(X) \leq c(X) \leq n+1$ by Theorems 5.2 and 5.10, establishing the implication from left to right. Conversely, if $c(X) = n+1$, then Sierksma's inequality $c \leq \max\{h, e-1\}$ yields that either $h(X) \geq n+1$ (which settles the result), or $e(X) - 1 \geq n+1$, contradicting that $e(X) \leq n+1$; cf. Theorem 5.10. ∎

The (standard median) Hilbert cube illustrates that this result does not hold in infinite dimensions. In the course of proving Theorem 5.4, we obtained an upper bound of $r(X)$ of type $2^{cind(X)} - 1$. The following provides a much better upper bound.

5.14. Corollary. *Let X be a closure stable, JHC and FS_{3+} space with connected convex sets and with compact polytopes. Then*

$r(X) \leq cind(X) \cdot (h(X) - 1) + 1$, *if* $h(X) \leq cind(X)$;
$r(X) \leq cind(X) \cdot (h(X) - 1) + 2$, *if* $h(X) = cind(X) + 1$.

Proof. If $cind(X) = \infty$, then $r(X) = \infty$ and $h(X) \geq 2$, so both formulas are valid. Assume $cind(X) < \infty$. If $h(X) \leq cind(X)$, then $c(X) = cind(X)$, and the first formula is a consequence of the Eckhoff-Jamison Inequality (cf. Theorem II§1.9(3)). If $h(X) = cind(X) + 1$ then $c(X) = cind(X)$, and the second formula is a consequence of the special Eckhoff-Jamison Inequality. ∎

5.15. Proposition. *Let X be a closure stable, JHC, and FS_{3+} space with connected convex sets. If X has the Ramification Property, then*

$$h(X) = r(X) = c(X) = e(X) = cind(X) + 1.$$

Proof. By Corollary 4.9, all segments of X are one-dimensional. It easily follows from this fact that all segments are decomposable:

$$\forall c \in ab: ab = ac \cup cb; \ \{c\} = ac \cap cb$$

(cf. Section I§7). Then, by Proposition II§1.3 and Theorem 5.2, all four classical

§5: Dimension and Convex Invariants   479

invariants are equal. We know already that the Helly number is at most $cind(X) + 1$. So it suffices to show that

(1) $cind(X) + 1 \leq c(X)$.

We may assume $c(X) < \infty$. Then $r(X) < \infty$ and by Theorem 5.4, we find $cind(X) < \infty$. The case $cind(X) \leq 0$ being easy, let (1) be valid in dimensions $< n$ and consider $cind(C) = n$. Let O be an open half-space with $cind(Bd(O)) = n-1$ and let $F \subseteq Bd(O)$ be a C-independent set with n points, say: $F = \{a_1,..,a_n\}$. We let $F_i = F \setminus \{a_i\}$. There is a point $u \in co(F)$ which is in no proper face $co(F_i)$ of the polytope. For each $i = 1,..,n$ there is a convex open set O_i with

$$u \in O_i; \quad O_i \cap co(F_i) = \emptyset.$$

Then $\cap_{i=1}^n O_i$ is a neighborhood of u and we find a point $t \in \cap_{i=1}^n O_i \cap O$. We verify that $F \cup \{t\}$ is C-independent.

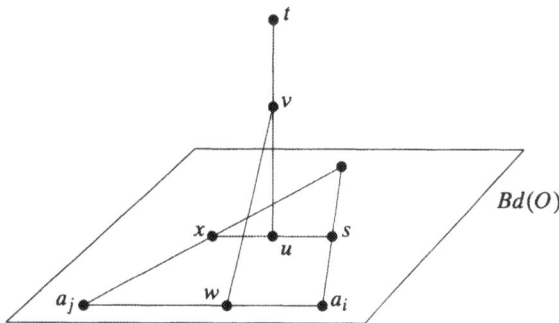

Fig. 3: Invariants under the Ramification Property

Suppose first that $u \in co(\{t\} \cup F_i)$ for some i. by JHC, there is a point $x \in co(F_i)$ such that $u \in xt$. By the Subdivision Property II§1.2, we have $co(F) = \cup_{j=1}^n co(\{x\} \cup F_j)$. So there is an index j and a point $s \in co(F_j)$ with $u \in xs$. Observe that $s \notin ut$ since $ut \subseteq \cap_{i=1}^n O_i$ and the latter is disjoint from $co(F_j)$. Also, $s \notin xu$ since $u \in xs$ and segments are geometric. Therefore,

$$s \notin xt = xu \cup ut.$$

On the other hand, $t \notin xs$ since $t \notin Bd(O)$. This contradicts the Ramification Property, showing that

(2) $u \notin \bigcup_{i=1}^n co(\{t\} \cup F_i)$.

Consider a point $v \in ut \setminus \{u,t\}$. We show that

$$v \in co(F \cup \{t\}) \setminus (\bigcup_{i=1}^n co(F_i \cup \{t\}) \cup co(F)),$$

to the effect that $F \cup \{t\}$ is C-independent. Clearly, $v \in co(F \cup \{t\})$ and $v \notin co(F)$. If $v \in co(\{t\} \cup F_k)$ for some k, then there is a point $w \in co(F_k)$ with $v \in wt$. However,

$w \notin ut$ and $u \notin wt$ by (2), contradicting the Ramification Property. ∎

We next concentrate on finding a lower bound for c and e in terms of *cind*. This will be done under the additional assumption of the Continuous Positions Property.

5.16. Lemma. *Let X be a space with connected convex sets and with the Continuous Positions Property. Let $C \subseteq X$ be convex and let $O_1,..,O_n$ be open half-spaces of X, such that for each $i = 1,..,n$, the set $\cap_{j<i} Bd(O_j) \cap C$ meets O_i and $X \setminus \overline{O}_i$. Then for each partition $\{F, G\}$ of $\{1,..,n\}$,*

$$\underset{i \in F}{\cap} O_i \cap \underset{j \in G}{\cap} Bd(O_j) \cap C \neq \emptyset.$$

Proof. We first show by induction on $p \leq n$ that

$(*_p)$ For all $i \leq p$, the set $\cap \{Bd(O_j) \mid j \leq p, j \neq i\} \cap O_i \cap C$ is not empty.

For $i = p$ this is an assumed property. so we only have to consider the case $i < p$. For $p = 1$ there is nothing left to be proved. Suppose $p > 1$ and let $(*_q)$ be valid for $q < p$. By (1) and (2) there exist points

$$x \in \underset{k<p}{\cap} Bd(O_k) \cap C \cap O_p; \quad y \in \underset{k<p}{\cap} Bd(O_k) \cap C \cap X \setminus \overline{O}_p.$$

Let $i < p$. By inductive assumption, O_i meets the convex set

$$D = \cap \{Bd(O_j) \mid j \leq p-1, j \neq i\}.$$

By CPP, this yields

$$\underset{j<p}{\cap} Bd(O_j) \cap C = Bd(O_i) \cap D = Cl_D(O_i \cap D) \setminus O_i.$$

Note that x and y are in the left hand set. Since O_p, resp., $X \setminus \overline{O}_p$, is a neighborhood of x, resp., y, we obtain two points

$$x' \in O_p \cap O_i \cap D; \quad y' \in X \setminus \overline{O}_p \cap O_i \cap D.$$

Since $x'y' \subseteq O_i \cap D$ and since the segment $x'y'$ is connected, there is a point $z \in x'y' \cap Bd(O_p)$, as required in $(*_p)$.

Having completed the induction, we conclude that $(*_n)$ holds, which is the conclusion of the lemma in case $\#F = 1$. Let $\#F = p > 1$ and suppose the conclusion holds for sets with less than p points. Take $k \in F$ and consider the sets

$$F' = F \setminus \{k\}; \quad G' = G \cup \{k\}.$$

By assumption, there are points

$$x \in \underset{i \in F'}{\cap} O_i \cap \underset{j \in G'}{\cap} Bd(O_j) \quad \text{and} \quad y \in O_k \cap \underset{l \neq k}{\cap} Bd(O_l) \cap C.$$

Then the segment xy is included in $\cap_{j \in G} Bd(O_j)$, whereas $x \in \overline{O}_k$ and $y \in O_k$. Hence by CPP, $x \in Cl_X(xy \cap O_k)$. As $\cap_{i \in F'} O_i$ is a neighborhood of x, we find

$$\emptyset \neq \underset{i \in F'}{\cap} O_i \cap xy \cap O_k \subseteq \underset{i \in F}{\cap} O_i \cap \underset{j \in G}{\cap} Bd(O_j). \quad ∎$$

§5: Dimension and Convex Invariants

5.17. Theorem. *Let X be a closure stable FS_3 space with compact polytopes and with the Continuous Positions Property. Let $Y \subseteq X$ be a connected convex subset. Then*
$$cind(Y) \leq c(Y); \quad cind(Y) + 1 \leq e(Y).$$

Proof. We will verify the following statements.
$$n \leq cind(Y) \Rightarrow n \leq c(Y); \quad n \leq cind(Y) \Rightarrow n+1 \leq e(Y).$$
Both are clear if $n \leq 1$; we assume $n > 1$. Let $\mathcal{O}$ denote the collection of all half-spaces $O \subseteq X$ such that $O \cap Y \neq \emptyset$ and $Y \not\subseteq O$. As a consequence of CPP, a set of type $Bd(O) \cap Y$, where $O \in \mathcal{O}$, is the relative boundary of $O \cap Y$ in Y. As X is FS_3, the collection of such boundaries separates relatively convex closed sets from points in Y. As $cind(Y) \geq n$, there is $O_1 \in \mathcal{O}$ such that
$$cind(Bd(O_1) \cap Y) \geq n-1 \geq 0.$$
If $n-1 > 0$, then this procedure can be repeated on the connected convex subspace $Y_1 = Bd(O_1) \cap Y$. This leads to a sequence $O_1,..,O_n$ of open half-spaces in X, such that for each $i \leq n$,
$$\bigcap_{j<i} Bd(O_j) \cap Y \cap O_i \neq \emptyset; \quad \bigcap_{j<i} Bd(O_j) \cap Y \not\subseteq \overline{O}_i;$$
$$cind(\bigcap_{j \leq i} Bd(O_j) \cap Y) \geq n-i \geq 0.$$
In particular, there is a point u in $\bigcap_{i=1}^{n} Bd(O_i) \cap Y$. By Lemma 5.16, there exists a point x_i in $\bigcap_{j \neq i} Bd(O_j) \cap Y \cap O_i$ for each $i = 1,..,n$. Consider the sets
$$F = \{x_1,..,x_n\}; \quad F_i = F \setminus \{x_i\}.$$
We have $F \subseteq Y$ and $co(F_i) \subseteq co(\{u\} \cup F_i) \subseteq Bd(O_i)$ for each i. As O_i meets $co(F)$ (e.g., in x_i), CPP yields that
$$co(F) = \overline{O}_i \cap co(F) = Cl(O_i \cap co(F)).$$
Consequently, $co(F_i)$ is a closed and nowhere dense subset of $co(F)$. It follows that $\bigcup_{i=1}^{n} co(F_i)$ is a proper subset of $co(F)$: the set F is C-independent. On the other hand, $\bigcap_{i=1}^{n} O_i \cap co(F)$ is a dense subset of $co(F)$ disjoint from $co(\{u\} \cup F_i)$. It follows that $co(F) \not\subseteq \bigcup_{i=1}^{n} co(\{u\} \cup F_i)$: the set $\{u\} \cup F$ is E-independent. Summarizing, we have $c(Y) \geq n$ and $e(Y) \geq n+1$. ∎

In Section II§4 we considered a few other invariants. One of them is the rank, which is loosely connected with the dimension of a space. Instead, it is closely connected with the dimension of the convex hyperspace. Let us verify these facts.

5.18. Proposition. *For a closure stable FS_3 space X with connected convex sets,*
$$d(X) \geq 2 \cdot cind(X).$$

Proof. If $cind(X) \geq n \geq 0$, then by Theorem 4.13 there is a CP map of X onto $[0,1]^n$. The rank of the n-cube equals $2n$ (cf. II§4.14.5), and surjective CP functions do not raise rank (cf. II§4.34). ∎

5.19. Theorem. *Let X be a uniform S_4 space with connected convex sets and with compact polytopes. Then*

$$cind(\mathcal{IC}_{comp}(X)) = d(X).$$

Proof. We first verify that

$$cind(\mathcal{IC}_{comp}(X)) \leq d(X).$$

We may assume that $d(X) < \infty$. Then the convex closure of the union of two compact convex sets is compact (cf. Topic 5.22.1), which is a necessary condition in the discussion of hyperspace properties presented in 3.10. We will show for each n ($-1 \leq n < \infty$) that

(*) $\quad n \leq cind(\mathcal{IC}_{comp}(X))$ implies $n \leq d(X)$.

This is clear if $n \leq 0$. Let $n > 0$ and $n \leq cind(\mathcal{IC}_{comp}(X))$. There is a CP map f of $cind(\mathcal{IC}_{comp}(X))$ onto $[0,1]^n$. For each $i = 1,..,n$ we have two closed half-spaces of $\mathcal{IC}_{comp}(X)$,

$$\mathcal{H}_i = f^{-1}\{y \mid \pi_i(y) = 0\}; \quad \mathcal{H}'_i = f^{-1}\{y \mid \pi_i(y) = 1\},$$

where π_i denotes the ith coordinate projection. By Lemma 2.25.2, each closed half-space of $\mathcal{IC}_{comp}(X))$ is of type

(i) $<H>$ or (ii) $<H,X>$,

where H is a closed half-space of X. Note that two half-spaces of type (ii) intersect. Therefore, $\mathcal{H}_i$ and $\mathcal{H}'_i$ cannot both have the second representation. Without loss of generality, $\mathcal{H}_i = <H_i> \cap \mathcal{IC}_{comp}(X)$ for some closed half-space H_i of X; we will no longer make use of the other half-space $\mathcal{H}'_i$. As f is surjective, for each $i = 1,..,n$ there is a compact convex set D_i such that

$$\pi_i f(D_i) = 1; \quad \pi_j f(D_i) = 0 \quad (j \neq i).$$

Consequently, $D_i \subseteq \cap_{j \neq i} H_j$ and $D_i \not\subseteq H_i$. Choosing $x_i \in D_i \setminus H_i$ for $i = 1,..,n$ yields an independent n-point set, showing that $n \leq d(X)$, establishing (*).

To prove the remaining inequality

$$d(X) \leq cind(\mathcal{IC}_{comp}(X)),$$

we verify for each n ($-1 \leq n < \infty$) that

(**) $\quad d(X) \geq n$ implies $cind(\mathcal{IC}_{comp}(X)) \geq n$.

We argue as follows ($n > 0$). Consider a polytope of X, spanned by n independent points. This yields a compact convex set $C \subseteq X$ with $d(C) \geq n$ and $cind(\mathcal{IC}_{comp}(X))$ is at least equal to $cind(\mathcal{IC}_{comp}(C))$. So, we only have to check that $n \leq cind(\mathcal{IC}_{comp}(C))$: for the remainder of the proof, we assume that X is a non-empty compact space of rank $\geq n$.

The result is evident if X is a one-point space. Let $cind(X) \geq 1$; in particular, $d(X) \geq 2$ by Proposition 5.18. So, we may assume $n \geq 2$. Let $x_1,..,x_n$ be n independent points of X. For each $i = 1,..,n$ there is a closed half-space $H_i \subseteq X$ such that

$$x_i \notin H_i; \quad co\{x_j \mid j \neq i\} \subseteq Int(H_i).$$

There is a convex neighborhood O_i of x_i, included in $\cap_{j \neq i} Int(H_j) \setminus H_i$. For each

§5: Dimension and Convex Invariants   483

$i = 1,..,n$, we consider a CP map $f_i: X \to [0,1]$ such that
$$f_i(x_i) = 0; \quad f_i(H_i) = \{1\}.$$
We can take care that $\sup f_i(O_i) = 1$. Consider the mapping
$$f: \mathcal{IC}_{comp}(X) \to [0,1]^n, \quad f(C) \mapsto (\inf f_1(C),.., \inf f_n(C)).$$
Each component mapping $C \mapsto \inf f_i(C)$ is continuous and CP (cf. III§4.21.3). Hence the mapping f is CP. If $(t_1,..,t_n) \in [0,1)^n$, then for each i we can find a point $u_i \in O_i$ such that $f_i(u_i) = t_i$. For each $j \neq i$ we have $u_i \in Int(H_j)$ and hence $f_j(u_i) = 1$. Therefore, f maps the "element" $co(\{u_1,..,u_n\})$ to $(t_1,..,t_n)$. This shows that f is surjective. By Theorem 4.13, we conclude that $cind(\mathcal{IC}_{comp}(X)) \geq n$, completing the proof. ∎

Under the assumptions of the last theorem, we can conclude that the dimension of a convex hyperspace is at least twice the dimension of the basic space.

Further Topics

5.20. A counterexample (van de Vel [1983g]). The next example illustrates the sharpness of Proposition 5.7. Consider the plane with an H-convexity which is symmetrically generated by the two coordinate projections f_1 and f_2, together with their sum f_0. Let
$$O_0 = f_0^{-1}(¼, \infty); \quad O_i = f_i^{-1}(0, \infty) \quad (i = 1, 2).$$
For each $i = 0, 1, 2$ this gives an open half-space
$$\mathcal{O}_i = <O_i> \cap \mathcal{IC}_{comp}(\mathbb{R}^2)$$
of the plane's convex hyperspace.

Let $X = \cap_{i=0}^{2} f_i^{-1}([0,1])$. Prove that the subspace
$$Bd(\mathcal{O}_0) \cap Bd(\mathcal{O}_1) \cap Bd(\mathcal{O}_2) \cap <X>$$
of $\mathcal{IC}_{comp}(\mathbb{R}^2)$ is three-dimensional, that its Carathéodory number equals 3, that its exchange number equals 4, and that it is not JHC. Hints. Use the equality of rank with convex hyperspace dimension, Theorem 5.19, in combination with the results on the dimension of a hyperplane, Proposition 4.7. To see that $c \leq 3$, use a characterization of closed half-spaces in convex hyperspaces. For non-JHC, use Topic I§2.26.6.

5.21. Spaces of arcs (van de Vel [1988]; compare II§3.2). Let S be a compact, arc-wise connected, and locally convex meet semilattice. Suppose S has a largest element and that S -- regarded as a lattice -- is distributive. Prove that the Helly number of the space of all maximal arcs in S equals the breadth of S. Hint: if $b(S) = n$, then there exists (meet) semilattice homomorphisms of S onto the n-cube and of a generalized arc-wise connected n-cube onto S. Use this to reduce the problem to an n-cube.

5.22. Rank and generating degree

5.22.1. (van de Vel [1984f]) Let X be an NS_3 space of finite rank d. Show that each compact convex set is a d-polytope. Hint: use the polytope characterization of I§1.6. Somewhat stronger results have been obtained in the above cited paper.

5.22.2. Note that the generating degree of a tcs need not be equal to the width of the family of all closed half-spaces, even if the space is compact and FS_4: consider a finite discrete space.

Problem. Under which circumstances is the width of the collection of all closed half-spaces of a tcs X equal to $gen(X)$? A reasonable guess is, that the space be connected, closure stable, FS_3, and satisfies CPP.

5.22.3. (van de Vel [1984f]) Let X be a closure stable and FS_3 space, with connected convex sets and with compact polytopes. Prove that if $gen(X) < \infty$, then X has the weak topology and a set $A \subseteq X$ is closed iff $A \cap P$ is closed in P for each polytope P of X.

5.22.4. Let G be a finite median graph. Show that G and its realization $|G|$ have the same generating degree and the same directional degree. Show that the rank and the generating degree of $|\lambda(n)|$ are equal. See II§4.20.2 and Topic II§4.37 for the computation of $gen(\lambda(n))$ and $d(\lambda(n))$.

5.22.5. Find conditions under which the rank and the generating degree of a space are equal. No counterexample is known to us in the class of FS_3 spaces with connected convex sets and with compact polytopes.

5.23. Problems

5.23.1. The upper bound for the Radon number, obtained in Corollary 5.14, is obtained with the aid of the (special) Eckhoff-Jamison inequality. Even in the context of set-theoretic convexity, this inequality is not known to be sharp. Is it true that (under conditions as in Corollary 5.14) $r \leq cind + 1$?

5.23.2. A lower bound for the Carathéodory and exchange number has been obtained under the additional assumption of the Continuous Positions Property; cf. Theorem 5.17. Can this condition be removed?

5.24. Metrizability of median spaces (van de Vel [1986b]).

Let X be a connected, finite-dimensional median space with compact polytopes. Assume further that X carries the weak topology. Prove that the topological weight and the density of X are equal. In particular, X is (topologically) metrizable iff it is separable. The result applies to compact, connected and distributive lattices of finite dimension (compare Stralka [1970]) as well as to connected and locally connected trees (compare Eberhart [1969]).

Hints. Concentrate first on (compact) polytopes P. Use the theory on directional and generating degree (cf. II§4) to show that P has a convex closed subbase which is the union of finitely many chains. Reduce each chain to a subcollection of cardinality equal to the density of P. Finally, embed X in a product of polytopes spanned by finite subsets of a dense set.

Notes on Section 5

Theorem 5.2, on the inequality $h \leq c$, is taken from van de Vel [1983c]. The same inequality was obtained by R. Hammer [1977] under the (a-topological) assumptions of decomposability and JHC; cf. Proposition II§1.3. Recently, Joó [1990] obtained the same conclusion for a class of convex structures on compact spaces X, where the convexity is determined by a parameterized segment operator $X \times X \times [0,1] \to X$, and where each convex closed set is the zero set of a continuous "convex" functional. Theorems 5.3, 5.4, and 5.5, on the relation between h resp., r with $cind$, are taken from the author's paper [1983c]. The sharpness of the last cited result (see Proposition 5.6) was demonstrated in our paper [1983e].

The upper bounds on c and e, presented in Theorem 5.10, were obtained by the author in [1983c] under the additional condition of CPP. We have been unable to remove this assumption from Theorem 5.17, where a lower bound of c and e is given. Proposition 5.15, on equality of the classical invariants with convex dimension, is based in part on the results of R. Hammer [1977], who showed that under certain conditions these invariants are all equal. Some of Hammer's conditions (e.g., interior stability) turned out to be superfluous.

The relationship between the rank of X and the dimension of $\mathcal{IC}_*(X)$ was studied by van de Vel in [1984d]. The present proof is new and more elementary. This result is close in nature to Lawson's theorems on the relation between breadth and dimension in certain classes of compact semilattices. See Lawson [1970] and [1971]; compare II§4.35.1.

6. Fixed Points

In this section we collect some results on fixed points and invariant subsets. The only common feature is that convexity is involved somehow. We present two results concerning metric spaces. First, an edge preserving function of a finite median graph into itself leaves some cube invariant. On the other hand, each non-expansive map of a weakly compact metric space leaves some diametrical closed set invariant. In many situations, such sets consist of only one point. Although graphic cubes are diametrical sets, the two results are formally unrelated.

Van Maaren's technique on complete crystals is an extension of the Sperner Lemma on completely labeled simplices. The proof can be transformed into an algorithm to find such crystals. A topological version of the result is included, leading directly to the classical Brouwer Fixed Point Theorem.

The Keimel-Wieczorek Theorem is a general result relating the Kakutani Fixed Point Property of a space with the same property of its polytopes. The Kakutani Property is obtained for a class of compact spaces with connected convex sets and a separation property. This leads to a Minimax Theorem and to a result on Invariant Arcs.

As in the previous sections, **all spaces are assumed point-convex.**

Our first result is on cubes which are invariant under edge preserving self-maps of a median graph. Two auxiliary results are needed.

6.1. Lemma. *Let G be a median graph.*

(1) *The median operator of G is edge preserving in each variable separately.*

(2) *If $H \subseteq G$ is connected, then the median stabilization med (H) is connected, and it is an induced median graph.*

Proof of (1). Let $a, b \in G$ and consider the function

$$f: G \to G, \quad x \mapsto m(a,b,x).$$

Application of the Five-point Transitive Law (cf. I§6.37.1) yields that

$$f(m(x,y,z)) = m(f(x), f(y), f(z)),$$

which expresses that f is median preserving. If x, y form an edge of G, then $xy = \{x,y\}$. As CP maps of median spaces map convex sets to convex sets, we conclude that $f(x)f(y) = \{f(x), f(y)\}$.

Proof of (2). The median stabilization of H is obtained as follows (cf. I§6.34). Let $H_0 = H$ and recursively $H_{n+1} = m(H_n^3)$. Then $med(H) = \cup_{n=0}^{\infty} H_n$. We verify by induction on $n \geq 0$ that H_{n+1} is connected. Let $x = m(a,b,c)$, where $a, b, c \in H_n$. It sufices to construct a path from x to a in H_{n+1}. To this end, fix a path $c_0 = c, c_1,.., c_k = a$ in H_n and define

$$x_0 = x; \quad x_i = m(a,b,c_i) \quad (i = 1,..,k).$$

By (1), this yields a path from x to $m(a,b,a) = a$ in H_{n+1}.

The underlying median graph of $med(H)$ is evidently an induced subgraph. ∎

Suppose $\pi: X \to \mathbb{N}$ is a weight function on a finite median space X. Recall that the weight of a subset Y of X is

$$\pi(Y) = \sum_{y \in Y} \pi(y).$$

The *weighted median of* (X, π) is the set

$$M(X, \pi) = \bigcap \{H \mid H \subseteq X \text{ a half-space, } \pi(H) > \pi(X \setminus H)\}.$$

6.2. Lemma. *Let π be a weight function on a finite median space X. Then the weighted median $M(X, \pi)$ of X is a graphic cube. If X has odd weight, then $M(X, \pi)$ is a singleton.*

Proof. It is clear that two sets $Y_1, Y_2 \subseteq X$ such that

$$\pi(Y_1) > \pi(X \setminus Y_1); \quad \pi(Y_2) > \pi(X \setminus Y_2),$$

have to intersect. As the Helly number of X is 2, we conclude that $M(X, \pi)$ is a non-empty convex set. Let $H_1, H_2 \subseteq M(X, \pi)$ be disjoint non-empty relative half-spaces. Let $p: X \to M(X, \pi)$ be the gate map and consider the half-spaces

$$\overline{H}_i = p^{-1}(H_i) \quad (i = 1, 2).$$

As both sets meet the weighted median without containing it, we find that

$$\pi(\overline{H}_1) = \pi(\overline{H}_2) = \tfrac{1}{2}\pi(X).$$

As each point has a positive weight, it follows that $\overline{H}_1 \cup \overline{H}_2 = X$.

This shows that two disjoint non-empty half-spaces of $M(X, \pi)$ cover this set. By I§6.35.3, $M(X, \pi)$ is a graphic cube. If the total weight of X is odd, then no half-space can have the same weight as its complement. The previous argument then shows that the weighted median consists of one point. ∎

6.3. Theorem (*Invariant Cube Theorem*). *Let G be a finite median graph and let $f: G \to G$ be an edge preserving function. Then there is a graphic cube of G which is mapped isomorphically onto itself by f. If the number of vertices of G is odd, then f has a fixed point.*

Proof. Take a point x at minimal distance $d(x, f(x))$ of its image $f(x)$. Since f is non-expansive, we find

$$\forall n \in \mathbb{N}: d(f^n(x), f^{n+1}(x)) = d(x, f(x)).$$

Since G is finite, the sequence $(f^n(x))_{n=0}^{\infty}$ has a period k. Without loss of generality, $f^k(x) = x$. Choose a geodesic α connecting x with $f(x)$. Then $f^k(\alpha)$ is another geodesic connecting these points. Hence we may assume $f^{kl}(\alpha) = \alpha$ for some $l > 0$. Let

$$Y = \cup_{j=1}^{kl} f^j(\alpha).$$

Note that Y is connected and that

(1) $f(Y) \subseteq Y$,
(2) f^{kl} is the identity map on Y.

The first statement is obvious. As to (2), $f^{kl}(\alpha) = \alpha$ and f^{kl} preserves edges, which shows that f^{kl} is the identity map on α.

We next consider the median subalgebra $\hat{Y}$ generated by Y. Explicitly, if $Y_0 = Y$ and (recursively) $Y_{n+1} = m(Y_n^3)$, then $\hat{Y} = \cup_{i=1}^{\infty} Y_n$. Suppose that for some $n \in \mathbb{N}$ we have

(3) $f(Y_n) \subseteq Y_n$.
(4) f^{kl} is the identity map on Y_n.

Let $x \in Y_{n+1}$, say: $x = m(a_1, a_2, a_3)$ with $a_1, a_2, a_3 \in Y_n$. Then

(5) $\quad \forall i \neq j \in \{1,2,3\}: d(a_i, x) + d(x, a_j) = d(a_i, a_j)$.

By (4), and as f^{kl} preserves edges, we find

$$d(a_i, f^{kl}(x)) + d(f^{kl}(x), a_j) = d(f^{kl}(a_i), f^{kl}(x)) + d(f^{kl}(x), f^{kl}(a_j))$$
$$\leq d(a_i, x) + d(x, a_j)$$
$$= d(a_i, a_j) = d(f^{kl}(a_i), f^{kl}(a_j)) \quad \text{(cf. (5))}.$$

By the triangle inequality we have equalities throughout. Hence

$$f^{kl}(x) = m(a_1, a_2, a_3) = x.$$

Having shown that f^{kl} is the identity on Y_{n+1} we infer that f preserves the distance between points of Y_{n+1}. By (5) we obtain

$$d(f(a_i), f(x)) + d(f(x), f(a_j)) = d(f(a_i), f(a_j)).$$

Hence

(6) $\quad f(x) = m(f(a_1), f(a_2), f(a_3))$,

which is in Y_{n+1} by (3). The induction being completed, we conclude that $f(\hat{Y}) = \hat{Y}$. By Lemma 6.1(2), $\hat{Y}$ is a median graph.

Formula (6) shows that f is median preserving on $\hat{Y}$. Since f is a bijection of $\hat{Y}$, it is even an isomorphism of the corresponding median convex structure. Assign weight 1 to each vertex, and consider the corresponding weighted median Q. Clearly, f maps Q isomorphically onto itself. By Lemma 6.2, Q is a graphic cube, and even a singleton if G has an odd number of vertices. ∎

In general, there need not be a fixed vertex, as is illustrated by the antipodal map of the n-cube ($n \geq 1$).

6.4. Some metric concepts. The next fixed point theorem requires some terminology involving metrics. Let X be a metric space with a metric ρ, let $A \subseteq X$ be non-empty, and let $a \in A$. The *diameter diam*(A) of A, the *radius of A at a*, and the *radius of A* are defined as follows.

$$\text{diam}(A) = \sup \{\rho(a_1, a_2) \mid a_1, a_2 \in A\};$$

$r_a(A) = \sup \{\rho(a,x) \mid x \in A\}$;

$r(A) = \inf \{r_a(A) \mid a \in A\}$.

A point $a \in A$ is *diametrical* provided $r_a(A) = diam(A)$. Note that for all $a \in A$,

$r(A) \le r_a(A) \le diam(A)$.

6.5. Examples.

6.5.1. Compact convex sets. Let X be a normed vector space and let $C \subseteq X$ be a compact convex set. Consider a countable dense subset $\{p_n \mid n \in \mathbb{N}\}$ of C and let $p = \sum_{n=1}^{\infty} 2^{-n} p_n$. Using an argument as in 2.29.1, it can be seen that each line segment joining a point $x \in C$, $x \ne p$, with p can be extended at p within C. In particular, the distance between p and x is strictly less than the diameter of C. We conclude that all points of C are diametrical iff C is a singleton.

6.5.2. Uniformly convex Banach spaces. In a normed vector space $(X, ||.||)$ a function $\delta: \mathbb{R}^+ \to [0,1]$ can be defined as follows.

$$\delta(\varepsilon) = \inf \{1 - \frac{||u+v||}{2} \mid ||u|| \le 1, ||v|| \le 1, ||u-v|| \ge \varepsilon\}.$$

The number $\delta(\varepsilon)$ measures how close the middle of two points in the unit disk can get to the unit sphere if the given points are at least ε away from each other. A normed vector space X is *uniformly convex* provided δ maps $(0,2]$ into $(0,1]$. Let $C \subseteq X$ be a closed, convex and bounded set of diameter $d > 0$, and let $u, v \in C$ be such that $||u-v|| > d/2$. If X is uniformly convex, then for each $x \in C$,

$$||x - \frac{u+v}{2}|| \le d(1 - \delta(\tfrac{1}{2})) < d.$$

Hence, C is diametrical iff it is a singleton. This yields a (partial) extension of the previous example.

6.5.3. Proposition. *A convex subset of a median graph is diametrical iff it is a graphic cube.*

Proof. It is evident that a cube is diametrical. We show that if a median graph G is diametrical, then it is a cube. To this end, let $a, b \in G$ be at maximal distance. Observe that if a' is a neighbor of a, then $d(a,b) \ne d(a',b)$ since G is bipartite. Hence all neighbors of a are in the interval ab. Let c be the b–infimum of these neighbors. By I§6.37.4, the segment ac is a graphic cube. Similarly, if $d \in G$ is diametrical to c, then all neighbors of c are in cd and (as a is the c–supremum of some set of c–neighbors) it follows that $a \in cd$. Then $a = d$, for otherwise some neighbor a' of a satisfies $a <_c a'$, whereas all a–neighbors are in ac. As $c \in ab$ and $d(a,c) = d(a,b)$, we conclude that $c = b$

To end the argument, we verify that $G = ab$. If this were false, some $c \in ab$ would have a neighbor $c' \notin ab$. However, if x is the antipode of c in the cube ab, then $d(c',x) = d(c,x) + 1$, a contradiction. ∎

§6: Fixed Points

We recall that a topological space is *countably compact* provided each countable open cover has a finite subcover. Equivalently, a space is countably compact iff each countable family of closed sets has a non-empty intersection provided each finite subcollection has. A function $f: X \to Y$ between metric spaces X and Y is *non-expansive* provided

$$\forall x_1, x_2 \in X: d(f(x_1), f(x_2)) \le d(x_1, x_2).$$

6.6. Lemma. *Let (M, ρ) be a non-empty bounded metric space with a convexity in which all metric disks are convex. If $f: M \to M$ is non-expansive, then for each $\varepsilon > 0$ there is a convex closed set C with $f(C) \subseteq C$ and $\operatorname{diam}(C) \le r(M) + \varepsilon$.*

Proof. We may assume that M has more than one point. Let

$$M_0 = \{x \in M \mid r_x(M) \le r(M) + \varepsilon\},$$

and consider

$$C_0 = \bigcap \{C \subseteq M \mid C \text{ convex closed}; M_0 \subseteq C; f(C) \subseteq C\}.$$

Note that the set $M_1 = M_0 \cup f(C_0)$ is included in C_0, whence $M_0 \subseteq \operatorname{co}^*(M_1) \subseteq C_0$ and $f(\operatorname{co}^*(M_1)) \subseteq f(C_0) \subseteq M_1 \subseteq \operatorname{co}^*(M_1)$. We conclude that $\operatorname{co}^*(M_1) = C_0$.

The desired convex set is defined as

$$C_1 = \{x \in C_0 \mid r_x(C_0) \le r(M) + \varepsilon\}.$$

This set is the intersection of closed disks, and hence it is convex closed. Fix a point $x \in C_1$. As $C_1 \subseteq C_0$, we find that both x, $f(x)$ are in C_0. To see that $M_1 \subseteq D(f(x), r(M) + \varepsilon)$, we have to consider two types of points. If $y \in C_0$, then

$$\rho(f(x), f(y)) \le \rho(x, y) \le r(M) + \varepsilon$$

since $x \in C_1$. If $z \in M_0$, then by the construction of M_0,

$$\rho(f(x), z) \le r(M) + \varepsilon.$$

We conclude that

$$C_0 = \operatorname{co}^*(M_1) \subseteq D(f(x), r(M) + \varepsilon),$$

and hence that $f(x) \in C_1$. By construction, $\operatorname{diam}(C_1) \le r(M) + \varepsilon$. ∎

6.7. Theorem. *Let (M, ρ) be a non-empty bounded metric space, equipped with a convexity in which all metric disks are convex. If M is weakly countably compact, and if $f: M \to M$ is a non-expansive map, then there is a non-empty closed convex set C such that $f(C) \subseteq C$ and each point of C is diametrical.*

Proof. For each non-empty convex closed set D such that $f(D) \subseteq D$, let

$$\delta_0(D) = \inf \{\operatorname{diam}(C) \mid C \text{ non-empty convex closed}; f(C) \subseteq C \subseteq D\}.$$

A decreasing sequence of non-empty convex closed sets C_n is constructed recursively as follows. Let $C_1 = M$ and (given C_n) choose $C_{n+1} \subseteq C_n$ such that

$$\operatorname{diam}(C_{n+1}) \le \delta_0(C_n) + 1/n; \quad f(C_{n+1}) \subseteq C_{n+1}.$$

The set $C = \bigcap_{n=1}^{\infty} C_n$ is convex closed, $f(C) \subseteq C$, and C is non-empty since M is weakly countably compact. Let $\varepsilon > 0$. By Lemma 6.6, there is a convex closed set $C' \subseteq C$ such that $f(C') \subseteq C'$ and $diam(C') \leq r(C) + \varepsilon$. Hence,

$$diam(C) - 1/n \leq diam(C_{n+1}) - 1/n \leq \delta_0(C_n) \leq diam(C') \leq r(C) + \varepsilon.$$

Letting $n \mapsto \infty$, we see that

$$diam(C) \leq r(C) + \varepsilon.$$

Considering that $\varepsilon > 0$ is arbitrary, we conclude that $diam(C) = r(C)$. ∎

A general method involving embedding into superextensions has been described in Topic III§4.33 to produce metrics on tcs's with the property that all metric disks are convex. We have no information, however, on the corresponding diametrical sets.

The previous theorem leads to a fixed point theorem for spaces in which singletons are the only closed convex sets consisting entirely of diametrical points. The next result provides one class of examples. For a fixed point theorem in bounded hyperconvex metric spaces, see Topic 6.26; for a discussion of (median) graphs, see Topic 6.23.

6.8. Corollary. *Let B be a uniformly convex Banach space and let C be a closed and bounded convex subset of B. Then each non-expansive map $C \to C$ has a fixed point.*

Proof. A bounded and closed convex set of a uniformly convex Banach space is weakly compact with respect to the standard convexity.[1] The disk convexity is coarser than the standard convexity, so the weak topology meant in Theorem 6.7 is compact as well. The result follows from Example 6.5.2. ∎

6.9. Multiply ordered sets, labels and crystals. We recall that a quasi-order is a reflexive, transitive relation. It is total provided $x \leq y$ or $y \leq x$ for all x, y in the domain. In this situation, two points x, y are indifferent provided $x \leq y$ and $y \leq x$; in symbols: $x \approx y$. Then $x < y$ can be defined by $x \leq y$ and not $x \approx y$, and is referred to as strict inequality.

Let X be a set, let P a non-empty index set, and for each $p \in P$ let $\leq_p$ be a total quasi-order on X. We regard the members of P as the names of the orderings, and we refer to the pair (X,P) as a *multiply ordered set*. The family P of orders is *(point) separating* provided for each pair of distinct points u, v there is $p \in P$ such that $u <_p v$.

Total quasi-orders on a set X correspond with functionals of X in the following manner. If L is a (genuine) totally ordered set and if $p: X \to L$, then the prescription "$u \leq_p v$ iff $p(u) \leq p(v)$" determines a quasi-order on X. Conversely, dividing X by the indifference relation of a total quasi-order yields a totally ordered quotient set. In this way, multiply ordered sets correspond to convex structures with a distinguished set of generating functionals.

A *crystal* in (X,P) is a pair (Q, σ) where $Q \subseteq P$ is non-empty and $\sigma: Q \to X$ is a function satisfying the following two conditions.

---

1. Köthe [1960, p. 358].

(CR-1) If $q \in Q$ then $\sigma(q)$ is a minimal element of $(\sigma(Q), \leq_q)$, that is, $\sigma(q) \leq_q x$ for all $x \in \sigma(Q)$.

(CR-2) The system of inequalities

$$\sigma(q) <_q x \quad (q \in Q)$$

has no solution $x \in X$.

The elements of $\sigma(Q)$ are the *vertices* and the elements of Q are the *faces* of the crystal. If $\#Q = k$ then (Q, σ) is referred to as a *k-crystal*.

A labeling of (X, P) is a function

$$l: X \to P;$$

the triple (X, P, l) is called a *labeled multiply ordered set*. A crystal (Q, σ) in a labeled multiply ordered set (X, P, l) is said to be complete provided $Q = l \circ \sigma(Q)$; it is said to be almost complete provided the sets Q and $l \circ \sigma(Q)$ differ in at most one point.

6.10. Theorem. *A non-empty, finite, and labeled multiply ordered set admits a complete crystal.*

Proof. Let (X, P, l) be a labeled multiply ordered space. For each $p \in P$ we consider a total proper (i.e., antisymmetric) order extending $\leq_p$. This is done by choosing a total order on each of the indifference sets $\{x \mid a \approx x\}$ for $a \in X$. It is clear that if each quasi-order is replaced this way, there will be less crystals. Therefore, it suffices to prove the result under the assumption that all total orders in consideration are proper orders.

We consider a graph $\mathcal{C}$, the vertices of which are the almost complete crystals. An edge is drawn between (Q_1, σ_1) and (Q_2, σ_2) provided one of the following holds.

(i) $Q_1 = Q_2$, and the symmetric difference of the sets $\sigma_1(Q_1)$ and $\sigma_2(Q_2)$ consists of one point.

(ii) $\sigma_1(Q_1) = \sigma_2(Q_2)$ and the symmetric difference of the sets Q_1 and Q_2 consists of one point.

The theorem will be derived from a detailed study of the graph $\mathcal{C}$. A sequence of seven preliminary statements is required.

6.10.1. *Let $(Q_i, \sigma_i) \in \mathcal{C}$ $(i = 1, 2)$ and let $q \in Q_1 \cap Q_2$ be such that $\sigma_1(Q_1) \subseteq \sigma_2(Q_2)$ and $\sigma_2(q) \in \sigma_1(Q_1)$. Then $\sigma_1(q) = \sigma_2(q)$.*

By (CR-1), we have $\sigma_2(q) = \min(\sigma_2(Q_2), \leq_q)$. As $\sigma_2(q) \in \sigma_1(Q_1) \subseteq \sigma_2(Q_2)$, another application of (CR-1) yields the result.

6.10.2. *If $(Q, \sigma) \in \mathcal{C}$ and if σ is not injective, then exactly two faces of the crystal share a vertex and all vertex labels are distinct.*

This follows directly from two cardinality considerations: $\#\sigma(Q) < \#Q$, and $\#(Q \setminus l\sigma(Q)) \leq 1$.

6.10.3. *A 1-crystal $(\{q\}, \sigma)$ is almost complete; it is complete iff $l \circ \sigma(q) = q$. Moreover, $\sigma(q) = \max(X, \leq_q)$.*

The first part is easy; the last part follows from (CR-2).

6.10.4. *A non-complete 1-crystal has precisely one neighbor in $\mathcal{C}$.*

Let q, resp., b be the face resp., vertex of the 1-crystal. The edge definition of $\mathcal{C}$ requires that either a face, or a vertex should be added to the crystal. The latter is evidently impossible (an image cannot have more points than a domain). So any neighbor is of type $(\{q,p\}, \sigma)$, with $\sigma(p) = \sigma(q) = b$. Taking $p = l(b)$ yields one neighbor. Considering that $l(b) \neq q$ (non-completeness) and $\#Q \setminus l \circ \sigma(Q) = 1$, it is clear that $p = l(b)$ is the only solution.

6.10.5. *Two injective crystals cannot be neighbors in $\mathcal{C}$. The same goes for two non-injective crystals.*

Just observe that in an injective crystal the number of vertices equals the number of faces. To produce an edge of $\mathcal{C}$, it is allowed to change only one type at the time. A similar reasoning works in the second case.

6.10.6. *For $k \geq 2$, a non-injective k-crystal $(Q, \sigma) \in \mathcal{C}$ has precisely two neighbors.*

By 6.10.2, there is one and only one pair of faces $p_1 \neq p_2$ in Q such that $\sigma(p_1) = \sigma(p_2)$. Let (Q', σ') be a neighbor of (Q, σ). By 6.10.5, σ' is injective. Also, Q' cannot have an additional face nor can $\sigma'(Q')$ have less vertices than $\sigma(Q)$. Two possibilities remain.

Case 1: $\sigma(Q) = \sigma'(Q')$ and $Q' = Q \setminus \{p\}$ for some $p \in Q$. By 6.10.1, $\sigma' = \sigma$ on Q'. As σ' is injective, p is one of p_1, p_2. Application of (CR-2) to (Q', σ') yields that the inequalities

$$\sigma(q) <_q x \quad (q \in Q \setminus \{p\})$$

have no solution $x \in X$. Summarized, Case 1 leads to two possible neighbors, and to a condition on (Q, σ) in either case.

Case 2: $Q = Q'$ and $\sigma'(Q') = \sigma(Q) \cup \{v\}$ for some $v \in X$ not in $\sigma(Q)$, say: $v = \sigma'(p)$, where $p \in Q$. Let $w = \sigma(p)$ and let $r \in Q$ be such that $w = \sigma'(r)$. Observe that $v \neq w$ and hence that $p \neq r$. If $q \in Q \setminus \{p\}$ then $\sigma'(q) \neq \sigma'(p)$ since σ' is injective. Hence $\sigma'(q) \in \sigma(Q)$. Application of 6.10.1 yields that $\sigma = \sigma'$ on $Q \setminus \{p\}$. Therefore, we have $\sigma(r) = \sigma'(r) = \sigma(p)$, whence $\{p, r\} = \{p_1, p_2\}$ by 6.10.2. This shows that p is one of p_i and that v is a solution of the inequalities

$$\sigma(q) <_q x \quad (q \in Q \setminus \{p\}).$$

If x is any other solution, then $v <_p x$ would conflict with (CR-2) on (Q', σ'). Hence, among all solutions of the above inequalities, v is the largest w.r.t. the order $<_p$. This completely determines the neighbor. Summarized, Case 2 leads to two possibilities, each

§6: Fixed Points   495

giving rise to a condition, which is the negation of a condition met in Case 1.

We conclude that there are at most two neighbors of (Q, σ) in $\mathcal{C}$. The above information is detailed enough to obtain an effective description of two different neighbors.

6.10.7. *For $k \geq 2$, a non-complete, injective k-crystal $(Q, \sigma) \in \mathcal{C}$ has precisely two neighbors.*

We know already that a neighbor (Q', σ') is a non-injective crystal, and that either it has one extra face, or one vertex less.

Case 1: $Q' = Q \cup \{p\}$ for some $p \notin Q$. Then $\sigma(Q) = \sigma'(Q')$ and by 6.10.1, we find that $\sigma = \sigma'$ on Q. Moreover, all vertex labels involved are distinct by 6.10.2. As (Q, σ) is not complete, there exists a face in

$$Q \setminus l \circ \sigma(Q) = Q \setminus l \circ \sigma'(Q').$$

Therefore, $p \in l \circ \sigma'(Q)$ since (Q', σ') is almost complete. Note that $l \circ \sigma$ is injective. Hence p is the unique point in $l \circ \sigma(Q) \setminus Q$ and $\sigma'(p)$ is the least element of $(\sigma(Q), <_r)$. This determines (Q', σ') completely, and effectively leads to the construction of one neighbor if the labeling is injective on the vertices of (Q, σ).

Case 2: $Q = Q'$ and $\sigma'(Q') = \sigma(Q) \setminus \{\sigma(p)\}$ for some $p \in Q$. This relation between two crystals has been considered before, yielding that $\sigma = \sigma'$ on $Q \setminus \{p\}$. Note that l is injective on $\sigma'(Q')$ since otherwise the number of labels of (Q, σ') would be two less than the number of faces. We distinguish two possibilities.

(i) $\#l\sigma(Q) = \#Q - 1$. There is a unique pair of vertices $w_1, w_2 \in \sigma(Q)$ with equal label. We obtain a unique face $p_i \in Q$ with $\sigma(p_i) = w_i$. As l is injective on $\sigma'(Q')$, we infer that p is one of p_i. A neighboring crystal obtains by taking $\sigma'(p)$ equal to the smallest point of $(\sigma(Q) \setminus \{w\}, <_p)$. So we effectively obtain two neighbors. Note that Case 1 does not apply to (Q, σ).

(ii) $\#l\sigma(Q) = \#Q$. Note that l is injective on the vertices of this crystal. As (Q, σ) is not complete, there is a unique point in $Q \setminus l\sigma(Q)$. This set is a subset of $Q' \setminus l\sigma'(Q')$, which therefore does not contain $l\sigma(p)$ since (Q', σ') is almost complete. We deduce that that $l\sigma(p) \notin Q' = Q$, which determines p uniquely. We can effectively construct such a neighbor by taking $\sigma'(p)$ equal to the smallest point of $(\sigma(Q) \setminus \{\sigma'(p)\}, <_p)$. Taking into account the result of Case 1, we arrive at precisely two neighbors.

A proof of the theorem goes as follows. Start at an almost complete 1–crystal $(\{p\}, \sigma)$. If it is not a complete one, consider a maximal path of crystals in the graph $\mathcal{C}$ starting at the 1–crystal. At the first moment when p is introduced as a label, or omitted as a face, we have arrived at a complete crystal. What makes the chain stop at the end point? We consider every possible next-step, incomplete crystal; in particular, p is a face and not a label. If the crystal has been met before, then by 6.10.4 it cannot be the original 1–crystal. So it is met "on the way", and it would have at least three neighbors. If the crystal is not on the path, we could have gone one further. If there is no next step at all, then we have arrived at a 1–crystal, which by 6.10.4 cannot be the original one. But then the initial face p must have been dropped somewhere. Therefore, the only possible next step is a complete crystal. ∎

6.11. An algorithm. The argument used in the proof of Theorem 6.10 essentially shows that the following algorithm leads to a complete crystal. We assume that all orders are total and antisymmetric.

1. Take the first order p_0 and take the maximum of $(X, \leq_{p_0})$ as $\sigma(p_0)$.
2. Having obtained an almost complete crystal (Q, σ), check first if σ is injective or not.
   - **2a.** Injective crystal: check if the labeling is injective.
     - **2ai.** Injective labeling: there is precisely one label p of the crystal which is not in Q. Introduce this face at the p-smallest point of $\sigma(Q)$. If this happens to undo the last step, take the unique face q with $l\sigma(q) \notin Q$ and remove the vertex $\sigma(q)$ by moving the face q up to the q-next vertex of the crystal.
     - **2aii.** Non-injective labeling: there is precisely one pair of faces leading to the same label. Pick the oldest face $p \in Q$ (to avoid undoing a recent step) and move it up to the p-next vertex of the crystal.
   - **2b.** Non-injective crystal: Consider the unique pair of faces sharing a vertex. Pick the oldest one, p, (to avoid undoing a recent step) and move it backward to the p-largest point enclosed by the other faces. If there is no such point, remove p.
3. If a new face has been introduced or an old one has been moved up or down, classify it as "most recent".

The algorithm stops at a complete crystal. In fact, it stops at the moment when the initial face p_0 is removed, or added as a label.

The following is a topological version of the preceding result. A quasi-order on a topological space X is topological provided its graph is closed in $X \times X$. A *multiply ordered space* is a topological space with a set of topological total quasi-orders. Equivalently, all corresponding functionals of the space are continuous. We consider a *topological multilabeling* $\mathcal{L}$ on (X,P), defined to be a set-valued function $X \multimap P$ such that if $x_n \mapsto x$ and $p \in \mathcal{L}(x_n)$ for all $n \in \mathbb{N}$, then $p \in \mathcal{L}(x)$. The definition of completeness of a crystal (Q, σ) extends in the obvious way: $Q \subseteq \cup_{q \in Q} \mathcal{L}(\sigma(q))$. We recall that a topological space is *sequentially compact* provided each sequence of points has a converging subsequence.

6.12. Theorem. *Let (X,P) be a separable, sequentially compact, multiply ordered space with a finite set of orders P, and let $\mathcal{L}$ be topological multilabeling. Then (X,P) has a complete crystal.*

Proof. Consider an increasing sequence of finite sets $X_n \subseteq X$ with a dense union. Let $l: X \to P$ be a selection of $\mathcal{L}$. For each $n \in \mathbb{N}$ we obtain a labeled multiply ordered set (X_n, P, l), and by Theorem 6.10, we obtain a crystal (Q_n, σ_n). Since the available

§6: Fixed Points

number of sets $Q_n \subseteq P$ is finite, we obtain a set $Q \subseteq P$ which can be used in a subsequence. Concretely, we assume that there is a crystal (Q, σ_n) (Q constant) with $l \circ \sigma_n(Q) = Q$. Observe that $l \circ \sigma_n$ is a permutation of Q. After passing to an appropriate subsequence, we obtain the following. There is a set $Q \subseteq P$ and a bijection τ of Q, such that for each $n \in \mathbb{N}$ there is a crystal (Q, σ_n) with $l \circ \sigma_n = \tau$.

For each $q \in Q$ we have a sequence of points $\sigma_n(q)$ in X. By the sequential compactness of X, we may assume that the sequence converges to a point $\sigma(q)$. As the multi-labeling $\mathcal{L}$ is topological, we find that $\tau(q) \in \mathcal{L}(\sigma(q))$ for $q \in Q$. As the order $\leq_q$ is topological, the inequalities $\sigma_n(q) \leq_q \sigma_n(p)$ for $p \in Q$ yield $\sigma(q) \leq_q \sigma(p)$ in the limit.

Suppose $x \in X$ satisfies

$$\sigma(q) <_q x \quad (q \in Q).$$

As the set of all such x is open, by the density of $D = \cup_{n \in \mathbb{N}} X_n$ in X, there is a point $d \in D$ with

$$\sigma(q) <_q d \quad (q \in Q).$$

For large enough N, we conclude that $d \in X_N$ and

$$\sigma_N(q) <_q d \quad (q \in Q),$$

conflicting with the fact that (Q, σ_n) is a crystal. We conclude that (Q, σ) is a crystal of the multi-labeled, multiply ordered space $(X, P, \mathcal{L})$. ∎

6.13. Corollary (*Brouwer's fixed point Theorem*). *Each continuous function of the standard n-simplex Δ^n has a fixed point.*

Proof. Each point $x \in \Delta^n$ has affine coordinates $x_0, ..., x_n$, such that $x_i \geq 0$ and $\sum_{i=0}^{n} x_i = 1$. We define the ith ordering as follows.

$$x \leq_i y \quad \Leftrightarrow \quad x_i \leq y_i.$$

Let (Q, σ) be a crystal. For $i, j \in Q$ we find that $\sigma(i)_i \leq \sigma(j)_i$, whence

$$\Sigma_{i \in Q} \sigma(i)_i \leq \Sigma_{i \in Q} \sigma(j)_i \quad (j \in Q).$$

If the left hand sum is strictly smaller than 1, then a point $x \in \Delta^n$ can be found such that $x_i > \sigma(i)_i$ for all $i \in Q$, contradicting that (Q, σ) is a crystal. Therefore, $\Sigma_{i \in Q} \sigma(i)_i = 1$, whence $\sigma(i)_i = \sigma(j)_i$ for $i, j \in Q$ and $\sigma(j)_k = 0$ for $k \notin Q$. We conclude that $\sigma(Q)$ consists of one point a of Δ^n, such that $a_k = 0$ for $k \notin Q$.

Let f be a continuous function $\Delta^n \to \Delta^n$. A topological multilabeling is obtained as follows.

$$\mathcal{L}(x) = \{i \mid x \leq_i f(x)\} \quad (x \in X).$$

By Theorem 6.12, we obtain a complete crystal (Q, σ) of $(\Delta^n, \{0, 1, ..., n\}, \mathcal{L})$. We conclude that $Q \subseteq \mathcal{L}(a)$ for some $a \in \Delta^n$. So $a_i \leq f(x)_i$ for $i \in Q$, whereas $a_k = 0$ for $k \notin Q$. It follows that $a = f(a)$. ∎

The next definition is inspired by a standard topic in fixed point theory. A tcs X has the **Kakutani Property** provided each USC multifunction $X \circ \rangle X$ with convex closed value sets has a fixed point.

6.14. Theorem. *Let X be a compact subspace of an NS_3 space, and let S be a dense subset of X such that for each non-empty finite set $F \subseteq S$ the polytope $co(F)$ has the Kakutani Property. Then X has the Kakutani Property.*

Proof. Let $F: X \multimap X$ be a USC, convex closed-valued multifunction. Being a compact subspace of an NS_3 space, each open cover of X has a finite refinement consisting of (relatively) convex closed sets. So it suffices to prove that if $\mathcal{D}$ is a finite convex closed cover of X, then $F(D) \cap D \neq \emptyset$ for some $D \in \mathcal{D}$. Assume that the latter fails to hold. For each $x \in X$ we take a convex closed set $C(x)$ such that $F(x) \subseteq Int\ C(x)$ and

$$\forall D \in \mathcal{D}: D \cap F(x) = \emptyset \text{ implies } D \cap C(x) = \emptyset.$$

We also take a neighborhood $N(x) \subseteq star(x, \mathcal{D})$ of x such that $F(x') \subseteq Int\ C(x)$ for each $x' \in N(x)$. There is a finite collection $G \subseteq X$ such that the sets $N(q)$ for $q \in G$ cover X. For each $G' \subseteq G$ with $\cap_{q \in G'} Int\ C(q) \neq \emptyset$ we take a point of S in the intersection. This leads to a finite collection of points in S, the convex hull of which is P.

Consider the multifunction

$$K: P \multimap P, \quad K(x) = \cap \{C(q) \mid x \in N(q), q \in G\} \cap P.$$

Note that each value $K(x)$ is a non-empty convex closed set by construction. If $x' \in \cap \{N(q) \mid x \in N(q)\}$ then $K(x') \subseteq K(x)$, and it easily follows that K is USC. Consider a fixed point $x \in P$ of K, a point $q \in G$ with $x \in N(q)$, and a set $D \in \mathcal{D}$ such that q and x are both in D. By assumption, $D \cap F(q) = \emptyset$ and hence $D \cap C(q) = \emptyset$. However, $K(x) \subseteq C(q)$ by construction, whence x is common to D and $C(q)$, a contradiction. ∎

6.15. Theorem. *Let X be a compact Hausdorff tcs with connected convex sets. If either X is properly locally convex, closure stable and S_4, or X is FS_4, then X has the Kakutani Property.*

Proof. We aim at an application of the following result[2].

(Leray's Fixed Point Theorem for convexoid Spaces) *Let X be a compact space and let $\mathcal{C}$ be a collection of closed acyclic sets, such that the intersection of two sets in $\mathcal{C}$ is in $\mathcal{C}$, and such that each point of X has a neighborhood subbase, consisting of sets in $\mathcal{C}$. Then each USC multivalued mapping of X into itself with values in $\mathcal{C}$ has a fixed point.*

Let C be a non-empty convex closed subset of X, and let $\mathcal{U}$ be an open (in X) cover of C. Assuming X to be properly locally convex, there is a convex open cover of C refining $\mathcal{U}$. This refinement has a contractible nerve on C by Theorem 3.2(1). Assuming X to be FS_4 instead, there is a finite convex closed refinement $\mathcal{D}$ of C refining $\mathcal{U}$, such that

---

2.   Leray [1959]; see also Begle [1950]. Acyclicity is understood relative to Čech homology over a field.

$C \subseteq \bigcup \{ Int\, D \mid D \in \mathcal{S} \}$. The polyhedron $|\mathcal{S}(C)|$ is contractible by Theorem 3.2(2). In either situation, we conclude that C has trivial Čech homology, and Leray's result can be applied. ∎

We note that a Lawson semilattice is FS_4 (cf. III§4.13.2) but not necessarily closure stable. We have no example of a compact tcs which is properly locally convex, closure stable and S_4, but not FS_4. The remainder of this section is concerned with the applications of the previous result. Let X be a convex structure and let $A \subseteq X$. By a *Knaster-Kuratowski-Mazurkiewicz function* (briefly, a *KKM function*) is meant a multifunction $G: A \multimap X$ such that

$$\forall F \in 2^A_{fin}: co(F) \subseteq G(F).$$

6.16. Corollary (*KKM Function Principle*). *Let X be a compact Hausdorff tcs with connected convex sets such that either X is properly locally convex, closure stable and S_4, or X is FS_4. If $A \subseteq X$ and if $G: A \multimap X$ is a KKM function with compact value sets, then*

$$\bigcap \{ G(a) \mid a \in A \} \neq \emptyset.$$

Proof. Suppose the contrary. Then there exist finitely many points $a_1,..,a_n \in A$ such that $\bigcap_{i=1}^n G(a_i) = \emptyset$. Let $P = co\{a_1,..a_n\}$ and consider the following multifunction

$$F: P \multimap P, \quad F(x) = co\{a_i \mid x \notin G(a_i)\}.$$

Clearly, F is USC and has convex closed value sets. Hence F has a fixed point x_0 by Theorem 6.15. However, $F(x)$ is covered by sets $G(a_i)$ with $x \notin G(a_i)$. Applying this to x_0 yields a contradiction. ∎

6.17. Lemma. *Let X and Y be compact Hausdorff tcs's with connected convex sets such that either X and Y are properly locally convex, closure stable and S_4, or X and Y are FS_4. Let $A, B: X \multimap Y$ be such that*

(i) $\forall x \in X: A(x)$ *is open and* $B(x)$ *is non-empty convex.*
(ii) $\forall y \in Y: B^{-1}(y)$ *is open and* $A^{-1}(y)$ *is non-empty convex.*

Then there is a point $x_0 \in X$ with $A(x_0) \cap B(x_0) \neq \emptyset$.

Proof. The product space $Z = X \times Y$ inherits the properties assumed on X and Y. Consider the function

$$G: Z \multimap Z, \quad G(x,y) = Z \setminus (B^{-1}(y) \times A(x)).$$

All value sets of G are compact. Moreover, there is no point common to all value sets. Indeed, if $(u,v) \in Z$, then take $x \in A^{-1}(v)$ and $y \in B(u)$. We find that $(u,v) \notin G(x,y)$. By the previous corollary, G is not a KKM function. Hence there exist n points $z_i = (x_i, y_i)$ for $i = 1,..,n$, together with a point

$$w_0 = (x_0, y_0) \in co\{z_1,..,z_n\} \setminus \bigcup_{i=1}^n G(z_i)$$
$$= \bigcap_{i=1}^n (B^{-1}(y_i) \times A(x_i)) \cap (co\{x_1,..,x_n\} \times co\{y_1,..,y_n\})$$

We find that $y_i \in B(x_0)$ and $x_i \in A^{-1}(y_0)$ for all i. As $B(x_0)$ and $A^{-1}(y_0)$ are convex, we

conclude that
$$y_0 \in co\{y_1,...,y_n\} \subseteq B(x_0);$$
$$x_0 \in co\{x_1,...,x_n\} \subseteq A^{-1}(y_0).$$
Consequently, $y_0 \in A(x_0) \cap B(x_0)$. ∎

6.18. Theorem (*Minimax Theorem*). *Let X and Y be compact Hausdorff tcs's with connected convex sets such that either X and Y are properly locally convex, closure stable and S_4, or X and Y are FS_4. Let $f: X \times Y \to \mathbb{R}$ be an upper semi-continuous and upper CP function in the first variable, and a lower semi-continuous and lower CP function in the second variable. Then*
$$\max_x \min_y f(x,y) = \min_y \max_x f(x,y).$$

Proof. As the assignment $x \mapsto f(x,y)$ is upper semi-continuous, $\max_x f(x,y)$ exists for each $y \in Y$ and the functional
$$Y \to \mathbb{R}, \quad y \mapsto \max_x f(x,y)$$
is lower semi-continuous. Hence the minimum over Y is taken at some point of Y. Similarly, the maximum of $\min_y f(x,y)$ is taken at some point of X. The inequality
$$\max_x \min_y f(x,y) \leq \min_y \max_x f(x,y).$$
is clearly valid. Suppose it a strict inequality, and fix a real number r strictly in between. Define
$$A: X \multimap Y, \quad A(x) = \{y \mid f(x,y) > r\};$$
$$B: X \multimap Y, \quad B(x) = \{y \mid f(x,y) < r\}.$$
Note that
$$A^{-1}(y) = \{x \mid f(x,y) > r\};$$
$$B^{-1}(y) = \{x \mid f(x,y) < r\}.$$
Then $A(x)$ and $B^{-1}(y)$ are open since $f(x,y)$ is lower semi-continuous in y and upper semi-continuous in x, whereas $A^{-1}(y)$ and $B(x)$ are convex since $f(x,y)$ is upper CP in x and lower CP in y. The set $B(x)$ is non-empty since $\min_y f(x,y) < r$, and the set $A^{-1}(y)$ is non-empty since $r < \max_x f(x,y)$. By Lemma 6.17, there exist $x_0 \in X$ and $y_0 \in A(x_0) \cap B(x_0)$, which is impossible. ∎

We now concentrate on spaces of arcs. Let S be a compact join semilattice with universal bounds $\mathbf{0}, \mathbf{1}$. We recall that an arc in S is a totally ordered subcontinuum. The set of all arcs joining $\mathbf{0}, \mathbf{1}$ is given the relative hyperspace topology and is equipped with the convexity described in I§2.4.2. The resulting tcs is denoted by $\Lambda(S)$. See 3.14 for detailed information.

§6: Fixed Points

6.19. Lemma. *Let S, T be bounded Lawson join semilattices, and let $q: S \to T$ be a continuous surjective homomorphism. Then the function*

$$\Lambda(q): \Lambda(S) \to \Lambda(T), \quad A \mapsto q(A),$$

is well-defined and continuous, and it inverts convex sets of $\Lambda(T)$ into convex sets of $\Lambda(S)$. If all fibers of q are connected, then $\Lambda(q)$ is surjective.

Proof. The universal bounds of S and T are distinguished by the subscripts S and T, respectively. The assumptions on q imply that $q(\mathbf{0}_S) = \mathbf{0}_T$, that $q(\mathbf{1}_S) = \mathbf{1}_T$, and that an arc $A \subseteq S$ is transformed into an arc $q(A) \subseteq T$. Hence $\Lambda(q)$ is well-defined. Note that this function is a restriction of a function between topological hyperspaces,

$$\mathcal{T}_*(S) \to \mathcal{T}_*(T), \quad A \mapsto q(A),$$

which is easily seen to be continuous provided the "basic" function q is.

To see that $\Lambda(q)$ inverts convex sets into convex sets, it is sufficient to verify that

$$\Lambda(q)(\operatorname{co}\{A_1,..,A_n\}) \subseteq \operatorname{co}\{\Lambda(q)(A_1),..,\Lambda(q)(A_n)\}$$

for each finite set of arcs $A_1,..,A_n \in \Lambda(S)$. To this end, let A be an arc in between $A_1,..,A_n$. Then $A \subseteq \vee_{i=1}^n A_i$ and as q is a homomorphism, we find that $q(A) \subseteq \vee_{i=1}^n q(A_i)$, as desired.

Finally, assume that q has connected fibers. The **Monotone Mapping Theorem** in general topology[3] asserts that a closed map with connected fibers inverts connected sets to connected sets. Therefore, if $B \in \Lambda(T)$, then the compact subsemilattice $q^{-1}(B)$ of S is connected. It contains the elements $\mathbf{0}_S$ and $\mathbf{1}_S$. By Koch's Arc Theorem,[4] there is an arc $A: \mathbf{0}_S \mapsto \mathbf{1}_S$ in $q^{-1}(B)$. Evidently, $q(A) = B$. ∎

Along the same lines, it is possible to prove that if all fibers of q are connected, then $\Lambda(q)$ maps convex sets of $\Lambda(S)$ to convex sets of $\Lambda(T)$.

6.20. Theorem (*Invariant Arc Theorem*). *Let S, T be connected Lawson join semilattices with respective lower bounds $\mathbf{0}_S$, $\mathbf{0}_T$, let $q: S \to T$ be a continuous surjective homomorphism with connected fibers, and let $f: S \to T$ be a continuous function such that if $x \leq y$ in S then $f(x)$, $f(y)$ are comparable in the order of T. Suppose that there is an arc in T joining $\mathbf{0}_T$ with $f(\mathbf{0}_S)$. Then there exists an arc $A \in \Lambda(S)$ such that*

$$f(A) \vee f(\mathbf{0}_S) \subseteq q(A) \vee f(\mathbf{0}_S).$$

In particular, if $f(\mathbf{0}_S) = \mathbf{0}_T$, then $f(A) \subseteq q(A)$ for some $A \in \Lambda(S)$.

Proof. Let $T_1 = \{t \in T \mid t \geq f(\mathbf{0}_S)\}$ and consider the mapping

$$q_1: T \to T_1, \quad t \mapsto t \vee f(\mathbf{0}_S).$$

Note that T_1 is a subsemilattice of T and that q_1 is a homomorphism. As there exists an arc joining $\mathbf{0}_T$ and $f(\mathbf{0}_S)$ in T, each fiber of q_1 is connected. So the composed

---
3. Engelking [1977, Thm. 6.1.29]
4. Full citation is given in III§2.29.3.

homomorphism
$$q_1 \circ q : S \to T_1$$
is continuous, surjective, and has connected fibers. Similarly, we transform f into a mapping
$$q_1 \circ f : S \to T_1$$
preserving comparability of points. These preliminary observations show that it is sufficient to establish the result in case f maps $\mathbf{0}_S$ to $\mathbf{0}_T$.

Under this assumption, we construct an operator
$$h : \Lambda(S) \to \Lambda(T)$$
as follows. By Koch's Arc Theorem, there is an arc $A_0 \subseteq T$ joining $f(\mathbf{1}_S)$ with $\mathbf{1}_T$. For $A \in \Lambda(S)$ we define $h(A)$ to be the arc in T_1 composed of the arcs $f(A)$ and $\sup(f(A)) \vee A_0$. This h is easily seen to be well-defined and continuous.

The multivalued function
$$F : \Lambda(S) \to \Lambda(S), \quad A \mapsto \Lambda(q)^{-1} h(A),$$
is well-defined since q is surjective, it is upper semi-continuous since $\Lambda(q)$ is a closed mapping, and its value sets are convex by Lemma 6.19. By Theorem 6.15, the multifunction F has a fixed "point" $A \in \Lambda(S)$. This arc is as desired. ∎

6.21. Some remarks and consequences

6.21.1. It is essential that there be an arc joining $\mathbf{0}_T$ and $f(\mathbf{0}_S)$ in T. For instance let $S = T$ be the (join) subsemilattice of $[0,1]^2$, consisting of $[½,1] \times [0,1]$, together with the

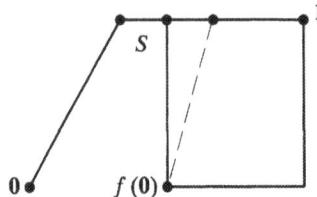

Fig. 1: A counterexample on invariant arcs

line segments joining $\mathbf{0}$ to $(^1/_3, 1)$ and $(^1/_3, 1)$ to $(½, 1)$ (Fig. 1). In addition, let q be the identity homomorphism and consider
$$f : S \to S; \quad f(x,y) = ((1+x)/2, y).$$
Then f preserves comparability of points (it is even a semilattice homomorphism), but the only arc A joining $\mathbf{0}$ and $\mathbf{1}$ is not left invariant by f.

6.21.2. Proposition. *Let $f : S \to S$ be an order preserving function, defined on a Lawson semilattice S, and let $x < y$ be two fixed points of f such that the interval $[x,y]$ is connected. Then there exists a point z such that $x < z < y$ and z, $f(z)$ are comparable.*

§6: Fixed Points 503

Proof. The function f maps $[x,y]$ into itself and preserves the lower and upper bounds of this interval. Then take z on some arc $x \mapsto y$ which is mapped into itself by f. ∎

It is not difficult to see that this result in turn allows to reprove the Invariant Arc Theorem in case $S = T$, $q = $ identity, and f is order-preserving.

6.21.3. The Invariant Arc Theorem can be applied to general continua as follows. Let Y, Z be continua and let f, $g: Y \to Z$ be maps of which g is surjective and has connected fibers. We consider the subspace $C(Y)$ of $\mathcal{T}_*(Y)$, consisting of all subcontinua. Suppose for a moment that there is a collection $\mathcal{C}$ of continua $C \subseteq Y$ with $f(C)$, $g(C)$ incomparable and such that $\mathcal{C}$ separates between the "point" $Y \in C(Y)$ and the collection of singletons of Y. Then f, g *cannot have a point of coincidence*, in other words: $f(y) \neq g(y)$ for all $y \in Y$.

To see this, suppose to the contrary that $f(p) = g(p) = q$, and consider the arc-wise connected Lawson semilattices

$$C_p(Y) = \{C \mid p \in C \in C(Y)\}; \quad C_q(Z) = \{C \mid q \in C \in C(Z)\};$$

(the union operator serves as the join). We have induced mappings

$$C(f), C(g): C_p(Y) \to C_q(Z),$$

where $C(h)(A) = h(A)$ for $h = f$, g. Both are homomorphisms. As all fibers of g are connected, the pre-image under g of a continuum is a continuum. It follows that $C(g)$ is surjective. By appealing to the Invariant Arc Theorem, we can consider an arc $\mathcal{A}$ of continua joining $\{p\}$ to Y, such that $C(f)(\mathcal{A}) \subseteq C(g)(\mathcal{A})$. The arc $\mathcal{A}$ must meet the separating family $\mathcal{C}$, yielding a $C \in \mathcal{C}$ such that $f(C)$ and $g(C)$ are comparable, a contradiction.

For an example of a collection $\mathcal{C}$, consider all subcontinua of Y of diameter r, where $0 < r < diam(Y)$. The corresponding result is somewhat non-intuitive as it actually states that the non-existence of a coincidence point (a fixed point, if $g = $ identity) can be established by inspecting the behavior of functions "in the large".

6.22. Theorem. *Let X be a compact space with a uniform S_4 convexity in which convex sets are connected, and let $F: X \multimap X$ be a LSC multifunction with convex values. For each pseudo-metric ρ on X there is a point $x \in X$ with $\rho(x, F(x)) = 0$.*

Proof. Let $q: X \to \tilde{X}$ be the topological quotient induced by the pseudo-metric ρ. By the Factorization Theorem III§3.19, there is a uniform CP quotient

$$q^*: X \to X^*$$

with the following properties.
(1) q factors through q^*, that is: there is a map $g: X^* \to \tilde{X}$ such that $g \circ q^* = q$.
(2) The weight of X^* equals the weight of $\tilde{X}$; in particular, X^* is metrizable as a topological space and hence also as a tcs.

By the Uniform Quotient Theorem III§3.17, q^* maps convex sets to convex sets as well. This leads us to an LSC convex-valued multifunction

$$X \to X^*, \quad x \mapsto q^*F(x)^-.$$

By (2) we can take a continuous selection f. Now we have two maps

$$f, q^* : X \to X^*,$$

giving rise to a USC multifunction with compact convex value sets,

$$X \multimap X, \ x \mapsto q^{*-1}(f(x)).$$

By Theorem 6.15, it has a fixed point x. By (1) we find that

$$q(x) = gq^*(x) = gf(x) \in gq^*F(x)^- = qF(x)^-,$$

which is as desired. ■

It is not known whether, in circumstances of the previous theorem, there has to be a fixed point of F. As to compact convex subsets of a locally convex vector space, it is not known how a uniform quotient looks like. A one-dimensional quotient can be a tree; cf. III§3.26.

Further Topics

6.23. Convex metric disks in a graph. Let G be a finite connected graph with geodesic distance d. We consider closed metric disks in G,

$$D(p,r) = \{x \in G \mid d(p,x) \le r\} \quad (p \in G; r > 0).$$

6.23.1. Verify that all neighborhoods $D(p, 1)$ ($p \in G$) of G are convex if and only if G has no isometric 4–cycle. Note that a median graph without 4–cycles is a tree.

6.23.2. Show that if *all* metric disks $D(p,r)$ ($p \in G$, $r \in \mathbb{N}$) are convex, then there are no isometric n–cycles except for $n = 3, 5$.

Note. Farber and Jamison [1987] showed that all metric disks about all convex sets of G are convex iff G is bridged.

6.23.3. Problem. Observe that an EP function $G \to G$ is non-expansive. If G is a median graph, then the diametric convex sets are precisely the graphic cubes. However, the preceding discussion shows that the median convexity of G will rarely include metric disks. Nevertheless, the similarities between the Invariant Cube Theorem 6.3 and Theorem 6.7 are too detailed to be pure coincidence. Is there a common generalization of both results?

6.24. Invariant simplices in dismantlable graphs (Bandelt and van de Vel [1987]; cf. Poston [1971] and Quilliot [1985]). Let G be a dismantlable graph and let $f : G \to G$ be edge preserving. Show that there is a simplex σ of G mapped isomorphically onto itself by f (*Invariant Simplex Property*). Observe that there need not be an invariant vertex.

6.25. Problems on multivalued functions. Formulate and prove an "invariant cube" property (resp., an "invariant simplex" property) for multivalued functions in a median graph (resp., a dismantlable) graph.

6.26. Hyperconvex spaces (Sine [1979], Soardi [1979]). A metric ρ (or, the corresponding metric space) is *hyperconvex* provided any family of closed disks $(D(p_i, r_i))_{i \in I}$ satisfying $\rho(p_i, p_j) \leq r_i + r_j$ for all i, j has a non-empty intersection. Verify that a hyperconvex metric space is weakly compact with respect to the disk convexity, and that each diametric convex closed set is a singleton. Deduce that a bounded hyperconvex metric space has the fixed point property for non-expansive mappings.

Note. Nachbin [1950] has characterized the (real) hyperconvex Banach spaces as function spaces of continuous real functionals on an extremally disconnected compact space.

6.27. Recursively contractible complexes (Wieczorek [1992]). A class $\mathcal{R}$ of finite simplicial complexes, called *recursively contractible complexes*, can be determined as follows.

(a) A simplicial complex, reduced to a simplex and its faces, is in $\mathcal{R}$.
(b) A simplicial complex, which is the union of two proper subcomplexes in $\mathcal{R}$ having an intersection in $\mathcal{R}$, belongs to $\mathcal{R}$.
(c) No finite simplicial complex is in $\mathcal{R}$ but those described in (a) and (b).

6.27.1. Show that the realization of a recursively contractible complex is a compact contractible space.

6.27.2. (compare 3.2) Let $\mathcal{F}$ and $\mathcal{O}$ be families of sets. Then $\mathcal{F}$ is said to screen $\mathcal{O}$ provided for each $F_0 \in \mathcal{F}$ and $O, O' \in \mathcal{O}$ with $F_0 \cap O \cap O' = \emptyset$ there exist $F, F' \in \mathcal{F}$ such that the pair $F \cap F_0, F' \cap F_0$ screens $O \cap F_0, O' \cap F_0$. On the other hand, $\mathcal{O}$ is said to penetrate $\mathcal{F}$ provided that if $F, F' \in \mathcal{F}$ and $F \cup F' \in \mathcal{F}$, then each $O \in \mathcal{O}$ meeting F and F' also meets $F \cap F'$.

If $\mathcal{F}$ and $\mathcal{O}$ are families of sets, closed under finite intersections, such that $\mathcal{F}$ screens $\mathcal{O}$ and $\mathcal{O}$ penetrates $\mathcal{F}$, then for each finite cover of $F \in \mathcal{F}$ by members of $\mathcal{O}$ the subcomplex on F is recursively contractible.

6.27.3. Let S be a simplicial complex which is recursively contractible, and let Φ be a function assigning to each simplex σ of S a recursively contractible subcomplex $\Phi(\sigma)$ of S, such that if τ is a non-empty face of σ then $\Phi(\tau) \supseteq \Phi(\sigma)$. Use the classical *Eilenberg-Montgomery Theorem* to show that Φ leaves some simplex invariant in the sense that $\sigma \in \Phi(\sigma)$ for some simplex σ.

Problem. Is there a direct (combinatorial) method to prove this invariance property? What is the relationship between recursively contractible complexes and dismantlable graphs?

6.28. Browder Fixed Point Theorem (compare Browder [1968]). Let X be a compact Hausdorff tcs with connected convex sets such that either X is properly locally convex, closure stable and S_4, or X is FS_4. Show that if $F: X \multimap X$ is a multivalued function and $F^{-1}(x)$ is open for each $x \in X$, then there is a point $x_0 \in X$ such that $x_0 \in co(F(x_0))$. The result is known as *Browder's Fixed Point Theorem*.

6.29. Helly number and Lebesgue covering dimension. Let X be a normal tcs with connected convex sets and with compact polytopes. In addition, either X is properly locally convex, closure stable and S_4, or each polytope of X is FS_4. Use the KKM Function Principle to show that $h(X) \leq dim(X) + 1$. Compare with Theorem 5.3, where *cind* is used instead of the Lebesgue dimension *dim*, and where somewhat different hypotheses are involved.

Notes on Section 6

Theorem 6.3, on invariant cubes in median graphs, is taken from Bandelt and van de Vel [1987]. It extends an earlier result of Nowakowski and Rival [1979] on invariant edges in graphic trees. Fixed point theorems for non-expansive mappings were obtained in Banach spaces by Browder [1965] and Kirk [1965]. See Corollary 6.8. The extension to general metric spaces, Theorem 6.8, is due to Kirk [1981]. A slightly weaker result was obtained before by Penot [1979].

The concept of a multiply ordered set is taken from van Maaren [1985], where the result 6.10 on completely labeled crystals is obtained. The topological version 6.12 of the result is also taken from van Maaren's paper, as is the application to the classical fixed point theorem of Brouwer, 6.13.

The Kakutani Property was defined for topological convex structures by Keimel and Wieczorek [1988]. This paper contains the elegant reduction of the property to polytopes, Theorem 6.14. Our treatment of the Minimax Theorem 6.18 via the KKM Function Principle 6.16 is fairly standard; cf. Dugundji and Granas [1982]. Komiya [1981] and Bielawski [1987] obtained a minimax theorem for convexities with a parameterization. The Invariant Arc Theorem, 6.20, and its applications are taken from the author's paper [1991].

Bibliography[1]

I. Anderson
[1987] *Combinatorics of finite sets,* Oxford Science Publications, Clarendon Press, Oxford, (1987), xv+250pp. <235>

N. Aronszajn, P. Panitchpakdi
[1956] *Extension of uniformly continuous transformations and hyperconvex metric spaces,* Pacific J. Math., 6, (1956), 405-439. <140, 141>

P. Assouad, M. Deza
[1982] *Metric subspaces of L^1,* Publications Mathematiques D' Orsay, Université de Paris-Sud, Bat. 425, 82.03, (1982), 47pp. <139>

G. Aumann
[1971] *Kontaktrelationen,* Verlag Bayer. Akad. Wiss., Math.- Natur. Kl., 1970, (1971), 67-77. <28, 44, 48>

S.P. Avann
[1948] *Ternary distributive semilattices,* Bull. Amer. Math. Soc., 54, (1948), 79.
<29, 141>
[1961] *Metric ternary distributive semi-lattices,* Proc. Amer. Math. Soc., 12, (1961), 407-414. <138, 141>

H.-J. Bandelt
[1983] *Tolerances on median algebras,* Czechoslovak Math. J., 33 (108), (1983), 344-347. <138>
[1989] *Graphs with intrinsic S_3 convexities,* J. Graph Theory, 13, (1989), 215-228.
<29, 64>

H.-J. Bandelt, V.D. Chepoi
[1991] *A Helly theorem in weakly modular space,* Preprint, (1991). <90, 112>

H.-J. Bandelt, V.D. Chepoi, M. van de Vel
[1993] *Pasch-Peano spaces and graphs,* Preprint, (1993). <88, 90, 218>

H.-J. Bandelt, A.W.M. Dress
[1992] *A canonical decomposition theory for metrics on a finite set,* Advances Math., 92, (1992), 47-105. <112>

H.-J. Bandelt, J. Hedlíková
[1983] *Median algebras,* Discrete Math., 45, (1983), 1-30. <25, 29, 63, 70, 140, 141>

H.-J. Bandelt, H.M. Mulder
[1990] *A Helly theorem for dismantlable graphs and pseudo-modular graphs,* in: Topics in Combinatorics, Physica Verlag Heidelberg, (1990), 65-71. <175>

H.-J. Bandelt, E. Pesch
[1989] *A Radon theorem for Helly graphs,* Archiv Math. (Basel), 52, (1989), 95-98.
<176>

---

1. Citation page numbers are enclosed with braces <..>

H.-J. Bandelt, M. van de Vel
[1987] *A fixed cube theorem for median graphs*, Discrete Math., 62, (1987), 129-137.
< 138, 504, 506 >
[1989] *Embedding topological median algebras in products of dendrons*, Proc. London Math. Soc., (3), 58, (1989), 439-453. < 242 >
[1991] *Superextensions and the depth of median graphs*, J. Combinatorial Theory, Series A, 57, (1991), 187-202. < 248 >
[1992] *Discrete spaces of arcs*, Preprint, (1992). < 48, 65, 69, 90, 218 >
[1992b] *The median stabilization degree of a median algebra*, Report WS-405, Vrije Universiteit Amsterdam, (1992). < 137 >

H.-J. Bandelt, M. van de Vel, E. Verheul
[1993] *Modular interval spaces*, Math. Nachrichten, to appear, (1993).
< 112, 132, 133, 134, 135, 138, 140, 141 >

Zs. Baranyai
[1975] *On the factorization of the complete uniform hypergraph*, in: Infinite and finite sets, Coll. Math. János Bolyai, 10 (Keshtely, Hungary, 1973), (1975), 91-108.
< 198, 203 >

J.-P Barthelémy
[1989] *From copair hypergraphs to median graphs with latent vertices*, Discrete Math., 76, (1989), 9-28. < 141 >

H. Bauer
[1960] *Minimalstellen von Funktionen und Extremalpunkte, II*, Archiv Math., 11, (1960), 200-205. < 433 >
[1961] *Šilovscher Rand und Dirichletsches Problem*, Ann. Inst. Fourier, Grenoble, 11, (1961), 89-136. < 433 >

P.W. Bean
[1974] *Helly and Radon type theorems in interval convexity*, Pacific J. Math., 51, (1974), 363-368. < 157, 175, 263, 264 >

G. Beer
[1983] *Approximate selections for upper semicontinuous convex valued multifunctions*, J. Approximation Theory, 39, (1983), 172-184. < 452 >

E.G. Begle
[1950] *A fixed point theorem*, Ann. of Math., 51, (1950), 544-550. < 498 >

D.E. Bell
[1977] *A theorem concerning the integer lattice*, Stud. Appl. Math., 56, (1977), 187-188. < 176 >

M. Benado
[1955] *Les ensembles partiellement ordonnés et le théorème de raffinement de Schreier. II*, Czechoslovak Math. J., 5 (80), (1955), 308-344. < 140 >

C. Berge, P. Duchet
[1975] *Une généralisation du théorème de Gilmore*, Cohiers Centre Etud. Rech. Opér., 17, (1975), 117-123. < 179 >

C. Bessaga, A. Pelczynski
[1975] *Selected Topics in infinite-dimensional Topology*, Monografie Matematyczne, PWN - Polish Scientific Publishers, Warszawa, 58, (1975), 353pp. < 67 >

R. Bielawski
[1987] *Simplicial convexity and its applications,* J. Math. Anal. Appl., 127, (1987), 155-171. <324, 451, 452, 506>

R.H. Bing
[1949] *Partitioning a set,* Bull. Amer. Math. Soc., 55, (1949), 1101-1110. <347>

G. Birkhoff
[1967] *Lattice Theory,* AMS Coll. Publ., Providence, R.I., XXV, (1967), 418pp.
<23, 28, 29, 48, 132, 133, 138, 140, 230>

G. Birkhoff, O. Frink
[1948] *Representations of lattices by sets,* Trans. Amer. Math. Soc., 64, (1948), 229-315. <28>

G. Birkhoff, S.A. Kiss
[1947] *A ternary operator in distributive lattices,* Bull. Amer. Math. Soc., 53, (1947), 749-752. <23, 29, 141>

G. Birkhoff, S. Mac Lane
[1958] *A survey of modern algebra,* MacMillan, New York, (1958), xi + 472pp.
<106>

L.M. Blumenthal
[1953] *Theory and Applications of Distance Geometry,* Clarendon Press, Oxford, (1953), 512pp. <29, 85, 157, 402, 403>

J. Ch. Boland, G. Sierksma
[1974] *The least-upper-bound for the Radon number of an Eckhoff space,* Report, University of Groningen, (1974). <202>
[1983] *On Eckhoff's conjecture for Radon numbers or how far the proof is still away,* J. Geometry, 20, (1983), 116-121. <264>

V.G. Boltyanskii
[1976] *Helly's theorem for H-convex sets,* Soviet Math. Dokl., 17, (1976), 78-81.
<249>

V.G. Boltyanskii, P.S. Soltan
[1978] *Combinatorial geometry and convexity classes,* Russian Math. Surveys, 33, (1978), 1-45. <29, 136, 249, 302>

J.A. Bondy, U.S.R. Murty
[1976] *Graph Theory with Applications,* MacMillan, London, (1976), 264pp. <197>

K. Borsuk
[1967] *Theory of retracts,* PWN - Polish Scientific Publishers, Warszawa, (1967).
<436>

O. Bottema, R.Ž. Djordjević, R.R. Janić, D.S. Mitrinović, P.M. Vasić
[1968] *Geometric Inequalities,* Wolters-Noordhoff, Groningen, the Netherlands, (1968), 151pp. <85>

D.G. Bourgin
[1943] *Linear topological spaces,* Amer. J. Math., 65, (1943), 637-659. <348>
[1952] *Restricted separation of polyhedra,* Portugaliae Math., 11, (1952), 133-136.
<70>

R.C. O'Brien
(cf. R.E. Jamison)

F.E. Browder
[1965] *Nonexpansive nonlinear operators in a Banach space,* Proc. Nat. Acad. Sc. U.S.A., 54, (1965), 1041-1044.   <506>
[1968] *The fixed point theory of multivalued mappings in topological vector spaces,* Math. Ann., 177, (1968), 283-301.   <506>

V.W. Bryant
[1974] *Independent axioms for convexity,* J. Geometry, 5, (1974), 95-99.   <158>
[1975] *Topological convexity spaces,* Proc. Edinburgh Math. Soc., 19, (1975), 125-132.   <285, 376>

V.W. Bryant, H. Perfect
[1980] *Independence theory in combinatorics,* Chapman and Hall, London, (1980).
<48>

V.W. Bryant, R.J. Webster
[1969] *Generalizations of the theorems of Radon, Helly, and Carathéodory,* Monatsh. Math., 73, (1969), 309-315.   <157, 179>
[1972] *Convexity spaces. I. The basic properties,* J. Math. Anal. Appl., 37, (1972), 206-213.   <157, 158, 376>
[1973] *Convexity spaces. II. separation,* J. Math. Anal. Appl., 43, (1973), 321-327.
<154, 158>
[1977] *Convexity spaces. III. Dimension,* J. Math. Anal. Appl., 57, (1977), 382-392.
<158, 466>

S. Burris
[1971] *The structure of closure congruences,* Coll. Math., 24(1), (1971), 3-5.   <25>
[1972] *Embedding algebraic closure spaces in 2-ary closure spaces,* Portugal. Math., 31, (1972), 183-185.   <87>

H. Busemann
[1955] *The geometry of geodesics,* Academic Press, New York, (1955).   <29, 403>

J. Calder
[1971] *Some elementary properties of interval convexities,* J. London Math Soc., (2), 3, (1971), 422-428.   <48, 90, 157, 158, 174, 179, 250, 262, 263, 264>

J. Cantwell
[1974] *Geometric convexity. I,* Bull. Inst. Math. Acad. Sinica, 2, (1974), 289-307.
<155, 158, 401>
[1978] *Geometric convexity. II: Topology,* Bull. Inst. Math. Acad. Sinica, 6, (1978), 303-311.   <371, 376, 401>

J. Cantwell, D.C. Kay
[1978] *Geometric convexity. III: Embedding,* Amer. Math. Soc., 246, (1978), 211-230.
<158, 399, 403>

C. Carathéodory
[1907] *Über den variabilitätsbereich der Koeffizienten von Potenzreihen, die gegebene Werte nicht annehmen,* Math. Ann., 64, (1907), 95-115.   <179>

J.H. Carruth
[1968] A note on partially ordered compacta, Pacific J. Math., 24, (1968), 229-231.
<347>

M. Changat, A. Vijayakumar
[1992] On order and metric convexities in $\mathbb{Z}^n$, Composition Math., 82, (1992), 57-65.
<176>

V.D. Chepoi
(cf. H.-J. Bandelt)
[1986] Geometric properties of d-convexity in bipartite graphs, Modelirovanie informacionnych sistem, (1986), 88-100. <64, 70, 88, 90>
[1986b] Some properties of domain finite convexity, Issled. po obščey algebre, geometrii, i ich priloženia, (1986), 142-148. <90>
[1991] Separation of two convex sets in convexity structures, Preprint, (1991).
<63, 70, 132, 244, 282>

G. Choquet, P.A. Meyer
[1963] Existence et unicité des représentations intégrales dans les convexes compacts quelconques, Ann. Inst. Fourier, Grenoble, 13, (1963), 139-154. <342>

M. Cochand, P. Duchet
[1983] Sous les Pavés .., Ann. of Discrete Math., 17, (1983), 191-202. <203>

R.P. Coelho
[1970] Axiomes de séparation dans les structures de convexité, Univ. Lisboa Revista Fac. Ci., A(2), 13, (1970), 83-108. <70>

H. Cohen
[1954] A cohomological definition of dimension for locally compact Hausdorff spaces, Duke Math. J., 21, (1954), 209-224. <462>

P.M. Cohn
[1965] Universal Algebra, Harper and Row, New York, (1965), xv + 333pp. <28>

H.S.M. Coxeter
[1947] Non-Euclidean Geometry, University of Toronto Press, Toronto, (1947).
<399, 403>
[1961] Introduction to geometry, Wiley & Sons, New York, (1961), 443pp.
<155, 156, 157>

D.W. Curtis
[1985] Application of a selection theorem to hyperspace contractibility, Canadian J. Math., 37, (1985), 747-759. <48, 324, 452>

L. Danzer
[1961] Über Durchschnittseigenschaften n-dimensionaler Kugelfamilien, J. Reine Angew. Math., 208, (1961), 181-203. <29>

L. Danzer, B. Grünbaum, V.L. Klee
[1963] Helly's theorem and its relatives, Proc. Symp. Pure Math., 7 (convexity), (1963), 101-180. <28>

E.B. Davies
[1967] A generalized theory of convexity, Proc. London Math. Soc., 17, (1967), 644-652. <28, 433>

M.M. Day
[1962] *Normed linear spaces*, Ergebnisse der Mathematik und ihrer Grenzgebiete, Springer, Berlin, 21, (1962). <22, 400>

E. Deak
[1966] *Eine Verallgemeinerung des Begriffs des linearen Raumes und der Konvexität*, Ann. Univ. Sc. Budapest, 9, (1966), 45-59. <285>

E. Degreef
[1981] *Some results in generalized convexity*, Dissertation, Vrije Universiteit Brussel, Brussel, Belgium, (1981), 125pp. <158, 261, 263, 264>
[1982] *Glued convexity spaces, frames, and graphs*, Compositio Math., 47, (1982), 217-222. <24, 45>

M. Deza
(cf. P. Assouad)

R.P. Dilworth
[1950] *A decomposition theorem for partially ordered sets*, Ann. of Math., 51, (1950), 161-166. <227>

R.Ž. Djordjević
(cf. O. Bottema)

J.-P. Doignon
[1973] *Convexity in cristallographic lattices*, J. Geometry, 3, (1973), 71-85. <176>
[1976] *Caractérisations d'espaces de Pasch-Peano*, Bull. Acad. Roy. Belgique, (5), 62, (1976), 679-699. <29, 155, 158, 403>

J.-P. Doignon, J.-C. Falmagne
[1985] *Spaces for the assessment of knowledge*, Internat. J. Man-Machine studies, 23, (1985), 175-196. <28>

J.-P. Doignon, J.R. Reay, G. Sierksma
[1981] *A Tverberg-type generalization of the Helly number of a convexity space*, J. Geometry, 16, (1981), 117-125. <261, 264>

J.-P. Doignon, G. Valette
[1977] *Radon partitions and a new notion of independence in affine and projective spaces*, Mathematika, 24, (1977), 86-96. <264>

C.H. Dowker
[1951] *On countably paracompact spaces*, Canadian J. Math., 3, (1951), 219-224.
<452>

A.W.M. Dress
(cf. H.-J. Bandelt)

A.W.M. Dress, R. Scharlau
[1987] *Gated sets in metric spaces*, Aequationes Math., 34, (1987), 112-120. <112>

P. Duchet
(cf. C. Berge; M. Cochand)
[1987] *Convexity in combinatorial structures*, Circ. Math. Palermo, (2) Suppl. No 14, (1987), 261-293. <28, 179>
[1988] *Convex sets in graphs, II: minimal path convexity*, J. Combinatorial Theory, Ser. B, 44, (1988), 307-316. <29, 173, 244>

P. Duchet, H. Meyniel
[1983] *Ensembles convexes dans les graphes. I. Théorèmes de Helly et de Radon pour graphes et surfaces*, Europ. J. Combinatorics, 4, (1983), 127-132.   < 29 >

J. Dugundji
[1966] *Topology*, Allyn and Bacon, Boston, (1966).   < 443 >

J. Dugundji, A. Granas
[1982] *Fixed point theory*, PWN - Polish Scientific Publishers, Warszawa, (1982).
   < 506 >

C. Eberhart
[1969] *Metrizability of trees*, Fund. Math., 65, (1969), 43-50.   < 484 >

J. Eckhoff
[1968] *Der Satz von Radon in konvexen Produktstrukturen I*, Monatsh. Math., 72, (1968), 303-314.   < 29, 201, 203 >
[1969] *Der Satz von Radon in konvexen Produktstrukturen II*, Monatsh. Math., 73, (1969), 17-30.   < 28, 202, 203 >
[1979] *Radon's theorem revisited*, in: Contributions to geometry, Proc. Geom. Symp., Siegen, 1978, Birkhäuser, Basel, (1979), 164-185.   < 28, 262, 264 >

P.H. Edelman
[1980] *Meet distributive lattices and the anti-exchange law*, Alg. Universalis, 10, (1980), 290-299.   < 45 >

P.H. Edelman, R.E. Jamison
[1985] *The theory of convex geometries*, Geometriae Dedicata, 19, (1985), 247-270.
   < 28, 45, 48, 109 >

J.W. Ellis
[1952] *A general set-separation theorem*, Duke Math. J., 19, (1952), 417-421.
   < 28, 48, 69, 90 >

R. Engelking
[1977] *General Topology*, PWN - Polish Scientific Publishers, Warszawa, (1977), 626pp.
   < 126, 219, 279, 289, 303, 305, 309, 312, 345, 347, 438, 441, 454, 460, 462, 463, 501 >

P. Erdös, D.J. Kleitman
[1970] *Extremal problems among subsets of a set*, Proc. Sec. Chapel Hill Conf. on Comb. Math. and its Appl., (1970), 146-170.   < 246 >

J.-C. Falmagne
(cf. J.-P. Doignon)

K. Fan
[1963] *On the Krein-Milman Theorem*, Proc. Symp. Pure Math., 7 (convexity), (1963), 211-221.   < 28, 433 >

M. Farber, R.E. Jamison
[1986] *Convexity in graphs and hypergraphs*, SIAM J. Algebraic Discrete Methods, 7, (1986), 433-444.   < 29, 45 >
[1987] *On local convexity in graphs*, Discrete Math., 66, (1987), 231-247.   < 504 >

C. Franchetti
[1971]   *Admissible sets and Kuratowski's number* α, Atti Accad. naz. Lincei, Rend sci. fis. mat. e nat., 50 (5), (1971), 550-554. < 22, 323 >

S.P. Franklin
[1962]   *Some results on order convexity*, Amer. Math. Monthly, 69, (1962), 357-359.
< 64, 90 >

O. Frink
(*cf.* G. Birkhoff)

B. Fuchssteiner
[1970]   *Verallgemeinerte Konvexitätsbegriffe und der Satz von Krein-Milman*, Math. Ann., 186, (1970), 149-154. < 285 >
[1971]   *Extrempunkte und Minimumsätze bei Hüllenbildungen*, Archiv Math., 22, (1971), 523-527. < 433 >

M.R. Garey, D.S. Johnson
[1977]   *The rectilinear Steiner problem is NP complete*, SIAM J. Appl. Math., 32, (1977), 826-834. < 141 >

G. Gierz, K.H. Hofmann, K. Keimel, J.D. Lawson, M. Mislove, D.S. Scott
[1980]   *A compendium of continuous lattices*, Springer Verlag, Berlin, (1980), xx+371pp. < 26, 246, 277, 285, 286, 301, 319, 320, 368, 375, 450, 452 >

J.R. Giles, D.A. Gregory, B. Sims
[1978]   *Characterization of normed linear spaces with Mazur's intersection property*, Bull. Austral. Math. Soc., 18, (1978), 105-123. < 22 >

V. Glivenko
[1936]   *Géométrie des systèmes de choses normées*, Amer. J. Math., 58, (1936), 799-828. < 141 >

S. Gottwald
[1980]   *Bemerkungen zur axiomatischen Konvexitätstheorie*, Beitrage zur Algebra und Geometrie, 10, (1980), 105-110. < 158 >

A. Granas
(*cf.* J. Dugundji)

G. Grätzer, E.T. Schmidt
[1962]   *On congruence relations of lattices*, Acta Math. Acad. Sci. Hungar., 13, (1962), 179-185. < 47 >

D.A. Gregory
(*cf.* J.R. Giles)

J. de Groot
[1969]   *Supercompactness and superextension*, in: Contributions to extension theory of topological structures, (Symp. Berlin 1967), Deutsche Verlag der Wissenschaften, Berlin, (1969), 89-90. < 29 >

B. Grünbaum
(*cf.* L. Danzer)

M.D. Guay
[1978]   *Introduction to the theory of convexity-topological spaces*, Coll. Math. Soc. János Bolyai, 23 (Topology), (1978), 521-545. < 29, 285 >

M.D. Guay and S.A. Naimpally
[1975] *Characterization of a convex subspace of a linear topological space,* Math. Japonicae, 20, (1975), 37-41. <401>
[1988] *Finite dimensional convexity topological spaces,* Bull. Math. Soc. Sci. Math. R.S. Roumanie, 32, (1988), 219-226. <283>

A. Guénoche,
[1986] *Graphical representation of a Boolean array,* Computers Humanities, 20, (1986), 277-281. <141>

P.C. Hammer
[1955] *General topology, symmetry, and convexity,* Trans. Wisc. Acad. Sci. Arts Lett., 44, (1955), 221-255. <28>
[1955b] *Maximal convex sets,* Duke Math. J., 22, (1955), 103-106. <64>
[1963] *Extended topology: domain finiteness,* Indag. Math., 25, (1963), 200-212. <28>
[1963b] *Semispaces and the topology of convexity,* Proc. Symp. Pure Math., 7 (convexity), (1963), 305-316. <28>
[1965] *Extended topology: Carathéodory's theorem on convex hulls,* Rendiconti del Circolo del Matematico di Palermo, Ser. II, 14, (1965), 34-42. <178>

R. Hammer
[1977] *Beziehungen zwischen den Sätzen von Radon, Helly und Carathéodory bei axiomatischen Konvexitäten,* Abh. Math. Sem. Univ. Hamburg, 46, (1977), 3-24. <158, 179, 216, 485>

M. Hanan
[1966] *On Steiner's problem with Rectilinear distance,* SIAM J. Appl. Math., 14, (1966), 255-265. <141>

O. Hanner
[1952] *Retraction and extension of mappings of metric and non-metric spaces,* Arkiv Mat., 2, (1952), 315-360. <446>
[1956] *Intersection of translates of convex bodies,* Math. Scand., 4, (1956), 65-87.
<136, 140, 241, 249>

W.R. Hare, G. Thompson
[1975] *Tverberg-type theorems in convex product structures,* Proceedings of a Conference held at Michigan State University, East Lansing, June 1974), Lecture Notes in Mathematics, 490, (1975), 212-217. <262, 263>

J. Hedlíková
(*cf.* H.-J. Bandelt)
[1977] *Chains in modular ternary latticoids,* Math. Slovaca, 27, (1977), 249-256.
<133>
[1983] *Ternary spaces, media and Chebyshev sets,* Czechoslovak Math. J., 33 (108), (1983), 737-389. <87, 90, 133, 140>

E. Helly
[1923] *Über Mengen konvexer Körper mit gemeinschaftlichen Punkten,* Jber. Deutsch. Mat. Verein., 32, (1923), 175-176. <179>

K.H. Hofmann
(*cf.* G. Gierz)
[1970] *A general invariant metrization theorem for compact spaces,* Fund. Math., 68,

(1970), 281-296. <347>

K.H. Hofmann, P. Mostert
[1966] *Elements of compact semigroups,* Merill, Columbus, Ohio, (1966). <285>

E. Howorka
[1981] *Betweenness in graphs,* Abstracts AMS, 2, (1981), *783-06-5. <140>

J.R. Isbell
[1980] *Median algebra,* Trans. Amer. Math. Soc., 260, (1980), 319-362.
<112, 138, 140, 141, 248>

R.E. Jamison
(cf. P.H. Edelman; M. Farber)

R.E. Jamison, R.C. O'Brien, P.D. Taylor
[1976] *On embedding a compact convex set into a locally convex vector space,* Pacific J. Math., 64, (1976), 193-205. <286>

R.E. Jamison (R.E. Jamison-Waldner)
[1974] *A general theory of convexity,* Dissertation, University of Washington, Seattle, Washington, (1974).
<28, 29, 48, 70, 84, 248, 282, 285, 298, 299, 301, 302, 348>
[1975] *A general duality between the theorems of Helly and Carathéodory,* Clemson University Technical report, 205, (1975). <178>
[1977] *Some intersection and generation properties of convex sets,* Compositio Math., 35, (1977), 147-161. <242, 249>
[1978] *Tietze's convexity theorem for semilattices and lattices,* Semigroup Forum, 15, (1978), 357-373. <29, 282>
[1980] *Copoints in antimatroids,* (Proc. 11th S.E. Conf. Comb. Graph Th. and Computing), Congressus Numerantium, 29, (1980), 535-544. <45, 64>
[1981] *Convexity and block graphs,* Congressus Numerantium, 33, (1981), 129-142.
<174>
[1981b] *Partition numbers for trees and ordered sets,* Pacific J. Math., 96, (1981), 115-140. <24, 112, 174, 175, 176, 248, 250, 261, 263, 264>
[1982] *A perspective on abstract convexity: classifying alignments by varieties,* in: Convexity and related Combinatorial Geometry, Proc. 2nd Univ. of Oklahoma Conf., Dekker, New York, (1982), 113-150. <28, 67, 68, 90, 177, 241, 246>

R.R. Janić
(cf. O. Bottema)

W. Jantosciak
(cf. W. Prenowitz)

D.S. Johnson
(cf. M.R. Garey)

I. Joó
[1990] *An inequality between the Helly and Carathéodory number (H≤C),* Ann. Univ. Sci. Budapest. Eötvös Sect. Math., 33, (1990), 143-146. <485>

S. Kakutani
[1937] *Ein Beweis des Sätzes von Edelheit über konvexe Mengen,* Proc. Imp. Acad. Tokyo, 13, (1937), 93-94. <90>

S. Kakutani, V.L. Klee
[1963] *The finite topology of a linear space,* Archiv Math., 14, (1963), 55-58.
 < 372, 376 >
E. Kamke
[1965] *Mengenlehre,* De Gruyter, Berlin, (1965), 194+31pp. < 381 >
D.C. Kay
(cf. J. Cantwell; W. Meyer)
[1977] *A nonalgebraic approach to the theory of linear topological spaces,* Geometriae Dedicata, 6, (1977), 419-433.
 < 29, 154, 157, 158, 281, 285, 298, 301, 376 >
[1977b] *Operators in convexity and a geometric solution to the linearization problem,* Preprint, (1977). < 44, 48, 88, 111, 158 >
[199*] *Dimension theory for general convexity,* Preprint, (199*). < 48 >
D.C. Kay, E.W. Womble
[1971] *Axiomatic convexity theory and the relationship between the Carathéodory, Helly and Radon numbers,* Pacific J. Math., 38, (1971), 471-485.
 < 48, 69, 179 >
K. Keimel
(cf. G. Gierz)
K. Keimel, A. Wieczorek
[1988] *Kakutani property of the polytopes implies Kakutani property of the whole space,* J. Math. Anal. Appl., 130, (1988), 97-109. < 348, 506 >
W.A. Kirk
[1965] *A fixed point theorem for mappings which do not increase distances,* Amer. Math. Monthly, 72, (1965), 1004-1006. < 506 >
[1981] *An abstract fixed point theorem for nonexpansive mappings,* Proc. Amer. Math. Soc., 82(4), (1981), 640-642. < 506 >
S.A. Kiss
(cf. G. Birkhoff)
V.L. Klee
(cf. L. Danzer; S. Kakutani)
[1951] *Convex sets in linear spaces,* Duke Math. J., 18, (1951), 443-466.
 < 282, 336, 348, 376 >
[1956] *The structure of semispaces,* Math. Scand., 4, (1956), 54-64. < 242, 249, 433 >
[1960] *Leray-Schauder theory without local convexity,* Math. Annalen, 141, (1960), 286-296. < 452 >
[1972] *Unions of increasing and intersections of decreasing sequences of convex sets,* Israel J. Math., 12, (1972), 70-78. < 28 >
D.J. Kleitman
(cf. P. Erdös)
M. Kolibiar, T. Marcisová
[1974] *On a question of J. Hashimoto,* Mat. Časopis, 24, (1974), 179-185. < 133, 138 >
K. Kołodziejczyk
[1985] *Starshapedness in convexity spaces,* Compositio Math., 56, (1985), 361-367.
 < 89 >

[1987] *Two constructional problems in aligned spaces*, Appl. Math., 19, (1987), 479-484. <216>
[1991] *Generalised Helly and Radon numbers*, Bull. Austral. Math. Soc., 43, (1991), 429-437. <260, 261>

K. Kołodziejczyk, G. Sierksma
[1990] *The semirank and Eckhoff's conjecture for Radon numbers*, Bull. Soc. Math. Belgique, 42, (1990), 383-388. <261>

H. Komiya
[1981] *Convexity on a topological space*, Fund. Math., 111, (1981), 107-113. <506>

G. Köthe
[1960] *Topologische Lineare Räume*, Springer verlag, Berlin, (1960), xii+456pp.
<50, 268, 285, 307, 492>

M. Lassak
[1975] *The dimension of Helly, Carathéodory and Radon and the metric independence of points in the theory of d-convex sets*, (1975). <179>
[1975b] *The Helly dimension for Cartesian products of metric spaces*, Mat. Issled., 10(2), (1975), 159-167. <90>
[1977] *On metric B-convexity for which diameters of any set and its hull are equal*, Bull. Acad. Pol. Sci., Sér. Sci. Math. Astr. et Phys., 25, (1977), 969-975.
<22, 29>
[1982] *Carathéodory's and Helly's dimension of products of convex structures*, Colloquium Math., 46, (1982), 215-225. <200, 202>
[1983] *The rank of product closure systems*, Archiv Math. (Basel), 40, (1983), 186-191. <245>
[1984] *Families of convex sets closed under intersections, homotheties, and uniting increasing sequences of sets*, Fund. Math., 120, (1984), 15-40. <283, 286>
[1986] *A general notion of extreme set*, Compositio Math., 57, (1986), 61-72. <433>

J.D. Lawson
(*cf.* G. Gierz)
[1970] *The relation of breadth and codimension in topological semilattices*, Duke Math. J., 37, (1970), 207-212. <466, 485>
[1971] *The relation of breadth and codimension in topological semilattices, II*, Duke Math. J., 38, (1971), 555-559. <466, 485>
[1973] *Comparison of taut cohomologies*, Aequationes Math., 9, (1973), 201-209.
<462>
[1973b] *Intrinsic topologies in topological lattices and semilattices*, Pacific J. Math., 44, (1973), 593-602. <362, 376, 449>
[1976] *Embeddings of compact convex sets and locally compact cones*, Pacific J. Math., 66, (1976), 443-453. <308>
[1979] *The duality of continuous posets*, Houston J. Math., 5, (1979), 357-386. <376>
[1980] *Algebraic conditions leading to continuous lattices*, Proc. Amer. Math. Soc., 78, (1980), 477-481. <276>

J. Leray
[1959] *Théorie des points fixes: indice total et nombre de Lefschetz*, Bull. Soc. Math. France, 87, (1959), 221-233. <498>

F.W. Levi
[1951] *On Helly's theorem and the axioms of convexity*, J. Indian Math. Soc., 15, Part A, (1951), 65-76. < 28, 154, 157, 177, 179 >

N. Lindquist, G. Sierksma
[1981] *Extensions of set partitions*, J. Combinatorial Theory, (A), 31, (1981), 190-198. < 202, 262 >

B. Lindström
[1972] *A theorem on families of sets*, J. Combinatorial Theory, (A), 13, (1972), 274-277. < 177, 262 >

W. Luxemburg, A. Zaanen
[1971] *Riesz spaces*, North-Holland Publishing Company, Amsterdam, (1971), 514pp. < 85 >

H. van Maaren
[1979] *On concepts of closure and generalizations of convexity*, Dissertation, University of Utrecht, Utrecht, Netherlands, (1979), 138pp. < 29 >
[1985] *Generalized Pivoting and Coalitions*, in: The Computation and Modelling of Economic Equilibria, North Holland, (1985), 155-176. < 506 >

S. Mac lane
(*cf.* G. Birkhoff)

P. Mah, S.A. Naimpally, J.H.M. Whitfield
[1976] *Linearization of a convexity space*, J. London Math. Soc., (2), 13, (1976), 209-214. < 400 >

T. Marcisóva
(*cf.* M. Kolibiar)

E. Marczewski
[1966] *Independence in abstract algebras; results and problems*, Colloquium Math., 14, (1966), 169-188. < 250 >

K. Menger
[1928] *Untersuchungen über allgemeine Metrik*, Math. Ann., 100, (1928), 75-163. < 29 >

P.A. Meyer
(*cf.* G. Choquet)

W. Meyer, D.C. Kay
[1973] *A convexity structure admits but one real linearization of dimension greater than one*, J. London Math. Soc., 7, (1973), 124-130. < 372, 400, 401 >

H. Meyniel
(*cf.* P. Duchet)

E. Michael
[1951] *Topologies on spaces of subsets*, Trans. Amer. Math. Soc., 71, (1951), 152-182. < 341 >
[1956] *Continuous selections I*, Ann. Math., 63 (2), (1956), 361-382. < 452 >
[1959] *Convex structures and continuous selection*, Canadian J. Math., 11, (1959), 556-575. < 28, 324, 452 >

J. van Mill

[1977] *Supercompactness and Wallman spaces,* Math. Centre Tracts, Mathematisch Centrum, Amsterdam, 85, (1977). < 112, 284 >

[1980] *Superextensions of metrizable continua are Hilbert cubes,* Fund. Math., 107, (1980), 201-224. < 432 >

[1988] *Infinite-dimensional topology: Prerequisites and introduction,* North-Holland, Amsterdam, (1988), 401pp. < 298, 454, 463 >

J. van Mill, M. van de Vel

[1978] *Convexity preserving mappings in subbase convexity theory,* Proc. Kon. Ned. Akad. Wet., A, 81, (1978), 76-90.
< 29, 140, 141, 285, 300, 301, 322, 346, 348 >

[1978b] *Path connectedness, contractibility and LC properties of superextensions,* Bull. Acad. Polon. Sc., Sér. Sci. Math. Astr. et Phys., 26, (1978), 261-269. < 112 >

[1979] *On an internal property of absolute retracts,* Topology Proceedings, 4, (1979), 193-200. < 284 >

[1979b] *On superextensions and hyperspaces,* in: Topological structures II, Math. Centre tracts, Mathematisch Centrum, Amsterdam, 115, (1979), 169-180.
< 285, 300, 302, 345 >

[1981] *Subbases, convex sets, and hyperspaces,* Pacific J. of Math., 92, (1981), 385-402. < 29, 301 >

[1986] *Equality of the Lebesgue and the inductive dimension functions for compact spaces with a uniform convexity,* Colloquium Math., 50, (1986), 187-200.
< 29, 70, 323, 324, 467 >

J. van Mill, E. Wattel

[1978] *An external characterization of spaces which admit binary normal subbases,* Amer. J. Math., 100, (1978), 987-994. < 29 >

C.F. Mills, W.H. Mills

[1980] *The calculation of* $\lambda(8)$, preprint, (1980). < 250 >

W.H. Mills

(*cf.* C.F. Mills)

M. Mislove

(*cf.* G. Gierz)

D.S. Mitrinović

(*cf.* O. Bottema)

A. Mitschke, R. Wille

[1973] *Freie modulare Verbände FM* $(_DM_3)$, Proc. Lattice Theory Conf. Houston, (1973), 383-396. < 130 >

B. Monjardet

[1974] *Problèmes de transversalité dans les hypergraphes,, les ensembles ordonnés et en théorie de la décision collective,* Doctorat d' état de mathématiques, Université de Paris VI, (1974). < 250 >

[1975] *Eléments ipsoduaux du treillis distributif libre et familles de Sperner ipso-transversales,* J. Combinatorial Theory, (A), 19, (1975), 160-176. < 137, 345 >

E.H. Moore
  [1910] *Introduction to a general form of analysis*, AMS Coll. Publ., New Haven, 2, (1910). <28>

P. Mostert
  (*cf.* K.H. Hofmann)

H.M. Mulder
  [1978] *The structure of median graphs*, Discrete Math., 24, (1978), 197-204. <63>
  [1980] *The interval function of a graph*, Math. Centre Tracts, Mathematisch Centrum, Amsterdam, 132, (1980). <29, 70, 112, 138, 140, 141>

H.M. Mulder, A. Schrijver
  [1979] *Median graphs and Helly hypergraphs*, Discrete Math., 25, (1979), 41-50. <249>

J.A. Murtha, E.R. Willard
  [1969] *Linear Algebra and Geometry*, Holt, Rinehart and Winston, New York, (1969). <47>

U.S.R. Murty
  (*cf.* J. A. Bondy)

L. Nachbin
  [1950] *A theorem of Hahn-Banach type*, Trans. Amer. Math. Soc., 68, (1950), 28-46. <505>
  [1965] *Topology and Order*, Van Nostrand, (1965). <285, 286, 348>

S.B. Nadler, Jr.
  [1978] *Hyperspaces of sets*, Monographs in Pure and Appl. Math., Dekker, New York, 49, (1978), 707pp. <452>

S.A. Naimpally
  (*cf.* M.D. Guay; P. Mah)

L. Nebeský
  [1970] *Graphic algebras*, Comment. Math. Univ. Carolinae, 11, (1970), 533-544. <141>

J. Nieminen
  [1978] *The ideal structure of simple ternary algebras*, Colloquium Math., 40, (1978), 23-29. <140, 141>
  [1979] *Betweenness and ternary operations on partial lattices*, Serdica, 5, (1979), 374-377. <133>
  [1988] *Boolean graphs*, Comment. Math. Univ. Carolinae, 29, (1988), 387-392. <90>

R. Nowakowski, I. Rival
  [1979] *Fixed-edge theorem for graphs with loops*, J. Graph Theory, 3, (1979), 339-350. <506>

R.C. O'Brien
  [1976] *On the openness of the barycenter map*, Math. Ann., 223, (1976), 207-212. <281>

S. Onn
  [1990] *On the geometry and computational complexity of Radon partitions in the integer lattice*, Preprint, (1990). <176, 177>

S.V. Ovchinnikov
[1980]   *Convexity in subsets of lattices*, Stochastika, 4, (1980), 129-140.   <112>

L. Pasicki
[1987]   *On continuous selections*, Opuscula Math., 3, (1987), 65-71.   <324, 451, 452>
[1990]   *A fixed point theory and some other applications of weeds*, Opuscula Math., 7, (1990), 96pp.   <324>

A. Pelczynski
(cf. C. Bessaga)

J.P. Penot
[1979]   *Fixed point theorems without convexity*, Bull. Soc. Math. France, 60, (1979), 129-152.   <506>

H. Perfect
(cf. V.W. Bryant)

E. Pesch
(cf. H.-J. Bandelt)

R.R. Phelps
[1960]   *A representation theorem for bounded convex sets*, Proc. Amer. Math. Soc., 11, (1960), 976-983.   <22>
[1966]   *Lectures on Choquet's Theorem*, Van Nostrand Mathematical Studies, Van Nostrand, Princeton, 7, (1966), 130pp.   <433>

E. Pitcher, M.F. Smiley
[1942]   *Transitivities of betweenness*, Trans. Amer. Math. Soc., 52, (1942), 95-114.
   <90>

T. Poston
[1971]   *Fuzzy geometry*, Dissertation, University of Warwick, (1971).   <504>

R. Precup
[1981]   *Sur l'axiomatique des espaces à convexité*, Anal. Numer. Theor. Approx., 9 (2), (1981), 255-260.   <155>

W. Prenowitz
[1946]   *Descriptive geometries as multigroups*, Trans. Amer. Math. Soc., 59, (1946), 333-380.   <28, 157>
[1961]   *A contemporary approach to classical geometry*, Amer. Math. Monthly, 68, (1961), 1-67.   <157>

W. Prenowitz, W. Jantosciak
[1972]   *Geometries and Join spaces*, J. Reine angew. Math., 257, (1972), 100-128.
   <28, 157, 158>

A. Quilliot
[1985]   *On the Helly property working as a compactness criterion on graphs*, J. Combinatorial Th., A, 40, (1985), 186-183.   <504>

R. Rado
[1949]   *Axiomatic treatment of rank in infinite sets*, Canadian J. Math., 1, (1949), 337-343.   <48>

J. Radon
[1921] *Mengen konvexer Körper, die einen gemeinsamen Punkt enthalten,* Math. Ann., 83, (1921), 113-115. < 179 >

G.N. Raney
[1953] *A subdirect-union representation for completely distributive complete lattices,* Proc. Amer. Math. Soc., 4, (1953), 518-522. < 367 >

J.R. Reay
(cf. J.-P. Doignon)
[1968] *An extension of Radon's Theorem,* Illinois J. Math., 12, (1968), 184-189.
< 264 >
[1982] *Open problems around Radon's theorem,* Convexity and related combinatorial geometry (Norman, Okla.), Lecture Notes in Pure and Applied Mathematics, 76, (1982), 151-172. < 264 >

I. Rival
(cf. R. Nowakowski)

J.W. Roberts
[1977] *A compact convex set with no extreme points,* Studia Math., 60, (1977), 255-266. < 277 >
[1978] *The embedding of compact convex sets in locally convex spaces,* Canadian J. Math., 30, (1978), 449-454. < 308 >

G.S. Rubinstein
[1970] *On an intrinsic characterization of relatively open convex sets,* Soviet Math. Dokl, 11, (1970), 1076-1079. < 399, 403 >

R. Scharlau
(cf. A.W.M. Dress)

E.T. Schmidt
(cf. G. Grätzer)

J. Schmidt
[1953] *Einige grundlegende Begriffe und Sätze aus der Theorie der Hüllenoperatoren,* Bericht über die Mathematiker-Tagung in Berlin, (1953), 21-48. < 24, 28 >

A. Schrijver
(cf. H.M. Mulder)

D.S. Scott
(cf. G. Gierz)

M. Sholander
[1952] *Trees, lattices, order, and betweenness,* Proc. Amer. Math. Soc., 3, (1952), 369-381. < 29, 141 >
[1954] *Medians and betweenness,* Proc. Amer. Math. Soc., 5, (1954), 801-807.
< 138, 141 >
[1954b] *Medians, lattices and trees,* Proc. Amer. Math. Soc., 5, (1954), 808-812.
< 47, 112, 134, 141 >

G. Sierksma
(cf. J. Ch. Boland; J.-P. Doignon; K. Kołodziejczyk; N. Lindquist)
[1975] *Carathéodory and Helly-numbers of convex-product-structures,* Pacific J.

Math., 61, (1975), 275-282. <179>
[1976] *Axiomatic convexity theory and the convex product space,* Dissertation, University of Groningen, Groningen, Netherlands, (1976), 113pp.
<28, 48, 174, 179, 199, 201, 202, 203>
[1977] *Relationships between Carathéodory, Helly, Radon and Exchange numbers of convexity spaces,* Nieuw Archief Wisk., (3), 25, (1977), 115-132. <179>
[1981] *Convexity on unions of sets,* Compositio Math., 42, (1981), 391-400.
<24, 216, 218>
[1982] *A cone condition for convexity spaces,* Bull. Soc. Math. Belgique, Sér. B, 34, (1982), 41-48. <46>
[1982b] *Generalizations of Helly's theorem,* Convexity and related combinatorial geometry (Norman, Okla.), Lecture Notes in Pure and Applied Mathematics, 76, (1982), 173-192. <264>
[1982c] *Generalized Radon partitions in convexity spaces,* Archiv Math. (Basel), 39, (1982), 568-576. <203, 262>
[1984] *Exchange properties of convexity spaces,* Annals of Discrete Math., 20, (1984), 293-305. <202>
[1984b] *Extending a convexity space to an aligned space,* Indag. Math., 46, (1984), 429-435. <29>

B. Sims
(*cf.* J.R. Giles)

R. Sine
[1979] *On nonlinear contractions in sup norm spaces,* Nonlinear Analysis, 3, (1979), 885-890. <505>

M.F. Smiley
(*cf.* E. Pitcher)
[1947] *A comparison of algebraic, metric and lattice betweenness,* Bull. Amer. Math. Soc., 49, (1947), 246-252. <24>

R.E. Smithson
[1972] *Multifunctions,* Nieuw Archief Wiskunde, (3), 20, (1972), 31-53. <289>

P. Soardi
[1979] *Existence of fixed points of nonexpansive mappings in certain Banach lattices,* Proc. Amer. Mat. Soc., 73, (1979), 25-29. <505>

E.L.N. Soetens
[1973] *On locally Desarguesian planes,* Bull. Sc. Acad. Roy. Belgique, 59, (1973), 725-734. <403>

P.S. Soltan
(*cf.* V.G. Boltyanskii)
[1972] *Helly's Theorem for d-convex sets,* Soviet Math. Dokl., 13, (1972), 975-978.
<90>

V.P. Soltan
[1976] *Some questions in the abstract theory of convexity,* Soviet Math. Dokl., 17 (3), (1976), 730-733. <179, 202>
[1979] *Star-shaped sets in the axiomatic theory of convexity,* Soobshch. Akad. Nauk Gruzin. SSR, 96 (1), (1979), 45-48. <89>

[1981] *Substitution numbers and Carathéodory numbers of the Cartesian product of convexity structures,* Ukrain Geom. Sb., 24, (1981), 104-108. <202>
[1982] *Some questions of the axiomatic theory of convex sets,* Soviet Math. Dokl., 25(1), (1982), 244-248. <23>
[1983] *d–convexity in graphs,* Soviet Math. Dokl., 28(1), (1983), 419-421. <29, 45>
[1984] *Introduction to the axiomatic theory of convexity,* Shtiinca, Kishinev, (1984), 224pp. <28, 248>
[1987] *Some properties of d-convex functions. II,* Amer. Math. Soc. Transl., (2) 134, (1987), 45-51. <429>

E. Spanier
[1966] *Algebraic topology,* McGraw-Hill, New York, (1966), 528+xiv pp.
<436, 438, 439>

E. Sperner
[1938] *Zur Begründung der Geometrie im begrenzten Ebenenstück,* Schriften der Königsberg Gelehrten Gesellschaft, 6, (1938), 121-143. <399, 403>

F.W. Stevenson
[1972] *Projective spaces,* W.H. Freeman & Co., (1972), x + 416pp.
<47, 158, 399, 403>

A.R. Stralka
[1970] *Locally convex topological lattices,* Trans. Amer. Math. Soc., 151, (1970), 629-640. <484>

D.P. Strauss
[1968] *Topological lattices,* Proc. London Math. Soc., 18, (1968), 217-230. <375>

D.A. Szafron, J.H. Weston
[1976] *An internal solution to the problem of linearization of a convexity space,* Canad. Math. Bull., 19, (1976), 487-494. <401>

A. Tarski
[1930] *Fundamentale Begriffe der Methodologie der deduktiven Wissenschaften,* Monatsh. Math. Phys., 37, (1930), 360-404. <28>

P.D. Taylor
(*cf.* R.E. Jamison)

G. Thompson
(*cf.* W.R. Hare)

H. Tietze
[1928] *Über Konvexheit im kleinen,* Math. Z., 28, (1928), 697-707. <282>

H. Tong
[1952] *Some characterizations of normal and perfectly normal spaces,* Duke Math J., 19, (1952), 289-292. <452>

H. Torunczyk
[1980] *On CE images of the Hilbert cube and characterization of Q-manifolds,* Fund. Math., 106, (1980), 31-40. <447>

H. Tverberg
[1966] *A generalization of Radon's theorem,* J. London Math. Soc., 41, (1966), 123-128. <28, 70, 264>

[1971]   *On equal union of sets,* Studies in Pure Math., (1971), 249-250.   < 177, 262 >
[1981]   *A generalization of Radon's Theorem, II,* Bull. Austral. Math. Soc., 24, (1981), 321-325.   < 264 >

A. Tychonov
[1935]   *Ein Fixpunktsatz,* Math. Ann., 111, (1935), 767-776.   < 376 >

F.A. Valentine
[1964]   *Convex sets,* McGraw-Hill, (1964), 238pp.   < 28, 90 >

G. Valette
(cf. J.-P. Doignon)
[1982]   *Addendum to "A cone condition for convexity spaces" by Sierksma,* Bull. Soc. Math. Belge, Sér. B, 34, (1982), 68.   < 46 >

J.C. Varlet
[1975]   *Remarks on distributive lattices,* Bull. Acad. Pol. Sci., Sér. Sci. Math., Astr., et Phys., 23, (1975), 1143-1147.   < 29 >

P.M. Vasić
(cf. O. Bottema)

M. van de Vel
(cf. H.-J. Bandelt; J. van Mill)
[1979]   *Superextensions and Lefschetz fixed point structures,* Fund. Math., 104, (1979), 27-42.   < 29, 112 >
[1982]   *Finite dimensional convex structures I: general results,* Topology Appl., 14, (1982), 201-225.   < 154, 464, 465, 467 >
[1983]   *Dimension of convex hyperspaces: non-metric case,* Compositio Math., 50, (1983), 95-108.   < 67, 246, 285 >
[1983b] *Euclidean convexity cannot be compactified,* Math. Ann., 262, (1983), 563-572.   < 48, 324 >
[1983c] *Finite dimensional convex structures II: the invariants,* Topology Appl., 16, (1983), 81-105.   < 246, 485 >
[1983d] *Pseudo-boundaries and pseudo-interiors for topological convexities,* Dissertationes Math., 210, (1983), 1-72.
         < 24, 67, 69, 70, 112, 286, 342, 344, 348, 430, 431, 432, 433, 467 >
[1983e] *Matching binary convexities,* Top. Appl., 16, (1983), 207-235.
         < 110, 112, 138, 177, 218, 248, 301, 376, 485 >
[1983f] *Dimension of convex hyperspaces: nonmetric case,* Compositio Math., 50, (1983), 95-108.   < 247 >
[1983g] *Two-dimensional convexities are join-hull commutative,* Top. Appl., 16, (1983), 181-206.   < 46, 48, 217, 245, 250, 299, 433, 483 >
[1984]   *Binary convexities and distributive lattices,* Proc. London Math. Soc., (3), 48, (1984), 1-33.   < 29, 112, 141, 249, 373, 376 >
[1984b] *Continuity of convexity preserving functions,* J. London Math. Soc, (2), 30, (1984), 521-532.   < 465 >
[1984c] *Dimension of binary convex structures,* Proc. London Math. Soc., (3), 48, (1984), 34-54.   < 29, 246, 248, 465, 466 >
[1984d] *Dimension of convex hyperspaces,* Fund. Math., 122, (1984), 11-31.
         < 29, 67, 70, 285, 324, 485 >

[1984e] *Lattices and semilattices: a convex point of view,* in: Continuous lattices and their applications, Lecture notes in pure and applied mathematics, Dekker, New York, 101, (1984), 279-302. < 28, 29, 112, 286, 301, 467 >

[1984f] *On the rank of a topological convexity,* Fund. Math., 119, (1984), 17-48.
< 250, 484 >

[1986] *Convex Hilbert cubes in superextensions,* Topology Appl., 22, (1986), 255-266. < 432 >

[1986b] *Metrizability of finite dimensional spaces with a binary convexity,* Canadian J. Math., 37, (1986), 1-18. < 247, 250, 484 >

[1988] *A Helly property of arcs,* Archiv Math. (Basel), 52, (1988), 298-306.
< 218, 483 >

[1988b] *Continuous position and existence sets,* Math. Ann., 282, (1988), 629-636.
< 432 >

[1990] *Collapsible Polyhedra and median spaces,* report WS-360, Vrije Universiteit Amsterdam, (1990). < 218 >

[1991] *Invariant arcs, Whitney levels, and Kelley continua,* Trans. Amer. Math. Soc., 326, (1991), 749-771. < 48, 90, 323, 324, 506 >

[1993] *A selection theorem for topological convex structures,* Trans. Amer. Math. Soc., 336(2), (1993), 463-496. < 29, 281, 322, 324, 449, 450, 452 >

M. van de Vel, E. Verheul

[1990] *Decomposing modular Banach Spaces into a rigid and an $l_1(I)$ part,* report WS-373, Vrije Universiteit Amsterdam, (1990). < 136, 283, 286 >

[1990b] *Isometric embedding of median algebras into $L_1(\mu)$ spaces,* report WS-384, Vrije Universiteit Amsterdam, (1990). < 139 >

[1991] *Completeness and topology of modular spaces,* preprint, (1991).
< 112, 135, 141, 284, 286 >

A. Verbeek

[1972] *Superextensions of topological spaces,* Math. Centre Tracts, Mathematisch Centrum, Amsterdam, 41, (1972), 155pp. < 109, 250, 345, 348 >

E. Verheul

(cf. H.-J. Bandelt; M. van de Vel)

[1993] *Multimedians in metric and normed spaces,* CWI Tract, Amsterdam, Netherlands, (1993), to appear. < 90, 111, 112, 134, 135, 138, 139, 140, 218 >

[199*] *Modular metric spaces,* Preprint, (199*). < 135, 141 >

A. Vijayakumar

(cf. M. Changat)

H.-J. Voss

(cf. H. Walther)

H. Walther, H.-J Voss

[1974] *Über Kreise in Graphen,* VEB Deutscher Verlag der Wiss., Berlin, (1974).
< 243 >

E. Wattel

(cf. J. van Mill)

R.J. Webster
   (cf. V.W. Bryant)
D.J.A. Welsh
   [1976]   *Matroid theory*, Academic Press, New York, (1976). <48>
J.H. Weston
   (cf. D.A. Szafron)
J.H.M. Whitfield
   (cf. P. Mah)
J.H.M. Whitfield, S. Yong
   [1981]   *A characterization of line spaces*, Canad. Math. Bull, 24(3), (1981), 273-277.
   <158>
H. Whitney
   [1935]   *On the abstract properties of linear dependence*, Amer. J. Math., 57, (1935), 509-533. <48>
G.T. Whyburn
   [1968]   *Cut points in general topological spaces*, Proc. Nat. Acad. Sc. USA, 61, (1968), 380-387. <430>
A. Wieczorek
   (cf. K. Keimel)
   [1983]   *Fixed points of multifunctions in general convexity spaces*, ICS PAS reports, 508, (1983), 32pp. <324>
   [1989]   *Spot functions and peripherals: Krein-Milman type theorems in abstract setting*, J. Math. Anal. Appl., 138, (1989), 293-310. <428, 433, 451>
   [1991]   *Compressibility: a property of topological spaces related with abstract convexity*, J. Math. Anal. Appl., 161, (1991), 9-19. <451>
   [1992]   *Kakutani and fixed point properties of topological spaces with abstract convexity*, J. Math. Anal. Appl., 168, (1992), 483-499. <505>
E.R. Willard
   (cf. J.A. Murtha)
R. Wille
   (cf. A. Mitschke)
E.W. Womble
   (cf. D.C. Kay)
S. Yong
   (cf. J.H.M. Whitfield)
A. Zaanen
   (cf. W. Luxemburg)

Index of Terms

A

absolute retract, 284, 446
absolutely dispensable, 24
Absorption Law (of medians), 8, 284
Additive Law
– (of closure), 4
– (of *cind*), 454
admissible set (of a convex system), 19
affine
– basis, 152
– combination, 34
– dimension → dimension
– function, 16
– hull, 151
– set (in a BW space), 151, 152
– – (in a vector space), 34
Affine Dimension Formula, 43, 44, 152, 154, 170
Affine Hull Formula, 151
affinely dependent set → dependence
affinely independent set → independence
algebra
– Boolean, 60
– – topological, 282
– Borel, 409
– Heyting, 366
– median, 8, 17, 23, 123, 178, 342, 343, 345
– – free, 137
– – topological, 269, 273, 285, 370
– modular, 133
algebraic closure operator
  → closure operator, domain finite
aligned space → convex structure
alignment → convexity
alternating set function, 201
amalgam → gated
angle, 156
anti-Exchange Law → Exchange Law
anti-homomorphism → homomorphism
anti-matroid → convex geometry
AR → absolute retract
arc, 33, 318

– space of ˜s, 33, 41, 51, 82, 318, 445
Archimedian property, 381
Arithmetic Laws (of joining sets), 146
arity
– (of a convex structure), 5, 15, 72, 248
– (of a convex system), 19
Associative Law (of a join operator), 146
attaching point, 64

B

base (of a convex structure), 10
base-point (quasi-)order, 91, 109, 110, 134, 143, 154, 288, 352, 457
basis (of a matroid), 38, 152
Bauer Maximum Principle, 417
Beer Approximation Theorem, 444
betweenness, 32
– (in cones), 32
– (in spaces of arcs), 33, 318
– Fan, 406
– strict, 428
biconvex set → half-space
boundary (of a join space), 147
Brace-Daykin Theorem, 235
branch (of a tree), 7
breadth (of a semilattice), 171, 177, 246
Brouwer's Fixed Point Theorem, 497
Browder Fixed Point Theorem, 506
Browder Selection Theorem
      → Selection Theorem
Brunn's property, 89
Bryant-Webster space, 149, 155, 170, 351, 379
bundle of lines, 399
BW space → Bryant-Webster space

C

C-dependent set → dependence
C-independent set → independence
Cantor cube → cube
Carathéodory
– number, 167, 174, 182, 185, 187, 288, 295, 469

– Theorem, 168
Cartesian
– metric → metric
– order, 7, 85, 86
– product
– – (of graphs), 87, 243
– – (of metric spaces), 86
– – (of ordered sets), 86
CC function → convex-to-convex function
center of perspectivity, 379
Choquet extreme point/set
  → extreme point/set
Choquet-Krein-Milman Theorem, 411
chromatic number, 242
cind → dimension
CIP → Copoint Intersection Property
clique, 175
– number, 175
close (w.r.t a cover)
– functions which are ~, 445
– sets which are ~, 304
closed set (of a closure space), 4
Closedness (of a join operator), 146
closure
– (of a set), 3
– operator, 4, 271
– – convex, 271, 287, 288, 327
– – domain finite, 4, 5
– – Kuratowski, 4
– space → closure structure
– stability, 272
– structure, 4, 10
Commutative Law (of a join operator), 146
compact element, 26
Compact Intersection Theorem, 220
– (for rank), 246
compatible
– metric → metric
– set, 230
– uniformity → uniformity
complementary pair (in a lattice), 60, 95
complete
– (interval space), 105, 107
– (join space), 146

– (metric space), 108
– (poset), 105
– monotonely ~ (metric space), 108, 129
– strong ~ (interval space), 111, 361
completeness (of a join operator), 146
completion
– (of a modular metric space), 126
– (of a uniform convex structure)
  → Uniform Completion Theorem
– Dedekind, 325
compressible space, 451
concave set, 3
conditionally distributive semilattice
  → semilattice, conditionally distributive
conditionally modular semilattice
  → semilattice, conditionally modular
cone (of a convex structure), 33, 45, 46, 59, 217, 281, 316, 328, 335, 431
Cone-union Property, 39, 163, 217, 328
Continous positions property, 420, 471
continuous
– functor, 346
– tcs, 288, 332
continuous position (of a set), 418, 432
contractible, 285
convex
– base → base
– closure operator → closure operator
– geometry, 45, 64, 153, 295, 429
– hull operator, 32, 354
– – (of a convex system), 19
– hyperspace → hyperspace
– invariant, 167, 174
– – classical, 167, 469
– metric → metric
– set, 3
– – interval, 71
– – strongly, 342, 343, 415
– structure, 3
– – metric/metrizable, 304, 443
– – topological, 267
– – uniform/uniformizable, 304
– subbase
  → subbase of a convex structure

Index   531

– succession, 148
– system, 19
– – metric/metrizable, 312, 452
– – topological, 298
– – uniform/uniformizable, 312
convex-to-convex function, 15, 170
convexity, 3
– (consisting of subsemilattices), 24, 45, 67, 177
– (of a lattice), 7, 16, 45, 51, 57, 105, 172, 178, 269, 273, 277
– (of a poset) → convexity, order
– (of a semilattice), 6, 11, 16, 45, 47, 50, 56, 93, 105, 165, 171, 176, 177, 229, 269, 274, 276, 278, 292, 328, 333
– (of a subspace), 170
– affine ˜ (of a vector space), 34, 154
– coarse, 10
– coarser, 9
– disk, 22, 179, 505
– finer, 10
– free, 10, 188, 235, 297
– k-free, 44
– generated by a base, 10
– generated by a subbase, 10, 62, 267
– generated by functionals, 11, 335, 406
– – symmetrically, 11, 335
– geodesic, 8, 17, 24, 64, 83, 86, 110, 136, 157, 282
– H-convexity, 11, 17, 25, 42, 50, 56, 64, 223, 229, 232, 238, 241, 244, 268, 273, 292, 299, 478
– induced ˜ (of a convex system), 19
– interval, 71
– linear, 34
– lower, 6, 25, 45, 325
– median, 9, 176, 235, 291, 300, 336, 346
– order, 6, 16, 24, 45, 64, 81, 86, 109, 174, 178, 188, 268, 273, 276, 334
– partial ˜ (of a convex system), 19
– quotient, 18, 68
– relative, 13, 87
– standard median ˜ (of a solid cube), 210
– standard ˜ (of a vector space), 6, 11, 16, 24, 45, 80, 92, 106, 149, 188, 191, 228, 232, 242, 268, 272, 300, 328, 333
– upper, 6, 25, 45, 325
– Vietoris, 12, 58, 66, 140, 309
convexity preserving function, 15, 25, 66, 170
convexity space → convex structure
convexly dependent set → dependence
convexly homogeneous space, 431
convexly independent set → independence
copoint, 64, 178, 220, 231, 241
Copoint Intersection Property, 263
core (of a set), 350
– intrinsic, 258, 273, 350, 407
corner points → complementary pair
Countable Intersection Theorem
– (for rank), 246
– I, 226
– IIA, 424
– IIB, 425
counting measure, 114
CP function
   → convexity preserving function
CP isomorphism → isomorphism
CPP → Continous positions property
crystal, 492
cube, 7
– Cantor, 59, 60, 219, 240
– general, 243, 246
– graphic, 51, 138, 176, 229, 244, 247, 248
– Tychonov, 229, 247
cubical polyhedron → polyhedron
CUP → Cone-union property
Curtis Selection Theorem
   → Selection Theorem
cut point
– between two points, 413
– order, 430

D

decomposable segment → interval
decreasing sequence (w.r.t a base-point order), 108
degree

– directional, 245, 247
– – (at a point), 258
– generating, 228, 245, 247
Density Property (of a join operator), 145, 146
dependence
– affine, 35, 164
– Carathéodory, 161, 163, 164, 173
– convex, 35, 227
– exchange, 161, 163, 164, 173
– Helly, 161, 163, 164, 173
– linear, 35
– minimal linear (of functionals), 222, 223, 244
– Radon, 161, 163, 164, 173
– Tverberg, 251
depth
– (of a convex structure), 248
– (of a poset), 177, 248
Desargues Property, 381
– dual, 383
– modified, 388
diametric points, 490
Dilworth Theorem, 227
dimension
– affine, 152
– convex, 453
dir → degree, directional
direction, 230, 242, 247
– covering with ˜s, 231, 242
directional degree, 230, 245 → degree
discrete
– interval space → interval space
– poset → poset
disjoint sum (of convex structures), 17, 24, 46, 65
distributive lattice → lattice
divisable set → dependence
down-complete subset (of a poset), 105
dual ideal → filter
Dyadic midpoint formula, 393
dyadic point (in a join segment), 392

E

E-dependent set → dependence
E-independent set → independence

Eckhoff conjecture, 262
Eckhoff-Jamison inequality, 169, 237
– (special), 174, 216
Eilenberg-Montgomery Theorem, 505
embedding (of convex structures), 15, 60, 154
– into products of trees, 243, 246, 248
Embedding Theorem, 396
end point (of a tree), 94
enlargement (of a multifunction), 443
equi-perspective segments, 390
exchange
– function, 185, 199
– number, 167, 186, 187, 235, 246, 473
Exchange Law, 34, 48
– anti ˜ (of convex geometry), 45, 295
exchangeable set → dependence
expansion
– convex, 63, 176, 218
Extensibility (of a join operator), 146
extension
– (of a convex structure), 13
– at a point → ray
– away from a set, 145
Extension Lemma, 189, 262
extension operator, 54, 71, 145
Extensive Law
– (of betweenness), 32
– (of closure), 4
– (of an interval operator), 71
externally convex metric → metric
extremality structure, 405
extreme point/set, 23, 45
– Choquet, 409
– δ_X ˜, 415
– Fan, 406
– standard, 226
– strict, 428
– synthetic, 405

F

$\mathcal{F}$-betweenness → betweenness
$\mathcal{F}$-closed set, 406
$\mathcal{F}$-protopology, 406
facet, 45
Fan extreme point/set → extreme point/set

Fan-Krein-Milman Theorem, 408
field
 – totally ordered, 5
filter, 51
 – prime, 51
Finitary Law (of independence), 35
finite topology (of a vector space), 372
flat → affine set
 – (of a matroid), 36
Fréchet vector space → vector space
Frattini-Neumann Intersection Theorem, 24
free
 – convex structure → convexity, free
 – median algebra → algebra
 – set, 175, 294
Fuchssteiner's property, 416, 420
fully independent set → independence
function space with supremum norm, 115
functional, 325, 341
 – ρ- convex, 429
 – equivalent, 325
 – lower CP, 325
 – lower semi-continuous, 326
 – separating collection of ˜s, 60, 329, 342, 400
 – separating two sets, 329
 – strictly convex, 428
 – symmetric collection of ˜s, 11, 409
 – upper CP, 325
 – upper semi-continuous, 326
functionally convex set, 343
functionally generated convex structure
    → convexity generated by functionals

G

gate, 98, 138
 – map, 98, 100, 139
 – mutual, 98
gated amalgam, 102, 104, 105, 110, 125, 135, 208, 370
gated set (in an interval space), 98
Gaussian integers, 176, 260
gen → degree, generating
general position, 190

geometric interval space → interval space
gluing (of convex structures), 24, 46
 – normal (of convex structures), 25
graph, 8
 – bipartite, 64, 242, 243
 – dismantlable, 175
 – Helly, 175
 – median, 243, 248, 371, 488
 – simplex, 242, 248
 – Steiner → Steiner tree
 – weighted, 129
 – – connecting a set, 129
graphic cube → cube
graphic interval space → interval space

H

H-convexity → convexity
H-dependent set → dependence
H-independent set → independence
half-open (co)sector, 178
half-space, 49, 221, 277
 – non-trivial, 49
 – separating two sets, 53
Hanner's Problem, 136
Hasse diagram, 96
Hausdorff
 – metric → metric
 – uniformity → uniformity
Helly
 – number, 167, 175, 220, 221, 223, 225, 231, 241, 244, 469
 – – sigma finite, 226
Helly graph → graph
Helly Theorem, 168
hemispace → half-space
hereditary property, 161, 173
homomorphism
 – (of lattices), 16
 – (of median algebras), 17, 123
 – (of semilattices), 16
 – anti ˜ (of lattices), 16
homothety, 17
 – center, 17
 – coefficient, 17
 – positive/negative, 17, 50
hull operator → convex hull operator

hyperflat (of a matroid), 45, 355
hyperoctahedron, 89
hyperplane (of a tcs), 456, 470
hyperspace
 − convex, 12, 66, 140, 217
 − − (of a tcs), 269, 301, 334, 341, 442, 482
 − topological, 267, 300
hyperspace convexity
   → convexity, Vietoris
Hyperspace Polytope Formula, 12

I

ideal, 51
 − prime, 51
Idempotent Law
 − (of closure), 4
 − (of an extension operator), 146
 − (of gates), 99
 − (of an interval operator), 74
 − (of a join operator), 146
II function → interval-to-interval function
increasing sequence (w.r.t a base-point order), 108
independence
 − affine, 35
 − Carathéodory, 161
 − convex, 35, 36, 227
 − exchange, 161
 − full, 252
 − Helly, 161
 − linear, 35
 − Radon, 161
 − Tverberg, 251
independence structure, 36, 44
independent set (of an independence structure), 36
indexed chain, 325
inner point, 24
interior stability, 272, 432
Intermediacy Property, 200
Interpolation Property, 350
intersectional subbase, 64
interval, 71
 − decomposable, 143, 148, 478
 − geodesic, 72, 83

 − lattice, 72
 − median, 73
 − modular, 72
 − order, 72
 − semilattice, 72
 − standard, 72
interval operator, 71, 136, 147, 173, 323
 − geometric, 74
 − modular, 113
 − relative, 73
interval preserving function, 74
interval space, 71, 87, 109
 − dense, 143, 148
 − discrete, 96
 − geometric, 74
 − graphic, 97, 110
 − modular, 113
 − semimodular, 115
 − straight, 143, 148
interval-to-interval function, 74
Invariant Arc Theorem, 501
Invariant Cube Theorem, 138, 488
Invariant Simplex Theorem, 504
Inversion Law (of an interval operator), 74
IP function → interval preserving function
ipsodual point → self-dual point
irreducible
 − element, 24
 − set, 165
isomorphism
 − (of convex structures), 15, 66
 − (of convex systems), 20
IW topology → topology, intrinsic weak

J

JHC → Join-hull commutativity
join
 − (of convex structures), 14, 109
 − (of two sets), 145
 − operator
 − − open, 145, 147, 148
 − − (in a meet semilattice), 278
 − space (open), 147
Join-hull commutativity, 39, 43, 144, 148, 163, 328, 342, 473

- weak, 40, 43, 152
Jordan-Hölder Theorem, 117

K

K-convexity, 241
k-Radon number → partition number
Kakutani Property, 497
Kakutani separation property
　　→ separation axioms
Keller Theorem, 448
KKM function
　　→ Knaster-Kuratowski-
　　Mazurkiewicz
KKM Function Principle, 499
Knaster-Kuratowski-Mazurkiewicz function, 499
Koch's Arc Theorem, 301

L

L_1-space
lattice, 7, 23, 26
- algebraic, 26, 240
- Boolean, 60, 69
- distributive, 94, 246
- - infinitely, 366
- modular
- opposite, 17
- topological, 269
- vector
Lawson semilattice → semilattice
Lemma on Roofs, 399
Leray Fixed Point Theorem (for convexoid Spaces), 498
Levi inequality, 169, 216
lexicographic order, 399
limit
- direct ˜ (of convex structures), 26, 46, 65
- inverse ˜ (of convex structures), 26, 65, 240, 345
line
- (in an interval space), 154, 400
- long, 399
- metric, 157
line space, 155, 401
linear combination, 34

linear set, 34
linear variety → affine set
linearization family, 400, 401
linearly independent set → independence
linked system (of sets), 12, 68
locally a convex set, 282
locally convex, 275
- at a point, 275
- proper, 275
long line → line
lower
- convexity → convexity
- CP → functional
- set, 6
lsc functional
　　→ functional, lower semi-
　　continuous
LSC multifunction
　　→ multifunction, lower semi-
　　continuous

M

Mac Lane Steinitz axiom
　　→ Exchange Law
Majority Law → Absorption Law
matroid, 34, 36, 43, 44, 153, 190, 252, 261
- affine ˜ (of a BW space), 152
- - (over a field), 35
- linear ˜ (over a field), 34
- projective, 42, 47, 154, 253
- - (over a field), 35
md → dependence
median
- (of three points), 8, 113
- algebra → algebra
- convex structure, 9
- graph → graph
- operator, 8
- - multi˜, 113, 135
- preserving function
　　→ homomorphism of median algebras
- space, 333, 364, 366, 367, 369, 471
- stable subset, 121, 130, 176
median stabilization, 130, 241

metamorphism
    → convexity preserving function
metathetism → betweenness
metric
 – adapted, 322, 346
 – Cartesian, 86, 115
 – compatible, 304, 436
 – convex, 115, 157
 – – externally, 157
 – Hausdorff, 126
 – hyperconvex, 505
 – Manhattan, 73, 115
 – path, 111
 – Ptolemaic, 85, 282
 – radially convex ˜ (in a pospace), 347
metric contraction, 346
metric nearest point function, 346
metric radius, 489
Michael Selection Theorem
    → Selection Theorem
midpoint, 389
van Mill Theorem, 448
minimal linear dependence → dependence
Minimax Theorem, 500
Minkowski functional, 326
mixer, 284
mls → linked system
modular
 – algebra → algebra
 – interval space → interval space
 – lattice → lattice
 – operator, 133
 – semilattice → semilattice
Monotone Law
 – (of an interval operator), 74
 – (of closure), 4
Monotone Mapping Theorem, 501
monotonely complete metric space
    → complete metric space
Moore family → protopology
Moulton plane, 150, 155, 382, 398
MP function
    → homomorphism of median algebras
multifunction, 287

 – continuous, 287
 – lower semi-continuous, 287, 436, 503
 – upper semi-continuous, 287, 445
multimedian
 – operator → median
 – w.r.t. a point (set which is ˜), 135
multiply ordered set, 492
 – labeled, 493
 – topological, 496
 – – multilabeled, 496

N

nerve (of a cover), 435
neutral element
 – (of a lattice), 133
 – (of a modular space), 133
non-expansive function, 491
Nondegeneracy Law (of independence), 35
norm
 – Cartesian, 115, 136
 – Manhattan, 114, 178
normal family, 62
Normal Law (of closure), 4
normalization (of betweenness), 32

O

OP function → order preserving function
opposite side (of a vertex in a triangle), 381
order dense, 143, 151
order convex set, 6
order preserving function, 16
ordered geometry, 155
outer point, 24, 45

P

parallel lines, 156
partial function, 379
partially ordered set, 6, 174, 261, 263, 349
 – completely sequential, 242
 – continuous, 349, 352
 – discrete, 40
 – semimodular, 40
partially ordered space, 268, 347

Index   537

– arc-wise connected, 318
partition number, 251, 260, 262
– restricted, 263
Pasch-Peano space, 79
Pasch Property, 77, 79
Pasch property (of a join operator), 146
path metric → metric
Peano Property, 77
– weak, 88
permutahedron, 65
perspective correspondence, 379
Petersen graph, 249
point-convex space
    → separation axioms, S_1
point-separating
– family of functionals
    → separation (with functionals)
– family of orders
    → separation (with orders)
pointwise directional degree
    → degree, directional
polyhedron
– cubical, 210, 472
– recursively contractible, 505
polytope, 3, 9, 55, 68, 267, 354
Polytope Screening Characterization, 55
poset → partially ordered set
pospace → partially ordered space
power set, 4
PP-space → Pasch-Peano space
pre-hull operator, 451
prime
– dual ideal → filter, prime
– filter → filter, prime
– ideal → ideal prime
product
– (of convex structures), 14, 45, 46, 49, 59, 86, 245, 262, 270, 274, 293, 350
– (of convex systems), 27
– (of interval spaces), 73, 96, 105
– (of topological convex structures), 270
product convexity, 14, 86, 87, 181, 346, 353
Product Polytope Formula, 15
projective
– pairs, 117

– set, 35
– space → matroid, projective
Projective Dimension Formula, 43, 44, 154
protopology, 4
– generated by a subbase, 10
– generated by functionals, 406
pseudo-boundary absolute/relative, 411
pseudo-boundary point, 411
pseudo-interior –, 411
Ptolemaic metric space → metric
pure subspace → subspace

Q

quadrangle (in a join space), 386
Quadrangle Lemma, 386
quasi-ordered set, 91, 492
quotient
– (of a convex structure), 18
– (of a convex system), 20

R

R-dependent set → dependence
R-independent set → independence
radially convex metric → metric
Radon
– function, 189, 190, 191, 195
– number, 167, 176, 177, 192, 193, 197, 244, 246, 247, 251, 259, 470
– partition, 161, 189
– probability measure, 409
– Theorem, 168, 177
ramification order (of a point), 110
ramification point, 7
Ramification Property, 478
– (of a join operator), 146
– (of an interval space), 143, 163
rank, 227, 245, 246, 247, 481
– (of a matroid), 39, 43, 152
– sigma finite, 246
Rank Formula, 43, 152, 154, 155
ray, 145, 379
– closed, 156
– open, 156
– parallel, 156
ray operator → extension operator

realization, 215
recursively contractible polyhedron
    → polyhedron
reducible set, 165
redundant point, 165
refinement
  – associated
       → refinement, corresponding
  – corresponding, 304
  – function (between covers), 435
  – of a functional, 325
  – of covers, 435
relative convexity
       → convexity of a subspace
Relative hull formula, 13, 96
relatively complete sets (in a convex
    structure), 353
relatively convex set, 13
Replacement Law (of independence), 36
  – strong ˜ (of independence), 36
representing measure, 409
Riesz space → vector lattice
rotund norm
       → vector space, normed, rotund

S

Sand-glass Property, 79, 80
Sand-glass property extended, 90
screening, 54, 55, 62, 344
  – convex closed, 332, 336, 344, 474
Screening Lemma, 99
segment, 5, 78, 235
  – decomposable
       → interval, decomposable
selection continuous, 436
Selection Theorem, 440
  – Browder, 450
  – Curtis ˜ (in spaces of arcs), 445
  – Michael ˜ (in Fréchet spaces), 442
  – – (in general convexity), 452
  – Nadler ˜ (in trees), 442
self-dual point, 137
semilattice, 6, 165, 171
  – conditionally distributive, 47, 134
  – conditionally modular, 132
  – join-continuous, 278

  – Lawson, 276, 318, 445, 449
  – median, 134, 246
  – modular, 132
  – topological, 268
semimodular
  – interval space → interval space
  – poset → poset
semispace → copoint
separating
  – functionals → functionals
  – orders, 492
separation axioms
  – S_1, 53, 54
  – S_2, 53, 225
  – S_3, 53, 80, 81, 83, 221, 225, 231
  – S_4, 53, 79, 80, 81, 144, 148, 245
  – functional, 331, 343
  – – (FS_2 , FS_3 , FS_{3+} , FS_4), 331
  – neighborhood, 327, 341
  – – (NS_2 , NS_3 , NS_4), 327
separator, 453
set-valued function → multifunction
Shilov
  – boundary, 429
  – subset, 429
Shrinking Property, 431
Sierksma
  – inequalities, 169, 216
  – number → exchange
simplex
  – (of a graph), 242
  – (of a simplicial complex)
simplex graph → graph
simplicial
  – complex, 21, 435
  – function, 22
Singleton Intersection Principle, 386
smooth norm
       → vector space, normed, smooth
smooth point, 22
social affinity, 44
solid cube → cube
space
  – → convex structure
  – → topological convex structure
  – (of measurable functions), 67

spanning set, 4, 271, 328
Sperner Theorem, 235
spread, 230, 241
 – at a point, 258
star-center, 89
star-shaped
 – set, 89
 – space, locally ~, 283
Steiner
 – point, 138
 – tree, 130
Stirling number of the second kind, 262
Stone Representation Theorem, 61
straightening, 318
subbase
 – (of a closure space), 11
 – (of a convex structure), 10, 56, 64, 231, 241, 337, 340
Subdivision Property, 162, 295, 351
subspace
 – (of a convex structure), 13, 59, 86
 – (of a convex system), 19
 – (of an interval space), 73, 87, 96, 105
 – (of a topological convex structure), 270
 – pure ~ (of a topological convex structure), 294, 338
substructure
 → subspace, of a convex structure
subsystem
 → subspace, of a convex system
superextension, 13, 24, 51, 69, 109, 137, 237, 238, 240, 246, 248, 249, 260, 279, 284, 337, 344, 448
superspace
 → extension of a convex structure
support
 – (of a measure), 409
 – point, 412
symmetrically generated convex structure
 → convexity
Symmetry Law
 – (of an interval operator), 71
 – (of medians), 8
system
 – direct ~ (of convex structures), 26
 – inverse ~ (of convex structures), 26

T

T_k-dependent set → dependence, Tverberg
T_k-independent set
 → independence, Tverberg
T_1 family, 13
tcs → topological convex structure
Tietze Property, 282
topological
 – convex hyperspace → hyperspace
 – convex structure → convex structure
 – convex system → convex system
 – hyperspace → hyperspace
 – multilabeling → multiply ordered space
 – space
 – – countably compact, 491
 – – sequentially compact, 496
topology
 – compatible ~ (of a convex structure), 267
 – core, 353
 – inclusion-exclusion, 219, 220, 240
 – intrinsic weak, 278, 280, 328, 335, 360
 – Lawson, 240
 – weak ~ (of a tcs), 278
Torunczyk Theorem, 447
totally ordered set, 11, 197, 228, 325
Transitive Law
 – (of betweenness), 32
 – (of dependence), 36
 – (five-point), 138
 – (of gates), 99, 212
 – (of medians), 8
transposed pairs, 117
transversal set, 269, 300, 345
tree, 7, 23, 45, 56, 93, 110, 174, 178, 188, 233, 239, 242, 247, 281, 323, 372
 – Steiner → Steiner tree
 – topological, 269, 309, 412, 430, 442, 464
Tri-Spherical Intersection Property, 115
triangle
 – (in a join space), 381
 – axially perspective ~s, 381
 – centrally perspective ~s, 381

– dual ˜ (in a join space), 381
Triangle Inequality, 8
Triangle Property, 97, 110
trihedron, 399
 – face, 399
Tverberg dependent set → dependence
Tverberg independent set → independence
Tverberg number → partition number
Tverberg partition, 251
Tverberg Theorem, 255, 263
Two-triples Property, 157

U

ultrametric, 347
underlying graph
 – (of a discrete space), 96, 110
 – (of a polyhedron), 210
Uniform Completion Theorem, 312
Uniform Factorization Theorem, 317
Uniform Separation Theorem, 311
uniform space, 303
uniformity, 303
 – compatible, 304
 – exotic, 322
 – Hausdorff ˜ (of a hyperspace), 305, 309
Union Property, 174
 – weak, 174
universal bound (of a semilattice), 33
up-complete subset (of a poset), 105
up-directed set, 4
upper
 – convexity → convexity
 – CP → functional
 – semi-continuous functional
   → functional, upper semi-continuous
 – – multifunction
   → multifunction, upper semi-continuous
 – set, 6
Urysohn Theorem, 330
Urysohn-Carruth Metrization Theorem, 347
usc functional
  → functional, upper semi-continuous

USC multifunction
  → multifunction, upper semi-continuous

V

valuation (on a lattice), 119
variety, 68, 241
vector convexity, 283
 – symmetric, 283
vector lattice, 85
vector space, 92, 106, 149, 164, 168
 – Fréchet, 426, 442
 – normed
 – – smooth, 22
 – – rotund, 24
 – – uniformly convex, 490
 – punctured, 34
 – topological, 268, 272
vertex
 – (of a crystal), 493
 – (of a polyhedron), 210
 – dominated, 175
Vietoris convexity
  → convexity, hyperspace

W

way-below relation, 349
weak
 – Euclidean Four-point Property, 157
 – JHC → Join-Hull Commutativity
 – Join-Hull Commutativity
   → Join-Hull Commutativity
 – Peano Property → Peano
 – topology → topology
 – Union Property → Union Property
weakly continuous function, 279
wedge, 20, 308
weighted
 – graph → graph
 – median, 488
 – set, 167, 488
width (of a poset), 175, 227, 241

www.ingramcontent.com/pod-product-compliance
Ingram Content Group UK Ltd.
Pitfield, Milton Keynes, MK11 3LW, UK
UKHW020657050526
12271UKWH00003B/7